E. Möbius
ıd NIST Space Science Center and
Department of Physics
University of New Hampshire
Durham, NH, USA
ınd
R. von Steiger
ıersity International Space Science Institute
Bern, Switzerland

ustration is a schematic view extending from the heliosphere via the bubble on an approximate logarithmic distance scale. The heliosphere—es, the solar wind termination shock (TS) and the heliopause (HP)—is gion that may be a transition between the Local Interstellar Cloud (LIC) ıutward trajectories and current positions of the most distant spacecraft, re also indicated. Also shown schematically are other interstellar clouds heliosphere around other stars).

insky, and E. Möbius, adapted from Fig. 1 of "From the Heliosphere to Ve Learned?" by E. Möbius, this volume

ımber: 2009925997

'-4

e-ISBN-978-1-4419-0247-4

ness Media, BV

AF813690

J.L. Li
R. von
Editors

From
to the

Compa
with Th

Previously pu
Issues 1–4, 20

Spring

J.L. Linsky
JILA, University of Colorado a
Boulder, CO, USA

V.V. Izmodenov
Department of Aeromechanics
Gas Dynamics
Lomonosov Moscow State Uni
Moscow, Russia

Cover illustration: The cover il
Local Bubble to the Loop I Supe
shown with its boundary structu
moving at ≈26 km/s through a r
and the so-called G-Cloud. The
the two Voyagers (V1 and V2),
and astrospheres (analogs of the

Illustration by K. Donahue, J. L
the Local Bubble—What Have

Library of Congress Control N

DOI: 10.1007/978-1-4419-024

ISBN-978-1-4419-0246-7

Printed on acid-free paper.

1

springer.com

Contents

Group Photograph

Front row, left to right:

Vladimir Baranov
Elena Provornikova
Burkhard Fuchs
Vladimir Florinski
Merav Opher
Hans-Reinhard Müller
André Balogh
Steve Snowden
George Gloeckler
Robin Sheldon
Johannes Geiss
Steve Spangler
Ed Stone
Jonathan Slavin
Nick Pogorelov
Dimitra Koutroumpa
Barry Welsh
Eric Quémerais
Brian Wood
Maciej Bzowski

Back row, left to right:

Eberhard Möbius
Vlad Izmodenov
Alex Lazarian
Ed Jenkins
Jeff Linsky
John Richardson
Dave McComas
Rosine Lallement
Seth Redfield
Rudolf von Steige
Randy Jokipii

Foreword

Jeffrey Linsky · Vlad Izmodenov · Eberhard Möbius

Originally published in the journal Space Science Reviews, Volume 143, Nos 1–4, 1–3.
DOI: 10.1007/s11214-008-9426-0 © Springer Science+Business Media B.V. 2008

For many years, studies of the outer heliosphere and the interstellar medium were two interesting research fields developing in near isolation from each other. Theoreticians knew that interstellar gas pressure provides the outer boundary condition for models of the solar wind, but little was known about the properties of interstellar gas either inside or outside of the heliosphere. There were no *in situ* plasma measurements of the distant solar wind to guide their studies of the outer heliosphere and beyond.

Since the 1980s, this situation has changed dramatically as a consequence of several developments: (i) measurements of pickup ions of the interstellar gas flowing through the heliosphere and direct observations of the neutral helium flow, (ii) first the prediction (in 1991) and then the detection (in 1996) of the hydrogen wall in the heliopause, (iii) the development of kinetic-fluid and multifluid theoretical models of the interaction of the solar wind with interstellar plasma including charge exchange, (iv) studies of local interstellar gas using ultraviolet absorption-line spectroscopy of nearby stars, and (v) plasma and magnetic-field measurements from Voyagers 1 and 2 as they crossed the termination shock. These developments have enriched both research fields enormously and demonstrated their inherent interrelations.

The primary drivers for this abrupt change in our understanding of the outer heliosphere and local interstellar medium (LISM) are the critical data provided by space missions. Launched in 1984, the Active Magnetospheric Particle Tracer Explorer Ion Release Module (AMPTE IRM) carried the SUprathermaL Energy Ionic Charge Analyzer (SULEICA) experiment that enabled the discovery of the interstellar pickup ions. Beginning in 1990 and still operating successfully, Ulysses carries the Solar Wind Ion Composition Experiment (SWICS) instrument that measures elemental charge states and composition in the solar wind and the pickup ions that result from the charge exchange of solar wind protons and ions with incoming neutrals. The Hubble Space Telescope (HST), also launched in 1990, initially carried the Goddard High Resolution Spectrometer (GHRS) and since 1997, the

J. Linsky (✉) · V. Izmodenov · E. Möbius
Boulder, CO 80309-0440, USA
e-mail: jlinsky@jila.colorado.edu

Space Telescope Imaging Spectrograph (STIS). Both instruments have measured absorption lines of hydrogen and other elements in the LISM. The Solar and Heliospheric Observatory (SoHO), launched in 1995 and still operating successfully, carries many instruments, including the Solar Wind Anisotropies (SWAN) instrument that measures backscattered-solar Lyman-α radiation that traces the flow of neutral hydrogen in the heliosphere. The Advanced Composition Explorer (ACE), launched in 1997, continues to provide information on solar wind composition and pickup ions. And, we must not forget the intrepid Voyager spacecraft, launched in 1977, that are providing unique information on the plasma and magnetic fields on both sides of the termination shock.

George Gloeckler, to whom this volume is dedicated on the occasion of his seventieth birthday, played a major role in the analysis of data from these space experiments and in our understanding of physical processes in the outer heliosphere and LISM.

The convenors of this ISSI workshop desired a format that would summarize these developments, identify and critically assess the important physical processes, and foster interdisciplinary research. Our specific objectives were to:

- Cross-fertilize the two scientific disciplines by bringing together active researchers with wide expertise in both fields in a setting that provides opportunities for detailed interactions in both the scientific sessions and in informal small groups.
- Discuss the rapidly advancing developments in both theory and observations, mainly due to space experiments.
- Exploit the rich archive of calibrated data and sophisticated theories tested by comparison with observations of nearby interacting plasmas to develop physical insights and models that could be applied to understanding plasma interactions elsewhere in the universe where the data are sparse and compromised by line of sight averages and the theories are less mature and tested.
- Stimulate modelers to test the accuracy of their various numerical codes by computing models with the same input parameters.
- Stimulate the development of broader perspectives.

With these objectives in mind, the convenors formulated a workshop program based on seven questions that the speakers were asked to address:

- What are the dominant physical processes in the termination shock and inner heliosheath?
- What are the three-dimensional shape and structure of the dynamic heliosphere?
- How are the interstellar plasmas and dust located inside and outside of the heliosphere related?
- What are the origin and physical properties of the very local ISM?
- What are the energy and pressure equilibria in the Local Bubble?
- What are the important physical processes in the multiphase interstellar medium located inside the Local Bubble?
- What are the roles that magnetic fields play in the outer heliosphere and Local Bubble?

During the course of the meeting, it became clear that magnetic fields play important roles in most of the phenomena discussed at the workshop. Therefore, the final question may be the most basic of the seven questions to answer.

To place these questions into a broader context, the convenors requested that several speakers give "big picture" talks on the time evolution of the heliosphere, the origin and evolution of the Local Bubble, the ISM beyond the Local Bubble, and challenges in modeling the heliospheric/ISM interface.

New data and more sophisticated theoretical models with predictive power raised new questions and identified those areas where new observations, both remote sensing and *in situ* measurements, are needed. For example, the new data from Voyager 1's crossing of the termination shock at 94 AU (on December 16, 2004) and Voyager 2's crossing at 84 AU (on August 30, 2007, conveniently just before the workshop), provided critical data for use in new MHD models for inferring the orientation and strength of the interstellar magnetic field near the Sun.

To place these new data in context, one often needs new terminology, but it is critical not to confuse the nonexperts and even the experts. For example, what is the best name for the partially ionized plasma that surrounds the heliosphere — now that there is evidence that the heliosphere is located just outside of the Local Interstellar Cloud. After much discussion, the participants decided on the term "circum-heliospheric interstellar medium" or CHISM and proposed that this term be used in the future.

Looking to the near future, the participants hoped for the successful launch and operation of the Interstellar Boundary Explorer (IBEX) satellite and the repair of the high-resolution ultraviolet spectrometers on HST/STIS. They also looked forward to the analysis of both new and archival data for use with 3D kinetic MHD models of the solar wind interaction with the CHISM. The rapid increase in the power of modern computers should facilitate calculations with such complex models.

Looking further into the future, the participants saw the need for a fully instrumented space probe that would measure the plasma and magnetic-field properties through the termination shock and heliopause into the bow shock region and beyond into the CHISM, which is unaltered by the Sun's influence.

We conclude by thanking all of those who made this workshop an extraordinary success. We thank the Directors of ISSI for their generous sponsorship and, in particular, Ruedi von Steiger for his advice in planning and editorial matters. We also thank the local organization team led by Brigitte Fasler for its professional work in planning and supporting the workshop. We thank all of the workshop participants for their excellent presentations, insightful discussions, and well-written papers that constitute the legacy of the workshop. Finally, we thank our summary speaker for his excellent summary and insightful comments on all that transpired.

Dedication to Dr. George Gloeckler

Eberhard Möbius

Originally published in the journal Space Science Reviews, Volume 143, Nos 1–4, 5–6.
DOI: 10.1007/s11214-008-9476-3

The ISSI Workshop "From the Heliosphere to the Local Bubble" was dedicated to Dr. George Gloeckler on the occasion of his 70th birthday. This workshop, which has brought together space scientists and astrophysicists to talk about the galactic neighborhood of our solar system, could not have been better suited for this purpose.

Starting his career with the study of energetic particles from the Sun, George Gloeckler has contributed to this emerging field with roots in both scientific communities through visionary instrument development, effective team building and strategic placement, meticulous data analysis and innovative ways to extract previously inaccessible information. Recognizing that interplanetary space was full of energetic particles and that their generation and acceleration happened at lower energies, he pushed instrument development to new horizons, by injecting novel instrument designs into space physics. He was one of the first to realize the potential of time-of-flight mass spectrometers for particle populations from the solar wind to low energy cosmic rays.

This quest included the energy range of pickup ions that are implanted into the solar wind upon their ionization, thus acquiring between zero and four times the solar wind energy. After having built a time-of-flight spectrometer within the long-standing Maryland-Garching collaboration forged between George Gloeckler and Dieter Hovestadt, the author was fortunate to discover the first interstellar pickup ions. Being convinced of their universal importance, George Gloeckler exploited the capabilities of the Ulysses SWICS instrument, the concept and design of which he was responsible for, to land coup after coup in revealing parameters of the Sun's immediate neighborhood in the galaxy. Here are just a few key examples: He used the unique properties of He^{2+} pickup ions that had been predicted by the late Daniel Rucinski to derive the most precise interstellar He density, without the need of absolute sensor calibration. Together with Johannes Geiss, he extracted the cosmologically important $^3He/^4He$ ratio for the local interstellar gas from pickup ions. In the heliophysical

E. Möbius (✉)
Space Science Center and Department of Physics, University of New Hampshire,
Morse Hall, 8 College Road, Durham, NH 03824, USA
e-mail: Eberhard.Moebius@unh.edu

DOI: 10.1007/978-1-4419-0247-4_2

triumvirate with Johannes Geiss and Lennard Fisk, he derived the density of interstellar H in the heliosphere and subsequently constrained the ionization state of the interstellar medium and the strength of the surrounding magnetic field. They also demonstrated that pickup ions lend themselves much more easily to further acceleration than ions of the solar wind proper. At this Workshop, George Gloeckler surprised the audience once more with a very interesting new result. He showed how the ubiquitous suprathermal tails as seen by the Voyagers can be used to connect the well measured abundance ratios of several interstellar species with the precisely determined interstellar He density to deduce improved densities for these species.

An account of George Gloeckler's contributions to the scientific community and scientific advancement would be rather incomplete without noting his essential involvement in important committees, his foresight with the development of new methods, definition of new missions, and tireless politicking at agencies to get things into motion. Without his effort to repeatedly propose and push for a small and innovative mission to study the interstellar medium from within the inner heliosphere, the implementation of the Interstellar Boundary Explorer (IBEX) that has just been launched on October 19, 2008, would be unthinkable.

At this juncture, space exploration has reached the point when we truly start to venture into interstellar space and turn our attention to the connection with the wider neighborhood of our solar system. Therefore, it couldn't be more appropriate to connect the name of George Gloeckler with the topical ISSI Workshop that brought together scientists from space physics and astrophysics to start a serious collaboration. His contributions in the field have pointed into this direction for many years. Yet, this workshop has started such an endeavor in earnest.

Let me add a few personal notes. In the late 1970s and early 1980s when I had just switched from laboratory plasma to space physics, working with Dieter Hovestadt, visits to George Gloeckler's University of Maryland group were like coming to a buzzing hub in space science and a place where it felt like home. The group had attracted young scientists, such as Glenn Mason, Doug Hamilton, and Fred Ipavich, who flourished in this nurturing environment and subsequently established important programs of their own. This almost self-running success can be attributed in large part to George's inviting, friendly, and supportive personality, which makes it a pleasure to spend time with him also beyond professional encounters. There are fond memories of a hike in the Austrian Alps during a weekend of the COSPAR meeting in Innsbruck, which shaped ideas in my early career, and of a refreshing walk through Muir Woods, during which some of the concepts that ultimately led to IBEX were discussed. Finally, I must confess, George also had his helping hand in my career when he dropped my name at the Institute for the Study of Earth, Oceans, and Space of the University of New Hampshire. As a consequence, I was asked and then decided to cross the Atlantic. Joined by Lynn Kistler, one of Fred Ipavich's students, and later by Toni Galvin, one of George Gloeckler's own students, we often jokingly referred to our space plasma composition effort as "Maryland North".

At this point I would like to speak for all workshop participants and for the scientific community in our field in conveying our congratulations. We wish George many more years of fruitful scientific work and many more important results in this field whose golden age has just begun.

Section I: What Are the Three-Dimensional Shape and Structure Of The Dynamic Heliosphere?

The Solar Wind in the Outer Heliosphere

J.D. Richardson · E.C. Stone

Originally published in the journal Space Science Reviews, Volume 143, Nos 1–4, 7–20.
DOI: 10.1007/s11214-008-9443-z

Abstract The solar wind evolves as it moves outward due to interactions with both itself and with the circum-heliospheric interstellar medium. The speed is, on average, constant out to 30 AU, then starts a slow decrease due to the pickup of interstellar neutrals. These neutrals reduce the solar wind speed by about 20% before the termination shock (TS). The pickup ions heat the thermal plasma so that the solar wind temperature increases outside 20–30 AU. Solar cycle effects are important; the solar wind pressure changes by a factor of 2 over a solar cycle and the structure of the solar wind is modified by interplanetary coronal mass ejections (ICMEs) near solar maximum. The first direct evidences of the TS were the observations of streaming energetic particles by both Voyagers 1 and 2 beginning about 2 years before their respective TS crossings. The second evidence was a slowdown in solar wind speed commencing 80 days before Voyager 2 crossed the TS. The TS was a weak, quasi-perpendicular shock which transferred the solar wind flow energy mainly to the pickup ions. The heliosheath has large fluctuations in the plasma and magnetic field on time scales of minutes to days.

Keywords Solar wind · Termination shock · Heliosheath · Heliopause · Pickup ions · Interstellar neutral atoms · Anomalous cosmic rays

1 Introduction

We give a brief introduction to the solar wind in the outer heliosphere (HS) and discuss the interaction of the solar wind with the interstellar medium. Figure 1 shows an overview of the heliosphere, the bubble blown into the circum-heliospheric interstellar medium (CHISM) by

J.D. Richardson (✉)
Kavli Institute for Astrophysics and Space Research, Massachusetts Institute of Technology,
77 Massachusetts Avenue, Cambridge, MA 02139, USA
e-mail: jdr@space.mit.edu

E.C. Stone
Space Radiation Laboratory, California Institute of Technology, 1200 E. California Blvd., Pasadena,
CA 91125, USA

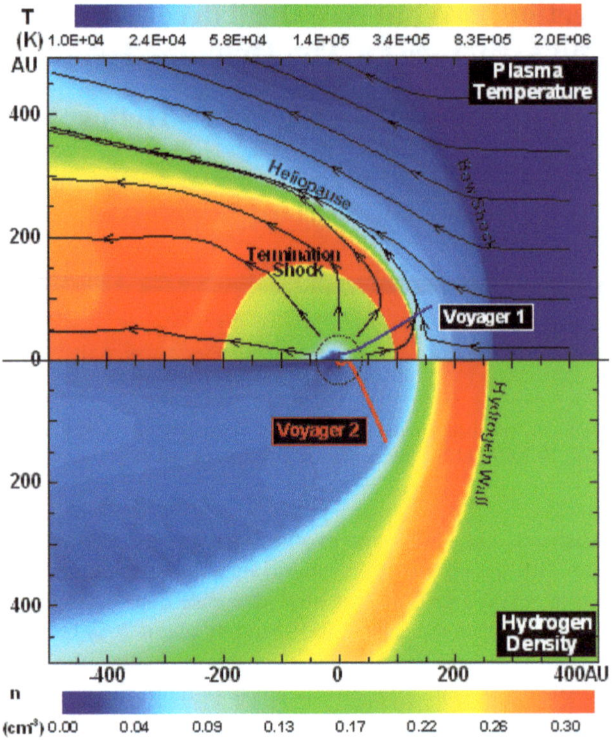

Fig. 1 A plot of the equatorial heliosphere from a plasma (*top*) and neutral (*bottom*) perspective. The *color bar* on the top panel shows the plasma temperature. The *lines* show the plasma flow. The main boundaries, the termination shock, heliopause, and bow shock are labeled. The *color bar* on the *bottom panel* shows the H density; the hydrogen wall in front of the heliopause is labeled and the trajectories of the Voyager spacecraft are shown. Figure courtesy of H. Müller

the solar wind which contains the Sun. Parker (1958) developed a theory for a supersonic solar wind at the dawn of the space age. His hypothesis was verified when the solar wind was first detected in the early 1960s (Gringauz 1961; Neugebauer and Snyder 1962).

The solar wind is seen in Fig. 1 flowing radially outward from the Sun. A corollary to Parker's theory is that the solar wind goes through a transition from supersonic to subsonic flow at the termination shock, where the solar wind senses the CHISM and deflects down the heliospheric tail (Weymann 1960). Figure 1 shows the CHISM moving to the left relative to the Sun. The size of the heliospheric bubble in the CHISM is determined by where the solar wind pressure is equal to the CHISM pressure (Parker 1963). The boundary between these two plasmas is called the heliopause (HP) and is analogous to the magnetopauses of Earth and other planets. The CHISM flow in this model is also supersonic (we don't know if the real CHISM is super- or sub-sonic), therefore it also goes through a shock so that the flow can divert around the heliosphere. The shock in the CHISM is called the bow shock. The region of shocked CHISM material which flows around the HS is called the outer heliosheath. The region of shocked solar wind where the solar wind diverts down the HS tail is called the inner heliosheath or commonly just the heliosheath.

The neutrals in the CHISM are not affected by the magnetic fields and flow into the heliosphere. Neutral He has few interactions with the plasma, so essentially pristine CHISM He flows into the inner HS where it can be directly measured by spacecraft (Witte 2004). The CHISM H interacts with the plasma both in the heliosheath (HSH) and in the solar wind via charge exchange. A proton gains an electron from a neutral; the neutral formed from the proton has a speed equal to the plasma speed. The newly created ion, called a pickup ion, is accelerated to the plasma speed and has an initial gyro-energy equal to the plasma flow

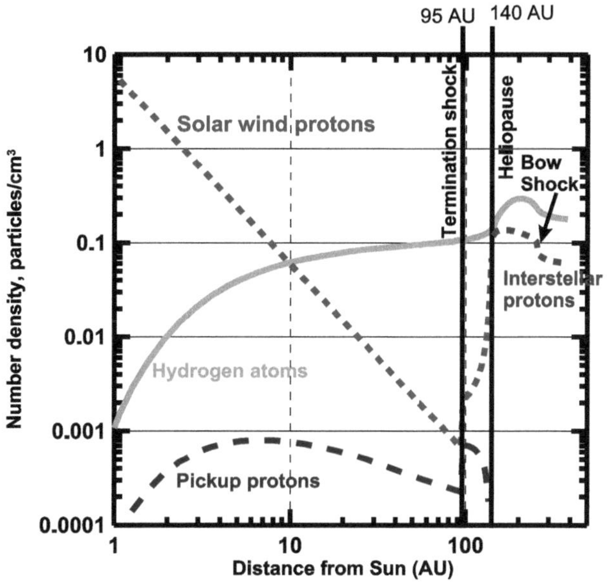

Fig. 2 The density of the plasma and neutral components from 1 to 1000 AU. The solar wind ions come from the Sun. The pickup ions are interstellar neutrals which have been ionized in the solar wind. The densities of the solar wind and of the pickup ions jump at the termination shock. Outside the heliopause, the ions are part of the CHISM. Both the ion and neutral density increase in front of the heliopause (in the hydrogen wall). The interstellar neutrals dominate the mass density outside 10 AU. Figure courtesy of V. Izmodenov

energy (about 1 keV in the solar wind). The energy for this acceleration comes from the plasma flow energy, so the plasma slows down (Richardson et al. 2008b).

One effect of charge exchange in the outer HSH is the formation of the hydrogen wall shown in Fig. 1. The CHISM plasma slows down as it approaches the heliopause. The CHISM H is coupled to this plasma via charge exchange, so it also slows down and thus has a higher density (Baranov and Malama 1993; Baranov and Malama 1995; Zank et al. 1996). This dense region upstream of the nose of the HS is called the hydrogen wall; similar walls are observed at other astrospheres (Linsky and Wood 1996).

Figure 2 shows the densities of ion and neutral populations in the heliosphere. The CHISM ion density increases at the bow shock where the flow speed decreases, but these ions do not enter the HS. The neutral density also increases at the bow shock, forming the hydrogen wall. The CHISM neutrals flow into the HS and are the HS population with the highest density outside 10 AU. The solar wind density decreases as R^{-2} out to the TS, increases at the TS, and again increases approaching the HP. The pickup ions, the ionized interstellar neutrals, make up an increasingly large fraction of the solar wind with distance and comprise about 20–30% of the solar wind at the TS (Richardson et al. 2008b). Since the pickup ions are hot, they dominate the thermal ion pressure outside 30 AU. Thus the CHISM not only stops the solar wind and diverts it down the tail, but also penetrates deep into the HS. This chapter discusses the outward flow of the solar wind and how it is affected by the CHISM.

2 Solar Wind Evolution

The slow solar wind reaches an asymptotic speed of about 400 km/s and, to first order, maintains that speed until the TS. Figure 3 shows 101-day averages of the solar wind speed, density, temperature and dynamic pressure at V2. The top panel also shows 101-day averages of the solar wind speed at 1 AU. Near the Sun (out to 30 AU) the speeds at Earth and

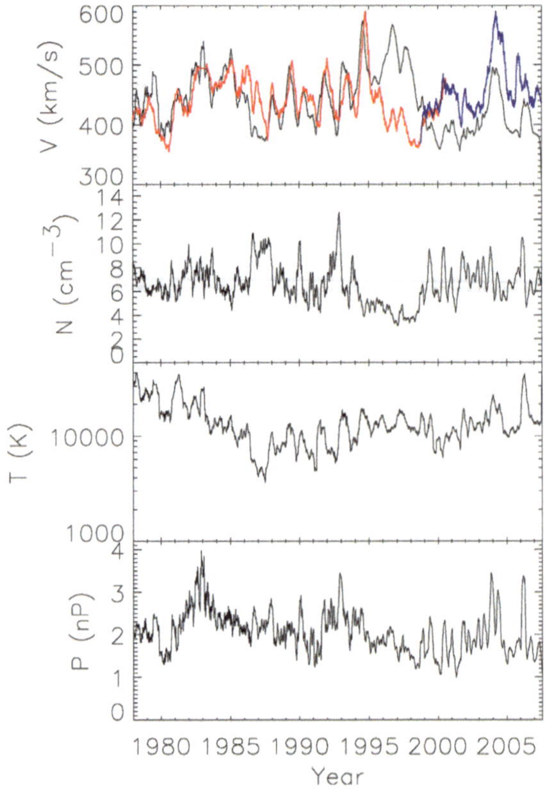

Fig. 3 Running 101-day averages of the solar wind speed, density, temperature and dynamic pressure observed by Voyager 2. The *top panel* also shows speeds at 1 AU from IMP 8 (*red*) and ACE (*blue*)

those at V2 are very similar. The solar wind parameters have a lot of variation, but to first order the speed is constant, the density decreases as R^2, and the temperature decreases out to 20–25 AU and then increases.

The solar wind changes over a solar cycle (Lazarus and McNutt 1990). The dynamic pressure, which determines the distance to the TS and HP, is least near solar maximum, increases for 2–3 years after solar maximum, then decreases to the next solar maximum (Richardson and Wang 1999). At solar maximum, the solar wind is slow and dense at all heliolatitudes. At solar minimum, the solar wind is slow and dense near the equator but fast and tenuous near the poles, with a transition region near 20–30° heliolatitude. This gradient in speed with heliolatitude at solar minimum causes the difference in solar wind speeds at Earth and V2 in 1986–87 and 1995–97. In 1986–87, V2 was at a lower average heliolatitude than Earth and observed lower speeds whereas from 1995–97 V2 was at a higher heliolatitude than Earth and observed much higher speeds. Variations also exist from solar cycle to solar cycle; the current solar minimum has a smaller solar wind dynamic pressure than in the previous two cycles (McComas et al. 2008). Other shorter scale features are also observed, such as the speed variations with a 1.3-year period observed from 1987–1998 (Richardson et al. 1995). This variation in speed was observed throughout the heliosphere and has been an occasional feature observed in historic solar wind data (Gazis et al. 1995; Szabo et al. 1995). A similar period has been observed in convection patterns in the Sun and may be related (Howe et al. 2000).

Solar activity varies over a solar cycle, with many more ICMEs at solar maximum than at solar minimum (Cane and Richardson 2003). As the ICMEs move outward, they expand until they reach 10–15 AU (Richardson et al. 2006); during solar maximum as much as 40% of the SW observed by V2 is from ICMEs. At times of high solar activity the Sun sometimes emits a series of ICMEs over time periods of days to months. The latter ICMEs catch up to earlier ICMEs and merge, compressing the solar wind ahead of them to form regions of high magnetic field and (often) density called merged interaction regions (MIRs) (Burlaga et al. 1984; Burlaga 1995; Richardson et al. 2002). Near solar maximum these structures dominate the solar wind profile; from 2001–2005 the MIRs observed in the outer solar system evolved so that the magnetic field, speed, density, and dynamic pressure were all correlated, resulting in large pressure pulses which reached the TS roughly twice each year and pushed it outward (Richardson et al. 2003).

3 Effects of the Neutral CHISM on the Solar Wind

The first effects of the CHISM on the solar wind are those of the neutral H and He which penetrate into the heliosphere. The solar wind contains stationary structures called pressure-balanced structures across which the pressure (thermal plus magnetic) is constant. By 20 AU, the observed structures could only be in balance if the plasma had a hot pickup ion component (Burlaga et al. 1994). The next observed effect of the CHISM neutrals was an increase in the thermal proton temperature starting between 20 and 30 AU, which overwhelmed the adiabatic cooling which would occur without a heat source. The pickup ions are the heat source; they are formed with a ring distribution (the particle trajectories are perpendicular to the field). These distributions are unstable and generate magnetic fluctuations which isotropize the distributions. The waves transfer a small amount (4%) of their to the thermal protons, which is enough to heat them as observed (Smith et al. 2006; Isenberg et al. 2005).

The slowdown of the solar wind due to pickup of CHISM neutrals increases with distance and was first reported near 30 AU (Richardson et al. 1995), although the slowdown was not observed in Pioneer 11 data at similar distances (Gazis 1995). By the time V2 neared the TS in 2007, the solar wind speed was about 83% of its speed at Earth, so the pickup ions made up about 19% of the total solar wind density (Richardson et al. 2008b). We note that the 17% decrease in speed represents a roughly 30% decrease in flow energy. Thus the neutrals from the interstellar medium acquire a substantial fraction of the solar wind energy well before the TS crossing.

4 The Termination Shock

The first direct sign of the approaching TS was a fairly sudden increase in tens of keV to MeV ions and electrons in mid-2002 observed at V1 but not at V2, which was 18 AU closer to the Sun (Krimigis et al. 2003; McDonald et al. 2003). Figure 4 shows that these ions streamed along the magnetic field lines in the outward direction, as if they were generated at the Sun. Although the approach of the TS seemed a likely source for these particles, the outward direction was initially confusing. An MIR passed V1 and V2 in early 2003, at which time the V1 particle fluxes decreased. The increased dynamic pressure associated with the MIR pushed the TS outward so the field lines at V1 became disconnected from the TS. The MeV particles were again observed starting in mid-2004, but the keV ions and electrons did not return until the start of 2004. This difference in particle energies suggests the connection to the TS was more distant, so that only the more energetic particles had time to make it to V1 before they were convected back to the TS with the solar wind flow.

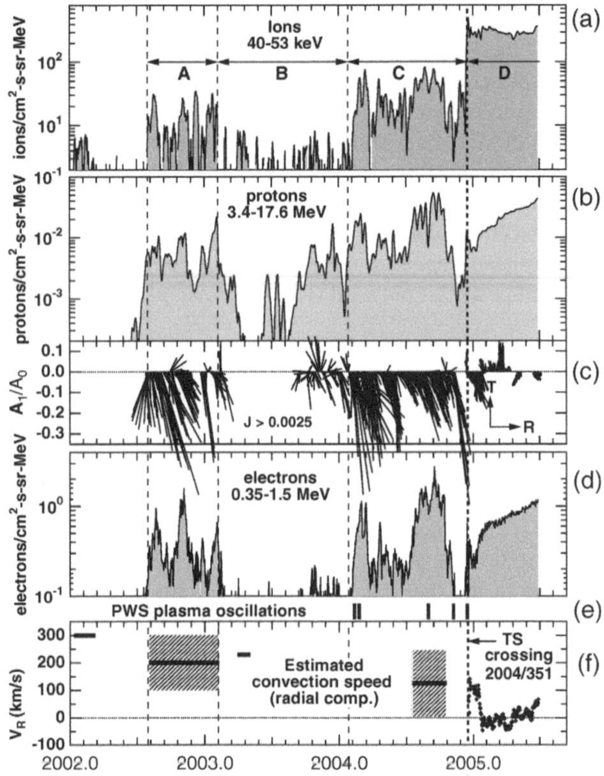

Fig. 4 Panels *1*, *2*, and *4* show 5-day smoothed V1 40–53 ion, 3.4–17.6 MeV proton, and 0.35–1.5 MeV electron intensities. Panel *3* shows the direction that particles are traveling. The *bottom panel* shows speeds estimated from the low-energy ions (since the plasma instrument is not working)

Another decrease, again likely due to an MIR, was observed in late 2004. Throughout this time period the particle intensities were highly variable and highly anisotropic (moving along the magnetic field away from the Sun). During a data gap on day 316, 2004, at 94 AU, V1 crossed the termination shock. The particle intensities jumped, became steady and isotropic, and the magnetic field magnitude increased (Burlaga et al. 2005; Decker et al. 2005; Stone et al. 2005).

The V1 TS crossing revealed the scale size of the heliosphere, roughly 90 AU for the TS and (based on models) 120–140 AU for the HP. Since V1 crossed the TS in a data gap, the TS strength (the ratio of downstream to upstream density and magnetic field, which is 4 for a strong shock) could only be estimated and was of order 2–3 (Burlaga et al. 2005; Stone et al. 2005). The HSH speeds derived from the particle data were very low just after the TS, −50 to 50 km/s (Decker et al. 2005), consistent with an inward-moving TS (Jokipii 2005). The keV particles seem to have their source at the TS, with a peak just outside the TS. However, the anomalous cosmic rays (ACRs), which were thought to be accelerated at the TS (Pesses et al. 1981), had intensities at the TS less than the peak values observed upstream in the solar wind and the flux of ACRs continued to increase after the TS crossing (Decker et al. 2005; Stone et al. 2005).

Let us return to the foreshock particles which were streaming the wrong way. They were telling us that the TS is blunt, with a larger radius of curvature than the Parker spiral field lines (Jokipii et al. 2004; Stone 2004). Figure 5 shows one such field line which has moved outward from the Sun. If the TS were blunt, the field line would first intersect the TS near where the TS is closest to the Sun, but would be in the solar wind on both sides. V1, as shown in the figure, would then see particles flowing from the TS in the direction outward

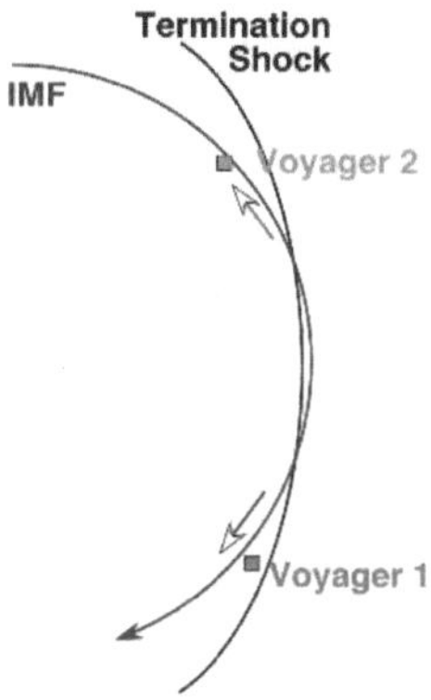

Fig. 5 Schematic diagram of a circular magnetic field line intersecting a blunt TS in two place. The Voyager spacecraft are on opposite sides of the nose of the heliosphere and see streaming in opposite directions as suggested by this figure

along the field line because this field line is connected to the TS. This hypothesis predicted that V2, on the other side of the closest point of the TS, would see particles streaming in the opposite direction. V2 entered the TS foreshock region in late 2004 and particles were streaming in the opposite sense (sunward) from those at V1, consistent with the blunt shock hypothesis (Decker et al. 2006).

V2 entered the foreshock region at 75 AU, about the same time V1 crossed the TS in late 2004. Since V1 entered this region at 85 AU, 10 AU further out, either the HS was asymmetric or the foreshock was much thicker in the V2 direction than the V1 direction. Models do suggest the foreshock is thicker at the V2 location because of the geometry of the crossing locations (Opher and Stone 2006), but only by a few AU. Models also show that, if the CHISM magnetic field were tilted from the CHISM flow direction, the heliosphere would be asymmetric (Linde et al. 1998; Ratkiewicz et al. 1998). Similar asymmetries are observed in Earth's magnetosphere (Paularena et al. 2001; Dmitriev et al. 2004) and in ICME models (Siscoe et al. 2007). The difficulty for global heliospheric models is to determine the actual CHISM field direction. Observations of a difference between flow directions of H and He coming into the HS provide constraints on this direction (Lallement et al. 2005; Izmodenov et al. 2005). Use of these field direction constraints in models gives HS asymmetries with the TS and HP closer in the south than the north, although the amount of the asymmetry varies between models (Opher et al. 2007; Pogorelov et al. 2007). V2 crossed the TS on day 242 of 2007 at 84 AU; calculations of the TS motion based on V2 solar wind data upstream of the TS and a 2-D model indicate the TS moved inward 2–3 AU between the V1 and V2 TS crossing, giving an asymmetry of 7–8 AU in the TS locations in the V1 and V2 directions (Burlaga et al. 2008; Decker et al. 2008; Richardson et al. 2008a; Stone et al. 2008).

Figure 6 shows daily averages of the plasma and magnetic field parameters before the TS crossing. The TS crossing is very obvious in the plasma data, marked by a sharp decrease in speed, an increase in temperature, and a change in flow angle. The density and magnetic field strength both increase, but these increases are comparable to increases observed in the solar wind. At the TS the solar wind begins to turn into a flow down the heliotail. The directions are as expected for this deflection, in the T and $-N$ directions (the RTN coordinate system has R radially outward, T parallel to the plane of the solar equator and positive in the direction of solar rotation, and N completes a right-handed system).

The solar wind begins to slow at about day 160, eighty days before the TS crossing, when a step-like decrease was observed. Two more downward speed steps were observed at days 190 and day 232, reducing the speed to near 300 km/s just before the TS. SW speeds of 300 km/s are sometimes observed near 1 AU, but V2 had not observed speeds this low since

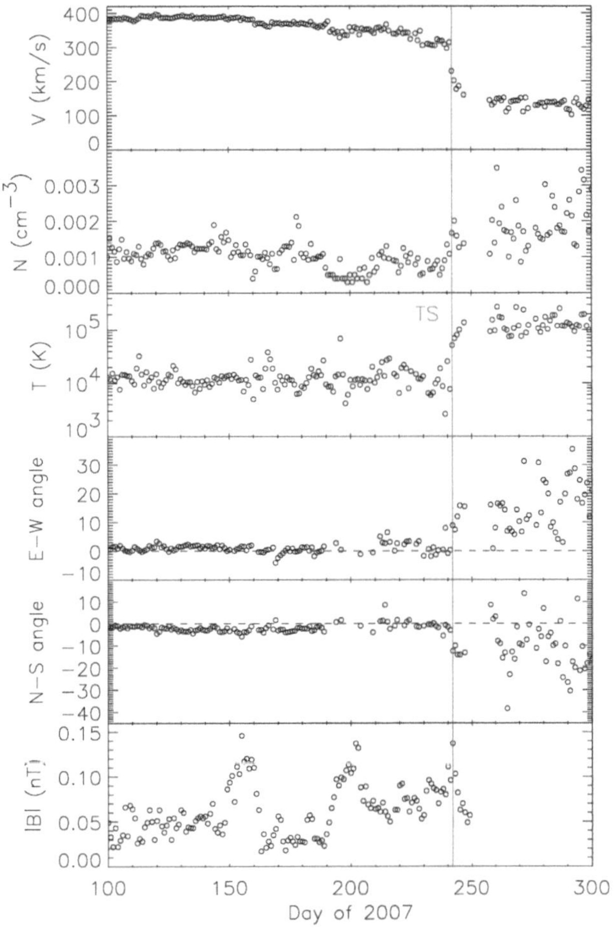

Fig. 6 Daily averages of the plasma speed, density, temperature, flow angles, and magnetic field magnitudes near the TS

1978, which suggests that these decreases are associated with the TS. The speed decrease from 380 to 300 km/s corresponds to a loss of about 40% of the solar wind flow energy before the TS. The first two speed decreases are associated with large increases in B. We do not yet understand the physics behind these speed decreases.

Figures 7 and 8 show the three crossings of the TS which occurred while V2 was being tracked. The first and fifth crossings occurred in data gaps. The third and fourth crossings at 244.0 and 244.11 have the classic foot ramp structure of a quasi-perpendicular, supercritical shock (Burlaga et al. 2008). At the foot, the magnetic field increases and the speed decreases; this region formed by ions which reflect from the shock. At the ramp the speed decreases and the temperature, density and field all increase. The second TS crossing at 243.84, a few hours earlier, looks very different. The speed steadily increases over about 30 minutes. The density does not have a clear increase but the temperature does jump. The magnetic field decreases across the foot region and two increases in the field were observed, both of which look like ramps. These data may show the TS in the process of reforming with the ramp moving upstream by an ion gyroradius (Burlaga et al. 2008).

Another surprise was the low thermal proton temperature in the heliosheath. Magnetosheaths of the outer planets all have proton temperatures of a few million degrees K and

Fig. 7 The plasma speed, density, thermal speed, and magnetic field magnitudes near the TS

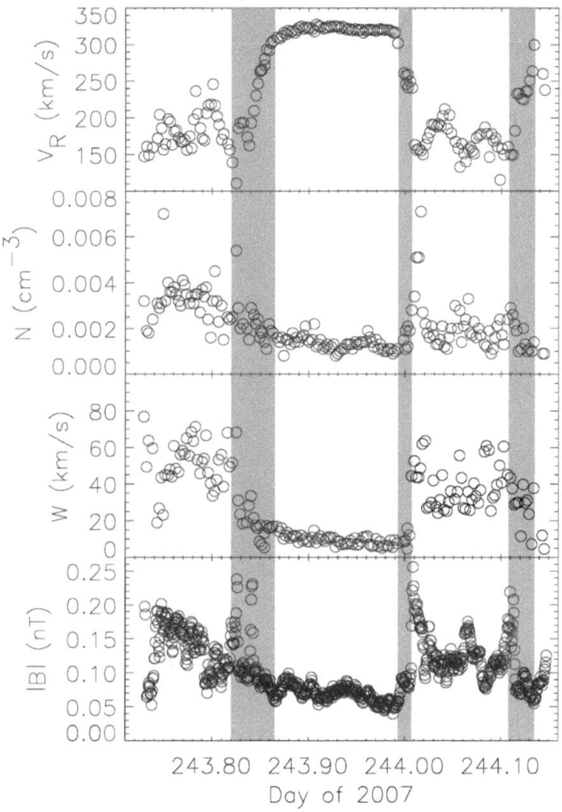

electron temperatures which are a factor of roughly 2 hotter. Models also predicted the HSH temperature would be a few million degrees. The models which predict these temperatures assume that almost all the SW flow energy goes into heating the thermal plasma, which is the case at the planetary magnetospheres. The observed HSH temperatures are about an order of magnitude less than these predictions, a few hundred thousand degrees K for protons (Richardson et al. 2008a). Electron temperatures are below the 10 eV instrument threshold; occasionally tails of the electron distributions are observed which allow us to estimate that the electron temperature is 3–4 hundred thousand degrees K.

What happens to the flow energy? We discussed above that about 40% of the flow energy is lost before the TS, probably heating energetic particles. The drop in speed at the TS is not a factor of 4 as at the planetary bow shocks but a factor of 2, so less energy is available for heating. But the major difference seems to be that most of the energy at the TS is going not to the thermal plasma but to the pickup ions. Based on V1 observations, Gloeckler et al. (2005) inferred that 80% of the flow energy transfered at the TS went into the pickup ions, which is consistent with the V2 plasma observations.

In addition to the average energy of the ions being low, some spectra look like they are not heated at all by the TS. Figure 9 shows examples of two ion spectra from the heliosheath which were fit to convected isotropic Maxwellian distributions. The first has a temperature of 22,000 K and the second a temperature of 228,000 K. The density and magnetic field go up by a factor of two at the shock, so the temperature should increase by a factor of two by conservation of the first adiabatic invariant. As shown in Fig. 9, some spectra in

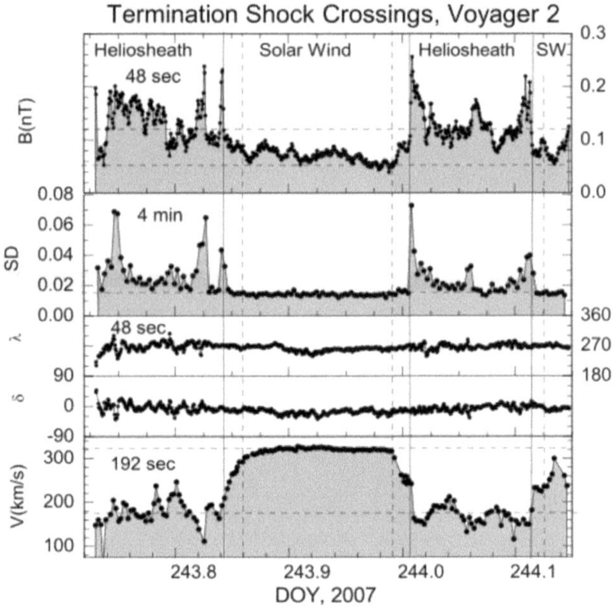

Fig. 8 The magnetic field magnitude, standard deviation of the magnetic field over 4-minute intervals, the azimuthal field angle, the elevations angle, and the plasma speed

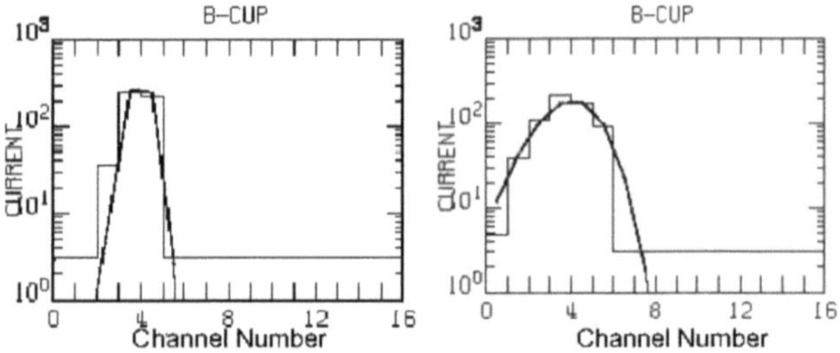

Fig. 9 Two-ion spectra from the heliosheath. Best fits of convected isotropic Maxwellian distributions to the histogram of observations of current versus energy are shown by the *curves*. For the *left* spectrum the temperature is 22,000 K and for the *right* spectrum it is 228,000 K

the HSH have temperatures of only 20,000 K, twice the average solar wind temperature. These protons were essentially unheated at the shock, they gained only the energy from compression of the plasma as it slowed at the shock. These cold spectra are seen sporadically throughout the heliosheath data encountered through August 2008. They could be plasma which has passed through the TS when it is in the process of reforming, or could have passed through the shock at a time when only hotter ions were effected by the shock (Zank et al. 1996). The TS is strongly time dependent based on the variations observed in the HSH.

Since data are available across the TS, the Rankine-Hugoniot equations can be solved to find the shock speed and normal (Richardson et al. 2008a). The shock normal angle is consistent with zero in the *N* direction but slightly off-radial in the *T* direction suggesting the shock has a smaller radius of curvature than a circle, opposite of the blunt shock hypothesized above. But this measurement is for one crossing of a very dynamic surface and so is not in real conflict with having a blunt TS. The TS was quasi-perpendicular and had compression ratios of 2.2–2.5 at TS-2 and 1.0–2.3 at TS-3. The shock moved at speeds of 50–100 km/s, similar to the speeds of planetary bow shocks. The upstream Mach numbers are 4.9 and 8.8 and the downstream Mach numbers are 1.1 and 2.8. These downstream Mach numbers highlight the lack of heating of the thermal plasma; this component of the plasma remains supersonic downstream of the TS. Thus the waves that convey information

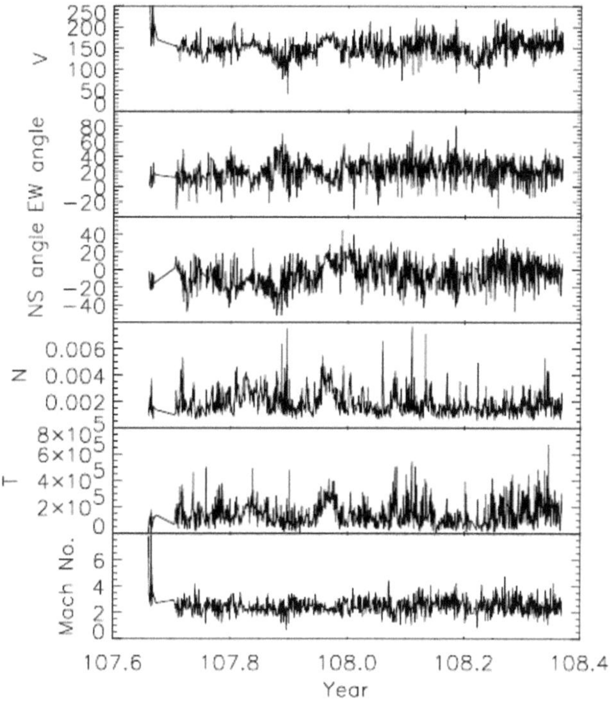

Fig. 10 The plasma speed, EW flow angle, NS flow angle, density, temperature,and Mach number of the thermal plasma

about the interstellar plasma upstream must propagate at speeds determined by the pickup ion sonic speed.

Figure 10 shows the plasma properties in the HSH. The average velocity components are 140 km/s in the radial direction, 47 km/s in the T direction, and -15 km/s in the $-N$ direction. The average speed is 144 km/s, density is 0.002 cm^{-3}, and $T = 144{,}000$ K. All the parameters are highly variable. These variations probably arise both from fluctuations in the shock speed, normal, and structure and from changes in the upstream solar wind. One feature which may be a solar wind feature is the increase in speed, density, and temperature beginning at about day 350 and lasting roughly 15 days. This increase is qualitatively similar to those observed in MIRs in the solar wind and is associated with a decrease in energetic particle fluxes, also analogous to a solar wind MIR. The N component of the speed changes from southward to northward at this time and remained northward for almost 40 days. This direction change suggests that a transient in the south pushed the flow ahead of it northward as it expanded.

5 Anomalous Cosmic Rays

Before the TS encounters, the TS was thought to be the source of anomalous cosmic rays (ACRs), singly ionized particles with energies of 20–100 MeV. These source of the ACRs are neutrals from the CHISM which are ionized, form pickup ions, and then are accelerated to high energy. The TS seemed a likely place for this acceleration to occur and pre-crossing expectations were that ACR intensities would peak at the TS and that the spectra would be power laws. The data from the TS in Fig. 4 shows that the low energy, tens of keV particle intensities peak at the TS but that the MeV particle intensities do not. Inspection of the particle spectra shows that the ACRs are modulated at the TS at both V1 and V2, so the ACRs are not accelerated where these spacecraft crossed the TS (Stone et al. 2005; Stone et al. 2008). Subsequent suggestions have been that the acceleration occurs in the flanks of the heliosphere (McComas and Schwadron 2008; Schwadron et al. 2008) or further out in the heliosheath (Fisk 2005).

6 Summary

The Voyager spacecraft are providing in situ observations of the interaction between the solar wind and the interstellar medium. The interstellar neutrals penetrate deep into the heliosphere and slow and heat the solar wind, removing about 35% of the solar wind energy before the TS. Upstream of the TS, a foreshock region of field-aligned streaming particles was observed. Ahead of the TS the solar wind slowed substantially starting 0.7 AU before the TS crossing. At the TS, little heating of the thermal plasma occurred with most of the energy going into the pickup ions. The HSH is a highly variable region due probably both to changes in the upstream solar wind and to motion of the TS.

Acknowledgements This work was supported by NASA; at MIT by NASA contract 959203 from JPL to MIT and NASA grant NAG5-8947.

References

V.B. Baranov, Y.G. Malama, Model of the solar wind interaction with the local interstellar medium–numerical solution of self-consistent problem. J. Geophys. Res. **98**, 15157–15163 (1993)

V.B. Baranov, Y.G. Malama, Effect of local interstellar medium hydrogen fractional ionization on the distant solar wind and interface region. J. Geophys. Res. **100**, 14755–14762 (1995)

L.F. Burlaga, F.B. McDonald, N.F. Ness, R. Schwenn, A.J. Lazarus, F. Mariani, Interplanetary flow systems associated with cosmic ray modulation in 1977–1980. J. Geophys. Res. **89**, 6579–6588 (1984)

L.F. Burlaga, N.F. Ness, J.W. Belcher, A. Szabo, P.A. Isenberg, M.A. Lee, Pickup protons and pressure-balanced structures: Voyager 2 observations in merged interaction regions near 35 AU. J. Geophys. Res. **99**, 21,511 (1994)

L.F. Burlaga, *Interplanetary Magnetohydrodynamics* (Oxford Univ. Press, London, 1995)

L.F. Burlaga, N.F. Ness, M.H. Acuna, R.P. Lepping, J.E.P. Connerney, E.C. Stone, F.B. McDonald, Crossing the termination shock into the heliosheath, magnetic fields. Science **309**, 2027–2029 (2005)

L.F. Burlaga, N.F. Ness, M.H. Acuna, R.P. Lepping, J.E.P. Connerney, J.D. Richardson, Observations of magnetic fields at the termination shock by Voyager 2. Nature **454**, 75–77 (2008)

H.V. Cane, I.G. Richardson, Interplanetary coronal mass ejections in the near-Earth solar wind during 1996–2002. J. Geophys. Res. **108**, 1156–1168 (2003). doi:10.1029/2002JA009817

R.B. Decker, S.M. Krimigis, E.C. Roelof, M.E. Hill, T.P. Armstrong, G. Gloeckler, D.C. Hamilton, L.J. Lanzerotti, Voyager 1 in the foreshock, termination shock, and heliosheath. Science **309**, 2020–2024 (2005). doi:10.1126/science.1117569

R.B. Decker, S.M. Krimigis, E.C. Roelof, M.E. Hill, Low-energy ions near the termination shock. In *Physics of the Inner Heliosheath: Voyager Observations, Theory, and Future Prospects*, AIP Conference Proceedings, vol. 258 (2006), pp. 73–78

R.B. Decker et al., Shock that terminates the solar wind is mediated by non-thermal ions. Nature **454**, 67–70 (2008)

A.V. Dmitriev, A.V. Suvorova, J.K. Chao, Y.-H. Yang, Dawn-dusk asymmetry of geosynchronous magnetopause crossings. J. Geophys. Res. **109**, A05203 (2004)

L.A. Fisk, Journey into the unknown beyond. Science **309**, 2016–2017 (2005)

P.R. Gazis, Limits on deceleration and asymmetry of solar wind speed. Geophys. Res. Lett. **22**, 2441–2444 (1995)

P.R. Gazis, J.D. Richardson, K.I. Paularena, Long term periodicity in solar wind velocity during the last three solar cycles. Geophys. Res. Lett. **22**, 1165–1168 (1995)

G. Gloeckler, L.A. Fisk, L.J. Lanzerotti, Acceleration of solar wind and pickup ions by shocks, in *Solar Wind 11/SOHO 16 Programme and Abstract Book (pdf file)*, European Space Agency (2005), p. 52

K.I. Gringauz, Some results of experiments in interplanetary space by means of charged particle traps on Soviet space probes. Space Res. **2**, 539–553 (1961)

R. Howe, J. Christensen-Dalsgaard, F. Hill, R.W. Komm, R.M. Larsen, J. Schou, M.J. Thompson, J. Toomre, Dynamic variations at the base of the solar convection zone. Science **287**, 2456–2460 (2000)

P.A. Isenberg, C.W. Smith, W.H. Matthaeus, J.D. Richardson, Turbulent heating of the distant solar wind by interstellar pickup protons with a variable solar wind speed, in *Proceedings of Solar Wind 11: Connecting Sun and Heliosphere*, ed. by B. Fleck, T.H. Zurbuchen. ESA SP-592 (European Space Agency, The Netherlands, 2005), pp. 347–350

V.V. Izmodenov, Y.G. Malama, M. Ruderman, Solar cycle influence on the interaction of the solar wind with Local Interstellar Cloud. Astron. Astrophys. **429**, 1069–1080 (2005). doi:10.1051/004-6361:20041348

J.R. Jokipii, J. Giacalone, J. Kota, Transverse streaming anisotropies of charged particles accelerated at the solar wind termination shock. Astrophys. J. **611**, L141–L144 (2004)

J.R. Jokipii, The magnetic field structure in the heliosheath. Astrophys. J. **631**, L163–L165 (2005)

S.M. Krimigis, R.B. Decker, M.E. Hill, T.P. Armstrong, G. Gloeckler, D.C. Hamilton, L.J. Lanzerotti, E.C. Roelof, Voyager 1 exited the solar wind at a distance of 85 AU from the Sun. Nature **426**, 45–48 (2003). doi:10.1038/nature02068

R. Lallement, E. Quemerais, J.L. Bertaux, S. Ferron, D. Koutroumpa, R. Pellinen, Deflection of the interstellar neutral hydrogen flow across the heliospheric interface. Science **307**, 1447–1449 (2005). doi:10.1126/science.1107953

A.J. Lazarus, R.L. McNutt, Jr., Plasma observations in the distant heliosphere: A view from Voyager, in *Physics of the Outer Heliosphere*, ed. by S. Grzedzielski, D.E. Page (Pergamon, New York, 1990), pp. 229–234

T.J. Linde, T.I. Gombosi, P.L. Roe, K.G. Powell, D.L. DeZeeuw, Heliosphere in the magnetized local interstellar medium: Results of a three-dimensional MHD simulation. J. Geophys. Res. **103**, 1889–1904 (1998)

J.L. Linsky, B.E. Wood, The Alpha Centauri line of sight: D/H ratio, physical properties of local interstellar gas, and measurement of heated hydrogen (the 'hydrogen wall') near the heliopause. Astrophys. J. **463**, 254–270 (1996)

D.J. McComas, N.A. Schwadron, An explanation of the Voyager paradox: particle acceleration at a blunt termination shock. Geophys. Res. Lett. **33**, L04102 (2006). doi:10.1029/2005GL025437

D.J. McComas et al., Weaker solar wind from the polar coronal holes and the whole Sun. Geophys. Res. Lett. **35**, L18103 (2008)

F.B. McDonald et al., Enhancements of energetic particles near the heliospheric termination shock. Nature **426**, 48–51 (2003)

M. Neugebauer, C.W. Snyder, Mariner 2 observations of the solar wind, 1, average properties. J. Geophys. Res. **71**, 4469–4484 (1962)

M. Opher, E.C. Stone, P.C. Liewer, The effects of a local interstellar magnetic field on Voyager 1 and 2 observations. Astrophys. J. **640**, L71–L74 (2006)

M. Opher, E.C. Stone, T.I. Gombosi, The orientation of the local interstellar magnetic field. Science **316**, 875–878 (2007). doi:10.1126/science.1139480

E.N. Parker, Dynamics of the interplanetary gas and magnetic fields. Astrophys. J. **128**, 664–676 (1958)

E.N. Parker, *Interplanetary Dynamical Processes* (Interscience, New York, 1963)

K.I. Paularena, J.D. Richardson, M.A. Kolpak, C.R. Jackson, G.L. Siscoe, A dawn-dusk density asymmetry in Earth's magnetosheath. J. Geophys. Res. **106**, 25377–25394 (2001)

M.E. Pesses, J.R. Jokipii, D. Eichler, Cosmic ray drift, shock wave acceleration, and the anomalous component of cosmic rays. Astrophys. J. **246**, L85 (1981)

N.V. Pogorelov, E.C. Stone, V. Florinski, G.P. Zank, Termination shock asymmetries as seen by the Voyager spacecraft: The role of the interstellar magnetic field and neutral hydrogen. Astrophys. J. **668**, 624 (2007)

R. Ratkiewicz, A. Barnes, G.A. Molvik, J.R. Spreiter, S.S. Stahara, M. Vinokur, S. Venkateswaran, Effect of varying strength and orientation of local interstellar magnetic field on configuration of exterior heliosphere: 3D MHD simulations. Astron. Astrophys. **335**, 363–369 (1998)

J.D. Richardson, C. Wang, The global nature of solar cycle variations of the solar wind dynamic pressure. Geophys. Res. Lett. **26**, 561–564 (1999)

J.D. Richardson, K.I. Paularena, A.J. Lazarus, J.W. Belcher, Evidence for a solar wind slowdown in the outer heliosphere? Geophys. Res. Lett. **22**, 1469–1472 (1995)

J.D. Richardson, K.I. Paularena, C. Wang, L.F. Burlaga, The life of a CME and the development of a MIR: From the Sun to 58 A. J. Geophys. Res. **106** (2002). doi:10.1029/2001JA000175

J.D. Richardson, C. Wang, L.F. Burlaga, Correlated solar wind speed, density, and magnetic field changes at Voyager 2. Geophys. Res. Lett. **30**, 2207–2210 (2003). doi:10.1029/2003GL018253

J.D. Richardson, Y. Liu, C. Wang, L.F. Burlaga, ICMES at very large distances. Adv. Space Res. **38**, 528–534 (2006)

J.D. Richardson, J.C. Kasper, C. Wang, J.W. Belcher, A.J. Lazarus, Termination shock decelerates upstream solar wind but heliosheath plasma is cool. Nature **454**, 63–66 (2008a)

J.D. Richardson, Y. Liu, C. Wang, D.J. McComas, Determining the LIC H density from the solar wind slow-down, Astron. Astrophys. (2008b, in press)

N.A. Schwadron, M.A. Lee, D.J. McComas, Diffusive acceleration at the blunt termination shock. Astrophys. J. **675**, 1584–1600 (2008)

G. Siscoe, P.J. MacNeice, D. Odstrcil, East–west asymmetry in coronal mass ejection geoeffectiveness. Space Weather **5**, S04002 (2007). doi:10.1029/2006SW000286

C.W. Smith, P.A. Isenberg, W.H. Matthaeus, J.D. Richardson, Turbulent heating of the solar wind by newborn interstellar pickup protons. Astrophys. J. **638**, 508–517 (2006)

E.C. Stone, A.C. Cummings, F.B. McDonald, B. Heikkila, N. Lal, W.R. Webber, Voyager 1 explores the termination shock region and the heliosheath beyond. Science **309**, 2017–2020 (2005)

E.C. Stone, Voyager in the vicinity of the termination shock: Observations and questions, Eos Trans. AGU, 85(47), Fall Meet. Suppl. (2004), Abstract SH41B-01

E.C. Stone et al., Voyager 2 finds an asymmetric termination shock and explores the heliosheath beyond. Nature **454**, 71–74 (2008)

A. Szabo, R. Lepping, J. King, Magnetic field observations of the 1.3-year solar wind oscillation. Geophys. Res. Lett. **22**, 1845–1848 (1995)

R. Weymann, Coronal evaporation as a possible mechanism for mass loss in red giants. Astrophys. J. **132**, 380 (1960)

M. Witte, Kinetic parameters of interstellar neutral helium—Review of results obtained during one solar cycle with the Ulysses/GAS-instrument. Astron. Astrophys. **426**(3), 835–844 (2004). doi:10.1051/0004-6361:20035956

G. Zank, H. Pauls, I. Cairns, G. Webb, Interstellar pickup ions and quasi-perpendicular shocks: Implications for the termination shock and interplanetary shocks. J. Geophys. Res. **101**, 457 (1996)

Constraints on the Structure of the Heliospheric Interface Based on Lyα Absorption Spectra

Brian E. Wood · Vladislav V. Izmodenov · Yury G. Malama

Originally published in the journal Space Science Reviews, Volume 143, Nos 1–4, 21–30.
DOI: 10.1007/s11214-008-9369-5

Abstract Spectroscopic observations of the hydrogen Lyα lines from nearby stars taken by the *Hubble Space Telescope* (HST) sometimes show absorption signatures from the heliosphere. This absorption is a unique diagnostic of material in the outermost parts of our heliosphere. We summarize how the HST data have been used to test various models of the heliosphere, particularly new 3-dimensional MHD models that have recently become available. We also focus on new detections of heliospheric absorption in very downwind directions, which can only be modeled using heliospheric model codes with extended grids in the downwind direction. We illustrate a couple attempts to reproduce this absorption using a couple of these extended tail models.

Keywords Solar wind · Interstellar medium · Heliosphere · Ultraviolet spectroscopy

1 Introduction

Ever since the first detection of Lyα absorption from the outer heliosphere in 1996 (Linsky and Wood 1996), the absorption has been used to test models of the heliosphere. The absorption comes from hot hydrogen that permeates the heliosphere due to charge exchange processes that occur during the collision of the solar wind and ISM. Detection of the absorption requires high resolution UV spectroscopy, and the only instruments capable of such observations are spectrometers on board the *Hubble Space Telescope* (HST); first the Goddard

B.E. Wood (✉)
JILA, University of Colorado, 440 UCB, Boulder, CO 80309-0440, USA
e-mail: woodb@colorado.edu

V.V. Izmodenov
Dept. of Aeromechanics and Gas Dynamics, Lomonosov Moscow State University, Moscow 119899, Russia
e-mail: izmod@ipmnet.ru

V.V. Izmodenov · Y.G. Malama
Space Research Institute (IKI), Russian Academy of Science, and Institute for Problems in Mechanics, Moscow 117526, Russia

J.L. Linsky et al. (eds.), *From the Outer Heliosphere to the Local Bubble.*
DOI: 10.1007/978-1-4419-0247-4_4

High Resolution Spectrograph (GHRS) and then its 1997 replacement the Space Telescope Imaging Spectrograph (STIS), which unfortunately has been inoperable since August 2004.

The Lyα absorption has confirmed the existence of the "hydrogen wall" predicted by models to exist between the heliopause and bow shock (Baranov et al. 1991). Attempts have been made to use the absorption to constrain ISM properties around the Sun, but with limited success due to model dependence and the sensitivity of the absorption to too many ISM parameters (Wood et al. 2000, 2007a; Izmodenov et al. 2002). Finally, analogous astrospheric absorption detected in the HST spectra have been used to study the solar-like winds of nearby stars and determine how the properties of such winds vary with stellar age and activity (Wood et al. 2005a).

2 The Heliospheric Absorption Detections

Only a small fraction of the Lyα spectra of nearby stars in the HST archive show direct evidence of heliospheric absorption (Wood et al. 2005b). There are two primary factors that determine whether an observed line of sight will have a detectable heliospheric signal. One is the ISM hydrogen column density. Detecting heliospheric absorption requires that the absorption is separable from interstellar absorption. The higher the interstellar H column density, the broader the interstellar absorption, and the more likely that the ISM absorption will hide any heliospheric absorption.

The second factor is the direction of the line of sight with respect to the orientation of the heliosphere. This is quantified by the angle θ between the line of sight and the upwind direction of the ISM flow seen by the Sun, so $\theta = 0°$ is upwind and $\theta = 180°$ is downwind. Figure 1 plots the logarithmic interstellar H column density versus θ for all HST-observed lines of sight, with the boxes being lines of sight with detected heliospheric absorption and the diamonds being nondetections. As expected, the detections tend to have low ISM columns. It is also clearly easier to detect heliospheric absorption in upwind directions than in downwind directions. This is due to the stronger deceleration of heliospheric H in upwind directions, which helps to shift the heliospheric absorption away from the ISM absorption, making it easier to detect.

Most of the absorption detections are in directions where the hydrogen wall dominates the absorption, meaning the absorption is a diagnostic of neutral H properties in between

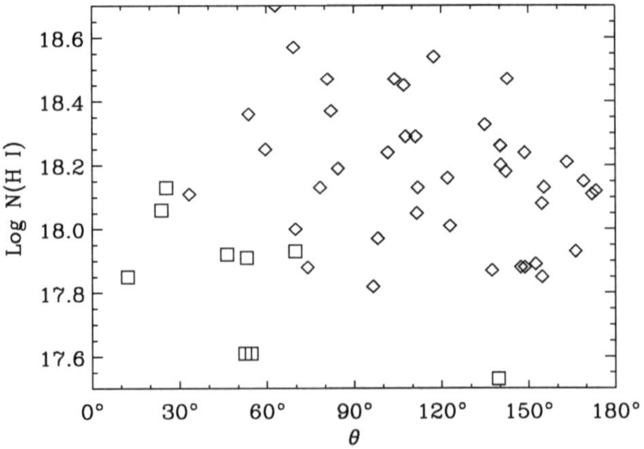

Fig. 1 The ISM H I column densities measured for all HST-observed lines of sight are plotted versus the angle of the lines of sight relative to the upwind direction of the ISM flow seen by the Sun. The *boxes* and *diamonds* indicate lines of sight that yield detections and nondetections of heliospheric absorption, respectively. From Wood et al. (2005b)

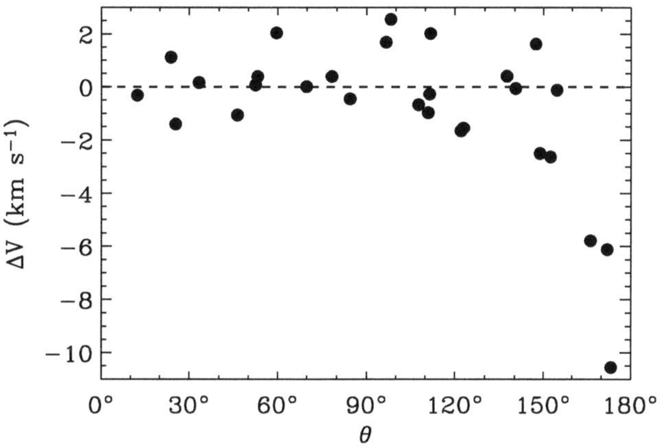

Fig. 2 Velocity difference between the bisectors of stellar Lyα lines and the stellar rest frame plotted as a function of θ, the angle between the line of sight and the ISM flow vector. The Lyα profiles of the most downwind lines of sight ($\theta > 160°$) are systematically blueshifted with respect to the stellar rest frame, suggesting the presence of heliosheath absorption. From Wood et al. (2007b)

the heliopause and bow shock. However, very downwind lines of sight with extended path lengths through the heliotail can potentially be dominated by absorption from heliosheath neutrals, neutrals formed by charge exchange in the inner heliosheath between the termination shock and heliopause. There is only one truly downwind heliospheric absorption detection in Fig. 1 where heliosheath absorption will dominate. The heliosheath absorption is much broader than the hydrogen wall absorption, but is also much shallower, which makes it harder to detect.

However, a less direct method of detecting heliosheath absorption has recently been devised (Wood et al. 2007b). The broad, shallow heliosheath absorption will absorb some of the flux from the red side of the observed stellar Lyα emission line. Therefore, the heliosheath absorption will cause the wings of the stellar emission line to be blueshifted relative to their original centroid, which would correspond to the stellar radial velocity. Figure 2 plots the velocity shifts of reconstructed stellar Lyα profiles as a function of θ. In most directions the profiles are within 2 km s^{-1} of the stellar radial velocity. But not in the most downwind directions with $\theta > 160°$. For the three lines of sight within 20° of the downwind direction the profiles are blueshifted relative to the star. This is the expected signature of broad heliosheath absorption, and the most downwind directions are where it is expected to be seen, so these three lines of sight can now be considered to represent heliospheric absorption detections, though the manner of detection is very different from the detections depicted in Fig. 1. The θ angle of 160° where the heliosheath absorption starts to become apparent may prove to be a useful diagnostic of the nature of the heliotail.

Figure 3 shows a sky map of the HST-observed lines of sight with Lyα spectra that have been searched for evidence of heliospheric absorption. The boxes are the lines of sight with detections of absorption, while the diamonds and plus signs are nondetections. As noted above, most of the detections are in the upwind direction, but the figure also shows the 3 new detections that are very close to the downwind direction. Even the nondetections can potentially be used to constrain models, as they place limits on the amount of heliospheric absorption that can be present in those directions. The 20 numbered lines of sight scattered around Fig. 3 indicate the sight lines that have been chosen in the past to test models (Wood et al. 2007a). These include the 11 sight lines with detected heliospheric absorption, and 9 sight lines with nondetections chosen to sample directions not covered by the detections.

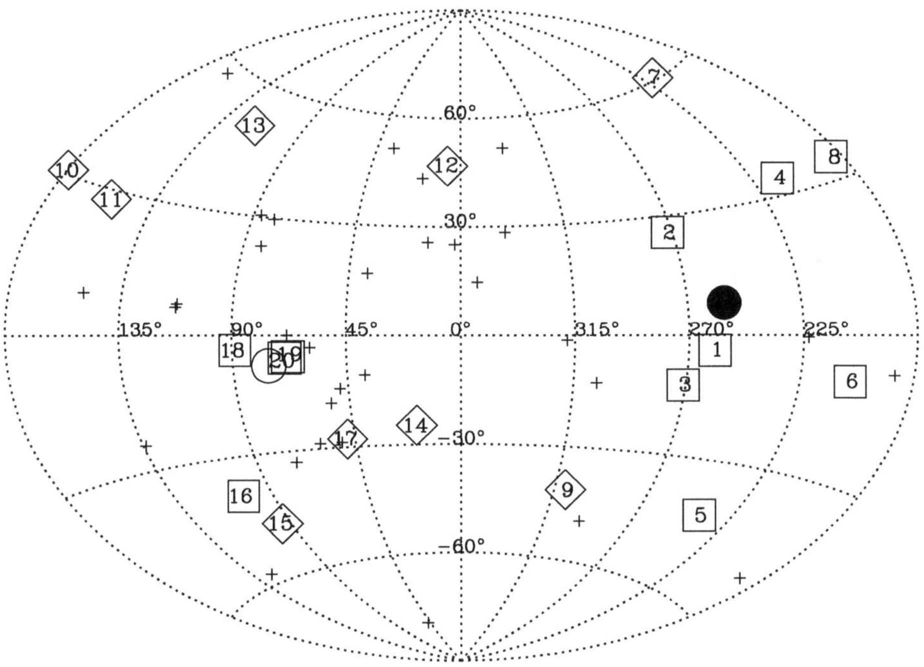

Fig. 3 Sky map in ecliptic coordinates of all HST-observed lines of sight with analyzed Lyα spectra. Numbered symbols indicate spectra that have been used to test model predictions of Lyα absorption (see Figs. 4, 5, and 7). *Boxes* indicate lines of sight with detected heliospheric absorption, while the *diamonds* and *plus signs* are nondetections. The *diamonds* indicate lines of sight with nondetections selected to provide upper limits for absorption in those directions. The *filled* and *open circles* indicate the upwind and downwind directions of the local ISM flow vector, respectively. From Wood et al. (2007a)

3 Absorption Dependence on the ISM Magnetic Field

Evidence for an asymmetric heliosphere has recently been provided by the *Voyager* satellites' observations near the termination shock, and by the different flow vectors measured for interstellar H and He within the solar system (Lallement et al. 2005; Opher et al. 2006; Stone and Richardson 2008). The principle cause of heliospheric asymmetry is expected to be an ISM magnetic field that is skewed with respect to the ISM flow seen by the Sun. The observations have stimulated the development of 3D MHD heliospheric models that can model the observed asymmetries and explore how heliospheric asymmetry depends on ISM field strength and orientation (Izmodenov et al. 2005; Opher et al. 2006; Pogorelov and Zank 2006; Opher et al. 2008; Izmodenov 2008). Some of these models are able to model the neutrals and can also be used to assess what effects the ISM magnetic field should have on the Lyα absorption diagnostic.

Figure 4 compares observed absorption with that predicted by models towards 4 of the lines of sight indicated in Fig. 2 (Wood et al. 2007a). The 3 models assumed different magnetic field strengths: $B = 0$, 1.25, and 2.5 μG. Field strengths higher than this tend to push the termination shock inside the 94 AU distance found by *Voyager 1* (Stone et al. 2005). The orientation of the magnetic field is assumed to be at an angle of $\alpha = 45°$ from the upwind direction of the ISM flow, in the plane suggested for the ISM field by the deflection of H atoms relative to He atoms within the solar system (Lallement et al. 2005). The

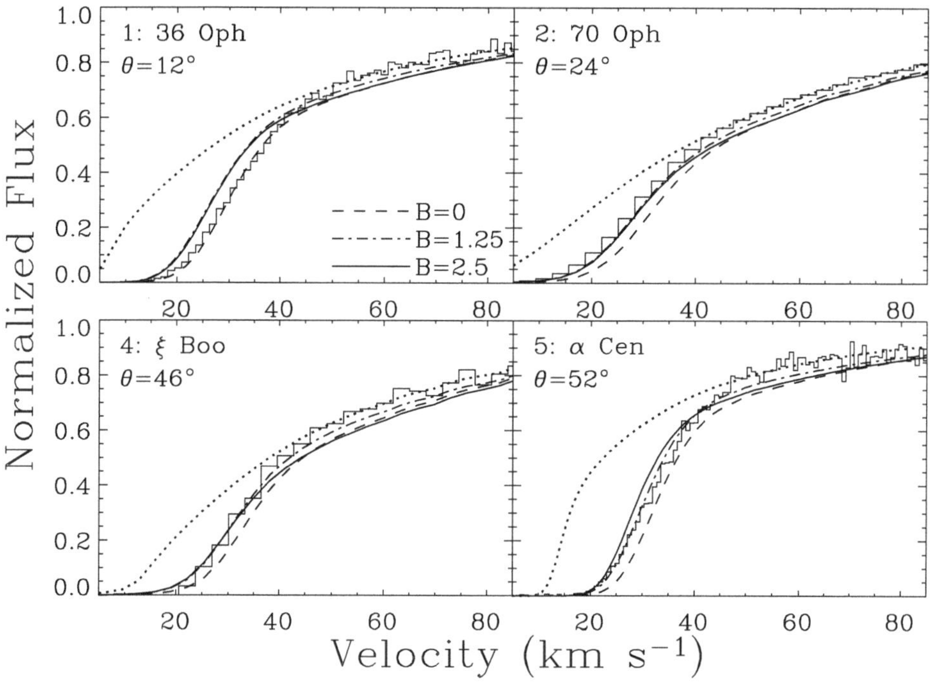

Fig. 4 The red side of the H I Lyα absorption line (histogram) for four stars with detection heliospheric absorption, with the number of the star indicating its location in Fig. 3. The stars are in order of increasing angle from the upwind direction of the ISM flow (θ). In each panel, the *dotted line* is the ISM absorption alone. Absorption predictions are shown for heliospheric models computed assuming ISM magnetic field strengths of 0, 1.25, and 2.5 μG (Wood et al. 2007a)

ISM hydrogen and proton densities assumed in these models are $n_\infty(\mathrm{HI}) = 0.18\ \mathrm{cm}^{-3}$ and $n_\infty(\mathrm{H}^+) = 0.06\ \mathrm{cm}^{-3}$, the ISM flow speed is $V_\infty = 26.4\ \mathrm{km\,s}^{-1}$, and the temperature is $T_\infty = 6400$ K (Izmodenov et al. 2005).

The panels in Fig. 4 show only the redshifted side of the Lyα absorption profiles where the heliospheric absorption will be located. Each panel shows the ISM absorption alone, and the excess absorption from the heliosphere, which the models are attempting to reproduce. Full descriptions of the Lyα data and its analysis can be found in Wood et al. (2005b).

Increasing the magnetic field generally decreases the amount of absorption, though there is little difference in the absorption predicted by the $B = 1.25$ μG and $B = 2.5$ μG models. The decrease is because a higher ISM field weakens the bow shock, leading to lower densities in the hydrogen wall and less absorption, despite the hydrogen wall being somewhat thicker (Wood et al. 2007a). The $B = 1.25$ μG and $B = 2.5$ μG models seem to fit the data particularly well in Fig. 4, though the $B = 0$ μG model works best for the most upwind line of sight (36 Oph). The lines of sight chosen in Fig. 4 are all upwind lines of sight with detected heliospheric absorption. There is little point in comparing the data and model predictions for downwind lines of sight because these particular models do not extend far enough downwind to capture all the absorption in such directions.

Not only does the absorption depend on the strength of the ISM field, it also depends on the orientation of the field. Figure 5 compares the observed absorption with that predicted by various models, for the same 4 lines of sight as in Fig. 4. The models are like the $B = 2.5$ μG model from Fig. 4, but with different α orientation angles. Rather than show the direct

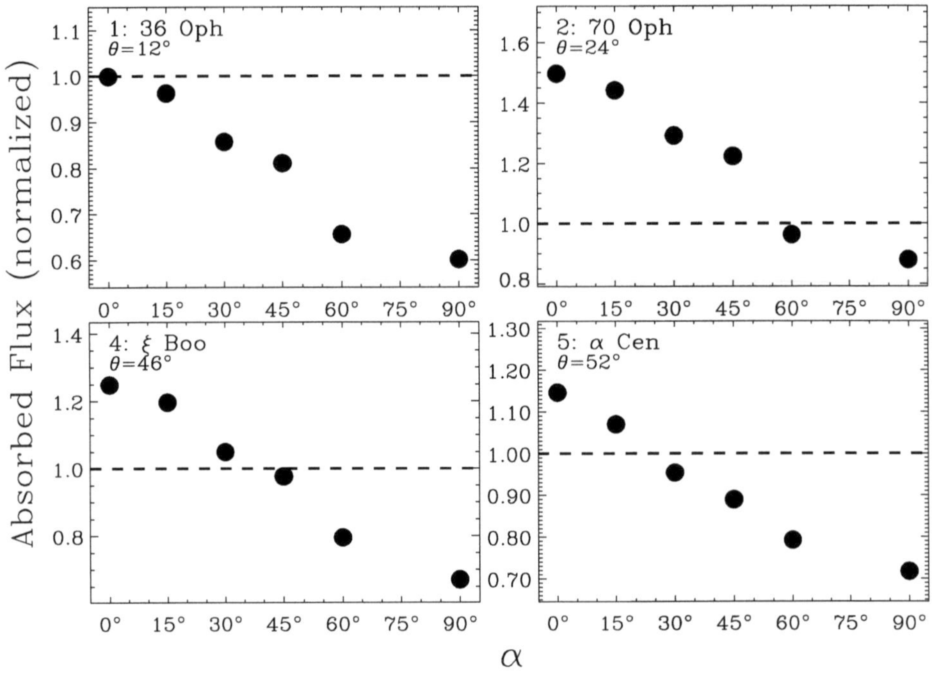

Fig. 5 For the same four upwind lines of sight as in Fig. 4, the predicted wavelength-integrated Lyα flux absorbed by heliospheric H I beyond that absorbed by the ISM is computed for six models with different ISM field orientation, where the α orientation angle is the angle of the field relative to the upwind direction of the ISM flow. The fluxes are normalized to the observed amount of flux absorbed, so in each panel a flux of 1 (*dashed lines*) corresponds to agreement with the data (Wood et al. 2007a)

data/model comparison like in Fig. 4, Fig. 5 computes the absorbed flux predicted by the models divided by the absorbed flux observed in the data, and plots this normalized flux versus α. Absorption decreases with increasing α. There is certainly some scatter in the results, but on average the $\alpha = 15°$–$45°$ models fit the data best, consistent with the $\alpha = 30°$–$60°$ range suggested by Opher et al. (2006) on the basis of termination shock asymmetries observed by the *Voyager* satellites.

For the upwind lines of sight that are the focus of Figs. 4 and 5, the absorption will be dominated by hydrogen wall neutrals. The ISM magnetic field results in a hydrogen wall that is asymmetric, thicker in one hemisphere than the other. However, the models predict that these asymmetries surprisingly do not lead to significant asymmetries in the Lyα absorption. The reason is that densities are lower in parts of the hydrogen wall that are thicker. The hydrogen wall thickness and density effects nearly cancel each other out, and as a consequence there is little azimuthal dependence in the hydrogen wall absorption to act as a signature of heliospheric asymmetries induced by the magnetic field.

However, this is not the case for the heliosheath absorption that dominates in downwind directions. It turns out that unlike the hydrogen wall, the asymmetries in the heliosheath induced by the ISM magnetic field *will* be accompanied by asymmetries in the heliosheath Lyα absorption. But it is only in downwind directions that heliosheath absorption is detectable, and there are only 4 stars with detected absorption in that direction (stars 16 and 18–20 in Fig. 3). This is really not enough to search for azimuthal asymmetries in the absorption that might be diagnostic of the ISM field orientation. Another problem is that most models, in-

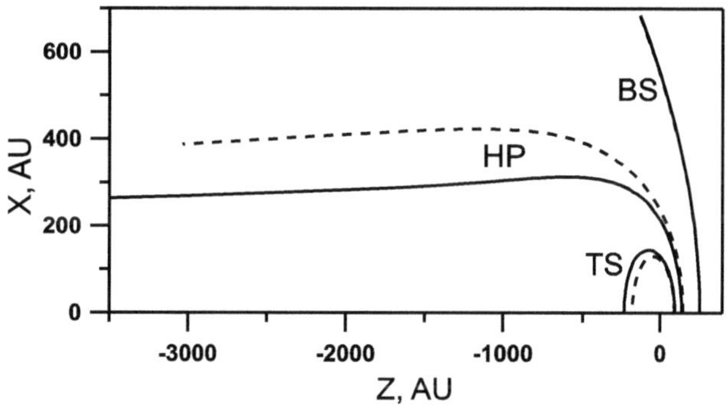

Fig. 6 The locations of the termination shock (TS), heliopause (HP), and bow shock (BS) according to the single-fluid (*dashed lines*) and multi-component (*solid lines*) plasma models (see text for model details). The Sun is at the origin

cluding the ones used in Figs. 4 and 5, do not extend far enough downwind to allow for a comparison between the data and models. Thus, it is not entirely clear from existing models what the extent of the absorption asymmetry should be (Wood et al. 2007a).

4 Extended Tail Models

There are only a few heliospheric models that have been computed that can provide a meaningful data/model absorption comparison for very downwind directions. Such a comparison requires that the model have a proper self-consistent treatment of the neutrals, and the model grid must extend many thousands of AU downwind to be able to capture all the absorption in those directions. Izmodenov and Alexashov (2003) and Alexashov et al. (2004) report on 2-dimensional models of the Baranov and Malama (1993, 1995) type that have been computed with extended tails and a full kinetic treatment of the neutrals. The dashed lines in Fig. 6 show the heliospheric structure of one of these long tail models, with input parameters identical to those of the models in Sect. 3, but with no ISM magnetic field.

The solid lines in Fig. 6 show the heliospheric structure of a model with identical input parameters, but with a much more sophisticated treatment of the plasma in the heliosphere. Although the traditional Baranov and Malama (1993, 1995) type models include a fully kinetic treatment of the neutrals in the heliosphere, they use a single-fluid treatment of the plasma. This is simplistic, however, as the example of pickup protons indicates, which are found not to be thermalized with the ambient solar wind protons within the solar system (e.g., Gloeckler and Geiss 2004). Malama et al. (2006) have modified the Baranov and Malama code to allow for a complex multicomponent treatment of the plasma. The solid lines in Fig. 6 indicate how the multicomponent treatment affects the heliospheric structure predicted by the model. The two models in Fig. 6 are both extended-tail models that can be used to assess whether the multicomponent treatment affects the Lyα absorption predicted by the models. It turns out that the hydrogen wall absorption is completely unchanged. However, the heliosheath absorption that dominates in downwind directions is affected (Wood et al. 2007b).

The model grid for these models extends 5000 AU in the downwind direction. In order to see if this is far enough to capture the absorption towards even the most downwind directions in Fig. 3, we compute the absorption towards these stars using both the full model grid and one that is truncated at shorter distances. At a distance of 3000 AU we see little difference in the absorption, but discrepancies become more significant at shorter truncation distances. Thus, we conclude that modeling the most downwind HST-observed directions requires grids that extend at least 3000 AU downwind.

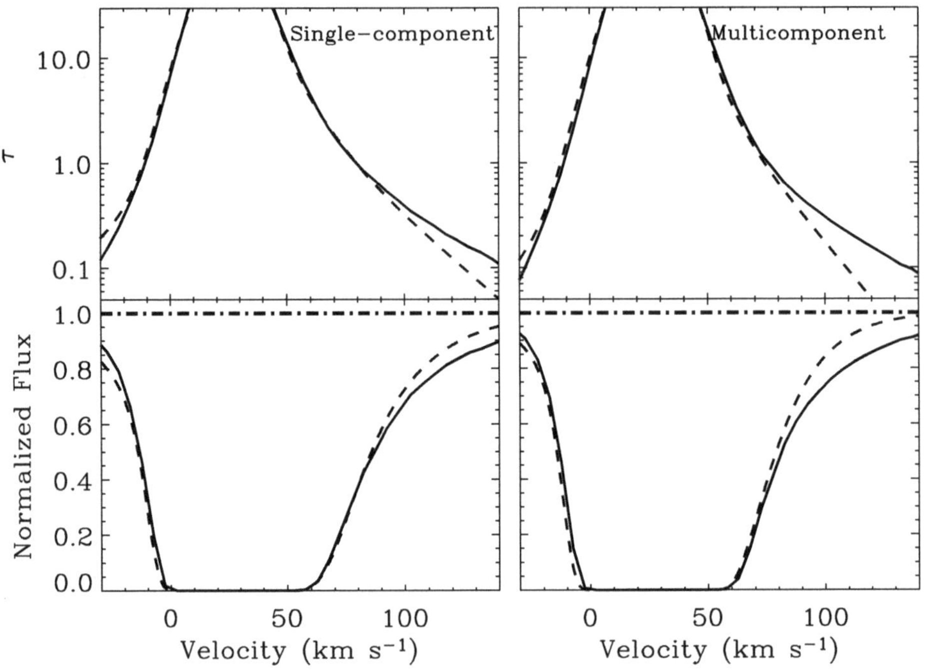

Fig. 7 The *top panels* show the opacity profiles for the absorption towards χ^1 Ori (star #18 from Fig. 3) predicted by the extended-tail single-component and multicomponent models, where the *solid lines* are the exact profiles derived from the kinetic codes and the *dashed lines* are profiles computed using local Maxwellian approximations to the velocity distributions of individual particle components. The *bottom panels* show the absorption profiles that correspond to these opacity functions. There are significant differences, particularly in the red wing of the absorption, emphasizing the inadequacy of the local Maxwellian approximation

The original method used to compute absorption profiles from Baranov and Malama (1993, 1995) style models used local Maxwellian approximations of individual populations of neutrals within the heliosphere (Izmodenov et al. 2002). This may be a good approximation for the hydrogen wall neutrals that dominate the absorption in upwind and sidewind directions, but we find that it does not work well at all for the heliosheath neutrals that dominate in downwind directions. This is shown explicitly in Fig. 7, which compares opacity and absorption profiles computed directly from line-of-sight integrated velocity distributions to ones computed using the local Maxwellian approximations. This is done for a downwind line of sight for both the single-component and multicomponent extended-tail models. There are significant differences, especially in the red wing of the line, which is where the detectable heliosheath absorption lies. Using the local Maxwellian approximation would lead to significant underestimation of heliosheath absorption in the red wing.

Figure 8 shows the absorption predicted by both the single-component and multicomponent models shown in Fig. 6, for four of the downwind lines of sight indicated in Fig. 3 (sight lines 15–19). The Sirius panel of Fig. 8 is analogous to those in Fig. 4, with detected heliospheric absorption. The ϵ Eri panel shows the absorption data for a line of sight with no detected heliospheric absorption. The dashed line in the panel indicates an upper limit for the amount of absorption that can be present while still being consistent with the data. A successful model should predict an absorption profile above the dashed line. The very downwind χ^1 Ori and HD 28205 lines of sight have heliospheric absorption detections, but

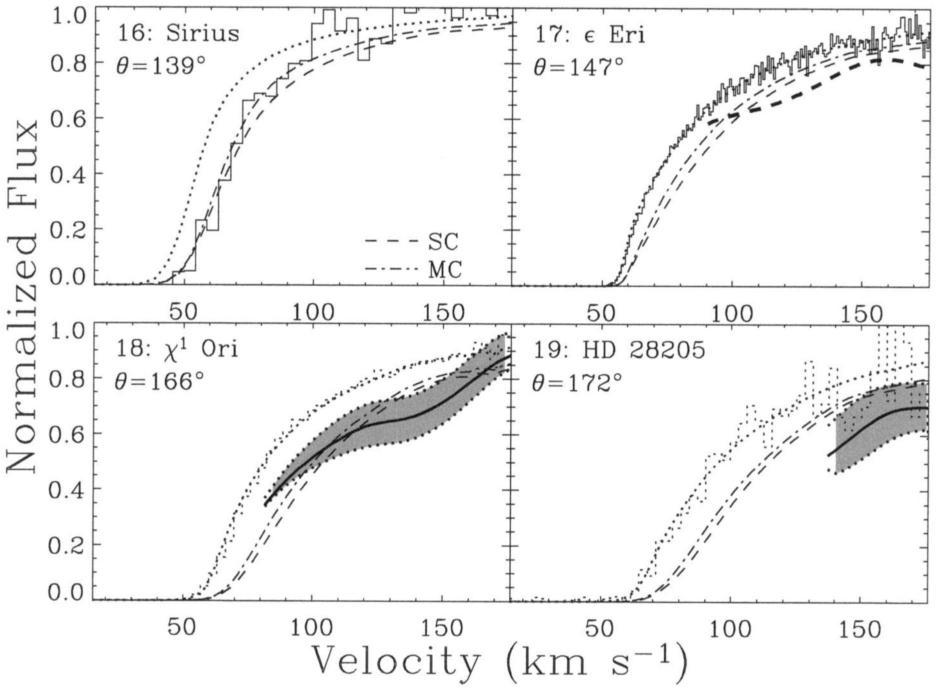

Fig. 8 The red side of the H I Lyα absorption line (histogram) for four of the downwind lines of sight indicated in Fig. 3, with θ indicating the angle from the upwind direction of the ISM flow (θ). The Sirius panel is like those from Fig. 4, with detected heliospheric absorption beyond that from the ISM (the *dotted line*). The ϵ Eri line of sight is a nondetection, where *the thick dashed* line in the panel provides an upper limit for how much absorption a model can predict towards ϵ Eri while still being consistent with the data. Model absorption profiles should lie above that line. The χ^1 Ori and HD 28205 lines of sight are heliospheric absorption detections inferred from line shifts of the reconstructed stellar Lyα profiles. The *shaded regions* in those panels indicate the amount of absorption that models *should* predict if the real stellar Lyα profile is centered on the stellar rest frame. For these lines of sight, the absorption predicted by the models should not fit the data (which are *dotted* histograms in these cases) but should instead fall within the shaded regions. The *dashed* and *dot-dashed lines* are the predicted heliospheric Lyα absorption of models with a single component (SC) plasma and a multi-component (MC) plasma, respectively

only on the basis of the line shift arguments from Sect. 2, and illustrated by Fig. 2. The line shift means that the original reconstructed stellar Lyα profile is inaccurate and the normalized fluxes in Fig. 8 (dashed histograms) are therefore inaccurate. The shaded regions in the figure indicate where the normalized fluxes *should* be if the background stellar profile is forced to be centered on the rest frame of the star. A successful model should predict absorption that lies within these shaded regions for those two lines of sight (Wood et al. 2007a).

The multicomponent model predicts less heliosheath absorption than the single component model. This is due to the overemphasis of the effects of minority high-temperature constituents on the plasma velocity distributions in the simplistic single-component model. The effect is to artificially broaden the distributions, a problem that is then transmitted to the neutral distributions via charge exchange. The reduced absorption of the multicomponent model provides improved agreement with the data for the Sirius and χ^1 Ori lines of sight in Fig. 8. The absorption predicted by the single-component model may be in conflict with the absorption limit towards ϵ Eri near 100 km s^{-1}, but the conflict is lessened for the multicomponent model. Both models may slightly underestimate the amount of absorption in the most downwind direction (HD 28205).

It is worth noting that the heliosheath neutrals that are responsible for the heliosheath absorption observed and modeled in Fig. 8 are essentially the same population of neutrals that the *Interstellar Boundary Explorer* (IBEX) will observe after it is launched in 2008 (McComas et al. 2005; McComas 2008). The effects that the multicomponent treatment has on the heliosheath absorption, which seems to improve agreement with the data, suggest that the heliosheath particle distributions that IBEX will observe locally may also be better modeled with a multicomponent treatment of heliospheric plasma. Combining IBEX and Lyα absorption constraints on such multicomponent models will be an interesting exercise for the future.

Acknowledgements This work was supported by NASA grant NNG05GD69G to the University of Colorado. V.I. was supported in part by RFBR grants 07-02-01101 and 05-02-22000.

References

D.B. Alexashov, V.V. Izmodenov, S. Grzedzielski, Adv. Space Res. **34**, 109 (2004)
V.B. Baranov, Y.G. Malama, J. Geophys. Res. **98**, 15157 (1993)
V.B. Baranov, Y.G. Malama, J. Geophys. Res. **100**, 14755 (1995)
V.B. Baranov, M.G. Lebedev, Y.G. Malama, Astrophys. J. **375**, 347 (1991)
G. Gloeckler, J. Geiss, Adv. Space Res. **34**, 53 (2004)
V.V. Izmodenov, Space Sci. Rev. (2008, this issue)
V.V. Izmodenov, D.B. Alexashov, Astron. Lett. **29**, 58 (2003)
V.V. Izmodenov, B.E. Wood, R. Lallement, J. Geophys. Res. **107**, 1308 (2002)
V. Izmodenov, D. Alexashov, A. Myasnikov, Astron. Astrophys. **437**, L35 (2005)
R. Lallement, E. Quémerais, J.L. Bertaux, S. Ferron, D. Koutroumpa, R. Pellinen, Science **307**, 1447 (2005)
J.L. Linsky, B.E. Wood, Astrophys. J. **463**, 254 (1996)
Y.G. Malama, V.V. Izmodenov, S.V. Chalov, Astron. Astrophys. **445**, 693 (2006)
D.J. McComas, Space Sci. Rev. (2008, this issue)
D. McComas et al., in *Solar Wind 11/SOHO 16: Connecting the Sun and Heliosphere*, ed. by B. Fleck, T.H. Zurbuchen (ESA, Noordwijk, 2005), p. 689
M. Opher, E.C. Stone, P.C. Liewer, Astrophys. J. **640**, L71 (2006)
M. Opher, E.C. Stone, J.C. Richardson, G. Toth, T. Gombosi, Space Sci. Rev. (2008, this issue)
N.P. Pogorelov, G.P. Zank, Astrophys. J. **636**, L161 (2006)
E.C. Stone, J.C. Richardson, Space Sci. Rev. (2008, this issue)
E.C. Stone, A.C. Cummings, F.B. McDonald, B.C. Heikkila, N. Lal, W.R. Webber, Science **309**, 2017 (2005)
B.E. Wood, H.-R. Müller, G.P. Zank, Astrophys. J. **542**, 493 (2000)
B.E. Wood, H.-R. Müller, G.P. Zank, J.L. Linsky, S. Redfield, Astrophys. J. **628**, L143 (2005a)
B.E. Wood, S. Redfield, J.L. Linsky, H.-R. Müller, G.P. Zank, Astrophys. J. Suppl. Ser. **159**, 118 (2005b)
B.E. Wood, V.V. Izmodenov, J.L. Linsky, D. Alexashov, Astrophys. J. **659**, 1784 (2007a)
B.E. Wood, V.V. Izmodenov, J.L. Linsky, Y.G. Malama, Astrophys. J. **657**, 609 (2007b)

Influence of the Interstellar Magnetic Field and Neutrals on the Shape of the Outer Heliosphere

N.V. Pogorelov · J. Heerikhuisen · G.P. Zank · S.N. Borovikov

Originally published in the journal Space Science Reviews, Volume 143, Nos 1–4, 31–42.
DOI: 10.1007/s11214-008-9429-x

Abstract Formed as a result of the solar wind (SW) interaction with the circum-heliospheric interstellar medium (CHISM), the outer heliosphere is generically three-dimensional because of the SW asphericity and the action of the interstellar and interplanetary magnetic fields (ISMF and IMF). In this paper we show that charge exchange between neutral and charged components of the SW–CHISM plasmas plays a dominant role not only in determining the geometrical size of the heliosphere, but also in the modulation of magnetic-field-induced asymmetries. More specifically, charge exchange between SW and CHISM protons and primary neutrals of interstellar origin always acts to decrease the asymmetry of the termination shock and the heliopause, which can otherwise be very large. This is particularly pronounced because the ionization ratio of the CHISM plasma is rather low. To investigate the deflection of the CHISM neutral hydrogen flow in the inner heliosphere from its original orientation in the unperturbed CHISM, we create two-dimensional neutral H velocity distributions in the inner heliosphere within a 45-degree circular conical surface with the apex at the Sun and the axis parallel to the interstellar flow vector. It is shown that the distribution of deflections is very anisotropic, that is, the most probable orientation of the H-atom velocity differs from its average direction. We show that the average deflection of the H-atom flow, for reasonable ISMF strengths, occurs mostly in the plane formed by the ISMF and CHISM velocity vectors at infinity. The possibility that the ISMF orientation may

This work is supported by NASA grants NNG05GD45G, NNG06GD48G, NNG06GD43G, and NNX08AJ21G, and NSF award ATM-0296114. Supercomputer time allocations are provided by DOE's INCITE project PSS001 and NCSA project MCA07S033.

N.V. Pogorelov (✉) · J. Heerikhuisen · G.P. Zank · S.N. Borovikov
Institute of Geophysics and Planetary Physics, University of California, Riverside, CA 92521, USA
e-mail: nikolaip@ucr.edu

J. Heerikhuisen
e-mail: jacobh@ucr.edu

G.P. Zank
e-mail: zank@ucr.edu

S.N. Borovikov
e-mail: sergeyb@ucr.edu

DOI: 10.1007/978-1-4419-0247-4_5

influence the 2–3 kHz radio emission, which is believed to originate in the outer heliosheath, is discussed.

Keywords ISM: kinematics and dynamics · Solar wind

1 Introduction

It has been known for about two decades (Fahr et al. 1988; Pogorelov and Matsuda 1998; Ratkiewicz et al. 1998) that the interstellar magnetic field (ISMF) of strength $\mathbf{B}_\infty$, by exerting pressure on the surface of the heliopause (HP), affects its shape and position with respect to the Sun and the direction of the CHISM plasma velocity vector $\mathbf{V}_\infty$. The asymmetry in the HP position is immediately seen in the shape of the heliospheric termination shock (TS). Following Linde et al. (1998) who considered a number of cases with $\mathbf{B}_\infty$ either parallel or perpendicular to $\mathbf{V}_\infty$, Pogorelov et al. (2004) revisited the problem of the ISMF influence of the heliospheric interface by analyzing the ISMF coupling with the interplanetary magnetic field (IMF) at the heliopause in an ideal MHD formulation. A range of different orientations of $\mathbf{B}_\infty$ with respect to $\mathbf{V}_\infty$ and the ecliptic plane was considered, and for all of them the changes in the HP and TS shape and orientations were analyzed. These studies remained rather academic until Lallement et al. (2005) suggested that the asymmetry in the plasma distribution on the interstellar side of the heliopause is responsible for the difference between the H-atom and the He-atom flow directions observed in the inner heliosphere in the *SOHO* SWAN experiment. This idea was confirmed by the calculations of Izmodenov et al. (2005), who estimated the direction of the H-atom flow in the inner heliosphere by analyzing statistically averaged trajectories on neutral hydrogen atoms. This was done in the absence of the IMF for a spherically-symmetric SW, which made the heliosphere symmetric with respect to the plane formed by $\mathbf{B}_\infty$ and $\mathbf{V}_\infty$ (a BV-plane). This means that the average neutral trajectory which starts at the CHISM point belonging to the BV-plane will remain in this plane. Meanwhile, trajectories that start at two points lying symmetrically with respect to the BV-plane will acquire out-of-plane velocity components oriented in opposite directions. Thus, the average out-of-plane deflection will be zero. Opher et al. (2006), remaining in the framework of the ideal MHD model, made the first attempt to quantify the TS asymmetries in the directions of the *Voyager 1* (*V1*) and *Voyager 2* (*V2*) spacecraft. However, in the absence of neutral H this model tends to exaggerate the effect of the ISMF on the heliospheric asymmetries considerably. Pogorelov et al. (2006) and Pogorelov and Zank (2006) also noticed that the flow of neutral hydrogen never preserves its original orientation in the unperturbed CHISM, even for $\mathbf{V}_\infty \parallel \mathbf{B}_\infty$. Moreover, in the presence of the IMF the deflection inevitably takes place both within and perpendicular to the BV-plane. This means that the hydrogen deflection plane (HDP) and the BV-plane generically do not coincide. The deflections parallel to the BV-plane and perpendicular to it become comparable if the angle between $\mathbf{B}_\infty$ and $\mathbf{V}_\infty$ is not small (Pogorelov and Zank 2006). To acquire a $4° \pm 1°$ hydrogen deflection observed in the *SOHO* SWAN experiment, one would need to assume a strong ISMF (perhaps greater than 4 μG). Magnetic fields of such strength should not be summarily disregarded because of the presumption that the TS/HP heliocentric distances may be too small for such fields. Since the magnetic pressure acts perpendicularly to magnetic field lines, plasma can actually move unobstructed along the lines. As a result, for $\mathbf{B}_\infty$ directed at small angles with respect to $\mathbf{V}_\infty$, the increase in $\mathbf{B}_\infty$ results in higher TS and HP stand-off distances in the upstream CHISM directions (Baranov and Krasnobaev 1971; Florinski et al. 2004). On the other hand, the presence of CHISM neutrals allows for 2D

(Florinski et al. 2004) and 3D (Pogorelov et al. 2006) solutions with the HP at finite distances from the Sun for sub-Alfvénic CHISM flows. These are impossible in the absence of neutrals. We note here that one of the theories of the origin of the Local Bubble (Cox and Helenius 2003) suggests a strong ISMF nearly parallel to the CHISM velocity vector. Since a multi-fluid analysis of the H flow deflection performed by Pogorelov and Zank (2006) and Pogorelov et al. (2007) assumes the fluid approximation for primary (born in the unperturbed CHISM) and secondary (born in the thermodynamically distinct SW and CHISM regions), one would want to see a kinetic analysis of neutral H trajectories. This is the main subject of the current paper. Additionally, we perform a comparison of numerical results obtained with the five-fluid (one plasma and four neutral fluids) and MHD-kinetic approaches. As suggested by Gurnett et al. (2006) and shown by Pogorelov et al. (2007), who followed the idea of Mitchell et al. (2008), the assumption of the BV-plane being parallel to the observed HDP results in band-like distributions of the ISMF strength in the outer heliosheath. Such bands are elongated nearly perpendicular to the BV-plane. Another approach, based on topological considerations to analyze the distribution of radio emission sources, was used by Opher et al. (2007). We explore this effect using our newly developed MHD-kinetic model (Heerikhuisen et al. 2006, 2007; Pogorelov et al. 2008a).

2 Heliospheric Asymmetry Induced by the ISMF

We performed numerical calculations in a Cartesian coordinate system (grid resolution is about 1.5 AU near the TS and 2.5 AU near the HP) with the origin at the Sun (Fig. 1). The x-axis is oriented along the Sun's rotation axis, which we assume to be perpendicular to the ecliptic plane (yz-plane). The z-axis belongs to the plane defined by the x-axis and $\mathbf{V}_\infty$, and is directed upstream into the CHISM. The y-axis completes the right coordinate system. The direction of the CHISM velocity is known (Moebius et al. 2004; Witte 2004) to be aligned with the vector $\mathbf{l}_{\mathrm{He}} = (-0.087156, 0, -0.9962)$. The HDP is defined by $\mathbf{l}_{\mathrm{He}}$ and the vector $\mathbf{l}_{\mathrm{H}}$ with the coordinates $(-0.1511, -0.04049, -0.9877)$ at

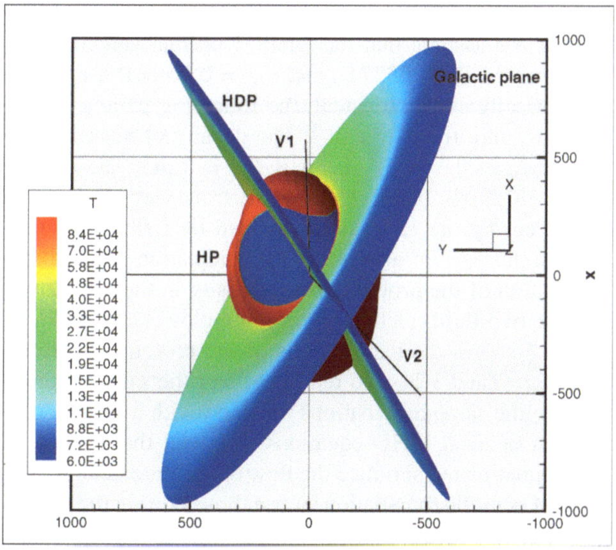

Fig. 1 Frontal view of the HP, HDP, Galactic plane, and *Voyager 1* and *2* trajectories. *Blue* and *red* colors on the surface of the HP correspond to the regions of negative and positive values of the ISMF radial component B_R, respectively. The planes have colors corresponding to the plasma temperature distributions in them. The HP is clearly asymmetric with respect to the BV-plane

which CHISM neutrals enter the inner heliosheath, according to the observations of Lallement et al. (2005). Throughout this paper we assume that $\mathbf{B}_\infty$ is aligned with the vector $\mathbf{l}_B = (-0.5, -0.2678, -0.82356)$. Thus, $\mathbf{B}_\infty$ belongs to the observed HDP and is directed into the southern hemisphere at an angle of 30° to the ecliptic plane. This means that the ISMF enters the heliosphere at the following ecliptic longitude and latitude: $\lambda \approx 238.8°$, $\beta \approx 22.9°$. This direction gave one of the largest *V1*–*V2* asymmetries of the TS in the two-fluid (one ion fluid and one neutral fluid) calculations of Pogorelov et al. (2007). Pogorelov et al. (2004) considered a case with $\mathbf{B}_\infty$ belonging to the plane tilted at 60° to the ecliptic plane, which was unwittingly identified with the Galactic plane, but in fact is very close to the observed HDP. As noticed by Stone et al. (2005), Opher et al. (2006), this orientation introduces east-west asymmetry of the TS which is consistent with the direction of energetic proton fluxes observed by *V1* and *V2*.

Voyager 2 observations (Stone et al. 2005) in the supersonic SW indicated that *V2* started measuring energetic charged particles at about 10 AU closer to the Sun than *V1*. This was demonstrably confirmed by the actual crossing of the TS by *V2* at ~84 AU compared to ~94 AU for *V1* (Stone et al. 2008). As can be seen from the model of Pogorelov et al. (2004), the ISMF with $\mathbf{B}_\infty$ at an angle of about 45° to $\mathbf{V}_\infty$ can indeed introduce a *V1*–*V2* asymmetry of this magnitude in the TS heliocentric distances. The figure showing this effect was adapted from Pogorelov et al. (2004) for the 2005 *Voyager* Senior Review Proposal. However, Pogorelov (2006) and Pogorelov et al. (2007) showed that this is an artifact of neglecting charge exchange between charged and neutral particles. Furthermore, the presence of neutrals does not allow *V2* to be directly connected by magnetic field lines to the TS at distances greater than 3 AU even for B_∞ as large as 3 μG (Pogorelov et al. 2007), while the corresponding fluxes of energetic protons were measured for at least two years (Stone et al. 2008). This implies a possible importance of time-dependent processes.

This is in distinction from the results of the ideal MHD models of Pogorelov et al. (2004) and Opher et al. (2006). The latter model, to arrive at some quantitative results, uniformly scales the ideal MHD solution to the known distance of the TS crossing by *V1*. The importance of this can be seen from the fact that the TS heliocentric distances can be 1.5 times larger if neutral particles are neglected and the scaling is nonuniform (Baranov and Malama 1993; Pauls et al. 1995; Zank et al. 1996; Zank 1999; Müller et al. 2008).

Here we assume that the CHISM plasma velocity, temperature and density are $V_\infty = 26.4\ \mathrm{km\,s^{-1}}$, $T_\infty = 6527\,\mathrm{K}$, and $n_\infty = 0.06\ \mathrm{cm^{-3}}$, respectively. It is assumed that the SW is spherically symmetric with the following parameters at 1 AU: $V_\mathrm{E} = 450\ \mathrm{km\,s^{-1}}$, $T_\mathrm{E} = 51\,100\,\mathrm{K}$, and $n_\mathrm{E} = 7.4\ \mathrm{cm^{-3}}$. The density of neutral hydrogen is $n_{\mathrm{H}\infty} = 0.15\ \mathrm{cm^{-3}}$. The magnitude of the ISMF vector is $\mathbf{B}_\infty = 3$ μG. The radial component of the IMF at 1 AU is set to 37.5 μG. Figure 1 shows the front view of the HP, obtained with our MHD-kinetic model, cut by the HDP as determined by Lallement et al. (2005), as well as the *V1* and *V2* trajectories. We also show the orientation of the Galactic plane. Figure 2 shows the distribution of the proton number density in the *V1*–*V2* plane. It is seen that, in agreement with the two-fluid calculations of Pogorelov et al. (2007), the asymmetry of the TS is minor. Figure 3 shows the distributions of the proton temperature in the directions of *V1* (solid black lines) and *V2* (solid red lines). For the sake of comparison, we also use dashed lines to show the same distributions obtained with a five-fluid model. The latter is based on the solution of ideal MHD equations to model the flow of protons and four coupled sets of Euler equations to simulate the flow of separate neutral H fluids. These consist of the parent CHISM neutrals (population 0) and those born in the outer heliosheath (population 1), inner heliosheath (population 2), and supersonic SW (population 3). It is interesting to see that if

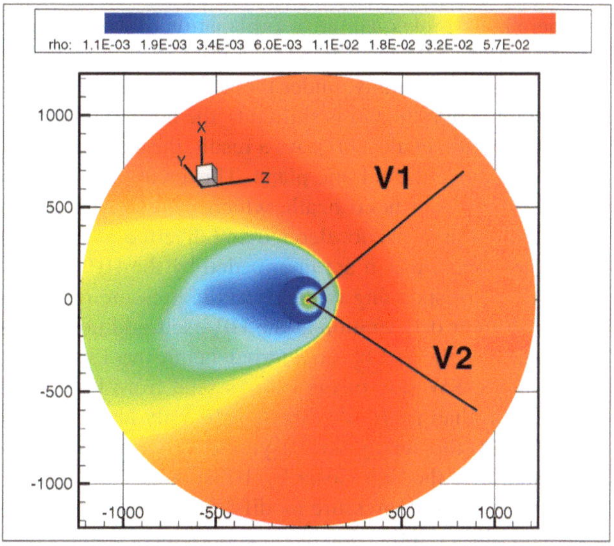

Fig. 2 Plasma number density distribution in the *V1–V2* plane. The *straight lines* show the *V1* and *V2* trajectories

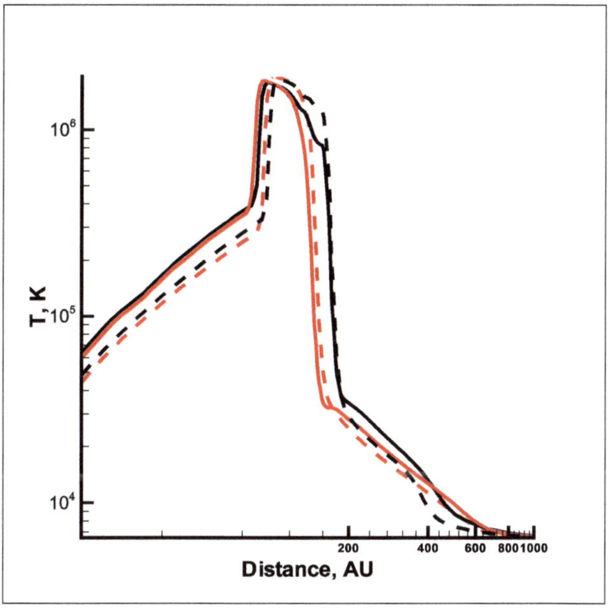

Fig. 3 The distributions of plasma temperature in the *V1* (*black lines*) and *V2* (*red lines*) directions. The results shown with *solid* and *dashed lines* are obtained with MHD-kinetic and five-fluid models, respectively (Pogorelov et al. 2008a)

charge exchange is treated kinetically, the bow shock, intrinsic to an equivalent ideal MHD model, disappears. This is because hot secondary neutral atoms created in the inner and outer heliosheaths have the ability to propagate upstream into the distant CHISM, where they heat the plasma ahead of the bow shock. One can notice that the introduction of population 1 neutrals considerably improves the solution in the outer heliosheath. On the other hand, the difference in the effective SW plasma temperature (the combination of the core and pick-up ion distributions) remains substantial because the filtration of neutral H into the HP is greater in the multi-fluid model than in the MHD-kinetic model.

There is a simple qualitative explanation of the symmetrizing effect of charge exchange. For the chosen SW and CHISM parameters, the neglect of neutral particles results in the HP rotating considerably, under the action of magnetic pressure, so that its nose is shifted to the south while the tail is shifted to the north with the appropriate shift of the nose westward (Pogorelov et al. 2007). As a result, the stagnation point of the CHISM plasma on the HP moves above the ecliptic plane. This creates an asymmetric distribution of the CHISM plasma in the outer heliosheath, which results in an enhanced charge exchange in this region. CHISM protons, decelerated and diverted at the HP, by exchanging their charge with CHISM neutrals, which are unimpeded by the HP, produce secondary (pick-up) ions with a pressure corresponding to the dynamic pressure of the parent atoms. These PUI's, in the process of their deceleration at the HP, exert additional pressure which acts to decrease the asymmetry of the HP by counterbalancing the ISMF pressure effect. A similar scenario works on the SW side of the HP. As seen from Fig. 3, the TS is closer to the Sun in the *V2* direction than in the *V1* direction by only about 3 AU in our MHD-kinetic simulation. In any event, the steady-state asymmetry is too small to ensure a *V2* crossing of the TS at a distance to the Sun closer by 10 AU than *V1*. Instead, it is likely that temporal variations in the SW ram pressure modify the TS location significantly (Scherer and Fahr 2003; Zank and Müller 2003; Borrmann and Fichtner 2005; Pogorelov et al. 2007; Washimi et al. 2007), thus supplementing the ISMF-pressure effect.

The MHD-neutral analysis of Pogorelov et al. (2007) shows that the east-west asymmetry of the TS due to the action of the ISMF lying in the observed HDP is also insufficiently large to allow *V2* to be directly (by less than a full winding of the IMF spiral) connected to the TS at distances larger than about 3 AU ahead of the TS. This is 2 AU smaller than the corresponding distance previously found in ideal MHD calculations by Opher et al. (2006) for a smaller ISMF (1.8 μG vs. 3 μG in this paper). The difference between ideal MHD and MHD-neutral results would, of course, have been larger for the same ISMF magnitude. Pogorelov et al. (2007) showed that energetic protons with energies below 7 MeV cannot reach *V2* by being connected to the TS indirectly. It is possible, in principle, to increase the TS asymmetry to 8 AU by increasing B_∞ to the surprisingly large value of 4 μG (Pogorelov et al. 2008b). This may result, however, in the H flow deflection being greater than that observed in the *SOHO* SWAN experiment (see the deflection results below). That is, these two effects are mutually related. Moreover, for magnetic fields of this strength, the radio emission conditions (Cairns and Zank 2002) will be satisfied in every point beyond the HP.

In Fig. 4 we compare the distributions of plasma density and magnetic field magnitude obtained with our five-fluid and MHD-kinetic models. Black, red, blue, and purple lines correspond to the directions $\phi = 180°, \theta = 35°$, $\phi = 0, \theta = 0$, $\phi = 0, \theta = 90°$, and $\phi = 0$, $\theta = 180°$, respectively. As usual, the angles ϕ and θ are measured from the x-axis in the xy-plane and from the z-axis, respectively. It is evident that the results obtained with these two substantially different models are in a very good qualitative and reasonable quantitative agreement.

Figure 5 shows the distributions of the neutral hydrogen atom density along the z-axis. Blue, purple, green, and red lines correspond to populations 0, 1, 2, and 3, respectively. Solid and dashed curves show the results obtained with our MHD-kinetic and five-fluid models, respectively. One will notice a discrepancy in the distributions of population 1 at large z, which is the result of a certain arbitrariness in our subdivision of populations 0 and 1 caused by the absence of a clearly defined bow shock. As noticed by Pogorelov et al. (2008a), hot neutrals born in the inner heliosheath propagate upstream into the CHISM, decelerate and heat it, thus eliminating a bow shock for the set of parameters used in this calculation. However, this discrepancy is of minor importance, since upstream propagating population 1 atoms have very low number density. It is notable that the results show

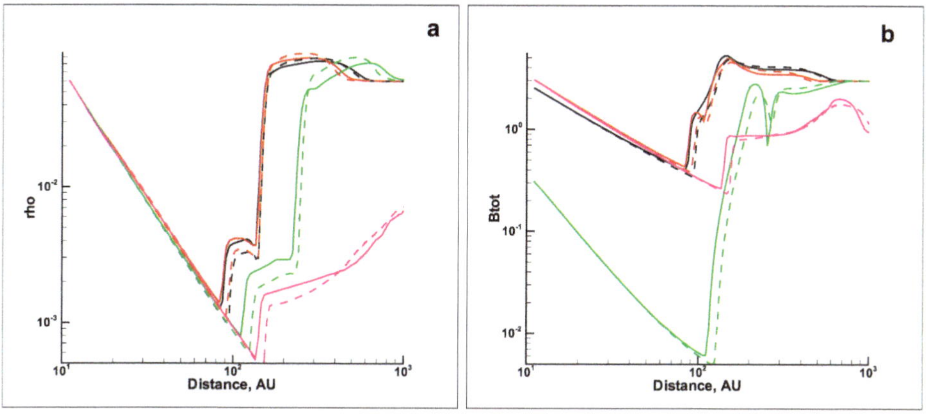

Fig. 4 The distributions of (**a**) plasma number density (in cm $^{-3}$) and (**b**) magnetic field magnitude (in μG) along the rays $\phi = 180°, \theta = 35°$ (*black lines*), $\phi = 0, \theta = 0$ (*red lines*), $\phi = 0, \theta = 90°$ (*green lines*), and $\phi = 0, \theta = 180°$ (*purple lines*). The results shown with *solid* and *dashed lines* are obtained with MHD-kinetic and five-fluid models, respectively

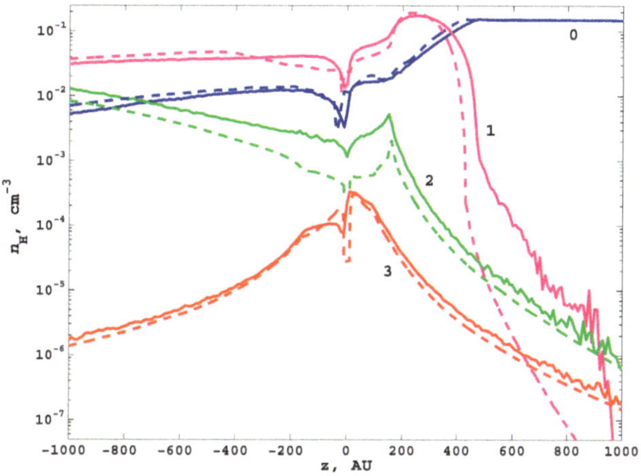

Fig. 5 The distributions of different populations of neutral hydrogen atoms along the z-axis. Blue, purple, green, and red lines correspond to populations 0, 1, 2, and 3, respectively. The results shown with solid and dashed lines are obtained with MHD-kinetic and five-fluid models, respectively

a very good qualitative agreement. As expected (see Heerikhuisen et al. 2006), the largest quantitative disagreement is in the distribution of population 2 neutrals. This is an intrinsic drawback of multi-fluid models which assume a Maxwellian distribution for these neutrals. This quantitative disparity can partially be leveled by the uncertainty in our knowledge of such properties of the CHISM as its ionization ratio and ion density.

To quantify the effect of the neutral H flow deflection, we run our kinetic neutral-atom code and collect statistics on the H-atom velocity distribution in the SW. We record the deflection from $\mathbf{V}_\infty$ of all H-atoms within a 45-degree cone about $\mathbf{V}_\infty$ out to 80 AU, both in the BV-plane and perpendicular to it, thus creating a two dimensional distribution of deflections. In Fig. 6 we show these for primary (population 0) CHISM H-atoms (top panel), secondary (population 1) H-atoms (middle panel), and the total (weighted) distribution (bottom panel) in the plane perpendicular to $\mathbf{V}_\infty$. Although the primary CHISM distribution starts out as Maxwellian, its interaction with the heliosphere results in a distribution of deflections that is obviously not isotropic. This is because charge-exchange losses may preferentially

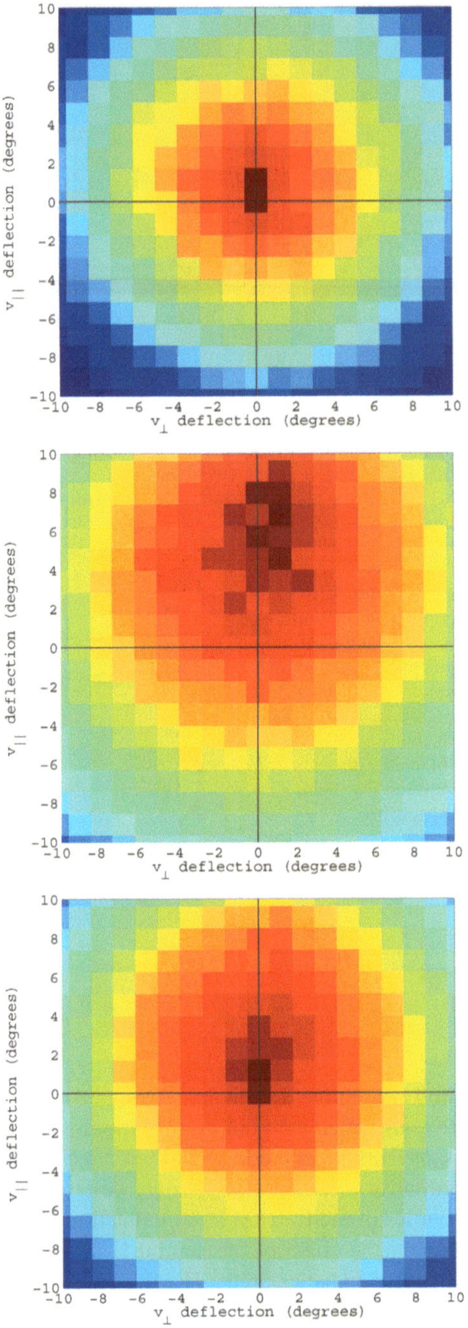

Fig. 6 Two-dimensional distribution of H-atom deflections from $\mathbf{V}_{\infty}$ in the plane perpendicular to the CHISM BV-plane (the interstellar perspective). *Top* are primary interstellar H-atoms, middle are secondary (i.e. last charge-exchange occurred in the outer heliosheath), while on the *bottom* is the combined distribution. The normal is determined by the vector product $\mathbf{l}_{\mathrm{H}} \times \mathbf{l}_{\mathrm{He}}$. The *darkest red* color corresponds to a particle count twice larger than that of the *darkest blue* color. (Pogorelov et al. 2008a)

cull a particular part of the distribution, due to asymmetric plasma flow and the dependence of the charge-exchange rate on the relative plasma flow speed. Secondary H-atoms (and the combined distribution, by extension) are clearly not isotropic, and the mean of the distribution does not coincide with its center, making it more difficult to quantify the overall deflection. We find that the average deflection of primary neutrals is about 1.8° in the BV-plane and $-0.18°$ perpendicular to this plane. For secondary neutrals, the corresponding values are 4.7° and 0.15°, while for the combined population these are 3.8° and 0.05°. We point out, however, that the peaks of the distributions are not at these locations. Instead, the primary population shows a peak close to zero deflection and the most common deflection of the secondary neutrals is around 7° in the BV-plane and 1° out of it. It appears in this particular example, that the average deflection takes place almost entirely in the BV-plane. Thus, the actual angle between the BV-plane and the HDP is determined by the accuracy of measuring the H-flow direction in the *SOHO* SWAN experiment. Although we assume that the BV-plane is parallel to the average HDP, an additional deflection of the order of $\pm 1°$ perpendicular to the average HDP cannot be excluded. This gives us an estimate for the angle between the HDP and BV-plane at $\sim 15°$ (Pogorelov et al. 2007).

3 Sources of the 2–3 kHz radio emission

Radio emission in the 2–3 kHz range is thought to be generated when a global merged interaction region (GMIR) enters the outer heliosheath, where plasma is primed with an enhanced level of superthermal electrons (Cairns and Zank 2002). The origin of the hot electrons is due to their energization by lower-hybrid waves generated by pick-up ions, created from hot secondary neutrals born in the inner heliosheath as they propagate into the outer heliosheath and charge-exchange with the shocked plasma. These pick-up ions have a ring-beam distribution. Since ring-beam driven lower hybrid waves propagate almost perpendicularly to the magnetic field vector $\mathbf{B}$ in the outer heliosheath, Gurnett et al. (2006) suggested that regions which satisfy this property ahead of GMIRs might preferentially radiate in the 2–3 kHz range. If $\mathbf{B}_\infty$ lies in the observed HDP, the distribution of magnetic field strength in the outer heliosheath suggests that radio-emission-source distribution will be elongated in the direction almost perpendicular to it (Pogorelov et al. 2007). If one assumes, as was done for the purpose of identifying the distribution of radio emission sources by Opher et al. (2007), that a GMIR is spherical initially and preserves its sphericity after crossing the TS and the HP, then the regions on the HP and in the outer heliosheath where the radial component B_R of the magnetic field is equal to zero might be candidate regions for radio emission. The HP shown in Fig. 1 is painted with two colors, red and blue, which correspond to $B_R > 0$ and $B_R < 0$, respectively. The boundary between them is where $B_R = 0$. Here, like Pogorelov et al. (2007, 2008a), we see that there is a line that contains possible radio emission sources on the HP surface and is nearly perpendicular to the BV-plane. Note that *V1* detected radio emission sources distributed along the Galactic plane and also additional sources not aligned with the Galactic plane, called "ambiguous" by Kurth and Gurnett (2003). One can see, however, that the approach based on determining regions where $B_R = 0$ can exhibit the distributions of radio emission sources, which have many different orientation in space. From this viewpoint, the use of the space orientation of possible-radio-emission source distributions as a test bed (Opher et al. 2007) for the HDP definition is questionable. Of course, this is not necessary if we model the transport of neutral H atoms self-consistently with the plasma flow.

There is another issue that might be important if we suppose that the $B_R = 0$ condition can determine the distribution of radio emission sources. The line (or lines) defined by this equation on the surface of the HP is ambiguous because the exact location of the HP is not known with enough precision. It would be much more relevant to look at the shape of the surface $B_R = 0$ in the outer heliosheath. This is shown in Fig. 7, which is produced from

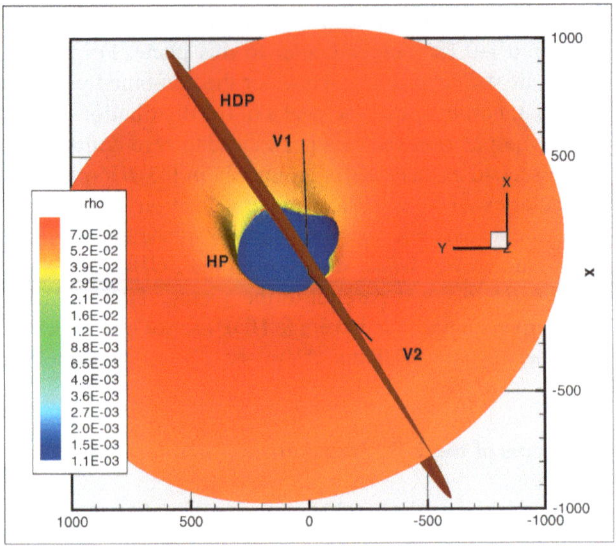

Fig. 7 The iso-surface $B_R = 0$ crossing the heliopause (compare with Fig. 1). The visible part of the HP surface has *blue* color indicating the region where $B_R < 0$. The *straight lines* corresponding to *V1* and *V2* trajectories and the HDP are also shown. 2–3 kHz radio emission sources are supposed to be distributed favorably on the $B_R = 0$ surface in the vicinity of the heliopause. This region is slightly elongated in the direction perpendicular to the HDP, although the lines of possible radio emission have no single particular orientation

Fig. 1 by adding the above-mentioned iso-surface and removing the Galactic plane. The blue color here shows negative values of B_R on the surface of the HP. The iso-surface $B_R = 0$ is colored according to the plasma density value on it. It is clear that the visible part of the boundary of the blue region and its vicinity belonging to the shown iso-surface are the best candidates for radio emission. It is seen that the domain is slightly elongated in the direction perpendicular to the HDP. On the iso-surface, there seems to exist a region containing the sources of possible radio emission, which is perpendicular to the HDP in the vicinity of the z-axis crossing the HP. There is also a region which is nearly parallel to the Galactic plane. However, there are also other regions that can be excluded only after analyzing whether physical conditions for radio emission are satisfied. It is therefore clear that one would need to use MHD-kinetic simulations to combine the condition $B_n = 0$, where B_n is the magnetic field component normal to a (nonspherical) GMIR shock surface, and utilize the physical conditions for generation of radio waves, if we are to use radio emission data as an additional constraint on the orientation of $\mathbf{B}_\infty$. As the angle between the HDP and the BV-plane can be as large as 15°, the ISMF orientation can presumably be adjusted to be consistent with the radio emission observations.

4 Conclusions

We have discussed the importance of charge exchange between charged and neutral particles for quantifying heliospheric asymmetries induced by the ISMF. It is clear, both from the qualitative analysis and the numerical modeling, that charge exchange tends to decrease the ISMF effect on the HP and, finally, on the TS. On the basis of a physical model that treats the motion of charged particles using ideal MHD equations and the transport of neutral particles kinetically, by solving the Boltzmann equation, we created distributions of neutral H velocity deflections from their original direction in the unperturbed CHISM, which is conventionally assumed to coincide with the neutral He velocity direction. We found that these distributions are highly anisotropic, with the most probable deflection not coinciding with the average deflection. It is however remarkable that, for the case considered in this paper, the velocity deflection in the BV-plane is substantially larger than that in the direction perpendicular to the BV-plane. Since the deflection observed by Lallement et al. (2005) is $4^\circ \pm 1^\circ$, the meaning of the word "larger" is defined by the accuracy to which the deflection is measured. That is, any deflection of less than 1° in the direction perpendicular to the BV-plane is of no practical interest when comparing with the *SOHO* SWAN measurements. Being cautiously optimistic, we can conclude that, for reasonable ISMF strengths of less than 3 μG, the angle between the BV-plane and the observed HDP should not be larger than $\arctan(\tan 1^\circ / \tan 4^\circ) \approx 14^\circ$ (see also Pogorelov et al. 2007, 2008a, 2008b). For $B_\infty = 3$ μG, the angle of 30° between $\mathbf{V}_\infty$ and $\mathbf{B}_\infty$ is sufficient to introduce a 4°-deflection of neutral H, which is consistent with the *SOHO* observations. For $B_\infty = 2.5$ μG, one might need to increase the angle to 45°. We showed that naive geometrical considerations do not allow us to determine the distribution of 2–3 kHz radio emission sources in the outer heliosheath. Moreover, ideal MHD simulations based on the models of Pogorelov et al. (2004), Opher et al. (2007) do not determine all geometrically-possible radio emission source distributions. To perform meaningful comparison of theoretical results with *Voyager* radio observations, one needs to combine numerical simulations of physical processes underlying the radio emission with the theoretical conditions ensuring its generation.

We performed a comparison of numerical results from different (MHD-kinetic and five-fluid) theoretical models of the heliospheric interface. The flow of charged particles is governed by the ideal MHD equations in both cases. The difference is in the way we treat the transport of neutral particles. The results for the distribution of protons obtained from our five-fluid and MHD-kinetic models are very consistent qualitatively. There are some quantitative differences, which are very well understood from a physical viewpoint, because kinetic transport of neutrals throughout the heliosphere differs from that based on the continuum description of multiple neutral fluids. Although the kinetic approach is physically relevant for the processes occurring in the outer heliosphere, one can easily see that the plasma distributions can be matched by modifying the CHISM properties to within the accuracy we know them. As far as the distribution of neutral particles is concerned, it is known (Heerikhuisen et al. 2006) that populations 0, 1, and 3 of neutrals are fairly well represented by Maxwellian distribution functions, while population 2 is not. As a result, the difference between our five-fluid and MHD-kinetic models is mostly due to the action of this population. While kinetic modeling of neutrals is unquestionably superior to a multi-fluid approach, the averaged results obtained with both approaches are very similar. The application of multi-fluid approaches might prove to be especially useful for realistic time-dependent problems addressing the propagation of SW transients through the heliosphere. Stochastic Monte Carlo approaches become very inefficient for problems where very small time (of the order of a day) and length (of the order of a fraction of AU) scales should be resolved. The question of physical validation of different SW–CHISM interaction models has been addressed recently by Müller et al. (2008).

Acknowledgements The authors are grateful to Horst Fichtner and Edward C. Stone for invaluable comments.

References

V.B. Baranov, K.V. Krasnobaev, Cosm. Res. **9**(4), 568–571 (1971)
V.B. Baranov, Y.G. Malama, J. Geophys. Res. **98**, 15157–15163 (1993)
T. Borrmann, H. Fichtner, Adv. Space Res. **35**, 2091–2101 (2005)
I.H. Cairns, G.P. Zank, Geophys. Res. Lett. **29**(7), 47-1–47-4 (2002)

D.P. Cox, L. Helenius, Astrophys. J. **583**, 205–228 (2003)

H.-J. Fahr, S. Grzedzielski, R. Ratkiewicz, Acta Geophys. **6**, 337–354 (1988)

V. Florinski, N.V. Pogorelov, G.P. Zank, B.E. Wood, D.P. Cox, Astrophys. J. **604**, 700–706 (2004)

D.A. Gurnett, W.S. Kurth, I.H. Cairns, J.J. Mitchell, in *Physics of the Inner Heliosheath: Voyager Observations, Theory, and Future Prospects*, ed. by J. Heerikhuisen et al. American Institute of Physics Conf. Proc., vol. 858 (AIP, New York, 2006), pp. 129–134

J. Heerikhuisen, V. Florinski, G.P. Zank, J. Geophys. Res. **111**, A06110 (2006)

J. Heerikhuisen, N.V. Pogorelov, G.P. Zank, V. Florinski, Astrophys. J. **655**, L53–L56 (2007)

V. Izmodenov, D. Alexashov, A. Myasnikov, Astron. Astrophys. **437**, L35–L38 (2005)

W.S. Kurth, D.A. Gurnett, J. Geophys. Res. **108**(A10), 8027 (2003)

R. Lallement, E. Quémerais, J.L. Bertaux, S. Ferron, D. Koutroumpa, R. Pellinen, Science **307**, 1447–1449 (2005)

T.J. Linde, T.I. Gombosi, P.L. Roe, K.G. Powell, D.L. DeZeeuw, J. Geophys. Res. **103**(A2), 1889–1904 (1998)

J.J. Mitchell, I.H. Cairns, N.V. Pogorelov, G.P. Zank, J. Geophys. Res. **113**(A4), A04102 (2008)

E. Moebius et al., Astron. Astrophys. **426**, 897–907 (2004)

H.-R. Müller, V. Florinski, J. Heerikhuisen, V.V. Izmodenov, K. Scherer, D. Alexashov, H.-J. Fahr, Astron. Astrophys. (2008, in press)

M. Opher, E.C. Stone, P.C. Liewer, Astrophys. J. **640**, L71–L74 (2006)

M. Opher, E.C. Stone, T.I. Gombosi, Science **316**, 875–878 (2007)

H.L. Pauls, G.P. Zank, L.L. Williams, J. Geophys. Res. **100**, 21595–21604 (1995)

N.V. Pogorelov, in *Physics of the Inner Heliosheath: Voyager Observations, Theory, and Future Prospects*, ed. by J. Heerikhuisen et al. American Institute of Physics Conf. Proc., vol. 858 (AIP, New York, 2006), pp. 3–13

N.V. Pogorelov, T. Matsuda, J. Geophys. Res. **103**(A1), 237–245 (1998)

N.V. Pogorelov, G.P. Zank, Astrophys. J. **636**, L161–L164 (2006)

N.V. Pogorelov, G.P. Zank, T. Ogino, Astrophys. J. **614**, 1007–1021 (2004)

N.V. Pogorelov, G.P. Zank, T. Ogino, Astrophys. J. **644**, 1299–1316 (2006)

N.V. Pogorelov, E.C. Stone, V. Florinski, G.P. Zank, Astrophys. J. **668**, 611–624 (2007)

N.V. Pogorelov, J. Heerikhuisen, G.P. Zank, Astrophys. J. **675**, L41–L44 (2008a)

N.V. Pogorelov, S.N. Borovikov, J. Heerikhuisen, I.A. Kryukov, G.P. Zank, in *Particle Acceleration and Transport in the Heliosphere and Beyond*, ed. by G. Li et al. American Institute of Physics Conf. Proc., vol. 1039 (AIP, New York, 2008b), pp. 410–417

R. Ratkiewicz, A. Barnes, G.A. Molvik, J.R. Spreiter, S.S. Stahara, M. Vinokur, Astron. Astrophys. **335**, 363–369 (1998)

K. Scherer, H.J. Fahr, Geophys. Res. Lett. **30**(2), 1045 (2003)

E.C. Stone, A.C. Cummings, F.B. McDonald, B. Heikkila, N. Lal, W.R. Webber, Science **309**, 2017–2020 (2005)

E.C. Stone, A.C. Cummings, F.B. McDonald, B. Heikkila, N. Lal, W.R. Webber, Nature **454**, 71–74 (2008)

H. Washimi, G.P. Zank, Q. Hu, T. Tanaka, K. Munakata, Astrophys. J. **670**, L139–L142 (2007)

M. Witte, Astron. Astrophys. **426**, 835–844 (2004)

G.P. Zank, Space Sci. Rev. **89**, 413–688 (1999)

G.P. Zank, H.-R. Müller, J. Geophys. Res. **108**(A6), 1240 (2003)

G.P. Zank, H.L. Pauls, L.L. Williams, D.T. Hall, J. Geophys. Res. **101**, 21639–21656 (1996)

Confronting Observations and Modeling: The Role of the Interstellar Magnetic Field in Voyager 1 and 2 Asymmetries

M. Opher · J.D. Richardson · G. Toth · T.I. Gombosi

Originally published in the journal Space Science Reviews, Volume 143, Nos 1–4, 43–55.
DOI: 10.1007/s11214-008-9453-x

Abstract Magnetic effects are ubiquitous and known to be crucial in space physics and astrophysical media. We have now the opportunity to probe these effects in the outer heliosphere with the two spacecraft Voyager 1 and 2. Voyager 1 crossed, in December 2004, the termination shock and is now in the heliosheath. On August 30, 2007 Voyager 2 crossed the termination shock, providing us for the first time in-situ measurements of the subsonic solar wind in the heliosheath. With the recent *in-situ* data from Voyager 1 and 2 the numerical models are forced to confront their models with observational data. Our recent results indicate that magnetic effects, in particular the interstellar magnetic field, are very important in the interaction between the solar system and the interstellar medium. We summarize here our recent work that shows that the interstellar magnetic field affects the symmetry of the heliosphere that can be detected by different measurements. We combined radio emission and energetic particle streaming measurements from Voyager 1 and 2 with extensive state-of-the art 3D MHD modeling, to constrain the direction of the local interstellar magnetic field. The orientation derived is a plane ~60°–90° from the galactic plane. This indicates that the field orientation differs from that of a larger scale interstellar magnetic field, thought to parallel the galactic plane. Although it may take 7–12 years for Voyager 2 to leave the heliosheath and enter the pristine interstellar medium, the subsonic flows are immediately sensitive to the shape of the heliopause. The flows measured by Voyager 2 in the heliosheath indicate that the heliopause is being distorted by local interstellar magnetic field with the same orientation as derived previously. As a result of the interstellar magnetic field the solar system is asymmetric being pushed in the southern direction. The presence of hydrogen atoms tend to symmetrize the solutions. We show that with a strong interstellar magnetic field with our

M. Opher (✉)
Department of Physics and Astronomy, George Mason University, 4400 University Drive, Fairfax, VA 22030, USA
e-mail: mopher@physics.gmu.edu

J.D. Richardson
Massachusetts Institute of Technology, 77 Massachusetts Ave., Cambridge, MA, USA

G. Toth · T.I. Gombosi
Center for Space Environment Modeling, University of Michigan, Ann Arbor, MI, USA

J.L. Linsky et al. (eds.), *From the Outer Heliosphere to the Local Bubble*.
DOI: 10.1007/978-1-4419-0247-4_6

most current model that includes hydrogen atoms, the asymmetries are recovered. It remains a challenge for future works with a more complete model, to explain all the observed asymmetries by V1 and V2. We comment on these results and implications of other factors not included in our present model.

Keywords Solar magnetic field · Solar system: Solar wind · Heliosphere · Termination Shock · Heliosheath

1 Introduction

Due to the motion of the solar system through the interstellar medium with a velocity of 26.3 ± 0.4 km/s, the solar wind is deflected by an interstellar wind, producing a comet-like shape with an extended tail. We observe the collision of winds in astrophysical media via *remote sensing*, such as the bow shocks formed ahead of R Hya (Wareing et al. 2006). Through the study of the interaction of the solar system and the interstellar medium, we can learn with *in-situ data*, on how two magnetized winds collide.

Several boundaries characterize this interaction: the termination shock, where the solar wind become subsonic, and the heliopause, where the two ionized winds are deflected with respect to each other. If the interstellar wind is supersonic, there is a third boundary, the bow shock, where the interstellar wind becomes subsonic. Beyond the termination shock, the solar wind is gradually deflected tailward. As the Sun rotates, the solar magnetic field is carried outward by the solar wind and forms a spiral, becoming almost completely azimuthal in the outer heliosphere.

The size and shape of the heliosphere will depend on the properties of both the solar wind and the local interstellar medium. From all the quantities that affect the boundaries of this interaction, the least known quantity is the interstellar magnetic field intensity and direction.

The interaction of the solar system with the interstellar medium is a highly complex system. Not only are the two winds are ionized winds, and carry magnetic fields, but they also carry neutral H atoms, neutral He atoms; cosmic rays and pickup ions. Another complication is the solar cycle. The solar cycle affects the solar wind and the magnetic field embedded in it. The heliospheric current sheet (HCS) is known to change its inclination in respect to the solar rotation axis as the solar cycle progresses. The flows produced by the HCS and the solar cycle in the heliosheath has not yet been studied in detail. Also, with the solar cycle there will be also more global merger interaction regions (that result of coronal mass ejections) that will disturb this interaction as well.

There have been several different techniques trying to model this interaction (Suess 1993; Zank 1999; Izmodenov and Baranov 2006). The challenge is to swift through all the different effects and isolates the more crucial ones. A special difficulty arises from the fact that a lot of the effects go hand in hand.

We are fortunate to have the twin spacecraft Voyager 1 (V1) and 2 (V2) probing in-situ the northern and southern hemisphere. Jointly with IBEX (McComas et al. 2004) that will remotely image the three-dimensional structure of the heliosphere, we now have a new view of the outer boundaries of the solar system. V1 crossed the termination shock in December 2004 (Decker et al. 2005; Burlaga et al. 2005; Stone et al. 2005). V2 is the only spacecraft that has the capability to measure in-situ the plasma flows. In August 2007 V2 crossed the Termination Shock (TS) providing us for the first time with in-situ measurements of the subsonic flows in the heliosheath.

There is several recent observational evidence from V1 and 2 that there are asymmetries of the solar system. The observational evidence comes from different instruments and locations (termination shock, heliopause and heliosheath), which indicates that V1 and 2 are detecting the signs of a global distortion of the heliosphere.

In several recent papers (Opher et al. 2006, 2007, 2008) we proposed that the asymmetries are due to a *global* factor distorting the solar system that is the interstellar magnetic field. Although it is intuitive to expect that interstellar magnetic field will have an effect on heliospheric asymmetries, it is not clear how strong this effect is and that the interstellar magnetic field solely is responsible for them (and not due to temporal or solar cycle effects). We analyzed the effect that the interstellar magnetic field direction has in the termination shock, in the heliopause and in the heliosheath. We found that the orientation of the interstellar magnetic field that matches the observations is in a plane 60°–90° from the galactic plane. This direction differs from the direction in large scale of the magnetic field that is the plane of the galaxy.

There are few previous modeling studies that include both the solar magnetic field and the interstellar magnetic field (Linde et al. 1998; Linde 1998; Ratkiewicz and Ben-Jaffel 2002; Pogorelov and Matsuda 1998; Pogorelov and Zank 2006; Pogorelov et al. 2004, 2007, 2008a, 2008b; Washimi and Tanaka 2001; Washimi et al. 2007). Linde (1998) and Linde et al. (1998) included the effects of neutral hydrogen atoms treated in a fluid way along with the solar magnetic field and an interstellar magnetic field either parallel or perpendicular to the interstellar wind velocity. Pogorelov and Zank (2006) and Pogorelov et al. (2008a) investigated models with inclined magnetic fields with neutral atoms treated as a separate fluid, while the model by Izmodenov et al. (2005b) included neutrals with a kinetic treatment, but did not include the solar magnetic field.

Until recently, results from computer modeling were studies showing *qualitative* general asymmetries in the solar system due to interstellar magnetic field (such as in Pogorelov and Matsuda 1998; Ratkiewicz et al. 1998; Ratkiewicz and Ben-Jaffel 2002). Izmodenov et al. (2005b) show that an interstellar magnetic field can produce the 4° degrees deflection of H atoms from the direction of the pristine interstellar gas flow (with a model that include kinetic neutrals and no interplanetary magnetic field); in agreement with Lallement et al. (2005).

In models such as Pogorelov and Matsuda (1998), Pogorelov and Zank (2006), Zank (1999) it can be seen that the interstellar magnetic field affect the termination shock; although these pioneer works didn't analyze the asymmetries at the V1 and V2 locations.

Works such as Izmodenov et al. (2003), Izmodenov and Alexashov (2006) and used observations of interstellar H atoms, and modeling of kinetic H and He to constrain the crossing of the termination shock by Voyager 1 and 2.

Opher et al. (2006) was the first study to compare quantitative the computer models with the observations of V1 and V2 both for the TS distances; as with the east-west asymmetry; and the particles produced at the TS (TSP particles). The model described in Opher et al. (2006) explained the direction of streaming of the TSPs detected by V1 and predicted that V2 will detect TSPs streaming from the opposite direction than V1 what was later observed! Opher et al. (2007) showed that the interstellar magnetic field can produce the observed asymmetries also in the radio emissions; and that an interstellar magnetic field in the galactic plane wont be consistent with either the TSPs streaming directions or the radio emissions. After these works, Pogorelov et al. (2007, 2008a, 2008b) followed with a more complete model (including a self-consistent neutral H atoms) and showed that asymmetries are diminished in comparison with pure MHD calculations.

In this paper we reviews our previous work giving special attention to the observations that need to be explained. It remains an open challenge to follow our previous works and with a more complete model reproduce all the heliospheric asymmetries seen. In Sect. 2, we describe in detail the observational evidence of asymmetry and in Sect. 3 the possible directions for the interstellar magnetic field. In Sect. 4, we describe our model and in Sect. 5 we discuss our results including the orientation of the interstellar magnetic field direction. In Sect. 6, we conclude with future work and discussions.

2 Observational Evidences for Asymmetries

A. *Position of the Termination Shock*: V1 crossed the termination shock in 2004 at a distance of 94 AU (Stone et al. 2005). V2 on August 30, 2007 crossed the shock at 83.7 AU (Stone et al. 2008), indicating that the termination shock is 10 AU closer to the Sun in the southern hemisphere.

B. *Anisotropic streaming of Low Energy Particles*: In mid 2002, V1 began observing strong beams of energetic termination shock particles (TSPs) streaming *outward* along the spiral magnetic field. The strong upstream TSP beams were observed much of the time until V1 crossed the shock at 94 AU in December 2004. Jokipii et al. (2004) and Stone et al. (2005) suggested that the upstream beaming resulted from a non-spherical shock. For a spherical shock, V1 would have observed upstream TSPs streaming inward along the magnetic field. With a non-spherical shock, V1 could be connected to the termination shock along magnetic field lines that crossed the termination shock (the source of TSPs) and then crossed back into the supersonic solar wind. The north-south asymmetry was also seen by the observations (Stone et al. 2005), where V2 started detecting TSPs 10 AU before V1 started. V2 observed TSPs streaming *inward* (Cummings et al. 2005).

C. *The distance of the spacecraft to the shock when starting to detect the low-energy particles from the shock:* Although there was no direct indication how distant V1 was from the shock when the upstream episodes of TSPs were observed, it is possible to estimate the distance from time-dependent models of the heliosphere (e.g., Izmodenov et al. 2008c). MHD models based on V2 solar wind pressure measurements (Richardson and Wang 2005), indicate that the distance was less than 3 to 4 AU; and at the time when V2 (was at 75 AU) started measuring the TSPs it was at a distance 5–7 AU from the shock (Washimi et al. 2007), indicating that the termination shock is further pushed in, in the southern hemisphere, by 9 AU.

D. *The unrolling of the energetic particle energy spectra:* Far from the TS the energetic particle spectra form a peak, essentially corresponding to the classical anomalous cosmic rays. Above the peak the source spectrum is observed since the high energy particles are hardly disturbed in route from the source region to the spacecraft. Below the peak the intensities fall sharply because these lower energy particles are prevented from reaching the spacecraft by transport effects such as scattering or the geometry of the magnetic connection and the relative velocity of the particle and the speed at which it is convected through the shock. As the spacecraft and shock approach one another the peak shifts to lower energies and higher intensities as fewer particles at progressively lower energies are prevented from reaching the spacecraft. This is seen at both Voyager 1 and 2, however for V1 the process is apparent early while at V2 it is only just before crossing the shock that this particular spectral evolution (or "unrolling") takes place (Decker et al. 2005). This indicates that the point of magnetic connectivity is farther away from V2 than from V1.

E. *Radio Emission at the Heliopause*: In the last 20 years, V1 and 2 have been detecting radio emissions in the outer heliosphere at frequencies from 2 to 3 kHz (Kurth et al. 1984;

Gurnett et al. 1993; Gurnett et al. 2003). The radio emissions were detected each solar cycle: first in 1983–84 during the solar cycle 21 (Kurth et al. 1984), second in 1992–94 during solar cycle 22 (Gurnett et al. 1993), and most recently in solar cycle 23 (Gurnett et al. 2003). The currently accepted scenario is that the radio emissions are generated when a strong interplanetary shock produced by a period of intense solar activity reaches the vicinity of the heliopause and move into the interstellar plasma beyond (Gurnett and Kurth 1995; Gurnett et al. 2003). Radio direction-finding measurements from V1 and V2 have been used to determine the positions near the heliopause at which the radio emission are generated (Kurth and Gurnett 2003). The sources lie along a line that passes near the nose of the heliosphere that roughly parallels the galactic plane. The GAL plane is 120° from the ecliptic plane. Based on the fact that the galactic magnetic field is oriented nearly parallel to the galactic plane, Kurth and Gurnett (2003) suggested the local interstellar magnetic field (in the local neighborhood of the sun) was also parallel to the galactic plane.

F. *Flows in the Heliosheath*: The termination shock separates the supersonic solar wind from the subsonic solar wind. Subsonic flows are sensitive to the obstacle ahead (Opher et al. 2008). Therefore, beyond the termination shock the flows will immediately be sensitive to the shape of the heliopause and start deflecting in response. The shape of the heliopause is affected by the pressure of the local interstellar magnetic field. The measured flows by V2 can probe additional asymmetries.

These observations indicate that V1 and 2 are detecting the signs of a global distortion of the heliosphere by the interstellar magnetic field.

3 Interstellar Magnetic Field Direction

As mentioned above, among the several physical quantities that describe the interaction of the solar system with the interstellar medium, the least known are the direction and intensity of the local interstellar magnetic field ($B_{\rm ISM}$). Models suggest that the strength of $B_{\rm ISM}$ is around few μG (Cox and Helenius 2003). Based on the polarization of light from nearby stars, Frisch (1990) suggested that the magnetic field direction is parallel to the galactic plane (and directed toward $l \sim 70°$) (GAL). Voyager 3 kHz radio emission data also show preferred source locations in a plane parallel to the galactic plane (Kurth and Gurnett 2003). On the other hand, Lallement et al. (2005), mapping the solar Lyman-α radiation that is resonantly backscattered by interstellar hydrogen atoms, found that the neutral hydrogen flow direction differs from the helium flow direction by 4°. The plane of the H deflection is tilted from the ecliptic plane by ~60° and is consistent with an interstellar magnetic field parallel to the H-deflection plane (Izmodenov et al. 2005b). We refer to this plane as the H-deflection plane (HDP).

However, Gurnett et al. (2006) recently pointed out that at the earth's bow shock and interplanetary shocks, the radio emission occurs where the magnetic field lines are tangential to the shock surface and suggested that heliospheric radio emissions occur where the local interstellar magnetic field is tangential to the surface of the shock that excites the plasma (or $\mathbf{B} \cdot \mathbf{n} = 0$, where **B** is the magnetic field and **n** is the shock normal). They conclude that the condition $\mathbf{B} \cdot \mathbf{n} = 0$ combined with the observed source location by Voyager spacecrafts implies that the local interstellar magnetic field is perpendicular to the galactic plane. This direction differs from the earlier suggestion (Gurnett and Kurth 1995) and is within 16° of the HDP plane.

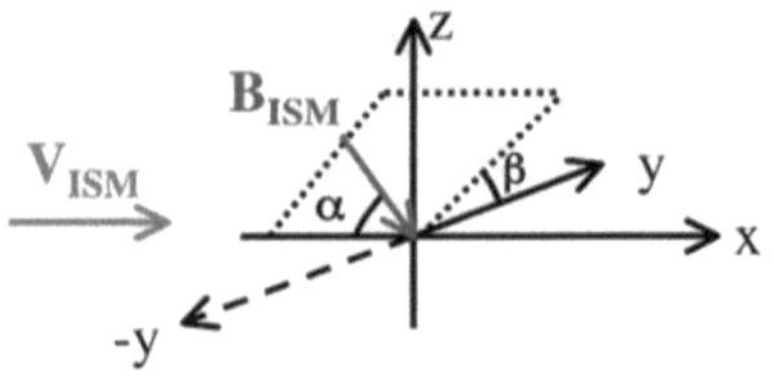

Fig. 1 Coordinate system used in the model

4 Model

The model that we used in Opher et al. (2006, 2007, 2008) is based on the BATS-R-US code, a three-dimensional magnetohydrodynamic (MHD) parallel, adaptive grid code developed by University of Michigan (Powell et al. 1999) and adapted by Opher et al. (2003, 2004) for the outer heliosphere problem. Since we were interested in the global properties of the heliosphere, we used a coarse grid with cells sizes ranging from 1.5 AU to 20 AU. The inner boundary was set at 30 AU and the outer boundary was from −1500 AU to 1500 AU in the y and z direction; and from −800 AU to 800 AU in the x direction. The solar magnetic field axis was aligned with the solar rotation axis with a 26 days solar rotation period. The solar wind was taken as uniform 450 km/s; only the ionized component was included. In Opher et al. (2008) neutral H atoms are included as well in a 5 multi-fluid description. The parameters of the solar wind at the inner boundary (30 AU) are $n = 7.8 \times 10^{-3}$ cm^{-3}, $T = 1.6 \times 10^3$ K and a Parker spiral magnetic field with $B = 2$ μG at the equator. For the interstellar wind, we used $n = 0.07$ cm^{-3}, and $T = 10^4$ K (Frisch 1996). The interstellar magnetic field ($B_{\rm ISM}$) magnitude is taken to be $B_{\rm ISM} = 1.8$ μG to 2.5 μG (and in Opher et al. 2008 we used 4.375 μG) (with the y component of $B_{\rm ISM}$, $B_{{\rm ISM},y} < 0$). The coordinate system has the interstellar velocity direction in the $+x$ direction and the z-axis is parallel to the solar rotation axis, with y completing the right-handed coordinate system. In this coordinate system, V1 is at 29.1° in latitude and 178.4° in longitude and V2 is at −31.2° and 213.4° in longitude.

Figure 1 shows the plane for the $B_{\rm ISM}$ in the model coordinate system. The coordinate system has the z-axis as the solar rotation axis of the sun, the interstellar velocity direction in the x direction, with y completing the right handed coordinate system. In this coordinate system β is the angle between the interstellar magnetic field and the solar equator and α the angle is the angle between the interstellar magnetic field and interstellar wind velocity.

5 Squashed Solar System: Global Asymmetries

The interstellar magnetic field is frozen into interstellar plasma that is deflected around the heliopause, causing the field to drape over the heliopause. If the angle between the interstellar magnetic field and interstellar velocity is non zero, it should break the axial symmetry of the heliosphere. This should be seen in the distortion of the heliopause and the termination shock. The shape of the heliopause, is distorted by pressure of the local interstellar magnetic field. For intensities around a few microgauss, the ambient interstellar magnetic pressure is comparable to the gas pressure, with the magnetic pressure increasing further in those regions where the interstellar flow decreases as it approaches the heliopause.

The heliopause surface will vary with the orientation and strength of the local interstellar magnetic field. Figure 2a indicates that the heliopause is strongly influenced by the interstellar magnetic field direction; the heliopause is asymmetric both north/south and east/west and has a plane of symmetry approximately parallel to the plane of the local interstellar magnetic field. The orange field lines are the interstellar magnetic field lines. The yellow

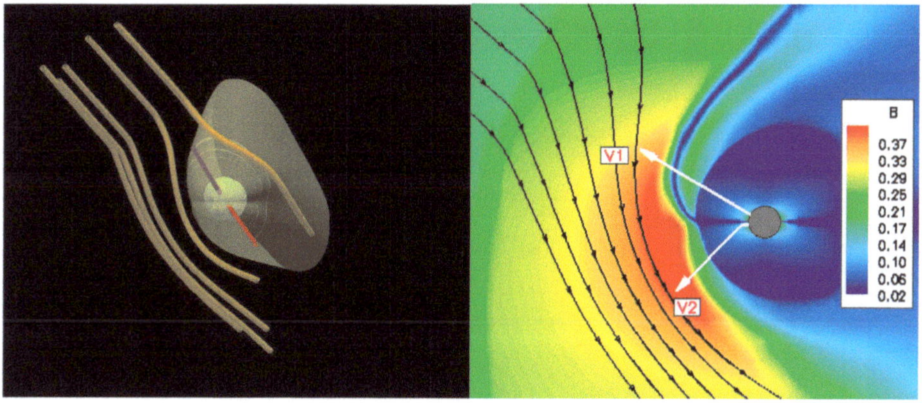

a) b)

Fig. 2 (**a**) Iso-surface of the Heliopause at $\log T = 11.5$ (*yellow*). The *green* iso-surface is the termination shock and the *pink* and *red* are, respectively the trajectory of V1 and V2. The *orange* field lines are the interstellar magnetic field in the HDP plane. (**b**) Meridional cut of the case (**a**) showing the contours of the magnetic field intensity. The *black* streamlines are the interstellar magnetic field lines

surface is the heliopause and the green iso-surface is the termination shock. The trajectory of V1 and 2 are shown in magenta and red, respectively.

At the heliopause due to the slow down of the plasma flow and piling up of the interstellar magnetic field the intensity of the magnetic field outside the heliopause is larger at the southern hemisphere rather that at the northern hemisphere (Fig. 2b).

In Opher et al. (2006, 2007) we investigated the effect of different interstellar magnetic field directions on the shape of the heliosphere. We considered several directions of interstellar magnetic field: the hydrogen deflection plane (HDP) ($\beta = 60°$); the galactic plane (GAL) ($\beta = 120°$); and the plane perpendicular to the galactic plane (PPG) ($\beta = 44°$).

In Opher et al. (2006) we predicted that an interstellar magnetic field parallel to the HDP, distort the TS such that the TSPs will stream inward at V2; which was later confirmed (Cummings et al. 2005; Decker et al. 2008; Stone et al. 2008). The model explained the outward streaming of the TSPs at V1 (Fig. 3a and 3b) (Asymmetry B). It also predicted the distance of the TS being closer in the southern hemisphere by 10 AU (Fig. 1a) (Asymmetry A).

We show (Opher et al. 2007) that for an interstellar magnetic field in the GAL the TS is distorted in the opposite direction, such that the TSPs stream inward towards V1 opposite to what was observed (Cummings et al. 2005) (Fig. 3c and 3d). Additionally the model also predicted that V1 will be 2–3 AU from the shock and V2 will be 7–10 AU. This is in agreement with Asymmetry C.

The model also explained the difference in unrolling of the energetic particles (Asymmetry D). Figure 4a shows that for $B_{\rm ISM}$ parallel to the HDP, the longitude of the MD (MD stands for the position at which the shock has the minimum radial distance to the Sun) at the northern hemisphere is closer to V1 than at the southern hemisphere where MD is farther away to V2 (in agreement with Asymmetry D). The model also explains the location of the radio sources (Opher et al. 2007) (Asymmetry E). Assuming a spherical interplanetary shock, the tangential field condition for the radio sources translates to $B_r = 0$ with B_r being the radial component of the interstellar magnetic field. For each modeled direction of the interstellar magnetic field, we compared the expected location of the radio sources ($B_r = 0$ at the heliopause) with the observed location of the radio sources detected by V1

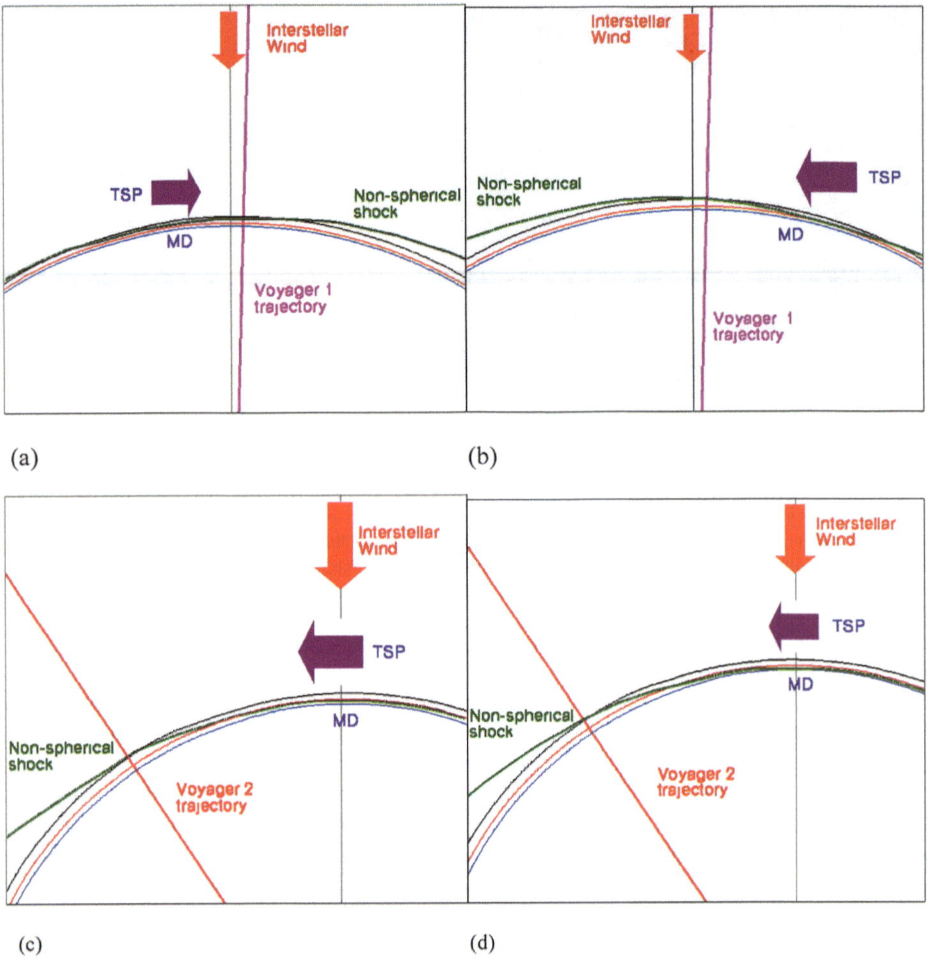

Fig. 3 Streaming of termination shock particles (TSPs) (*magenta arrows*) from the minimum distance of the termination shock to the Sun (MD) to Voyager 1 (V1) and 2 (V2), for the interstellar magnetic field in the (**a**) hydrogen deflection plane (with $\alpha = 45°$) and (**b**) galactic plane (with $\alpha = 45°$). The interplanetary magnetic field is carried radially outward by the solar wind, forming a spiral on a conical surface. The conical surfaces coinciding with the V1 and V2 trajectories are shown. V1 and V2 are first connected to the shock along the spiral magnetic field lines that contact the shock at the point of its minimum distance from the Sun (labeled MD). In the (**a**)–(**d**) panels the *green line* indicates the non-spherical termination shock. The (**a**) and (**b**) panels show the solar magnetic field lines that intersect Voyager 1; the field line intersecting the shock where V1 crosses the shock is labeled 0 AU (*black*) with *red* and *blue* indicating, respectively, magnetic field lines 2.0 AU and 3.0 AU upwind from the 0 AU line. The *magenta arrow* indicates the streaming direction of the termination shock particles from the shock along the field line to V1. Panels (**c**) and (**d**) show similar plot for V2, showing field lines 3.0 and 5.0 AU upwind of the 0 AU line. Note that in both views the solar magnetic field spirals clockwise with increasing distance outward

and 2. With $B_{\rm ISM}$ parallel to GAL (with $\alpha = 45°$, Fig. 5b), the region where $B_r = 0$ is almost perpendicular to the galactic plane, which is inconsistent with the radio observations. An interstellar magnetic field perpendicular to the galactic plane (PPG plane, with $\alpha = 30°$) produces the best agreement with the radio observations by Voyager, as suggested by Opher

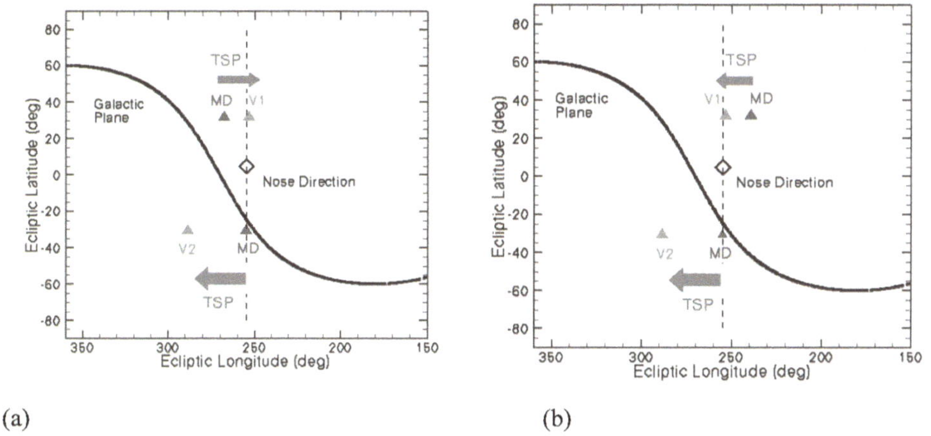

Fig. 4 Same as Fig. 3. The panels (**a**) and (**b**) summarize the streaming of TSPs from MD back to V1 and V2. The nose direction (*diamond*) and the galactic plane are indicated

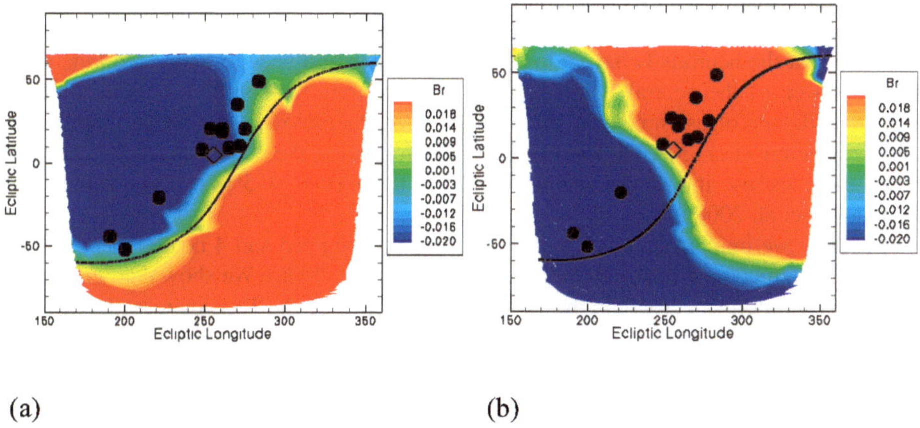

Fig. 5 The radio source location as a function of the interstellar magnetic field (B_{ISM}) direction for B_{ISM}, in the (**a**) HDP plane and (**b**) GAL plane (with $\alpha = 45°$). Panels (**a**) and (**b**) show the surface of the heliopause converted to ecliptic coordinates. The direction of the nose of the heliosphere (*diamond*) and the galactic plane (*black lines*) are indicated for reference. The radio sources detected by V1 and V2 are shown (*black points*) as viewed from upwind from the interstellar wind point of view. The contours show the strength of the radial component of the interstellar magnetic field, B_r, on the heliopause. The *green* band is the location of the radio sources (at $B_r = 0$)

et al. (2007). The HDP orientation differs from PPG by only 16° and is also in general agreement (Fig. 5a). The offset of ~15° between the observations and the region with $B_r = 0$ for the model in best agreement indicates that the accuracy of the model is not adequate to distinguish between the PPG and HDP field orientations.

Finally, the model is also in agreement with the heliosheath flows (Opher et al. 2008) (Asymmetry F). We used the heliosheath flows from days 277–320 of 2007, when the heliosheath flows were relatively quiet before a transient arrived on day 320. The flows are mainly radial with the ratios $V_N/V_R = -0.30$ and $V_T/V_R = 0.35$. We perform an unweighted average calculation so as to equally weigh each day. The mean angle $\theta = \tan^{-1}(V_N/V_T)$ for the period is $-38.7°$ and an uncertainty of 2.1°. For B_{ISM} parallel to PPG or HDP and small α(30°–45°), the angle agrees with the observed flows (−28.6° and −37.4° respectively)

6 Discussion and Conclusions

In this paper we discussed the consequences of the interstellar magnetic field on the heliosphere. In particular, we show that the orientation of the local interstellar magnetic field in a plane inclined 60°–90° from the large scale plane of the interstellar magnetic field (with $\alpha = 30°$–45°) can explain the different observational indication that exist a global asymmetry (A-F asymmetries outlined in Sect. 2).

Although the model explains these different observations, future work needs to be done to include and access the effect of additional factors not included in the model. Below we comment briefly on some of these aspects.

Neutral H Atoms The model of Opher et al. (2006, 2007) did not include the neutral hydrogen atoms that interact with the ionized component by charge exchange. This can affect the quantitative amount of asymmetry (Pogorelov et al. 2007, 2008a, 2008b). An important aspect is that the neutrals H have a long mean free path and need to be treated kinetically. The inclusion of the neutral H atoms tend to symmetrize the solution and quantitatively affect the degree of asymmetry as seen recently by Pogorelov et al. (2008a, 2008b) (see also Izmodenov et al. 2005b). The asymmetry in respect to the TS distances are recovered when more intense interstellar magnetic field is included (∼4–5 μG) (Izmodenov 2008; Izmodenov et al. 2008b).

Recently we implemented a 5 multi-fluid model (1 ionized and 4 neutral H fluids) and reproduced Izmodenov et al. (2008c) results (Opher et al. 2008). We show that with the inclusion of an interplanetary field the asymmetry in respect to the TS is increased (∼12 AU). Figure 6 shows the same view as Fig. 3a and 3c for the case with the interstellar magnetic field in the HDP plane with $\alpha = 20°$ and intensity $B = 4.375$ μG. It can be seen, in a case

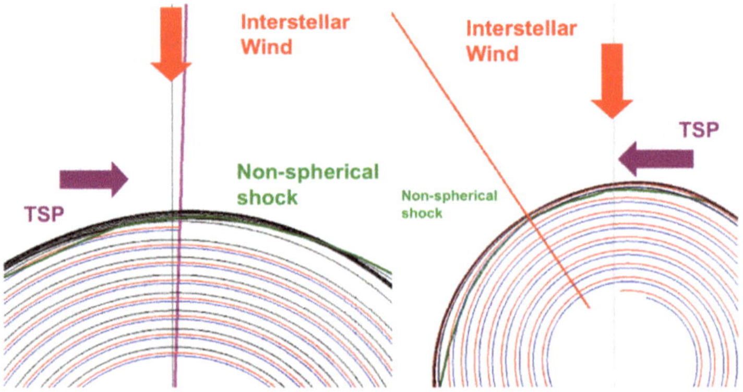

(a) (b)

Fig. 6 Same as Fig. 3a and 3c for the case of 5 fluid MHD model with B_{ISM} in the HDP plane with $\alpha = 20°$ with $B = 4.375$ μG

with strong interstellar field, that even with a presence of neutral H atoms, the east-west asymmetry remains (Opher et al. 2008).

It remains a challenge for future works to, with a more complete model, explain quantitatively all the observed asymmetries by V1 and V2 (including the deflection of 4° of the H flow in respect to the main interstellar medium flow as seen by the analysis of backscattered solar Lyman-alpha radiation).

Solar Cycle The models discussed above are stationary models. Several previous works investigated also the effect of time-varying heliosphere (e.g., Florinski et al. 2005; Karmesin et al. 1995; Whang et al. 2005; Zank and Muller 2003; Borrmann and Fichtner 2005; Izmodenov et al. 2005a, 2008a). Recently Washimi et al. (2007) investigated with a 3D-MHD model the effect of the solar cycle in the termination shock position using V2 plasmas data to drive a time-varying inner boundary. This model was a pure MHD model. As noted above, the solar cycle affects the solar wind and the magnetic field imbedded in it. The heliospheric current sheet (HCS) is known to change its inclination in respect to the solar rotation axis as the solar cycle progresses. The flows produced by the HCS and the solar cycle in the heliosheath has not yet been studied in detail. Opher et al. (2003, 2004, 2005) had investigated the effect of the current sheet in the heliosheath. This requires a very high numerical resolution, a challenge for future studies. With the solar cycle there will be also more global merger interaction regions (that result of coronal mass ejections) that will disturb this interaction as well. However, we expect that although details of the heliosheath structure will be affected, a global asymmetry still will remain produced by an external agent, the interstellar magnetic field.

Energetic-Particle Mediated Shock Recent observations indicate (Richardson et al. 2008) that the TS is a particle mediated shock. The non-thermal pressure produced by the energetic ions in the heliosheath compare (or exceeds) the thermal and magnetic pressure. The high intensity of the electrons implies that the electron impact ionization can play a comparable (or greater) role in the generation of suprathermal pickup hydrogen than charge exchange ionization (Richardson et al. 2008). The temperature at the sheath was significantly lower than the value expected. The heliosheath plasma has only about 20% of the pre-shock solar wind energy. The remaining energy must be transmitted to some other components of the heliosheath, possible the pick up protons or other particles and waves. These aspects should be included in future modeling of the interaction of the solar system with the interstellar medium (see works of Malama et al. 2006; Heerikhuisen et al. 2008).

Our result that the orientation of the local magnetic field differs from the large scale magnetic field could be a result of turbulence in the interstellar medium (Jokipii 2007) of a consequence of a local distortion at the Local Bubble. The study of Redfield and Linsky (2008) that mapped the three-dimensional structure of the warm local interstellar medium within the Local Bubble, indicated that the heliosphere lies in a transition region between the LIC and the G cloud. The interface orientation of the two clouds is also 60° to 90° from the galactic plane; what could indicate the local interstellar magnetic field is parallel to the edges of this interface.

Further upcoming missions such as IBEX will further probe this interface with the energetic neutral atoms (ENAs). Preliminary studies show the effect of the interstellar magnetic field (Prested et al. 2008; Izmodenov et al. 2008c) will be able to be detected.

Acknowledgements I would like to thank the conference conveners for organizing an extremely productive conference. I would like to acknowledge the generosity of the International Space Science Institute (ISSI), NASA-Voyager Guest Investigator grant #201351, National Science Foundation CAREER Grant #201467 for support. We thank the staff at NASA Ames Research Center for the use of the Columbia supercomputer. It is a pleasure to thank E.C. Stone for comments on the manuscript.

References

T. Borrmann, H. Fichtner, Adv. Space Res. **35**, 2091 (2005)

L.F. Burlaga, N.F. Ness, M.H. Acuna, R.P. Lepping, J.E.P. Connerney, E.C. Stone, F.B. McDonald, Science **23**, 2027 (2005)

D.P. Cox, L. Helenius, Astrophys. J. **583**, 205 (2003)

A.C. Cummings, E.C. Stone, F.B. McDonald, B.C. Heikkila, N. Lal, W.R. Webber, AGUFMSH51A-1184 (2005)

R.B. Decker, S.M. Krimigis, E.C. Roelof, M.E. Hill, T.P. Armstrong et al., Science **309**, 2020 (2005)

R. Decker et al., Nature **454**, 67 (2008)

V. Florinski, G.P. Zank, N.V. Pogorelov, J. Geophys. Res. **110**(A7), A07104 (2005)

P.C. Frisch, in *Physics of Outer Heliosphere, Proceedings of the 1st COSPAR Colloquium* (Pergamon Press, Elmsford, 1990), p. 19

P.C. Frisch, Space Sci. Rev. **78**, 213 (1996)

D.A. Gurnett, W.S. Kurth, Adv. Space Sci. **16**, 279 (1995)

D.A. Gurnett, W.S. Kurth, S.C. Allendorf, R.L. Poynter, Science **262**, 199 (1993)

D.A. Gurnett, W.S. Kurth, E.C. Stone, Geophys. Res. Lett. **30**, SSC8-1 (2003)

D.A. Gurnett, W.S. Kurth, I.H. Cairns, J. Mitchell, in *5th Annual IGPP International Astrophysics Conference*. AIP Conference Proceedings, vol. 858 (2006), p. 129

J. Heerikhuisen, N.V. Pogorelov, G.P. Zank, J.A. le Roux, Astrophys. J. **682**, 679 (2008)

V.V. Izmodenov, Space Sci. Rev. (2008), this issue

V.V. Izmodenov, D.B. Alexashov, in *American Institute of Physics Conference Proceedings*, vol. 858 (2006), pp. 14–19

V.V. Izmodenov, V.B. Baranov, in *The Physics of the Heliospheric Boundaries*, ed. by V. Izmodenov, R. Kallenbach. ISSI Scientific Report 5, ESA-ESTEC, Paris (2006), pp. 67–135. (Available online at: http://www.issibern.ch/PDF-Files/SR-005.pdf)

V.V. Izmodenov, G. Gloeckler, Y. Malama, Geophys. Res. Lett. **30**(7), 3–1 (2003). CiteID 1351, doi:10.1029/2002GL016127

V.V. Izmodenov, Y.G. Malama, M. Ruderman, Astron. Astrophys. **429**, 1069–1080 (2005a)

V.V. Izmodenov, D. Alexashov, A. Myasnikov, Astron. Astrophys. **437**, L35 (2005b)

V.V. Izmodenov, Y.G. Malama, M.S. Ruderman, J. Adv. Space Res. **41**(2), 318–324 (2008a)

V.V. Izmodenov, D.B. Alexashov, S.V. Chalov, O.A. Katushkina, Y.G. Malama, E.A. Provornikova, Space Sci. Rev. (2008b), IBEX issue

V.V. Izmodenov et al., Adv. Space Res. (2008c)

J.R. Jokipii, Science **307**, 1424 (2007)

J.R. Jokipii, J. Giacalone, J. Kota, Astrophys. J. **611**, L141 (2004)

S.R. Karmesin, P.C. Liewer, J.U. Brackbill, Geophys. Res. Lett. **22**, 1153 (1995)

W.S. Kurth, D.A. Gurnett, J. Geophys. Res. **108**, LIS2-1 (2003)

W.S. Kurth, D.A. Gurnett, F.L. Scarf, R.L. Poynter, Nature **312**, 27 (1984)

R. Lallement, E. Quemerais, J.L. Bertaux, S. Ferron, D. Koutroumpa, R. Pellinen, Science **307**, 1449 (2005)

T. Linde, PhD thesis, Univ. of Michigan, 1998

T.J. Linde, T.I. Gombosi, P.L. Roe, K.G. Powell, D.L. DeZeeuw, J. Geophys. Res. **103**, 1889 (1998)

Y.G. Malama, V.V. Izmodenov, S.V. Chalov, Astron. Astrophys. **445**(2), 693 (2006)

D. McComas et al., in *Physics of the Outer Heliosphere*, ed. by V. Florinski, N.V. Pogorelov, G.P. Zank. AIP Conference Proceedings, vol. 19, Riverside, California, 8–13 February 2004 (American Institute of Physics, Melville, 2004), pp. 162–181

M. Opher, P.C. Liewer, T.I. Gombosi, W. Manchester, D.L. DeZeeuw, I. Sokolov, G. Toth, Astrophys. J. **591**, L61 (2003)

M. Opher, P.C. Liewer, M. Velli, L. Bettarini, T.I. Gombosi, W. Manchester, D.L. DeZeeuw, G. Toth, I. Sokolov, Astrophys. J. **611**, 575 (2004)

M. Opher, P. Liewer, M. Velli, T. Gombosi, W. Manchester, D. DeZeew, G. Toth, AGUSMSH23A (2005)

M. Opher, E.C. Stone, P.C. Liewer, Astrophys. J. Lett. **640**, L71 (2006)

M. Opher, E.C. Stone, T. Gombosi, Science **316**, 875 (2007)

M. Opher, E.C. Stone, G. Toth, V. Izmodenov, J. Richardson, T.I. Gombosi, (2008, in preparation)

N.V. Pogorelov, T. Matsuda, J. Geophys. Res. **10**, 237 (1998)

N.V. Pogorelov, G.P. Zank, Astrophys. J. **636**, L161 (2006)

N. Pogorelov, G.P. Zank, T. Ogino, Astrophys. J. **614**, 1007 (2004)

N.V. Pogorelov, E.C. Stone, V. Florinski, G.P. Zank, Astrophys. J. **668**, 611 (2007)

N.V. Pogorelov, J. Heerikhuisen, G. Zank, Astrophys. J. **675**, L41 (2008a)

N.V. Pogorelov, G. Zank, T. Ogino, Adv. Space Res. **41**, 306 (2008b)

K.G. Powell, P.L. Roe, T.J. Linde, T.I. Gombosi, D.L. DeZeeuw, J. Comput. Phys. **154**, 284 (1999)

C. Prested et al., J. Geophys. Res. **113**(A6), A06102 (2008). doi:10.1029/2007JA012758
R. Ratkiewicz, A. Barnes, G.A. Molvik, J.R. Spreiter, S.S. Stahara, M. Vinokur, S. Venkateswaran, Astron. Astrophys. **335**, 363 (1998)
R. Ratkiewicz, L. Ben-Jaffel, J. Geophys. Res. **107**, 2 (2002)
S. Redfield, J.L. Linsky, Astrophys. J. **673**, 383 (2008)
J.D. Richardson, C. Wang, AGUFMSH51A-1186 (2005)
J.D. Richardson et al., Nature **454**, 63 (2008)
E.C. Stone, A.C. Cummings, F.B. McDonald, B.C. Heikkila, N. Lal, W.R. Webber, Science **23**, 2017 (2005)
E.C. Stone et al., Nature **454**, 71 (2008)
S. Suess, J. Geophys. Res. **93**, 15147 (1993)
G. Zank, Space Sci. Rev. **89**, 413 (1999)
G. Zank, H.-R. Muller, J. Geophys. Res. **108**(A6) SSH7-1 (2003)
C.J. Wareing et al., Mon. Not. R. Astron. Soc.: Lett. **372**(1), L63–L67 (2006)
H. Washimi, T. Tanaka, Adv. Space Res. **27**, 509 (2001)
H. Washimi, G.P. Zank, Q. Hu, T. Tanaka, K. Munakata, Astrophys. J. **670**, 139L (2007)
Y.C. Whang, Y.-M. Wang, N.R. Sheeley, L.F. Burlaga, J. Geophys. Res. **110**(AR), A03103 (2005)

The Dynamic Heliosphere: Outstanding Issues

Report of Working Groups 4 and 6

V. Florinski · A. Balogh · J.R. Jokipii · D.J. McComas · M. Opher · N.V. Pogorelov · J.D. Richardson · E.C. Stone · B.E. Wood

Originally published in the journal Space Science Reviews, Volume 143, Nos 1–4, 57–83.
DOI: 10.1007/s11214-009-9488-7

Abstract Properties of the heliospheric interface, a complex product of an interaction between charged and neutral particles and magnetic fields in the heliosphere and surrounding Circumheliospheric Medium, are far from being fully understood. Recent Voyager spacecraft encounters with the termination shock and their observations in the heliosheath revealed multiple energetic particle populations and noticeable spatial asymmetries not accounted for by the classic theories. Some of the challenges still facing space physicists include the origin of anomalous cosmic rays, particle acceleration downstream of the termination shock, the role of interstellar magnetic fields in producing the global asymmetry of

V. Florinski (✉)
Department of Physics and Center for Space Plasma and Aeronomic Research, University of Alabama in Huntsville, 301 Sparkman Dr., Huntsville, AL 35806, USA
e-mail: vaf0001@uah.edu

A. Balogh
International Space Science Institute, Hallerstrasse 6, Bern 3012, Switzerland
e-mail: balogh@issibern.ch

J.R. Jokipii
Department of Planetary Sciences and Lunar and Planetary Laboratory, University of Arizona, 1629 E. University Blvd., Tucson, AZ 85721, USA
e-mail: jokipii@lpl.arizona.edu

D.J. McComas
Southwest Research Institute, 6220 Culebra Rd., San Antonio, TX 78238, USA
e-mail: dmccomas@swri.edu

M. Opher
Department of Physics and Astronomy, George Mason University, 4400 University Drive, Fairfax, VA 22030, USA
e-mail: mopher@physics.gmu.edu

N.V. Pogorelov
Department of Physics and Center for Space Plasma and Aeronomic Research, University of Alabama in Huntsville, 301 Sparkman Dr., Huntsville, AL 35806, USA
e-mail: np0002@uah.edu

the interface, the influence of charge exchange and interstellar neutral atoms on heliospheric plasma flows, and the signatures of solar magnetic cycle in the heliosheath. These and other outstanding issues are reviewed in this joint report of working groups 4 and 6.

Keywords Solar Wind · Termination shock · Heliosheath · Heliopause · Pickup ions · Interstellar neutral atoms · Anomalous cosmic rays

1 Introduction

The heliospheric termination shock (TS) and heliosheath represent the interface of the heliosphere with the surrounding circumheliospheric interstellar medium or CHISM. In the past few years we have experienced a phenomenal growth in our knowledge, with important observations, both remote and *in situ* coming in almost monthly. Many of our cherished views have undergone significant challenges and were discussed at this meeting.

One of the most important goals of heliospheric exploration has been to observe the solar-wind (SW) TS, where the supersonic plasma flow undergoes a sudden transition to subsonic flow, in response to the inward pressure of the CHISM. Beyond the shock is the heliosheath, a region of subsonic solar outflow, where the solar plasma is deflected by the incoming interstellar plasma to flow backward toward the tail of the heliosphere. The boundary between the plasma of solar origin and the interstellar plasma is called the heliopause (HP).

The Voyager 1 spacecraft crossed the termination shock in December of 2004 (Stone et al. 2005), and since then has been in the heliosheath. Voyager 2 crossed the shock in August of 2007 and is also now in the heliosheath (Decker et al. 2008; Stone et al. 2008). This region of subsonic outward flow and deflection of the postshock solar plasma has been predicted theoretically since the earliest discussions of the solar wind (Parker 1961). However, its structure and properties have been discussed only for large scales. The observed fluctuations and inward and outward motions of the termination shock can change the observations and make it difficult to interpret.

As one example the anomalous cosmic rays (ACR) revealed the complexity of the termination shock. Earlier ideas, based on a simple spherical shock were found to be wanting. A number of ideas have been proposed to explain this, ranging from a turbulent, fluctuating shock to doing away with shock acceleration altogether. This kind of analysis depends on knowing the large-scale structure.

Remote observations using backscattered radiations and energetic neutral atoms have provided new insights into the overall structure and asymmetries of the heliosphere and

J.D. Richardson
Kavli Institute for Astrophysics and Space Research, Massachusetts Institute of Technology, 77 Massachusetts Avenue, Cambridge, MA 02139, USA
e-mail: jdr@space.mit.edu

E.C. Stone
Space Radiation Laboratory, California Institute of Technology, 1200 E. California Blvd., Pasadena, CA 91125, USA
e-mail: ecs@srl.caltech.edu

B.E. Wood
Space Science Division, Naval Research Laboratory, 4555 Overlook Ave. S.W., Washington, DC 20375, USA
e-mail: brian.wood@nrl.navy.mil

hence provided important new constraints on the CHISM. In particular, the very local interstellar magnetic field just outside of the HP was essentially unknown. Recent observations (Lallement et al. 2005) used the deflection of neutral hydrogen to constrain this field for the first time. The combination of the *in situ* and remote observations are essential to the interpretation of the data.

This chapter describes some of the outstanding issues currently being discussed and some of the solutions. Section 2 discusses new theories of ACR origin. Dynamic effects of the solar cycle on the large-scale structure of the heliosphere are discussed in Sect. 3. Here we also mention the possibility of large-scale unstable structures developing near the HP. Section 4 connects *in situ* observations of energetic neutral atoms with remote sensing using Lyα absorption technique—an important issue for the *Interplanetary Boundary Exprorer* (IBEX) mission (McComas et al. 2004). In Sect. 5 the properties of the heliosheath are compared with those of other interplanetary sheaths, including the Jovian magnetosheath and the sheaths behind the interplanetary coronal mass ejection (ICME) shocks. Finally, Sect. 6 speculates on the properties of the heliopause and the surrounding region, which the Voyagers are predicted to reach by the year 2020.

2 On the Origin of Anomalous Cosmic Rays

2.1 Observations of ACRs Near the Termination Shock

Voyager 1 and 2 space probes have been monitoring anomalous cosmic ray activity in the heliosphere for over 30 years. Prior to the TS precursor events that started in 2002.5 (at V1) the only visible part of the ACR spectrum was the broad maximum at $\sim$10–20 MeV/n for He and a portion of the slope at higher energies (e.g., Stone et al. 1996; Hill et al. 2002), with lower energies being modulated by the SW. From 2002 on V1 began to observe the new "termination shock particle" (TSP) component at lower energies (1–10 MeV for He), but the combined spectra did not fit the concept of a single source of energetic particles produced by shock acceleration because ACRs were at a much lower intensity level than predicted by the models (McDonald et al. 2003). ACR intensities were highly variable during 2002–2005 mostly due to a modulation effect of merged interaction regions that were abundant in the outer heliosphere during that time period. The TS was crossed by V1 in December of 2004, however the measured spectra did not "unfold" into a single continuous power law, characteristic of diffusive acceleration at a stationary shock wave. Instead, ACRs remain modulated with intensities still below those measured during the previous solar minimum in 1998 (Stone et al. 2005).

In the following years V1 spectra in the heliosheath continued to fill in at low energies and by 2007.0 appeared to form a continuous power law with a spectral index of -1.6 (Cummings et al. 2008). Some examples of heliosheath spectra are shown in Fig. 1, taken from their paper. In 2007–2008 V2 ACR spectra still remained modulated and resembled V1 spectra observed in 2004–2005 (cf. the rightmost top and bottom panels in Fig. 1). ACR intensities at V1 rose substantially between 2005.0 and 2006.0, which was at the time interpreted as evidence of a large positive spatial gradient in the heliosheath. This trend is evident in Fig. 2 taken from McDonald et al. (2007). However, as also seen in that figure, V2 intensities grew by a similar amount in the SW, making it clear that the effect was largely a temporal one indicating either an increase in the source strength, or a reduced modulation (McDonald et al. 2007; Cummings et al. 2008). When V2 crossed the TS in September of 2007, energetic particle spectra did not completely unfold and a break between ACRs and TSPs was still evident (Stone et al. 2008).

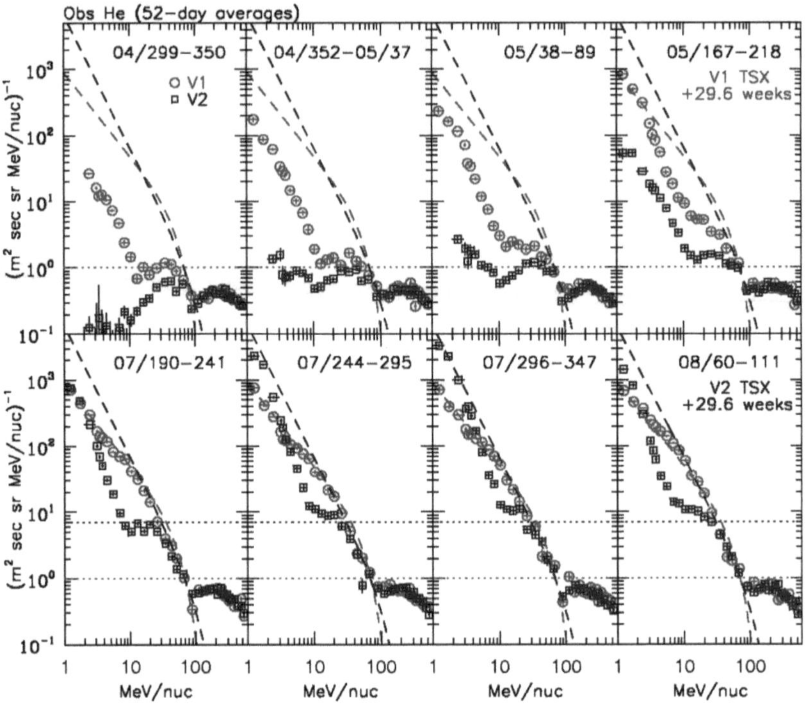

Fig. 1 Fifty-two day averaged energy spectra of He at V1 and V2 for 8 time periods. V1 crossed the TS on day 351 of 2004, between the *first two panels of the top row*. The ACR portion of the energy spectrum is from ~10–100 MeV/nuc. Examples of possible source spectra at the TS are shown by the *dashed lines*. The *bottom row* shows similar data near the time of the V2 shock crossing, which occurred between the *first two panels*. The ACR He spectrum remained in a modulated state during both shock crossings. Note that by 2008/60-111 the V1 energy spectrum may have unfolded to the ACR source spectrum. From Cummings et al. (2008)

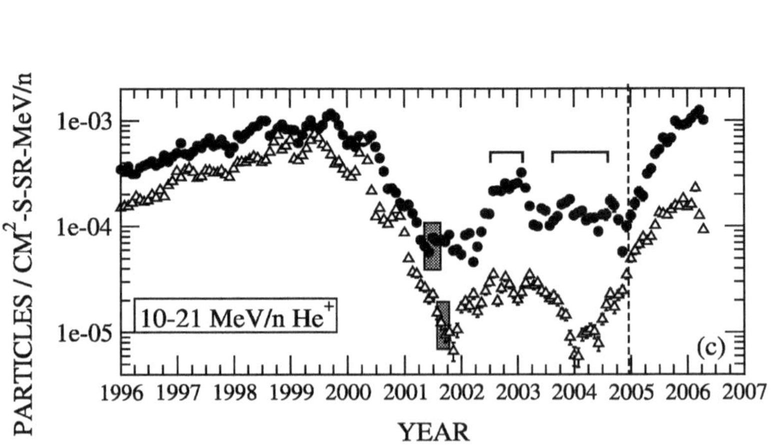

Fig. 2 10–21 MeV/n anomalous He intensity from Voyager 1 (*solid circles*) and Voyager 2 (*open triangles*). The *vertical dashed line* marks the V1 TS crossing. From McDonald et al. (2007)

The entirety of Voyager observations indicate that solar activity and the solar magnetic cycle have a profound effect on the efficiency of ACR acceleration. The broad maximum moved to a higher energy during the last $A < 0$ cycle compared with the previous $A > 0$ period, with the transition completed within a year of the magnetic polarity reversal (McDonald et al. 2007). This result implies that drifts play an important role in ACR acceleration and transport even during periods of high solar activity. Intensities in the outer heliosphere have not fully completed their recovery at the time of writing. Further increases following the complete recovery past the 2008 solar minimum would be indicative of spatial variations in the heliosheath.

2.2 Classic (Drift) Paradigm

The first "modern" model of ACR acceleration and transport was developed by Jokipii (1986). The model was two-dimensional and featured both the termination shock, where ACRs were accelerated, and the heliosheath region, bounded by a free escape boundary, approximating the heliopause. It was shown that positively charged particles drifted from high latitudes toward the ecliptic along the shock front during the $A < 0$ solar cycle. Because particles gained energy at the shock in the process, their intensities were enhanced near the ecliptic compared with polar regions. In $A > 0$ cycles the direction of drifts was reversed and polar intensities started to dominate.

Despite a number of improvements made in later work the basic paradigm remains the same today (see, e.g., Steenberg and Moraal 1996; Jokipii and Giacalone 1998; Langner and Potgieter 2004). Some of the more notable improvements include the self-consistent effects of ACR pressure on the TS (Florinski et al. 2004), and the more realistic structure of the solar wind and heliosheath from global heliospheric models (Ferreira and Scherer 2006; Ferreira et al. 2007). All models of this kind are based on the Parker transport equation for a nearly isotropic distribution function $f(\mathbf{r}, p)$, where $\mathbf{r}$ is spatial position and p is the magnitude of the particle momentum, which may be written in the most general form as

$$\frac{\partial f}{\partial t} + (\mathbf{u} + \mathbf{v}_d) \cdot \nabla f - \nabla \cdot (\kappa \cdot \nabla f) - \frac{\nabla \cdot \mathbf{u}}{3} \frac{\partial f}{\partial \ln p} - \frac{1}{p^2} \frac{\partial}{\partial p}\left(p^2 D \frac{\partial f}{\partial p}\right) = Q, \quad (1)$$

where $\mathbf{u}$ is the plasma flow velocity, $\mathbf{v}_d$ is the drift velocity, κ is the spatial diffusion tensor, D is the momentum diffusion coefficient, and Q is the source term. Typically, an injection source at the shock, possibly dependent on PUI intensity is specified. It should be noted that few of the "classic" models consistently included the process of preacceleration of PUIs required to reach the injection threshold.

Figure 3 shows typical spectra obtained with a drift ACR transport model (Florinski et al. 2004). Similar trends are present in other models (Langner and Potgieter 2004; Ferreira and Scherer 2006). Low-energy spectra are power laws with the energy slope dependent on the shock compression ratio, as $\sigma = (s/2 + 1)(s - 1)$. At the highest energies (above 100 MeV) diffusive acceleration timescale becomes longer than the timescale for adiabatic cooling in the SW and the spectra roll over exponentially. Between these two extremes, a "hump" may appear during the $A < 0$ periods (right panel). The hump is a drift effect present only when positive ions drift from high to low latitudes along the shock front. It may be seen that ions are accelerated to higher energies during $A < 0$ cycles, in agreement with the observations (McDonald et al. 2007). However, the separation between the low-energy part of the spectrum and the broad maximum is much less pronounced in theoretical models than in the observed spectra (cf. Fig. 1).

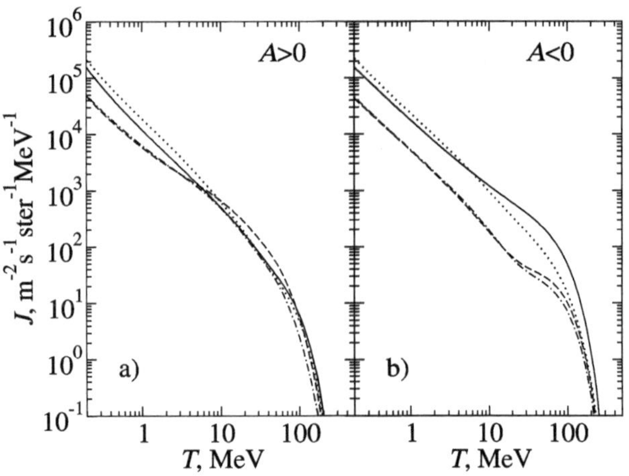

Fig. 3 Anomalous hydrogen spectra predicted by a 2D self-consistent model for an $A > 0$ solar minimum (**a**) and an $A < 0$ solar minimum (**b**). *Solid* and *dashed lines* are spectra at the TS in the ecliptic, and at 90° (polar direction). *Dotted* and *dot-dashed lines* are spectra at these respective latitudes, but 30 AU into the heliosheath. From Florinski et al. (2004)

In classic models the peak ACR intensity is in the ecliptic during the negative solar cycles (and at high latitudes in the opposite case). The maximum may lie some distance beyond the TS because ACRs experience additional compression in the heliosheath where their energy is boosted by another $\sim 40\%$ (Florinski et al. 2004; Ferreira et al. 2007). Compressive flow in the heliosheath results from charge exchange with interstellar hydrogen atoms producing $\nabla \cdot \mathbf{u} < 0$ with the associated acceleration timescale of the order of 8 years (Florinski et al. 2004), which is long compared with shock acceleration, but comparable to ACR storage time in the heliosheath. For the present heliospheric conditions ($A < 0$ minimum) drift models predict the main source of ACRs to lie at low latitudes, possibly some 10–20 AU beyond the TS in the heliosheath.

2.3 Shock Geometry and Longitudinal Effects

Most of the classic models discussed above were axially symmetric and included only latitudinal effects. The TS was typically approximated by a sphere centered on the Sun. This approximation, however, is highly oversimplified and contemporary global models of the heliosphere predict a shock with a more complicated geometry—blunt in the direction facing the interstellar flow, and somewhat elongated over the heliographic poles and extended in the direction of the heliotail (e.g., Pauls and Zank 1997, Pogorelov et al. 2008b). This implies that the spiral magnetic field intersects the shock at different angles, depending on longitude. Because the shock is expected to be less oblique in the flanks, PUI injection was expected to be more efficient there than near the "nose" (Chalov 1993; Chalov and Fahr 2000; McComas and Schwadron 2006).

In addition to examining the injection effect produced by a blunt termination shock, McComas and Schwadron (2006) further proposed that a realistic termination shock geometry should also lead to acceleration of particles up to classic ACR energies only along the flanks and tail of the shock. These authors pointed out that because of the solar winds radial outflow, the spiral IMF flux tubes must first pierce the shock near the nose and then are connected to the shock for progressively longer times as the flux tube moves out into the heliosheath and its connection points move back along the flanks (see Fig. 4). A typical diffusive shock acceleration time for ACRs is roughly a year, which is similar to the connection time far back along the flanks and tail of a realistic termination shock. This geometric effect provides a simple explanation for the unexpected dearth of higher energy particles observed by both Voyagers at their crossing points near the nose of the heliosphere.

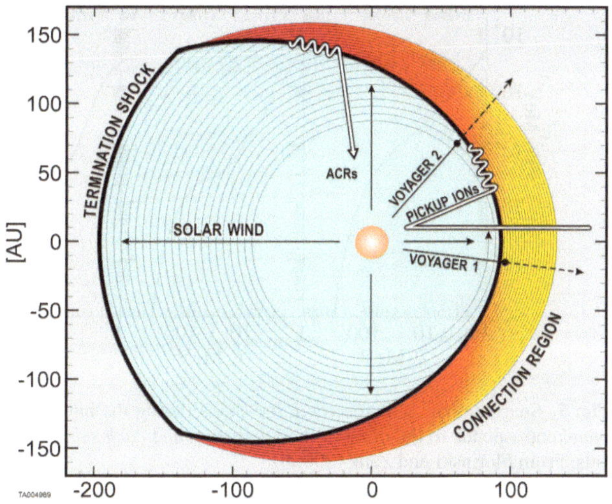

Fig. 4 Schematic diagram of an equatorial cut through the termination shock taken from McComas and Schwadron (2006). The expanding Archimedean spiral flux tubes first cross the termination shock near the nose and their connection point to the shock move back along the flanks as they continue to propagate out through the heliosheath. These long connection times for the flanks and tail can allow diffusive shock acceleration to energize particles up to observed ACR energies

Schwadron and McComas (2007) extended their previous work in this area and described the implications of the blunt shock geometry on the modulation of anomalous and galactic cosmic rays in the inner heliosheath. That study anticipated the strong correlation in the modulation of ACRs and GCRs observed by the Voyagers based on the magnetic connection of flux tubes from the flanks to a region in the heliosheath well out ahead of the nose of the shock (and well beyond the Voyagers current locations). Finally, Schwadron et al. (2008) developed analytically based models of diffusive acceleration along non-spherical shocks. With reasonable parameters for the scattering mean free path and the perpendicular to parallel diffusion ratio, their model produced energy spectra that did a good job of matching the power-law spectrum of energetic protons below several MeV and the more sharply falling spectrum above this energy, as observed by the Voyagers.

2.4 Dynamical Effects

Most of V1 TS-related observations were made during the declining phase of an $A < 0$ solar cycle. This period was especially abundant in large-scale SW transient structures, typically shocks and merged interaction regions (MIRs) with at least 7 MIRs detected by V2 between 2002 and 2005 (Richardson et al. 2005). The passage of MIRs was accompanied by marked drops in energetic particle intensity. McDonald et al. (2007) argued that MIRs were the cause of unexpectedly low ACR intensity during the V1 TS crossing. MIR activity has ceased since then, but smaller variations in SW dynamic pressure with an amplitude of $\sim 50\%$ on the timescales of months to a year remain (Richardson et al. 2008b).

Transients affect charged particle propagation and acceleration in several important ways: by (a) modifying the shock compression ratio, (b) by altering magnetic field connection and shock-normal angle, and (c) by changing the turbulence/scattering properties of the interplanetary medium. At low ion energies (but above the DSA injection threshold) spectral slope should closely track variations in the TS compression ratio. Conversely, highest energy ACRs are not significantly affected by the transients because their acceleration timescale is long (~ 1 year) compared with the typical width of a structure. Florinski and Zank (2006a) argued that collisions of SW structures with the TS could lead to a separation between the fast responding low-energy part of the spectrum and the slowly-varying high-energy ACR

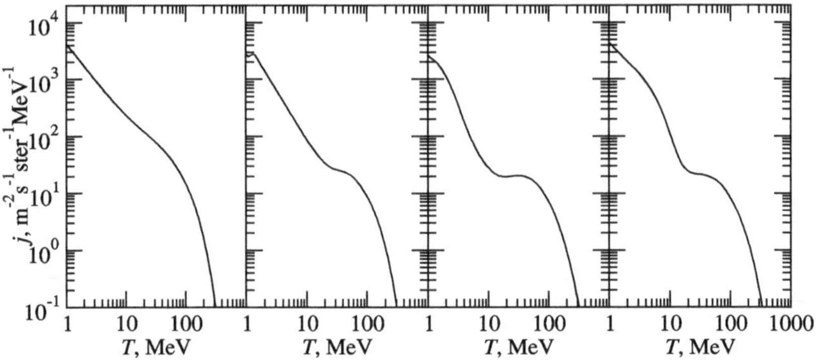

Fig. 5 Snapshots of ACR spectra at the shock during the interaction of a large MIR with the TS. The *first panel* corresponds to the onset of the interaction and *each subsequent panel* shows spectra at 3 month intervals. From Florinski and Zank (2006b)

component. A similar scenario was proposed by Jokipii (2006) who used rapid periodic motion of the TS and variable injection rate to obtain a break in the accelerated spectrum.

Spectra from the MIR collision scenario are shown in Fig. 5. Before the onset (panel 1) the spectrum is an ordinary power law with a cutoff. Three months after the leading edge of the MIR collided with the shock, the latter is pushed back and weakened. Low-energy particles become trapped between a pair of waves in the heliosheath and are removed from the acceleration region. Later, as the rarefaction following the MIR hits the shock, the latter becomes stronger again producing a harder power law distribution at low energies (panels 3 and 4). This is the mechanism by which low and high energy particles become spectrally separated.

The analysis of Florinski and Zank (2006a) was restricted to a single large MIR with a width (passage time for a stationary observer) of ~ 3 months. V2 spectra during its shock crossing in September of 2007 were probably not affected by MIRs in the SW because the last such large structure was observed in 2006. The TS however was moving with the speed of the order of 60–100 km/s at the time of Voyager 2 crossing (Richardson et al. 2008a) indicating that a source of variability was still present in 2008. One such potential source of variability is the MIRs reflected off the HP and reaching the TS from the downstream side (Washimi et al. 2007).

2.5 Acceleration in the Heliosheath

The inner heliosheath is expected to be a turbulent environment offering possibilities for second-order Fermi acceleration. In quasi-linear theory the momentum diffusion coefficient for Alfvén waves is $D \sim V_a^2 p^2/\kappa_\parallel$ and the associated timescale of ion acceleration is $\tau_d \sim \kappa_\parallel/V_a^2$. Momentum diffusion is negligible in the SW at ACR energies because V_a^2 is small and $\kappa_\parallel$ is large so that τ_d is much longer than the other relevant timescales. In the heliosheath these timescales become comparable (Kallenbach et al. 2005).

Stochastic acceleration of energetic particles in the heliosheath was first included in the model of Chalov et al. (2003). In addition to Alfvénic turbulence these authors also considered large scale compressive structures accelerating particles via a turbulent stochastic mechanism. Unlike resonant mechanisms, this mode of acceleration has an advantage of being essentially independent of particle velocity and consequently becomes dominant at high (ACR) energies. Some authors investigated the possibility that second

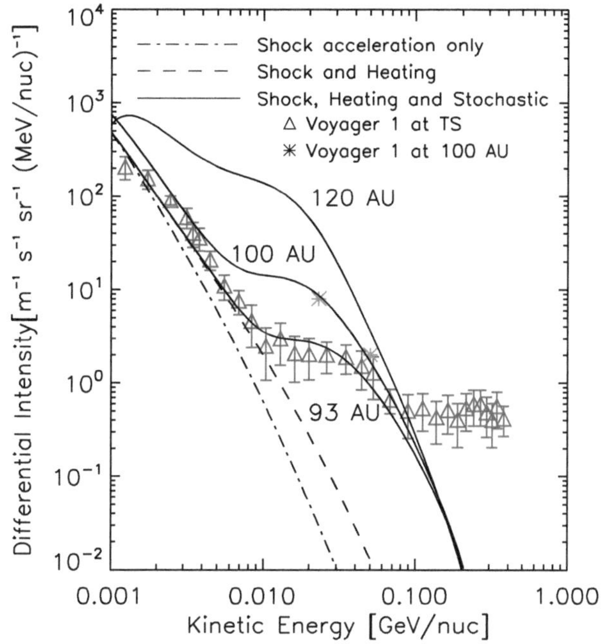

Fig. 6 Spectra of anomalous helium produced primarily by stochastic acceleration in the heliosheath. Shock acceleration and compression in the heliosheath are also included. From Ferreira et al. (2007)

order Fermi acceleration in the heliosheath, rather than DSA at the TS, was the dominant source of ACR particles (Moraal et al. 2006; Zhang 2006; Chalov et al. 2007; Ferreira et al. 2007). In these models the bulk of stochastic acceleration occurred close to the heliopause, where the field was compressed near the stagnation point. These ACRs then diffused inward to the TS where they appeared as a "hump" in the spectrum below the cutoff energy. This is illustrated in Fig. 6 taken from Ferreira et al. (2007).

Stochastic acceleration sets up a positive radial gradient of accelerated ions beyond the TS. Zhang (2006) and Ferreira et al. (2007) predict that 4–5 MeV/n helium intensity would increase by a factor of $\sim$ 10 some 10 AU into the heliosheath. There is no agreement between the models as to where ACRs would reach their peak intensity. As discussed in Sect. 3.1, the actual radial gradient, if present, was obscured by the large temporal increase as the intensities recovered after a passage several MIRs during 2004 (Fig. 2). It is also possible that the rate of momentum diffusion has declined as the wave/turbulence intensities decreased near the solar minimum. In that case intensities would be expected to rise again during the next 4–5 years. However, by that time V1 will be close to the HP, where ACR intensity may decline yet again because of their escape to the interstellar medium.

Gloeckler et al. (2008) argued that the observed energetic ion spectra are the sum of a stochastically-accelerated PUI tail at lower energies and a modulated ACR component at high energies. In their scenario ACRs are accelerated to higher energies in the heliosheath reaching their peak intensity near the HP.

3 Variability of the Termination Shock Region and Heliosheath: Solar Cycle and Transient Effects

Most of currently used numerical models of the heliospheric interface (Izmodenov et al. 2005a; Heerikhuisen et al. 2007; Heerikhuisen et al. 2008; Pogorelov et al. 2006; Pogorelov et al. 2007; Pogorelov et al. 2008a) are three-dimensional, in one way or another include the effect of charge exchange between ions and atoms, and are time-dependent. This is because the structure of the SW–CHISM interaction implies the presence of both subsonic and supersonic flow regions. A time-dependent system of ideal MHD equations is always hyperbolic, which allows us to apply a variety of sophisticated shock-capturing numerical schemes (Kulikovskii et al. 2001) to its solution. If the boundary conditions are stationary and the solution to be found is stable, one would expect that regardless of the choice of initial conditions a unique steady state should be found as $t \to \infty$. It is not surprising that, historically, steady solutions for the SW–CHISM interaction were found (see, e.g., Baranov and Malama 1993; Pauls et al. 1995). The reality is that neither SW nor the CHISM are time-invariant. Moreover, the HP is subject to various type instabilities, which results in the mixing of plasmas with considerably differing properties. The SW flow is characterized by the presence of various transients, such as coronal mass ejections (CME's), corotating interaction regions (CIR's), MIR's, and global MIR's (GMIR's) which affect the distribution of plasma and neutral flow properties in the distant SW and beyond the TS, in the inner heliosheath. Besides short-time transients, of great importance are phenomena related to the 11-year period of solar activity. During this period, a relatively narrow latitudinal extent of slow SW (McComas et al. 2000) characteristic of solar minima gradually increases so that all latitudes are filled with a non-uniform, high-speed flow. The angle α between the Sun's rotation and magnetic-dipole axes is small near the solar minimum. Within the solar activity period, the polarity of the Sun's magnetic field changes to the opposite, which can be loosely interpreted as the increase of α to 90° and further on to 180°. In this section we will briefly consider these three effects.

3.1 Propagation of a Solar Wind Velocity Pulse through the Heliosphere

The interstellar magnetic field (ISMF) is known to be of great importance in determining the shape and orientation of the HP (Linde et al. 1998; Pogorelov and Matsuda 1998; Ratkiewicz et al. 1998; Pogorelov et al. 2004; Opher et al. 2006). The interplanetary magnetic field (IMF) is an important observable quantity whose distribution in the inner heliosheath affects cosmic ray transport throughout the heliosphere (Florinski et al. 2003; Ball et al. 2005; Ferreira et al. 2007). Pogorelov et al. (2004) performed a detailed parametric investigation of the solar wind (SW) interaction with the CHISM and showed that the IMF–ISMF coupling at the HP results in a bending and rotation of the heliospheric current sheet (HCS), especially near the solar minimum (Pogorelov et al. 2006). The combined action of the ISMF and IMF creates asymmetries that can affect propagation of high-amplitude perturbations of the solar wind beyond the termination shock. Although ISMF-induced asymmetries are substantially smaller in the presence of neutral particles (Pogorelov et al. 2006; Pogorelov et al. 2007; Pogorelov et al. 2008a), the calculations of Pogorelov et al. (2004), which were obtained in the ideal MHD formulation of the problem, introduced some confusion and resulted in a considerable exaggeration of the importance of magnetic fields by some authors (e.g., Opher et al. 2006).

A GMIR was defined by Burlaga et al. (1993) as an extended region of intense magnetic field with a global shell-like structure. GMIR's are important for cosmic-ray modulation and their evolution as they move from the Sun to the HP has already been analyzed on the basis of spherically symmetric MHD models (e.g. Burlaga et al. 2003). Here we show some numerical results that address three-dimensional effects of the propagation of a spherically-symmetric pulse in the SW radial velocity throughout the heliosphere.

For simplicity, we will use a 2-fluid (Pauls et al. 1995) model of the SW–CHISM interaction here which has been revised by adding three-dimensional magnetic fields and improving the numerical scheme and performance of far-field boundary conditions (Pogorelov et al. 2006). It is assumed that the CHISM velocity, ionized and neutral hydrogen densities and temperatures in the distant LISM are equal to $V_\infty = 25\ \mathrm{km\,s^{-1}}$, $n_\infty = 0.07\ \mathrm{cm^{-3}}$, $n_{\mathrm{H}\infty} = 0.1\ \mathrm{cm^{-3}}$, and $T_\infty = 5679$ K, respectively. The ISMF strength is supposed to be equal to its average galactic value $B_\infty \approx 1.5\ \mu$G. In this particular calculation we neglect all hydrogen atoms that originate in the inner heliosheath. We choose the orientation of the ISMF vector $\mathbf{B}_\infty$ perpendicular to $\mathbf{V}_\infty$ and belonging to the plane tilted at 60° to the ecliptic plane. The orientation of this plane in space is very close to that of the hydrogen deflection plane (Lallement et al. 2005)—the plane formed by $\mathbf{V}_\infty$ and the velocity of neutral hydrogen flow at distances of about 10 AU from the Sun. The solar wind plasma is assumed to be spherically symmetric initially, with the following parameters: $U_{\mathrm{E0}} = 450\,\mathrm{km\,s^{-1}}$, $M_{\mathrm{E}} = U_{\mathrm{E0}}/c_{\mathrm{E}} = 10$, $n_{\mathrm{E}} = 7\,\mathrm{cm^{-3}}$. The subscript "E" corresponds here to the parameters evaluated at 1 AU. The magnetic and rotational axes of the Sun are assumed parallel. Although the magnetic field of the Sun is not dipolar at solar maximum, for simplicity, we suppose that it has the shape of Parker's spiral at the inner boundary located at 10 AU and that the radial component of the IMF at 1 AU is approximately equal to 37.5 μG. The spiral is directed from the Sun in the northern hemisphere and towards it in the southern hemisphere. It is interesting to note in this connection that the *Voyager 1* observations of the IMF strength agree with Parker's model from 1 to 81 AU and from 1978 to 2000, when one considers the solar cycle variations in the source magnetic field and the latitudinal/time variation of the solar wind speed (Burlaga et al. 2002). Since the IMF is discontinuous across the equatorial plane, a heliospheric current sheet (HCS) naturally forms in accordance with the chosen boundary conditions. To model a GMIR, we assume that velocity of the solar wind increases three times at $R = 10$ AU within 60 days and then returns to its original value within another 60 days. More specifically,

$$U_{\mathrm{E}}(t) = U_{\mathrm{E0}}\left[1 + 2e^{-\left(\frac{t-1.864}{0.57}\right)^2}\right], \qquad (2)$$

where time is dimensionless and its unit is approximately equal to 69.4 days. As shown later, this results in a region of higher than average magnetic field with a size of about 20–30 AU beyond $R = 40$ AU in the direction of *Voyager 1*.

Initial conditions are represented by a steady state corresponding to $U_{\mathrm{E}} = U_{\mathrm{E0}}$. In Fig. 7, we show the variation of the magnetic field strength distribution in the meridional (xz) plane. The snapshots were made at $t = 179$, 528, 644, and 1457 days (note that the peak velocity at the inner boundary appears at $t = 129$ days). Figure 7a shows that at an early stage of a GMIR evolution, the termination shock crosses the positive half of the x-axis at $R \approx 95$ AU. The distribution of $B_{\mathrm{tot}} = (B_x^2 + B_y^2 + B_z^2)^{1/2}$ shows that there is a rather thin region of small magnetic field behind the termination shock. This region bends to the southern hemisphere, which is the result of the HP rotating with respect to the meridional plane (Pogorelov et al. 2004).

One can see that a region of increased magnetic field is formed in the solar wind, which is initially nearly spherically symmetric, except for the polar regions, where the magnetic field is weaker (Fig. 7a). As the perturbation crosses the termination shock, the asymmetry of the inner heliosheath leads to an asymmetric distribution of magnetic field, which is seen in Figs. 7b and 7c. The largest increase in the magnetic field strength occurs earlier in the northern than in the southern hemisphere. After the GMIR reaches the heliopause (see

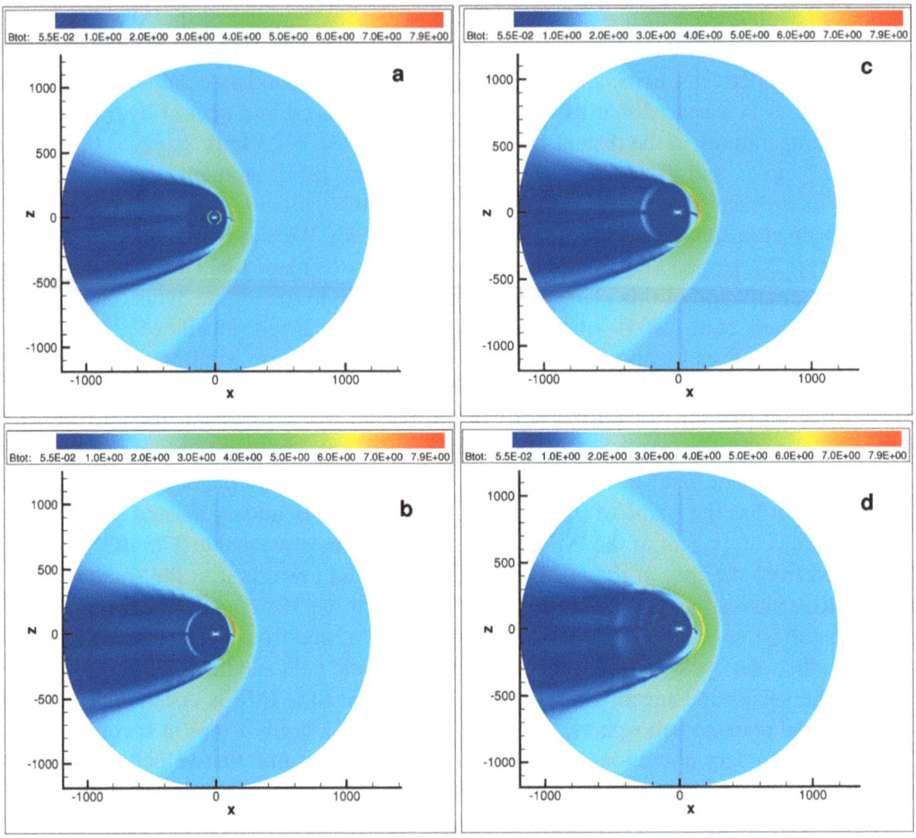

Fig. 7 Magnetic field strength variation in the meridional plane as a GMIR propagates towards the heliopause

Fig. 7d), it initiates a perturbation that results in an increase of the ISMF strength in the outer heliosheath. While unsteady phenomena take place, the termination shock can decrease its distance to the Sun to $R \approx 80$ AU at the upwind point in the ecliptic plane. The interaction pattern returns to a steady state after approximately 35 years.

To better understand the detailed variation during GMIR propagation, we present several one-dimensional plots that show radial distributions along the rays belonging to the half-plane $\phi = 0$ along the ray $\theta \approx 55°$, which is approximately in the direction of *Voyager 1*. Here θ is measured from the z-axis. Separate subsequent curves in Fig. 8 correspond to $t = 0$ (a thick solid line), 66.5, 121.7, 154.6, 179.5, 206.9, 242.7, 292.6, 346.2, 400, and 457.5 days. The right and left panels of Fig. 8 show the distributions of the magnetic field strength and radial velocity of plasma, respectively. One can see that the magnitude of the IMF vector can increase to ≈ 6.3 μG in the supersonic solar wind. Magnetic fields of this strength are not observed in the supersonic SW at the considered distances from the Sun and are completely due to our assumption of a very strong increase in the SW velocity at the inner boundary. After crossing the termination shock this GMIR increases the magnetic barrier at the inner side of the HP to 8.4 μG, which is considerably higher than the maximum ISMF strength. It appears that, after the ascending velocity profile breaks, the increase and further decrease of the magnetic field in each curve representing its distribution in the supersonic solar wind correspond to forward and reverse shock waves. Theoretical justification of their appearance

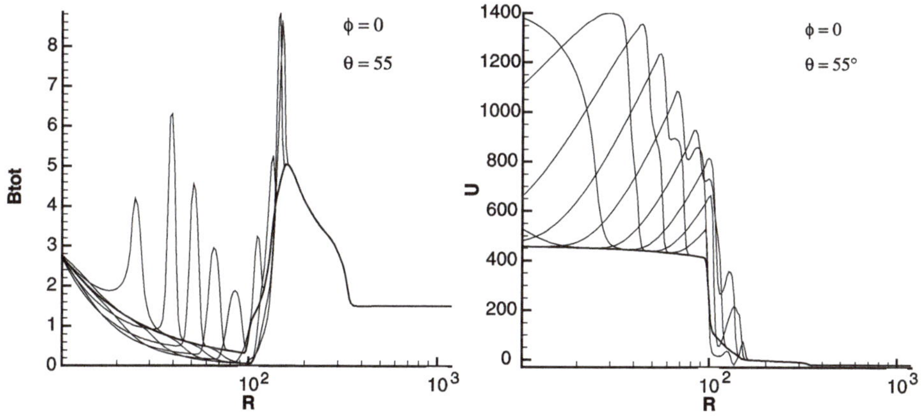

Fig. 8 The initial stage of GMIR propagation through the heliosphere: time evolution of the magnetic field strength (*left*) and radial velocity (*right*) distribution along the ray $\phi = 0$, $\theta = 55°$

Fig. 9 GMIR propagation into the outer heliosphere: time evolution of the magnetic field strength distribution along the ray $\phi = 0$, $\theta = 55°$

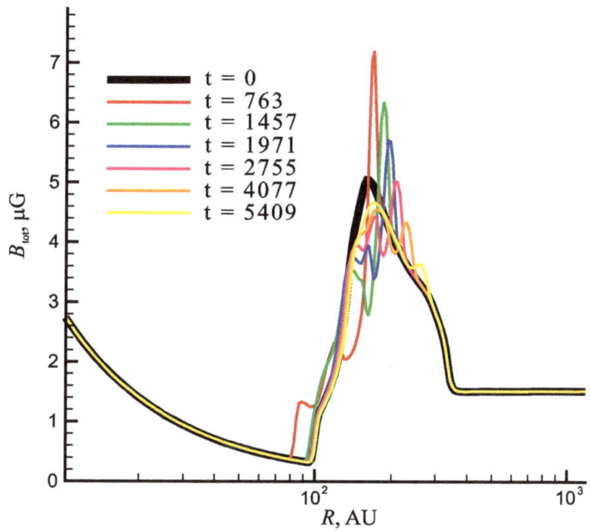

was given by Simon and Axford (1966). The shocks in this pair are both carried by the solar wind towards the termination shock. The splitting of the compression wave into two shocks is very well seen from the distribution of the radial velocity U in Fig. 8 (right panel). Note the regions of negative radial velocity in the distributions shown in this figure, which are produced by the motion of the HP.

In Fig. 9 we present the distributions of the magnetic field strength and the neutral hydrogen density along the ray $\phi = 0$, $\theta = 55°$ on a larger time scale. This allows us to analyze the propagation of disturbances beyond the HP into the outer heliosheath and their reflection from the HP. We see that there is a substantial ISMF increase in the outer heliosheath which is caused by the deformation and motion of the heliopause interacting with a GMIR. Although this increase fades away in the direction of the bow shock, it is capable of creating asymmetric ISMF configurations consistent with the distribution of radio emission sources (Kurth and Gurnett 2003).

3.2 Instability of the Heliopause

The HP is a tangential discontinuity that divides the interacting streams of the SW and CHISM. It has been known for some time that the HP is subject to Rayleigh–Taylor instabilities driven and mediated by interstellar neutral atoms (Zank et al. 1996; Florinski et al. 2005). Borovikov et al. (2008) identified a new form of instability on the flanks of HP driven by hot neutral hydrogen atoms created by charge exchange of interstellar neutrals with hot heliosheath plasma. To investigate the instabilities and the consequences of these on the heliospheric structure high-resolution, shock-capturing, adaptive-mesh-refined calculations of the SW–LISM interaction were performed. Low numerical dissipation allowed us to analyze the fine structure of the heliosheath resulting from the HP instability. We found that further increasing the grid resolution results in the absence of a dominant frequency in oscillations of the HP. It should be understood that by refining grids we add new degrees of freedom to the nonlinear process under consideration. That is, decreasing the cell size (until we reach the physical dissipation limit) allows us to resolve successively finer structures of the instability and, consequently, smaller time scales.

Borovikov et al. (2008) showed that secondary neutrals play an essential role in destabilizing the flanks of the HP. The time-dependent location of the HP and the termination shock and the influence of their excursions on the plasma distribution in the inner heliosheath were analyzed. It was found that perturbations generated by the HP instability can affect the distribution of plasma in the inner heliosheath at shorter time scales than the time-scale of the instability as it develops near the stagnation axis. The characteristic time scale t_{lin} for the linear stage of the instability development is related to the charge exchange frequency, which was estimated by Florinski et al. (2005) to be about $2 \times 10^{-8}\ \text{s}^{-1}$ in the heliosheath, corresponding to $t_{\text{lin}} \approx 1.6$ yrs. The slower nonlinear growth time scale was ($\sim$60 yrs) in our simulation, if we disregard the finer vortex structures. Observationally, this means that *Voyagers* are likely to travel through a structurally complex region with the width of about 20 AU in the vicinity of the HP. This complex structure will not be due to SW variability. Time-dependent simulations based on solar cycle variations do not introduce this type of structural, long-lived modifications to the HP (see Zank and Müller 2003; Scherer and Fahr 2003; Izmodenov et al. 2005b). Additionally, propagation of SW disturbances (GMIRs, etc.) will not create this kind of variability at the HP because (i) SW disturbances have latitudinal/longitudinal extent much greater than 20 AU, and (ii) these disturbances weaken progressively as they propagate from the TS to the HP, so their effect on the HP is not very dramatic.

By way of application, Borovikov et al. (2008) estimated the intensity of the soft X-ray emission generated in different regions of the HP and showed that its instability leads to a substantial enhancement in the X-ray emission which bears an imprint of the HP shape. Figure 10 (left panel) shows a snapshot in the plasma density distribution which illustrates the HP instability. The influence of this instability on the distribution of ion density in the inner heliosheath is illustrated in Fig. 10 (right panel). Here this distribution is shown at the *Voyager 1* location as a function of time after *V1* crossed the TS. The variations clearly occur on a time scale of about 1.5–2 months, which is notably similar to the time scale of the SW velocity distribution extracted from *V1* energetic ion data by Decker et al. (2006).

3.3 The Effects of the Solar Activity Cycle

Solar-cycle-dependent variation in the SW parameters also affect the distribution of plasma quantities in the inner heliosheath (e.g., Zank and Müller 2003; Pogorelov et al. 2006,

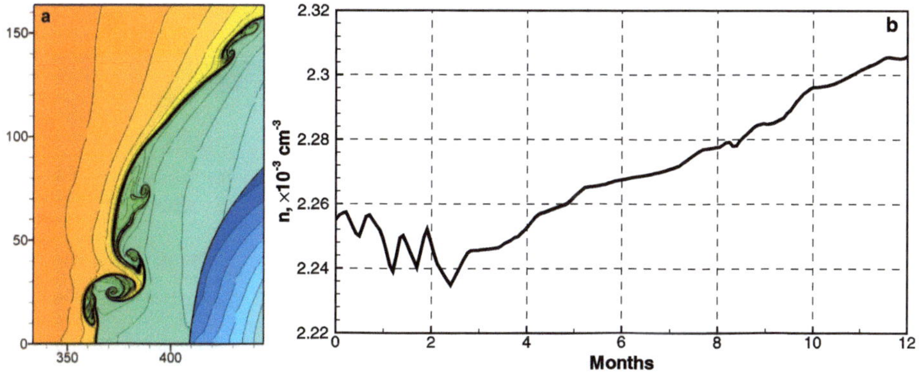

Fig. 10 Instability of the HP initiated by charge exchange between charged and neutral particles (**a**); the distribution of plasma density as if it were measured by *Voyager 1* as a function of time after crossing the termination shock (**b**)

2008b). We illustrate this here by assuming a rather simplified model for the solar cycle. Following McComas et al. (2000), we assume that at the solar minimum slow SW occupies heliolatitudes $-\theta_{0\min} \leq \theta \leq \theta_{0\min}$, where $\theta_{0\min} = 35°$. SW velocity, density, and temperature in the slow region are $V_{Es} = 4 \times 10^7$ cm s^{-1}, $n_{Es} = 8$ cm^{-3}, and 10^5 K, respectively. The parameters in the fast SW are $V_{Ef} = 8 \times 10^7$ cm s^{-1}, $n_{ES} = 3.6$ cm^{-3}, and 2.6×10^5 K. The IMF radial component at 1 AU is 28 μG. The vector **B** in the SW is assumed to be given by Parker's formulae (Parker 1961) at the inner boundary and throughout the SW at $t = 0$. This means that our initial conditions correspond to the solar activity minimum. We also assume that the angle between the Sun's rotation and magnetic-dipole axes is initially $\alpha = \alpha_{\min} = 9°$. To model solar activity cycle, we introduce the following dependence of the slow wind latitudinal extent angle

$$\theta_0(t) = \frac{\theta_{0\max} + \theta_{0\min}}{2} - \frac{\theta_{0\max} - \theta_{0\min}}{2} \cos \frac{2\pi t}{T}, \tag{3}$$

where $\theta_{0\max} = 80°$, and $T \approx 61.05$ is the dimensionless period of solar activity. A similar formula is used to specify $\alpha(t)$ ($\alpha_{\max} = 80°$), but at the maximum we allow the dipole axis to flip its orientation from one hemisphere to another. With this approach, we specify the HCS location on the inner boundary sphere (see Pogorelov et al. 2007).

Let the z-axis of our Cartesian coordinate system be aligned with the Sun's rotation axis. The x-axis belongs to the plane defined by the z-axis and $\mathbf{V}_\infty$ and is directed upstream into the LISM. The y-axis completes the right coordinate system. We choose the following parameters of the unperturbed LISM: velocity $v_\infty = 26.4$ km s^{-1}, proton density $n_\infty = 5$ cm^{-3}, proton temperature $T_\infty = 6527$ K, and neutral hydrogen density $n_{H\infty} = 0.15$ cm^{-3}. The LISM vector $\mathbf{V}_\infty$ is aligned with the vector $\mathbf{r}_{He} = (-0.087156, 0, -0.9962)$, that is, with the direction of the interstellar helium flow (Möbius et al. 2004). The ISMF vector is aligned with the vector $\mathbf{r}_B = (-0.5, -0.2678, -0.82356)$. Thus, $\mathbf{B}_\infty$ belongs to the hydrogen deflection plane (HDP) determined by Lallement et al. (2005) and is directed into the southern hemisphere at an angle of 30° to the ecliptic plane. We choose $B_\infty = 3$ μG.

Figure 11 shows the distribution of plasma temperature at the moments of time close to solar activity maximum (a) and minimum (b). It is clear that a large portion of the inner heliosheath will be consecutively occupied by slower and hotter or faster and colder SW plasma. Any spacecraft traveling through these regions will see periodic changes both in

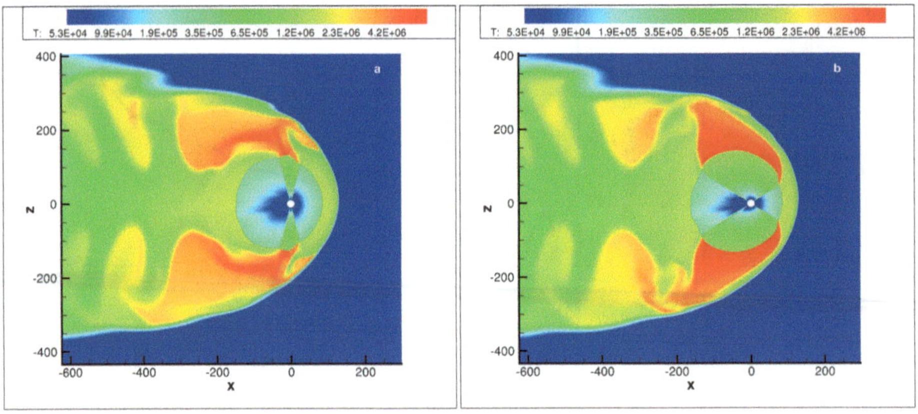

Fig. 11 Solar cycle induced variations of plasma temperature in the meridional plane during solar activity maximum (**a**) and minimum (**b**)

the plasma properties and the IMF strength. Changes in the SW ram pressure cause oscillations of the HP, which in turn generate waves propagating through the CHISM in the outer heliosheath. These waves can generate magnetic pressure enhancements, which can affect the process of 2–3 kHz radio emission generation (Cairns and Zank 2002; Mitchell et al. 2008; Pogorelov et al. 2008a, 2008b). MHD-kinetic modeling of Heerikhuisen et al. (2008) showed that the presence of fast and slow stream is readily seen on energetic neutral atom (ENA) maps. The IBEX mission, launched in October of 2008, will be measuring ENAs arriving from different regions of the heliosphere surrounding the Sun. Comparison of the observational and theoretically-predicted sky maps of ENA fluxes in different energy bands will be invaluable for determining truly three-dimensional properties of the SW–LISM interaction regions. The instruments on board the IBEX spacecraft will be able to resolve solar-cycle-related effects, which makes 3D, time-dependent modeling of the heliospheric interface one of the most challenging tasks for space physicists.

We gave a few examples of time-dependent phenomena that affect SW properties observable in the inner heliosheath and generation of soft X-ray and radio emission. The formulations of the problems considered in this section are oversimplified in order to separate one effect from another. Realistic SW flows bear imprints of the solar cycle, Sun's rotation, CME's, CIR's, MIR's, and GMIR's. To quantify the effect of these on the heliosheath flow one should use both the observational data provided by a fleet of spacecraft measuring data at distances from 1 to 5 AU and remote diagnostics of events occurring on the Sun.

4 Connecting ENAs with Lyα Absorption Measurements

Ultraviolet spectra from the *Hubble Space Telescope* (HST) have proved to be an unexpected source of information about the heliospheric structure. It turns out that neutral hydrogen in the outer heliosphere can produce detectable absorption in spectra of the chromospheric H I Lyman-α emission lines of nearby stars. There are two populations of heliospheric hydrogen that have been detected and studied using the HST data: "hydrogen wall neutrals" created by charge exchange in between the HP and bow shock, and "heliosheath neutrals" created by charge exchange in between the TS and the HP. The former produce detectable absorption in mostly upwind directions, and the latter are only detectable in very downwind directions (Wood et al. 2005, 2007).

As one of the few observational diagnostics of material in the outermost parts of the heliosphere, the Lyα absorption data have been useful for testing sophisticated hydrodynamic models that properly treat neutrals in a self-consistent manner with the protons. These models suggest that the absorption is sensitive to many physical properties of the surrounding CHISM, so the absorption can potentially be used to constrain these properties. A detailed description of these aspects of Lyα absorption studies is provided elsewhere in these proceedings (Wood and Izmodenov 2009).

We here address the progress that can be made with regards to the Lyα diagnostic in the near future. One obvious growth area, which will be emphasized here, involves connecting the Lyα diagnostic with the ENAs that IBEX will observe after it launches, currently scheduled for August 2008 (McComas et al. 2004, 2006, also McComas 2009)

The IBEX mission and the HST Lyα data are natural allies in that they are both used to study the same populations of heliospheric neutral hydrogen atoms. While IBEX will make detailed all-sky maps of local ENA fluxes and detailed measurements of their energy distributions from ~ 10 eV to 6 keV, the HST data provide information about line-of-sight integrated energy distributions for certain directions through the heliosphere. Heliosheath neutrals will dominate IBEX observations, so the most relevant HST data for comparison with IBEX are the heliosheath absorption detections in very downwind directions.

As an example of how Lyα data have been used to constrain purported ENA distributions, we consider a curious population of alleged ENAs detected first by the ASPERA-3 instrument on *Mars Express* and later by the identical ASPERA-4 instrument on *Venus Express*. These instruments were designed to study particles escaping from the atmospheres of Mars and Venus (Barabash and Lundin 2006; Barabash et al. 2007). But the ASPERA instruments appeared to detect ENAs even when not pointed at their respective planets, both en route to their destinations and while in orbit. It was initially argued that these observations were of the same population of heliospheric particles that IBEX will study (Galli et al. 2006; Wurz et al. 2006). More recent evidence, however, points to a local source involving the solar wind's interaction with the spacecraft (Grigoriev et al. 2006). In any case, we show here how constraints provided by Lyα data have been used as strong evidence against the heliospheric interpretation (Wood et al. 2007).

The logarithmic particle fluxes reported by ASPERA-3 for the non-planetary ENAs are in the range $\log F_{A3} = 3.7$–5.0 (in $\mathrm{cm^{-2}\,s^{-1}\,ster^{-1}}$ units) within ASPERA's energy range of 0.2–10 keV. Dotted lines in each panel of Fig. 12 illustrate this observed flux range, with the energy spectrum observed by the instrument. The lower bound of ASPERA's energy range, 0.2 keV, represents a particle velocity of 196 $\mathrm{km\,s^{-1}}$. This corresponds with the *upper* bound of the 50–200 $\mathrm{km\,s^{-1}}$ velocity range where the Lyα data provide the best absorption constraints for heliosheath neutrals (see Wood et al. 2007). Thus, before the ASPERA measurements can be used to predict Lyα absorption, the ASPERA energy spectra must be extrapolated to lower energies. There are three different extrapolations assumed in Fig. 12. The four panels of the figure show the resulting spectra for four different values of $\log F_{A3}$.

The biggest difficulty in connecting ENA and HST Lyα data lies in the issue of how to extrapolate a locally observed particle spectrum along an entire line of sight through the heliosphere. If the heliosheath is the source of the ENAs, then for our purposes the region inside the termination shock is in effect an empty cavity filled by a flux of heliosheath particles from all directions. Therefore, to first order the particle distributions can be assumed to be isotropic and the same throughout the region. This will not be the case within the heliosheath itself, so the absorption calculation illustrated in Fig. 12 considers only absorption

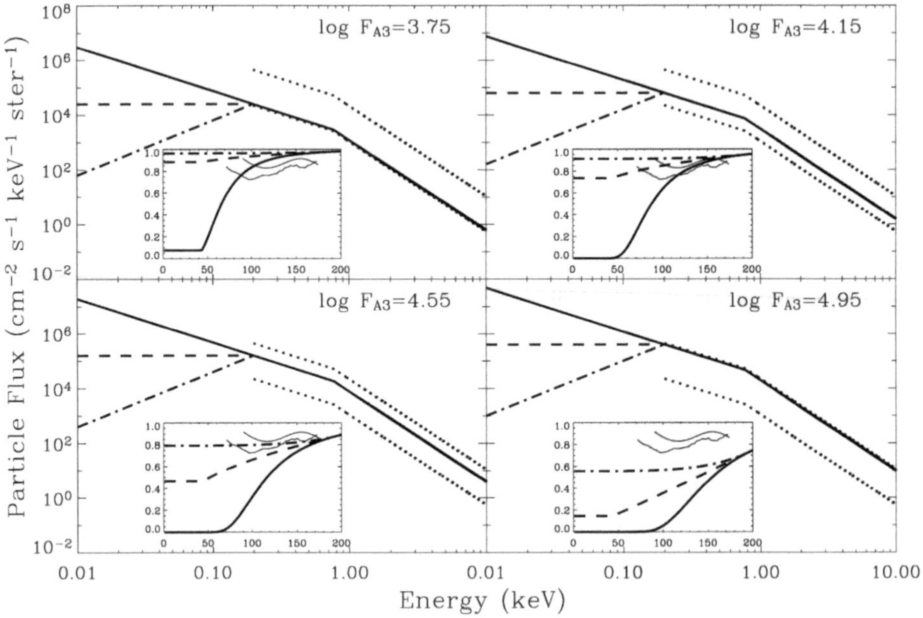

Fig. 12 The *dotted lines in each panel* encompass the range of ENA spectra observed by ASPERA-3, corresponding to different 0.2–10 keV particle fluxes (in cm^{-2} s^{-1} ster^{-1} units) of $\log F_{A3} = 3.7$–5.0. Assuming different fluxes within this range, *each panel* shows three different spectra, which use different extrapolations to low energies. The *inset figures* show the H I Lyα absorption from within the termination shock predicted by these spectra, plotted on a heliocentric velocity scale (in km s^{-1}). The *two thin solid lines* that stretch from ~ 70 to 175 km s^{-1} are the absorption limits derived for two lines of sight observed by HST (τ Cet and ϵ Eri). The absorption predicted by the energy spectra must lie above these limits to be consistent with the Lyα data. From Wood et al. (2007)

by heliosheath neutrals *within the termination shock.* In downwind directions, which are the most appropriate for heliosheath Lyα absorption, we assume a termination shock distance of ~ 200 AU. The inset figures in Fig. 12 show the Lyα absorption produced by heliosheath neutrals within this distance based on the ENA spectra. The upper limits on absorption derived for two lines of sight observed by HST (τ Cet and ϵ Eri; see Wood et al. 2007) are shown for comparison. The absorption predicted by the ENA spectra must lie above these limits to be consistent with the Lyα data.

The two highest flux values explored in Fig. 12, $\log F_{A3} = 4.55$ and $\log F_{A3} = 4.95$, clearly lead to too much Lyα absorption to be consistent with the limits from the HST data. A flux of $\log F_{A3} = 4.15$ is only consistent with the Lyα data if ENA fluxes decrease below 0.2 keV, and even the lowest flux of $\log F_{A3} = 3.75$ is only consistent with the absorption limits if the ENA fluxes are flat or decrease below 0.2 keV. The situation is actually worse than this, given that the absorption predictions in Fig. 12 only include absorption from within the termination shock. A downwind line of sight will have a path length through the heliosheath much longer than the 200 AU path length inside the termination shock, so including absorption from the heliosheath would likely increase the amount of absorption well above the predictions shown in Fig. 12. With this in mind, it is very doubtful whether *any* of the ENA spectra in Fig. 12 are consistent with the relative lack of heliospheric absorption observed in downwind directions. This is a strong argument against the ASPERA ENAs being heliosheath neutrals. However, a very recent recalibration of the ASPERA data has resulted in a downward shift of the ENA particle spectrum to closer to the lower bound of the flux range shown in Fig. 12 (Wurz et al. 2008). Partly due to this adjustment, Wurz et al. (2008) claim that a heliosheath origin for the ENAs is consistent with absorption constraints and is still the most likely interpretation of the observed ENAs.

The ENA spectrum shown as a dashed line in the upper left panel of Fig. 12 is just barely consistent with the Lyα constraints. The integrated particle flux of this spectrum in the 0.01–0.2 keV energy range most applicable to the Lyα absorption data is $F < 5000\ \mathrm{cm^{-2}\,s^{-1}\,ster^{-1}}$. Although the ASPERA instruments are not sensitive to these energies, IBEX will be, and $F < 5000\ \mathrm{cm^{-2}\,s^{-1}\,ster^{-1}}$ represents a conservative upper limit on the particle fluxes that IBEX can expect to see in the 0.01–0.2 keV range. The actual 0.01–0.2 keV flux is likely to be well below this limit since the limit was derived considering only absorption within the termination shock, ignoring the absorption from the heliosheath itself.

More precise constraints on ENA fluxes from the Lyα data will require the assistance of kinetic heliospheric models that properly model the neutral velocity distributions throughout the heliosphere (e.g., Izmodenov et al. 2005a; Malama et al. 2006; Heerikhuisen et al. 2007). Such models can provide a direct link between the ENA and absorption diagnostics by being able to predict both ENA properties within the inner solar system and heliospheric Lyα absorption for various directions. Once IBEX is launched it will be very interesting to see if heliospheric models are capable of simultaneously reproducing both the local ENA spectrum and the heliosheath Lyα absorption data. And it will be fascinating to see if this agreement can be achieved only with a unique set of parameters describing the ambient ISM.

5 A Comparison between the Heliosheath and Other Sheaths in Space

When a supersonic flow encounters an obstacle, a shock forms so that the flow can divert around the obstacle. At the shock the flow becomes slower, denser, hotter, and changes direction. Sheaths are the regions of subsonic flow between the shock and the obstacle. The supersonic SW encounters several types of obstacles. Magnetized planets have magnetospheres which deflect the SW flow; a bow shock forms upstream of the planet and the subsonic solar wind flow behind the shock is called the magnetosheath. Non-magnetized planets and comets with atmospheres form ionized layers where the neutral atmospheres are ionized by photons and/or charge exchange with solar wind protons. The SW magnetic field cannot penetrate this layer, resulting in the formation of an induced magnetopause (or cometopause). The SW passes through a bow shock in front of these obstacles and a sheath region forms as at a magnetized planet.

The largest obstacle for the SW is the local interstellar cloud (LIC). The LIC flows at about 26 km/s relative to the Sun; the solar wind forms a bubble of solar material in the CHISM, but as it moves outward and expands the dynamic pressure grows weaker and the SW flow turns and moves down the heliotail in the direction of the LIC flow. The TS is where the SW becomes subsonic and starts to divert toward the heliotail. The subsonic region is called the heliosheath. Another type of sheath is formed due to transient events in the SW. CMEs propel fast solar wind outward from the Sun. When these ICMEs encounter the slower ambient SW a shock forms at the boundary; the region between the shock and the ICME material is the ICME sheath. This chapter reviews similarities and differences between these various types of sheaths.

The topics we compare are 1) ion distributions in sheaths, 2) sheath size and structure, and 3) waves in sheaths.

5.1 Ion Distributions

Sheaths are composed of shocked fluids; for solar wind sheaths these fluids are plasmas. At the giant planets the proton distributions in the sheaths are often observed to have two components, with distributions fairly well-represented by two Maxwellian distributions with different temperatures (Richardson 1987; Szabo and Lepping 1995). For a high Mach number quasi-perpendicular shock, some portion of the ion distribution function does not have enough energy to pass through the shock potential and is reflected (Sckopke et al. 1983). These reflected ions are convected back to the shock by the solar wind and gain additional energy. The two proton distributions are formed from those protons which traverse the shock the first time and those that traverse it the second time.

Figure 13 shows an example of an ion distribution from Jupiter's magnetosheath which has two distinct proton components; a low temperature component with about 45% of the density at 100 eV and a high temperature reflected component with a temperature of 600 eV. These distributions are observed at all the giant planets except Uranus. They have also been observed in some (but not all) ICME sheaths (Richardson and Liu 2007); Fig. 13b shows an example of an ICME sheath where about 40% of the ions are in the reflected component, which has a temperature 10 times the colder component. Given that the TS is a quasi-perpendicular high-Mach number shock, expectations were that the distributions in the heliosheath would have a high percentage of heavy ions. Figure 13c shows a distribution in the heliosheath showing a good fit to a single Maxwellian; in this spectra the plasma is very cold, only twice the SW temperature. Not only are ions not reflected, but in this case the thermal plasma is essentially unheated by the TS; the energy gain is consistent with conservation of the first adiabatic invariant, p^2/B = constant, with B increasing by a factor of 2 at the TS. Not all heliosheath spectra are this cold, some spectra show evidence of significant heating, but the average heating is a factor of 10–20 less than in planetary magnetospheres. The difference between these shocks seems to be the presence of pickup ions upstream of the TS. Upstream of the TS, the pickup ions make up almost 20% of the solar wind density and dominate the thermal pressure since they are much hotter than the thermal ions (Richardson et al. 2008a).

In shocks with Maxwellian distributions in the upstream flow, energy conservation is maintained by having some of the ions reflect and gain extra energy (i.e., Fuselier and Schmidt 1994). At the TS, this extra energy all seems to go the pickup ions; both plasma and extrapolations of particle data to lower energy are consistent with 80% of

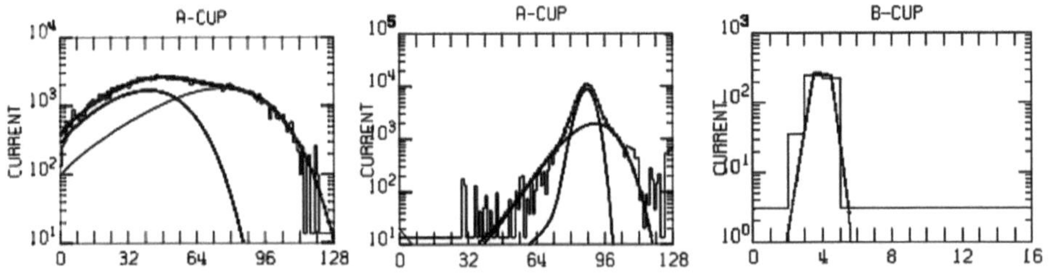

Fig. 13 Proton distributions in the magnetosheath of Jupiter, in an ICME sheath, and in the heliosheath. The *x axis* shows the energy channel of the cup; the channels range in energy from 10 eV to 5950 eV and are roughly logarithmically spaced. The magnetosheath and ICME sheath both show a thermal and a hot component; the heliosheath spectra does not have a hot component. The ratio of the hot/total density is 0.54 for the Jupiter spectrum and 0.4 for the ICME spectrum. The ratios of the hot/cold temperatures are 6 and 9, respectively

the solar wind flow energy being transferred to the pickup ions (Gloeckler et al. 2005; Richardson et al. 2008a). The only giant planet without these double distributions in the magnetosheath is Uranus (Richardson 1987). Uranus is surrounded by an extensive corona of neutral H; this similarity to the TS suggests that pickup ions from the Uranian H corona may be soaking up the shock energy at that planet.

5.2 Sheath Size and Structure

The size of the sheath regions varies with upstream conditions; lower Mach numbers generally give broader sheaths. In planetary magnetosheaths generally the sheath widths at the nose are 30–40% greater than the magnetopause standoff distance, or about 0.2 times the radius of curvature of the shock. The flow through magnetosheaths is comparable to the flow time of the solar wind past the planet, hours to a week. The distance between the TS and the HP is not known, since the heliopause boundary has not been reached. Estimates in the literature vary from 30 to 70 AU, with residence times of order tens of years. ICME sheaths are very different because ICMEs are evolving structures. ICMEs expand in size out to about 15 AU (Richardson et al. 2006); thus the material entering the sheath has a tangential speed comparable to the sheath expansion speed, so the lifetime of material in these sheaths is long. The sheath width is less, roughly 0.1 times the radius of curvature of the ICME. Since the ICME is expanding this sheath is also expanding.

Asymmetries also are common features of sheaths which can be created if the upstream magnetic field is tilted from the flow direction. The 45° average angle of the solar magnetic field to the solar wind flow direction at Earth results in asymmetries in the magnetosheath plasma (Paularena et al. 2001) and the magnetopause locations (Dmitriev et al. 2004). The draping of the magnetic field lines around the magnetopause gives a stronger field at dusk causing more plasma to move to the dawn side. A similar mechanism works at ICMEs. The ICME runs into the predicting solar wind and the spiral field lines drape around the ICME, giving stronger magnetic fields on the east flank of the ICME (Siscoe et al. 2007). The heliosphere is asymmetric, with the TS closer toward V2 than toward V1. Models suggest this asymmetry could arise from a CHISM magnetic field which is tilted out of the galactic plane and at an angle to the CHISM flow direction (Linde et al. 1998; Opher et al. 2006; Pogorelov et al. 2006).

A common structure in planetary magnetosheaths is the plasma depletion layer (PDL) which forms near the magnetopause. When the magnetic shear at the magnetopause boundary is small (no reconnection occurs), magnetic field lines pile up as they run into this boundary. The large magnetic field pressure causes the plasma to flow out of this region along the field, creating a high-field, low density, low-beta region in front of the magnetopause (Zwan and Wolf 1976). PDLs have been observed at Earth, Jupiter, Saturn (Crooker et al. 1979; Hammond et al. 1993; Violante et al. 1995). A similar geometry is present at ICMEs at the boundary of the ejecta and the shocked solar wind in the ICME sheath, and PDLs have been observed in ICME sheaths (Farrugia et al. 1997; Liu et al. 2006). A region of high field and low plasma has been hypothesized to exist in front of the HP and predicted by models (Cranfill 1971; Pogorelov et al. 2006); one precursor of the HP may be a heliospheric PDL.

5.3 Waves

The high beta, anisotropic plasmas behind quasi-perpendicular shocks are ideal environments for generating mirror mode waves. These waves have alternating regions of high field and low density and their structure can evolve with time. These waves are observed in most planetary magnetosheaths with periods ranging from tens of seconds at Earth to 5 minutes at Saturn (Kaufmann et al. 1970; Violante et al. 1995; Joy et al. 2006). The amplitudes of the field variations can be a factor of two in magnetosheaths, possibly affecting particle transport and acceleration. Mirror mode wave generation ends in the low-beta PDL, leaving behind magnetic holes which convect with the flow (Bavassano-Cattaneo et al. 1998). Mirror mode waves have also been identified in ICME sheaths and in the heliosheath (Tsurutani et al. 1992; Liu et al. 2006; Burlaga et al. 2006a). The ICME waves are weak, but the heliosheath waves have periods of hours and amplitudes of a factor of three so these waves may have an important affect on particles in the heliosheath.

Other aspects of sheaths are similar but not yet quantified. All have highly variable plasma and magnetic field due to changes of the upstream flow and movement of the shock. PDFs may change from power laws to Gaussians (Burlaga et al. 2006b). As the Voyagers move through the heliosheath we will have a unique opportunity to study the physics of these complex regions.

6 Outlook: Heliosheath Thickness and Processes at the Heliopause

6.1 Estimating the Thickness of the Heliosheath

Since both V1 and V2 are now in the inner heliosheath, the next question to ask is, how long until each spacecraft reaches the boundary of the solar system—the heliopause? The thickness of the inner heliosheath is generally different in the V1 and V2 directions and depends on the SW and CHISM parameters. A number of numerical computations (Pogorelov et al. 2004, 2008a; Heerikhuisen et al. 2008) were performed with boundary conditions which reasonably fit the observational data. It should be understood that ideal MHD calculations, like those of Pogorelov et al. (2004) or Opher et al. (2006), significantly overestimate the width of the heliosheath. This is because the geometrical size of the interaction region is considerably larger in the absence of the SW deceleration and heating by interstellar neutral particles. For example, choosing the following CHISM and SW parameters: $V_\infty = 25$ km s^{-1}, $n_\infty = 0.07$ cm^{-3}, $n_{H\infty} = 0.1$ cm^{-3}, $T_\infty = 5679$ K, $B_\infty = 3$ μG (the ISMF vector is at 45° to the CHISM velocity vector), $U_E = 450$ km s^{-1}, $M_E = U_E/c_E = 10$, and $n_E = 7$ cm^{-3}, and disregarding the IMF, the width of the heliosheath, according to Pogorelov (2008, unpublished) can be as large as 86 AU and 55 AU along the V1 and V2 trajectories, respectively. This results in substantial asymmetry of the HP and TS in these directions (Pogorelov et al. 2007). If one assumes that the heliosphere can be scaled uniformly to the distance of 94 AU, at which V1 crossed the TS in 2003, the heliosheath width becomes 53 AU and 34 AU in the V1 and V2 directions, respectively.

It is known, however, from various MHD-neutral calculations that such uniform scaling is wrong. If we use the results of the MHD-kinetic simulations of Pogorelov et al. (2008a), where $\mathbf{B}_\infty$ belongs to the hydrogen deflection plane (HDP) and forms an angle of 30° to $\mathbf{V}_\infty$, the width of the heliosheath becomes 79 AU and 55 AU in the *V1* and *V2* directions, respectively. The calculations of Pogorelov et al. (2008a) assumed a Maxwellian distribution for the SW ions, that is, pick-up ions are assumed to equilibrate with core protons. In reality, PUI's form a separate, non-thermal population. Heerikhuisen et al. (2008) suggested to model this effect by assuming a Lorentzian (kappa) distribution function instead, with kappa being as small as 1.63. In their calculation the width of the heliosheath becomes smaller by approximately a factor of 1.25. This results in the widths equal to 63 AU and 44 AU in the V1 and V2 directions, respectively. Per year, V1 covers the distance of about 3.5 AU, while the velocity of V2 is about 3.15 AU per year. From this we can conclude that V1 and V2 will reach the HP approximately in 2021 and 2020, respectively. This estimate does not take into account any time-dependent processes in the SW–CHISM interaction (see more discussion in Sect. 3).

6.2 Energetic Particles at the Heliopause

The HP is a tangential discontinuity, meaning there is a discontinuous change in plasma density, velocity, and magnetic field across its surface. There is no flow of plasma across the interface and the magnetic fields are parallel to it on both sides (although, not necessarily parallel to each other, i.e., rotation is permitted). Because $\nabla \cdot \mathbf{u}$ is zero across the interface, no first-order acceleration could take place according to (1). There are, however, certain second order effects that are driven by flow shear and that are traditionally omitted from the cosmic-ray transport equation.

Berezhko and Krymskii (1981) (see also Earl et al. 1988; Williams and Jokipii 1991) noted that if the diffusion approximation is carried out to second, rather than to a first order in p/mu, four additional terms appear in the transport equation that are related to momentum and energy transport conducted by the particles traveling across a flow shear. Under the assumption of isotropic diffusion (the approximation made by Earl et al. 1988) and ignoring second-order stochastic acceleration, the diffusive transport equation assumes the form

$$\frac{\partial f}{\partial t} + (\mathbf{u} + \mathbf{v}_d) \cdot \nabla f - \nabla \cdot (\kappa \cdot \nabla f) - \frac{\nabla \cdot \mathbf{u}}{3} \frac{\partial f}{\partial \ln p} - \frac{\Gamma}{p^2} \frac{\partial}{\partial p}\left(p^2 \kappa \frac{\partial f}{\partial p}\right) + \cdots = Q, \quad (4)$$

where we did not write explicitly the remaining drift terms associated with the flow inertia. The coefficient Γ describes viscous transfer of energy and is a function of the shear $\partial u_i / \partial x_j$. The viscous term in (4) is formally similar to the momentum diffusion term in (1) and provides the mechanism for second-order acceleration in a shear flow. Unlike momentum diffusion, however, viscous acceleration is $\sim \kappa$, rather than p^2/κ, and is potentially more efficient when κ is large (i.e., at higher energies).

The above analysis however, neglects the large anisotropy in the diffusion coefficients. Indeed, in weak turbulence transport is predominantly field-aligned. Perpendicular transport in space plasmas is provided by the wandering of magnetic field lines. However, such wandering is not permitted across the HP (the fields on the two sides are clearly separated) and perpendicular transport is thus strongly suppressed. Under such conditions it may be shown that the viscous term is effectively zero. Tangential discontinuities are routinely observed in the inner heliosphere, most notably as stream interfaces inside merged interaction regions. Observations show no evidence of acceleration at these structures and theoretical modeling shows that diffusion across the interface is small (Intriligator and Siscoe 1995). It appears therefore that viscous acceleration is not a viable process at the HP.

Energetic particles may be efficiently transported along the surface of the HP by gradient and curvature drifts. The direction of transport is dependent on both the direction of the interstellar field, that is expected to be relatively stable, but is poorly known, and of the heliospheric field which changes polarity with latitude and with time because of the solar cycle (see Sect. 2). More detailed modeling of the solar cycle effects in the heliosheath will be needed to understand the transport of energetic particles in this region.

Finally, energetic particles may be accelerated by the general second-order mechanism if the HP is unstable due to charge exchange with interstellar hydrogen (Sect. 3) producing enhanced levels of turbulence. This process could potentially offset the loss of energetic particles across the interface to the CHISM.

7 Conclusions

Our knowledge and understanding of the distant heliosphere, including the heliosheath region and the HP, are still very incomplete. Theoretical modeling was until recently the principal source of ideas, some of which are now being put to the test following the Voyager approach to and eventual crossing of the TS. This review discussed several outstanding problems of heliospheric research at the time when both Voyagers are finally in the heliosheath and just before the launch of IBEX.

Voyager observations of energetic particles in the vicinity of the TS, particularly their spectra in the foreshock region and the heliosheath, reopened the question of the origin of ACRs. Currently the two acceleration site candidates are the TS (the traditional view), and the heliosheath—possibly the region near the HP—although the HP itself is unlikely to be a source of accelerated particles. In addition to stochastic acceleration in the heliosheath, two scenarios have been proposed to explain unusual accelerated particles spectral features and temporal behavior near the TS, involving temporal changes (shock dynamic evolution) and spatial variations (acceleration at different longitudes), see Sect. 2. Several more years of additional observations in the heliosheath should narrow down the list of possibilities.

The second outstanding issue is the structure and properties of the heliosheath, a region that is highly dynamic in nature. Unlike the SW, at any given time several consecutive solar cycles are stored in the heliosheath in a highly compressed form. Current 3D MHD models are capable of predicting the coarser features of the magnetic field topology, such as the location of the envelopes of different field polarity, but lack the resolution to simulate the behavior of the neutral sheet. This is yet another challenge for the next generation of computer models of the global heliosphere.

Compared to other sheaths observed in space, the heliosheath is unique in that the core plasma is cold (supersonic) and most of the energy is deposited by the TS into the PUI component. IBEX data will provide a much more accurate measure of the PUI heating at the TS. The distance to the heliopause is estimated to be between 44 and 63 AU according to computer models; this distance is probably different along the V1 and V2 trajectories due to the global asymmetry of the HP induced by the tilt of the interstellar magnetic field. It is estimated that both Voyagers will reach the HP sometime near year 2020. It is anticipated that new phenomena will be recorded on approach to this boundary of the heliosphere. This includes a possible plasma depletion layer that is commonly observed in planetary magnetospheres, and dynamic effects associated with the possible instability of the HP from charge-exchange.

Acknowledgements V.F. was supported by NASA grants NNG05GD45G, NNG06GD43G, and NNG06GD48G. J.R.J. received support from NSF grant ATM0447354 and NASA grants NNG05GE83G and NNX07AH19G. D.J.M. was supported by NASA as a part of the Interstellar Boundary Explorer (IBEX) mission. N.V.P.'s work was supported by NASA grants NNG06GD48G, NNG06GD43G, and NNX08AJ21G. Supercomputer time allocations were provided by DOE's INCITE project PSS001, NSF Teragrid project MCA07S033 and NASA High-End Computing program award SMD-08-0763. J.D.R. was supported under NASA contract 959203 and NASA grants NAG5-8947 and NNX08AC04G. E.C.S. was supported by NASA contract NAS7-03001. B.E.W. acknowledges support from NASA grant NNG05GD69G to U of Colorado.

References

B. Ball, M. Zhang, H. Rassoul, T. Linde, Astrophys. J. **634**, 1116–1125 (2005)
S. Barabash, R. Lundin, Icarus **182**, 301–307 (2006)

S. Barabash, J.-A. Sauvaud, H. Gunell, H. Andersson, A. Grigoriev, K. Brinkfeldt, M. Holmström, R. Lundin, M. Yamauchi, K. Asamura, W. Baumjohann, T.L. Zhang, A.J. Coates, D.R. Linder, D.O. Kataria, C.C. Curtis, K.C. Hsieh, B.R. Sandel, A. Fedorov, C. Mazelle, J.-J. Thocaven, M. Grande, H.E.J. Koskinen, E. Kallio, T. Säles, P. Riihela, J. Kozyra, N. Krupp, J. Woch, J. Luhmann, S. McKenna-Lawlor, S. Orsini, R. Cerulli-Irelli, M. Mura, M. Milillo, M. Maggi, E. Roelof, P. Brandt, C.T. Russell, K. Szego, J.D. Winningham, R.A. Frahm, J. Scherrer, J.R. Sharber, P. Wurz, P. Bochsler, Planet. Space Sci. **55**, 1772–1792 (2007)
V.B. Baranov, Y.G. Malama, J. Geophys. Res. **98**, 15157–15163 (1993)
M.B. Bavassano-Cattaneo, C. Basile, G. Moreno, J.D. Richardson, J. Geophys. Res. **103**, 11961–11972 (1998)
E.G. Berezhko, G.F. Krymskii, Sov. Astron. Lett. **7**, 352–354 (1981)
S.N. Borovikov, N.V. Pogorelov, G.P. Zank, I.A. Kryukov, Astrophys. J. **682**, 1404–1415 (2008)
L.F. Burlaga, F.B. McDonald, N.F. Ness, J. Geophys. Res. **98**, 1–11 (1993)
L.F. Burlaga, N.F. Ness, Y.-M. Wang, N.R. Sheeley, J. Geophys. Res. **107**, 1410 (2002)
L.F. Burlaga, C. Wang, J.D. Richardson, N.F. Ness, Astrophys. J. **585**, 1158–1168 (2003)
L.F. Burlaga, N.F. Ness, M.H. Acuña, Geophys. Res. Lett. **33**, L21106 (2006a)
L.F. Burlaga, N.F. Ness, M.H. Acuña, J. Geophys. Res. **111**, A09112 (2006b)
I.H. Cairns, G.P. Zank, Geophys. Res. Lett. **29**, 1143 (2002)
S.V. Chalov, Planet. Space Sci. **41**, 133–136 (1993)
S.V. Chalov, H.J. Fahr, Astron. Astrophys. **360**, 381–390 (2000)
S.V. Chalov, H.J. Fahr, V. Izmodenov, J. Geophys. Res. **108**, 1266 (2003)
S.V. Chalov, H.J. Fahr, Y.G. Malama, Ann. Geophys. **25**, 575–580 (2007)
C.W. Cranfill, Flow problems in astrophysical systems. Thesis (Ph.D.), University of California, San Diego, CA (1971)
N.U. Crooker, T.E. Eastman, G.S. Stiles, J. Geophys. Res. **84**, 869–874 (1979)
A.C. Cummings, E.C. Stone, F.B. McDonald, B.C. Heikkila, N. Lal, W.R. Webber, in *Particle Acceleration and Transport in the Heliosphere and Beyond*, ed. by G. Li et al.. AIP Conf. Proc., vol. 1039 (AIP, Melville, 2008), pp. 343–348
R.B. Decker, E.C. Roelof, S.M. Krimigis, M.E. Hill, in *Physics of the Inner Heliosheath*, ed. by J. Heerikhuisen, V. Florinski, G.P. Zank, N.V. Pogorelov. AIP Conf. Proc., vol. 858 (AIP, Melville, 2006), pp. 73–78
R.B. Decker, S.M. Krimigis, E.C. Roelof, M.E. Hill, T.P. Armstrong, G. Gloeckler, D.C. Hamilton, L.J. Lanzerotti, Nature **454**, 67–70 (2008)
A.V. Dmitriev, A.V. Suvorova, J.K. Chao, Y.-H. Yang, J. Geophys. Res. **109**, A05203 (2004)
J.A. Earl, J.R. Jokipii, G. Morfill, Astrophys. J. **331**, L91–L94 (1988)
C.J. Farrugia, N.V. Erkaev, H.K. Biernat, L.F. Burlaga, R.P. Lepping, V.A. Osherovich, J. Geophys. Res. **102**, 7087–7094 (1997)
S.E.S. Ferreira, M.S. Potgieter, K. Scherer, Astrophys. J. **659**, 1777–1783 (2007)
S.E.S. Ferreira, K. Scherer, Astrophys. J. **642**, 1256–1266 (2006)
V. Florinski, G.P. Zank, Geophys. Res. Lett. **33**, L15110 (2006a)
V. Florinski, G.P. Zank, in *Physics of the Inner Heliosheath*, ed. by J. Heerikhuisen, V. Florinski, G.P. Zank, N.V. Pogorelov. AIP Conf. Proc., vol. 858 (AIP, Melville, 2006b), pp. 159–164
V. Florinski, G.P. Zank, N.V. Pogorelov, J. Geophys. Res. **108**, 1228 (2003)
V. Florinski, G.P. Zank, J.R. Jokipii, E.C. Stone, A.C. Cummings, Astrophys. J. **610**, 1169–1181 (2004)
V. Florinski, G.P. Zank, N.V. Pogorelov, J. Geophys. Res. **110**, A07104 (2005)
S.A. Fuselier, W.K.H. Schmidt, J. Geophys. Res. **99**, 11539–11546 (1994)
A. Galli, P. Wurz, S. Barabash, A. Grigoriev, R. Lundin, Y. Futaana, H. Gunell, M. Holmström, E.C. Roelof, C.C. Curtis, K.C. Hsieh, A. Fedorov, D. Winningham, R.A. Frahm, R. Cerulli-Irelli, P. Bochsler, N. Krupp, J. Woch, M. Fraenz, Astrophys. J. **644**, 1317–1325 (2006)
G. Gloeckler, L.A. Fisk, L.J. Lanzerotti, in *Proc. Solar Wind 11—SOHO 16, Connecting Sun and Heliosphere*, ed. by B. Fleck, T.H. Zurbuchen. ESA SP, vol. 592 (ESA, Noordwijk, 2005), pp. 107–112
G. Gloeckler, L.A. Fisk, G.M. Mason, M.E. Hill, in *Particle Acceleration and Transport in the Heliosphere and Beyond*, ed. by G. Li et al. AIP Conf. Proc., vol. 1039 (AIP, Melville, 2008), pp. 367–374
A. Grigoriev, T. Sotirelis, E.C. Roelof, P.C. Brandt, S. Barabash, J. Vandegriff, Eos Trans. AGU **87**(52). Fall Meet. Suppl., Abstract P51B-1193 (2006)
C.M. Hammond, J.L. Phillips, S.J. Bame, E.J. Smith, C.G. Maclennan, Planet. Space Sci. **41**, 857–868 (1993)
J. Heerikhuisen, N.V. Pogorelov, G.P. Zank, V. Florinski, Astrophys. J. **655**, L53–L56 (2007)
J. Heerikhuisen, N.V. Pogorelov, G.P. Zank, V. Florinski, J.A. le Roux, Astrophys. J. **682**, 679–689 (2008)
D.S. Intriligator, G.L. Siscoe, J. Geophys. Res. **100**, 21605–21612 (1995)
V.V. Izmodenov, D.B. Alexashov, A. Myasnikov, Astron. Astrophys. **437**, L35–L38 (2005a)
V.V. Izmodenov, Y.G. Malama, M.S. Ruderman, Astron. Astrophys. **429**, 1069–1080 (2005b)

J.R. Jokipii, J. Geophys. Res. **91**, 2929–2932 (1986)

J.R. Jokipii, in *Physics of the Inner Heliosheath*, ed. by J. Heerikhuisen, V. Florinski, G.P. Zank, N.V. Pogorelov. AIP Conf. Proc., vol. 858 (AIP, Melville, 2006), pp. 143–152

J.R. Jokipii, J. Giacalone, Space Sci. Rev. **83**, 123–136 (1998)

S.P. Joy, M.G. Kivelson, R.J. Walker, K.K. Khurana, C.T. Russell, W.R. Paterson, J. Geophys. Res. **111**, A12212 (2006)

R. Kallenbach, M. Hilchenbach, S.V. Chalov, J.A. le Roux, K. Bamert, Astron. Astrophys. **439**, 1–22 (2005)

R.L. Kaufmann, J.-T. Horng, A. Wolfe, J. Geophys. Res. **75**, 4666–4676 (1970)

A.G. Kulikovskii, N.V. Pogorelov, A.Y. Semenov, *Mathematical Aspects of Numerical Solution of Hyperbolic Systems* (Chapman & Hall/CRC, London, Boca Raton, 2001)

W.S. Kurth, D.A. Gurnett, J. Geophys. Res. **108**, 8027 (2003)

M.E. Hill, D.C. Hamilton, S.M. Krimgis, Astrophys. J. **572**, L169–L172 (2002)

R. Lallement, E. Quémerais, J.L. Bertaux, S. Ferron, D. Koutroumpa, R. Pellinen, Science **307**, 1447–1449 (2005)

U.W. Langner, M.S. Potgieter, Ann. Geophys. **22**, 3063–3072 (2004)

T. Linde, T.I. Gombosi, P.L. Roe, K.G. Powell, D.L. DeZeeuw, J. Geophys. Res. **103**, 1889–1904 (1998)

Y. Liu, J.D. Richardson, J.W. Belcher, J.C. Kasper, R.M. Skoug, J. Geophys. Res. **111**, A09108 (2006)

Y.G. Malama, V.V. Izmodenov, S.V. Chalov, Astron. Astrophys. **445**, 693–701 (2006)

D. McComas, Space Sci. Rev. (2009, this issue)

D.J. McComas, N.A. Schwadron, Geophys. Res. Lett. **33**, L04102 (2006)

D.J. McComas, B.L. Barraclough, H.O. Funsten, J.T. Gosling, E. Santiago-Muños, R.M. Skoug, B.E. Goldstein, M. Neugebauer, P. Riley, A. Balogh, J. Geophys. Res. **105**, 10419–10433 (2000)

D. McComas, F. Allegrini, P. Bochsler, M. Bzowski, M. Collier, H. Fahr, H. Fichtner, P. Frisch, H. Funsten, S. Fuselier, G. Gloeckler, M. Gruntman, V. Izmodenov, P. Knappenberger, M. Lee, S. Livi, D. Mitchell, E. Möbius, T. Moore, D. Reisenfeld, E. Roelof, N. Schwadron, M. Wieser, M. Witte, P. Wurz, G. Zank, in *Physics of the Outer Heliosphere*, ed. by V. Florinski, N.V. Pogorelov, G.P. Zank. AIP Conf. Proc., vol. 719 (AIP, Melville, 2004), pp. 162–181

D. McComas, F. Allegrini, L. Bartolone, P. Bochsler, M. Bzowski, M. Collier, H. Fahr, H. Fichtner, P. Frisch, H. Funsten, S. Fuselier, G. Gloeckler, M. Gruntman, V. Izmodenov, P. Knappenberger, M. Lee, S. Livi, D. Mitchell, E. Möbius, T. Moore, S. Pope, D. Reisenfeld, E. Roelof, H. Runge, J. Scherrer, N. Schwadron, R. Tyler, M. Wieser, M. Witte, P. Wurz, G. Zank, in *Physics of the Inner Heliosheath*, ed. by J. Heerikhuisen, V. Florinski, G.P. Zank, N.V. Pogorelov. AIP Conf. Proc., vol. 858 (AIP, Melville, 2006), pp. 241–250

F.B. McDonald, E.C. Stone, A.C. Cummings, B. Heikkila, N. Lal, W.R. Webber, Nature **426**, 48–51 (2003)

F.B. McDonald, E.C. Stone, A.C. Cummings, W.R. Webber, B. Heikkila, N. Lal, Geophys. Res. Lett. **34**, L05105 (2007)

J.J. Mitchell, I.H. Cairns, N.V. Pogorelov, G.P. Zank, J. Geophys. Res. **113**, A04102 (2008)

E. Möbius, M. Bzowski, S. Chalov, H.-J. Fahr, G. Gloeckler, V. Izmodenov, R. Kallenbach, R. Lallement, D. McMullin, H. Noda, M. Oka, A. Pauluhn, J. Raymond, D. Rucinski, R. Skoug, T. Terasawa, W. Thompson, J. Vallerga, R. von Steiger, M. Witte, Astron. Astrophys. **426**, 897–907 (2004)

H. Moraal, R.A. Caballero-Lopez, K.G. McCracken, F.B. McDonald, R.A. Mewaldt, V. Ptuskin, M.E. Wiedenbeck, in *Physics of the Inner Heliosheath*, ed. by J. Heerikhuisen, V. Florinski, G.P. Zank, N.V. Pogorelov. AIP Conf. Proc., vol. 858 (AIP, Melville, 2006), pp. 219–225

M. Opher, E.C. Stone, P.C. Liewer, Astrophys. J. **640**, L71–L74 (2006)

E.N. Parker, Astrophys. J. **134**, 20–27 (1961)

K.I. Paularena, J.D. Richardson, M.A. Kolpak, C.R. Jackson, G.L. Siscoe, J. Geophys. Res. **106**, 25377–25394 (2001)

H.L. Pauls, G.P. Zank, J. Geophys. Res. **102**, 19779–19787 (1997)

H.L. Pauls, G.P. Zank, L.L. Williams, J. Geophys. Res. **100**, 21595–21604 (1995)

N.V. Pogorelov, T. Matsuda, J. Geophys. Res. **103**, 237–245 (1998)

N.V. Pogorelov, G.P. Zank, T. Ogino, Astrophys. J. **614**, 1007–1021 (2004)

N.V. Pogorelov, G.P. Zank, T. Ogino, Astrophys. J. **644**, 1299–1316 (2006)

N.V. Pogorelov, E.C. Stone, V. Florinski, G.P. Zank, Astrophys. J. **668**, 611–624 (2007)

N.V. Pogorelov, J. Heerikhuisen, G.P. Zank, Astrophys. J. **675**, L41–L44 (2008a)

N.V. Pogorelov, G.P. Zank, T. Ogino, Adv. Space Res. **41**, 306–317 (2008b)

R. Ratkiewicz, A. Barnes, G.A. Molvik, J.R. Spreiter, S.S. Stahara, M. Vinokur, Astron. Astrophys. **335**, 363–369 (1998)

J.D. Richardson, J. Geophys. Res. **92**, 6133–6140 (1987)

J.D. Richardson, Y. Liu, in *Turbulence and Nonlinear Processes in Astrophysical Plasmas*, ed. by D. Shaikh, G.P. Zank. AIP Conf. Proc., vol. 932 (AIP, Melville, 2007), pp. 387–392

J.D. Richardson, F.B. McDonald, E.C. Stone, C. Wang, J. Ashmall, J. Geophys. Res. **110**, A09106 (2005)

J.D. Richardson, Y. Liu, C. Wang, L.F. Burlaga, Adv. Space. Res. **38**, 528–534 (2006)
J.D. Richardson, J.C. Kasper, C. Wang, J.W. Belcher, A.J. Lazarus, Nature **454**, 63–66 (2008a)
J.D. Richardson, Y. Liu, C. Wang, Adv. Space Res. **41**, 237–244 (2008b)
K. Scherer, H.J. Fahr, Geophys. Res. Lett. **30**, 1045 (2003)
N.A. Schwadron, D.J. McComas, Geophys. Res. Lett. **34**, L14105 (2007)
N.A. Schwadron, M.A. Lee, D.J. McComas, Astrophys. J. **675**, 1584–1600 (2008)
N. Sckopke, G. Paschmann, S.J. Bame, J.T. Gosling, C.T. Russell, J. Geophys. Res. **88**, 6121–6136 (1983)
G. Siscoe, P.J. MacNeice, D. Odstrcil, Space Weather **5**, S04002 (2007)
M. Simon, W.I. Axford, Planet. Space Sci. **14**, 901–908 (1966)
C.D. Steenberg, H. Moraal, Astrophys. J. **463**, 776–783 (1996)
E.C. Stone, A.C. Cummings, W.R. Webber, J. Geophys. Res. **101**, 11017–11025 (1996)
E.C. Stone, A.C. Cummings, F.B. McDonald, B.C. Heikkila, N. Lal, W.R. Webber, Science **309**, 2017–2020 (2005)
E.C. Stone, A.C. Cummings, F.B. McDonald, B.C. Heikkila, N. Lal, W.R. Webber, Nature **454**, 71–74 (2008)
A. Szabo, R.P. Lepping, J. Geophys. Res. **100**, 1723–1730 (1995)
B.T. Tsurutani, D.J. Southwood, E.J. Smith, A. Balogh, Geophys. Res. Lett. **19**, 1267–1270 (1992)
L. Violante, M.B. Bavassano-Cattaneo, G. Moreno, J.D. Richardson, J. Geophys. Res. **100**, 12047–12055 (1995)
H. Washimi, G.P. Zank, Q. Hu, T. Tanaka, K. Munakata, Astrophys. J. **670**, L139–L142 (2007)
L.L. Williams, J.R. Jokipii, Astrophys. J. **371**, 639–647 (1991)
B.E. Wood, V.V. Izmodenov, Space Sci. Rev. (2009, this issue)
B.E. Wood, S. Redfield, J.L. Linsky, H.-R. Müller, G.P. Zank, Astrophys. J. Suppl. Ser. **159**, 118–140 (2005)
B.E. Wood, V.V. Izmodenov, J.L. Linsky, Y.G. Malama, Astrophys. J. **657**, 609–617 (2007)
P. Wurz, A. Galli, A. Barabash, A. Grigoriev, in *Physics of the Inner Heliosheath*, ed. by J. Heerikhuisen, V. Florinski, G.P. Zank, N.V. Pogorelov. AIP Conf. Proc., vol. 858 (AIP, Melville, 2006), pp. 269–275
P. Wurz, A. Galli, A. Barabash, A. Grigoriev, Astrophys. J. **683**, 248–254 (2008)
G.P. Zank, H.L. Pauls, L.L. Williams, D.T. Hall, J. Geophys. Res. **101**, 21639–21655 (1996)
G.P. Zank, H.-R. Müller, J. Geophys. Res. **108**, 1240 (2003)
M. Zhang, in *Physics of the Inner Heliosheath*, ed. by J. Heerikhuisen, V. Florinski, G.P. Zank, N.V. Pogorelov. AIP Conf. Proc., vol. 858 (AIP, Melville, 2006), pp. 226–232
B.J. Zwan, R.A. Wolf, J. Geophys. Res. **81**, 1636–1648 (1976)

Section II: What Are the Dominant Physical Processes in the Termination Shock and Inner Heliosheath?

The Heliospheric Magnetic Field and Its Extension to the Inner Heliosheath

A. Balogh · J.R. Jokipii

Originally published in the journal Space Science Reviews, Volume 143, Nos 1–4, 85–110.
DOI: 10.1007/s11214-008-9485-2

Abstract The general structure of the heliospheric magnetic field is well known and has been extensively studied, mostly in the inner heliosphere, out to the orbit of Saturn. Beyond 10 AU, the Pioneer and now the Voyager spacecraft have provided a view of the outer heliosphere. Its structure is strongly affected by large-scale phenomena originating in the Sun's activity, such as the pattern of fast and slow solar wind streams around solar minimum that lead to Corotating Interaction Regions, and the increased frequency and strength of Coronal Mass Ejections around solar maximum. The large current sheet that separates the dominant magnetic polarities in the heliospheric medium, the Heliospheric Current Sheet, provides a variable structure that evolves from a relatively simple geometry close to the solar equatorial plane to what is likely to be a highly complex and dynamic surface reaching to high heliolatitudes at high levels of solar activity. The magnetic field observed in a fluctuating, dynamical heliosheath differs considerably from that in a static heliosheath. In particular, the time between current-sheet crossings (sectors) is quite sensitive to the radial speed of the solar-wind termination shock. If an inwardly moving termination shock moves past an observer on a slowly moving spacecraft, the time between current-sheet crossings in the heliosheath becomes larger, and can become very large, for reasonably expected inward shock speeds. This effect may help to explain recent observations of the magnetic field from the Voyager 1 spacecraft, where, in the heliosheath, the magnetic field remained directed outward from the Sun for several months without a current-sheet crossing. The crossings finally resumed and now occur somewhat regularly. In addition, the magnetic fluctuations in the heliosheath are observed to be quite different from those in the supersonic upstream solar wind.

Keywords Heliosphere · Heliospheric magnetic field · Corotating interaction regions · Interplanetary coronal mass ejections · Inner heliosheath

A. Balogh (✉)
International Space Science Institute, 3012 Bern, Switzerland
e-mail: balogh@issibern.ch

J.R. Jokipii
University of Arizona, Tucson, AZ 85721, USA

1 The Global Heliospheric Magnetic Field: the Parker Formulation

The heliospheric magnetic field (HMF) originates in the solar corona and is dragged from the corona into the heliosphere by the (nearly everywhere) infinitely conducting but collisionless solar wind plasma (for an extensive recent review see Smith 2007). The simplest model of the HMF (Parker 1958) assumes a radially oriented magnetic field at some (spherical) source surface of radius R_{SS} around the Sun where the solar wind entrains the magnetic field lines in its radial flow with velocity V_{SW}. Given a distribution of the magnetic field strength $B_R(R_{SS})$ that may be also a function of solar latitude (this is sufficient since the direction is defined as radial, normal to the source surface and it is also assumed that the magnetic field is axisymmetric), and the solar rotation rate Ω, assumed to be uniform on the source surface, the resulting HMF can be derived very simply as a geometric consequence of solar rotation and a uniform (spherically symmetric), constant speed radial solar wind flow.

The geometry of the HMF and its dependence on heliolatitude and heliocentric distance were derived by Parker (1958) at the same time when the concept and first model of the solar wind as a supersonic solution of the expanding solar atmosphere was also proposed. To describe the HMF, the generally used coordinate system, the RTN system, is used to describe the magnetic field vector. The radial coordinate axis, R, points away from the centre of the Sun; the polar or N axis is normal to the radial axis in the solar meridian plane and is towards the solar North pole; the azimuthal or T axis completes the right handed system. In the solar wind flow, the radial component of the magnetic field, $B_R(R)$ is given as

$$B_R(R) = B_R(R_{SS})\frac{R_{SS}^2}{R^2}. \tag{1}$$

The tangential or azimuthal component $B_T(R)$ of the magnetic field, at a colatitude θ measured from the solar rotation axis in the North, is

$$B_T(R,\theta) = B_R(R_{SS})\frac{R_{SS}^2}{R^2}\frac{\Omega R\sin\theta}{V_{SW}}. \tag{2}$$

In this equation for the azimuthal component, we assume that $R \gg R_{SS}$. Note that the radial component of the magnetic field $B_R(R)$ decreases as R^{-2}, while the azimuthal component $B_T(R)$ varies as R^{-1}. In this simple geometrical view of the solar wind flow and the HMF, the polar ("North–South") component of the magnetic field is zero, $B_N(R) = 0$. The magnitude of the magnetic field, assuming that $R \gg R_{SS}$, is

$$B(R) = \left(B_R^2 + B_T^2\right)^{1/2} = B_R(R_{SS})\frac{R_{SS}^2}{R^2}\left[1 + \left(\frac{\Omega R\sin\theta}{V_{SW}}\right)^2\right]^{1/2}. \tag{3}$$

In a somewhat more realistic case, the magnetic field at the source surface is a function also of heliolongitude ϕ and heliolatitude (or colatitude θ) while remaining in direction normal to the spherical source surface. In this case, the rotation of the point of origin of the solar wind parcel on the source surface also needs to be taken into account, so that $B_R(R_S)$ in (1) and (2) needs to be replaced by

$$B_R(R_{SS}) \Rightarrow B_R\left(R_{SS}, \theta, \phi + \frac{\Omega(R - R_{SS})\sin\theta}{V_{SW}}\right). \tag{4}$$

This formula provides the basis for "projecting" the magnetic field measured in interplanetary space back to the solar source surface, by taking into account the rotation of the Sun during the propagation time of the solar wind to the point of observation. Usually, differential rotation is not taken into account. Although the photosphere rotates at a latitude dependent rate, at periods of ~25.4 days at the equator and ~29.5 days close to the poles, these may not be representative of the rotation rate associated with the HMF. The HMF originates from open magnetic field lines in the corona, associated with coronal holes; at least some of these appear to rotate rigidly with the rotation rate characteristic of the equator (Timothy et al. 1975). However, the rotation rate of coronal holes is in general variable, and rigid rotation at the equatorial photospheric rotation rate cannot be normally assumed (Nash et al. 1988; Wang and Sheeley 1993). This variability introduces a time-dependent variability into the latitude dependence in (2) and (3)

We note that the HMF magnetic field lines form the so-called Parker spiral pattern, that is an Archimedes spiral in the solar equatorial plane, and a spiral wound at constant pitch on the surface of a cone, axially aligned with the solar rotation axis, and of half-angle equal to the co-latitude of the observations away from the equatorial plane. In the solar equatorial plane, the equation of the Parker spiral is

$$\phi = -\frac{\Omega(R - R_S)}{V_{SW}}. \tag{5}$$

For field lines away from the solar equatorial plane, (5) becomes

$$\phi = -\frac{\Omega(R \sin\theta_0 - R_S)}{V_{SW}} \tag{6}$$

where θ_0 is the co-latitude of the footpoint of the field line on the notional source surface.

Observations of the average magnetic field direction have confirmed that the field is oriented along the Parker direction with a significant, but close to symmetric spread (Thomas and Smith 1980; Forsyth et al. 2002).

2 Deviations from the Parker Model of the Heliospheric Magnetic Field

It is clear that effectively all the assumptions enumerated above need considerable modification to describe the solar wind and the HMF observed in the heliosphere. The solar wind is not uniform but is emitted with significant differences in speed and other properties (density, temperature, composition) from different regions of the corona. Fast solar wind (~750 $\mathrm{km\,s^{-1}}$) is emitted from coronal holes in which magnetic field lines are open to the heliosphere, while the qualitatively different, much more variable slow solar wind (at speeds ~300 to 450 $\mathrm{km\,s^{-1}}$) is associated with the magnetically closed regions in the corona.

The other properties of the solar wind are equally varying according to the coronal configuration of the source regions. The solar wind flow is not radial everywhere above the corona, super-radial expansion is due to the dominance of the magnetic pressure close to the Sun. As already noted above, the photosphere does not rotate uniformly, there is a significant slow-down in the rotation rate from the equator to the poles. However, the corona from which the solar wind and HMF originate does not rotate in the same way as the photosphere, in fact (the larger) coronal holes may rotate more or less independently of latitude at the equatorial rate of the photosphere (Wang and Sheeley 1993). The coronal magnetic field that is entrained by the solar wind also has a variable non-radial component. Quite apart from these spatial differences from those assumed in the simple model above, most of the boundary conditions that affect the expansion of the coronal magnetic fields into the heliosphere are highly time variable, on many scales.

In the light of these differences from the simple assumptions, it is remarkable how robust the simple Parker model of the HMF is and how well it can be used as the first order approximation to the actual HMF. The deviations from the simple model as observed in space are often quite considerable, yet these remain deviations rather than imposing the need for a new model.

Three aspects of the differences from the simple model need to be noted. The first of these, first put forward by Jokipii and Kóta (1989) has its origin in the earlier concept of the random motion of photospheric magnetic field lines (Jokipii and Parker 1968) which then leads to, when extended over the polar regions of the Sun, to a relatively increasing importance of the transverse (fluctuating) components of the magnetic field vector. The reason for this is that the radial component of the magnetic field decreases as the square of the heliocentric distance, while the transverse components decrease proportionally with heliospheric distance. Although the effect predicted by Jokipii and Kóta (1989) may become significant, even possibly dominant at some tens of AU from the Sun, the observations of Ulysses over the polar regions at 2 to 3 AU showed that the variances in the transverse components of the magnetic field were consistent with that picture (Jokipii et al. 1995; Forsyth et al. 1996). Comparison of the model based on the statistics of photospheric field line diffusion with the Ulysses data brought further confirmation of the relative increase in the transverse variance of the magnetic field with heliospheric distance over the polar regions (Giacalone and Jokipii 2004).

The second effect is related to the heliolatitude dependence of the radial component of the magnetic field. Ulysses observations as it first surveyed the heliosphere over the Sun's south polar region in late 1994 close to solar minimum showed that the strength of the radial component of the magnetic field, normalised to 1 AU, was independent of heliolatitude and in fact matched the measurements made simultaneously in the ecliptic plane (Smith and Balogh 1995). The complete heliolatitude survey in 1994–95 confirmed that the normalised value of the radial component of the magnetic field was indeed independent of heliolatitude both in the southern and northern hemispheres (Forsyth et al. 1996). This observation confirms that the magnetic field from polar coronal holes is transported equatorward close to the Sun, close to the source region of the solar wind; the solar wind itself must diverge from the polar coronal holes, as had been foreseen based on solar observations and modelling (Wang and Sheeley 1991; Wang 1993).

A potentially significant modification of the Parker geometry of the magnetic field was identified by Fisk (1996). This arises from the way the magnetic field line geometry is affected by the difference in rotation rates between the solar photosphere and the overlying coronal holes. The magnetic field lines are anchored in the photosphere; there is a significant slowing down of the rate at which the photosphere rotates from the equator to the polar regions (see, e.g. Snodgrass 1983). The fast solar wind from the polar coronal holes that transports the magnetic field into the heliosphere expands non-radially, as already mentioned above. Coronal holes tend to rotate faster than the underlying high latitude photosphere, approximately at the equatorial rotation rate of the photosphere. This rigid rotation rate is not completely general, however it appears to apply to the large scale coronal holes associated with the fast solar wind streams around solar minimum (Wang and Sheeley 1993). At the same time, the coronal holes rotate around an axis (notionally, the magnetic dipole axis) of the Sun that is offset by maybe 10 to 30 degrees from the rotation axis near solar minimum.

The compound effect of the difference in the rotation of the coronal holes and the footpoints of the magnetic field is that field lines originating at high latitudes are emerging in the solar wind at lower latitudes. This effect thus modifies the Parker configuration that implies that the heliolatitude of the magnetic field lines is constant. This structuring of the magnetic field lines away from high latitudes is based on the observation of the non-radial expansion of the solar wind and the rotation rate of the polar coronal holes that is faster than the underlying photosphere in which the footpoints of the magnetic field are embedded. The extent to which this effect occurs and can be measured in the observations of the magnetic field depends on the detailed geometry of the coronal holes and the stability of the source regions of the fast solar wind. This effect may be significant at times of low solar activity, although it is probably difficult to isolate it from other temporal changes as has been shown by Forsyth et al. (2002). More recently, Roberts et al. (2007) analysed the Ulysses data and concluded that the observations precluded the field predicted by Fisk (1996) and Zurbuchen et al. (1997) by at least a factor of approximately 3. Even in the absence of a steady state modification of the overall structure of the heliospheric magnetic field, this effect may introduce some magnetic flux transport away from the Sun's polar regions.

The HMF as observed in the heliosphere results from the solar boundary conditions and the dynamic effects that the non-uniform solar wind flow imposes on it. Any description of the HMF has to cover the effects that are dependent on heliocentric distance as well as heliolatitude and longitude. In the following, the HMF's most significant features are described as a function of the solar activity cycle. The dependence of the structure of the corona on the solar cycle is the dominant effect in the HMF that modifies the model based on the simple boundary conditions assumed in Parker's model.

3 Structures in the Heliospheric Magnetic Field: Solar Cycle Effects

The solar activity cycle manifests itself, as far as the solar wind is concerned, in two main effects. First of all, around solar minimum, coronal holes are large, cover both polar regions of the Sun, and change relatively slowly compared to the solar rotation period. This configuration gives rise to quite stable fast solar wind streams that fill the heliosphere polewards from about 15 to 30 degrees in heliolatitude. This relative stability of the solar wind flows on the scale of several solar rotations generates an equally stable dynamic pattern in the inner heliosphere. This pattern is the dynamic outcome of the interaction of fast and slow streams that appear to corotate with the Sun as the resultant of the radial propagation of the solar wind and the solar rotation, hence their designation as Corotating Interaction Regions (CIRs). Around solar maximum activity, this relative stability gives way to a great deal of variability, due to the source regions of the solar wind (coronal holes, and magnetically closed loop structures) being less well differentiated in the corona, while the corona itself becomes highly time variable on the scale of the solar rotation period.

Coronal mass ejections (CMEs) occur much more frequently near solar maximum activity, resulting in significant changes in the corona as well as the injection of solar material typical of the hot, closed coronal loops into the solar wind The magnetic flux that is carried in CMEs generally retains the complex structure characteristic of the closed loop in the corona.

The second effect of the solar cycle is the change that occurs in the Sun's large scale magnetic field. Around solar minimum, the solar magnetic field, from the outer corona, is dominated by the dipole term with an axis close to the Sun's rotational axis with only relatively small contributions from higher order or non-axisymmetric terms. In the heliosphere, what is observed is that the large-scale HMF is oriented on average away from the Sun along the Parker spiral in one hemisphere (northern or southern, depending on the solar cycle), and towards the Sun in the other hemisphere. The surface separating the two polarities is the Heliospheric Current Sheet (HCS), the largest recognisable plasma boundary inside the heliosphere (for a review, see Smith 2001). Given that the magnetic equator of the Sun is close to the rotational equator when the axial dipole dominates the large scale field, with generally only a relatively small deviation (up to 15 to 30 degrees) from it, the HCS is a surface that remains near the solar equatorial plane. For an observer near the ecliptic plane, the rotation of the HCS with the Sun generates the magnetic sector structure first identified by Ness and Wilcox (1965) and related to the solar magnetic field by Wilcox and Ness (1965). This consists of observing the two polarities of the HMF (away from the Sun or towards the Sun along the Parker spiral) in alternating intervals, depending on whether the observer is North or South of the HCS. In most cases, there are two or four magnetic sectors in each solar rotation.

3.1 The Solar Corona, the Heliospheric Magnetic Field and the Heliospheric Current Sheet

Around solar maximum activity, the magnetic complexity of the corona, as already mentioned, reflects the photospheric and the global magnetic fields that are significantly more complex than around solar minimum (Wang and Sheeley 2003a). Different descriptions have been used for the large scale solar magnetic field around solar maximum, such as a combination of two dipole terms, one axisymmetric, the other oriented in the solar equatorial plane or, alternatively, an axial dipole together with large, even dominant quadrupole and octupole terms (Wang and Sheeley 2003b). In the latter description, the dipole term first becomes small then reverses at solar maximum, to grow in magnitude and relative importance towards the next solar minimum, while the higher order poles decrease in relative magnitude. Despite the magnetic complexity on the Sun and in the corona near solar maximum activity, the geometry of the large scale HMF generally remains on average recognisably Parker-like, although the dominant (towards and away) polarities are separated by a HCS that is very complex in shape and reaches to high, near-polar heliolatitudes. The dynamically evolving shape of the HCS in the heliosphere at this time, when the neutral line in the corona nearly reaches the poles, is difficult to visualise and model.

A particularly successful way of linking the HMF with its solar origin, particularly its large scale structure and variations as a function of the phase of the solar cycle and of location in the heliosphere has been provided by the potential field, source surface (PFSS) model (Schatten et al. 1969; Hoeksema et al. 1982). In this model, the measured magnetic field in the photosphere is used as an input and, assuming that there are no currents in the corona, the magnetic field is calculated between the solar surface and a spherical surface at a radius of 2.5 R_{sun}. On this "source surface", magnetic field lines that intersect it are forced to be normal to the surface. Most of the magnetic field lines from the photosphere close in loop structures below the source surface; magnetic flux that is not closed at the surface is considered to be open and to extend into the heliosphere. While the current-free (potential field) assumption is clearly a restriction that may be unrealistic, there are also other restrictions to the model: the lack of precise measurements of the photospheric fields at high (from about 60° to 70°) latitudes, and the fact that temporal changes within a solar rotation period are averaged, are two of the basic limitations of the model.

Nevertheless, magnetic maps of the corona at the source surface, showing a calculated magnetic neutral line separating the two magnetic polarities have been produced since 1976 by the Wilcox Solar Observatory (WSO), Stanford University, now covering three solar activity cycles. The modelled neutral line has predicted with considerable success the large scale magnetic structure of the heliosphere, in particular the position of the HCS which corresponds to the extension of the coronal neutral line into the heliosphere. Three such coronal maps are shown in Fig. 1, illustrating the changes that occur in the coronal structure

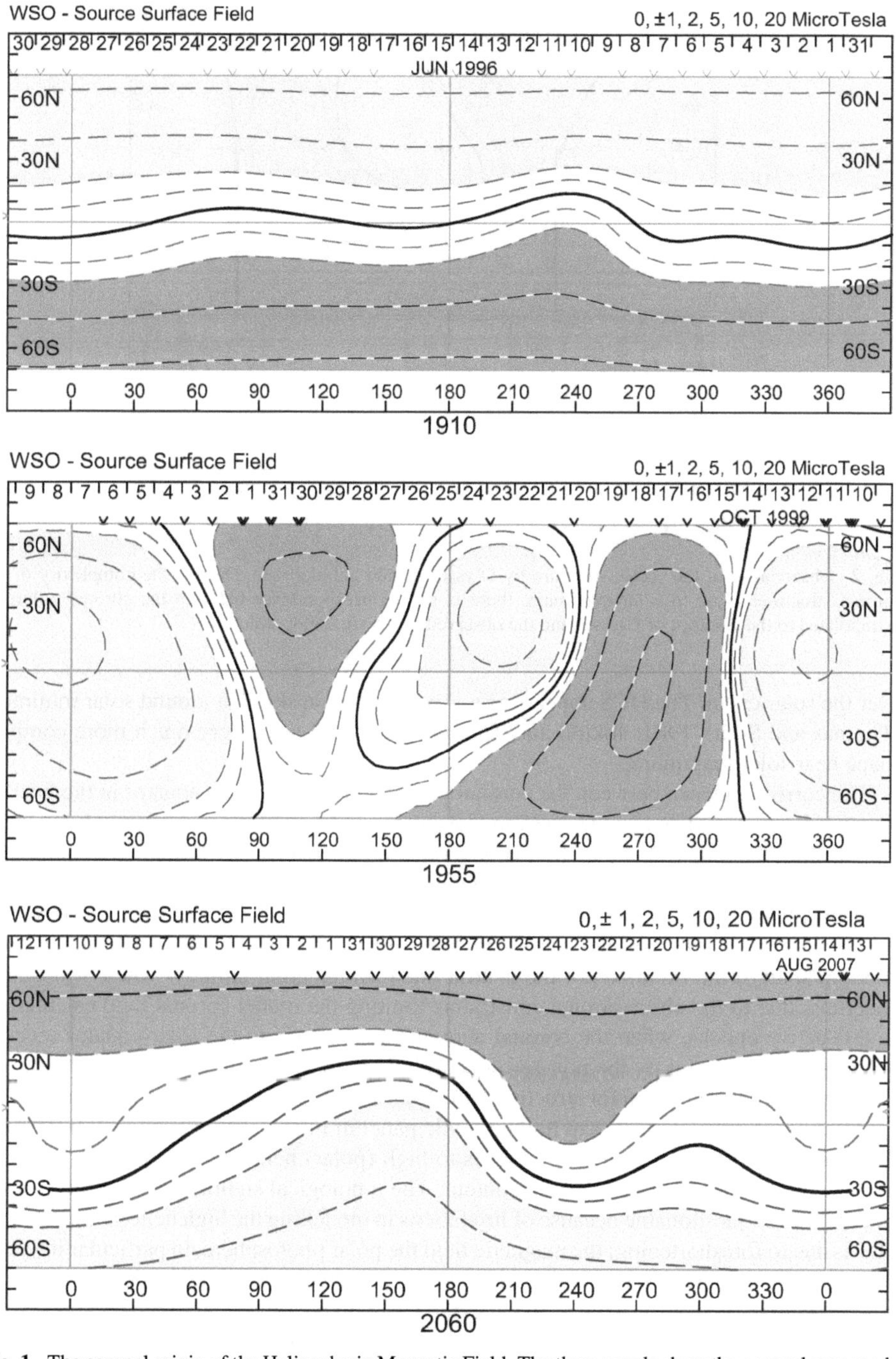

Fig. 1 The coronal origin of the Heliospheric Magnetic Field. The three panels show the coronal map model derived from the Source Surface Potential Field model of the Wilcox Solar Observatory. The *upper* and *lower panels* represent the state of the corona near solar minimum, with the magnetic neutral line separating the two dominant magnetic polarities close to the solar equator. The *middle panel* shows the model corona close to solar maximum activity, with a magnetic neutral line that may be topologically discontinuous, but reaching to high heliolatitudes where the SSPF model is unreliable

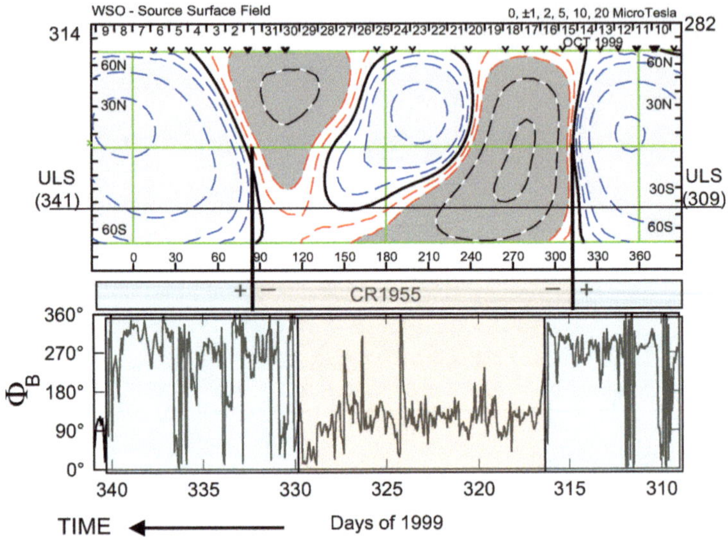

Fig. 2 Observation of the sector structure by Ulysses at mid-heliolatitude. Despite the complexity of the coronal structures close to solar maximum, there is good correspondence between the coronal polarities extrapolated to the distance of Ulysses and the observed magnetic sector structure

over the solar cycle. The HCS appears as a wavy "ballerina skirt" at around solar minimum (Thomas and Smith 1981; Jokipii and Thomas 1981), but has a very much more complex shape near solar maximum.

The correspondence between the coronal position of the magnetic equator in the form of the neutral line calculated by the PFSS model and the position of magnetic sector boundaries representing the passage of the HCS is generally very good (Suess 1993a). An example is shown in Fig. 2. Although the observations were made at mid-heliolatitudes, at about 45° south by Ulysses close to solar maximum, a clear two-sector polarity structure was observed. When compared to the expected sector structure based on the PFSS model, the match is quite good, in spite of the excursion of the neutral line towards polar latitudes where it becomes uncertain, due to the observational constraints limiting the model coronal field calculation. In less active epochs, when the coronal neutral line is closer to the solar equator (even if waviness is introduced by the non-zero quadrupole term in the global magnetic field), the match between observed sector structure and projected HCS position is generally very good.

As shown by the coronal map in the middle panel in Fig. 1, the magnetic neutral line in the corona around solar maximum reaches to high (polar) heliolatitudes and even appears to show topologically more than one contour. The topological significance of such coronal neutral lines is questionable because of limitations in modelling the high heliolatitude corona This is due to foreshortening; the magnetic field the polar photosphere, in particular its radial component, cannot be accurately determined. The potential field model is then limited in its resolution, and leads to the neutral line, if extended to high latitudes (above 70 degrees North or South of the solar equator), being largely uncertain. The complexity and, additionally, the time variability of the photospheric magnetic field near solar maximum make the SSPF model uncertain. Topologically unconnected neutral lines, with one (or occasionally two) "islands" of opposite polarity on the source surfacewould imply detached current sheets, but the Ulysses results are compatible with a single, topologically connected HCS at least in the inner heliosphere even around solar maximum (Jones et al. 2003).

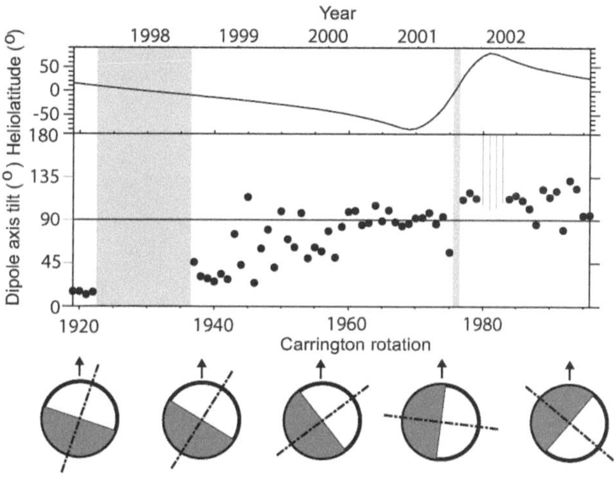

Fig. 3 Observation of the magnetic reversal of the Sun by Ulysses. Taking the ratios of the duration of the positive and negative sectors, the direction of the solar dipole axis can be estimated. The data are consistent, in a first approximation, with a rigid rotation of the dipole axis across the solar equator at the time of magnetic reversal. This implies a Heliospheric Current Sheet that is nearly in a solar meridian plane. (From Jones et al. 2003)

Projections of this neutral line at around solar maximum activity into the heliosphere, complicated by solar rotation, generate a shape for the HCS that is difficult to visualise. In addition to projecting a three-dimensional surface into the heliosphere, the non-uniform solar wind flows and the frequent Coronal Mass Ejections introduce dynamical effects, discussed in more detail below, that further distort the HCS. In any case, even during these disturbed times, the magnetic sector structure of the HMF continues to show two or four sectors at all latitudes within the heliocentric distance range covered by the Ulysses observations, at least to 5 AU. This has been interpreted as evidence that, while complex, and affected by the dynamics of transient solar wind streams, the HCS remains a single connected structure throughout the heliosphere even if largely inclined to the solar equator. In fact, there is evidence that at the time of magnetic reversal near solar maximum activity, the HCS rotates, in a first approximation, as the rigid equatorial plane of the solar dipole while the axis rotates from one pole to the other, as illustrated in Fig. 3 (Jones et al. 2003). These results have been derived from the high heliolatitude measurements of Ulysses during the last magnetic reversal, showing that the HCS remains an identifiable boundary between the two dominant polarities even at the highest latitudes, through the magnetic reversal process. There is, therefore, a magnetic neutral line that reaches to high heliolatitudes in the corona and a corresponding HCS at up to the heliocentric distances covered by Ulysses during the reversal (up to about 3 AU). This means that there may be a rotating current sheet close to a meridional plane that sweeps through the inner heliosphere at the period of the solar rotation. It is not known what happens dynamically to such a high latitude current sheet at larger heliocentric distances.

Voyager results in the late declining phase of the solar cycle in 1994 (Burlaga and Ness 1996) are compatible with a wavy HCS within a limited heliolatitude range around the solar equator out to 40 to 45 AU. During that year (that corresponds to the Ulysses observations of the sector structure in 1992–93, see Fig. 7 below) the two Voyager spacecraft were, respectively, at 32.5° north (V1) and 11.9° south (V2). Whereas Voyager 1 observed a unipolar magnetic field corresponding to the inward solar polarity of the northern solar hemisphere, Voyager 2 observed the two polarities, inward and outward, with almost equal probability. Clearly Voyager 1 was above the northernmost extent of the HCS, while Voyager 2 remained within the sector zone (Burlaga and Ness 1996).

Further results in the new solar cycle 23 at the increasing heliocentric distances of the two Voyager spacecraft are compatible with an increasing extent of the HCS in heliolatitude

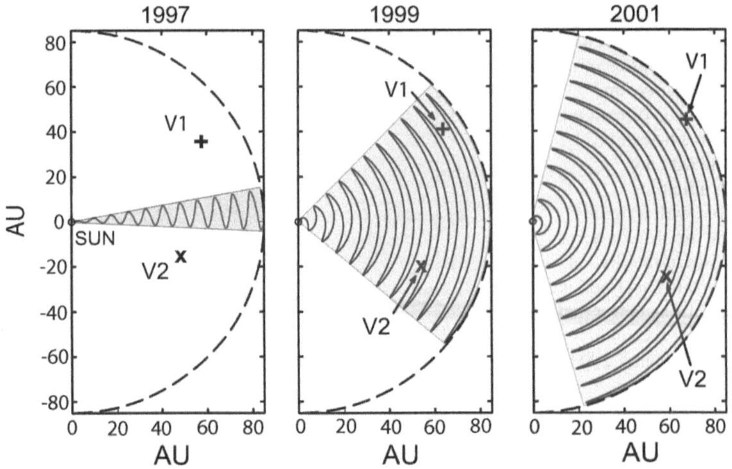

Fig. 4 The volume of the heliosphere in which alternating magnetic polarities (the sector structure) can be observed close to solar minimum (*left panel*), during the ascending phase of the solar activity cycle (*middle panel*) and close to solar maximum (*right panel*). These volumes, called the sector zones, contain the Heliospheric Current Sheet, represented schematically by the wavy line. The sketch is based on observations by the two Voyager spacecraft in the outer heliosphere. (From Burlaga et al. 2002)

with increasing solar activity. These results (Burlaga et al. 2002) are qualitatively compatible with the magnetic neutral line on the source surface in the SSPF model, as well as with observations in the inner heliosphere by Ulysses. A sketch of the evolution of the HCS from 1997 (close to solar minimum) to 2001 (around solar maximum) is shown in Fig. 4. While compatible with the periodic changes in the direction of the HMF, the sketch is best seen as indicative of the increasingly large volume of the heliosphere in which both dominant polarities of the Sun can be observed. To what extent the HCS remains stable enough to show rigid corotation with the Sun is unclear. A further level of complexity is introduced into this picture by the CMEs that affect heliospheric structure at all latitudes around solar maximum (see below, in Sect. 3.3). A significant fraction (variously estimated to be 30 to 50%) of the magnetic flux in the solar wind is associated with CMEs at that epoch and the general structure of the HCS and the heliospheric medium in general are considerably more complex than implied in the sketch in Fig. 4 for 2001.

3.2 Corotating Interaction Regions

During the declining phase of the solar activity cycle, relatively stable coronal holes form in the polar regions, but also with extensions towards the equator. One such period, during the Skylab epoch, with an equatorial extension of the evolving coronal hole, first led to the identification of coronal holes as the source regions of the fast solar wind (Krieger et al. 1973). The alternating slow and fast solar wind streams that are formed in this coronal configuration lead to the characteristic dynamic pattern of the Corotating Interaction Regions (CIRs) in the inner and middle heliosphere, foreseen by Parker (1963) and confirmed both at 1 AU (Gosling et al. 1972) and beyond (Smith and Wolfe 1976). CIRs normally form within a maximum of about 45° of the solar equator (for in-depth reviews and comprehensive references see (Balogh et al. 1999)); a sketch of the projection into the solar equatorial plane is shown in Fig. 5.

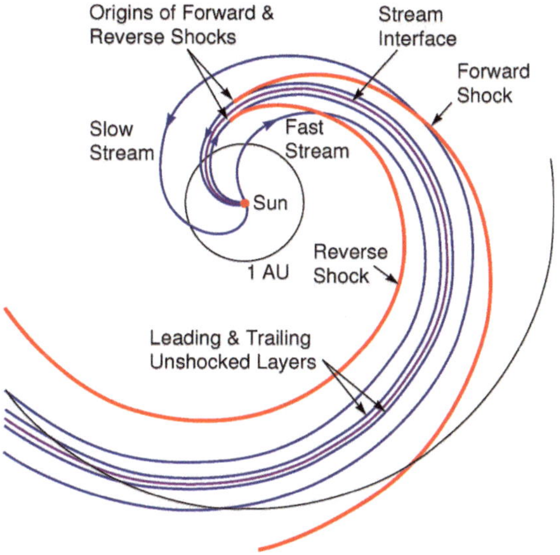

Fig. 5 A Corotating Interaction Region represented schematically in the solar equatorial plane. Shocks (*red lines*) form usually beyond 1 AU; the separation surface between the preceding slow stream and the following fast stream is the stream interface. The compressed interaction region is between the forward and reverse shocks. (After Crooker et al. 1999)

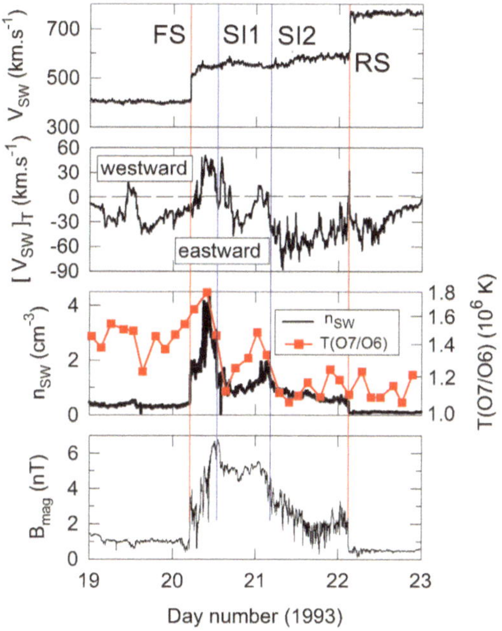

Fig. 6 Morphology of a CIR observed by Ulysses. The *upper panel* is the solar wind velocity; the *second panel* shows the transverse component of the soar wind velocity with its characteristic westward–eastward deflection across the CIR; the *third panel* shows the solar wind density (*black line*) with its compressive features and the ratio of O^{7+} to O^{6+} (*red squares*) that indicate the temperature of the corona from which the solar wind streams originated; the *lower panel* shows the compressed magnetic field intensity between the forward and reverse shocks

The time profile of a CIR, observed by the Ulysses spacecraft at a heliocentric distance of 5.0 AU and at a heliolatitude of 30° south, is shown in Fig. 6, typical of a fully formed structure at distances from about 1.5 AU out to about 10 AU. The interface between the preceding slow solar wind and the following fast solar wind is the stream interface SI; in the case illustrated in Fig. 6 the first slow stream was followed by a short-duration fast stream leading to the compression of the plasma at the leading edge of the CIR. In this event, a second, short-duration slower stream was observed that was compressed by a larger fast stream; as a result, two SIs have been identified in this CIR. The identification of the SI relies on the difference in ion composition and the charge state of ions between slow and fast streams. In particular, the origin of the solar wind plasma can be identified from the coronal freezing in temperature of the oxygen ion population that is determined from the ratio of the ionisation states of oxygen ions measured in the solar wind (Wimmer-Schweingruber et al. 1997). The kinetic parameters of the plasma (speed, density and temperature) do not provide an unambiguous identification as they are subject to the dynamic evolution at the interface. Such complexity, involving several identifiable streams involved in the interaction, is not unusual.

There are two important dynamic phenomena associated with the CIRs. The first is the formation of shock waves at the leading and trailing edges of the CIRs. These shock waves form because of the increase in the pressure of the compressed plasma within the CIR. At the leading edge, a forward shock wave is formed, while at the trailing edge a reverse shock wave is formed. Shocks usually evolve from compressive waves beyond 1 AU. The forward shock wave propagates away from the Sun in the solar wind frame and the reverse shock propagates sunward in the same frame. Because the solar wind is highly supersonic, both shock waves are seen by a stationary observer (such as a spacecraft) as propagating away from the Sun. The pressure is higher within the CIR bounded by the two shock waves than in the preceding or following solar wind plasma. At the forward shock, the solar wind speed and density and the magnetic field increase abruptly; at the reverse shock, the speed increases, but the density and the magnetic field intensity decrease.

The second important dynamic phenomenon is the deflection in the flow direction of the solar wind as the fast stream collides with the preceding slow stream. On the leading, slow-wind side of the stream interface, the flow is deflected westward, while the slow wind is accelerated by the momentum of the colliding fast wind. Behind the stream interface, the fast wind is decelerated and is deflected eastward. This is also illustrated in Fig. 6. In addition to the azimuthal, westward-eastward deflection, there is also a meridional, north–south deflection in the flow that depends on the hemisphere in which the CIR is observed. In both hemispheres, the deflection near the forward shock is towards the equator, so southward in the northern hemisphere and northward in the southern hemisphere. The 3D structure of CIRs is the result of a dynamic evolution of the interaction of the slow and fast streams and is dependent on their respective configurations near their solar source regions (Pizzo and Gosling 1994; Gosling and Pizzo 1999; Lee 2000).

The stability of the heliospheric pattern (comprising both the properties of the solar wind and the heliospheric magnetic field) that develops during the declining phase of solar activity is illustrated in Fig. 7. A long-lasting pattern of coronal holes extending to low heliolatitudes developed during the declining phase of solar cycle 22, in 1992–93. The Ulysses spacecraft began the first of its out-of-ecliptic orbits in February 1992, and progressed to about 45° south by the end of 1994.

As CIRs propagate from the inner heliosphere, they widen to cover a larger radial distance between their leading and trailing edges; thus CIRs catch up with preceding CIRs and merging occurs (Burlaga et al. 1984; Whang and Burlaga 1985; Smith 1989). This phenomenon manifests itself in the coalescence of the interaction regions, resulting in fewer, but more intense compression regions beyond 10 AU. A comparison between observations near the Earth and at Pioneer 11 at heliocentric distances between 15 and 17 AU in 1984 is shown in Fig. 8 (following Smith 1989). The solar wind is dynamically compressed into narrow merged interaction regions as the comparison between IMP-8 and Pioneer 11 measurements show; the magnetic field magnitude is similarly compressed as the comparison between ISEE-3 and Pioneer 11 measurements indicate. Similar results, with merged CIRs recurring at the rate of one per solar rotation were observed by the Voyager

Fig. 7 The stability of the coronal configuration of the sources of high and low speed solar winds over approximately 15 solar rotations during the declining phase of the solar activity cycle in 1992/93 observed by Ulysses. The *upper panel* is the solar wind speed, varying periodically between high and low speeds; the *second panel* represents the ratio of oxygen charge states 7 to 6, used as a proxy for the temperature of the coronal region from which the solar wind originated; the *third panel* is the magnitude of the magnetic field, showing the periodic compressions associated with CIRs; the *lower panel* shows the magnetic field azimuth angle as it varies periodically between outward and inward polarities

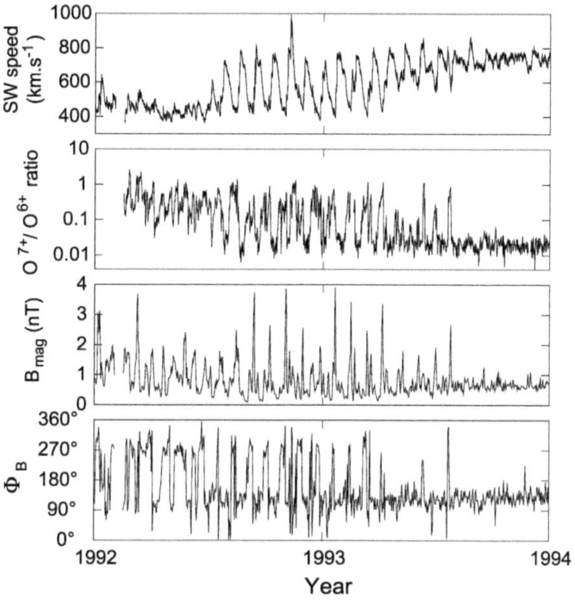

Fig. 8 Observations of the magnetic field magnitude during the declining phase of solar cycle 21 at 1 AU by ISEE-3 (*lower panel*) and between 15 and 17 AU by Pioneer 11. The highly variable field profile at 1 AU is consolidated into merged interaction regions in the middle heliosphere. (After Smith 1989)

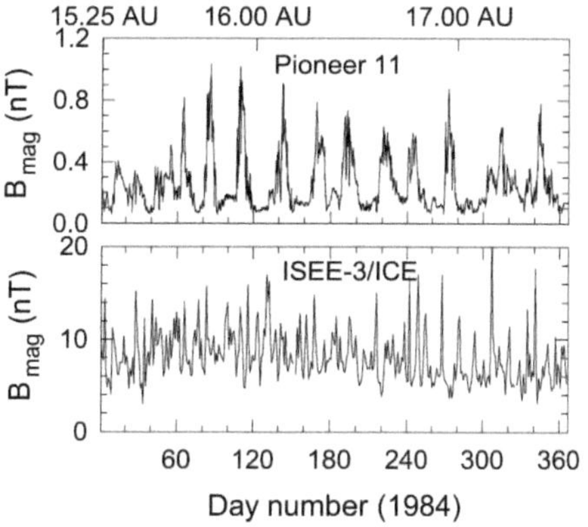

spacecraft at these heliocentric ranges (Burlaga et al. 1997). Voyager results at greater heliocentric ranges, in the range ~40 to 50 AU (Burlaga et al. 1997; Lazarus et al. 1999; Richardson et al. 2008) showed a structured heliospheric medium, but much lower correlation between magnetic field and solar wind compressive features than at medium (~10 to 20 AU) heliocentric distances. Modelling of interaction regions and the comparison with Voyager data has shown the evolution from periodic structures to a more disordered state (Burlaga et al. 2003). Shock waves that recurred apparently remained identifiable, in particular the forward shocks, at least in terms of very abrupt and significant increases in the solar wind speed. However, reverse shocks could not be unambiguously identified (Lazarus et al. 1999). Voyager 2 observations in 1994, during the decaying phase of the solar cycle, are shown in Fig. 9, at ~ 43 AU, illustrating the recurring nature of the fast solar wind streams, apparently led by forward shock waves. The density of the solar wind and the magnetic field intensity have a more complex profile in which the associated simultaneous increases that would be expected to accompany compressions cannot be easily distinguished.

Fig. 9 Solar wind and magnetic field parameters observed by Voyager 2 in the distant middle heliosphere at about 43 AU. This interval is comparable to the interval of Ulysses observations in Fig. 7. The solar wind speed shows a profile that is approximately periodic at close to the solar rotation period, but the solar wind density and the magnetic field intensity appear to be uncorrelated and show no evidence of periodic structuring of the interaction regions

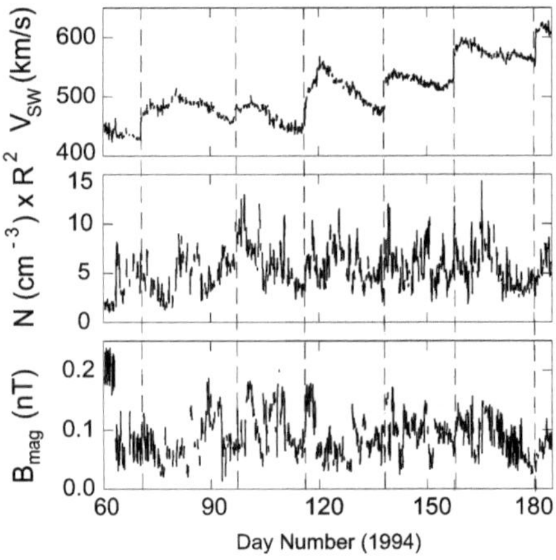

The overall picture around solar minimum at large heliocentric distances is less that of an organised and structured medium, but rather that of a medium in which the dynamics has fragmented the structures that had developed in the inner heliosphere. Contrary to the observations one solar cycle earlier at ~14 AU, in 1994 Voyager 2 observed, at 43 AU, a recurrent solar wind speed profile, but no recurrence pattern at around 26 days in the form of the previously highly structured compressive features, such as solar wind density and the magnetic field intensity (Burlaga et al. 1997). It would appear that the continuing dynamic evolution of the merging of interaction regions that is observed out to ~15 AU leads to this fragmentation of the recognisable interaction regions. A qualitative change occurs in the properties of the heliospheric medium; this change also affects the propagation of energetic particles, including, primarily galactic cosmic rays. The fragmentation of structures remains the dominant characteristic of the outer heliosphere.

3.3 "Interplanetary" Coronal Mass Ejections through the Heliosphere

Instabilities in the corona that develop in coronal loop structures associated with active regions result in large volumes of previously closed magnetic fields and the coronal plasma that they contained to be expulsed into the solar wind. For in-depth reviews on all aspects of Coronal Mass Ejections (CMEs), "Interplanetary" Coronal Mass Ejections (ICMEs) and comprehensive references, see the volume by Kunow et al. (2006). CMEs are recognised by their compositional signatures such as a large (~0.08 or greater) alpha particles to protons density ratio and a high ionisation level for Fe ions (greater than about 12, up to 16 or more, see Lepri and Zurbuchen 2004). For an important subset of them, CMEs are recognised through their magnetic structure. The most recognisable magnetic structure is a magnetic cloud (Burlaga et al. 1981; Gosling 1990), often considered to be a force free magnetic structure recognisable by the large scale rotation of the magnetic field. An indirect magnetic signature is the presence of the bi-directional streaming of supra-thermal electrons along magnetic field lines, interpreted as indicating that the field lines are connected at both their end points into the corona, thus corresponding to coronal loops that have expanded into the heliosphere. Magnetic loops recognised in the form of flux ropes are evidence of the origin of CMEs in the closed magnetic field regions of the corona. However, none of the signatures are either necessary or sufficient and identifying CMEs is often quite difficult. An example is shown in Fig. 10, where the rotation of the magnetic field is quite obvious, but the solar wind parameters are less clearly associated with CMEs. However bidirectional streaming of

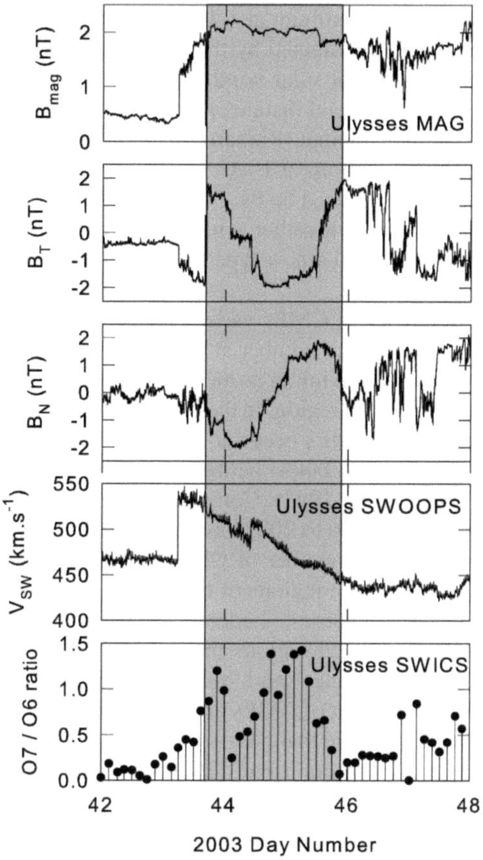

Fig. 10 Profile of a magnetic cloud corresponding to a Coronal Mass Ejection from the Sun. The *uppermost panel* shows the enhanced magnetic field, the *next two panels* show the transverse magnetic field components, showing the large scale rotation of the magnetic field. The *fourth panel* shows the solar wind velocity, while the *lowermost panel* shows the oxygen ion charge ratio as a proxy of the high temperature of the coronal region from which the CME originated

supra-thermal electrons (the shaded interval in Fig. 10) is a strong indication, as is the high freezing in temperature of oxygen, as evidenced by the increased ratio of O^{7+}/O^{6+} in the interval corresponding to the large scale rotation of the magnetic field.

As CMEs have their origin in the complex, variable and hot closed-loop corona, they introduce structures that are qualitatively and topologically different from the (mainly) Parker-like structure of the HMF. The flux ropes and other structures that come from the closed, hot corona obviously evolve dynamically in the solar wind, dependent on many contextual parameters. They often have complex structures, often with embedded current sheets apparently separating originally distinct loops in the corona. Multiple loops are not unusual (see, e.g. Rees and Forsyth 2004).

As CMEs propagate away from the Sun, the largest and most energetic ones retain a positive speed differential with respect to the solar wind in which they are embedded, but the majority of them effectively travel with the solar wind. Those that have been positively identified at large distances are those which were noted to be the largest in the inner heliosphere, such as the so-called Bastille Day event in July 2000 and the series of large solar outbursts collectively known as the Halloween event in 2003. The propagation of CMEs out to ~30 AU was reviewed and summarised by Richardson et al. (2006). The first characteristic they note is the expansion of the CMEs, due to their originally larger internal pressure than that of the ambient solar wind, from 1 AU to 10 to 15 AU, until pressure balance on a large scale is established. This has as a consequence that CME material fills up to 40% of space near solar maximum at ~10 AU; therefore the magnetic structure in the heliospheric medium is equally affected by the magnetic structure of the CMEs as these evolve dynamically in the ambient solar wind. At greater distances, there is further evolution, although tracking CMEs beyond distances of about 10 or more AU from the Sun becomes problematic. In one such attempt (Richardson et al. 2002) a CME was tracked from the Sun, through near-Earth observations at 1 AU, at Ulysses at 5.3 AU and at Voyager 2 at 58 AU. The CME material was recognised by the expected enhancement in the alpha to proton ratio. However, dynamic evolution, together with the proximity of another CME that apparently merged with the first created a new type of merged structure that appears to be more characteristic at large distances.

The frequency of CMEs varies greatly with solar activity, with a few only around solar minimum, but a large number at times of high solar activity; the largest ones usually occur after the activity maximum as measured by sunspot numbers. At solar minimum, CMEs occur in the equatorial region, in the band of closed field lines under the helmet streamers. Near solar maximum, CMEs occur at all heliolatitudes; their distribution is not related to active regions or sunspots, but to helmet streamers that reach high heliolatitudes at high activity levels (Hundhausen 1993). Comprehensive statistics of the CMEs observed by the SOHO spacecraft from 1996 to 2002 have been analysed by Yashiro et al. (2004). These data show the increase in the number of CMEs observed as solar activity increases, and also an extension of the source regions of the CMEs extending to high heliolatitudes. The frequently occurring CMEs evolve considerably in the context of the solar wind and interact with each other to form merged regions that contain CME material and structures. A schematic illustration of the "inner" and "middle" heliosphere is shown in Fig. 11. Given the almost isotropic distribution of CMEs around solar maximum, this sketch can represent any planar cut of the heliosphere that passes through the Sun. An example event that supports the large scale merging of several solar CMEs was analysed by McDonald et al. (1993), using observations in the outer heliosphere by Pioneer 10 (at ~52 AU), Voyager 1 (at ~46 AU) and Voyager 2 (~35 AU). Using magnetic field, plasma and energetic particle measurements, they concluded that the complex pattern of successive CMEs during the March and June 1991 large solar event series merged into an apparently global structure that affected the propagation of both solar energetic particles and cosmic rays, compatible with the sketch in Fig. 11.

The structure at the largest distances sampled by Voyager 2 was reviewed by Richardson et al. (2006). The fragmentation and decay seen in the corotating structures does not happen in the case of large structures of especially large Coronal Mass Ejections that are associated with the active Sun. It appears that around solar maximum, magnetic fields and the solar wind plasma coalesce into large scale correlated structures, each lasting about half a year or so, as shown in Fig. 12, from Richardson et al. (2006). These structures (that also affect the propagation of cosmic rays) arise from the merging of pressure waves that are the remnants of CMEs as they interact and evolve while travelling to the outer heliosphere.

Thus the picture of the outer heliosphere in terms of its magnetic structure and solar wind environment is responsiveness to gross solar conditions, but in a highly compound manner. The evolution of structures that arise from either stable/periodic or impulsive coronal structures and events passes through several stages of interaction and merging from the Sun to the boundary of the heliosphere. Structures that remain recognisable through most of the heliosphere are the largest CMEs (or a series of large CMEs that merge) and the magnetic sector structure. This latter implies that the HCS may retain its integrity through the heliosphere. However, given the complex dynamic effects that shape the HCS from the Sun to the outer boundary of the heliosphere, it is difficult to visualise it and its evolution to the termination shock.

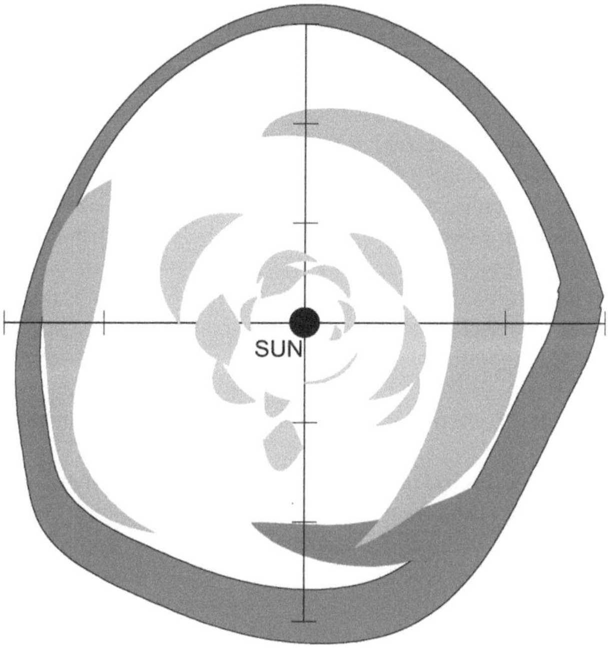

Fig. 11 A schematic sketch of a cut-through the inner and middle heliosphere near solar maximum, when a large number of CMEs is emitted from the Sun in all directions. The CMEs combine and merge as they propagate with their different velocities and may, for the largest of them, form a giant merged interaction region in the heliosphere. Because CMEs are emitted in all directions, this sketch can represent a projection on any plane that includes the Sun

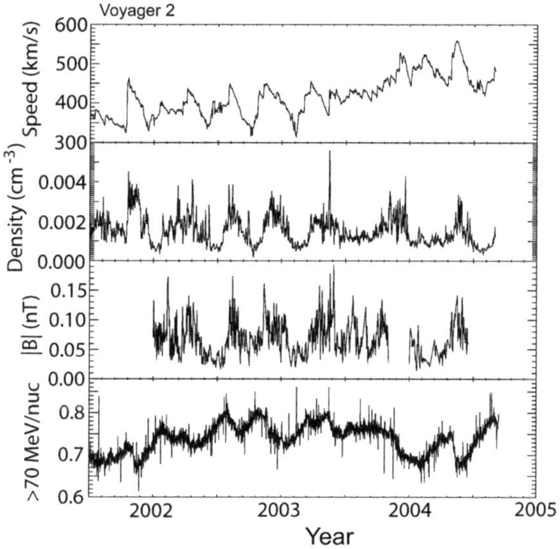

Fig. 12 Heliospheric conditions in the outer heliosphere observed by Voyager 2. The *upper panel* represents the solar wind speed, the *second panel* shows the solar wind density, and the *third panel* shows the magnetic field intensity. The structure of the solar wind and of the magnetic field in these years following the solar activity maximum are affected by the evolved products of continual interaction over the propagation of CMEs from the Sun out to >60 AU. The *bottom panel* represents the flux of cosmic rays that is clearly modulated by the structures in the solar wind and the HMF. (From Richardson et al. 2006)

4 The Magnetic Field in the Heliosheath

One of the most important goals of heliospheric exploration has been to observe the solar-wind termination shock, where the supersonic solar wind undergoes a sudden transition to subsonic flow, in response to the pressure of the local interstellar medium. Beyond the termination shock is the heliosheath, a region of subsonic solar outflow, where the solar plasma is deflected by the incoming interstellar plasma to flow backward toward the tail of the heliosphere. The boundary between the plasma of solar origin and the interstellar plasma is called the heliopause.

The Voyager 1 spacecraft crossed the termination shock in December of 2004 (Stone et al. 2005), and since then has been in the heliosheath. This region of subsonic outward flow and deflection of the postshock solar plasma has been predicted theoretically since the earliest discussions of the solar wind (Parker 1963). However, its structure and properties have been discussed only for large scales. It is the purpose of this paper to present a more-detailed analysis of possible properties of the heliosheath magnetic field as observed by a slowly moving spacecraft such as Voyager 1. In particular, expected inward and outward motions of the termination shock can change the observations and make it difficult or impossible to detect the sector structure from observations from a spacecraft. This may help understand recently reported observations.

4.1 The Magnetic-Field Sector Structure in the Heliosheath

The general spiral nature of the interplanetary magnetic field, predicted nearly 50 years ago by Parker (1958) was discussed above. In the mid 1970's, observations showed that the sector structure was the result of a general heliospheric current sheet which oscillated up and down in latitude about the heliographic equator and rotated with the Sun.

At any given time relative to the sunspot cycle, an observer moving slowly (as a spacecraft or planet) or at rest, in the supersonic solar wind will see the sector structure as a small, even number (2, 4, ...) of current sheet crossings during a solar rotation if located at a heliographic latitude closer to the equator than the maximum extent of the interplanetary current sheet at that time. If above the maximum latitudinal extent of the current sheet the observer will see the appropriate polar magnetic field with no changes in direction corresponding to sector structure.

The general magnetic structure described above for the supersonic solar wind, including the current sheet is expected to be convected into the heliosheath as well. We consider next what the nature of the sector structure or current sheet will be in the heliosheath. Since the termination shock is located far from the Sun, at a distance of 80 AU or greater, we may regard the current sheet at low latitudes as oriented essentially normal to the heliographic radius vector and the solar wind velocity. Our conclusions are valid for a spherical shock and will be approximately correct for a non-spherical shock if the local deviation of the shock normal from the radial direction is small.

For simplicity, assume that the solar wind speed, V_{sw} is a constant upstream of the termination shock. Hence, along a given radius vector in the solar wind, if the current sheet crosses a stationary observer in a time τ_1, the sector extends a radial distance $S_1 = V_{sw}\tau_1$, where V_{sw} is the solar wind velocity. If we observe from a spacecraft moving radially at V_{obs}, then $S = (V_{sw} - V_{obs})\tau_1$, where τ_1 is the time to cross the moving observer. Hence, for a fixed S and V_{sw}, τ_1 will depend on the motion of the observer. In this case, τ_1 is the solar rotational period divided by the number of sectors.

If the termination shock is stationary, the radial velocity downstream of the shock, in the heliosheath, is $U_2 = V_{sw}/r_{sh}$, where r_{sh} (≤ 4 for an ideal shock) is the shock compression ratio. Also, the plasma is compressed radially by the ratio r_{sh}. The end result is that the radial extent of the sector is reduced by the ratio r_{sh}. Since the radial velocity is also reduced by the same factor, the observed time for the sector to cross the observer is unchanged. Hence, not surprisingly, a stationary or slowly-moving observer, in this case, will continue to see approximately the same time between current sheet crossings, with the same temporal scale in the post-shock flow as in the solar wind.

However, the first crossing of the termination shock most likely occurred during rapid inward motion of the shock. If the shock is moving radially, the observed time between sector crossings can change dramatically because the post-shock flow speed depends on the shock speed relative to the upstream flow. If the shock is moving inward at the proper speed, so that the post-shock flow moves with the spacecraft, it is possible for the sector structure to disappear completely. The spacecraft observes the same magnetic sector for a very long time.

To see this, consider the following analysis. Suppose that the shock is moving in the *radial* direction at a constant speed V_{sh}. The shock ratio is the ratio of upstream to downstream speed in the shock frame. Denoting values in the shock frame with primes, with the upstream speed being U_1, we find immediately

$$U_1' = V_{sw} - V_{sh} \tag{7}$$

and

$$U_2' = \frac{V_{sw} - V_{sh}}{r_{sh}}. \tag{8}$$

And in the frame at rest with respect to the Sun

$$U_2 = U_2' + V_{sh} = \frac{V_{sw}}{r_{sh}} + \left(1 - \frac{1}{r_{sh}}\right) V_{sh}. \tag{9}$$

We note that U_2 becomes negative if the inward shock speed is such that

$$V_{sh} < -\frac{V_{sw}}{r_{sh} - 1}. \tag{10}$$

From this defining τ_1 and τ_2 as the times between current-sheet crossings in the upstream and the post-shock flow, respectively, we find that the times τ_1 and τ_2 are related by

$$\frac{\tau_2}{\tau_1} = \frac{V_{sw}}{r_{sh} U_2} = \frac{V_{sw}}{V_{sw} + (r_{sh} - 1) V_{sh}}. \tag{11}$$

The above equation applies to an observer at a fixed distance from the Sun. If the observer is moving at a radial velocity V_{obs}, Equation (4) becomes

$$\frac{\tau_2}{\tau_1} = \frac{V_{sw} - V_{obs}}{r_{sh}(U_2 - V_{obs})} = \frac{V_{sw} - V_{obs}}{V_{sw} + (r_{sh} - 1) V_{sh} - r_{sh} V_{obs}}. \tag{12}$$

Note that in this expression, *both* τ_1 and τ_2 are defined with respect to the observer, moving at velocity V_{obs}.

Figure 13 illustrates the ratios τ_2/τ_1 as a function of the shock speed, for an observer moving outward at 17 km/s (approximately the speed of Voyager 1). Two different scales are shown in the two panels. Clearly, for strong shocks and inward shock speeds in the range 60–100 km/s, the sector crossing times can become very large.

We also point out that the shock speed and the downstream flow are expected to change with time and distance behind the shock, so the above results are quantitatively correct only close to the shock. However, they are readily extended to any specified variation in a more-complicated analysis.

This simple calculation establishes the major point that the sector structure or time between successive current crossings in the heliosheath can differ significantly from that in

Fig. 13 Illustration of the ratio of the sector-crossing time downstream of the termination shock (τ_2) to that upstream (τ_1), for an observer moving outward from the Sun at 17 km/s and for a shock ratio $r_{sh} = 3.8$. The two panels show the same curves at different scales. In each panel, the *solid line* is for an upstream solar wind speed of 350 km/s, and the *dashed line* for 500 km/s. The singularities occur when the downstream flow speed relative to the observer is zero

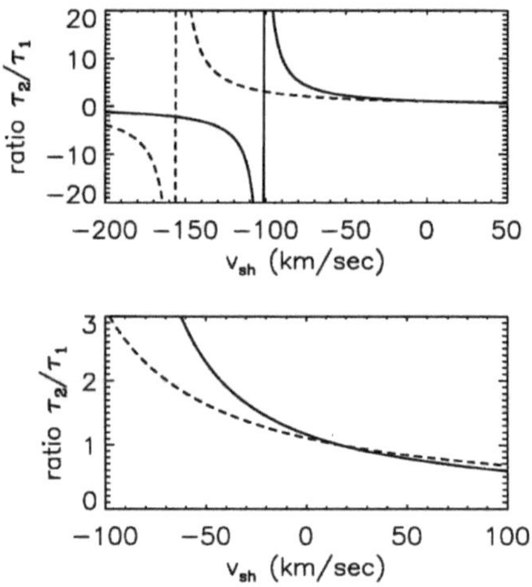

the upstream solar wind. Extension of this analysis to more general, or time varying, shock speeds is trivial, but clearly depends on the specific velocity profile. Similarly, if the shock is not spherical (Jokipii and Giacalone 2003) the sector structure will also be different downstream from the shock. These considerations should apply to any given radial direction since the solar wind flow and flow immediately beyond the termination shock are very nearly radial, and the magnetic field is frozen into the plasma.

These considerations should be relevant to the first crossing of the termination shock by Voyager1, when the shock was probably moving in toward the Sun. Hence the sector crossing time is expected to be larger than that in the solar wind, perhaps by a large factor. Various theoretical estimates of the shock propagation speed have been published. Barnes (1993) suggested that speeds of 100 km/s might be expected. Suess (1993b) suggested even larger speeds. Most relevant here is the paper of Whang et al. (2004) who considered Voyager 1 explicitly and suggested a shock speed $V_{sh} = 67 \pm 41$ km/s. The curves in Fig. 13 show that the sector structure observed from a spacecraft can be profoundly altered and perhaps even reversed for inward shock speeds which may be reasonably expected. Further analysis of the observations may be able to confirm or deny this possibility.

4.2 Application to Observations

The above discussion demonstrates that a dynamical heliosheath can alter the observed magnetic field and current sheet structure so as to make the sector crossing period much larger than that observed in the upstream solar wind. Voyager 1crossed the termination shock in December of 2004. After this crossing, a variety of observations of the heliosheath post-shock solar wind were reported. In particular, Burlaga et al. (2005) reported the puzzling fact that the magnetic field in the heliosheath did not show any evidence of the change in sign associated with the sector structure, beginning at the time of the crossing in December, 2004 until mid-May, 2005. This is illustrated in Fig. 14. The sector structure had been observed to be normal in the period prior to the crossing of the termination shock. Moreover, the sense of the field was outward during this entire time. In addition, Decker et al. (2005) reported that the radial anisotropy of the low-energy anomalous-cosmic-ray particles was very small, indicating a very low plasma flow speed of less than 50 km/s relative to Voyager 1.

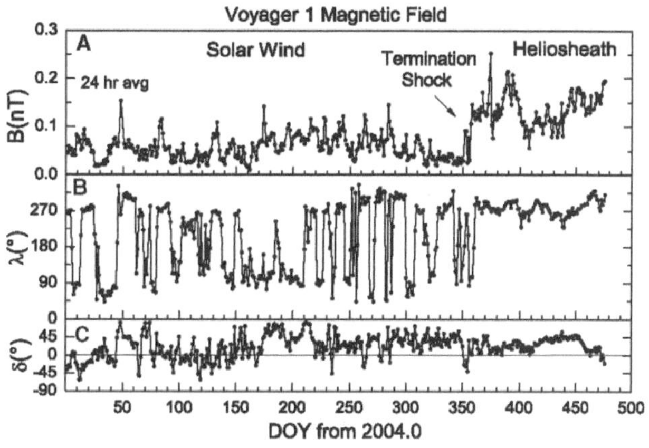

Fig. 14 Observations of Burlaga et al. (2005), showing the lack of sector structure observed on Voyager 1 after it crossed the termination shock

These observations are quite likely a consequence of the phenomena discussed above—that the shock was moving rapidly inward, so that the postshock flow at Voyager 1 was very small, and that this velocity continued for months. Since the flow of the plasma relative to Voyager 1 is slow, the current sheet structure takes a very long time to cross, resulting in the observed constant magnetic field sense. The outward direction of the field is simply a reflection of the fact that the sector which Voyager was in since the shock crossing contained outwardly directed magnetic field. If the field had happened to have the opposite sign at shock crossing, the observed field in the post-shock flow would have had that sign. In particular, in this picture, the sense of the observed magnetic field has little to do with the global magnetic field structure. A spacecraft (such as Voyager 2), crossing the termination shock at a different time would be expected to see quite different temporal behaviour.

It should be noted that at a speed of several tens of km/s, the shock may have moved several AU. At this distance from the shock, the subsonic postshock flow may already be deflected away from the radial direction, in the northward direction by the interstellar medium. This would deflect the outward magnetic field northward, as well, producing a northward component in the average magnetic field. This may account for the persistent northward latitudinal component of the field which was observed in the heliosheath (Burlaga et al. 2006, Fig. 5).

The analysis here is for an ideal shock, where the energy in the fluid is conserved across the shock. It is expected, for example, that a fraction of the flow energy goes into the acceleration of anomalous cosmic rays, and this loss of energy may change the quantitative conclusions, although the general nature of the effect should remain unchanged.

4.3 Magnetic Fluctuations in the Heliosheath

Burlaga et al. (2006) also presented an analysis of large-amplitude, irregular fluctuations in the heliosheath. One cause of these fluctuations may be the interaction of upstream solar-wind turbulence with the termination shock. This process, discussed by Giacalone and Jokipii (2007), robustly produces large-amplitude magnetic fluctuations downstream of a shock which is interacting with upstream density fluctuations. This is illustrated in Fig. 15. The physical cause of the large downstream magnetic-field fluctuations is large-amplitude transverse and vortical motions induced by the upstream density fluctuations.

Burlaga et al. (2005) also showed that the probability distribution of the magnetic fluctuations is quite different from that observed previously in the solar wind, upstream of the

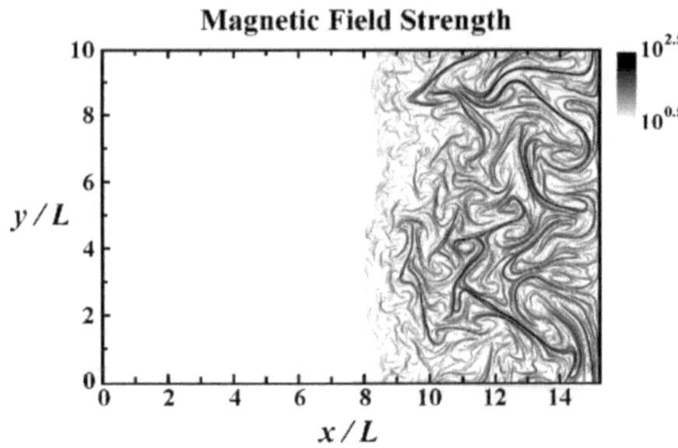

Fig. 15 A grayscale plot from a two-dimensional MHD simulation of a shock propagating to the left through a pre-shock medium containing density fluctuations, showing the enhanced magnetic fluctuations downstream of the shock. From Giacalone and Jokipii (2007)

termination shock. The probability distribution of the magnetic field magnitude in the supersonic, upstream solar wind is robustly a log-normal distribution, whereas downstream of the shock, in the heliosheath, the distribution is gaussian. A physical reason for this difference has yet to be published. It is possibly caused by the turbulent fluctuations in the upstream solar wind being processed by the shock, but simulations which we have carried out in Tucson (J. Giacalone, private communication) have been unable to reproduce the observed change.

5 Summary

A great deal is known about the heliospheric magnetic field, but mostly in the inner heliosphere, out from the Sun to about 10 AU, thanks to such space missions as HELIOS, ISSE3/ICE, Ulysses, ACE, WIND, Cassini-Huygens and the fleet of Earth-orbiting spacecraft. Vitally important missions, Pioneer 10 and 11, and Voyager 1 and 2 have explored beyond Saturn and have provided an insight into heliospheric processes that become dominant at large distances from the Sun. While the picture in the inner heliosphere is one of a succession of well understood and quite distinct dynamic states driven by coronal conditions that change in response to the 11-year activity cycle, the picture (or what we know of it) at larger distances is quite different.

The inner heliosphere, possibly out to about 20 to 30 AU (as modelled by Burlaga et al. 2003), is dominated around solar minimum, during the epoch of stable (on the timescale of many solar rotations) and large coronal holes, by corotating structures that develop from the interaction of the stable fast and slow flow patterns of the solar wind. The magnetic field becomes increasingly concentrated into the interaction regions by compression. The latitudinal extent of the corotating structures is maybe from about 30° south to 30° north of the solar equator. These structures contain the heliospheric current sheet that is the extension of the magnetic neutral line in the corona that separates the dominant polarities of the Sun. As the CIRs move outward, they coalesce into larger scale, merged structures (the so-called Merged Interaction Regions, or MIRs) from about 10 AU or so (e.g. Burlaga et al. 1985, 1996).

At higher heliolatitudes, the uniformly fast solar wind from the polar coronal holes carries a magnetic field that is on average Parker-like, but with large transverse fluctuations superimposed on the mean field (in fact, for the fluctuations, $\Delta B/B \sim 1$).

At the time of increasing solar activity, the coronal order and stability seen around activity minimum is replaced by increasingly important magnetically closed, hot plasma, with only transient and small coronal holes. The large-scale division between coronal holes and mostly equatorial active regions is replaced by smaller scale transient structures, inherently unstable and frequently generating Coronal Mass Ejections. CMEs are distributed at all heliolatitudes; they are frequently magnetically complex due to the underlying complexity of the ejected coronal structures. Only about a third of CMEs are magnetic clouds with a recognisable expanding flux rope; the majority are compound structures. Due to the frequency of CMEs and their widely different speeds, they interact with each other dynamically from close to the Sun throughout their propagation into the heliosphere. Already within the inner heliosphere there are repeated interactions and merging so that CME-related structures that strongly affect the dynamics, structure and composition of the solar wind and the heliospheric magnetic field cannot be correlated with the original events on the Sun. In the case of the largest of solar events, usually compounds consisting of more than one CME, these may form the so-called Global Merged Interaction Regions (GMIRs, Burlaga et al. 1993) and can be tracked to larger distances.

There is, however, an evolution towards the outer heliosphere that tends to destroy the orderly structures observed in the inner heliosphere. This applies equally to structures formed from the merging of CIRs and to those formed by the coalescing remnants of CMEs (le Roux and Fichtner 1999; Richardson et al. 2002). The transition to the more fragmented (although still evolving) structures takes place over the distance range of 20 to 40 AU. Despite this continual fragmentation and presumably partly reforming process, the global polarities and the sector structure of the magnetic field remain recognisable all the way to the termination shock. Although unlikely to be a simple surface that can be easily visualised, the Heliospheric Current Sheet apparently retains its integrity through the evolving solar wind, even at quite high heliolatitudes and high levels of solar activity.

The heliosheath is a fluctuating, dynamic environment. The radial speed of the termination shock as it moves past a slowly moving observer strongly influences the time between current-sheet crossings (the duration of the magnetic sectors). An inwardly moving termination shock leads to increasing the intervals between the current sheet crossings, in fact making them very large. This would explain the Voyager 1 observations that the magnetic field in the heliosheath remained directed outward for several months, while the observations just inside the termination shock showed a sustained, periodic alternation of the two magnetic polarities, consistent with previous observations throughout the supersonic solar wind. Also, the observed magnetic-field fluctuations in the heliosheath are quite different from those observed in the supersonic upstream flow.

Acknowledgements A. Balogh acknowledges useful discussions with G. Erdős. The work of J.R. Jokipii was supported by NASA and NSF. The authors thank the organisers of the ISSI Workshop for a stimulating program and environment for discussing the state of the heliosphere. The paper has greatly benefited from the very helpful comments and suggestions of the referee.

References

A. Balogh, J.T. Gosling, J.R. Jokipii, R. Kallenbach, H. Kunow, *Corotating Interaction Regions, Space Science Series of ISSI*, vol. 7 (Kluwer, Dordrecht, 1999)

A. Barnes, Motion of the heliospheric termination shock—A gas dynamic model. J. Geophys. Res. **98**, 15,137–15,146 (1993)

L.F. Burlaga, N.F. Ness, Magnetic fields in the distant heliosphere approaching solar minimum: Voyager 1 and 2 observations during 1994. J. Geophys. Res. **101**, 13,473–13,481 (1996)

L.F. Burlaga, E. Sittler, F. Mariani, R. Schwenn, Magnetic loop behind an interplanetary shock—Voyager, Helios, and IMP 8 observations. J. Geophys. Res. **86**, 6673–6684 (1981)

L.F. Burlaga, R.P. Lepping, K.W. Behannon, L.W. Klein, Large-scale interplanetary magnetic fields—Voyager 1 and 2 observations between 1 AU and 9.5 AU. J. Geophys. Res. **89**, 10659–10668 (1984)

L.F. Burlaga, M.L. Goldstein, F.B. McDonald, A.J. Lazarus, Cosmic ray modulation and turbulent interaction regions near 11 AU. J. Geophys. Res. **90**, 12,027–12,039 (1985)

L.F. Burlaga, C. Wang, J.D. Richardson, N.F. Ness, Evolution of magnetic fields in Corotating Interaction Regions from 1 to 95 AU: Order to chaos. Astrophys. J. **590**, 554–566 (1993)

L.F. Burlaga, N.F. Ness, J.W. Belcher, Radial evolution of corotating merged interaction regions and flows between ~14 AU and 43 AU. J. Geophys. Res. **102**, 4,661–4,671 (1997)

L.F. Burlaga, N.F. Ness, Y.-M. Wang, N.R. Sheeley Jr., Heliospheric magnetic field strength and polarity from 1 to 81 AU during the ascending phase of solar cycle 23. J. Geophys. Res. **107**(A11), 1410 (2002). doi:10.1029/2001JA009217

L.F. Burlaga, F.B. McDonald, N.F. Ness, Cosmic ray modulation and the distant heliospheric magnetic field. J. Geophys. Res. **98**, 1–11 (2003)

L.F. Burlaga, N.F. Ness, M.H. Acuña, R.P. Lepping, J.E.P. Connerney, E.C. Stone, F.B. McDonald, Crossing the termination shock into the heliosheath: Magnetic fields. Science **309**, 2027–2029 (2005)

L.F. Burlaga, N.F. Ness, M.H. Acuña, Magnetic fields in the heliosheath: Voyager 1 observations. Astrophys. J. **642**, 584–592 (2006)

N.U. Crooker, J.T. Gosling et al., CIR morphology, turbulence, discontinuities, and energetic particles. Space Sci. Rev. **89**, 179–220 (1993)

R.B. Decker, S.M. Krimigis, E.C. Roelof, M.E. Hill, T.P. Armstrong, G. Gloeckler, D.C. Hamilton, L.J. Lanzerotti, Science **309**, 2020–2024 (2005)

L.A. Fisk, Motion of the footpoints of heliospheric magnetic field lines at the Sun: Implications for recurrent energetic particle events at high heliographic latitudes. J. Geophys. Res. **101**, 15,547–15,554 (1996)

R.J. Forsyth, A. Balogh, T.S. Horbury, G. Erdoes, E.J. Smith, M.E. Burton, Astron. Astrophys. **316**, 287 (1996)

R.J. Forsyth, A. Balogh, E.J. Smith, The underlying direction of the heliospheric magnetic field through the Ulysses first orbit. J. Geophys. Res. **107**(A11), SSH19-1 (2002). doi:10.1029/2001JA005056

J. Giacalone, J.R. Jokipii, Magnetic footpoint diffusion at the Sun and its relation to the heliospheric magnetic field. Astrophys. J. **616**, 573 (2004)

J. Giacalone, J.R. Jokipii, Magnetic field amplification by shocks in turbulent fluids. Astrophys. J. **63**, 41 (2007)

J.T. Gosling, Coronal mass ejections and magnetic flux ropes in interplanetary space, in *Physics of Magnetic Flux Ropes*. (American Geophysical Union, Washington, 1990), pp. 343–364. A92-31201 12-75

J.T. Gosling, V.J. Pizzo, Formation and evolution of corotating interaction regions and their three-dimensional structure. Space Sci. Rev. **89**, 21–52 (1999)

J.T. Gosling, A.J. Hundhausen, V. Pizzo, J.R. Asbridge, Compressions and rarefactions in the solar wind: Vela 3. J. Geophys. Res. **77**, 5,442–5,454 (1972)

T.J. Hoeksema, J.M. Wilcox, P.H. Scherrer, Structure of the heliospheric current sheet in the early portion of sunspot cycle 21. J. Geophys. Res. **87**, 10,331–10,338 (1982)

A.J. Hundhausen, Sizes and locations of coronal mass ejections—SMM observations from 1980 and 1984–1989. J. Geophys. Res. **98**, 13,177–13,200 (1993)

J.R. Jokipii, J. Giacalone, Anomalous cosmic rays at a termination-shock crossing, in *Proc. 28th International Cosmic Ray Conference*, vol. 7, Tsukuba, Japan (2003), p. 3753

J.R. Jokipii, J. Kóta, The polar heliospheric magnetic field. Geophys. Res. Lett. **16**, 1–4 (1989)

J.R. Jokipii, E.N. Parker, Random walk of magnetic lines of force in astrophysics. Phys. Rev. Lett. **21**, 44–47 (1968)

J.R. Jokipii, B.T. Thomas, Effects of drift on the transport of cosmic rays. IV—Modulation by a wavy interplanetary current sheet. Astrophys. J. **243**, 1115–1122 (1981)

J.R. Jokipii, J. Kóta, J. Giacalone, T.S. Horbury, E.J. Smith, Interpretation and consequences of large-scale magnetic variances observed at high heliographic latitude. Geophys. Res. Lett. **22**, 3385–3388 (1995)

G.H. Jones, A. Balogh, E.J. Smith, Solar magnetic field reversal as seen at Ulysses. Geophys. Res. Lett. **30**, ULY2-1 (2003). doi:10.1029/2003GL017204

A.S. Krieger, A.F. Timothy, E.C. Roelof, A coronal hole and its identification as the source of a high velocity solar wind stream. Sol. Phys. **29**, 505–525 (1973)

H. Kunow, N.U. Crooker, J.A. Linker, R. Schwenn, R. von Steiger (eds.), *Coronal Mass Ejections*. Space Science Series of ISSI, vol. 21 (Springer, Dordrecht, 2006)

A.J. Lazarus, J.D. Richardson, R.B. Decker, F.B. McDonald, Voyager 2 observations of Corotating Interaction Regions (CIRs) in the outer heliosphere. Space Sci. Rev. **89**, 53–59 (1999)

M.A. Lee, An analytical theory of the morphology, flows, and shock compressions at corotating interaction regions in the solar wind. J. Geophys. Res. **105**, 10491–10500 (2000)

S.T. Lepri, T.H. Zurbuchen, Iron charge state distributions as an indicator of hot ICMEs: Possible sources and temporal and spatial variations during solar maximum. J. Geophys. Res. **109**, A01112 (2004). doi:10.1029/2003JA009954

J.A. le Roux, H. Fichtner, Global merged interaction regions, the heliospheric termination shock, and time-dependent cosmic ray modulation. J. Geophys. Res. **104**, 4709–4730 (1999)

F.B. McDonald, A. Barnes, L.F. Burlaga, P. Gazis, J. Mihalov, R.S. Selenick, The effects of the intense solar activity of March/June 1991 observed in the outer heliosphere, in *Proc. 23rd International Cosmic Ray Conference*, vol. 3 (1993), p. 350

A.G. Nash, Jr. N.R. Sheeley, Y.-M. Wang, Mechanisms for the rigid rotation of coronal holes. Sol. Phys. **117**, 359–389 (1988)

N.F. Ness, J.W. Wilcox, Structure of the quiet interplanetary magnetic field. Science **148**, 1592–1594 (1965)

E.N. Parker, Dynamics of interplanetary gas and magnetic field. Astrophys. J. **128**, 644–670 (1958)

E.N. Parker, *Interplanetary Dynamical Processes* (Interscience, New York, 1963)

V.J. Pizzo, J.T. Gosling, Three-dimensional simulation of high-latitude interaction regions: Comparison with Ulysses results. Geophys. Res. Lett. **21**, 2063–2066 (1994)

A. Rees, R.J. Forsyth, Two examples of magnetic clouds with double rotations observed by the Ulysses spacecraft. Geophys. Res. Lett. **31**, L06804 (2004). doi:10.1029/2003GL018330

J.D. Richardson, K.I. Paularena, C. Wang, L.F. Burlaga, The life of a CME and the development of a MIR: From the Sun to 58 AU. J. Geophys. Res. **107**, SSH1-1 (2002). doi:10.1029/2001JA000175

J.D. Richardson, Y. Liu, C. Wang, L.F. Burlaga, ICMEs at very large distances. Adv. Space Res. **38**, 528–534 (2006)

J.D. Richardson, Y. Liu, C. Wang, Solar wind structure in the outer heliosphere. Adv. Space Res. **41**, 237–244 (2008)

D.A. Roberts, J. Giacalone, J.R. Jokipii, M.L. Goldstein, T.D. Zepp, Spectra of polar heliospheric fields and implications for field structure. J. Geophys. Res. **112**, A8 (2007). CiteID A08103

K.H. Schatten, J.M. Wilcox, N.F. Ness, A model of interplanetary and coronal magnetic fields. Solar Phys. **6**, 442–455 (1969)

E.J. Smith, Interplanetary magnetic field over two solar cycles and out to 20 AU. Adv. Space Res. **9**(4), 159–169 (1989)

E.J. Smith, The heliospheric current sheet. J. Geophys. Res. **106**, 15,819–15,831 (2001)

E.J. Smith, The global heliospheric magnetic field, in *The Heliosphere through the Solar Activity Cycle*, ed. by A. Balogh, L.J. Lanzerotti, S.T. Suess (Springer, Chichester, 2007), pp. 79–150

E.J. Smith, A. Balogh, Ulysses observations of the radial magnetic field. Geophys. Res. Lett. **22**, 3317–3320 (1995)

E.J. Smith, J.H. Wolfe, Observations of interaction regions and corotating shocks between one and five AU: Pioneer 10 and 11. Geophys. Res. Lett. **3**, 137 (1976)

H.B. Snodgrass, Magnetic rotation of the solar photosphere. Astrophys. J. **270**, 288 (1983)

E.C. Stone, A.C. Cummings, F.B. McDonald, B.C. Heikkila, N. Lal, W.R. Webber, Voyager 1 explores the termination shock region and the heliosheath beyond. Science **309**, 2017–2020 (2005)

S.T. Suess, The relationship between coronal and interplanetary magnetic fields. Adv. Space Res. **13**(9), 31–42 (1993a)

S.T. Suess, Temporal variations in the termination shock distance. J. Geophys. Res. **98**, 15,147–15,155 (1993b)

B.T. Thomas, E.J. Smith, The Parker spiral configuration of the interplanetary magnetic field between 1 and 8.5 AU. J. Geophys. Res. **85**, 6861–6867 (1980)

B.T. Thomas, E.J. Smith, The structure and dynamics of the heliospheric current sheet. J. Geophys. Res. **86**, 11105–11110 (1981)

A.F. Timothy, A.S. Krieger, G.S. Vaiana, The structure and evolution of coronal holes. Sol. Phys. **42**, 135–156 (1975)

Y.-M. Wang, On the latitude and solar cycle dependence of the interplanetary magnetic field strength. Astrophys. J. **410**, L123 (1993)

Y.C. Whang, L.F. Burlaga, Evolution and interaction of interplanetary shocks. J. Geophys. Res. **90**, 10,765–10,778 (1985)

Y.-M. Wang, N.R. Sheeley Jr., Why fast solar wind originates from slowly expanding coronal flux tubes. Astrophys. J. **372**, L45 (1991)

Y.-M. Wang, N.R. Sheeley Jr., Understanding the rotation of coronal holes. Astrophys. J. **414**, 916–927 (1993)

Y.-M. Wang, N.R. Sheeley Jr., The solar wind and its magnetic sources at sunspot maximum. Astrophys. J. **587**, 818–822 (2003a)

Y.-M. Wang, N.R. Sheeley Jr., On the topological evolution of the coronal magnetic field during the solar cycle. Astrophys. J. **599**, 1404–1417 (2003b)

Y.C. Whang, L.F. Burlaga, Y.-M. Wang, N.R. Sheeley Jr., The termination shock near 35° latitude. Geophys. Res. Lett. **31**, L0385 (2004)

J.W. Wilcox, N.F. Ness, Quasi-stationary corotating structure in the interplanetary medium. J. Geophys. Res. **70**, 5793 (1965)

R.F. Wimmer-Schweingruber, R. von Steiger, R. Paerli, Solar wind stream interfaces in corotating interaction regions: SWICS/Ulysses results. J. Geophys. Res. **102**, 17407–17418 (1997)

S. Yashiro, N. Gopalswamy, G. Michalek, O.C.St. Cyr, P. Plunkett, N.B. Rich, R.A. Howard, A catalog of white light coronal mass ejections observed by the SOHO spacecraft. J. Geophys. Res. **109**, A07105 (2004). doi:10.1029/2003JA010282

T.H. Zurbuchen, N.A. Schwadron, L.A. Fisk, Direct observational evidence for a heliospheric magnetic field with large excursions in latitude. J. Geophys. Res. **102**(A11), 24175–24182 (1997)

Pickup Ion Acceleration at the Termination Shock and in the Heliosheath

V. Florinski

Originally published in the journal Space Science Reviews, Volume 143, Nos 1–4, 111–124.
DOI: 10.1007/s11214-008-9451-z

Abstract We discuss pickup ion acceleration and transport near the solar wind termination shock from the perspective of their spectral, spatial, and pitch-angle distributions. Our study is performed in the framework of a recently developed anisotropic transport model based on a Legendre polynomial expansion technique. Voyager 1 LECP angular distributions of 1 MeV protons, represented in the form of an expansion in spherical harmonics in the frame aligned with the measured interplanetary magnetic field, are used as benchmarks for our theory. We find the observed distributions consistent with our model predictions for particle acceleration and reflection at a highly oblique shock wave. It is shown that first-order (field aligned) anisotropy is a measure of shock obliquity while the second-order (transverse) anisotropy reflects the energy dependence of the particle scattering mean free path. We also discuss the role of enhanced scattering and momentum diffusion on the spectral properties of energetic charged particles.

Keywords Termination shock · Heliosheath · Pickup ions · Focused transport

1 Introduction

The two Voyager spacecraft have recently traversed the solar wind (SW) termination shock (TS) region and entered the heliosheath (HS)—in December 2004 for Voyager 1 (Stone et al. 2005; Decker et al. 2005; Burlaga et al. 2005) and in August 2007 for Voyager 2 (Decker et al. 2008; Stone et al. 2008). Prior to their respective crossings both spacecraft witnessed several periods of enhanced intensities of energetic ions, lasting up to half a year, that were termed "termination shock precursor" events. The ions appeared to stream along the interplanetary magnetic field (IMF) direction away from the TS in the azimuthal direction, on average. Their complex spectral features appeared to exclude

V. Florinski (✉)
Department of Physics and Center for Space Plasma and Aeronomic Research, University of Alabama in Huntsville, 301 Sparkman Dr., Huntsville, AL 35899, USA
e-mail: vaf0001@uah.edu

DOI: 10.1007/978-1-4419-0247-4_9

the possibility of them being accelerated at any one particular location along a stationary TS. At least two populations were observed, the newly detected low energy component called termination (fore)shock particles (TSP) and the previously observed high energy anomalous cosmic ray (ACR) component (Cummings et al. 2006; McDonald et al. 2006; Hill et al. 2006).

Ion intensities peaked sharply at the TS and subsequently dropped during the Voyager 1 crossing (Decker et al. 2005) in contrast to the standard diffusive theory where phase space density is conserved across a shock. After the crossing, intensities continued to rise gradually until levelling off two years later and remained essentially flat since then (Decker et al. 2007). Because a similar increase was observed at Voyager 2 which was upstream of the TS during that time, the increase was probably of a temporal origin, rather than being associated with a possible spatial intensity gradient in the HS (Cummings and Stone 2007). However, the alternative view of a positive radial gradient of energetic particles in the HS has gained more attention in the space physics community and several previously unexplored acceleration sites were suggested. McComas and Schwadron (2006) and Schwadron and McComas (2007) proposed acceleration on the flanks of a blunt TS, whereas Moraal et al. (2006), Zhang (2006), and Ferreira et al. (2007) advocated momentum diffusion in the HS as the mechanism for acceleration. At the present time the length of Voyager 1 record in the HS is not sufficient to identify the global scale spatial trend in particle intensity that will allow us to differentiate between these theories.

Many of the spectral features of energetic particle distributions observed near the TS are even less well understood. Power law segments at high (>1 MeV) energies appear to be products of diffusive acceleration by the SWTS (e.g., Jokipii 1986; Steenberg and Moraal 1996; Florinski et al. 2004), whereas the lower energy pickup ion (PUI) population may be primarily accelerated through momentum diffusion in the SW or the HS (Kallenbach et al. 2005; Fisk and Gloeckler 2006). Varying power-law slopes and spectral breaks may be related to temporal evolution of the TS produced by large scale transient structures in the SW (Florinski and Zank 2006; Jokipii 2006). Florinski et al. (2008b) also argued that the gap appearing in Voyager 1 spectra at ~ 1 MeV for protons (Cummings et al. 2006) may be related to adiabatic reflection off a highly oblique portion of the TS. This hypothesis is further discussed below.

In this paper we briefly identify modeling strategies for energetic ion acceleration and transport in the outer heliosphere, focusing on the TS region. We concentrate on low-energy particles (PUIs, rather than the more energetic ACRs) featuring highly anisotropic (beamed) distributions in the SW frame. Our analysis is performed in the framework of a model (Sect. 2) that contains sufficient pitch-angle information in the distribution function to allow for anisotropy of up to ~ 1, but which at the same time is closely connected with the conventional diffusive theories. In other words, the present models follows in the *pitch* of diffusive shock acceleration theories, but adopts a somewhat different *angle* to treat the particles. Section 3 reviews some of the spectral and spatial features predicted by the theoretical models and relates them to Voyager 1 observations. We analyse pitch-angle distributions, or PADs, of accelerated PUI to show how they relate to the shock compression ratio and SW turbulence spectral slope in Sect. 4. Effects of momentum diffusion in the SW and the HS are briefly discussed in Sect. 5 and some additional acceleration mechanisms are also mentioned there.

2 Modeling Anisotropic Distributions: Focused Transport

Observed PUI distribution are often highly beamed in the direction of the interplanetary magnetic field with anisotropies reaching 90% or more at energies below a few MeV. Diffusive formalism becomes inadequate when anisotropies are as large as these and a more general approach is called for to capture pitch-angle properties of the particle distributions. The natural approach to adopt is to use the Fokker-Planck equation describing the evolution of a gyrotropic distribution function $f(\mathbf{r}, w, \mu)$, where r is position, w is velocity, measured in the plasma frame, and μ is the pitch-angle cosine (also in the plasma frame). In a single spatial dimension this equation, often called focused transport equation (FTE) may be written (Skilling 1971; Isenberg 1997) as

$$\begin{aligned}\frac{\partial f}{\partial t} &+ (u + w\mu b)\frac{\partial f}{\partial x} + \left[(1-3\mu^2)b^2 - (1-\mu^2) - \frac{2\mu u b}{w}\right]\frac{1}{2}\frac{du}{dx}\frac{\partial f}{\partial \ln w} \\ &+ \frac{1-\mu^2}{2}\left[(1-3b^2)\mu\frac{du}{dx} + w\frac{db}{dx} - \frac{2ub}{w}\frac{du}{dx}\right]\frac{\partial f}{\partial \mu} \\ &- \frac{\partial}{\partial \mu}\left(D_{\mu\mu}\frac{\partial f}{\partial \mu}\right) - \frac{1}{w^2}\frac{\partial}{\partial w}\left(w^2 D_{ww}\frac{\partial f}{\partial w}\right) = 0. \end{aligned} \tag{1}$$

In the above equation u is the SW plasma velocity and $b = B_x/B = \cos\theta_{bn}$ is the cosine of the angle between the IMF ($\mathbf{B}$) and the flow, or the shock normal. The terms on line one of this equation describe convection and adiabatic energy change, on line two, change in pitch angle (related to adiabatic focusing), on line three, pitch-angle scattering and velocity diffusion. In quasi-linear theory (e.g., Jokipii 1966; Schlickeiser 1989) the Fokker-Planck coefficients are

$$\begin{aligned} D_{\mu\mu} &= \frac{\pi\Omega^2(1-\mu^2)}{B^2 w|\mu|}P_\perp\left(\frac{\Omega}{w|\mu|}\right), \\ D_{ww} &= V_a^2 D_{\mu\mu}, \end{aligned} \tag{2}$$

where Ω is the cyclotron frequency, $P_\perp \sim k^{-q}$ is the spectral power in one perpendicular (to $\mathbf{B}$) slab turbulence component, and V_a is the speed of hydromagnetic waves in the plasma frame (typically, the Alfvén speed). In the above, k is the wavenumber and the value of q is typically $5/3$ for inertial-range turbulence. Note that the "mixed" Fokker-Planck coefficients, such as $D_{\mu p}$, are also not included in (1). Typically, these may be regarded as small under the assumption of low cross-helicity, i.e., equal intensities of oppositely propagating waves (Schlickeiser 1989).

In (1) we omitted the cross-field transport coefficient D_{xx} related to field-line meandering. Our neglect of perpendicular diffusion is based on the estimate of Zank et al. (2006) who calculate the ratio of the perpendicular to parallel diffusion coefficients, $\kappa_\perp/\kappa_\parallel$ to be only about 0.001 in the outer heliosphere at all relevant energies. Larger values of perpendicular diffusion $\kappa_\perp/\kappa_\parallel > 0.01$ are commonly used in cosmic-ray transport studies (e.g., Jokipii 1986; le Roux et al. 1996). In these large-scale models, a nominal Parker spiral angle (89.4°) is assumed and perpendicular transport naturally plays a dominant role. Obviously, the real magnetic field is rarely purely azimuthal, but fluctuates in direction on timescales ranging from seconds to months (e.g., Burlaga et al. 2005). Our analysis of Voyager daily averaged magnetic field data (Florinski et al. 2008a) shows that the field direction can vary by tens of degrees between consecutive 1-day intervals. This timescale is long compared with the characteristic shock acceleration time of particles near the low end of the energy range of the LECP instruments on the Voyagers (tens of keVs). Whereas there is no fundamental difficulty in adding perpendicular transport to (1), such an approach would be equivalent to averaging the FTE over long timescales and consequently losing fine-scale details of the particle distribution. Instead, as in Florinski et al. (2008a) we use the measured magnetic field azimuthal angles, which are typically between 80 and 88°, so that parallel transport dominates. In the present version of our model we keep the azimuthal angle fixed. A future revision will include variations in the magnetic field azimuthal angle in a more realistic time-dependent fashion.

Several approaches have been developed to solve the FTE. To preserve the complete pitch-angle information, the FTE may be directly integrated on a grid in phase space using finite differences (Klappong et al. 2001; Kota and Jokipii 2004; le Roux et al. 2007) or by emplyoing a Monte-Carlo technique to follow stochastic trajectories of particles in phase space (Gieseler et al. 1999; Chalov and Fahr 2000). Alternatively, one can expand $f(\mu)$ in some appropriate functions of pitch angle and derive a set of equations for the harmonics of f by evaluating the moments of the FTE (Zank et al. 2000; Lu et al. 2002). This approach was used by Florinski et al. (2008a, 2008b) to investigate pitch-angle distributions of energetic ions accelerated by the SWTS. In their model f is expanded in Legendre polynomials and only terms of second order and below are retained, i.e.,

$$f = \sum_{n=0}^{2} (2n+1) P_n(\mu) f_n = f_0 + 3\mu f_1 + \frac{5}{2}(3\mu^2 - 1) f_2. \tag{3}$$

Evaluating moments of (1) one obtains a set of partial differential equations for the harmonics f_n. These equations are somewhat cumbersome and will not be repeated here—the reader is referred to Florinski et al. (2008a) instead.

Truncation at P_2 does restrict the application of our model to PADs that are relatively smooth. Fortunately, both observations (Krimigis et al. 2005) and particle numerical simulations (Decker 1988) show that ion distributions likely to be measured upstream of a fast MHD shock are relatively smooth. Florinski et al. (2008a) show that in about half of Voyager 1 observations during 2004 distributions of 1 MeV protons can be well fitted with a second degree polynomial in pitch angle. In a different publication (Florinski et al. 2008b) we have shown that in shock simulations the three harmonics are ordered in magnitude, i.e., $5f_2 \ll 3f_1 < f_0$. In other situations, for example where *narrow* particle beams are expected, solving the full FTE (or, alternatively, retaining more harmonics in the expansion) would be required.

The complete system may be conveniently written in a matrix form as

$$\frac{\partial \mathbf{U}}{\partial t} + \mathbf{A}^x \frac{\partial \mathbf{U}}{\partial x} + \mathbf{A}^w \frac{\partial \mathbf{U}}{\partial \ln w} - \frac{1}{w^2} \frac{\partial}{\partial w} \left(\mathbf{A}^{ww} \frac{\partial \mathbf{U}}{\partial w} \right) = \mathbf{C}\mathbf{U}. \tag{4}$$

Here $\mathbf{U} = (f_0, f_1, f_2)^T$ is the state vector and the 3×3 matrices $\mathbf{A}^x$, $\mathbf{A}^w$, $\mathbf{A}^{ww}$, and $\mathbf{C}$ describe spatial transport, adiabatic energy change, momentum diffusion, and pitch-angle diffusion/focusing, respectively. The system (4) is linear and can be readily solved using finite differences—see Florinski et al. (2008a) for the details of the numerical algorithms used.

Here the system (4) is solved on a simple plasma background with constant up- and downstream states separated by a stationary reverse shock wave. Strong beaming is a persistent feature of the observed distributions (Decker et al. 2007), which indicates that pitch-angle scattering is relatively inefficient, i.e., slab turbulence is weak. Here we use $B = 4 \times 10^{-7}$ G and $\langle \delta B^2 \rangle = 6.6 \times 10^{-15}$ G^2 as our upstream values. These values are representative of the outer heliosphere (Zank et al. 2006). For the downstream region, we assumed that the turbulent field is amplified by the same amount as the mean IMF, i.e., $\langle \delta B^2 \rangle / B^2$ is approximately conserved across the TS. It is not clear what happens to the turbulence correlation length l_c as the turbulent flow traverses the shock. Parallel propagating (slab) waves are not compressed because the shock is nearly perpendicular, but transverse (2D) fluctuations are. Because the upstream turbulence is mostly 2D, their correlation length should decrease across the TS. To cover both possibilities we consider two different scenarios: one with l_c conserved across the shock and the other with l_c reduced downstream by a factor of s, the shock compression ratio. This leads to a significant enhancement in the scattering and stochastic acceleration rates due to a relatively steep ($\sim k^{5/3}$) spectral slope of the SW turbulent fluctuations.

3 Distribution Functions of Energetic Particles Near a Quasi-Perpendicular Shock

In this section we review some of the fundamental properties of charged particle distributions accelerated by highly oblique collisionless shock waves. We will be guided by results obtained with the help of the adiabatic theory, as discussed in the previous section. More accurate models exist, most notably those based on particle orbit integration (e.g., Decker 1988; Giacalone 2005) that also keep track of the particle's gyrophase in addition to pitch angle, something the FTE lacks. However, the advantage of the simpler FTE formalism is that a wide range of possibilities, such as strong vs. weak scattering transport, momentum and/or perpendicular diffusion present or not, as well as more general plasma backgrounds can be readily studied using minimal computational resources. We are also not restricted to small regions near the shock ramp, but can readily model large volumes of space including the entire heliosphere.

Florinski et al. (2008a, 2008b) discussed shapes of distribution functions produced by shocks of varying obliquity in the weak scattering limit. In addition to repeating their basic result here we also consider a scenario featuring stronger scattering in the downstream region with the correlation length l_c decreasing across the TS by a factor of s. The parallel mean free path (MFP) $\lambda_\parallel \sim w/D_{\mu\mu} \sim w^{2-q}$ therefore the ratio $\lambda_\parallel / r_g \sim w^{1-q}$, where r_g is the gyroradius is decreasing with energy. In the first ("weaker" turbulence) case the ratio $\lambda_\parallel / r_g$ in our simulations was 98 upstream and 201 downstream of the shock at $T = 100$ keV. At $T = 10$ MeV the ratio is 21 upstream and 43 downstream assuming $q = 5/3$. In the "stronger" scattering case the ratios are approximately the same on both sides of the shock (and equal to the upstream values quoted above). For the moment we ignore momentum diffusion entirely by setting $D_{ww} = 0$. The effects of second-order acceleration on intensities and spectra will be discussed below in Sect. 5. The principal result we wish to emphasize here is that in the highly anisotropic regime the spectra are more complicated than the simple power law predicted by the diffusion theory.

In all simulations discussed here we used the minimum particle energy corresponding to a velocity $w_{\min} = 1.5 V_{\text{dHT}}$, where $V_{\text{dHT}} = u_1 \sec \theta_{bn}$ is the de Hoffmann-Teller (dHT) speed and u_1 is the upstream plasma speed. For typical $\theta_{bn} = 80$–$85°$ this is of the order of $10 u_1$. The low energy boundary serves as the "injection" energy in our simulations whereby the particle intensity below that energy is maintained at a constant level throughout the simulation domain. Injection occurs at spatial locations with a negative value of $\nabla \cdot \mathbf{u}$ through a

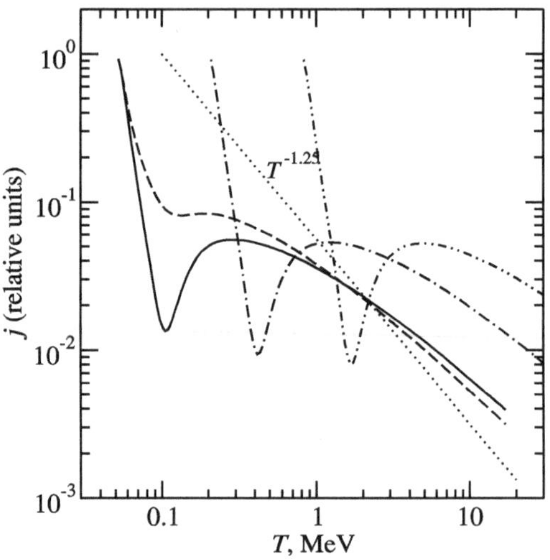

Fig. 1 Particle spectra immediately beyond the shock. The *solid line* is the spectrum at a $\theta_{bn} = 80°$ shock, corresponding to the weaker downstream scattering case (no wave compression across the TS). The *dashed line* is the particle spectrum obtained for the same shock obliquity under the assumption that the wave spectrum was shifted to higher wavenumbers. The *dash-dotted line* corresponds to $\theta_{bn} = 85°$ and the *dash-doubledotted line* to $\theta_{bn} = 87.5°$, both in the weak scattering case. The *dotted line* is the diffusive power law (slope -1.25)

positive flux along the velocity coordinate. It is assumed that the PUI distribution is isotropic below $w_{\min}$ in the upstream plasma frame, i.e., $f_1 = f_2 = 0$. Note that higher energy (ACR) population is not included in these simulations.

Figure 1 shows several characteristic spectral shapes calculated immediately downstream of a shock. Solid (dash-dotted) and dashed lines are spectra (i.e., the differential intensity $j = p^2 f_0$, p being the particle momentum) obtained in the weaker and the stronger scattering cases, respectively. The weak scattering spectrum features a prominent gap near the energy corresponding to about twice the dHT speed. This spectral dip and the following increase are a consequence of efficient acceleration by adiabatic reflection whereby a particle with an initial velocity w in the upstream plasma frame receives a maximum boost in energy in that frame given by (e.g., Decker 1988; Florinski et al. 2008b)

$$\Delta T = 2mV_{\mathrm{dHT}}^2\left(1-\frac{B_1}{B_2}\right)^{1/2}\left[\left(1-\frac{B_1}{B_2}\right)^{1/2}+\left(\frac{1}{\varepsilon^2}-\frac{B_1}{B_2}\right)^{1/2}\right], \tag{5}$$

where $\varepsilon = V_{\mathrm{dHT}}/w$ and

$$\frac{B_1}{B_2} = \frac{1}{\sqrt{\cos^2\theta_{bn} + s^2\sin^2\theta_{bn}}} \tag{6}$$

(MHD effects were neglected in deriving (6)). As an example, an $s = 3$ shock boosts the particle's energy by a factor of up to 5.3 in a single reflection for the initial velocity of $w = V_{\mathrm{dHT}}$ in the upstream plasma frame. After such a reflection a particle gains enough energy to propagate upstream against the flow of the SW. The dip is a common spectral feature found in simulations that retain pitch-angle information in the distribution function (Decker 1988; Chalov and Fahr 2000; Zank et al. 2001; Giacalone 2005; le Roux et al. 2007) and is a distinct signature of particles accelerated at an oblique shock. Far in the tail, the spectrum becomes softer with increasing energy. It is expected to approach the diffusive power law (shown with a dotted line in Fig. 1) at sufficiently high energy as the distribution approaches the isotropic limit.

In the second simulation featuring stronger downstream scattering, the appearance of the dip is significantly reduced compared with the weak scattering case. The overall spectrum

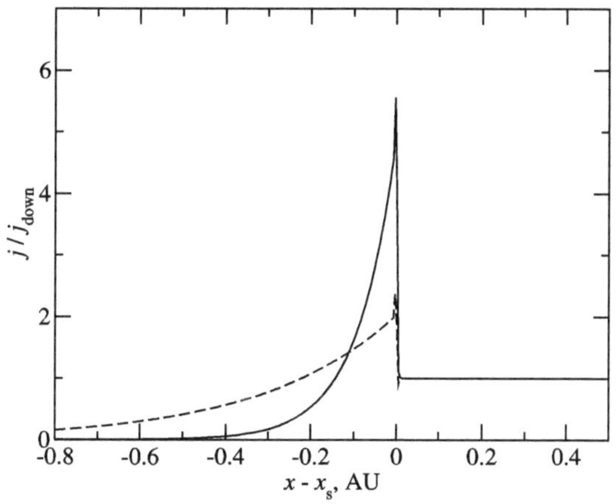

Fig. 2 Spatial dependence of ion intensity $j = p^2 f_0$ near the TS, normalized to the downstream value. A compression ratio of 3.0 and an obliquity of 85° are assumed for the TS. The *solid line* is the intensity at $T =$ 1 MeV and the *dashed line* is the intensity at $T = 5$ MeV

more closely resembles the diffusive power law. This trend appears to be qualitatively in agreement with the results of Giacalone (2005) who found, using an orbit integration model, that reducing the scattering rate led to a larger drop in intensity at energies of a few tens of the upstream flow velocity for a nearly perpendicular shock. His result, however, did not feature the same spectral recovery after the dip as did Chalov and Fahr (2000) and le Roux et al. (2007) and which is also present in our simulations. The difference may be due to a different treatment of the scattering process in the simulations based on adiabatic theory (FTE) and those based on particle orbit simulations as well as the neglect of perpendicular diffusion by the current generation of FTE-based models. Also shown in Fig. 1 are two spectra calculated in the weak scattering limit for the cases when the magnetic field makes an angle $\theta_{bn} = 85°$ and 87.5° with the shock normal. These spectra are shifted versions of the 80° spectrum and merely reflect the fact that the FTE is nearly invariant in almost perpendicular geometries, if velocity is measured in the units of V_{dHT} (Florinski et al., 2008a).

It should be noted that Voyager 1 did measure a spectral dip similar to the one depicted in Fig. 1 in several observation periods during the second half of 2004. The dip was centered at ~1 MeV and separated the low-energy power law PUI distribution from the more complicated higher-energy part of the spectrum (Cummings et al. 2006). Our theory predicts the same location for the spectral dip for the obliquity angle of just under 87°, which is reasonable for the TS. It appears possible that the observed feature is also a result of adiabatic reflection off a highly oblique segment of the TS.

Spatial dependence of the particle intensity $j = f_0 p^2$ is shown in Fig. 2 for a shock with $\theta_{bn} = 85°$ (the shock is located at $x = 0$ in this figure). Intensities are normalized to the downstream value, which is constant in this simulation in view of our neglect of the velocity diffusion. All intensities shown feature an exponential ramp upstream, which is similar to that in a diffusive model, but has a somewhat different slope. Florinski et al. (2008a) calculate the slope to be

$$j(x) = j(0) \exp\left(\frac{\alpha}{l_{\text{diff}}} x\right), \tag{7}$$

where

$$\alpha = \frac{\varepsilon + Q(3\varepsilon - \varepsilon^{-1}) + \sqrt{(3Q-1)^2\varepsilon^2 - 6Q^2 + 26Q/5 + Q^2\varepsilon^{-2}}}{6\varepsilon(3/5 - \varepsilon^2)}, \tag{8}$$

where $Q = 5q/(q+4)$ and $l_{\rm diff} = \kappa_{\parallel,1}\cos^2\theta_{bn,1}/u_1$ is the "classic" upstream diffusive length. From (8) we see that the e-folding distance is reduced compared with the diffusive limit, which corresponds to $\varepsilon \to 0$ (i.e., $\alpha \to 1$).

An important difference between the diffusive and the FTE results is the presence of an intensity drop across the shock ramp. Comparing the $T = 1$ MeV and $T = 5$ MeV profiles shows that the height of the drop decreases with increasing energy. The intensity structure in the immediate vicinity of the shock, sometimes called "shock spike", is sometimes observed at nearly perpendicular traveling interplanetary shocks in the inner heliosphere (Sarris and Krimigis 1985). It also features prominently in simulations dependent on pitch angle (Gieseler et al. 1999; Klappong et al. 2001; le Roux et al. 2007; Florinski et al. 2008b). The height of the drop behind the shock depends on both the compression ratio and the obliquity angle and disappears for purely parallel shocks (e.g., Florinski et al. 2008a). Physically, the drop is a consequence of adiabatic reflection at the shock: because not all upstream phase space is mapped into downstream phase space, Liouville's theorem predicts that the downstream intensity could be different from that immediately upstream (Webb 1983), depending on the specific form of $f(\mu)$. The downstream intensity may be larger than the upstream intensity due to the compression at the shock, conversely, the upstream intensity may be larger because it includes the reflected particles, which are missing in the downstream region. These two effects are precisely balanced when f is isotropic, and the intensity in then conserved (Drury 1983; Gieseler et al. 1999), but this is not true in general.

Voyager 1 did observe a distinct rise and a subsequent fall in energetic particle intensity lasting about 4 days and centered on the time of the shock crossing (Decker et al. 2005). Both le Roux et al. (2007) and Florinski et al. (2008b) argued that this event may be likewise interpreted as a shock spike at a nearly perpendicular TS. The observed increase in intensity of 1 MeV protons was of the order of 5–10, which is broadly consistent with the height of the peak shown in Fig. 2. The duration of the spike is consistent with the Voyager 1 traversal time of the shock precursor having a width of ~0.2 AU if the TS was moving inward with the speed of the order of 100 km/s (Jokipii 2005; Florinski and Zank 2006).

4 Pitch-Angle Distributions

The spatial and spectral distributions provide important clues to the origin and acceleration mechanisms of the TSPs and ACRs, but do not constitute a complete picture. The third and equally important piece of the puzzle are the pitch-angle distributions (PADs) and their associated anisotropies. These traditionally were regarded as less useful, in part because PAD measurements were not readily available and partly due to a relative difficulty in including the effects of large particle anisotropies in transport models. This perception is now changing since directional distributions became available for several low-energy channels from the analysis of the data from the 7 sectors of the Voyagers LECP instrument (Decker et al. 2007). As discussed in Florinski et al. (2008a), daily averaged PADs of 1 MeV protons in about half of all observations analyzed in their work may be adequately represented by a second-order polynomial in pitch angle, corresponding to the Legendre expansion discussed in Sect. 2. This makes the task of analyzing the observed PADs considerably simpler because we are allowed to use the relatively straightforward mathematics and the scaling relation discussed above. Florinski et al. (2008b) have derived a simple expression for the second-order harmonic in the Legendre expansion in the shock upstream region. Specifically, it may be shown that

$$f_1 = -\varepsilon f_0 \tag{9}$$

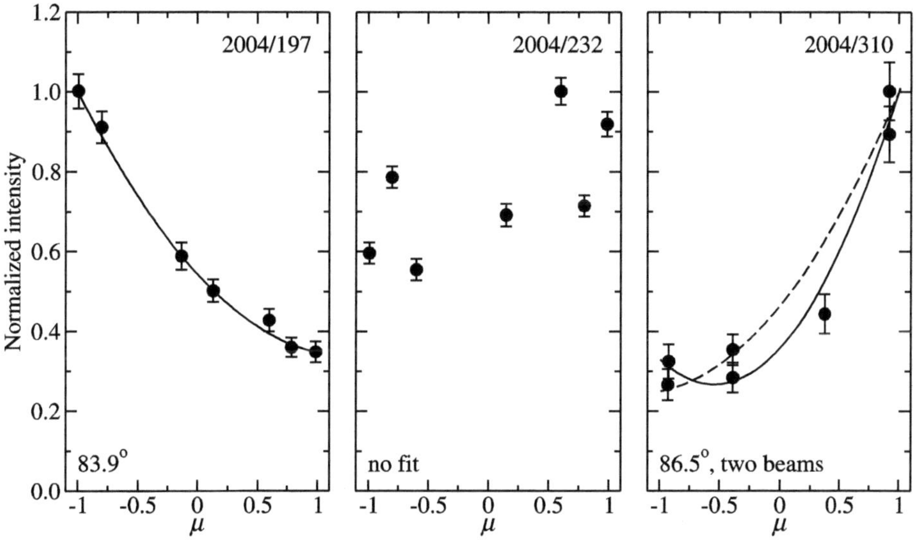

Fig. 3 PADs measured over three selected 1-day averaging periods. *Left panel* is an example of a good single beam quadratic fit. PAD in the *middle panel* cannot be fitted by a second-order polynomial. The *right panel* shows an example of a high-curvature measurement that may be explained by dual oppositely propagating beams. The *dashed line* in the *right panel* is a single beam fit

just as in diffusive theory and

$$f_2 = \frac{1+\alpha(3\varepsilon^2-1)}{2\alpha} f_0, \tag{10}$$

where α is given above. Because α is also a function of ε, we can only use the latter as a free parameter. This means it is not always possible to fit both first and second order harmonics to an observation.

Figure 3 shows three selected distributions observed by Voyager 1 during the second half of 2004. The PADs were initially measured in the spacecraft frame and rotated into the field-aligned frame using Voyager magnetometer data. On the left, the data points are exceptionally well fitted with a Legendre expansion to second order. About half of the observations analyzed by Florinski et al. (2008a) are of this type. The TS obliquity angle for that observation (83.9°) was derived using (9) by assuming $u_1 = 300$ km/s (velocity measurements are not available on Voyager 1 due to a nonfunctioning plasma instrument). For the PAD shown in the middle panel no quadratic fit was possible. In some cases such PADs are associated with weak or rapidly fluctuating magnetic fields and consequently high statistical errors in the directional data, while in other cases PADs appear to be genuinely too complicated and require more terms in a Legendre expansion.

The right panel in Fig. 3 is interesting because it depicts a rare event with a much larger second moment (i.e., curvature) than predicted by the theory. Florinski et al. (2008a) argued that this type of PAD indicates the presence of dual oppositely propagating beams at a location multiply-connected to the TS by the IMF lines. In this case 84% of particles were coming from the right, i.e., from the direction ($\mu = 1$) and the remaining 16% from the left ($\mu = -1$). The dual beam model allows to fit these high-curvature observations where single beam fits fail (cf. solid and dashed lines in the right panel of Fig. 3). Note that the majority of particles appear to be coming from opposite directions on days 2004/197 and 2004/310. This is not really the case, but merely indicates that Voyager 1 was in magnetic sectors with different polarities on these respective days.

In addition to providing us with information about the shock obliquity angle, PADs have another useful application. Comparing (8), (9) and (10) one can see that it is possible to derive the slope of the turbulent power spectrum q if both f_1 and f_2 are known. Using 1 MeV LECP data Florinski et al. (2008a) found that most of the data points in the f_1–f_2 regression plot lie between theoretical curves for the values of q between $1/2$ and $5/3$. It may be possible to narrow down the range of power law indices by using the actual measurements of u_1 from Voyager 2; this work is currently in progress.

5 The Heliosheath

The IMF increases across the TS by approximately the shock compression ratio and continues to grow with radial distance in the HS. A radially expanding wind yields $B \sim r$ beyond the TS in an incompressible limit. This translates to a radial gradient behind the TS of about 1% per AU, which is in fact similar to the Voyager 1 magnetometer results from 2005–2006 according to Burlaga et al. (2007). Global models of the heliosphere tend to predict a faster increase in B with distance, especially on approach to the heliopause where the field can pile up forming a "magnetic wall", where the IMF is up to 3 times stronger than just beyond the TS in the forward (upwind) part of the HS (Pogorelov et al. 2006). Large magnetic fields imply faster wave propagation and a possibility for more efficient momentum diffusion.

To simulate the effect of stochastic acceleration the D_{ww} term is now reintroduced into the transport equation (1) and the corresponding matrix equation for the moments (4). Again, we study both the weaker and the stronger scattering scenarios as described in Sect. 3. For simplicity we only consider a 80° shock and ignore the IMF increase with distance in the HS and the flow deflection and slowdown as both of these processes require treatment in a multidimensional model. The simulation domain used here only extends ~10–15 AU into the HS, which is similar to how far Voyager 1 would have traveled by the end of 2008. In our simulations we impose a condition of fixed intensity at the lowest energy, which is equivalent to assuming a preexisting distribution of low-energy particles with a sharp cutoff in energy. A better and more consistent approach, currently under development, is to start with an upstream spectrum in the shape of a power law tail (Gloeckler 1999). This can be followed through the TS with the low-energy particles having $w < V_{\rm dHT}$ being merely compressed by the shock (Fisk et al. 2006) and the higher energy particles being diffusively accelerated to attain a (perhaps) different power law slope.

Figure 4 illustrates the effect of stochastic acceleration on particle spectra and intensity profiles. Compared with Fig. 1 shock spectra are enhanced at low energies and the dip is less prominent. This result is to be expected because momentum diffusion tends to smooth out sharp momentum gradients in the distribution function. At $r_s + 9$ AU the spectra have filled in at low energy, more so in the high scattering limit, but the high-energy slope remains the same. In the right panel intensity at 200 keV shows a large increase in the high scattering simulation and a more modest increase in the low scattering case. 1 MeV intensity remain flat, however because the SW convection time from the TS to $r_s + 9$ is relatively short. We would like to emphasize that this result is valid for the resonant type of wave-particle interactions, whereby a pair of waves is required to accelerate the particles stochastically. The HS plasma also contains magnetoacoustic waves as well as large amplitude compressive structures (i.e., shock waves and rarefactions) which can accelerate ions via turbulent stochastic

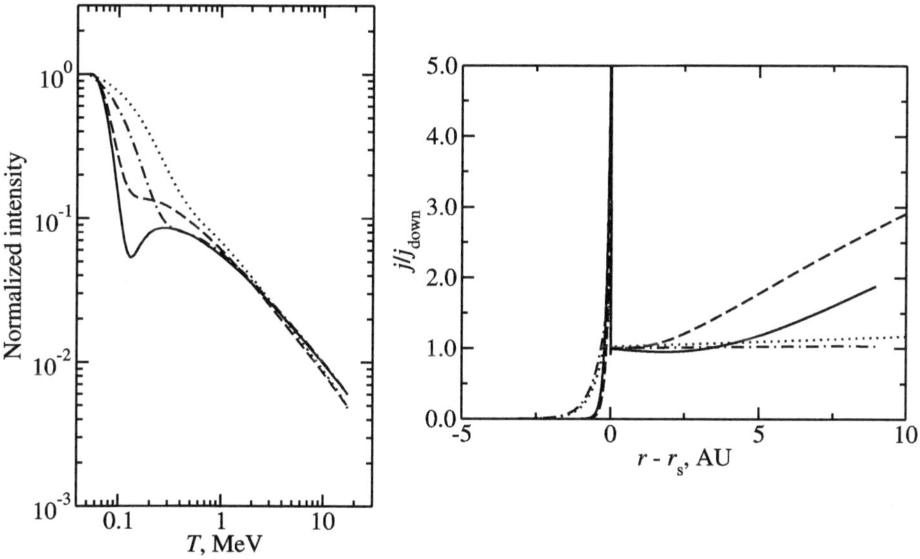

Fig. 4 *Left*: spectra of energetic protons at the TS subject to both first- and second-order acceleration in the low scattering case (*solid line*) and in the high scattering case (*dashed line*). *Dash-dotted* and *dotted lines* show spectra 9 AU beyond the TS for the two cases, respectively. *Right*: radial intensity profile of 200 keV protons in the low-scattering (*solid line*) and the high-scattering regimes (*dashed line*). *Dash-dotted* and *dotted lines* are radial profiles of 1 MeV protons for the two respective scattering cases

acceleration (Chalov et al. 1997, 2003). Unlike resonant diffusion, where the acceleration rate depends on velocity as w^{-q}, decreasing with energy, turbulent velocity diffusion is similar to ordinary adiabatic heating/cooling in that it is energy-independent and hence more efficient at high (ACR) energies.

It should be noted that in the simulations of Moraal et al. (2006) and Zhang (2006) second-order acceleration was concentrated near the heliopause where ACR energies were achieved with this mechanism. Even so, the gradient at low energies appears to be very high given a rather modest level of Alfvénic turbulence used in our simulation. As mentioned above there is presently no evidence of a sustained positive radial gradient of energetic particles in the HS (McDonald et al. 2006; Cummings and Stone 2007), therefore it is unknown if momentum diffusion is an efficient particle acceleration mechanism in the HS.

Another mode of acceleration that may be operating in the HS is adiabatic compression in a converging flow (Florinski et al. 2004). The HS flow is ordinarily nearly incompressible, i.e., plasma slowdown near the stagnation point on the heliopause is compensated by a spreading of the flow to other latitudes and longitudes, so that overall $\nabla \cdot \mathbf{u} \simeq 0$ is maintained. However, if charge exchange with interstellar atoms is taken into account, the plasma attains a certain degree of compressibility, such that

$$\nabla \cdot \mathbf{u} \simeq -\frac{\nu}{2\gamma}, \tag{11}$$

where ν is the charge-exchange rate in the HS and $\gamma = 5/3$ is the specific heat ratio. Florinski et al. (2004) estimated that the timescale for particle acceleration by this compression is of the order of 8 years, which is long compared with shock acceleration. The mechanism thus does not appear to be very efficient and some of the models confirm this (Zhang 2006;

Ferreira et al. 2007). It is also expected to compete with turbulent momentum diffusion since both are independent of the particle energy.

6 Conclusions

In this paper we briefly reviewed the adiabatic theory of PUI acceleration by the SWTS and incompressible turbulence in the SW and the HS. Our analysis was performed in the framework of a recently developed transport model based on spherical harmonic expansion of a pitch-angle dependent PUI distribution function. Contrary to the diffusive formalism, the new equations have finite signal propagation velocities, which is of particular advantage for the studies of time dependent particle transport. We have shown that three harmonics are sufficient to reproduce most essential features of particle spectral and pitch-angle distributions in about half of Voyager 1 observations in the 1 MeV energy channel.

The principal difference between our anisotropic results and those obtained in diffusive theory of shock acceleration is the presence of spectral gaps at particle energies near the dHT velocity and intensity spikes at the shock. Both of these features are produced by adiabatic reflection off a highly oblique (80–88°) segment of the TS. Both features also appear to be present in Voyager 1 data, although the spectral gap was diminished and eventually disappears in the HS. The last observation is also consistent with the expectation that the gap would be "filled in" due to velocity diffusion.

The anisotropic transport model appears to be a useful tool to analyze pitch-angle features of ion distribution. We proposed a technique to estimate the power spectral index of interplanetary turbulent fluctuations from the analysis of first- and second-order anisotropies. It was also argued that certain observations featuring abnormally large second-order anisotropies could be explained by dual oppositely propagating ion beams where an IMF line would be connected to the TS at two distinct locations. Some of our data analysis is currently impeded by the absence of plasma velocity measurements from Voyager 1. In the future we expect to significantly improve the accuracy of our calculations by using Voyager 2 particle, plasma, and magnetometer data. Extending the theoretical framework to include more Legendre harmonics would also improve the accuracy of the model and allow us to study a wider range of particle distributions, including those with narrow pitch-angle features.

The process of second-order Fermi acceleration is expected to be quite efficient in the HS at low energies, even under the most adverse conditions. Reaching higher (ACR) energies requires longer HS storage times and faster acceleration rates. Both of these conditions may be realizable only very close to the HP. At the time of writing of this report the availability of data from the HS was not sufficient to conclude that particle acceleration was taking place in that region, and it is presently not possible to rule out any of the suggested acceleration mechanisms.

Finally, we omitted a discussion of ion acceleration by shock surfing. In that process PUIs gain energy repeatedly in the SW motional electric field by being reflected off a cross shock electric potential, set up by a charge separation in the shock ramp (Lee et al. 1996; Zank et al. 1996). This process could lead to efficient acceleration of PUIs to ~50–100 keV (higher energy ions can overcome the shock electric potential and escape downstream) provided the width of the shock ramp is sufficiently narrow to be much less than an ion inertial length (le Roux et al. 2000). Whereas during Voyager 2 crossing the full extent of the TS ramp was measured to be wider than the ion inertial length (Burlaga et al. 2008), the ramp possessed complex structure on smaller scales. Multiple reflections at such structured shocks are also not ruled out.

Acknowledgements This work was supported, in part, by NASA grant NNG06GD43G. The author would like to thank R.B. Decker for providing Voyager 1 LECP data and J.A. le Roux and A.C. Cummings for helpful discussions.

References

L.F. Burlaga, N.F. Ness, M.H. Acuña, R.P. Lepping, J.E.P. Connerney, E.C. Stone, F.B. McDonald, Science **309**, 2027–2029 (2005)
L.F. Burlaga, N.F. Ness, M.H. Acuña, Astrophys. J. **668**, 1246–1258 (2007)
L.F. Burlaga, N.F. Ness, M.H. Acuña, R.P. Lepping, J.E.P. Connerney, J.D. Richardson, Nature **454**, 75–77 (2008)
S.V. Chalov, H.J. Fahr, Astron. Astrophys. **360**, 381–390 (2000)
S.V. Chalov, H.J. Fahr, V. Izmodenov, Astron. Astrophys. **320**, 659–671 (1997)
S.V. Chalov, H.J. Fahr, V. Izmodenov, J. Geophys. Res. **108**, 1266 (2003)
A.C. Cummings, E.C. Stone, in *Proc. 30th Intl. Cosmic Ray Conf. (Merida)*, paper 1118 (2007)
A.C. Cummings, E.C. Stone, F.B. McDonald, B.C. Heikkila, N. Lal, W.R. Webber, in *Physics of the Inner Heliosheath*, ed. by J. Heerikhuisen, V. Florinski, G.P. Zank, N.V. Pogorelov (Amer. Inst. Phys., Melville, 2006), p. 86
R.B. Decker, Space Sci. Rev. **48**, 195–262 (1988)
R.B. Decker, S.M. Krimigis, E.C. Roelof, M.E. Hill, T.P. Armstrong, G. Gloeckler, D.C. Hamilton, L.J. Lanzerotti, Science **309**, 2020–2024 (2005)
R.B. Decker, S.M. Krimigis, E.C. Roelof, M.E. Hill, in *Turbulence and Nonlinear Processes in Astrophysical Plasmas*, ed. by D. Shaikh, G.P. Zank. AIP Conf. Proc., vol. 932 (AIP, Melville, 2007), pp. 197–202
R.B. Decker, S. M Krimigis, E.C. Roelof, M.E. Hill, T.P. Armstrong, G. Gloeckler, D.C. Hamilton, L.J. Lanzerotti, Nature **454**, 67–70 (2008)
O'C. Drury, Rep. Prog. Phys. **46**, 973–1027 (1983)
S.E.S. Ferreira, M.S. Potgieter, K. Scherer, J. Geophys. Res. **112**, A11101 (2007)
L.A. Fisk, G. Gloeckler, Astrophys. J. **640**, L79–L82 (2006)
L.A. Fisk, G. Gloeckler, T.H. Zurbuchen, Astrophys. J. **644**, 631–637 (2006)
V. Florinski, G.P. Zank, Geophys. Res. Lett. **33**, L15110 (2006)
V. Florinski, G.P. Zank, J.R. Jokipii, E.C. Stone, A.C. Cummings, Astrophys. J. **610**, 1169–1181 (2004)
V. Florinski, R.B. Decker, J.A. le Roux, J. Geophys. Res. **113**, A07103 (2008a)
V. Florinski, G.P. Zank, J.A. le Roux, Adv. Space Res. **41**, 361–367 (2008b)
J. Giacalone, Astrophys. J. **624**, 765–772 (2005)
U.D.J. Gieseler, J.G. Kirk, Y.A. Gallant, A. Achterberg, Astron. Astrophys. **345**, 298–306 (1999)
G. Gloeckler, Space Sci. Rev. **89**, 91–104 (1999)
M.E. Hill, R.B. Decker, E.C. Roelof, S.M. Krimigis, G. Gloeckler, in *Physics of the Inner Heliosheath*, ed. by J. Heerikhuisen, V. Florinski, G.P. Zank, N.V. Pogorelov. AIP Conf. Proc., vol. 858 (AIP, Melville, 2006), pp. 98–103
P.A. Isenberg, J. Geophys. Res. **102**, 4719–4724 (1997)
J.R. Jokipii, Astrophys. J. **146**, 480–487 (1966)
J.R. Jokipii, J. Geophys. Res. **91**, 2929–2932 (1986)
J.R. Jokipii, Astrophys. J. **631**, L163–L165 (2005)
J.R. Jokipii, in *Physics of the Inner Heliosheath*, ed. by J. Heerikhuisen, V. Florinski, G.P. Zank, N.V. Pogorelov. AIP Conf. Proc., vol. 858 (AIP, Melville, 2006), pp. 143–152
R. Kallenbach, M. Hilchenbach, S.V. Chalov, J.A. le Roux, K. Bamert, Astron. Astrophys. **439**, 1–22 (2005)
K. Klappong, K. Leerungnavarat, P. Chuychai, D. Ruffolo, in *Proc. 27th Intl. Cosmic Ray Conf.* (Hamburg) (2001), pp. 3461–3464
J. Kota, J.R. Jokipii, in *Physics of the Outer Heliosphere*, ed. by V. Florinski, N.V. Pogorelov, G.P. Zank. AIP Conf. Proc., vol. 719 (AIP, Melville, 2004), pp. 272–278
S.M. Krimigis, R.B. Decker, E.C. Roelof, M.E. Hill, in *Solar Wind 11—SOHO 16*, ed. by B. Fleck, T.H. Zurbuchen. ESA SP-592 (ESA, Noordwijk, 2005), pp. 15–22
M.A. Lee, V.D. Shapiro, R.Z. Sagdeev, J. Geophys. Res. **101**, 4777–4789 (1996)
J.A. le Roux, M.S. Potgieter, V.S. Ptuskin, J. Geophys. Res. **101**, 4791–4803 (1996)
J.A. le Roux, G.P. Zank, H. Fichtner, V.S. Ptuskin, Geophys. Res. Lett. **27**, 2873–2876 (2000)
J.A. le Roux, G.M. Webb, V. Florinski, G.P. Zank, Astrophys. J. **662**, 350–371 (2007)
J.Y. Lu, G.P. Zank, R. Rankin, R. Marchand, Astrophys. J. **576**, 574–586 (2002)
D.J. McComas, N.A. Schwadron, Geophys. Res. Lett. **33**, L04102 (2006)

F.B. McDonald, W.R. Webber, E.C. Stone, A.C. Cummings, B.C. Heikkila, N. Lal, in *Physics of the Inner Heliosheath*, ed. by J. Heerikhuisen, V. Florinski, G.P. Zank, N.V. Pogorelov. AIP Conf. Proc., vol. 858 (AIP, Melville, 2006), pp. 79–85

H. Moraal, R.A. Caballero-Lopez, K.G. McCracken, F.B. McDonald, R.A. Mewaldt, V. Ptuskin, M.E. Wiedenbeck, in *Physics of the Inner Heliosheath*, ed. by J. Heerikhuisen, V. Florinski, G.P. Zank, N.V. Pogorelov. AIP Conf. Proc., vol. 858 (AIP, Melville, 2006), pp. 219–225

N.V. Pogorelov, G.P. Zank, T. Ogino, Astrophys. J. **644**, 1299–1316 (2006)

E.T. Sarris, S.M. Krimigis, Astrophys. J. **298**, 676–683 (1985)

R. Schlickeiser, Astrophys. J. **336**, 243–263 (1989)

N.A. Schwadron, D.J. McComas, Geophys. Res. Lett. **3**, L14105 (2007)

J. Skilling, Astrophys. J. **170**, 265–273 (1971)

C.D. Steenberg, H. Moraal, Astrophys. J. **463**, 776–783 (1996)

E.C. Stone, A.C. Cummings, F.B. McDonald, B.C. Heikkila, N. Lal, W.R. Webber, Science **309**, 2017–2020 (2005)

E.C. Stone, A.C. Cummings, F.B. McDonald, B.C. Heikkila, N. Lal, W.R. Webber, Nature **454**, 71–74 (2008)

G.M. Webb, Astron. Astrophys. **345**, 298–306 (1983)

G.P. Zank, H.L. Pauls, I.H. Cairns, G.M. Webb, J. Geophys. Res. **101**, 457–477 (1996)

G.P. Zank, J.Y. Lu, W.K.M. Rice, G.M. Webb, J. Plasma Phys. **64**, 507–541 (2000)

G.P. Zank, W.K.M. Rice, J.A. le Roux, I.H. Cairns, G.M. Webb, Phys. Plasmas **8**, 4560–4576 (2001)

G.P. Zank, G. Li, V. Florinski, Q. Hu, D. Lario, C.W. Smith, J. Geophys. Res. **111**, A06108 (2006)

M. Zhang, in *Physics of the Inner Heliosheath*, ed. by J. Heerikhuisen, V. Florinski, G.P. Zank, N.V. Pogorelov. AIP Conf. Proc., vol. 858 (AIP, Melville, 2006), pp. 226–232

ENA Imaging of the Inner Heliosheath—Preparing for the Interstellar Boundary Explorer (IBEX)

David J. McComas

Originally published in the journal Space Science Reviews, Volume 143, Nos 1–4, 125–138.
DOI: 10.1007/s11214-008-9410-8

Abstract The Voyager 1 and 2 spacecraft recently crossed the termination shock and are currently sending back groundbreaking and detailed observations at two locations in the inner heliosheath. Complementary global observations will soon be provided by the Interstellar Boundary Explorer—IBEX, which measures energetic neutral atoms (ENAs) produced via charge exchange with energetic protons in this region. While several data sets from instruments on other spacecraft have provided tantalizing observations that might be heliosheath ENAs, none has definitively shown that they are observing this source. In contrast, IBEX has been specifically designed and developed to make all-sky observations of inner heliosheath ENAs with very high sensitivity and signal to noise. These observations will provide the critical global perspective required to understand the three-dimensional heliospheric interaction with the Circum-Heliospheric Interstellar Medium (CHISM). This paper, written prior to the launch of IBEX, reviews previous observations and provides background on this important new mission.

Keywords IBEX · Heliosphere · Interstellar interaction · ENA

PACS 96.50.-e · 96.50.Ek · 96.50.Xy · 96.50.Zc

1 Introduction

Voyager 1 crossed the termination shock in December 2004 (Burlaga et al. 2005; Decker et al. 2005; Stone et al. 2005) at a heliocentric distance of 94 AU. More recently, in September 2007, Voyager 2 crossed this shock at ~84 AU (Burlaga et al. 2008; Richardson et al. 2008). These spacecraft are currently providing excellent local measurements and detailed information at two specific locations ~100 AU apart in the inner heliosheath (Richardson and Stone 2008). While these observations provide a wealth of knowledge about these two locations,

David J. McComas on behalf of the IBEX Science and Mission Teams.

D.J. McComas (✉)
Southwest Research Institute, 9503 W. Commerce, San Antonio, TX 78238, USA
e-mail: david.mccomas@swri.edu

they do not provide much information about the global configuration of the heliosphere's interaction with the interstellar medium. Remote imaging of this region is both critical and complementary to such detailed single point measurements, and global imaging is clearly required to understand the global interaction. Fortunately, a dedicated and optimized mission to globally image a broad range of plasma energy ENAs from the inner heliosheath is preparing for launch—IBEX, the Interstellar Boundary Explorer.

Energetic Neutral Atoms (ENAs) are produced by charge exchange between cold neutral atoms drifting in from the Circum-Heliospheric Interstellar Medium (CHISM) and hotter ions from the inner heliosheath. The supersonic solar wind carrying its population of suprathermal interstellar pickup ions (PUI) passes through the termination shock and into the inner heliosheath. Because it slows down, it becomes heated in passing through the shock. Flowing through this region are large numbers of interstellar neutrals atoms that drift into heliosphere from the CHISM. Recently, consensus results on the physical parameters in the CHISM have been compiled by Möbius et al. (2004) based on direct neutral gas, pickup ion, and UV backscattering observations of interstellar He, for which the heliospheric boundary is transparent. Direct neutral gas flow observations provide the most complete and least model dependent results (Witte 2004). The majority species, i.e. H, which is strongly affected by charge exchange in the boundary region, has been extensively mapped through backscattering of solar Ly-α (Quémerais et al. 1999). Also, its density has been derived from pickup ion observations (Gloeckler and Geiss 2001; Bzowski et al. 2008) and through the slowdown of the solar wind by PUI mass-loading (Richardson and Stone 2008). Interstellar PUIs also provide the seed population (when accelerated to much higher energies) for anomalous cosmic rays (ACRs). The hot solar wind protons charge exchange with interstellar neutrals to produce hydrogen ENAs whose flux (J_{ENA}) is calculated by the line-of-sight integral

$$J_{\mathrm{ENA}} = \int dl\, n_{\mathrm{H}} J_{\mathrm{ION}} \sigma(E) \tag{1}$$

of the local hydrogen density, n_{H}, local ion flux, J_{ION}, and energy dependent charge exchange cross section, $\sigma(E)$, between these neutrals and ions.

IBEX was proposed for the Small Explorer (SMEX) mission announcement of opportunity in February 2003. Table 1 lists the IBEX science team selected by NASA. After initial selection for a Phase A study, completion of that study, and its associated site visit, IBEX was finally selected for development in January 2005. Since then, progress has been extremely rapid with the Mission Preliminary Design Review (PDR) in January 2006, Critical Design Review (CDR) in September 2006, completion of the full science payload and its delivery to the spacecraft in September 2007. In December 2007, while still holding much more than adequate schedule reserves to reach our originally planned launch on 15 June 2008, the IBEX team was notified that NASA could not provide our launch on schedule. Instead, they announced the first in a series of delays that has pushed the IBEX launch back into the fall of 2008.

2 Early Efforts to Observe ENAs from the Heliosheath

Prior to the launch of IBEX, no space mission has been flown specifically to image ENAs propagating inward from the inner heliosheath. While several groups have made valiant efforts to scour existing data sets for directional measurements of these neutral atoms, and while some studies have claimed possible, limited detection of the heliospheric ENAs, the evidence has not been regarded as compelling by the community as a whole.

Table 1 The IBEX Science Team

Role	Team member	Organization
PI	David J. McComas	Southwest Research Institute
Co-Investigators	Peter Bochsler	University of Bern
	Maciej Bzowski	Space Research Centre, Polish Academy of Sciences
	Hans J. Fahr	University of Bonn
	Horst Fichtner	Ruhr-University Bochum
	Priscilla C. Frisch	University of Chicago
	Herbert O. Funsten	Los Alamos National Laboratory
	Stephan A. Fuselier	Lockheed Martin Advanced Technology Center
	Michael A. Gruntman	University of Southern California
	Vlad V. Izmodenov	Lomonosov Moscow State University
	Paul H. Knappenberger Jr.	Adler Planetarium and Astronomy Museum
	Martin A. Lee	University of New Hampshire
	Stefano A. Livi	Southwest Research Institute
	Donald G. Mitchell	Johns Hopkins University Applied Physics Laboratory
	Eberhard S. Möbius	University of New Hampshire
	Thomas E. Moore	NASA Goddard Space Flight Center
	Edmond C. Roelof	Johns Hopkins University Applied Physics Laboratory
	Nathan A. Schwadron	Boston University
	Peter Wurz	University of Bern
	Gary P. Zank	University of California, Riverside
Collaborators	Frederic Allegrini	Southwest Research Institute
	Michael R. Collier	NASA Goddard Space Flight Center
	George M. Gloeckler	University of Michigan
	Daniel Reisenfeld	University of Montana
	Martin Wieser	Swedish Institute of Space Physics
	Manfred Witte	Max-Planck Institute for Solar System Research

At higher energies, the CELIAS/HSTOF instrument on the SOHO spacecraft is capable of identifying H ENAs, and early analyses suggested the observation of ACR energy ENAs coming from the inner heliosheath (Hilchenbach et al. 1998). A re-analysis and recalibration of these data by the same authors (Hilchenbach et al. 2001) provided a new energy range and indicated that HSTOF was measuring neutral H atoms with energies from 58–88 keV. These authors examined 285 "quiet" days when the interplanetary charged-particle flux was low in 1996–1997 from two limited regions of the sky: the heliospheric apex (once) and anti-apex (twice). They interpreted the observations as ENAs from low-energy energetic ions (up to few hundred keV) in the outer heliosphere.

In another CELIAS/HSTOF study Czechowski et al. (2006) deduced ACR properties from ENA measurements by comparing HSTOF spectra with ACR measurements from Voyager 1. They argued that ENA fluxes measured were consistent with a heliosheath origin. These authors also found that the H/He ratio implied by HSTOF ENA data is close to the ACR value instead of the value typical of the termination shock particles. Assuming they were observing heliosheath ENAs, they calculated a heliosheath thickness of $\sim$75 AU toward the flank, a value that they suggest may be roughly three times thicker than what might be expected at the nose.

Finally, Hilchenbach et al. (2006) did a much broader study, examining HSTOF data from 1996–2005 for H and He ENAs. These authors concluded that the observations did not reveal a statistically significant pattern of variations in the apex direction sector of the heliosphere (observed averaged fluxes had only a $\sim 1\sigma$ significance). These authors went on to examine other possible ENA sources including solar energetic particles (SEPs), co-rotating interaction regions (CIRs), and pre-accelerated PUIs in the outer heliosphere. Finally, they concluded that the HSTOF observations at least set an upper bound on the higher energy ACR–ENAs that must constrain models.

At much lower energies, Galli et al. (2006) analyzed ASPERA-3 observations from Mars Express and described "Unknown Neutral Beam Events" (UNBEs). They argued that the energy spectra of these events were very similar and could be fitted by a two-component power law. They claimed to rule out ENAs of planetary and solar wind origin, contamination by UV from UV bright stars, and contamination by high-energy protons. Thus they concluded that the heliospheric termination shock is the most plausible source for these UNBEs.

Subsequently, the ASPERA-4 neutral particle detector (NPD) observed periods of UNBEs in cruise phase and in orbit on Venus-Express. Grigoriev et al. (2006) and Brandt et al. (2008) examined these events and found that the TOF spectra were even more intense than those observed by ASPERA-3/NPD on Mars-Express. However, these studies found no ordering of the UNBEs in inertial coordinates. Instead, they found that the strongest non-planetary UNBE signals in ASPERA-4 NPD were more likely associated with direction of spacecraft solar panels and deck surfaces. They indicated that the most likely explanation of these events was not heliospheric ENAs from the inner heliosheath, but rather solar wind ions trans-charging (neutralized) on solar panels and spacecraft surfaces.

Another criticism of a heliosheath origin for UNBEs on Mars Express was provided by Wood et al. (2007). These authors analyzed Ly-α absorption measurements from the Hubble Space Telescope and calculated an upper limit for heliosheath ENA fluxes in the 0.01–0.2 keV range of $<5000\ \mathrm{cm^{-2}\ s^{-1}\ sr^{-1}}$. They concluded that the claimed large ENA fluxes from ASPERA-3 on Mars Express could not be real ENA fluxes from the outer heliosphere. While the source of these measurements might be some hitherto unknown source of real ENAs in the space environment, a far more likely explanation seems to be some sort of instrumental or spacecraft-related background, possibly of the type indicated by Brandt et al. (2008) for ASPERA-4 on Venus-Express.

Thus, while the "jury is still out" on whether ENAs from the inner heliosheath, either at higher ACR energies or lower plasma energies, have already been marginally observed in one or a few directions in space, clearly no global imaging of the inner heliosheath has been carried out so far. On the other hand, the urgent need for global, energy resolved observations is absolutely clear.

3 The Interstellar Boundary Explorer

IBEX's sole, focused science objective is to discover the *global* interaction between the solar wind and the interstellar medium. We achieve this objective by taking a set of global ENA images that will answer four fundamental science questions:

I. What is the global strength and structure of the termination shock?
II. How are energetic protons accelerated at the termination shock?
III. What are the global properties of the solar wind flow beyond the termination shock and in the heliotail?
IV. How does the interstellar flow interact with the heliosphere beyond the heliopause?

In order to answer these questions, IBEX produces all-sky maps of the ENAs in 14 energy bands covering the energy range from ~10 eV to 6 keV. Details on how these data answer all four of these questions and provide closure on the science is provided in McComas et al. (2004). Briefly, the combination of all-sky maps at various energies and detailed energy spectra in ~1800 separate pixels covering all directions in the sky can be used to answer questions I, II, and III. Global ENA images easily differentiate gross differences such as asymmetries caused by the effects of the external magnetic field. More subtle asymmetries in global images further illuminate flow patterns beyond the termination shock. The energy spectra provide direct measures of solar wind ions, pickup protons, and energetic protons beyond the termination shock. Spectra as a function of direction will reveal the three dimensional configuration of the heating and energy partition of the ions by the shock. The Voyager 2 shock observations establish that only a fraction of the upstream free energy goes into heating the thermal plasma downstream (Richardson et al. 2008), while the energy density in the suprathermal ions >30 keV can exceed that of the downstream magnetic field (Decker et al. 2008). In other words, the shock is strongly mediated by suprathermal ions. The IBEX spectra will also provide information about how suprathermal energetic particle pressure modifies the shock, but more importantly of how it varies as a function of position along the shock.

Details of the global images and energy spectra also address question IV. In addition, IBEX will also make the first direct observations of interstellar neutral oxygen. By separately fitting the pristine oxygen drifting into the heliosphere from "filtered" oxygen, which has charge exchanged with the upstream plasma, IBEX will tell us about these upstream properties. In addition, by comparing these properties to those of the unfiltered He, IBEX should provide a very sensitive measure of the filtration process and the interstellar interaction further out, beyond the heliopause.

The IBEX team has a broadly formulated science strategy with three levels of exploration: Discovery, Exploration, and Understanding. At the Discovery level, fundamental properties of the global interstellar interaction should be easily found. Generally speaking, these results should be directly available from inspection of the IBEX data products, and thus are truly model independent. At the Exploration level, IBEX observations will let us explore the global fundamental properties of interstellar interaction by combining IBEX data products with simple physics-based calculations, theory and limited modeling. These results are therefore only slightly model dependent. Finally, at the level of full, detailed Understanding, IBEX observations will make it possible to understand the in depth global properties of the interstellar interaction. This will be done by iterative analysis using IBEX data to test and refine 3D models of the heliosphere. At the Understanding level, the results are in a sense model dependent, but more importantly, the IBEX observations will truly drive the models.

Table 2 shows examples of results for each of the levels of exploration and the derived measurement requirements needed to fully answer all questions at all three levels of exploration. Again, more detail is provided in McComas et al. (2004). Finally, the bottom portion of Table 2 provides the measurement requirements derived from the science goals above. These measurement requirements provided the Level 1 requirements that drove the entire IBEX mission, spacecraft, and instrument designs. They have remained unchanged from our original proposal. All have been fully met by the IBEX design and verified via calibrations and other verification techniques.

Figure 1 shows how IBEX produces global ENA maps. The charge exchange process is depicted schematically in the lower right corner. ENAs radiate in all directions from the hot inner heliosheath plasma with a tiny fraction just happening to have the trajectory to send

Table 2 IBEX science questions, science strategy, and derived measurement requirements

The IBEX approach and measurement requirements				
The sole IBEX objective to discover the global interaction between the solar wind and the interstellar medium is achieved by answering four fundamental science questions				
Levels / **Fundamental Science Questions**	**I:** *TS Strength/ Structure*	**II:** *Energetic Protons Near TS*	**III:** *Solar Wind Flow Patterns of Inner Heliosheath*	**IV:** *Interstellar Flow and Interaction*
[Raw Images] **Discover** (Straightforward interpretation of IBEX data products)	All-sky survey: strong vs. weak TS	Intensity and spectra of energetic protons near TS	Nose/tail asymmetries	First direct measurement of filtered interstellar neutral, O
Explore (Interpretation based on simple physics-based calculations and limited modeling)	All-sky map of shock strength	Modification of TS dynamics due to energetic particle pressure	Solar wind flow direction beyond TS vs. angle from the nose	Bulk speed, direction, and temperature of interstellar O inside TS
[Detailed Models] **Understand** (use data products to iteratively define, revise, and refine 3D models of the heliosphere)	-3D TS configuration incl. distance scale and strength -Energy partition of solar wind, pickup protons, and energetic protons	Bound injection processes at TS	-3D solar wind flow patterns beyond TS -LISM B	-Filtration -Ionized fraction of O in the LISM -Interstellar flow patterns beyond heliopause -Bow shock existence
Measurements Requirements				
Global heliospheric ENA imaging	7^ox7^o pixels	21^ox21^o pixels	7^ox7^o pixels	Determine O Direction to <2.0^o
H energy spectra: Solar wind through energetic particle ENAs	0.01-1 keV in 7 energy bands	1-6 keV in 4 energy bands	0.01-1 keV in 7 energy bands	
O fluxes and velocity direction vs. time of year				0.01-0.5 keV in 6 energy bands

them back in toward the inner solar system and ultimately to IBEX. The spacecraft spins at 4 RPM; during each spin, ENAs from near a great circle of the sky are imaged (mapped into crescents in the sky map in the upper left inset). At the end of each ∼8-day spacecraft orbit, the spin axis is repointed to track the Sun as the Earth moves around it. This repointing of the spin axis from one orbit to the next shifts the great circle being observed, producing an all-sky map each six months; in order to remain Sun-pointing, the spin axis has to be repointed 180° as the Earth orbits half a year around the sun. The image in the upper left corner of Fig. 1 is a simulated all-sky map of 300–600 eV ENAs from the inner heliosheath for a symmetric heliosphere, using the "strong shock" case from Gruntman et al. (2001) and including statistical counting noise.

Perhaps the greatest technical challenge of developing the IBEX mission within the cost constraints of NASA's SMEX program was to find a way to achieve a very high altitude orbit from NASA's least expensive launch option—the Pegasus launch vehicle. Pegasus is a small rocket dropped from the bottom of an L1011 airplane, capable of placing ∼470 kg in low Earth (∼200 km) orbit. Generally, Pegasus launches carry spacecraft to altitudes of only a few hundred km. In contrast, we needed to place IBEX into an orbit that would extend to at least 25 R_E, and preferably much higher, in order to spend most of the observing time well outside the magnetosphere. As a practical limit, we chose an apogee distance of 50 R_E; beyond this altitude lunar perturbations become exceedingly large and the orbital period becomes so long that we would have had to consider repointing maneuvers more than once per orbit.

Consequently, the IBEX team developed a new and highly innovative approach in order to achieve IBEX's very high altitude orbit from a Pegasus launcher (McComas et al. 2005, 2006). In order to accomplish this, first we added our own solid rocket motor (SRM) to the bottom of the IBEX spacecraft. This SRM lifts the IBEX spacecraft from low Earth orbit to a transfer orbit with apogee above ∼10 R_E. Then, we use our internal hydrazine thruster

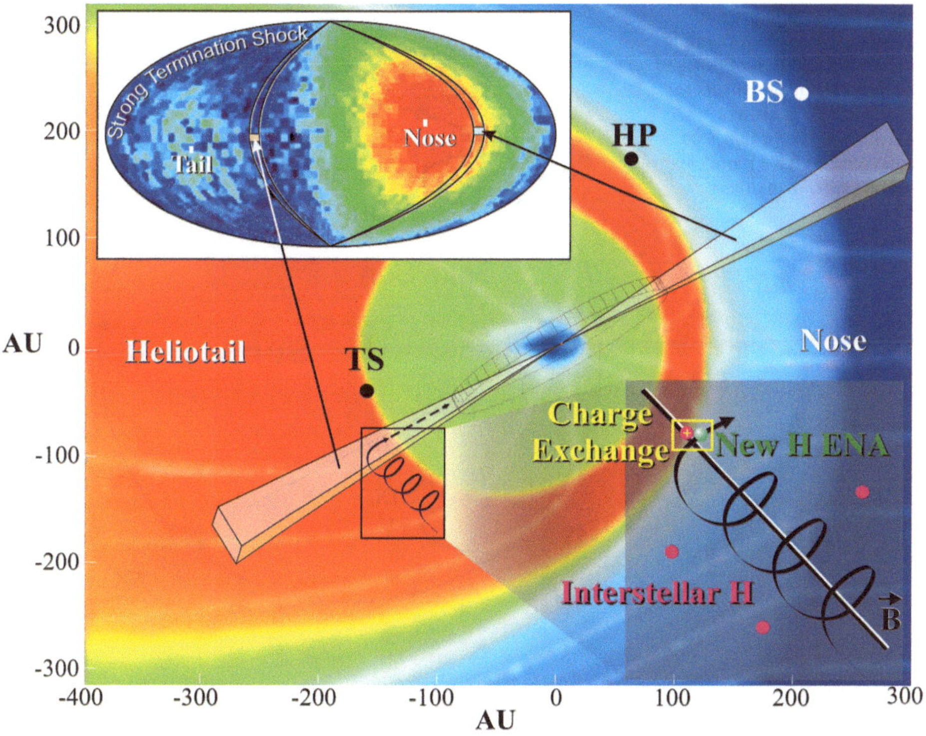

Fig. 1 Schematic representation of how IBEX images ENAs from the heliosphere's interstellar interaction, taken from McComas et al. (2006). ENAs produced by charge exchange (*lower right inset*) in the relatively hot and dense inner heliosheath radiate in all directions. Some of these travel inward toward Earth. Two instantaneous viewing *white shaded pixels* are shown looking outward from IBEX's location in the main image. These correspond to the two *white shaded pixels* in the all-sky map (*upper left inset*). As IBEX spins, these pixels sweep out a great circle in the sky (corresponding to the two crescents in this Aitoff-Hammer projection), allowing counts to be assigned to the appropriate pixels in the crescents by using spin phase. Finally, as the IBEX spacecraft repoints once each orbit in order to maintain its Sun-pointing attitude, different crescents are sampled such that an all-sky map is produced each six months

system to raise both perigee and apogee to our initial science orbit. At the end of orbit raising, IBEX is in a highly elliptical, ~8-day orbit with apogee at ~50 R_E and perigee at 7000 km altitude. Detailed Monte Carlo simulations, including all dispersions of both the Pegasus and SRM burns, statistically demonstrated an essentially 100% probability of being able to achieve this orbit. Owing to gravitational interactions with the moon, both perigee and apogee move up and down and the orbital inclination varies over time. This effect actually ends up helping us make the global images because the motion of the orbit, particularly its inclination, makes it easier to see around the "hole" where the magnetosphere obscures viewing of the heliosheath region behind it.

Because the IBEX mission was designed and developed solely to image ENAs from the inner heliosheath, all aspects of it were optimized for this unique purpose. The IBEX spacecraft is a simple Sun-pointed spinner with two very large aperture single pixel ENA cameras, pointed in opposite directions, perpendicular to the spacecraft spin axis. In order to keep on-orbit operations as simple as possible, we divide each orbit into two parts: high altitude science operations above ~10 R_E and low altitude housekeeping operations below

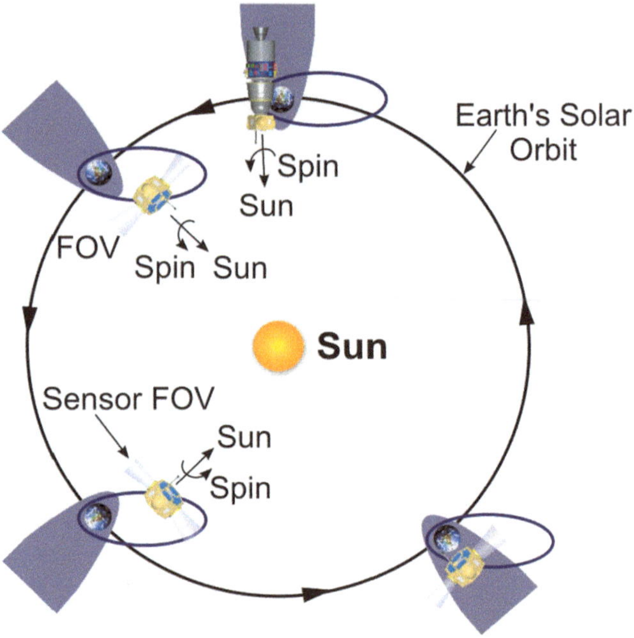

Fig. 2 Schematic diagram showing IBEX's Sun-pointing orientation at different times of the year. The magnetosphere (*blue shaded*) is always anti-sunward from Earth, so some seasons have more clear viewing time than others

this altitude. During the ~7.5 days at high altitude, the IBEX spacecraft is in science mode, collecting and storing data.

As the spacecraft descends below 10 R_E, stored commands reduce the high voltages on the sensors so that they can safely pass through the radiation belts. Then, generally near the start of the low altitude pass, we make contact through one of the USN ground antennas, download all the science data from the orbit and upload commands for the following orbits. In order to ensure that a bad contact cannot lose our science data or commands, the on-board memory always keeps the last two orbits worth of data, and commanding is uploaded for two orbits ahead. Shortly after the ~30-minute contact, stored commands execute the ~8° spin axis repointing; then, as IBEX again crosses above ~10 R_E, stored commands bring the sensor voltages back up for the next science orbit. This simple, repetitive process naturally produces nearly full-sky viewing each 6 months. Figure 2 schematically shows the IBEX orbit and pointing for several seasons.

Table 3 summarizes the instrument properties. The very large aperture areas and final geometric factors are needed to provide the large sensitivity required to observe the weak ENA emissions. The energy bands were chosen to optimize the trade space between energy resolution and fraction of the possible viewing time spent at each energy step. Importantly, the Hi and Lo sensors provide independent measurements of the overlap region from 300 eV to 2 keV, thus covering the critical region where much of the heliosheath ENA emission is expected to reside.

Because of the very low fluxes of heliosheath ENAs, the IBEX team has been especially careful to identify, quantify, and mitigate to the maximum extent possible, all potential sources of noise and background that we have been able to identify. We define noise sources as anything that generates uncorrelated (non-coincident) counts in the sensor detectors. Noise sources include UV light, X-rays, photoelectrons, secondary electrons, and penetrating radiation. In contrast, we use the term background to describe counts from ions or atoms that masquerade as a heliosheath ENA in the IBEX sensors. Background signals

Table 3 Nominal properties of IBEX sensors

	IBEX-Lo	IBEX-Hi
Energy range	~10 eV to 2 keV	~300 eV to 6 keV
Energy bands	8	6
Field-of-view	7° × 7° FWHM	7° × 7° FWHM
Coincidence	Triple with time-of-flight	Triple
Sensitivity	1 ENA $(cm^2\ s\ sr\ keV)^{-1}$	0.02 ENA $(cm^2\ s\ sr\ keV)^{-1}$
Geometric factor (triple coincidence)	$>3\times10^{-4}$ cm^2 sr keV/keV (average from 200 eV–1 keV)	$>1\times10^{-3}$ cm^2 sr keV/keV (for energies >1 keV)

Table 4 Potential noise and background sources for IBEX

Noise source	Background source
Diffuse UV, UV from stars	ENAs from planetary magnetospheres
X-rays from photoelectron acceleration toward, and impact with, biased collimator grids	Ions from magnetosheath and foreshock
Photoelectrons and secondary electrons generated at conversion surface	Charge exchange of plasma ions with outgassing spacecraft species
Penetrating radiation: radionuclide decay in detectors	Secondary ions generated in entrance subsystem
Penetrating radiation: cosmic rays	ENAs from CMEs, CIRs, and pickup ion charge exchange in the heliosphere
Penetrating radiation: SEP events	
Penetrating radiation: Magnetospheric energetic particles	

come from ions in the local environment and ENAs that are produced anywhere other than beyond the termination shock. Table 4 enumerates the various noise and background sources that we have identified and explicitly worked to minimize.

In the IBEX proposal, we estimated the ENA signal for Maxwellian distributions assuming both strong and weak termination shocks as described by Gruntman et al. (2001). Since then, there has been considerable work showing that higher-energy pickup-ion distributions have a high-energy power-law tail that extends well beyond the core solar wind Maxwellian distribution (Gloeckler et al. 2005; Fisk and Gloeckler 2006). A recent study by Prested et al. (2008) reanalyzed the expected IBEX signal assuming a κ-distribution with the power law index of 1.5 as suggested by Ulysses and Voyager observations. Figure 3 shows the calculated signal to noise ratio for IBEX assuming Maxwellian (dashed) and kappa (solid) distributions. Also shown for comparison is the signal to noise for the all-sky COBE observations of the 3 K background radiation.

Another significant improvement in the expected IBEX measurements compared to our original proposal comes from calibrated geometric factors that are substantially higher than the design requirements. As an example, the preliminary triple coincidence geometric factor from the IBEX-Hi flight calibration is typically two to three times larger than baseline requirement. Similarly, preliminary analysis of the calibration results for IBEX-Lo indicate that triple coincidences have roughly twice and double coincidences have three to four times larger efficiencies than the baseline requirements. Finally, cross-calibration between the two

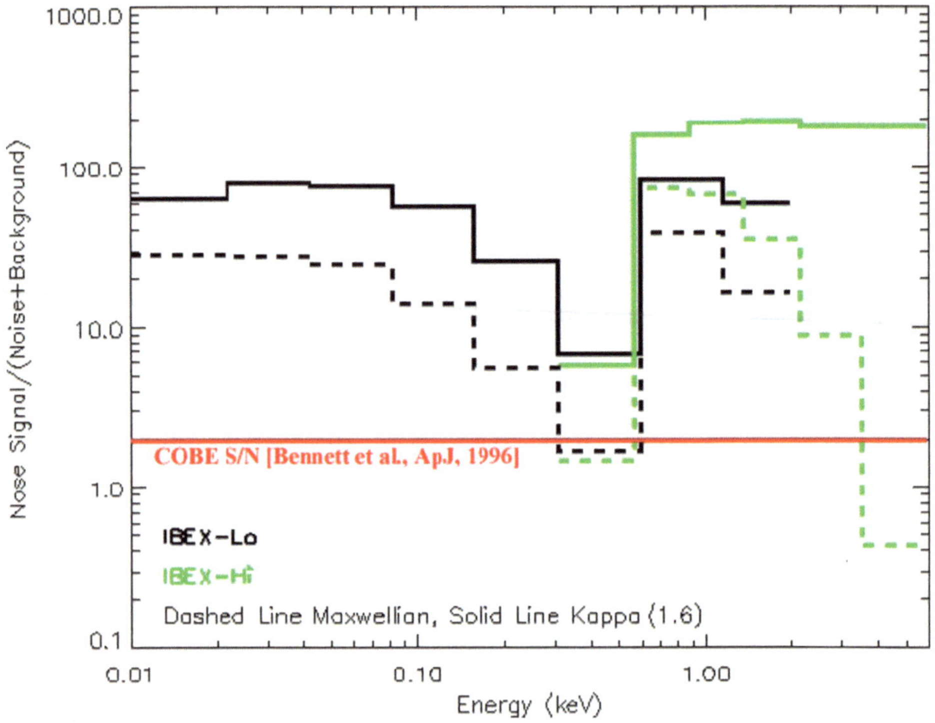

Fig. 3 Expected signal to noise ratio for IBEX, taken from Prested et al. (2008). IBEX-Hi (*green*) and Lo (*black*) are compared for Maxwellian distributions in the inner heliosheath (*dashed lines*) and more realistic κ-function distributions (*solid lines*)

IBEX sensors in the same chamber and at the same time demonstrated highly consistent results within the range of energy overlap.

4 Discussion

The new results from Voyager are bringing broad excitement to our scientific field and showing how little we really know about this critical region of space. IBEX will be launching at this most opportune time to provide the first global, energy-resolved images of the heliosphere's interstellar interaction. Just to show one example of how IBEX's global images will play a critical role in resolving the mysteries of this region, we consider the question of a large heliospheric asymmetry. The very large difference in radial distance between the Voyager 1 and 2 termination shock crossing distances (94 vs 84 AU), has ruled out a simple, symmetric, time-steady heliosphere, as assumed in most models prior to the last couple of years. Recently, Opher et al. (2007) provided an MHD simulation with a similarly large asymmetry by assuming an external magnetic field in the CHISM, tilted to be consistent with the external magnetic field direction suggested by Ly-α observations from SOHO (Lallement et al. 2005). This simulation, however, did not include neutrals, which have a significant impact on the properties of the inner heliosheath (Zank 1999 and references therein) and act to significantly reduce these sorts of asymmetries (Pogorelov et al.

2007; Heerikhuisen et al. 2008). Global observations provide the key to measuring any such asymmetry.

The results of a very recent simulation (Heerikhuisen et al. 2008), which includes neutrals, the more realistic κ-function ion distributions (these authors assumed $\kappa = 1.63$) and the external magnetic field, are shown in Fig. 4. The maps plot color-coded ENA fluxes in energy bands of 8–12 eV, 40–60 eV, 180–220 eV, 900–1100 eV, 2200–2600 eV and 5600–6400 eV from top left to bottom right. Note that the maps have been re-oriented from the image shown in Fig. 1; the upstream (nose) direction here is at the center of each plot. Clearly, these sorts of energy-resolved maps have the detailed all-sky observations needed to understand the true global configuration.

Another of the exciting things that IBEX should provide is a map of the thickness of the inner heliosheath as a function of direction in the sky. Nearly all of the ENA emission at IBEX energies comes from this region between the termination shock and heliopause, so we plan to extrapolate Voyager's observations of the suprathermal tails out to the heliopause and convolve them into the line of sight integral. The ENA intensities measured by IBEX would therefore yield an estimate of the thicknesses along these directions. In addition to being interesting in their own right, such thickness measurements along the Voyager 1 and 2 trajectories should provide the best estimates of how long it will take for these spacecraft to actually reach and cross the heliopause.

In the final analysis, IBEX data and 3-D models will need to be fully integrated to provide the best, ultimate understanding of the heliosphere's global interaction. This is why the original IBEX team (Table 1) contained many of the theorists and modelers working in this area. We were, however, mindful that others outside the formal team would also have much to contribute to the analysis, theory, and modeling. Therefore we set aside $2M (in FY03$—now worth ~$2.4M) from the SMEX cost cap when we proposed and asked NASA to assign them to a Guest Investigator (GI) program that would specifically enhance the IBEX mission science. The IBEX team looks forward to welcoming new colleagues into our team and efforts. In addition, the IBEX team is also delighted to welcome non-US "Team Members," who can bring their own support from their national governments. Please contact us if you need a formal letter of invitation to become a Team Member.

This paper has provided a very brief summary of some of the most important elements of the IBEX science and mission—clearly a complete description would fill a volume. The IBEX team is currently preparing just such a volume, which will be published in the summer or fall of 2008 in *Space Science Reviews* and as a separate bound book. This issue will contain the definitive pre-launch papers, fully describing the IBEX sensors, mission and science, and providing the information needed for interested scientists to understand and hopefully use the IBEX observations in their research.

In closing, ENA imaging is the only way to globally measure the interaction between the solar wind and interstellar medium. Imaging across multiple energy passbands provides critical energy spectral information needed for detailed analysis. IBEX launches in the fall of 2008 to make the first global, energy-resolved observations of the outer heliosphere. Copious new global observations are just around the corner, suggesting that we may need another ISSI workshop on this topic in a few years.

Acknowledgements I am eternally gratefully to the many scientists, engineers, technicians and other team members who have made phenomenal efforts and contributions to make this extraordinary mission a reality! I also thank Ed Roelof for reading and providing valuable comments on this manuscript and Wendy Mills for help with editing it. IBEX is funded by NASA through the GSFC Explorers Office.

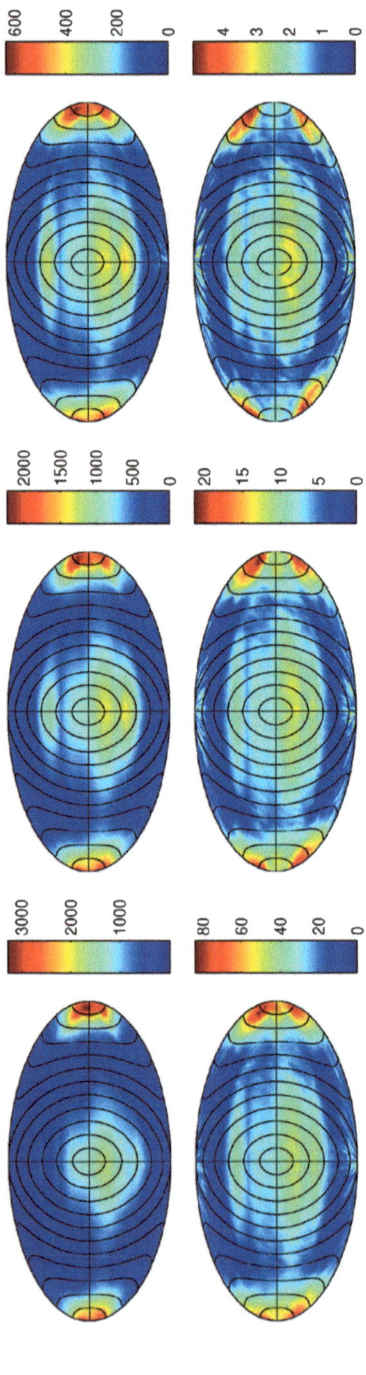

Fig. 4 Simulated all-sky maps taken from Heerikhuisen et al. (2008). ENA fluxes are color-coded in units of $(\mathrm{cm}^2\ \mathrm{sr\ s\ keV})^{-1}$

References

P.C. Brandt, T. Sotirelis, E.C. Roelof et al., Energetic neutral atom analysis of the solar wind interaction with Mars and Venus observed by ASPERA-3 and ASPERA-4, in *Proceedings of the AGU Chapman Conference: Solar Wind Interaction with Mars*, San Diego, January 2008

L.F. Burlaga, N.F. Ness, M.H. Acuña et al., Crossing the termination shock into the heliosheath: magnetic fields. Science **309**(5743), 2027–2029 (2005)

L.F. Burlaga, N.F. Ness, M.H. Acuña et al., Observations of magnetic fields at the termination shock by Voyager 2. Science (2008, submitted)

M. Bzowski, E. Möbius, S. Tarnopolski, V. Izmodenov, G. Gloeckler, Ulysses observations of pickup ions and their interpretation. Space Sci. Rev. (2008, this issue)

A. Czechowski, M. Hilchenbach, K.C. Hsieh, R. Kallenbach, J. Kota, Estimating the thickness of the heliosheath from CELIAS/HSTOF and Voyager 1 Data. Astrophys. J. **647**(1), L69–L72 (2006)

R.B. Decker, S.M. Krimigis, E.C. Roelof et al., Voyager 1 in the foreshock, termination shock, and heliosheath. Science **309**(5743), 2020–2024 (2005)

R.B. Decker, S.M. Krimigis, E.C. Roelof et al., Voyager 2 encounter with an energetic-particle-mediated termination shock. Science (2008, submitted)

L.A. Fisk, G. Gloeckler, The common spectrum for accelerated ions in the quiet-time solar wind. Astrophys. J. **640**(1), L79–L82 (2006)

A. Galli, P. Wurz, S. Barabash et al., Direct measurements of energetic neutral hydrogen in the interplanetary medium. Astrophys. J. **644**(2), 1317–1325 (2006)

G. Gloeckler, L.A. Fisk, L.J. Lanzerotti, Acceleration of solar wind and pickup ions by shocks, in *Proceedings of the solar wind 11/SOHO 16: Connecting Sun and Heliosphere Conference*, Whistler, Canada, June 2005. ESA Special Publication, vol. 592 (2005), p. 107

G. Gloeckler, J. Geiss, Heliospheric and interstellar phenomena deduced from pickup ion observations. Space Sci. Rev. **97**(1/4), 169–181 (2001)

A. Grigoriev, T. Sotirelis, E.C. Roelof et al., Energetic neutral atom observations at Venus from the ASPERA-4 experiment on Venus Express, in *American Geophysical Union Fall Meeting 2006*, San Francisco, Abstract No. P51B-1133, 2006

M. Gruntman, E.C. Roelof, D.G. Mitchell et al., Energetic neutral atom imaging of the heliospheric boundary region. J. Geophys. Res. **106**(A8), 767–781 (2001)

J. Heerikhuisen, N.V. Pogorelov, V. Florinski, G.P. Zank, J.A. le Roux, The effects of a κ-distribution in the heliosheath on the global heliosphere and ENA flux at 1 AU. Astrophys. J. (2008, submitted)

M. Hilchenbach, K.C. Hsieh, D. Hovestadt et al., Detection of 55–80 keV hydrogen atoms of heliospheric origin by CELIAS/HSTOF on SOHO. Astrophys. J. **503**, 916 (1998)

M. Hilchenbach, K.C. Hsieh, D. Hovestadt et al., Energetic neutral hydrogen of heliospheric origin observed with SOHO/CELIAS at 1 AU, in *The Outer Heliosphere: The New Frontiers*, ed. by K. Scherer, H. Fichtner, H.J. Fahr, E. Marsch. COSPAR Colloquium Series, vol. 11 (Pergamon, Amsterdam, 2001), p. 273

M. Hilchenbach, A. Czechowski, K.C. Hsieh, R. Kallenbach, Observations of energetic neutral atoms and their implications on modeling the heliosheath, in *Physics of the Inner Heliosheath: Voyager Observations, Theory, and Future Prospects, 5th Annual IGPP International Astrophysics Conference Proceedings*, vol. 858 (2006), pp. 276–281

R. Lallement, E. Quemerais, J.L. Bertaux et al., Deflection of the interstellar neutral hydrogen flow across the heliospheric interface. Science **307**(5714), 1447–1449 (2005)

D.J. McComas, F. Allegrini, P. Bochsler et al., The Interstellar Boundary Explorer (IBEX), in *Physics of the Outer Heliosphere, AIP Conference Proceedings*, vol. 719, ed. by V. Florinski, N.V. Pogorelov, G.P. Zank (2004) pp. 162–181

D.J. McComas, F. Allegrini, L. Bartolone et al., The interstellar boundary explorer (IBEX) mission, in *Proceedings of the Solar Wind 11—SOHO 16: Connecting Sun and Heliosphere*, Whistler, Canada, June 2005, ed. by B. Fleck, T. Zurbuchen, H. Locoste, ESA Special Publication, vol. 592 (2005), pp. 689–692

D.J. McComas, F. Allegrini, L. Bartolone et al., The Interstellar Boundary Explorer (IBEX): Update at the end of phase B, in *Physics of the Inner Heliosheath: Voyager Observations, Theory, and Future Prospects, 5th Annual IGPP International Astrophysics Conference Proceedings*, vol. 858 (2006), pp. 241–250

E. Möbius, M. Bzowski, S. Chalov et al., Synopsis of the interstellar He parameters from combined neutral gas, pickup ion and UV scattering observations and related consequences. Astron. Astrophys. **426**, 897–907 (2004)

M. Opher, E.C. Stone, T.I. Gombosi, The orientation of the local interstellar magnetic field. Science **316**(5826), 875 (2007)

N.V. Pogorelov, E.C. Stone, V. Florinski, G.P. Zank, Termination shock asymmetries as seen by the voyager spacecraft: The role of the interstellar magnetic field and neutral hydrogen. Astrophys. J. **668**(1), 611–624 (2007)

C. Prested, N. Schwadron, J. Passuite et al., Implications of universal kappa distributions for IBEX ENA flux maps. J. Geophys. Res. (2008, submitted)

E. Quémerais, J.-L. Bertaux, R. Lallement et al., Interplanetary Lyman α line profiles derived from SWAN/SOHO hydrogen cell measurements: Full-sky velocity field. J. Geophys. Res. **104**(A6), 12585–12604 (1999)

J. Richardson, E.C. Stone, The solar wind in the outer heliosphere. Space Sci. Rev. (2008, this issue)

J.D. Richardson, J.C. Kasper, C. Wang, J.W. Belcher, A.J. Lazarus, Plasma observations of the termination shock by Voyager 2. Science (2008, submitted)

E.C. Stone, A.C. Cummings, F.B. McDonald et al., Voyager 1 explores the termination shock region and the heliosheath beyond. Science **309**(5743), 2017–2020 (2005)

M. Witte, Kinetic parameters of interstellar neutral helium. Review of results obtained during one solar cycle with the Ulysses/GAS-instrument. Astron. Astrophys. **426**, 835–844 (2004)

B.E. Wood, V.V. Izmodenov, J.L. Linsky, Y.G. Malama, Lyman-α absorption from heliosheath neutrals. Astrophys. J. **657**(1), 609–617 (2007)

G.P. Zank, Interaction of the solar wind with the local interstellar medium: a theoretical perspective. Space Sci. Rev. **89**(3/4), 413–688 (1999)

Section III: How Are the ISM Inside and Outside of the Heliosphere Related?

Local Interstellar Parameters as They Are Inferred from Analysis of Observations Inside the Heliosphere

Vladislav V. Izmodenov

Originally published in the journal Space Science Reviews, Volume 143, Nos 1–4, 139–150.
DOI: 10.1007/s11214-008-9444-y

Abstract This paper provides a brief summary on the current knowledge of the properties of the Circum-Heliospheric Interstellar Medium (CHISM). It discusses what can be learnt on the parameters of CHISM's components from analysis of measurements performed inside the heliosphere. The analysis is based on the kinetic-gasdynamic models of the solar wind/interstellar medium interaction. We focus the analysis on three types of diagnostics: 1) interstellar H atom number density at the heliospheric termination shock inferred from pickup ion measurements, 2) the location and time of the Voyager 1 and 2 termination shock crossings, 3) the deflection of the interstellar H atom flow inside the heliosphere as been measured by SOHO/SWAN. From these results estimations of the unknown local interstellar parameters are deduced. The parameters are the number densities of interstellar H^+ and H and the magnitude and direction of the interstellar magnetic field in the vicinity of the solar system.

Keywords Heliospheric interface · Interstellar H atoms · Voyager

1 Introduction

The goal of this paper is to provide a brief summary on our current knowledge on the parameters of the circumsolar heliospheric interstellar medium (CHISM) as they can be determined from observations inside the heliosphere and numerical modelling. The heliosphere and CHISM are separated by the heliospheric interface region that is formed in the interaction of the solar wind with the interstellar gas (Fig. 1). The interstellar charged component

V.V. Izmodenov
Lomonosov Moscow State University, Moscow, Russia

V.V. Izmodenov (✉)
Space Research Institute (IKI) Russian Academy of Sciences, Moscow, Russia
e-mail: izmod@iki.rssi.ru

V.V. Izmodenov
Institute for Problems in Mechanics Russian Academy of Sciences, Moscow, Russia

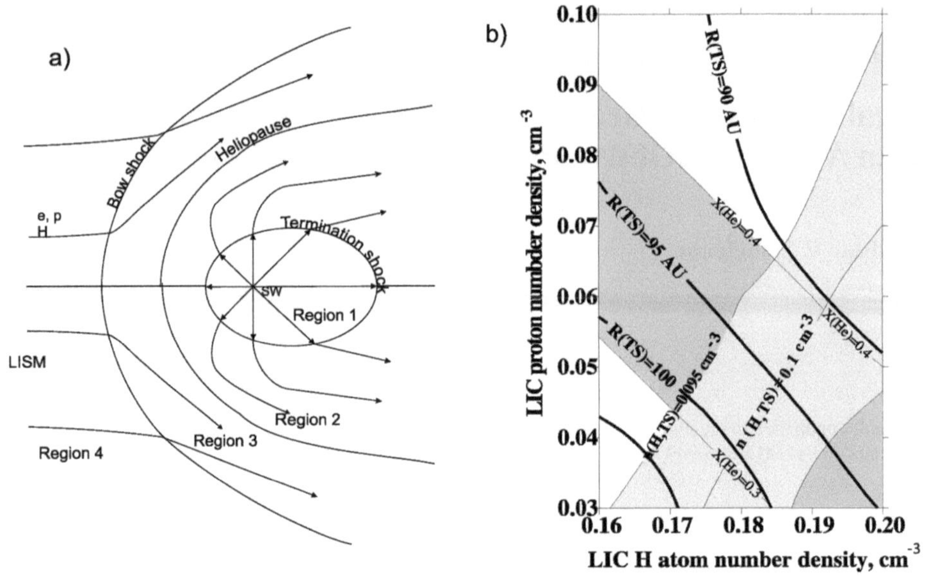

Fig. 1 **a**) Schematic view of the heliospheric interface structure. The Heliopause is the tangential discontinuity that separates the CHISM plasma from the solar wind. The termination shock and the bow shock decelerate the supersonic solar wind and the interstellar gas respectively. **b**) Results of a parametric study in the frame of a modified Baranov-Malama model (Izmodenov et al. 2003)

(electrons and protons) as well as the interstellar magnetic field are diverted around the heliopause by the interaction with the solar wind. So parameters of these components can not be measured from inside the heliosphere.

Luckily the CHISM is partly ionized. The CHISM neutral gas component consists of hydrogen ($\sim$90%), helium ($\sim$9%), oxygen, neon, nitrogen, and other minor components. The mean free path of the interstellar atoms is larger or similar to the size of the heliospheric interface region. Therefore, the neutrals penetrate the heliosphere where they can be measured.

The CHISM neutral component and its products (pickup ions and ACRs) are the major sources of information on the CHISM properties. Interstellar helium atoms penetrate the heliospheric interface undisturbed, because of a very weak coupling to protons. Indeed, due to the small cross sections of elastic collisions and charge exchange with protons, the mean free path of these atoms is much larger than the size of the heliospheric interface. Measurements of interstellar helium allow us to determine: 1) direction and velocity of the local interstellar flow, 2) local interstellar temperature. The most precise determination of these parameters was performed by using Ulysses/GAS direct measurements of interstellar helium atoms. Analysis of Ulysses/GAS observations (Witte 2004) results in a velocity $V_{\rm CHISM} = 26.3 \pm 0.4$ km/s with ecliptic longitude $\lambda = 74.7 \pm 0.5°$ and elevation $\beta = -5.2 \pm 0.2°$ of the CHISM inflow and a temperature $T_{\rm CHISM} = 6300 \pm 340$ K. The velocity and temperature of interstellar helium inside the heliosphere can also be obtained from measurements of backscattered solar radiation (Vallerga et al. 2004; Lallement et al. 2004a, 2004b). Independently, the velocity and temperature in the local interstellar cloud (CHISM) can be derived from the analysis of absorption features in stellar spectra (Lallement and Bertin 1992; Lallement 1996). However, this method provides only mean values along lines of sight toward nearby stars in the local interstellar medium. Comparison of the interstellar helium parameters obtained with the different methods are given in Möbius et al. (2004).

The mean free paths of interstellar hydrogen and oxygen are comparable with or larger than the characteristic size of the heliospheric interface. Therefore, similar to helium these components penetrate inside the heliosphere and can be measured. However, contrary to interstellar helium, interstellar hydrogen and oxygen are significantly disturbed in the heliospheric interface due to their coupling with protons by resonant charge exchange. The fact that these populations of interstellar atoms are disturbed in the interface allows to infer some properties of the heliospheric interface. Since the structure of the interface is determined by properties of the CHISM these properties can potentially be constrained from analysis of interstellar atom distributions inside the heliosphere. Such an analysis requires theoretical models of the solar wind interaction with the CHISM. In this paper we will combine current knowledge of the interstellar hydrogen parameters measured inside the heliosphere and theoretical models of the heliospheric interface in order to derive constraints on the remaining unknown CHISM parameters such as proton and H atom number densities, interstellar magnetic field magnitude and direction.

2 Observations from Inside the Heliosphere

Table 1 summarizes current major available diagnostics of the heliospheric interface. Most of the diagnostics are indirect and require special analysis and interpretation that is inherently model dependent. One of the most direct diagnostics of the heliospheric interface is provided by the crossings of the heliospheric termination shock (TS) by Voyager 1 and 2

Table 1 Summary of available diagnostics of the heliospheric interface and local interstellar parameters

Type of diagnostics	What is measured?	What can be determined?
Backscattered Ly-α		
SOHO/SWAN, HST	Spectral characteristics	Properties of H atom velocity distribution inside the heliosphere
Voyager, Pioneer, etc.	intensity	H number density inside the heliosphere
Pickup ions measured on board Ulysses and ACE	energy spectra	H atom number density inside the TS local interstellar composition
ACR and energetic particles	energy spectra	plasma properties in outer heliosphere local interstellar composition
Solar wind parameters measured by Voyager 2	density, velocity temperature	H atom number density inside the heliosphere
Voyager 1 and 2 crossing the TS		effective interstellar pressure; TS asymmetry
Absorption spectra toward nearby stars (HST)	Ly-alpha line spectra	constraints on the heliospheric flow (both inner and outer heliosheath)
2–3 kHz emission Voyagers		distance to the HP; plasma density gradient in the inner heliosheath

Table 2 Summary of parameters determined by using spacecraft data inside the heliosphere

Type of diagnostics	Determined parameters	References
Backscattered Ly-α		
SOHO/SWAN, HST	$V(\mathrm{H, TS}) = 22$–23 km/s,	Costa et al. (1999)
	$T(\mathrm{H, TS}) = 12000$ K	
	deflection of H atoms	Lallement et al. (2005)
	time-variation of H atom properties	Quemerais et al. (2008a, 2008b)
Voyager, Pioneer, etc.	$n(\mathrm{H, TS}) = 0.085$–$0.095\ \mathrm{cm}^{-3}$	Pryor et al. (2008)
Pickup ions measured	$n(\mathrm{H, TS}) = 0.11 \pm 0.022\ \mathrm{cm}^{-3}$	Bzowski et al. (2008)
by Ulysses and ACE	$n(\mathrm{H, TS}) = 0.10 \pm 0.01\ \mathrm{cm}^{-3}$	Geiss et al. (2006)
Solar wind parameters	$n(\mathrm{H, TS}) = 0.09 \pm 0.01\ \mathrm{cm}^{-3}$	Richardson et al. (2008)
measured by Voyager 2		
Crossing the TS by	94.1 AU in Dec 2004	Burlaga et al. (2005), Stone et al. (2005),
Voyager 1		Decker et al. (2005)
and by Voyager 2	83.7 AU in August 2007	Richardson and Stone (2008)
Absorption spectra		Wood et al. (2007, 2008)
toward nearby stars (HST)		
2–3 kHz emission measured	$n(\mathrm{e, CHISM}) = 0.04\ \mathrm{cm}^{-3}$	1.8 kHz emission cut-off
by Voyagers	$\sim$158 AU	Gurnett et al. (2006), and ref. therein

at 94 AU in December 2004 and at 83.7 AU in August–September 2007, respectively. As it will be seen later in this paper the knowledge of the TS distances in two directions already provides significant constraints on the CHISM parameters. However, the heliosphere is essentially three dimensional and time-dependent, so two locations of the TS are not enough to determine the global 3D time-dependent structure of the heliospheric interface.

Among other diagnostics of the heliospheric interface are:

- determination of the interstellar hydrogen properties from measurements of backscattered solar Lyman-alpha radiation with SOHO, Hubble Space Telescope, Voyager 1 and 2, Pioneer 10, Cassini, Ulysses, Galileo, etc.,
- observation of pickup ion spectra on board Ulysses and ACE,
- measurement of the solar wind speed and density in the distant heliosphere on board Voyager 2,
- observation of ACRs and energetic particles on boards of the two Voyagers and other spacecraft,
- measurements of Lyman-alpha absorption spectra toward nearby stars by Hubble Space Telescope (HST)
- measurements of the 2–3 kHz emission from the distant heliosphere on board the Voyagers.

In this paper we focus on three specific results that can be inferred from remote diagnostics of the heliospheric interface (Table 2). They are 1) the location of the TS at 94 AU in Dec 2004 in direction toward Voyager 1, and at 84 AU in Aug.–Sep. 2007 in direction toward Voyager 2; 2) the number density of the interstellar H atoms in the outer heliosphere; 3) the direction of motion of the interstellar H atoms in the outer heliosphere.

As it is seen from Table 2, the H atom number density in the outer heliosphere (i.e. near the heliospheric termination shock) is established with rather good accuracy. Different diagnostic methods provide similar values for the number density.

In addition, the direction of the H atom flow and its velocity and temperature can be determined from analysis of spectral properties of the backscattered solar Lyman-alpha radiation by using hydrogen-cells that are parts of SOHO/SWAN instrument. In particular, from analysis of SOHO/SWAN data it was obtained (Lallement et al. 2005) that the inflow direction of interstellar H atoms at the TS is few degrees deflected as compared with the direction of interstellar helium. This deflection is explained by an asymmetry of the heliospheric interface structure due to influence of the interstellar magnetic field (Lallement et al. 2005; Izmodenov et al. 2005b). The H-cell technique allows also to determine the bulk speed and effective temperature of the interstellar hydrogen in the outer heliosphere (e.g. Costa et al. 1999) and their variations with the solar cycle (Quemerais et al. 2006, 2008b). Recent development in the analysis of backscattered Lyman-alpha radiation is reported in Quemerais et al. (2008a).

The analysis of Lyman-alpha absorption spectra toward nearby stars is presented in Wood et al. (2007, 2008) and will not be touched here.

3 Multi-Component Models of the Global Heliosphere

In order to put constraints on the CHISM parameters one needs to employ a theoretical model of the heliospheric interface. This section briefly reviews current kinetic-gasdynamic and kinetic-MHD models of the heliospheric interface. We restrict this discussion to the kinetic-gasdynamic models developed in the Moscow group because results of these models will be used later in this paper. Models developed by other groups are described elsewhere. Most of other group models are based on multi-fluid approach for the neutral component (Fahr et al. 2000; Zank and Mueller 2003) or ignore or oversimplified the neutral component description so far (Ratkiewicz et al. 2002; Opher et al. 2004). The exception is recent models by Heerikhuisen et al. (2006, 2008), Pogorelov et al. (2008), where the component is treated kinetically as it is done in the Moscow group models reviewed here. Note, also, that comparisons between the kinetic and multi-fluid models are given in Alexashov and Izmodenov (2005), Heerikhuisen et al. (2006), Muller et al. (2008).

Table 3 summarizes the kinetic-gasdynamic models of the solar wind interaction that includes different components (or effects) of the solar wind or CHISM. Following Baranov and Malama (1993) model all further models take into account interstellar plasma and H atom components. The H atom component is treated kinetically, while the plasma component is treated as a fluid. In addition to protons and electrons, the plasma component includes solar wind alpha particles and interstellar helium ions. These components have minor effects on the interstellar H atoms in the heliosphere but noticeable influence on the location of the heliopause (e.g. Izmodenov et al. 2003). The entire heliosphere, including the heliospheric termination shock, shows significant fluctuations over the solar cycle (Izmodenov et al. 2005a, 2008a). The interstellar magnetic field makes the heliospheric interface asymmetric and influences both shape and location the heliospheric termination shock (Izmodenov et al. 2005b). Effects of the galactic and anomalous cosmic rays on the global distribution of the H atoms and plasma in the interface are insignificant (Myasnikov et al. 2000a, 2000b; Alexashov et al. 2004), but may change the location of the TS by a few AU. A recent model by Malama et al. (2006) makes a step toward a realistic description of the heliospheric plasma, treating pickup ions as a separate kinetic component co-moving with

Table 3 Modern multi-component kinetic-gasdynamic and kinetic-MHD models of the heliospheric interface developed by Moscow group

Component or effect	Reference
Interstellar H atoms (kinetic description)	Baranov and Malama (1993), Izmodenov et al. (2001) in all model below in the table
Interstellar plasma: protons, electrons + helium ions	Baranov and Malama (1993) Izmodenov et al. (2003)
Interstellar magnetic field	Izmodenov et al. (2005b)
Galactic cosmic rays	Myasnikov et al. (2000a, 2000b)
Anomalous cosmic rays	Alexashov et al. (2004)
Solar wind (protons, electrons) + alpha particles	Baranov and Malama (1993); Izmodenov et al. (2003)
Pickup ions (kinetic description)	Malama et al. (2006)
Solar cycle variations of the solar wind	Izmodenov et al. (2005a, 2008a)
Latitudinal variations of the solar wind	Izmodenov and Alexashov, Astron. Lett., in preparation

the solar wind protons and electrons. The different distribution of the heliospheric H atoms obtained in the model leads to some changes in the shape and location of the termination shock and the heliopause as compared with the Baranov-Malama model.

4 CHISM Parameters Inferred from Data and Theory

In this section we establish constraints on the CHISM H atom and proton number densities as well as magnitude and direction of the interstellar magnetic field on the basis of the data and numerical models of the heliospheric interface reviewed above.

As it was said above we examine three particular results of the heliospheric interface models: 1) filtration of interstellar atoms in the interface; 2) position of the TS in the Voyager 1 and 2 directions; 3) deflection of the interstellar H atom flow as compared to the interstellar helium direction.

We start this analysis with the standard stationary axisymmetric Baranov-Malama model modified by Izmodenov et al. (2003) and then consider the influence of additional components or physical phenomena. A parametric study performed in the frame of the standard model for rather large ranges of interstellar proton and H atom number densities (0.032 $\text{cm}^{-3} < n_{\text{p,CHISM}} < 0.07\ \text{cm}^{-3}$, $0.16\ \text{cm}^{-3} < n_{\text{H,CHISM}} < 0.20\ \text{cm}^{-3}$) has shown that the filtration (penetration) factor, which is defined as $F_H = n_{\text{H,TS}}/n_{\text{H,CHISM}}$, varies insignificantly and is equal to 0.54 ± 0.04. Assuming according to Geiss et al. (2006) and references therein that $n_{\text{H,TS}} = 0.100 \pm 0.01\text{cm}^{-3}$ we get $n_{\text{H,CHISM}} = 0.185 \pm 0.018\text{cm}^{-3}$.

To put constraints on $n_{\text{p,CHISM}}$ one can assume, for example, that the ratio of hydrogen to helium in the CHISM is 10. Then using that the ionization of helium in the CHISM is $35 \pm 5\%$ (Wolff et al. 1999) and $n_{\text{He,CHISM}} = 0.015 \pm 0.0015\ \text{cm}^{-3}$, one gets an estimate for the proton number density in the CHISM: $n_{\text{p,CHISM}} = 0.05 \pm 0.015\ \text{cm}^{-3}$ (see, Izmodenov et al. 2003).

Table 4 Results of parametric calculations

#	$n_{H,CHISM}$ cm^{-3}	$n_{p,CHISM}$ cm^{-3}	R(TS) AU	$F_{H,TS}$[a]	$F_{O,TS}$	$F_{N,TS}$
1	0.16	0.032	109	0.58	0.72 (0.84)	0.80 (0.90)[b]
2	0.16	0.05	102	0.55	0.70 (0.83)	0.80 (0.90)
3	0.16	0.06	99	0.54	0.70 (0.82)	0.80 (0.90)
4	0.16	0.07	96	0.53	0.69 (0.81)	0.80 (0.90)
5	0.18	0.032	101	0.57	0.69 (0.82)	0.77 (0.90)
6	0.18	0.05	96	0.54	0.68 (0.81)	0.79 (0.89)
7	0.18	0.06	93	0.53	0.68 (0.81)	0.79 (0.89)
8	0.18	0.07	88	0.52	0.66 (0.80)	0.79 (0.89)
9	0.20	0.032	94	0.55	0.68 (0.82)	0.76 (0.89)
10	0.20	0.04	93	0.54	0.67 (0.81)	0.77 (0.89)
11	0.20	0.05	90	0.53	0.67 (0.79)	0.78 (0.89)
12	0.20	0.06	88	0.52	0.67 (0.80)	0.78 (0.89)
13	0.20	0.07	86	0.51	0.67 (0.79)	0.78 (0.88)

[a] $F_{A,TS} = n_{A,TS}/n_{A,CHISM}$ (A = H, O, N) are the filtration factors of interstellar H, O, N atoms, respectively

[b] In parentheses we present filtration factors calculated under an assumption of reduced (by factor of 3) the electron temperature in the inner heliosheath between the TS and HP (Izmodenov et al. 2004)

In fact, detailed consideration of the filtration factors for H atoms (see Table 4) allows to establish range of possible pairs of ($n_{p,CHISM}$, $n_{H,CHISM}$) a little better than was shown above. This range is shown in Fig. 1 as the intersection of two shaded areas.

Additional constraints on $n_{p,CHISM}$ and $n_{H,CHISM}$ can be obtained from the location of the heliospheric termination shock. For example, if in the frame of a stationary axisymmetric model the TS is located at 95 AU (see Fig. 1) then the relation between $n_{p,CHISM}$ and $n_{H,CHISM}$ can be easily established: $n_{p,CHISM} + 1.15\, n_{H,CHISM} = 0.26$. (Here both $n_{p,CHISM}$ and $n_{H,CHISM}$ in cm^{-3}.) Assuming $n_{H,CHISM} = 0.185 \pm 0.018$ cm^{-3} the ratio gives $n_{p,CHISM} = 0.05 \pm 0.016$ cm^{-3}. This additional estimation of $n_{p,CHISM}$ is independent of the H/He ratio and the helium ionization in the CHISM, but depends on the H atom number density inferred from pickup observations and on the model of the heliospheric interface.

In fact it is possible to determine $n_{p,CHISM}$ independently from pickup observations. One of the possible additional diagnostics consists of spectral observations of backscattered Lyman-alpha radiation (e.g. Quemerais et al. 2006 and references therein) that allow to infer the temperature and velocity of H atoms at the termination shock (e.g. Costa et al. 1999).

To choose a final subset of possible ($n_{p,CHISM}$, $n_{H,CHISM}$) values some additional diagnostics are needed. This could be, for example, the temperature and velocity of the interstellar H atoms at the TS inferred from spectral measurements of backscattered Lyman-alpha radiation (e.g. Quemerais et al. 2006).

The method for obtaining $n_{p,CHISM}$ and $n_{H,CHISM}$ from remote observations, as presented above, would be straightforward in the case of an axisymmetric and stationary heliospheric interface. However, this is not a realistic situation. In fact, the heliosphere is breathing due to solar cycle variations of the solar wind dynamic pressure (e.g. Izmodenov et al. 2005a, 2008a). In addition, the interstellar magnetic field and heliolatitudinal asymmetry of the solar wind break the axis symmetry of the flow in the heliospheric interface (Lallement et al. 2005; Izmodenov et al. 2005b; Pogorelov et al. 2008).

Figure 2 shows fluctuations of the heliospheric termination shock in a time-dependent axisymmetric model of the heliospheric interface that takes into account the solar cycle variations of the solar wind parameters (Izmodenov et al. 2008a). The figure presents variations of the heliospheric termination shock distance for the upwind direction (black curve),

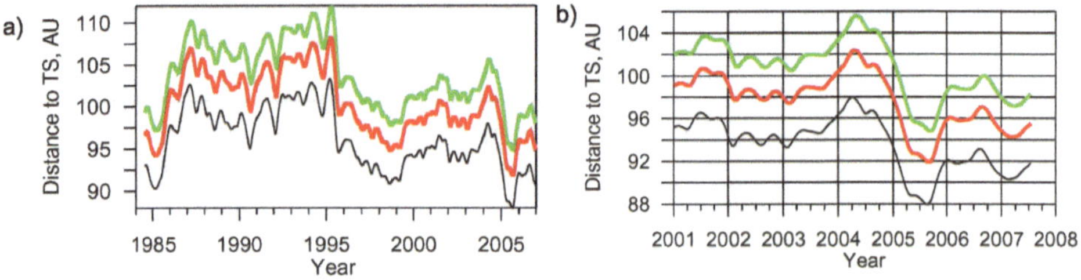

Fig. 2 **a** Distance to the heliospheric termination shock as function for the upwind direction (*black curve*), and directions of Voyager 1 and 2 (*red* and *green curves*, respectively). **b** Same as **a** zoomed for 2001–2008

and directions of Voyager 1 and 2 (red and green curves, respectively). The calculations were performed with the CHISM velocity and temperature that corresponds to Ulysses measurements of interstellar helium. Proton and H atom number densities were adopted as $n_{\text{p,CHISM}} = 0.06\ \text{cm}^{-3}$, $n_{\text{H,CHISM}} = 0.18\ \text{cm}^{-3}$. The solar wind parameters at the Earth orbit were taken from ONMI-2 Web site. Details for the model formulation and the boundary conditions are given in Izmodenov et al. (2008a).

As can be seen from Fig. 2, the axisymmetric time-dependent kinetic-gasdynamic model predicts that the distances to the heliospheric termination shock toward Voyager 1 in 2004 and toward Voyager 2 in 2007 are equal to 98 AU. For comparison, the actual crossing of the TS by Voyager 1 was at 94 AU in Dec 2004 and by Voyager 2 at ~84 AU in August–September 2007. Note, that the same distances to the termination shock in the Voyager 1 and 2 directions obtained in the time-dependent model is a coincidence (Fig. 2) that appears to come from a combination of spatial and temporal effects.

In view of this result, what are possible physical reasons for the difference between the theoretical predictions and observed distances to the heliospheric termination shock? There are many physical effects that may be responsible for the difference of ~4 AU in direction toward Voyager 1. For example, effects of galactic cosmic rays (Myasnikov et al. 2000a, 2000b) or interstellar magnetic field (Izmodenov and Alexashov 2006) may move the termination shock closer to the Sun by several AU. Conversely, the influence of anomalous cosmic rays (Alexashov et al. 2004) leads to an increase of the distance toward the termination shock.

Consideration of the multi-component nature of the heliospheric plasmas as it was introduced by Malama et al. (2006) also leads to an outward shift of the TS by 5 AU as compared with the one-fluid plasma component model by Izmodenov et al. (2003). In fact, the latter model is more realistic than a one-component plasma model because many observations show that pickup protons are not assimilated into the plasma component. The different (from Baranov-Malama model) distribution of solar wind and pickup protons in the heliosphere results in a different distribution of H atoms throughout the heliosphere. As a result of this process more energy is removed (by energetic neutrals that have a long mean free path) from the inner heliosheath as compared with models with one plasma component. This results in the displacement of the TS outward and of the heliopause inward. Therefore, under the same boundary conditions the multi-component model predicts the heliospheric termination shock at 103 AU. The situation can be improved by increasing the interstellar pressure in the models by increasing proton and/or H atom number densities. Another way to increase the total interstellar pressure is to take into account an interstellar magnetic field in the model.

Apparently, a need to introduce an *inclined* interstellar magnetic field becomes evident after considering the distance at which Voyager 2 crossed the heliospheric termi-

Table 5 Positions of the TS, HP and thickness of the inner heliosheath towards Voyager 1, Voyager 2 and upwind

	R_{TS}			R_{HP}			Heliosheath thickness		
	$\theta = 0$	V1	V2	$\theta = 0$	V1	V2	$\theta = 0$	V1	V2
no MF	96.0	99.0	104.0	157	167.2	181.0	61.0	68.2	77.0
$\alpha = 0$	98.5	100.9	104.7	169	174.6	183.3	70.5	73.7	78.6
$\alpha = 30$	88.5	92.7	92.8	149	162.3	156.6	60.5	69.6	63.8
$\alpha = 45$	83	86.9	87.6	136	147.1	144.8	53.0	60.2	57.2
$\alpha = 60$	81.6	85.2	86.6	130	139.8	140.5	48.4	54.6	53.9
$\alpha = 90$	80	82.5	87.1	128.6	133.6	148.1	48.6	51.5	61

nation shock. It is 84.7 AU, i.e. ~9 AU closer compared with the Voyager 1 crossing, while the axisymmetric time-dependent model predicts similar distances toward both directions. Any adjustments of the termination shock distance made within axisymmetric models can not explain the ~9 AU difference in the termination shock distance toward Voyager 1 and 2 as was measured. Therefore, an asymmetry of the heliospheric interface region is needed. Moreover, as it was mentioned above, the asymmetry is needed also to explain the deflection of the interstellar H atom flow in the heliosphere relative to the direction of interstellar helium (Lallement et al. 2005). The deflection is mainly associated with the effects of the interstellar magnetic field (Izmodenov et al. 2005b; Izmodenov and Alexashov 2006).

Table 5 presents distances to the heliospheric termination shock and to the heliopause and the thickness of the inner heliosheath obtained with our 3D kinetic-MHD models that include effects of the interstellar magnetic field (IsMF) (Izmodenov and Alexashov 2006). The distances are shown toward the Voyager 1 and 2 directions. To calculate the cases with the interstellar magnetic field inclined to the direction of the relative Sun/CHISM motion we assume that the interstellar magnetic field is in the plane determined by the velocity vectors of interstellar H and interstellar He (see, Lallement et al. 2005).

For an axisymmetric stationary heliosphere with no IsMF the TS is ~5 AU farther toward Voyager 2 as compared with Voyager 1 because Voyager 2 is farther from the upwind direction. In the case when the direction of the interstellar magnetic field coincides with the direction of CHISM's flow the interaction region remains axisymmetric. However, the difference of the distances to the termination shock in the Voyager 1 and 2 direction is smaller as compared to the model with no IsMF. This effect is connected with the shape of the heliopause, which is more distant from the Sun in the upwind direction and closer sidewind. In the presence of IsMF the heliopause is closer to a pencil-like shape.

In the 3D MHD case for $\alpha = 30, 45°$ (α is the angle between CHISM velocity vector and vector of the IsMF) the distances to the TS in the V1 and V2 direction become similar (Table 5). The difference of the heliospheric termination shock distances toward Voyager 1 and 2 increases with increasing angle α. In the case of $\alpha = 90°$ the heliospheric interface becomes axisymmetric. Therefore, the comparison of the axisymmetric kinetic-gasdynamic stationary model with a 3D kinetic-MHD stationary models suggests that the interstellar magnetic field may help to reduce the TS distance in the direction of Voyager 2 as compared with its distance toward the direction of Voyager 1. For $B_{CHISM} = 2.5$ μG (under the assumption that the vector of the IsMF lies in the plane determined by the directions of the interstellar helium and hydrogen flow) the TS distance at Voyager 2 can be close to the distance at Voyager 1 for $\alpha = 30$–$45°$. For $\alpha = 30°$ the TS is ~6 AU closer to the Sun toward

Voyager 1 direction as compared with a corresponding model without the interstellar magnetic field and in the Voyager 2 direction the TS is closer by $\sim$11 AU. Combining this result with time-dependent modelling of the heliosphere we obtain 92 AU for the TS distance toward Voyager 1 in 2004 and 87 AU toward Voyager 2 in 2007. For $\alpha = 45°$ the agreement with the model gets worse, $\sim$86 AU for the TS distance toward Voyager 1, i.e. significantly closer than 94 AU. The model's distance in the direction toward Voyager 2 is 82 AU in 2007. Therefore, we conclude that the model with $\alpha = 30°$ provides better agreement with actual crossings. In fact, to get perfect agreement between the model and data the TS should be 2 AU further in the Voyager 1 direction and 3 AU closer to the Sun in the Voyager 2 direction. This may require a somewhat stronger magnetic field and smaller value of α. Such possible solutions were demonstrated by Izmodenov et al. (2008b) already after the paper was prepared. Note, however, that such solutions are possible if employed cross section is different from the cross section used in this paper (see, Izmodenov et al. 2008b).

In conclusion, the magnitude and direction of the interstellar magnetic field can indeed be inferred from the deflection of the interstellar H atom flow as compared with the direction of the interstellar helium flow. Estimations performed by Izmodenov et al. (2005b), Izmodenov and Alexashov (2006) have shown that models with $B_{\mathrm{CHISM}} = 2.5$ μG, and $\alpha = 30$–$45°$ provide values for the deviation that are in agreement with the amount obtained from analysis of the SOHO/SWAN data.

5 Conclusions and Perspectives

In this paper we performed an analysis of data obtained from space experiments inside the heliospheric termination shock to obtain the local interstellar parameters such as proton and H atom number densities, magnitude and direction of interstellar magnetic field. We used the H number density at the termination shock determined from Ulysses pickup measurements, the Voyager 1 and Voyager 2 crossings of the heliospheric termination shock, and the deflection of the H atom flow direction inside the heliosphere from analysis of SOHO/SWAN data in comparison with the pristine interstellar wind flow as obtained from the observations. The study was based on results of 2D time-dependent kinetic-gasdynamic models (Izmodenov et al. 2005a) and stationary 3D kinetic-MHD models (Izmodenov et al. 2005b).

We conclude that $n_{\mathrm{H,CHISM}} = 0.185 \pm 0.018$ cm^{-3}, $n_{\mathrm{p,CHISM}} = 0.05 \pm 0.016$ cm^{-3}, $B_{\mathrm{CHISM}} = 2.5$–3.5 μG, $\alpha_{\mathrm{CHISM}} = 15$–$30$ deg.

These conclusions are model dependent and further advancement of the heliospheric interface models will provide even better constraints on the CHISM parameters. Tightening the constraints will require extended parametric calculations, including magnitudes of the interstellar magnetic field larger than 2.5–3 μG and angles between the CHISM flow and the IsMF direction smaller than 30 deg. In fact new model calculations by Izmodenov et al. (2008b) show that $B_{\mathrm{CHISM}} = 4.4$ and μG, $\alpha_{\mathrm{CHISM}} = 15$ deg is also possible solution.

Perspectives. The heliospheric interface is a great example of exploration by using a combination of different experimental diagnostics and intensive multi-component modelling.

In fall 2008, the Interstellar Boundary Explorer (IBEX) will be launched. The instrumentation implemented on the spacecraft will provide maps of fluxes and energy spectra of the heliospheric energetic neutral atoms (ENAs). ENAs originate in the region of the inner heliosheath between the heliospheric termination shock and the heliopause. 3D MHD-kinetic calculations show an asymmetry of the inner heliosheath and in the fluxes of heliospheric ENAs. Therefore, the ENA maps will provide new diagnostics of the global asymmetry of the heliospheric interface. Future expectations are also connected with the planned measurement of interstellar oxygen atoms with the IBEX/Lo sensor (Möbius et al. 2008). Interstellar oxygen is coupled to the plasma component in the heliospheric interface (Izmodenov et al. 1997, 1999, 2004; Izmodenov 2007) and may serve as an additional tracer of the heliospheric interface properties and local interstellar parameters.

In addition a large amount of information has been collected over the solar cycle by SOHO/SWAN instrument and new direct information from the inner heliosheath is continuously provided by Voyager 1 and 2.

Existing and new observations challenge theoreticians to develop more detailed and precise models of the heliospheric interface. With modern computers and previous modeling expertise complete 3D non-stationary and self-consistent multi-component kinetic-MHD models of the heliospheric interface are possible in the near future. However, since heliospheric hydrogen atoms have mean free paths comparable with or larger than the size of the heliospheric interface high-energy tails (e.g. Fisk and Gloeckler 2007) may become dynamically important. Therefore, a kinetic description should be employed in the models for both neutral and plasma components. Importance of the kinetic description for the plasma component is seen in the results of the first self-consistent model of the heliospheric interface that employs a kinetic description for both neutral and plasma components (Malama et al. 2006).

Acknowledgements This work was supported in part by RFBR grants 07-02-01101, 06-02-72557, Dynastia Foundation, and by Fundamental Research Program of OEMMPU RAS.

References

D. Alexashov, V. Izmodenov, Astron. Astrophys. **439**, 1171–1181 (2005)
D.B. Alexashov, S.V. Chalov, A.V. Myasnikov, V.V. Izmodenov, R. Kallenbach, Astron. Astrophys. **420**, 729–736 (2004)
V.B. Baranov, Y.G. Malama, J. Geophys. Res. **98**, 15157–15163 (1993)
L.F. Burlaga, N.F. Ness, M.H. Acuna, R.P. Lepping, J.E.P. Connerney, E.C. Stone, F.B. McDonald, Science **309**, 2027–2029 (2005)
M. Bzowski, E. Moebius, S. Tarnopolski, V. Izmodenov, G. Gloeckler, Astron. Astrophys. (2008, in press). arXiv:0710.1480
J. Costa, R. Lallement, E. Quemerais, J.-L. Bertaux, E. Kyrola, W. Schmidt, Astron. Astrophys. **349**, 660–672 (1999)
R.B. Decker, S.M. Krimigis, E.C. Roelof et al., Science **309**, 2020–2024 (2005)
H.J. Fahr, T. Kausch, H. Scherer, Astron. Astrophys. **357**, 268–282 (2000)
D.A. Gurnett, W.S. Kurth, I.H. Cairns, J.J. Mitchell, AIP Conf. Proc. **858**, 129 (2006)
J. Geiss, G. Gloeckler, L. Fisk, in *The Physics of the Heliospheric Boundaries*, ed. by V. Izmodenov, R. Kallenbach. ISSI Scientific Report 5, ESA-ESTEC, Paris, 2006, (pp. 137–181)
L. Fisk, G. Gloeckler, Proc. Nat. Acad. Sci. **104**, 5749–5754 (2007)
J. Heerikhuisen, V. Florinski, G.P. Zank, J. Geophys. Res. **111**, A6 (2006). CiteID A06110
J. Heerikhuisen, N.V. Pogorelov, V. Florinski, G.P. Zank, J.A. le Roux, Astrophys. J. **682**, 679–689 (2008)
V.V. Izmodenov, Space Sci. Rev. **130**, 377–387 (2007)
V.V. Izmodenov, D.B. Alexashov, AIP Conf. Proc. **858**, 14–19 (2006)
V.V. Izmodenov, Y.G. Malama, R. Lallement, Astron. Astrophys. **317**, 193–202 (1997)
V.V. Izmodenov, R. Lallement, J. Geiss, Astron. Astrophys. **344**, 317–321 (1999)
V.V. Izmodenov, M. Gruntman, Y.G. Malama, J. Geophys. Res. **106**, 10681–10690 (2001)
V.V. Izmodenov, Y.G. Malama, G. Gloeckler, J. Geiss, Astrophys. J. **594**, L59–L62 (2003)
V.V. Izmodenov, Y.G. Malama, G. Gloeckler, J. Geiss, Astron. Astrophys. **414**, L29–L32 (2004)
V.V. Izmodenov, Y.G. Malama, M.S. Ruderman, Astron. Astrophys. **429**, 1069–1080 (2005a)
V.V. Izmodenov, D.B. Alexashov, A.V. Myasnikov, Astron. Astrophys. **437**, L35–L38 (2005b)
V.V. Izmodenov, Y.G. Malama, M.S. Ruderman, Adv. Space Res. **41**, 318–324 (2008a)
V.V. Izmodenov, D.B. Alexashov, S.V. Chalov, O.A. Katushkina, Y.G. Malama, E.A. Provornikova, Space Sci. Res. (2008b, submitted)
R. Lallement, Space Sci. Rev. **78**, 361–374 (1996)
R. Lallement, P. Bertin, Astron. Astrophys. **266**, 479–485 (1992)

R. Lallement, J.C. Raymond, J. Vallerga, M. Lemoine, F. Dalaudier, J.-L. Bertaux, Astron. Astrophys. **426**, 875–884 (2004a)
R. Lallement, J.C. Raymond, J.-L. Bertaux, E. Quémerais, Y.-K. Ko, M. Uzzo, D. McMullin, D. Rucinski, Astron. Astrophys. **426**, 867–874 (2004b)
R. Lallement, E. Quemerais, J.L. Bertaux, S. Ferron, D. Koutroumpa, R. Pellinen, Science **307**, 1447–1449 (2005)
Y.G. Malama, V.V. Izmodenov, S.V. Chalov, Astron. Astrophys. **445**, 693–701 (2006)
E. Möbius, M. Bzowski, S. Chalov, H.-J. Fahr, G. Gloeckler, V. Izmodenov, R. Kallenbach, R. Lallement, D. McMullin, H. Noda, Astron. Astrophys. **426**, 897–907 (2004)
E. Möbius, H. Kucharek, L. Petersen, M. Bzowski, L. Saul, P. Wurz, S.A. Fuselier, V.V. Izmodenov, D. McComas, H. Muller, D.B. Alexashov, Space Sci. Rev. (2008, in press)
H.-R. Muller, V. Florinski, J. Heerikhuisen, V.V. Izmodenov, K. Scherer, D. Alexashov, H.-J. Fahr, Astron. Astrophys. (2008, in press). arXiv:0804.0125
A.V. Myasnikov, V.V. Izmodenov, D.B. Alexashov, S.V. Chalov, J. Geophys. Res. **105**, 5179–5188 (2000a)
A.V. Myasnikov, D.B. Alexashov, V.V. Izmodenov, S.V. Chalov, J. Geophys. Res. **105**, 5167–5178 (2000b)
M. Opher, P.C. Liewer, M. Velli et al., Astrophys. J. **611**, 575–586 (2004)
N.V. Pogorelov, J. Heerikhuisen, G.A. Zank, Astrophys. J. **675**, L41–44 (2008)
W. Pryor, P. Gangopadhyay, B. Sandel et al., Astron. Astrophys. (2008, in press). doi:10.1051/0004-6361:20078862
E. Quemerais, R. Lallement, J.-L. Bertaux, D. Koutroumpa, J. Clarke, E. Kyrola, W. Schmidt, Astron. Astrophys. **455**, 1135–1142 (2006)
E. Quemerais, R. Lallement, B. Sandel, J.T. Clarke, Space Sci. Rev. (2008a, this issue). doi:10.1007/s11214-008-9379-3
E. Quemerais, V. Izmodenov, D. Koutroumpa, Y. Malama, Astron. Astrophys. **488**, 351–359 (2008b)
R. Ratkiewicz, A. Barnes, H.-R. Muller, G.P. Zank, G.M. Webb, Adv. Space Res. **29**, 433–438 (2002)
J.D. Richardson, E. Stone, Space Sci. Rev. (2008, this issue)
J. Richardson, Y. Liu, C. Wang, D. McComas, Astron. Astrophys. (2008, in press). doi:10.1051/0004-6361:20078565
E.C. Stone, A.C. Cummings, F.B. McDonald, B.C. Heikkila, N. Lal, W.R. Webber, Science **309**, 2017–2020 (2005)
J. Vallerga, R. Lallement, M. Lemoine, F. Dalaudier, D. McMullin, Astron. Astrophys. **426**, 855–865 (2004)
M. Witte, Astron. Astrophys. **426**, 835–844 (2004)
B. Wolff, D. Koester, R. Lallement, Astron. Astrophys. **346**, 969–978 (1999)
B. Wood, V. Izmodenov, J. Linsky, D. Alexashov, Astrophys. J. **659**, 1784–1791 (2007)
B.E. Wood, V.V. Izmodenov, Y.G. Malama, Space Sci. Rev. (2008, this issue). doi:10.1007/s11214-008-9369-5
G.P. Zank, H.-R. Mueller, J. Geophys. Res. **108**, 1240 (2003)

Interplanetary Lyman α Observations: Intensities from Voyagers and Line Profiles from HST/STIS

Eric Quémerais · Rosine Lallement · Bill R. Sandel · John T. Clarke

Originally published in the journal Space Science Reviews, Volume 143, Nos 1–4, 151–162.
DOI: 10.1007/s11214-008-9379-3

Abstract We present an analysis of Voyager UVS data obtained between 1993 and mid-2007. These data are used to study the interplanetary background and the hydrogen number density in the outer heliosphere. Two types of observations are studied, first the heliospheric scans performed until 2003 and then the fixed line of sight observations, close to the upwind direction, which are still performed at the end of 2007.

We make comparisons with models including multiple scattering and hydrogen distributions derived from self-consistent modeling of the interface region. It is found that there is a remaining discrepancy between models and data. The origin of this difference is unknown but it may be linked to a possible tilting of the heliospheric interface due to the presence of an interstellar magnetic field.

We should also estimate alternate sources of emission which are not backscattering of solar photons like collisional excitation of hydrogen in the heliosheath and emission after charge transfer or recombination of proton and electron in HII regions.

Line profiles from HST/STIS are also presented.

Keywords Outer heliosphere · Interplanetary background · Hydrogen distribution

1 Introduction

Hydrogen atoms present in the interplanetary medium originate from the local insterstellar medium. The relative motion between the solar system and the local cloud in which it is embedded creates a flow of neutral particles, the interstellar wind. This flow was first observed

E. Quémerais (✉) · R. Lallement
Service d'Aéronomie, Université Versailles-St Quentin, Verrières-le-Buisson 91371, France
e-mail: eric.quemerais@aerov.jussieu.fr

B.R. Sandel
Lunar and Planetary Laboratory, University of Arizona, Tucson, USA

J.T. Clarke
Boston University, Boston, MA, USA

by the resonance scattering of the solar Lyman α photons which creates the interplanetary background (Thomas and Krassa 1971; Bertaux and Blamont 1971).

The interplanetary hydrogen distribution is not only defined by the parameters in the interstellar medium. In fact, it is strongly affected when hydrogen atoms cross the heliospheric interface (Baranov and Malama 1993; Izmodenov et al. 2001) which is the region where the expanding solar wind is decelerated and stopped by the ionized component of the interstellar medium. Therefore, studying the interplanetary background gives information on the hydrogen distribution and the heliospheric interface. According to model computations performed by Izmodenov et al. (2001), two hydrogen atoms out of three undergo a charge transfer when crossing the interface region.

Because the charge exchange cross section of helium atoms is much smaller than for hydrogen, the helium velocity distribution is essentially unchanged when crossing the interface region. As a consequence, the parameters of the helium flow reflect those of the local interstellar cloud.

In the outer heliosphere, beyond about 50 AU from the Sun, the hydrogen number density reflects the existence of the hydrogen wall (Baranov and Malama 1993). This wall is due to the accumulation of the heated and decelerated component of the hydrogen atoms created after charge exchange. As both Voyager spacecraft get closer to this wall, the interplanetary background they observe is more and more affected by the pile-up of hydrogen atoms.

In the inner heliosphere, the varying influence of the Sun is the dominant factor on the interplanetary background. On the other hand, because the flow of hydrogen atoms in the heliosphere is collisionless, it is still possible to make the distinction between the two hydrogen populations, the one which underwent a charge transfer and the one which didn't, by studying their velocity distribution. Therefore, the line profiles of the interplanetary background also bring information on the heliospheric interface.

2 Intensities in the Outer Heliosphere

In the outer heliosphere, the interplanetary background at Lyman α is still clearly visible (Quémerais et al. 2003). At 100 AU from the Sun, the upwind intensity observed by Voyager 1-UVS is around 30 R. The intensity pattern is controlled by two effects, first the hydrogen number density distribution which is influenced by the hydrogen wall, second by radiative transfer effects between the Sun and the observer. For a constant density close to 0.1 cm^{-3} and a gas temperature close to 12000 K, a distance of 10 AU roughly corresponds to an optical thickness at line center equal to 1. The coordinates for the upwind direction used in this analysis correspond to the values derived by Quémerais et al. (1999), i.e. 252.3° and 8.7° in ecliptic coordinates.

2.1 Heliosphere Scans with Voyager-UVS

The Voyager Outer Heliosphere scans were first implemented at the beginning of 1993. The original design of the scans is shown in Fig. 1. This design is based on two principles. First, scan platform movements are minimized which means that either the azimuth or the elevation of the platform is kept constant and that the scan follows a great circle in the sky. The plane containing the scan directions gets as close to the upwind and downwind direction as possible.

The spectrometers on Voyager 1 and 2 have a total bandwidth covering a range between 550 Å and 1600 Å. The pixel resolution is about 10 Å but the instrumental spectral response

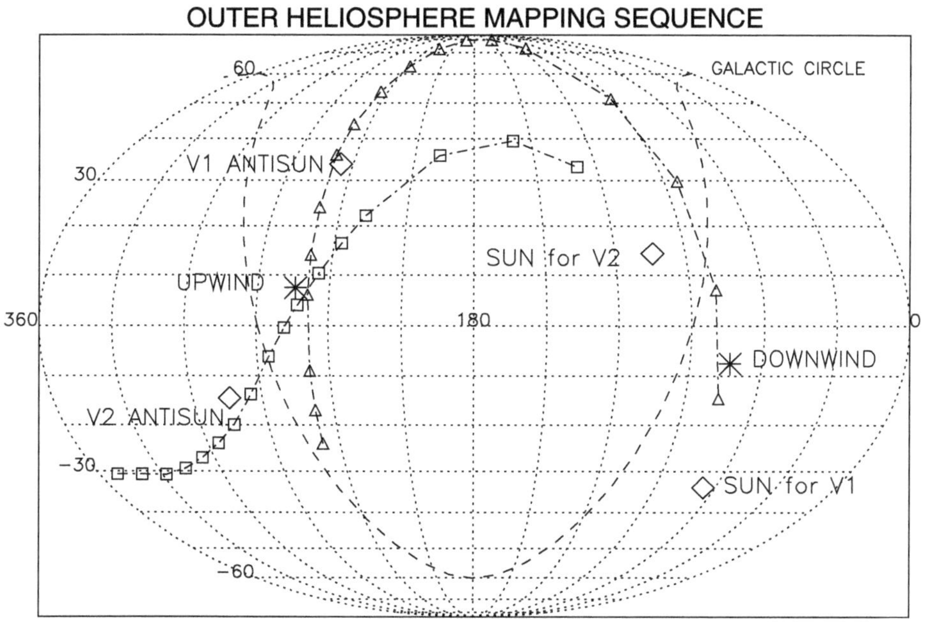

Fig. 1 Map of the celestial sphere as seen from Voyager 1-UVS or Voyager 2-UVS in ecliptic coordinates. Note that the positions of the Sun and anti-sun are quite different for each spacecraft. The *great circles* that guided the design of the H Lyman α mapping sequences are shown, along with the upwind and anti-solar directions that define them. The *discrete lines* of sight used in the mapping sequences are shown by the *triangles* for Voyager 1-UVS and the *squares* for Voyager 2. The *dashed line* shows the trace of the galactic plane

is roughly 9 pixel wide which means that a single line is spread over 9 pixels. This means that the IP line width cannot be studied and that intensities are derived by integrating over 9 channels.

The scans were performed regularly between 1993 and 2003 for Voyager 1 and 1993 and 1998 for Voyager 2. Between 1993 and 2003, the distance between the Sun and Voyager 1 increased from 55 AU to 85 AU. From 1993 to 1998, the distance between the Sun and Voyager 2 changed from 39 AU to 55 AU. To perform each scan, the total time for the maneuver varied between 10 days and 20 days, depending on the integration time chosen for each point, which was half a day or a full day. For this analysis, the variations of the solar illuminating flux have been taken into account by interpolating between daily measurements of the solar flux and correcting for solar rotation and for the different directions between the Sun, the Voyager spacecraft and the Earth. In most cases, these corrections correspond to a few percent due to the large distance between the Sun and the spacecraft.

Detailed descriptions of the data are given in Quémerais et al. (2003). The big advantage of the scan is that it allows to determine the ratio between upwind and downwind directions, thus removing the need to correct for solar Lyman α flux and instrumental calibration variations. Due to the position of the spacecraft in the heliosphere, the downwind direction is rather close to the direction of the Sun. Data obtained for lines of sight close to downwind can then be used to follow the temporal variations of the solar Lyman α flux with the 11-year solar cycle. The upwind to downwind intensity ratio is then used to study the hydrogen number density gradient in the upwind direction.

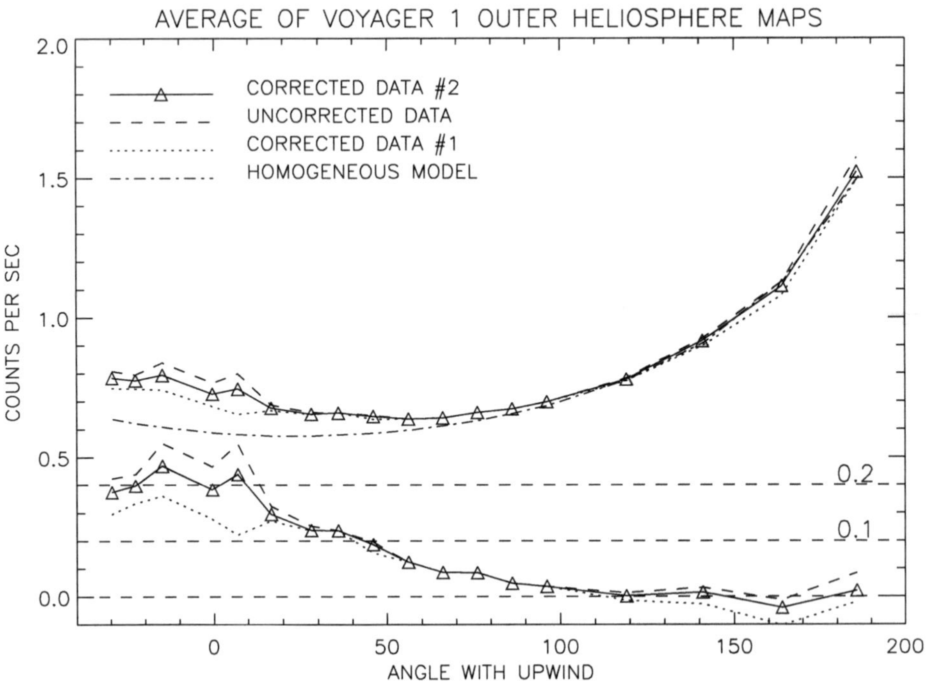

Fig. 2 Average of the Voyager 1-UVS brightnesses as a function of the angle with upwind. The model is a constant density model which fits the data in the downwind direction close to the Sun. The *lower curves* correspond to the relative variations between the data and the constant density model (no shock). The best estimate gives an increase of $17 \pm 3\%$ in the upwind direction. The *bottom curves* (difference between data and model) have been multiplied by a factor of 2. This is shown on the *right hand scale* where lines for 0.1 counts per second and 0.2 counts per sec are shown

Figures 2 and 3 show averages of the Voyager 1-UVS and Voyager 2-UVS outer heliosphere maps. See Quémerais et al. (2003) for details of the processing. The lower curves in each figure correspond to the relative variation according to the expected results from a homogeneous model (no interface) with $D_\infty = 0.16$ cm^{-3}. The best estimate gives an increase of intensity by $10 \pm 3\%$ for Voyager 2-UVS and $17 \pm 3\%$ for Voyager 1-UVS in the direction of the incoming interstellar hydrogen atoms. There is a small uncertainty due to a small contamination of some spectra by stellar light. However, the minimum values reported here are sufficient to show unambiguously that the observed intensity pattern departs from the prediction of an homogeneous model.

Comparisons with a model including the interface effects (Quemerais et al. 1996; Quémerais and Izmodenov 2002) do not completely fit the data. There is always a small excess about 20° from the upwind direction. Recently, Lallement et al. (2005) reported that the direction of the hydrogen flow was significantly shifted from the direction of the helium flow (≈ 4°). This was interpreted as a distortion of the heliospheric interface by the interstellar magnetic field. This could also result in a break of the symmetry of the hydrogen distribution around the wind axis. The excess intensity found when comparing the UVS data with models of the interplanetary background might be due to such an asymmetry of the real distribution which is not taken into account in the model. Such a study will require a self-consistent model of the interface including neutrals and the interstellar magnetic field.

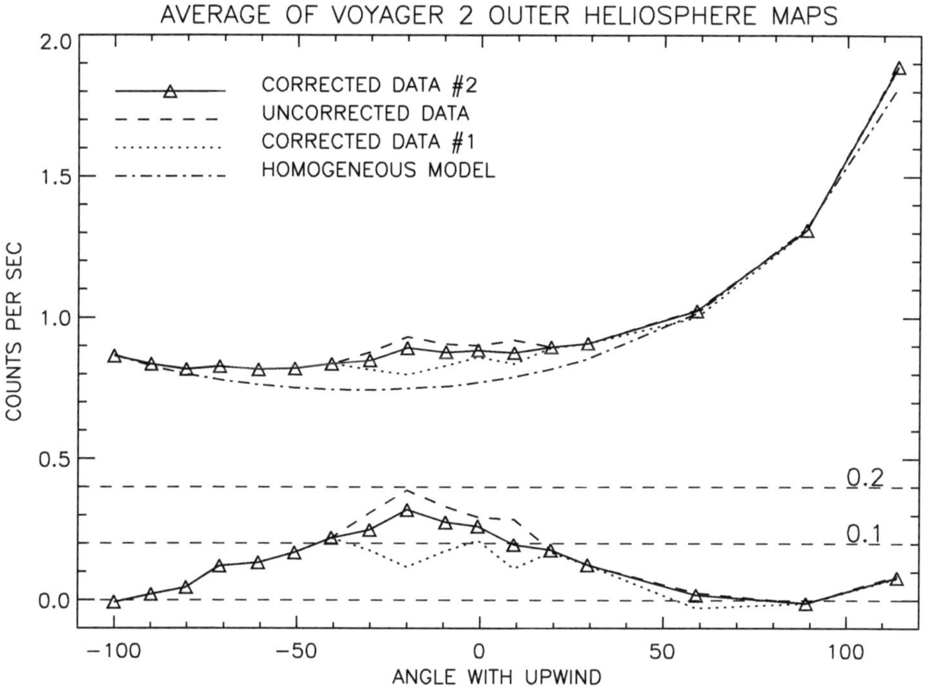

Fig. 3 Average of the Voyager 2-UVS brightnesses as a function of the angle with upwind. The *lower curves* correspond to the relative variations between the data and the homogeneous model with constant density (no shock). The best estimate gives an increase by 10 ± 3% in the upwind direction. This value is obtained by considering only the corrected points, following the two procedures given in the text. The *bottom curves* (difference between data and model) have been multiplied by a factor of 2. This is shown on the *right hand scale* where lines for 0.1 counts per second and 0.2 counts per sec are shown

2.2 Radial Variation of the Upwind Intensity

At the end of 2003, because of power constraints on the Voyager 1 spacecraft, it was decided to stop movements of the scan platform. The platform was put in a position that ensured that the line of sight of the main port of the UVS was pointing as close to the upwind direction as possible. Since then, UVS has been recording spectra in the same direction. Every day, the data which are received on Earth are cumulated to make a daily spectrum with excellent statistics.

Figure 4 shows a model computation of the variation of the upwind intensity as a function of the distance between the Sun and the observer. It is worth noting that multiple scattering effects ensure that, at 100 AU from the Sun, the IP glow is still important (30 R) while single scattering computations predict intensities more than an order of magnitude lower.

Hall et al. (1993) were the first to present a comprehensive study of the variation of the upwind intensity measured by UVS with solar distance. They introduced a radial coefficient defined by

$$\alpha = \frac{r}{I}\left(\frac{\partial I}{\partial r}\right)$$

where r is the solar distance and I the upwind intensity. It measures the slope of the intensity as a function of distance. In the case of constant density and neglecting radiative transfer

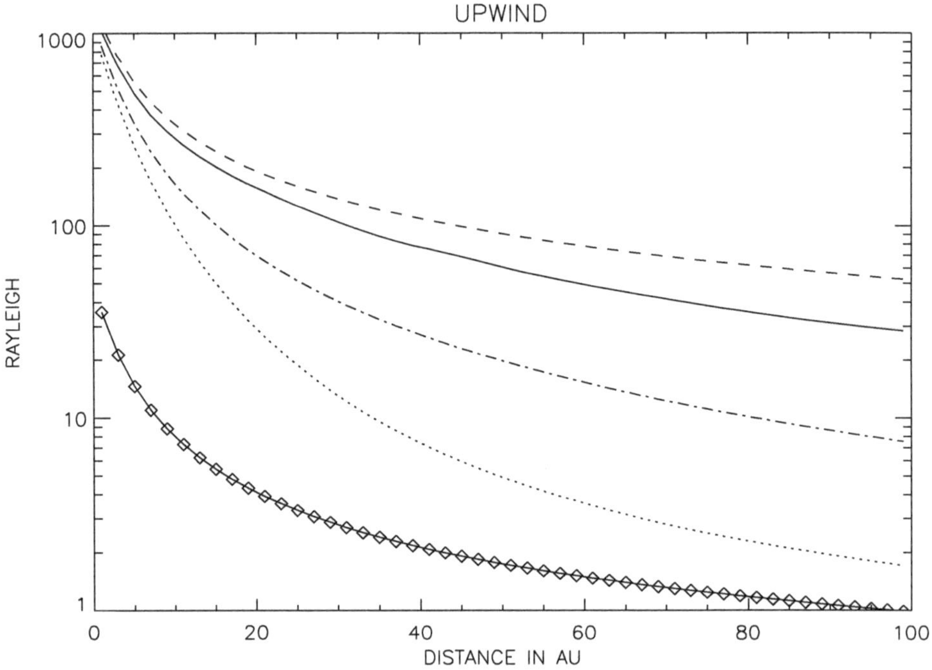

Fig. 4 Intensity as a function of solar distance in the Upwind direction. The full radiative transfer calculation is shown by the *solid line*. The optically thin calculation is shown by the *dotted line*. We have added the self-absorbed result (*dash-dot*) and the primary term (*dotted line*). The *diamonds* show the intensity due to the population created in the heliosheath

(extinction and multiple scattering), the emissivity varies with $1/r^2$ and the intensity as $1/r$, i.e. $\alpha = -1$. When extinction is considered, the contribution to the intensity gets closer to the spacecraft because of extinction and α tends asymptotically toward -2.

Hall et al. (1993) reported a value of -0.35 for the UVS data obtained between 15 and 35 AU. Such a value was considered to be a proof of the existence of the hydrogen wall. Later works by Quémerais et al. (2003) found that this value, between 15 and 35 AU is inconsistent with data obtained after 1993. Moreover, computations based on the Izmodenov et al. (2001) model distributions show that effects on the number density due to the wall are not yet visible at such distances. According to our assessment, this early detection was caused by an erroneous correction of the solar Lyman α flux. The flux corrections used by Hall et al. (1993) were based on the SME dataset. These values have been corrected after the publication of the SOLSTICE-UARS dataset (Woods et al. 2005). This correction not only changed the absolute values of the solar flux but also the ratio between solar minimum flux and solar maximum flux. Figure 5 shows the solar flux variations obtained from the latest data sets (Woods et al. 2005). These values have been used to correct the data for solar cycle flux variations. The diamonds show when UVS data in the upwind direction are available.

Figure 6 shows the data used in this analysis. The spectra have all the same line of sight within 20° of the upwind direction. The spectra are averages of all available individual spectra recorded in a day. Each individual spectrum has an integration time of 240 s. Each day, roughly 8 to 12 hours worth of data are recorded on the ground and used in the daily averages. The Y-axis unit is in counts per second. The statistical accuracy is very good, the number of counts for 240 s is about 100 counts, and over 10 hours this gives about

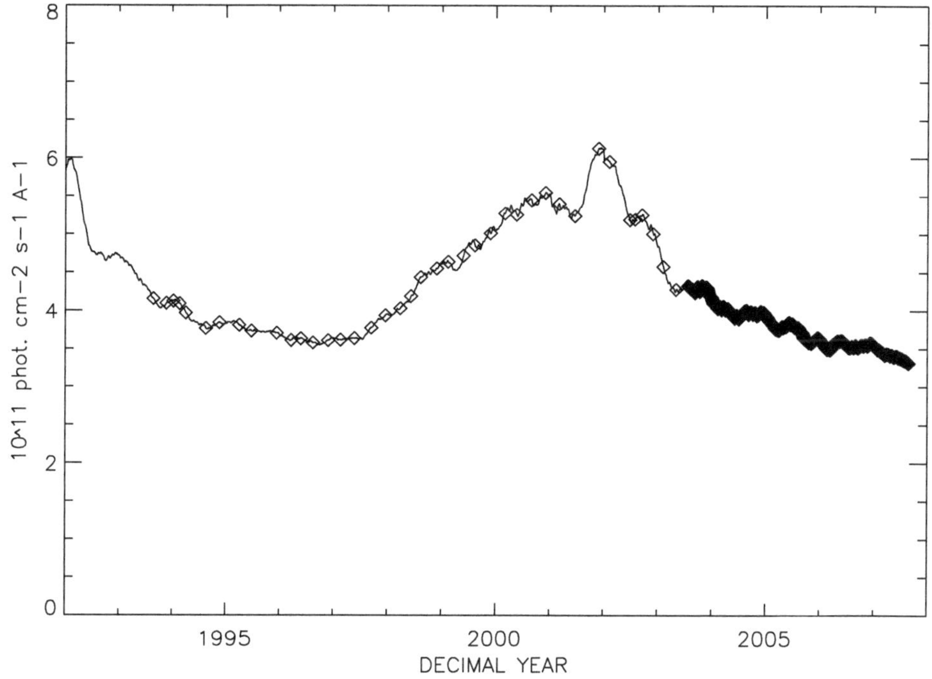

Fig. 5 Solar Lyman α flux variations based on the SOLSTICE-UARS and SOLSTICE-SORCE data sets (Woods et al. 2005). The values are smoothed over a solar rotation to account for multiple scattering averaging effects. The *diamonds* show the times when we have a UVS spectrum in the upwind direction

15000 counts, i.e. better than a 1% accuracy. Unfortunately, dark counts and fixed pattern noise corrections increase the uncertainty of the spectra to a few percent. The measurements cover the 1993–2007 period, more than a solar cycle. The solar cycle variations of the solar Lyman α flux clearly appear on the data. This plot shows how this correction is important for this analysis. During the period of time corresponding to these data, Voyager 1 covered a distance of more than 50 AU.

Figure 7 shows the data of Fig. 6 once the correction given by Fig. 5 is applied. The curves are normalized to the first values of the data set. We have added two model computations for comparison. The dotted line is the result for a model with constant density in the outer heliosphere (hot model) including multiple scattering. For such a model the radial coefficient is around -1.5 (Fig. 8 of Quémerais et al. 2003). The dashed line shows the multiple scattering computation in the case of a model with interface. The input hydrogen distribution was computed by Izmodenov et al. (2001). Similar results were found with distributions obtained from self-consistent models using time-dependent solar parameters (Izmodenov et al. 2008). In the case of the interface model, the radial intensity does not decrease as fast as in the case of constant density because of the positive gradient of hydrogen in the outer heliosphere. Still, the model cannot fit the data curve which appears nearly constant after 70 AU. At the present time, we do not have a model which has the same radial dependence as the data. To fit the data, it would be necessary to either bring the hydrogen wall closer or increase its peak value. It should be noted that the models used here do not include the deflection of the hydrogen flow. A full computation including this effect may improve the fit.

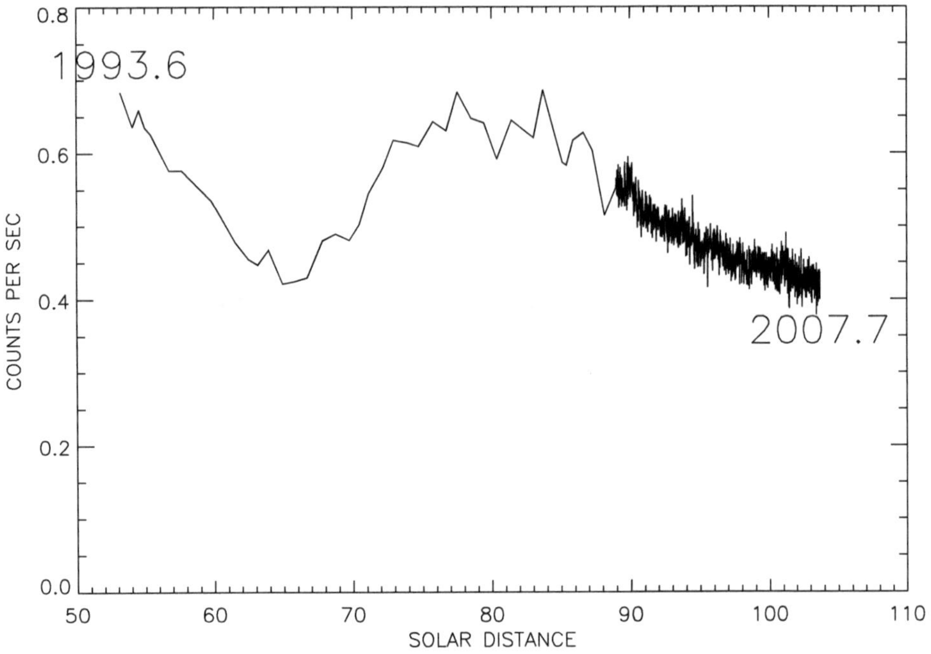

Fig. 6 V1-UVS upwind intensities as a function of solar distance. The values are in counts per second, daily averaged. The 11-year cycle variations of the solar illuminating flux are clearly visible. The values are shown in counts per second. Each individual spectrum has an integration time of 240 seconds and the average corresponds to a total integration of time of 8 to 12 hours in most cases

3 Line Profiles in the Inner Heliosphere

In the inner heliosphere, the hydrogen atoms are strongly affected by ionization processes from the Sun (charge exchange and photo-ionization), radiation pressure and the gravitational force of the Sun. This means that the IP background, as seen from one AU, is mainly affected by the Sun and its variability. It is therefore very difficult to study effects of the heliospheric interface by studying IP background intensities in the inner heliosphere.

As mentioned above, the flow of hydrogen atoms in the heliosphere is collisionless. This means that the different populations do not mix and that the changes in the distributions created at the interface do propagate inside the heliosphere. This is shown in Fig. 8 where we display a model computation of the upwind line profile which takes into account three populations of hydrogen atoms (Izmodenov et al. 2005a). The unperturbed interstellar H population is shown by the dash-dot line. This component crosses the heliospheric interface without interaction. In the vicinity of the Sun, however, slower atoms have a larger probability to get ionized which results in an apparent acceleration and cooling of the population by selection of the faster atoms. The second component is created by charge exchange in the outer heliosheath, i.e. by charge exchange between neutrals and the compressed interstellar plasma. This results in the slower and hotter component shown by the dashed line in Fig. 8. Finally, the third component is shown by the dotted line. This component is created by charge exchange with the solar wind protons beyond the termination shock, i.e. after the supersonic solar wind has been decelerated through the termination shock and strongly heated. The local temperature in this region is of the order of 10^5 K.

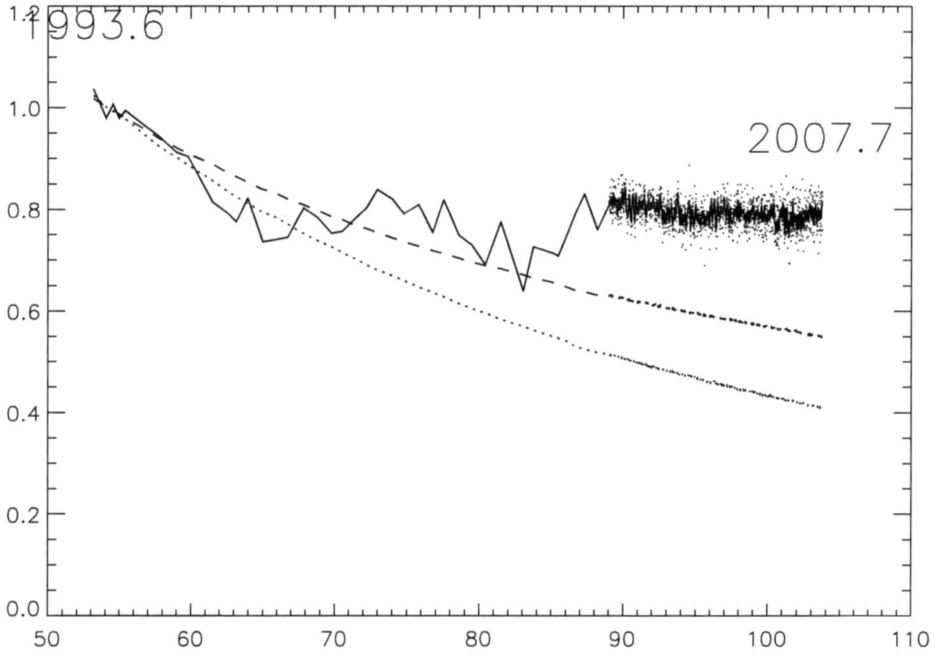

Fig. 7 V1-UVS upwind intensities corrected for flux variations as a function of solar distance. The x-axis shows distances in AU. The values are normalized to the values obtained at 55 AU (1993). The *dotted line* shows a multiple scattering model for a constant density in the outer heliosphere (hot model). The *dashed line* shows a multiple scattering model using the hydrogen distribution derived from an interface model. In this second case, the hydrogen wall creates a positive gradient of hydrogen density in the heliosphere which explains why the intensity is not decreasing as rapidly as in the case of constant density. Both models fail to represent the almost constant value obtain from the data after 70 AU from the Sun

Figure 9 shows an actual measurement of the upwind IP line profile obtained by STIS aboard Hubble Space Telescope in March 2001. In these data, it is not possible to identify the heliosheath component because the signal to noise ratio is not good enough. Note also that, according to the model in Fig. 8, the peak value of the heliosheath component is two orders of magnitude lower than the peak emission of the total line. Actually observing the hot component in the heliosheath would bring a strong constraint on the conditions in the inner heliosheath. However, it requires a very good signal to noise ratio which is not achievable by existing spaceborn UV spectrographs. This observation may be achieved by cumulating a large number of SWAN hydrogen cell observations (Quémerais et al. 1999).

4 Conclusion and Discussion

The analysis of the latest V1-UVS data confirms that the hydrogen distribution in the outer heliosphere is strongly affected by the heliospheric interface. However, we have not yet found a model which fits the data satisfyingly.

We see that, if we scale the data and model at the same value at about 50 AU from the Sun, then, at 100 AU from the sun, the data show an excess over the interface distribution model by about 10 R. The same exercise with a no shock model (constant density in the outer heliosphere) yields an excess of about 20 R.

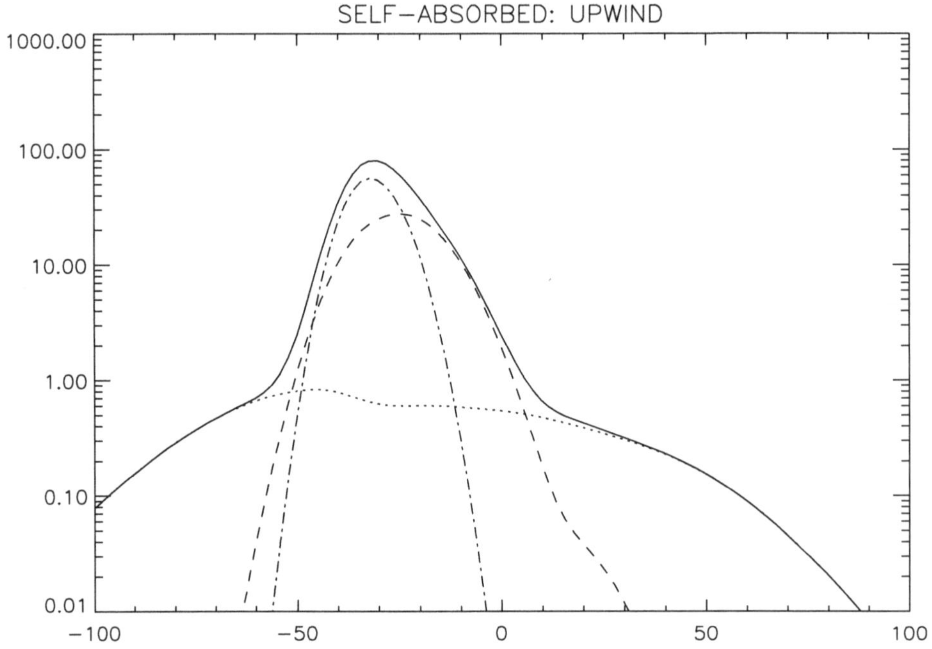

Fig. 8 Model of the Upwind intensity line profile showing the three hydrogen populations. The *dash-dot line* corresponds to the unperturbed interstellar hydrogen population. The *dashed line* shows the profile backscattered by the population created in the outer heliosheath. The *dotted line* shows the profile backscattered by the hot component created in the inner heliosheath. The *thick line* shows the sum with extinction on the line of sight taken into account. X-axis units are Doppler shifts in the solar rest frame expressed in km/s

To reconcile the data and our model, it would be necessary to increase the positive gradient of hydrogen number density in the outer heliosphere. This means either increase the peak value of the hydrogen wall or bring it closer to the Sun.

Our models do not include the deflection of the hydrogen flow at the interface. This may change our results in a positive way. For instance, the heliosphere scans performed before 2003 showed that the maximum intensity was not directly in the upwind direction but was shifted by about 20°. This may be a sign that the hydrogen distribution in the outer heliosphere is not symmetric around the mean hydrogen flow direction as assumed. To continue this analysis, we need detailed computations including a self-consistent model of the interface with the influence of an interstellar magnetic field. The direction of the magnetic field was defined by Lallement et al. (2005).

Our model considers only backscattering of solar photons as a significant source for the interplanetary background. However, there may be other sources. For instance, we estimated how much Lyman α photons may be produced after a charge transfer in the heliospheric interface region. The actual number depends on how probable a Lyman α photon is during the radiative cascade of the excited electron to the fundamental level. Just assuming the maximum number, i.e. one Lyman α photon for each charge transfer, we find a total intensity of 0.3 R which is an order of magnitude lower than the needed value. This value is overestimated because the actual probability to have a Lyman α photon is lower than one.

Emissions from outside the heliosphere, for instance emission after proton–electron recombination in HII regions, could also provide some answers. This should be investigated.

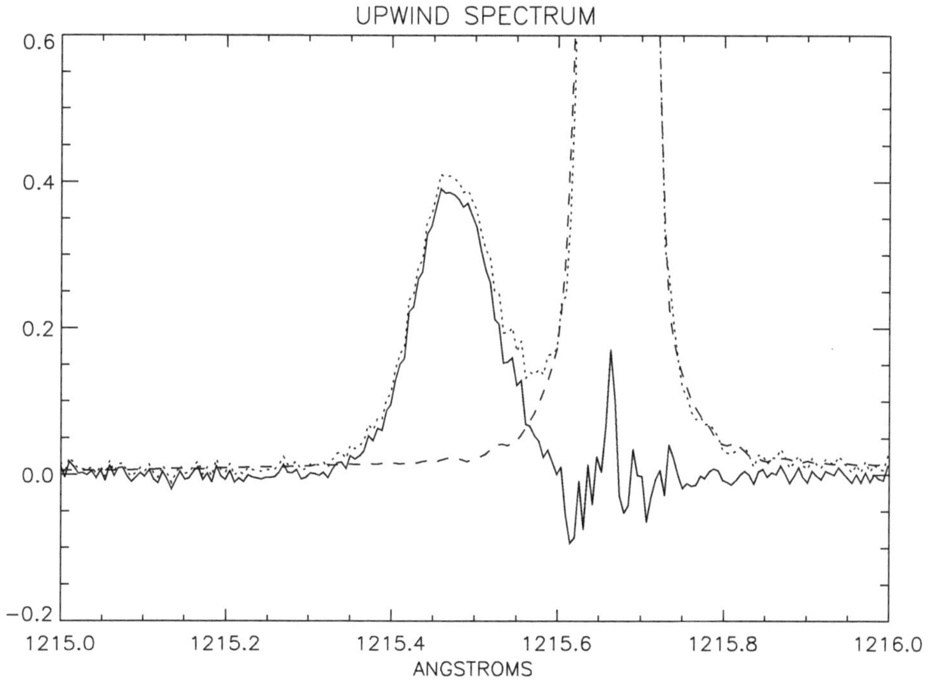

Fig. 9 Upwind line profile observed by STIS/HST on March 2001. The IP background data are shown by the *solid line*. The *dashed line* shows the geocoronal line which has been removed from the data. The original profile is shown by the *dotted line*. The geocoronal and interplanetary emissions are easily separated because of their large doppler shift due to the relative motions of the Earth and interplanetary hydrogen atoms. This shift is largest in March each year due to the direction of the Earth velocity vector. The *abscissa* is in wavelength unit in the Earth rest frame. Therefore, the geocoronal line is centered at 1215.66 Å

Finally, collisional excitation of H atoms in the heliosheath and beyond the heliopause should also be investigated.

Model computations of the interplanetary line profiles as seen from 1 AU suggest that there is a neutral "hot" component created in the inner heliosheath (Izmodenov et al. 2005a). Although the intensity of this component is two orders of magnitude lower than the core component, it is spectrally much wider than the core components and therefore is less absorbed at large distance with the Sun. The HST-STIS spectrum of the upwind IP background presented here doesn't show this hot component but uncertainties are larger than one percent. In future works, we plan to use SWAN-SOHO H cell measurements (Bertaux et al. 1995) to try to better constrain this "hot" component.

References

V.B. Baranov, Yu.G. Malama, Model of the solar wind interaction with the local interstellar medium: numerical solution of self-consistent problem. J. Geophys. Res. **98**(A9), 15157–15163 (1993)

J.L. Bertaux, J.E. Blamont, Evidence for an extra-terrestrial Lyman-alpha emission, the interstellar wind. Astron. Astrophys. **11**, 200 (1971)

J.L. Bertaux, E. Kyrölä, E.R. Quémerais, R. Lallement, W. Schmidt et al., SWAN: A study of solar wind anisotropies on SOHO with Lyman alpha sky mapping. Sol. Phys. **162**, 403–439 (1995)

D.T. Hall, D.E. Shemansky, D.L. Judge, P. Gangopadhyay, M.A. Gruntman, Heliospheric hydrogen beyond 15 AU—Evidence for a termination shock. J. Geophys. Res. **98**(A9), 15,185–15,192 (1993). (ISSN 0148-0227)

V. Izmodenov, M. Gruntman, Y.G. Malama, Interstellar hydrogen atom distribution function in the outer heliosphere. J. Geophys. Res. **106**(A6), 10681–10690 (2001)

V. Izmodenov, Y.G. Malama, M.S. Ruderman, Solar cycle influence on the interaction of the solar wind with Local Interstellar Cloud. Astron. Astrophys. **429**, 1069–1080 (2005a)

V.V. Izmodenov, Y.G. Malama, M.S. Ruderman, Modeling of the outer heliosphere with the realistic solar cycle. Adv. Space Res. **41**(2), 318–324 (2008)

R. Lallement, E. Quémerais, J.L. Bertaux, S. Ferron, D. Koutroumpa, R. Pellinen, Deflection of the interstellar neutral hydrogen flow across the heliospheric interface. Science **307**(5714), 1447–1449 (2005)

E. Quemerais, Y.G. Malama, B.R. Sandel, R. Lallement, J.-L. Bertaux, V.B. Baranov, Outer heliosphere Lyman alpha background derived from two-shock model hydrogen distributions: application to the Voyager UVS data. Astron. Astrophys. **308**, 279–289 (1996)

E. Quémerais, J.-L. Bertaux, R. Lallement, M. Berthé, E. Kyrölä, W. Schmidt, Interplanetary Lyman alpha line profiles derived from SWAN/SOHO hydrogen cell measurements: Full-sky velocity field. J. Geophys. Res. **104**(A6), 12585 (1999)

E. Quémerais, V. Izmodenov, Effects of the heliospheric interface on the interplanetary Lyman alpha glow seen at 1 AU from the Sun. Astron. Astrophys. **396**, 269–281 (2002)

E. Quémerais, J.-L. Bertaux, R. Lallement, B.R. Sandel, V. Izmodenov, Voyager 1/UVS Lyman a glow data from 1993 to 2003: Hydrogen distribution in the upwind outer heliosphere. J. Geophys. Res. **108**(A10), LIS4–1 (2003)

G.E. Thomas, R.F. Krassa, OGO 5 measurements of the Lyman alpha sky background. Astron. Astrophys. **11**, 218 (1971)

T.N. Woods, F.G. Eparvier, S.M. Bailey, P.C. Chamberlin, J. Lean, G.J. Rottman, S.C. Solomon, W.K. Tobiska, D.L. Woodraska, Solar EUV Experiment (SEE): Mission overview and first results. J. Geophys. Res. 110(A1) (2005). CiteID A01312

Composition of Interstellar Neutrals and the Origin of Anomalous Cosmic Rays

G. Gloeckler · L.A. Fisk · J. Geiss · M.E. Hill · D.C. Hamilton · R.B. Decker · S.M. Krimigis

Originally published in the journal Space Science Reviews, Volume 143, Nos 1–4, 163–175.
DOI: 10.1007/s11214-008-9482-5

Abstract Knowledge of the elemental composition of the interstellar gas is of fundamental importance for understanding galactic chemical evolution. In addition to spectroscopic determinations of certain element abundance ratios, measurements of the composition of interstellar pickup ions and Anomalous Cosmic Rays (ACRs) have provided the principal means to obtain this critical information. Recent advances in our understanding of particle acceleration processes in the heliosphere and measurements by the *Voyagers* of the energy spectra and composition of energetic particles in the heliosheath provide us with another means of determining the abundance of the neutral components of the local interstellar gas. Here we compare the composition at the termination shock of six elements obtained from measurements of (a) pickup ions at ~5 AU, (b) ACRs in the heliosphere at ~70 AU, and (c) energetic particles as well as (d) ACRs in the heliosheath at ~100 AU. We find consistency among these four sets of derived neutral abundances. The average interstellar neutral densities at the termination shock for H, N, O, Ne and Ar are found to be 0.055 ± 0.021 cm^{-3}, $(1.44 \pm 0.45) \times 10^{-5}$ cm^{-3}, $(6.46 \pm 1.89) \times 10^{-5}$ cm^{-3}, $(8.5 \pm 3.3) \times 10^{-6}$ cm^{-3}, and $(1.08 \pm 0.49) \times 10^{-7}$ cm^{-3}, respectively, assuming the He density is 0.0148 ± 0.002 cm^{-3}.

Keywords Interstellar medium: elemental composition · Anomalous cosmic rays: origin

G. Gloeckler (✉) · L.A. Fisk
Department of Oceanic, Atmospheric and Space Sciences, University of Michigan, Ann Arbor, MI 48109-2143, USA
e-mail: gglo@umich.edu

G. Gloeckler · D.C. Hamilton
Department of Physics, University of Maryland, College Park, MD 20742, USA

J. Geiss
International Space Science Institute, 3012 Bern, Switzerland

M.E. Hill · R.B. Decker · S.M. Krimigis
Applied Physics Laboratory, The Johns Hopkins University, Laurel, MD 20723, USA

DOI: 10.1007/978-1-4419-0247-4_13

1 Introduction

To obtain the composition of the local interstellar gas from *in situ* measurements of pickup ions in the inner heliosphere and energetic particles in the inner heliosheath requires a reasonable understanding of how these particles are created from the interstellar gas. The general scenario has been known for some time (Axford 1972; Fahr 1974). The neutral components of the interstellar gas, after some modifications due to charge exchange in the heliosheath (e.g. Izmodenov et al. 2005, 2004; Izmodenov 2009), penetrate deep into the heliosphere where they become progressively more and more ionized as they approach the Sun. This ionization process gradually reduces the density of the neutrals within several AU of the Sun and produces pickup ions. Möbius et al. (1985) discovered He^+ and later Gloeckler et al. (1993) observed pickup H^+ for the first time. From the velocity distributions and composition of interstellar pickup ions measured with the SWICS instrument on Ulysses at about 5 AU, the density of interstellar neutral H, ^{4}He, ^{3}He, N, O, and Ne near the termination shock ($\sim$100 AU) was estimated (Gloeckler et al. 1993, 2001; Gloeckler and Geiss 1998, 2004). Model calculations (e.g. Izmodenov et al. 2004) are then required to deduce (a) the likely composition of the neutral component in the local interstellar cloud (LIC), and finally (b) the composition of the ionized components in the LIC (e.g. Slavin and Frisch 2007) to provide the total elemental abundances of the local interstellar gas.

The composition of the Anomalous Cosmic Rays (see reviews by e.g. Klecker 1995; Jokipii 1998) has also been used to determine the relative density ratios of interstellar neutrals at the termination shock. Assuming that the Anomalous Cosmic Rays (ACRs) are pickup ions accelerated to high energies at the termination shock, Cummings et al. (2002) found that the He/O and especially H/O ratios were far below the values obtained from pickup ion measurements, and concluded that the diffusive shock acceleration mechanism they invoked required a mass-dependent injection and/or acceleration. However, when the *Voyagers* crossed the termination shock (TS), no evidence was found that diffusive shock acceleration produced ACRs (Stone et al. 2005, 2008), and alternative mechanisms and/or locations were suggested for the origin of these highly energetic pickup ions or ACRs (e.g. McComas and Schwadron 2006; Fisk and Gloeckler 2008a; Florinski 2009).

To estimate density ratios of interstellar neutrals near the termination shock using measurements of low energy particles in the heliosheath and of ACRs, we will use the Fisk and Gloeckler (2008a) model for the origin of ACRs. This model is based on the theory of Fisk and Gloeckler (2006, 2007, 2008b) which explains the extensive observations in the inner heliosphere of the commonly observed suprathermal -5 power law tails (e.g. Gloeckler et al. 2008 and references therein). Suprathermal power law tails with spectral index of -5 and a gentle rollover at an e-folding energy of $\sim$0.05–5 MeV/nuc on the velocity distributions of the solar wind and interstellar pickup ions are observed repeatedly not only during quiet times and in polar coronal holes at solar minimum, but also during quasi-steady or spatially homogeneous disturbed periods (e.g. downstream of shocks or in corotating interaction regions). For example, the proton distribution function (Fig. 1), averaged over many such disturbed time periods during all of 1998, has an extended power law tail above $W \approx 2$ with spectral index -5 (in the solar wind frame) and a gentle rollover at about 50 keV. The only time the -5 spectra are not observed is during rapidly changing (spatially inhomogeneous) solar wind flows, and for field-aligned, highly anisotropic beams upstream of shocks (e.g. the termination shock, see Gloeckler and Fisk 2006).

Fisk and Gloeckler (2006, 2007, 2008b) found that -5 power law tails are formed by compressional turbulence in the bulk solar wind acting on a thermodynamically isolated system, or equivalently a spatially homogeneous system of core (pickup plus halo solar wind

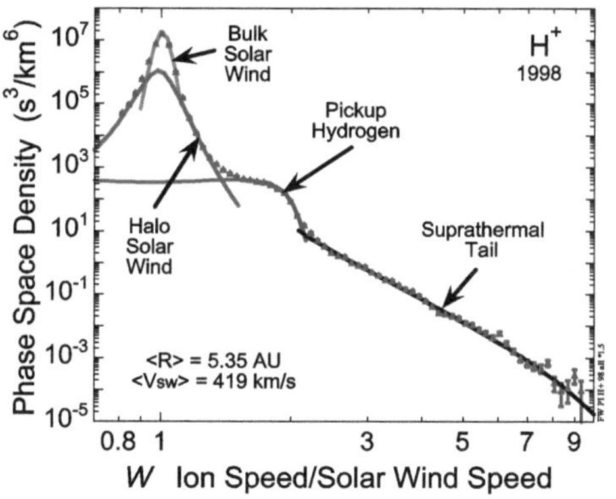

Fig. 1 Time-averaged (all of 1998) velocity distribution of H^+ (*filled symbols*), measured by SWICS on *Ulysses* in the slow, near-ecliptic solar wind at ~5 AU, consisting of the bulk (cold) and halo (hot) solar wind, pickup protons with a cutoff at $W \approx 2$, and the -5 power law suprathermal tail with a gentle roll-over at $W \approx 7$. The *curves* are model distributions

ions, see Fig. 1 caption for definitions) and tail particles. Energy flows from core to tail to produce a -5 power law spectrum (in the solar wind frame) when equilibrium is reached. The rollover energy depends on ambient solar wind conditions but in the heliosphere is limited to ~10 MeV/nuc because of adiabatic cooling in the expanding solar wind. Here we will apply these concepts to show that the ACRs are most likely an evolution of suprathermal -5 power law tails, convected from the TS to the heliopause, where they extend to energies of several hundred MeV/nuc. By fitting modulated -5 tail spectra to the ACR spectra measured both in the heliosheath and in the outer heliosphere by *Voyager 1*, we then derive the relative abundances of interstellar neutrals at the TS and compare them with updated abundances obtained from pickup ion measurements by SWICS on *Ulysses*.

2 Interstellar Neutral Densities at ~90–100 AU

The best estimates of absolute number densities of interstellar neutrals near the termination shock are obtained from direct measurements of these low energy (few eV/nuc) neutrals. So far this has been done only for He at several AU from the Sun, giving density values (after loss rate corrections) at the TS of 0.012–0.016 cm^{-3} (Witte et al. 2004). Interstellar neutral densities for H, He, N, O and Ne have up to now been obtained primarily from pickup ion measurements (e.g. Gloeckler and Geiss 2004). Estimates of absolute neutral densities obtained from pickup ions require knowledge of the total loss rates of neutrals, the total production rates of pickup ions and instrument specific parameters such as geometrical factor and counting efficiencies. The average total loss rate can be deduced from the measured pickup ion spectral shapes. The best estimate for the He number density (0.015 ± 0.0015) cm^{-3} at the TS was obtained by Gloeckler and Geiss (2004) using pickup He^{++} and a production rate determined from the solar wind He flux. Because both the solar wind alpha particles and the He^{++} pickup ions were measured simultaneously with the same instrument (SWICS on *Ulysses*), systematic geometrical factor uncertainties (approximately ±25%) cancelled out. This method is unfortunately not possible for other species. (See also Möbius et al. (2004) and references therein for consensus values of interstellar He parameters, such as number density and temperature.)

2.1 Abundance of Interstellar Neutrals at the TS from Pickup Ions

The process of finding interstellar neutral abundances from interstellar pickup ions begins with measurements of their distribution functions averaged over sufficiently long time periods to obtain reasonable statistical accuracy and to average out fluctuations produced by

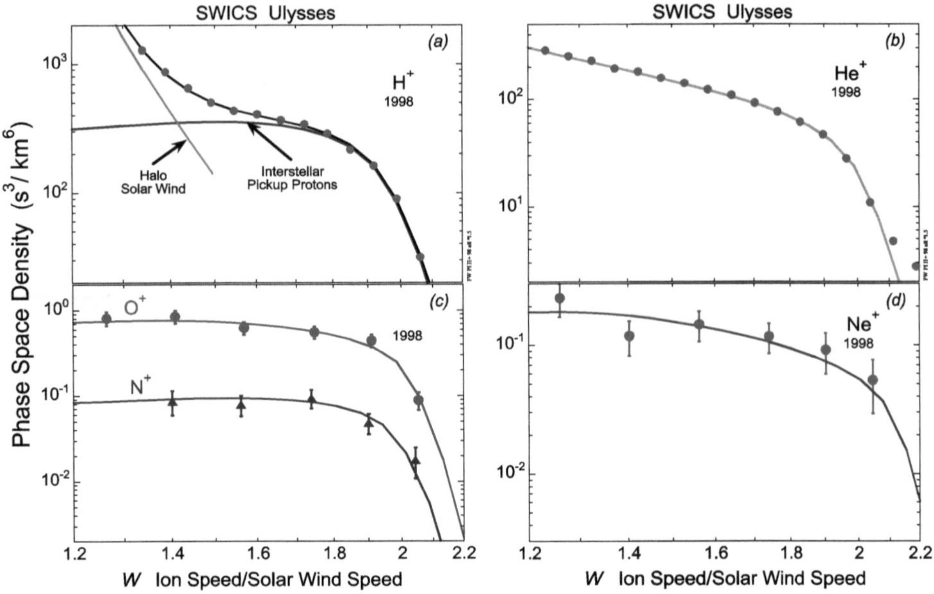

Fig. 2 Velocity distributions (symbols) of (**a**) H^+, (**b**) He^+, (**c**) N^+ and O^+, and (**d**) Ne^+ averaged over the entire one-year period in 1998, with no time periods excluded, along with model curves computed as described in the text. For H^+ (panel **a**) the halo solar wind (see Fig. 1) contributed significantly to the phase space density below $W \approx 1.6$. The pickup ion number densities, obtained by integrating the solar wind frame pickup ion velocity distributions, are listed in row 4 of Table 1

compressions and shocks in the solar wind. We show in Fig. 2 one-year averaged velocity distributions for H^+, He^+, N^+, O^+ and Ne^+ measured with SWICS on *Ulysses* in the slow solar wind at about 5 AU near the ecliptic. Model curves were computed using the standard hot model to find the variation of neutral densities as a function of radial distance from the Sun, with $\mu = 0$ ($\mu = 1.0$ for H) and values of 22, 26.3, 26.3, 24 and 26.3 km/s for the average relative speed, and of 12000, 6300, 6300, 8000 and 6300 K for the average gas temperature for H, He, N, O and Ne, respectively. For each element we assumed the same values for the production ionization rate ($\beta_{\rm prod}$) as for the loss ionization rate ($\beta_{\rm loss}$), and calculated the spacecraft-frame phase space densities as described in Gloeckler et al. (2001). For each element, $\beta_{\rm loss}$ and the neutral density at the TS are then varied until the best fit (minimum χ^2) to the pickup ion portions of the respective measured velocity distributions is obtained. The best estimates of the absolute interstellar gas densities at the TS, $N_{\rm PI}$, and the corresponding values of $\beta_{\rm prod} = \beta_{\rm loss}$ are listed in rows 2 and 3 of Table 1, respectively. With the assumption that $\beta_{\rm prod} = \beta_{\rm loss}$ the changes in densities ($\Delta N_{\rm P}$) of all elements (except He) with changes ($\Delta\beta$) in $\beta_{\rm prod} = \beta_{\rm loss}$ are small, ($\Delta N_{\rm P}/N_{\rm P} \ll \Delta\beta/\beta_{\rm loss}$), as demonstrated for H by Bzowski et al. (2008). However, because our assumption that $\beta_{\rm prod} = \beta_{\rm loss}$ may not hold, we included the estimated errors in $\beta_{\rm prod}$ listed in row 3 of Table 1 in our computation of uncertainties of the interstellar densities listed in row 2 of Table 1. We note that the values for $\beta_{\rm loss}$ obtained from fits to the spectra are the total rates (at 1 AU), the sum of charge exchange, photo-ionization and electron impact ionization, and that the one-year averaged velocity distributions used here included many disturbed time periods (e.g. downstream of shocks) during which electron impact ionization becomes important (Isenberg and Feldman 1995), especially for He, N and Ne, and thus contributes to the total loss rates.

Table 1 Interstellar neutral densities near the termination shock derived from pickup ions

Quantity	Element				
	H	He	N	O	Ne
Interstellar density[a] (cm^{-3})	0.067 ± 0.016	0.018 ± 0.003	$(1.48 \pm 0.49) \times 10^{-5}$	$(7.85 \pm 1.7) \times 10^{-5}$	$(1.26 \pm 0.43) \times 10^{-5}$
Interstellar density ratios	3.72 ± 1.08	$=1.000$	$(8.2 \pm 3.1) \times 10^{-4}$	0.0044 ± 0.0012	$(7.0 \pm 2.7) \times 10^{-4}$
$\beta_{prod} = \beta_{loss}$ (s^{-1})	$(6.5 \pm 1.5) \times 10^{-7}$	$(1.5 \pm 0.25) \times 10^{-7}$	$(10 \pm 2) \times 10^{-7}$	$(6 \pm 1) \times 10^{-7}$	$(3 \pm 0.7) \times 10^{-7}$
Pickup ion density (cm^{-3})	$(3.5 \pm 0.6) \times 10^{-4}$	$(1.2 \pm 0.2) \times 10^{-4}$	$(9.8 \pm 3.2) \times 10^{-8}$	$(6.5 \pm 1.4) \times 10^{-7}$	$(1.1 \pm 0.4) \times 10^{-7}$

[a]Systematic errors of $\sim \pm 25\%$ due to uncertainties in the SWICS geometrical factor are not included in the density errors in row 2

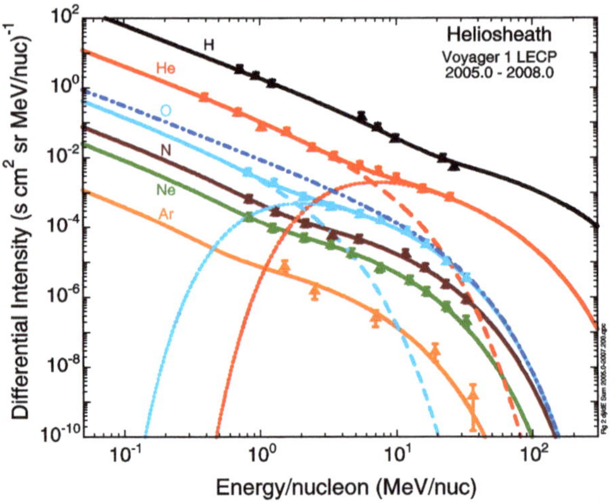

Fig. 3 Three-year averaged differential intensities of H, He, O, N, Ne and Ar measured with the Low Energy Particle Telescope (LEPT) of the LECP experiment on *Voyager 1*. LEPT is a dE/dx versus E telescope with ultra-thin front solid-state detectors (Krimigis et al. 1977). The ion spectrum (mostly protons, not shown here but see Decker et al. 2005, 2008, and Fig. 2 in Gloeckler et al. 2008), measured down to the low energy limit of ~40 keV of LECP, is a −1.5 power law. *Dotted curves* are the modulated ACR spectra and *dashed curves* the suprathermal tail spectra (shown for He and O). The *dashed-dotted curve* is the predicted shape of the unmodulated ACR oxygen spectrum near the heliopause. The *solid curves* are the sum of the tail and modulated ACR spectra. Equations used to fit all of the data are given in the text

2.2 Tails and ACRs in the Inner Heliosheath

It is now well established that ACRs are accelerated interstellar pickup ions, as first proposed by Fisk et al. (1974). Clues on how and where these pickup ions are accelerated are provided by *Voyager* measurements of spectra in the inner heliosheath shown for six elements in Fig. 3. We have used these data to guide us in developing our model for formation of suprathermal tails (Fisk and Gloeckler 2007) and the acceleration of ACRs (Fisk and Gloeckler 2008a). In the framework of these models, we will derive here the relative abundances of interstellar neutrals at the termination shock from *Voyager 1* measurements of suprathermal tails and ACRs in the heliosheath as well as ACR spectra obtained in the outer heliosphere.

Observations with the Low Energy Charged Particle (LECP) experiment (Krimigis et al. 1977) on *Voyager* revealed the steady, low-energy differential energy spectra that are power laws with index −1.5 down to the ~30 to 40 keV lower limit of LECP (e.g. Decker et al. 2005, 2008; Hill et al. 2006). (Such spectra when expressed as velocity distributions are power laws with index −5.) Fisk and Gloeckler (2008b, 2008a) argued that these spectra are formed by the same mechanism that produces suprathermal −5 power law tails observed consistently in the heliosphere. These are simply the suprathermal −5 tails with a gentle rollover starting at several MeV formed from the heliosheath pickup ion distributions heated by the TS, and have the form

$$dj/dE = j_{At} E^{-1.5} \exp\left[-A^{\alpha}(E/E_{ot})^{(\alpha+1)/2}\right], \tag{1}$$

where A is the mass number and α and E_{ot} are parameters determined by fitting (1) to the six spectra shown in Fig. 3. The only difference between the heliosheath proton tail in Fig. 3.

and that shown in Fig. 1 is that the intensity (j_t), is a factor of ~15 to 30 times higher, and thus easy to observe. Fisk and Gloeckler (2007) have shown that the pressure P_{At} in the heliosheath tails of particles with mass number A is related to the pressure P_{Ac} of pickup or core ions by

$$P_{At}/P_{Ac} = 0.4\beta/(1+\beta) = \varepsilon, \tag{2}$$

where $\beta = \delta P_c/P_c \approx 1$, and δP_c is the rms pressure of the core (pickup) particles. It can be shown that (2) holds for all elements. The pressure P_{At} of heliosheath tail particles is obtained by integrating the -5 power law velocity distributions from v_l, the low energy threshold to v_2, the high energy roll over e-folding speed. From (1) $v_2 \propto A^{-\alpha/(\alpha+1)}$, and Fisk and Gloeckler (2008b) showed that $v_l \propto A^{-\alpha/(\alpha+1)}$, with both v_l and v_2 depending on the spatial diffusion coefficient which may be expressed as $\kappa = \kappa_o A^\alpha E^{(\alpha+1)/2}$. Thus, $P_{At} \propto A j_{At} \ln(v_2/v_1) \propto A j_{At}$, and since from (2) $P_{At} = \varepsilon P_{Ac}$ and $P_{Ac} \propto A N_A \beta_{\text{Aprod}}$, where N_A is the neutral density near the TS, $N_A \propto j_{At}/\beta_{\text{Aprod}}$. However, if the index α dependents on energy and changes to γ at some lower energy, then the threshold v_l is lowered and additional pressure is injected into the tail. In that case, with $\delta = [\alpha/(\alpha+1) - \gamma/(\gamma+1)]/C$, where $C = \ln(v_2/v_1)$,

$$N_A = C \times j_{At} \times (1 - \delta \ln A)/\beta_{\text{Aprod}}. \tag{3}$$

Our interpretation of the high energy portions of the *Voyager* spectra shown in Fig. 3 is based on a model (Fisk and Gloeckler 2008b, 2008a) where we envision the rollover of the tail spectra to move to higher energies, particularly close to the heliopause (at ~140 AU) where additional energy is injected into the tail and E_o reaches several hundred MeV/nuc. This can occur in the heliosheath because adiabatic cooling is small, unlike in the supersonic solar wind where the cooling limits the rollover energy to less than ~10 MeV/nuc. Indeed, Hill et al. (2006) found evidence for such an increase of E_o in their *Voyager* data. Thus, the differential intensity spectra near the heliopause at ~140 AU are -1.5 power laws with a rollover e-folding energy of ~150 to 200 MeV/nuc (1). These are the ACRs. Observed closer in, such as in the heliosheath at ~100 AU or in the outer heliosphere, these ACR spectra will be modulated with ever stronger decrease in intensity with decreasing particle rigidity.

In Fig. 3 are shown the differential intensities of H, He, N, O, Ne and Ar, elements that have significant neutral components of the interstellar gas. The curves are fits to the observed spectrum of each element. For the low-energy component, the suprathermal tail spectrum (dashed curve in Fig. 2), we use (1). The high-energy components of the spectra are the modulated ACR spectra, i.e., the tail spectra that the *Voyagers* should measure near the heliopause (see dashed oxygen spectrum of Fig. 3) multiplied by a rigidity (mass/charge) dependent modulation function of the form $\exp[-\lambda A^{-\xi} E^{-(\xi+1)/2}]$, with the energy E expressed in MeV/nucleon. This model for the acceleration and modulation of ACRs in the heliosheath is described in more detail in Fisk and Gloeckler (2008a). All of the model ACR modulated spectra shown and used in Fig. 3 (the dotted curves) are expressed by

$$dj/dE = j_{\text{AACR}} E^{-1.5} \exp\left[-A^\alpha (E/E_{o\text{ACR}})^{(\alpha+1)/2}\right] \times \exp\left[-\lambda A^{-\alpha} E^{-(\alpha+1)/2}\right]. \tag{4}$$

The solid curves that are the sums of (1) and (4) fit all the spectra well. Values of j_A (subscript A denotes the mass/charge of the various elements) are given in rows 2 and 3 of Table 2. We note that the tail spectrum for Ar and the modulated heliosheath ACR spectrum for H are not well determined because they fall mostly outside the range of the LECP measurements. A single value for parameter $\alpha = 0.87$ was used for all spectra, $E_{o\text{tail}} = 17$ MeV/nuc

Table 2 Values for j_A used in (1) and (4) to fit the spectra shown in Figs. 3 and 4

Region/Parameter	j_{H}	j_{He}	j_{N}	j_{O}	j_{Ne}	j_{Ar}
Heliosheath tails	2.0 ± 0.1	0.140 ± 006	$(9.0 \pm 0.8) \times 10^{-4}$	$(5.0 \pm 0.35) \times 10^{-3}$	$(3.0 \pm 0.27) \times 10^{-4}$	$(1.5 \pm 1.1) \times 10^{-5}$
Heliosheath ACRs	4.0 ± 2.8	0.28 ± 0.015	$(1.8 \pm 0.08) \times 10^{-3}$	$(1.0 \pm 0.04) \times 10^{-2}$	$(6.0 \pm 0.3) \times 10^{-4}$	$(3.0 \pm 0.6) \times 10^{-5}$
Heliosphere ACRs	0.95 ± 0.07	0.10 ± 0.004	$(7.3 \pm 0.4) \times 10^{-4}$	$(4.5 \pm 0.25) \times 10^{-3}$	$(2.2 \pm 0.2) \times 10^{-4}$	$(9.5 \pm 0.4) \times 10^{-6}$

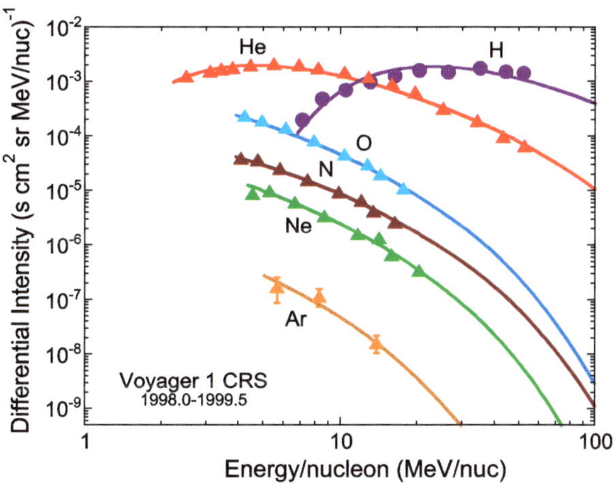

Fig. 4 One-and-a-half-year averaged differential intensities of H, He, O, N, Ne and Ar measured with the CRS instrument on *Voyager 1* in the outer heliosphere at a mean distance of 72 AU (*filled symbols*). The highest energy H intensities may have contributions from galactic cosmic rays. Data are taken from Fig. 4 of Cummings et al. (2002)

and $E_{o\text{ACR}} = 160$ MeV/nuc were used for the local heliosheath tail near the TS and ACR spectra respectively, and for the modulation parameter $\lambda = 38$ was used for the heliosheath ACRs.

2.3 ACRs in the Outer Heliosheath

The ACR spectra measured by the CRS instrument on *Voyager 1* (Cummings et al. 2002) are displayed in Fig. 4 along with model fits using (4) multiplied by an additional modulation function, $\exp[-65A^{-2}E^{-3/2}]$, that approximates modulation of the spectra in the heliosphere. Good fits to the measured spectra are obtained for all six elements using values for j_A given in row 4 of Table 2. The value used for parameters α, $E_{o\text{ACR}}$, and the modulation parameter was λ 0.87, 160 MeV/nuc and 15, respectively. The only parameter value that was different from those used to fit the heliosheath ACRs was λ. This is not surprising because the two measurements were taken during different time periods and thus different phases of the solar cycle.

2.4 Abundance of Interstellar Neutrals from Tails and ACRs

Accurate absolute densities of interstellar neutrals cannot be obtained at this time from measured abundances of ACRs or suprathermal tails in the heliosheath because we do not yet know with precision such parameters as the particle speed of the threshold between the core and the tail or exactly how much the pressure of the core or pickup ions increases when crossing the TS, or the exact fraction of the pickup ion pressure that flows into the tail. However, the relative abundance can be determined accurately because the constant C in (3) does not depend on the particle mass A (Fisk and Gloeckler 2007, 2008b, 2008a).

In Table 3 we list values for the interstellar neutral density ratios near the TS derived from composition measurements (j_A values from Table 2) of suprathermal tails measured in the heliosheath at ~100 AU (column 3), ACRs measured in the heliosheath at ~100 AU (column 4) and ACRs measured in the heliosphere at ~72 AU (column 5), using (3) with values for the respective production rates, $\beta_{A\text{prod}}$ (normalized to 1 AU) listed in column 2. The production rates for H and He were taken from row 3, columns 2 and 3, respectively of Table 1. For all other elements we used the 21-year averaged charge exchange plus photo-ionization rates (Cummings et al. 2002) to which we added the respective electron impact ionization

Table 3 Interstellar neutral density ratios near the termination shock derived from *Voyager 1* composition measurements, and average densities (normalized to the He density of 0.0148 ± 0.002 cm^3) computed from these and pickup-ion-derived density ratios

Element	β_{prod} (s^{-1})	Density ratios of neutrals at the termination shock			Average
		100 AU tails	100 AU ACRs	72 AU ACRs	density[a]
H	$(7 \pm 1) \times 10^{-7}$	3.67 ± 0.84	–	–	0.055 ± 0.021
He	$(1.5 \pm 0.25) \times 10^{-7}$	=1.000	=1.000	=1.000	0.0148 ± 0.002
N	$(7.5 \pm 0.22) \times 10^{-7}$	$(1.05 \pm 0.37) \times 10^{-3}$	$(1.05 \pm 0.36) \times 10^{-3}$	$(9.7 \pm 3.4) \times 10^{-4}$	$(1.44 \pm 0.45) \times 10^{-5}$
O	$(9.8 \pm 0.26) \times 10^{-7}$	$(4.4 \pm 1.4) \times 10^{-3}$	$(4.4 \pm 1.4) \times 10^{-3}$	$(4.4 \pm 1.4) \times 10^{-3}$	$(6.46 \pm 1.89) \times 10^{-5}$
Ne	$(4.3 \pm 0.22) \times 10^{-7}$	$(5.8 \pm 3.2) \times 10^{-4}$	$(5.8 \pm 3.2) \times 10^{-4}$	$(4.5 \pm 2.5) \times 10^{-4}$	$(8.5 \pm 3.3) \times 10^{-6}$
Ar	$(1.2 \pm 0.36) \times 10^{-6}$	–	$(9.3 \pm 3.8) \times 10^{-6}$	$(5.2 \pm 2.8) \times 10^{-6}$	$(1.08 \pm 0.49) \times 10^{-7}$

[a]Computed by first weight-averaging the respective *Voyager 1* derived density ratios of columns 3, 4 and 5, and the pickup ion derived ratios (column 3 of Table 1), and then normalizing to the helium density of 0.0148 ± 0.002 cm^{-3} of Gloeckler and Geiss (2004) and Möbius et al. (2004). Uncertainties include all errors listed for the ratios as well as the 10% error in the He density

rates computed using the Voronov (1997) formulae with electron density of 1.3 cm^{-3} and temperature of 5×10^6 K. The δ values of 0.12, 0.12 and 0.189 for tails at 100 AU, ACRs at 100 AU and ACRs at 72 AU, respectively, were obtained using (3), and requiring that the O/He interstellar density ratios derived from all three heliosheath and heliosphere local tail and ACR spectra be equal to the pickup ion derived O/He ratio of 4.4×10^{-3}.

3 Discussion and Summary

We derived the average number densities of interstellar neutrals at the termination shock (~90–100 AU) for six elements, including Ar, from four different samples of processed interstellar gas. The first sample comes from interstellar pickup ions measured during all of 1998 by *Ulysses* at ~5 AU. Interstellar abundances at the TS are obtained using the standard hot model with loss ionization rates derived from best fits to the pickup ion spectral shapes and values of the production ionization rates that are assumed to be equal to the respective loss rates. To eliminate the large systematic errors due to uncertainties of the SWICS instrument geometrical factor, we then calculate density ratios relative to He (row 3 of Table 1).

The remaining dominant source of uncertainties that limits the accuracy of the derived interstellar neutral abundances from pickup ion measurements is our poor knowledge of both the loss rates of interstellar neutrals as well as the production rates of pickup ions from these neutrals in inner heliosphere. By assuming that the production rate is equal to the loss rate we were left only with uncertainties in the loss rates, which we determine from the spectral shapes of the measured pickup ion velocity distributions. We emphasize that the loss and production rates obtained from spectral shapes include contributions from electron impact ionization, which may be significant during the many disturbed periods in 1998 that included turbulent regions downstream of shocks (Isenberg and Feldman 1995). This leads to values for ionization rates that are larger (especially for He) than those obtained using only charge exchange and photo-ionization. In computing model pickup ion reduced velocity distributions at the mean radial distance of Ulysses in 1998, we make the standard assumption that the pickup ions are convected with the solar wind and that the distributions in the solar wind frame are isotropic. Under different assumptions concerning the nature of the pickup ion distribution functions (e.g. asymmetric distributions) the shapes of the computed velocity distributions would be different as would be respective values of the total ionization rates necessary to fit the observed spectra. Because this systematic error in the ionization rates would affect all of the derived neutral densities in the same way (either increase all or decrease all, but not necessarily by the same factor), the density ratios relative to He would be less affected.

The other three samples were taken in the outer heliosphere and the inner heliosheath by *Voyager 1*. We estimated neutral interstellar densities from measured relative abundances in the three *Voyager 1* samples in the framework of the theory of Fisk and Gloeckler (2006, 2007, 2008b, 2008a) for the origin of suprathermal tails and ACRs. Our model for the origin of ACRs, build on concepts developed to explain the −5 power law tails commonly observed in the inner heliosphere, starts with the interstellar pickup ion distributions upstream of the TS whose pressure is increased by the Rankine-Hugoniot jump in crossing the termination shock. Suprathermal power law tails with spectral index −5 (−1.5 in the differential intensity representation) with a rollover at a few to tens of MeV/nucleon are then established due to the strong compressional turbulence of the plasma in the heliosheath. As these tails are convected outward, the rollover e-folding energy increases, most strongly near the heliopause. Additional energy is injected into the tail, primarily near the heliopause, where particles are accelerated to energies of several hundred MeV/nucleon. These -1.5 power law spectra near the heliopause are the ACRs, and we predict that the *Voyagers* will measure -1.5 ACR spectra when they reach ~130–150 AU. The density ratios of the neutral interstellar gas at the TS are then obtained from the composition measured in the tails in the heliosheath near the TS, and in the modulated ACRs in the outer heliosphere and heliosheath. We note that our theory is able to relate the composition of all elements measured in the ACRs and in the tails to that of pickup ions without invoking an unknown mass dependent shock acceleration/injection mechanism.

The dominant source of uncertainties that limits the accuracy of the derived interstellar neutral abundances from heliosheath suprathermal tails and ACRs is poor knowledge of the production rates of pickup ions from interstellar neutrals upstream of the termination shock. It is difficult to determine these rates from measurements near 1 AU. However, future 3-D and time-dependent modeling of spatial and temporal variations of the total ionization rates should reduce these uncertainties.

The H density obtained using two different methods and data sets, i.e. pickup ions and the suprathermal tails in the heliosheath (0.056 ± 0.023 and 0.054 ± 0.02 cm^{-3} respectively, assuming the He density to be 0.0148 ± 0.002 cm^{-3}) are almost identical to one another but lower than previously published values derived from pickup ions, which ranged from 0.077 ± 0.015 cm^{-3} (Gloeckler et al. 1993) to 0.115 ± 0.015 cm^{-3} (Gloeckler et al. 2001), and those obtained from analysis of the solar wind slowdown, which ranged from 0.075 ± 0.015 cm^{-3} to 0.11 ± 0.015 cm^{-3} (see Richardson et al. 2008). The reason for the difference is that, unlike in the present analysis, listed errors of previously published pickup-ion-derived H densities were primarily statistical and did not include the much larger systematic uncertainties in the loss and production rates used in the hot model nor systematic errors (estimated to be roughly ±25%) in the geometrical factor of the SWICS instrument. Including these systematic errors in the older values puts our present lower values within the error limits of previously published absolute interstellar neutral hydrogen densities at the TS.

We conclude by noting that our new lower value for the interstellar neutral hydrogen at the TS could have consequences for pressure balance between the heliosphere and the local interstellar medium and may require stronger interstellar magnetic fields to place the location of the termination shock at distances observed by the Voyager spacecraft.

Acknowledgements We are most grateful to E. Möbius for carefully reading the manuscript and for his numerous excellent suggestions for improving the paper. This work was supported, in part, by NSF Grant ATM 0632471, by the *Voyager* Guest Investigator Grant NNX07AH, and by *ACE* data analysis contract, 44A-1080828.

References

W.I. Axford, 1972, in *Solar Wind*, ed. by E.P. Sonnett, P.J. Coleman, J.M. Wilcox, NASA **SP-308**, 609
M. Bzowski, E. Möbius, S. Tarnopolski et al., Astron. Astrophys. (2008, in press)
A.C. Cummings, E.C. Stone, C.D. Steenberg, Astrophys. J. **578**, 194 (2002)
R.B. Decker, S.M. Krimigis, Roelof et al., Science **309**, 2020 (2005)
R.B. Decker, S.M. Krimigis, Roelof et al., Nature **454**, 67–70 (2008)
H.J. Fahr, Space Sci. Rev. **15**, 483 (1974)
L.A. Fisk, B. Kozlovsky, R. Ramaty, Astrophys. J. **190**, L35–L37 (1974)
L.A. Fisk, G. Gloeckler, Astrophys. J. **640**, L79 (2006)
L.A. Fisk, G. Gloeckler, Proc. Nat. Acad. Sci. **104**, 5749 (2007)
L.A. Fisk, G. Gloeckler, Adv. Space Res. (2008a, submitted)

L.A. Fisk, G. Gloeckler, Astrophys. J. **686**, 1466–1473 (2008b)
Florinski, Space Sci. Rev. (2009, this issue)
G. Gloeckler, L.A. Fisk, Astrophys. J. **648**, L63 (2006)
G. Gloeckler, J. Geiss, Space Sci. Rev. **86**, 127–159 (1998)
G. Gloeckler, J. Geiss, Adv. Space Res. **34**, 53–60 (2004)
G. Gloeckler, J. Geiss, H. Balsiger et al., Science **261**, 70–73 (1993)
G. Gloeckler, J. Geiss, L.A. Fisk, in *The Heliosphere near Solar Minimum: the Ulysses Perspectives*, ed. by A. Balogh, E.J. Smith, R.G. Marsden (Springer-Praxis, Berlin, 2001), p. 287
G. Gloeckler, L.A. Fisk, G.M. Mason, M.E. Hill, AIP Conf. Proc. **CP1039**, 367–374 (2008)
M.E. Hill, R.B. Decker, E.C. Roelof et al., AIP Conf. Proc. **CP858**, 98–103 (2006)
P.A. Isenberg, W.C. Feldman, Geophys. Res. Lett. **22**, 873–875 (1995)
V. Izmodenov, Space Sci. Rev. (2009, this issue)
V. Izmodenov, Y. Malama, G. Gloeckler, J. Geiss, Astron. Astrophys. **414**, L29 (2004)
V. Izmodenov, Y. Malama, M.S. Ruderman, Astron. Astrophys. **429**, 1069 (2005)
J.R. Jokipii, Space Sci. Rev. **86**, 161 (1998)
B. Klecker, Space Sci. Rev. **72**, 419 (1995)
S.M. Krimigis, C.O. Bostrom, T.P. Armstrong et al., Space Sci. Rev. **21**, 329–354 (1977)
D.J. McComas, N.A. Schwadron, Geophys. Res. Lett. **33**, L04102 (2006)
E. Möbius, D. Hovestadt, B. Klecker et al., Nature **318**, 426–429 (1985)
E. Möbius, M. Bzowski, S. Chalov et al., Astron. Astrophys. **426**, 897–907 (2004)
J.D. Richardson, Y. Liu, C. Wang, D.J. McComas, Astron. Astrophys. **491**, 1–5 (2008)
J.D. Slavin, P.C. Frisch, Space Sci. Rev. **130**, 409–414 (2007)
E.C. Stone, A.C. Cummings, F.B. McDonald et al., Science **309**, 2017–2020 (2005)
E.C. Stone, A.C. Cummings, F.B. McDonald et al., Nature **454**, 71–74 (2008)
G.S. Voronov, At. Data Nucl. Data Tables **65**(1), 1–30 (1997)
M. Witte, M. Banaszkiewicz, H. Rosenbauer, D. McMullin, Adv. Space Res. **34**, 61 (2004)

Neutral H Density at the Termination Shock: A Consolidation of Recent Results

M. Bzowski · E. Möbius · S. Tarnopolski · V. Izmodenov · G. Gloeckler

Originally published in the journal Space Science Reviews, Volume 143, Nos 1–4, 177–190.
DOI: 10.1007/s11214-008-9479-0

Abstract We discuss a consolidation of determinations of the density of neutral interstellar H at the nose of the termination shock carried out with the use of various data sets, techniques, and modeling approaches. In particular, we focus on the determination of this density based on observations of H pickup ions on Ulysses during its aphelion passage through the ecliptic plane. We discuss in greater detail a novel method of determination of the density from these measurements and review the results from its application to actual data. The H density at TS derived from this analysis is equal to $0.087 \pm 0.022\ \mathrm{cm}^{-3}$, and when all relevant determinations are taken into account, the consolidated density is obtained at $0.09 \pm 0.022\ \mathrm{cm}^{-3}$. The density of H in CHISM based on literature values of filtration factor is then calculated at $0.16 \pm 0.04\ \mathrm{cm}^{-3}$.

Keywords Heliosphere · Local interstellar medium · Neutral interstellar gas · Heliospheric interface · Pickup ions · Heliospheric UV glow

1 Introduction

With the plethora of detailed measurements available from heliospheric missions such as Voyager, Ulysses, SOHO, Cassini, ACE, Wind, and many others and in anticipation of the first mission dedicated to remote-sensing studies of the heliospheric interface IBEX, a consolidation of heliospheric parameters is an important task. Key parameters in this quest are

M. Bzowski (✉) · S. Tarnopolski
Space Research Centre PAS, Bartycka 18A, Warsaw, Poland
e-mail: bzowski@cbk.waw.pl

E. Möbius
Space Science Center and Department of Physics, University of New Hampshire, Durham, NH, UK

V. Izmodenov
Moscow State University and Space Research Institute RAS, Moscow, Russia

G. Gloeckler
Department of Atmospheric, Oceanic, and Space Science, University of Michigan, Ann Arbor, MI, USA

the density of neutral interstellar hydrogen and helium at the nose of the termination shock. While a consensus was reached on the parameters of interstellar helium already a few years ago (Möbius et al. 2004, 2005: Witte 2004; Gloeckler et al. 2004a; Vallerga et al. 2004; Lallement et al. 2004b), the hydrogen density remained more uncertain until very recently due to more complex effects on this species within the heliosphere. A dedicated team supported by the International Space Science Institute from Bern, Switzerland, has recently reached a consolidation of the density of neutral interstellar hydrogen at the nose of the termination shock based on different data sets and modeling. We discuss this consolidation, focusing mostly on the determination of the density at the termination shock based on hydrogen pickup ion (H PUI) observations by Ulysses during its passage through the ecliptic plane at aphelion of its orbit, performed using a novel analysis method.

2 Neutral Gas Density from Pickup Ion Observations

2.1 Past Attempts to Derive the Density at the Termination Shock

Pickup ions are neutral interstellar atoms which become ionized and picked up by the frozen-in magnetic field of the solar wind. A simple model of the transport of pickup ions in the solar wind was developed by Vasyliunas and Siscoe (1976) under assumption that (1) the atoms of the seed (neutral) population are stationary with respect to the Sun at the moment of ionization; (2) the solar wind expands radially away from the Sun with a velocity V_{SW}, is spherically symmetric and stationary, and the frozen-in magnetic field is perpendicular to the expansion direction—thus there is no net parallel streaming of the PUI population (the PUI fluid expands radially); (3) the population of newly-injected pickup ions undergoes an immediate isotropization of the pitch angle with respect to the local magnetic field direction, forming a spherical shell distribution; (4) the evolution of the pickup ion distribution function with distance is adiabatic; and (5) the ionization rate of the seed population is spherically symmetric and invariable with time.

With these assumptions, the distribution function of pickup ions at a heliocentric distance r at the upwind axis is described by the formula:

$$f_{\mathrm{pui}}(r, w) = \frac{3}{8\pi V_{\mathrm{SW}}^4} \beta_{\mathrm{ion}} \left(\frac{r_{\mathrm{E}}}{r}\right)^2 \left(r w^{-3/2}\right) n_{\mathrm{H}}\left(r w^{-3/2}\right), \tag{1}$$

where $w = v / V_{\mathrm{SW}}$, v is the specific velocity of a pickup ion in the co-moving frame of the solar wind, and

$$\beta_{\mathrm{ion}} = \beta_{\mathrm{ion,E}} \left(r / r_{\mathrm{E}}\right)^2, \tag{2}$$

where $\beta_{\mathrm{ion,E}}$ is the production rate of pickup ions at the reference distance r_{E}, usually adopted as equal to 1 AU. The production rate is assumed to be equal to the loss rate of the interstellar gas, which implies that n_{H} is also a function of β_{ion}.

It is evident from (1) that the distribution function observed at a location (r, θ) for a given velocity v corresponds to the density of the seed population at a distance from the Sun equal to $r w^{-3/2}$. Thus, the absolute spectrum of pickup ions in velocity at a given location in space yields the density profile of the interstellar gas seed population with distance from the Sun if the PUI production rate and the speed of the solar wind are known. With the local density of the neutral gas in hand, one can extrapolate this quantity to the termination shock providing a sufficiently accurate model of the depletion of neutral interstellar gas in the inner heliosphere is available. Such a method has been used to infer the density of interstellar H at the termination shock by Gloeckler (1996) and Gloeckler and Geiss (2001). A similar method was used by Möbius et al. (1988) to determine the termination shock density of interstellar He from observations at 1 AU.

In some cases, an alternative method can be used. Since He^{+} pickup ions are solely created by charge exchange between the neutral He atoms and solar wind alphas, and since Ulysses SWICS is able to simultaneously observe the He^{2+} PUI and solar wind fluxes at ~5 AU, where neutral interstellar He is not significantly depleted, Gloeckler et al. (1997) could obtain the local He density with an uncertainty that depends only on the accuracy of the knowledge of charge exchange cross section and does not depend on the absolute calibration of the instrument. This method, however, is not directly applicable to H because the production of pickup ions depends on two comparable ionization channels, i.e. charge exchange with solar wind and EUV ionization. Consequently, the rate cannot be easily eliminated from the analysis by simultaneous measurements. The remaining uncertainty in the total ionization rate translates into the uncertainty in the density and its significance is illustrated by the fact that Gloeckler and Geiss (2001) had to invoke a high electron ionization rate of 2.4×10^{-7} s^{-1} at 1 AU to explain the pickup ion velocity distribution, while current estimates for the time interval of the observations suggest that it is equal to about 0.8×10^{-7} s^{-1} (Bzowski 2008).

2.2 Modeling Uncertainties in Pickup Ion Interpretations

Because the aforementioned method to deduce the density at the termination shock depends on the knowledge of the neutral gas density inside the observer location and interplanetary parameters greatly influence the relation between the local density and the density at the shock, it is rather sensitive to modeling choices and the knowledge of parameters, such as ionization rates and radiation pressure. The transport of PUI from their birthplace to the detector is quite complex (see, e.g. Isenberg 1987; Chalov and Fahr 1998; Chalov 2006; Malama et al. 2006) and the attenuation processes of interstellar gas, including ionization and radiation pressure, are strongly anisotropic and time-dependent on time scales from days to decades (see, e.g., Bzowski 2008). Thus, for an instantaneous distribution of PUI (1) one cannot assume the local production rate of PUI is equal to the loss rate of the interstellar gas, which is one of the factors shaping $n_{\mathrm{H}}(r, \theta)$, (2) one cannot be sure where exactly the boundary of the cavity is located because it changes with time, and (3) one can use neither the cold model, nor even the classical hot model (Fahr 1978, 1979; Thomas 1978; Wu and Judge 1979; Lallement et al. 1985) for the local density of interstellar gas. The reasons for this, partly discussed already by Ruciński and Bzowski (1996), are as follows. Firstly, H is subject to a complex behavior of the radiation pressure, which depends on the radial velocity of an atom (Tarnopolski and Bzowski 2008; Bzowski 2008) and is proportional to the net solar flux in the Lyman-α line. Its magnitude varies with the phase of the solar cycle and over a solar rotation (e.g., Bzowski 2001b, 2008). Secondly, the ionization rate is composed of charge exchange with the solar wind protons, photoionization (Bzowski 2001b; Auchère et al. 2005; Bzowski et al. 2008; Bzowski 2008), and electron-impact ionization (Bzowski et al. 2008; Bzowski 2008). All three contributors vary both with time (Bzowski 2001b, 2008) and with heliolatitude (McComas et al. 1999; Bzowski et al. 2003; Quémerais et al. 2006). Gloeckler and Geiss (2001) have determined the total ionization rate that includes all of these components from the slope of the observed PUI distribution. However, this method works under the stated assumption that at least over the observation period the PUI production rate is equal to the loss rate of the interstellar neutrals. Also it should be noted, the modeling

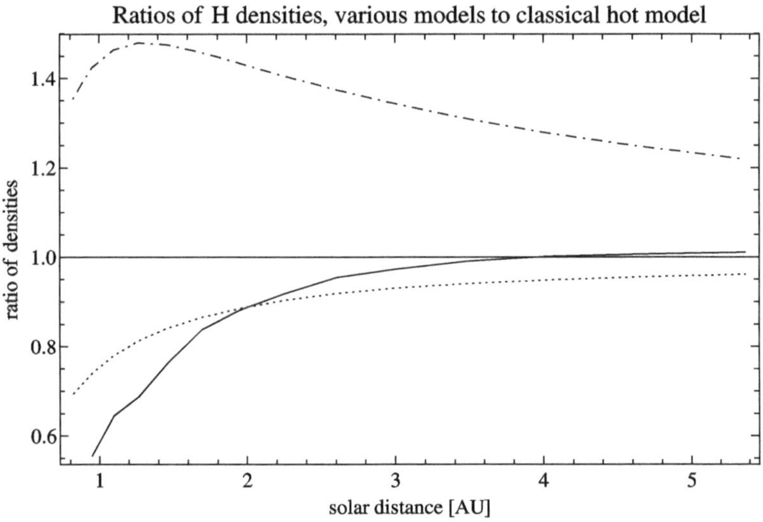

Fig. 1 Ratios of densities of neutral interstellar H computed using various models relative to the results of the classical hot model. Shown is the profile along the Ulysses-Sun line in the solar equatorial plane towards the Ulysses aphelion. *Solid line*: Moscow MC model (i.e., conditions at the termination shock non-Maxwellian and non-homogeneous); *dotted line*: radiation pressure sensitive to the radial velocity of the atoms (non-flat solar Lyman-α line shape adopted); *dash-dot*: full time-dependent treatment of the radiation pressure and the ionization rate

of the PUI distributions for all comparisons with observations, except if just the PUI flux at the observer location, i.e. at the PUI cut-off is used, is tied to the reasonable assumption that the full adiabatic cooling in the expanding solar wind (Vasyliunas and Siscoe 1976) is valid, an assumption that has not been independently tested with observations. Finally, there is a strong departure of the hydrogen distribution function at the termination shock from a shifted-Maxwellian due to the filtering processes in the heliospheric interface region (Baranov et al. 1998; Izmodenov 2001; Izmodenov et al. 2005a; Heerikhuisen et al. 2006; Izmodenov and Baranov 2006).

An illustration how strongly including or excluding different physical processes in the modeling can influence the simulated density profile is shown in Fig. 1. It shows simulated densities as a function of distance from the Sun in the equatorial plane towards the aphelion position of Ulysses obtained with three different modeling approaches normalized to the density profile from the classical hot model of the interstellar gas.

As can be seen in Fig. 1, at heliospheric locations relevant to the Ulysses in-ecliptic measurements, various modeling strategies lead to differences in the predicted gas density, reaching a 30% level at ~5 AU and increasing towards the Sun. The magnitude of variations shown here should be regarded as a lower boundary because the model used does not take into account the variations in the distribution function in the source region, i.e. at the termination shock, which were shown to amount to about 10% by Izmodenov et al. (2005b) and Izmodenov et al. (2008). Therefore, the classical hot model can only be regarded as a qualitative representation of the true density, and the quality of the approximation is reduced with a decreasing distance from the Sun. The model distribution function derived from the classical hot model is also affected by the inherent deficiencies of this model, of course, and its direct use to the interpretation of PUI measurements is not recommended.

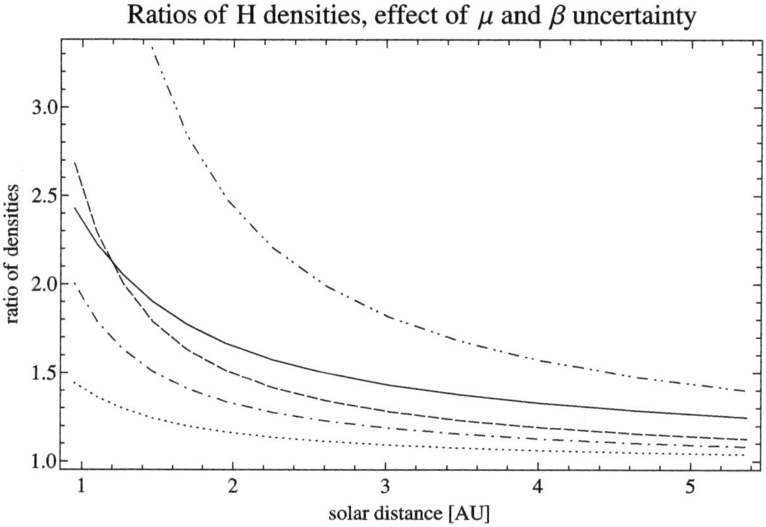

Fig. 2 Effects of changes in the absolute calibration of the ionization rate and solar Lyman-α flux on the neutral gas density profile. Shown are ratios of densities computed using the hot model for reduced μ and β values over nominal values obtained from recent measurements. *Solid line*: β reduced by 25%; *dots, dash-dot, dashes*: μ reduced by 10%, 20%, and 30%, respectively; *dash-dot-dot*: both μ and β reduced, respectively, by 25% and 30%

Another factor relevant for the calculation of the absolute value of the H PUI distribution function is the accuracy of the heliospheric parameters such as the solar wind flux, solar EUV flux, solar Lyman-α flux, and the shape of the solar Lyman-alpha line. The absolute calibration of the solar wind flux measured by various instruments has been known with an accuracy of about 25% (cf. the OMNI-2 data base, King and Papitashvili (2005), and an early review by Bzowski (2001a)). Similarly, the absolute calibration of the flux in the solar Lyman-alpha line has changed within the past 30 years by about 30% (see Vidal-Madjar 1975; Woods and Rottman 1997; Tobiska et al. 2000; Woods et al. 2000; Floyd et al. 2002) and even now is known to no better than 10%. The influence of uncertainties in these parameters on the absolute densities of neutral interstellar hydrogen inside the Ulysses orbit near ecliptic is shown in Fig. 2.

As evident from Fig. 2, uncertainties in the absolute ionization rate of 25% translate into uncertainties of ~25% in the density of interstellar gas at Ulysses at ~5 AU, and an uncertainty of 30% in radiation pressure leads to a density uncertainty of 10% at 5 AU. Both uncertainties increase dramatically towards the Sun. All these uncertainties have to be included in the determination of the H density at the termination shock from PUIs in the inner heliosphere.

2.3 Robust Method to Determine the Density at the Termination Shock

Recently, Bzowski et al. (2008) have presented a novel method to determine the density of interstellar H at the nose of the termination shock using measurements performed at the boundary of the cavity of neutral interstellar gas in the inner heliosphere.

2.3.1 Simplified analytic derivation

Based on the distribution function of H PUI as proposed by Vasyliunas and Siscoe (1976) and repeated in (1), the phase space density at $w = 1$ (i.e., for locally injected ions) is directly proportional to the local source function of pickup ions $S(r)$, defined as:

$$S(r) = \beta_{\text{ion}}\, n_{\text{H}}(r, \beta_{\text{ion}}(r)). \tag{3}$$

Using a cold model of neutral interstellar hydrogen (Fahr 1968; Axford 1972) and assuming perfect compensation of the solar radiation pressure and gravity, the gas density $n(r)$ on the upwind axis varies as:

$$n(r) = n_0 \exp\left[-\beta_{\text{ion}} \frac{r}{V_{\text{ISM}}}\right]. \tag{4}$$

The local PUI production rate is equal to:

$$S(r) = \beta_{\text{ion}} n_0 \exp\left[-\frac{\beta_{\text{ion}} r_{\text{E}}^2}{V_{\text{ISM}} r}\right] \tag{5}$$

and scales linearly with the density at the termination shock. Now we seek a distance r_0 such that the source function does not depend on β_{ion}, requiring that:

$$\frac{\partial S(r_0)}{\partial \beta_{\text{ion}}} = 0 = n(r_0, \beta_{\text{ion}}(r_0)) + \beta_{\text{ion}}(r_0) \frac{\partial n(r_0, \beta_{\text{ion}})}{\partial \beta_{\text{ion}}}. \tag{6}$$

Combining (4) and (6) we get:

$$n(r_0) = -\beta_{\text{ion}}(r_0) \frac{\partial n(r_0, \beta_{\text{ion}})}{\partial \beta_{\text{ion}}}. \tag{7}$$

With

$$\frac{\partial n}{\partial \beta_{\text{ion}}} = -n \frac{r}{V_{\text{ISM}}}, \tag{8}$$

the distance r_0 becomes:

$$r_0 = \frac{V_{\text{ISM}}}{\beta_{\text{ion}}} \text{ or } r_0 = \frac{\beta_{\text{E}} r_{\text{E}}^2}{V_{\text{ISM}}} \tag{9}$$

in connection with (2). Inserting this result to (4), we find that the exact location where the source function is insensitive to the ionization rate is found precisely at the boundary of the heliospheric cavity, i.e. where the density is decreased to $1/e$ of the value at the termination shock. To calculate n_0 from the locally observed PUI source strength $S(r_0)$ the ionization rate must be determined. However, as long as the observations are made close to r_0 moderate uncertainties in the knowledge of the ionization rate only result in rather small uncertainties in the determination of the termination shock density n_0. In a comparison between different simulations and Ulysses PUI data Bzowski et al. (2008) showed that this method is also robust against uncertainties in the radiation pressure and against variations in the modeling approaches.

2.3.2 Simulations of the H PUI Production Rate at Ulysses

Bzowski et al. (2008) took a subset of the Ulysses H PUI data obtained over 13 months in 1997 and 1998 during the Ulysses passage through the ecliptic plane at aphelion and averaged the observed distribution function to eliminate fluctuations of the local heliospheric conditions. The resulting local production rate of H PUI at Ulysses was equal to 7.26×10^{-10} cm^{-3} s^{-3}.

The simulations to calculate the density of neutral hydrogen in the inner heliosphere were performed with the use of a combination of two models. First, the density of interstellar hydrogen at the nose of the termination shock was calculated using the Moscow MC model Baranov and Malama (1993) with solar wind alphas and interstellar He^+ ions included (Izmodenov et al. 2003). This density was split into two populations, primary and secondary, and their respective bulk velocities and temperatures were registered. The parameters of the two populations were subsequently handed over to the Warsaw kinetic time- and latitude-dependent model (Ruciński and Bzowski 1995; Bzowski et al. 1997, 2002; Tarnopolski and Bzowski 2008), which calculated a time series of densities and PUI production rates from the primary and secondary populations at 16 evenly-spaced moments during the observation interval at distances and offset angles coinciding with the Ulysses positions. The production rates were averaged over the appropriate time periods so that they could be compared with the value obtained from observations.

Bzowski et al. (2008) started with a test whether the handover of the calculations from the Moscow MC to the Warsaw kinetic model is robust. The point was that the distribution function of the neutral gas at the termination shock is not independent of the offset angle from the upwind direction, but the Warsaw model makes this assumption. This check was made by running both models with identical parameters, assuming that radiation pressure and ionization rate are averaged over the solar cycle, spherically symmetric, and invariable with time. The version of the MC model used in the simulations is stationary and axially symmetric. The Warsaw model was de-rated to similar assumptions and run with the parameters of the heliospheric gas populations as predicted by the Moscow model at the nose of the termination shock. Finally, the densities predicted by both models along a radial line connecting Ulysses with the Sun were compared. It was discovered that for this geometry, discrepancies between the results begin at $\sim$3 AU from the Sun and increase inward, but at the Ulysses locations, the two models agree within $\sim$1%. Hence, using the static and axially symmetric MC model to determine the parameters of the gas at the termination shock and handing these parameters to a time-dependent, full 3D Warsaw kinetic model was a reasonable and unbiased approach.

Bzowski et al. (2008) studied in detail variations in the solar wind density and velocity as a function of time and heliolatitude as well as radiation pressure, photoionization and electron-impact rates for the observations time. They developed a time- and heliolatitude-dependent model of the charge exchange ionization rate and a spherically symmetric, time-dependent models of the photoionization and electron-ionization rates, as well as a time- and radial velocity-dependent model of radiation pressure, covering a large time span straddling the interval of Ulysses measurements, and implemented them all in the Warsaw code. With those in hand, they calculated the PUI production rates at Ulysses for the 16 observation periods and averaged the result.

The robustness of this result was verified in a multitude of ways. First, tests of the robustness against modeling strategies were performed. The reaction of the simulated PUI production rate to a down-grading of the model from its full 3D time dependence was tested. First, the fully time-dependent ionization rate and radiation pressure were replaced with static values relevant for the 16 observation points. Then, the sensitivity of the radiation pressure to radial velocities of the atoms was switched off. Tests of the influence of the latitude dependence in the ionization rate were performed.

With this result in hand, the gas density at the termination shock was computed from the relation:

$$n_{\mathrm{TS}} = n_{\mathrm{TS,sim}} \frac{S_{\mathrm{obs}}}{S_{\mathrm{sim}}} \tag{10}$$

where the index "sim" refers to a simulated quantity and "obs" to an observed value. The magnitudes of the combined effects are shown in Fig. 3 (from Bzowski et al. 2008), which presents the combined variation of all modeling parameters in comparison with the observations. It is evident that the calculated PUI production rate at Ulysses only weakly depends on these variations.

Each of the simulations was repeated with departures of the ionization rate and radiation pressure from their nominal (measured) values. In particular, the radiation pressure was reduced by a factor of 0.9, 0.8, and 0.7 (i.e., by 10%, 20%, and 30%), and the ionization rate by 25%. The local PUI production rate changed just by a few percent in the case of the ionization rate variation. The effect of radiation pressure reduction was more pronounced, but still within ~15%.

Finally, it was checked whether the conclusions hold when the parameters in the LIC are changed. The entire simulation program was repeated with a different set of parameters, which introduced the simulated density at the termination shock increased by ~25%; the most important difference, however, was a totally different breakdown between the primary and secondary populations. The simulated production rates increased almost exactly by the same percentage (as can be seen from the fact that the pairs of lines between the two groups shown in Fig. 3 are almost perfectly parallel), proving that indeed, the PUI source function at Ulysses scales linearly with the density at the termination shock.

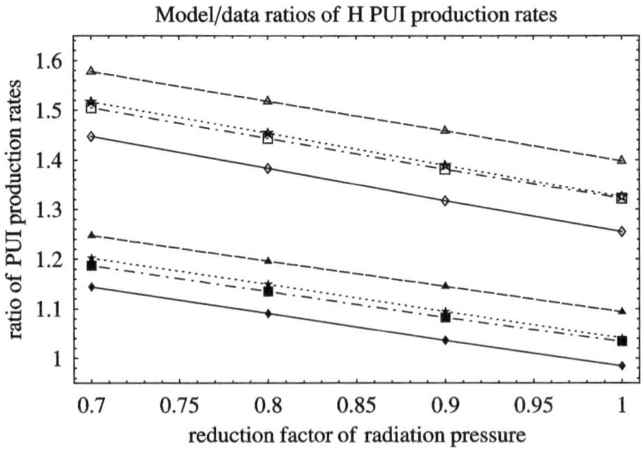

Fig. 3 Model-to-data ratios of the source function of the H pickup ions at Ulysses as a function of reduction factor of the net solar Lyman-alpha flux used in the simulation. The group of lines with *filled symbols* corresponds to the simulations performed with the density at the nose of the termination shock equal to 0.095 cm^{-3}, the *open symbols* correspond to 0.117 cm^{-3}. *Diamonds* correspond to the nominal values of the ionization rate and radiation pressure dependent on v_r, *stars* to the values of the ionization rate reduced by 25% and radiation pressure dependent on v_r, and *squares* and *triangles* to radiation pressure independent on v_r with the ionization rate, respectively, on the nominal level and reduced by 25%

In a final test, the simulations were repeated with a different coupling factor between the protons and neutral atoms. It could be shown that replacing the charge exchange cross-section formula from Maher and Tinsley (1977) with that by Lindsay and Stebbings (2005) does not visibly affect the results.

3 Determination of Hydrogen Density at the Termination Shock

To determine the density of neutral H at the nose of the termination shock, Bzowski et al. (2008) took the simulated values of the production rate at Ulysses relevant for the current calibration of the ionization rate and radiation pressure with their uncertainty bars (which correspond to the points for the reduction factor equal to 0.9 or 1.0 in Fig. 3) and calculated n_{TS} from the observed PUI fluxes at the Ulysses aphelion. Since the modeling uncertainty was just a few percent, almost the entire error in the result is due to the uncertainty of determination of the PUI production rates from Ulysses observations (mostly the geometric factor of the instrument), which is known to about 20%. This analysis brought the density of interstellar H at the termination shock to 0.087 $cm^{-3} \pm 25\%$.

4 Consolidation of the Interstellar H Density

4.1 Density of H at the Termination Shock

Determinations of the neutral H density at the termination shock, obtained using several different observation methods, were recently compared and consolidated in the framework of an International Space Science Institute Team. These values were then used to infer the H density in the CHISM based on a global heliosphere modeling.

As discussed in the previous section, a re-analysis of the Ulysses SWICS H PUI observations resulted in a H density at the termination shock of $n_{H,TS} = 0.087 \pm 0.022$ cm^{-3}, where all observational, modeling, and external parameter uncertainties are included in the overall uncertainty of 25% Bzowski et al. (2008). The neutral H density at the TS was also determined using two independent methods and data sources.

One of them was an analysis of the solar wind slowdown resulting from its mass-loading due to the accumulation of H pickup ions. Richardson et al. (2008) took the solar wind velocity measured by Ulysses at ~5 AU from the Sun, when it was at the same heliolatitude as Voyager 2 at ~80 AU, and using a 1D MHD model of the solar wind propagation, they calculated an expected solar wind velocity at Voyager assuming there are no pickup ions. The difference between the modeled and actually measured values, equal to 67 km/s, was adopted as the slowdown of the solar wind due to the "friction" of the solar wind against the inflowing neutral interstellar gas. Subsequently, they repeated the calculations of the solar wind speed at Voyager, but this time assuming three values of neutral interstellar hydrogen density at the termination shock: 0.07, 0.9, and 0.11 cm^{-3}. Having compared the measured slowdown value of 67 km/s with modeling results, they arrived at a value of $n_{H,TS} = 0.09 \pm 0.01$ cm^{-3}. With the observational uncertainty including potential variations in parameters such as the ionization rate and the modeling approach, the overall uncertainty for this derivation can be placed at ± 0.02 cm^{-3}.

A third method was applied by Pryor et al. (2008), who used UV backscattering observations in an approach that does not require any absolute calibration of the observed UV intensities. They used data from Cassini located at 10 AU, and Voyager at ~90 AU from the Sun, obtained during the same time interval 2003–2004. Both spacecraft observed about monthly-periodic variations in the backscattering signal. These were attributed to the scattering of the Lyman-alpha radiation from active regions, which rotate with the Sun and thus illuminate the surrounding interstellar gas like a "lighthouse". The amplitudes of the variations observed at Voyager were systematically lower than at Cassini, which was interpreted as attenuation by a factor of 0.2, resulting from multiple scattering of the Lyman-alpha photons in the ambient neutral interstellar gas in the inner heliosphere. Using the Moscow MC model of the heliosphere and the radiation transport code by Gangopadhyay et al. (2006), they simulated the attenuation for two sets of interstellar parameters: ($n_p = 0.06\ \text{cm}^{-3}, n_\text{H} = 0.18\ \text{cm}^{-3}$) and ($n_p = 0.05\ \text{cm}^{-3}, n_\text{H} = 0.15\ \text{cm}^{-3}$), which brought the interstellar gas density at the termination shock equal to 0.085 and 0.095 cm^{-3}, respectively. While it was impossible to exactly reproduce all observed variations, they concluded that the attenuation is consistent with a density at the termination shock of $\sim 0.09\ \text{cm}^{-3}$. However, the value derived is model-dependent. Therefore, an overall uncertainty of about 25% has been placed on the derived value of $n_{\text{H,TS}} = 0.09 \pm 0.024\ \text{cm}^{-3}$ (Pryor et al. 2008).

Although aimed primarily at a determination of the ionization rate in the inner heliosphere, another recent study of heliospheric backscatter glow by Quémerais et al. (2006), which is based on multi-year photometric all-sky observations with SOHO SWAN, has also returned a value of 0.09 cm^{-3} for the H density at the termination shock.

Hence, all three results obtained using independent methods and independent data sets are consistent with each other within their uncertainty bands, and with the result of the PUI analysis by Bzowski et al. (2008). Computing a weighted average from all these values as done for He by Möbius et al. (2004), the density of neutral interstellar hydrogen at the termination shock is found equal to $0.089 \pm 0.022\ \text{cm}^{-3}$.

As an alternative attempt, Gloeckler et al. (2008, this issue) combined the CHISM density of neutral He, based on the combined He parameter determination with a density $n_\text{He} = 0.015 \pm 0.0015 \pm 0.002\ \text{cm}^{-3}$ (Möbius et al. 2004), with the abundance ratios obtained with Voyager from the suprathermal tails of interstellar pickup ion distributions and from anomalous cosmic rays in the heliosheath. They present a noticeably lower H density at the termination shock of $0.055 \pm 0.021\ \text{cm}^{-3}$, but with a rather large uncertainty due to a number empirical modeling steps involved in the method.

4.2 Extrapolating the Termination Shock Density to the CHISM

The bulk velocity and temperature of the gas in the CHISM are now well established from interstellar helium observations (Witte 2004; Gloeckler et al. 2004b; Vallerga et al. 2004; Lallement et al. 2004a, compiled by Möbius et al. 2004), but the density, ionization, and magnetic field still require further studies. While the determinations of the density of interstellar H at the termination shock based on various observations, discussed earlier in this section, depend relatively weakly on the modeling of the heliosphere as a whole, extrapolating this density to the CHISM is certainly model-dependent and requires a detailed model of the interaction of the solar wind with the local interstellar medium. Since the mean free path of the interstellar H atoms is comparable with the size of the heliosphere, kinetic models are required. In fact, due to modifications in the interaction region the flow of H atoms at the termination shock varies with the angle off the upwind direction. This dependence should be taken into account in the analyses. In this paper we assumed that the interstellar hydrogen gas at the termination shock consists of two populations (primary and secondary) which flow homogeneously at the TS. It was verified (see Bzowski et al. 2008) that such an assumption does not introduce an error greater than 1% at the location of Ulysses observations discussed in this paper.

Under these assumptions we can estimate the H atom number density in CHISM by introducing the filtration factor (e.g. Izmodenov 2007) as: $F = n_{\mathrm{H,TS}}/n_{\mathrm{H,CHISM}}$, where $n_{\mathrm{H,TS}}$ is equal to the density of neutral interstellar hydrogen at the nose of the termination shock and $n_{\mathrm{H,CHISM}}$ to the H density in the unperturbed interstellar gas. Then, knowing $n_{\mathrm{H,TS}}$ upwind and the filtration factor from the models, we can obtain an estimate for $n_{\mathrm{H,CHISM}}$. The filtration factor should be calculated in the framework of a solar wind/CHISM interaction model.

Izmodenov et al. (2004) present an axisymmetric stationary kinetic-gasdynamic model of the solar wind/CHISM interaction. The model differs from the original (Baranov and Malama 1993) by taking into account the solar wind alpha particles and the ionized interstellar helium component. The solar wind parameters in this parametric study were similar to those used by Bzowski et al. (2008). Table 1 in Izmodenov et al. (2004) presents the filtration factors obtained for different interstellar proton and H atom number densities. It is shown that the filtration factor varies from 0.51 to 0.58 for a wide range of densities. Applying this filtration to the density at the nose of the termination shock obtained by Bzowski et al. (2008) we obtain $n_{\mathrm{H,CHISM}} = 0.16 \pm 0.04\ \mathrm{cm}^{-3}$. However, since the error bars for $n_{\mathrm{H,TS}}$ are large, a wider range of ($n_{\mathrm{H,CHISM}}$, $n_{\mathrm{p,CHISM}}$) is needed than available from Izmodenov et al. (2004).

In addition, filtration factors also depend on the interstellar magnetic field, which was not taken into account by Izmodenov et al. (2004), and on variations in the solar wind ram pressure (McComas et al. 2008; for model, see, e.g., Izmodenov et al. 2005a, 2008). Also, the dependence of the solar wind on heliolatitude may be important. Including these effects may extend the range of the filtration factor beyond the values mentioned above. To establish the interstellar parameters based on $n_{\mathrm{H,TS}}$, additional constraints, such as the location of the TS in Voyager 1 and 2 directions (see, e.g., Izmodenov, this issue), Lyman-alpha absorption (Wood et al., this issue), and solar backscattered Lyman-alpha radiation observations (Quemerais et al., this issue), should be added.

It is worthwhile to mention that the interaction of interstellar gas with the solar wind is also described with hydrodynamic multi-fluid models (e.g. Müller et al. 2006; Scherer and Fahr 2003; Florinski et al. 2005, 2003). Alexashov and Izmodenov (2005) compared results of their kinetic model with different multi-fluid models. Results relevant to present studies are listed in Table 2 of this paper and show that the parameters of the primary and secondary populations of interstellar atoms obtained from hydrodynamic models differ significantly from the results of the kinetic models.

Müller et al. (2008) compared the results from five different models of the heliosphere evaluated with identical boundary conditions. Two of these models are fully kinetic in their treatment of neutrals (Baranov and Malama 1993; Heerikhuisen et al. 2006), and three are hydrodynamic multi-fluid (Florinski et al. 2003, 2005; Müller et al. 2006, extended from Pauls et al. 1995, and Scherer and Fahr 2003, extended from Fahr et al. 2000). The parameter set used by Müller et al. (2008) was close to one of the sets of interstellar parameters used in the simulations by Bzowski et al. (2008) with the important exception that the solar wind speed was essentially lower. The filtration factors obtained in this comparison varied from 0.52 (for the two-fluid hydrogen model) to 0.69–0.74 (for the kinetic models).

Müller et al. (2008) did not present the parameters of the primary and secondary populations at the termination shock separately and offered only the mean values of the density, bulk velocity, and temperature. While all models reproduced the mean bulk velocity at the termination shock within a narrow range from 19.2 to 23.4 km/s, differences in the temperature were larger, i.e. from 12 000 to 30 900 K.

Since the goal of the comparison made by Müller et al. (2008) was checking the results of different calculating schemes, all models had to include the same physical effects, which required switching off some of their advanced features (different in different models). As a result, none of them was run in its full mode. Nevertheless, both of the comparison studies (Alexashov and Izmodenov 2005; Müller et al. 2008) seem to suggest that an uncritical use of hydrodynamic multi-fluid models for quantitative analysis of heliospheric observations may be misleading.

From the discussion above it follows that without performing a dedicated modeling, the H density in the CHISM can be assessed from the density at the termination shock by dividing this value by the filtration ratio being in the (0.51, 0.59) range, which brings the density of neutral H in the CHISM equal to $0.16 \pm 0.04\ \mathrm{cm}^{-3}$.

5 Conclusions

The density of neutral interstellar gas at the termination shock seems to be well constrained by observations performed using different techniques and appropriate modeling. The datasets used include Ulysses observations of H pickup ions, Voyager and Ulysses observations of solar wind slowdown, Cassini and Voyager observations of modulation in the heliospheric Lyman-alpha glow, and multi-year all-sky photometric observations of the glow by SWAN/SOHO. The densities obtained from analysis of these observations group in the range of 0.08–0.09 cm^{-3} and the uncertainty of this determination is about 20%. Extrapolation of this value to the CHISM based on analysis of heliospheric filtration factors from the literature brings the density of neutral interstellar hydrogen equal to $0.16 \pm 0.04\ \mathrm{cm}^{-3}$.

Acknowledgements The authors gratefully acknowledge the hospitality and excellent ambiance within the International Space Science Institute (ISSI) in Bern, Switzerland, where the workshop "From the Outer Heliosphere to the Local Bubble: Comparison of New Observations with Theory" was held. This research was performed within the framework of an International Space Science Institute (Bern, Switzerland) Working Group Neutral Interstellar Hydrogen. The authors are grateful for helpful discussions during the manuscript preparations with Martin Lee and Stan Grzedzielski. Support for this study from NASA Grant NNG06GF55G and Grant NAG 5-12929 through a subcontract from the California Institute of Technology is gratefully acknowledged. V.I. was supported part by RFBR grants 07-02-01101-a, 07-01-00291-a and Dynastia Foundation.

References

D. Alexashov, V. Izmodenov, Astron. Astrophys. **439**, 1171–1181 (2005)

F. Auchère, J.W. Cook, J.S. Newmark, D.R. McMullin, R. von Steiger, M. Witte, Astrophys. J. **625**, 1036–1044 (2005)

W.I. Axford, in *The Solar Wind*, ed. by J.M.W.C.P. Sonnet, P.J. Coleman. *NASA Spec. Publ.*, vol. 308 (1972), pp. 609–660

V.B. Baranov, Y.G. Malama, J. Geophys. Res. **98**, 15,157–15,163 (1993)

V.B. Baranov, V.V. Izmodenov, Y.G. Malama, J. Geophys. Res. **103**, 9575–9585 (1998)

M. Bzowski, Space Sci. Rev. **97**, 379–383 (2001a)

M. Bzowski, in *The Outer Heliosphere: The Next Frontiers*, ed. by K. Scherer, H. Fichtner, H.J. Fahr, E. Marsch. COSPAR Colloquia Series, vol. 11 (Elsevier, Pergamon, 2001b), pp. 69–72

M. Bzowski, Astron. Astrophys. **488**, 1057–1068 (2008)

M. Bzowski, H.J. Fahr, D. Ruciński, H. Scherer, Astron. Astrophys. **326**, 396–411 (1997)

M. Bzowski, T. Summanen, D. Ruciński, E. Kyrölä, J. Geophys. Res. **107**, 141 (2002). doi:10.1029/2001JA00

M. Bzowski, T. Mäkinen, E. Kyrölä, T. Summanen, E. Quèmerais, Astron. Astrophys. **408**, 1165–1177 (2003)

M. Bzowski, E. Möbius, S. Tarnopolski, V. Izmodenov, G. Gloeckler, Astron. Astrophys. **491**, 7–19 (2008)

S. Chalov, in *The Physics of the Heliospheric Boundaries*, ed. by V.V. Izmodenov, R. Kallenbach (2006), pp. 245–282

S.V. Chalov, H.J. Fahr, Astron. Astrophys. **335**, 746–756 (1998)
H.J. Fahr, Astrophys. Space Sci. **2**, 474–495 (1968)
H.J. Fahr, Astron. Astrophys. **66**, 103–117 (1978)
H.J. Fahr, Astron. Astrophys. **77**, 101–109 (1979)
H.J. Fahr, T. Kausch, H. Scherer, Astron. Astrophys. **357**, 268–282 (2000)
V. Florinski, G.P. Zank, N.V. Pogorelov, J. Geophys. Res. (Space Phys.) **108**, 1228 (2003)
V. Florinski, G.P. Zank, N.V. Pogorelov, J. Geophys. Res. (Space Phys.) **110**(A9), 7104 (2005)
L.E. Floyd, D.K. Prinz, P.C. Crane, L.C. Herring, Adv. Space Res. **29**, 1957–1962 (2002)
P. Gangopadhyay, V.V. Izmodenov, M. Gruntman, D.L. Judge, Astrophys. J. **637**, 786–790 (2006)
G. Gloeckler, Space Sci. Rev. **78**, 335–346 (1996)
G. Gloeckler, J. Geiss, Space Sci. Rev. **97**, 169–181 (2001)
G. Gloeckler, L.A. Fisk, J. Geiss, Nature **386**, 374–376 (1997)
G. Gloeckler, F. Allegrini, H.A. Elliott, D.J. McComas, N.A. Schwadron, J. Geiss, R. von Steiger, G.H. Jones, Astrophys. J. Lett. **604**, L121–L124 (2004a)
G. Gloeckler, E. Möbius, J. Geiss, M. Bzowski, S. Chalov, H. Fahr, D.R. McMullin, H. Noda, M. Oka, D. Ruciński, R. Skoug, T. Terasawa, R. von Steiger, A. Yamazaki, T. Zurbuchen, Astron. Astrophys. **426**, 845–854 (2004b)
G. Gloekler, L.A. Fisk, J. Geiss, M.E. Hill, D.C. Hamilton, R.B. Decker, S.M. Krimigis, Space Sci. Rev. (2008), this issue
J. Heerikhuisen, V. Florinski, G.P. Zank, J. Geophys. Res. **111**(A10), 6110 (2006)
P.A. Isenberg, J. Geophys. Res. **92**, 1067–1073 (1987)
V.V. Izmodenov, Space Sci. Rev. **97**, 385–388 (2001)
V.V. Izmodenov, Space Sci. Rev. **130**, 377–387 (2007)
V.V. Izmodenov, V.B. Baranov, in *The Physics of the Heliospheric Boundaries*, ed. by V. Izmodenov, R. Kallenbach. ISSI Sci. Rep., vol. SR-005 (2006), pp. 67–136
V. Izmodenov, G. Gloeckler, Y. Malama, Geophys. Res. Lett. **30**, 1351 (2003)
V. Izmodenov, Y. Malama, G. Gloeckler, J. Geiss, Astron. Astrophys. **414**, L29–L32 (2004)
V. Izmodenov, S. Alexashov, A. Myasnikov, Astron. Astrophys. **437**, L35–L38 (2005a)
V. Izmodenov, Y. Malama, M.S. Ruderman, Astron. Astrophys. **429**, 1069–1080 (2005b)
V.V. Izmodenov, Y.G. Malama, M.S. Ruderman, Adv. Space Res. **41**, 318–324 (2008)
J.H. King, N.E. Papitashvili, J. Geophys. Res. **110**(A9), 2104–2111 (2005)
R. Lallement, J.L. Bertaux, F. Dalaudier, Astron. Astrophys. **150**, 21–32 (1985)
R. Lallement, J.C. Raymond, J.-L. Bertaux, E. Quémerais, Y.-K. Ko, M. Uzzo, D. McMullin, D. Rucinski, Astron. Astrophys. **426**, 867–874 (2004a)
R. Lallement, J.C. Raymond, J. Vallerga, M. Lemoine, F. Dalaudier, J.L. Bertaux, Astron. Astrophys. **426**, 875–884 (2004b)
B.G. Lindsay, R.F. Stebbings, J. Geophys. Res. **110**, 12,213 (2005)
L.J. Maher, B.A. Tinsley, J. Geophys. Res. **82**, 689–695 (1977)
Y. Malama, V.V. Izmodenov, S.V. Chalov, Astron. Astrophys. **445**, 693–701 (2006)
D.J. McComas, R.W. Ebert, H.A. Elliot, B.E. Goldstein, J.T. Gosling, N.A. Schwadron, R.M. Skoug, Geophys. Res. Lett. **35**, L18,103 (2008)
D.J. McComas, H.O. Funstein, J.T. Gosling, W.R. Pryor, Geophys. Res. Lett. **26**, 2701–2704 (1999)
E. Möbius, B. Klecker, D. Hovestadt, M. Scholer, Astrophys. Space Sci. **144**, 487–505 (1988)
E. Möbius, M. Bzowski, S. Chalov, H.-J. Fahr, G. Gloeckler, V. Izmodenov, R. Kallenbach, R. Lallement, D. McMullin, H. Noda, M. Oka, A. Pauluhn, J. Raymond, D. Ruciński, R. Skoug, T. Terasawa, W. Thompson, J. Vallerga, R. von Steiger, M. Witte, Astron. Astrophys. **426**, 897–907 (2004)
E. Möbius, M. Bzowski, H.-J. Fahr et al., in *Solar Wind 11/SOHO 16, Connecting Sun and Heliosphere*. ESA Special Publication, vol. 592 (ESA, Noordwijk, 2005), pp. 363–366
H.-R. Müller, P.C. Frisch, V. Florinski, G.P. Zank, Astrophys. J. **647**, 1491–1505 (2006)
H.R. Müller, V. Florinski, J. Heerikhuisen, V.V. Izmodenov, K. Scherer, D. Alexashov, H.J. Fahr, Astron. Astrophys. **491**, 43–51 (2008)
H.L. Pauls, G.P. Zank, L.L. Williams, J. Geophys. Res. **100**, 21,595–21,604 (1995)
W. Pryor, P. Gangopadhyay, B. Sandel, T. Forrester, E. Quemerais, E. Möbius, L. Esposito, I. Stewart, W. McClintock, A. Jouchoux, J. Colwell, V. Izmodenov, Y. Malama, K. Tobiska, D. Shemansky, J. Ajello, C. Hansen, M. Bzowski, Astron. Astrophys. **491**, 21–28 (2008)
E. Quémerais, R. Lallement, S. Ferron, D. Koutroumpa, J.-L. Bertaux, E. Kyrölä, W. Schmidt, J. Geophys. Res. **111**, 9114–9131 (2006)
J.D. Richardson, Y. Liu, C. Wang, D. McComas, Astron. Astrophys. **491**, 1–5 (2008)
D. Ruciński, M. Bzowski, Astron. Astrophys. **296**, 248–263 (1995)
D. Ruciński, M. Bzowski, Space Sci. Rev. **78**, 265–276 (1996)
K. Scherer, H.-J. Fahr, Astron. Astrophys. **404**, L47–L50 (2003)

S. Tarnopolski, M. Bzowski, Astron. Astrophys. (2008). doi:10.1051/0004-6361:20077
G.E. Thomas, Annu. Rev. Earth Planet. Sci. **6**, 173–204 (1978)
W.K. Tobiska, T. Woods, F. Eparvier, R. Viereck, L. Floyd, D. Bouwer, G. Rottman, O.R. White, J. Atmos. Terr. Phys. **62**, 1233–1250 (2000)
J. Vallerga, R. Lallement, M. Lemoine, F. Dalaudier, D. McMullin, Astron. Astrophys. **426**, 855–865 (2004)
V. Vasyliunas, G. Siscoe, J. Geophys. Res. **81**, 1247–1252 (1976)
A. Vidal-Madjar, Sol. Phys. **40**, 69–86 (1975)
M. Witte, Astron. Astrophys. **426**, 835–844 (2004)
T.N. Woods, G.J. Rottman, J. Geophys. Res. **102**, 8769–8779 (1997)
T.N. Woods, W.K. Tobiska, G.J. Rottman, J.R. Worden, J. Geophys. Res. **105**, 27,195–27,215 (2000)
F.M. Wu, D.L. Judge, Astrophys. J. **231**, 594–605 (1979)

Is the Sun Embedded in a Typical Interstellar Cloud?

Connecting Interstellar Gas in and out of the Heliosphere

Priscilla C. Frisch

Originally published in the journal Space Science Reviews, Volume 143, Nos 1–4, 191–204.
DOI: 10.1007/s11214-008-9394-4

Abstract Interstellar material around the Sun is typical for our neighborhood of the Galaxy. The physical properties and kinematics of the partially ionized interstellar material (ISM) near the Sun are typical of warm diffuse clouds in the solar vicinity. The direction of the interstellar magnetic field at the heliosphere, the polarization of light from nearby stars, and the kinematics of nearby clouds are naturally explained in terms of the S1 superbubble shell. The interstellar radiation field at the Sun appears to be harder than the field ionizing ambient diffuse gas, which may be a consequence of the low opacity of the tiny cloud surrounding the heliosphere.

Keywords ISM · Heliosphere

1 Introduction

Observations of interstellar gas in the Milky Way Galaxy span over ten orders-of-magnitude in spatial scales, and over six orders-of-magnitude in temperature. Interstellar material (ISM) is observed at the Earth's orbit, where interstellar He^0 has been counted by the Ulysses GAS detector and is detected through fluorescence of solar 584 Å (Weller and Meier 1981; Flynn et al. 1998; Salerno et al. 2003; Witte 2004; Möbius et al. 2004). Interstellar H^0 and other neutral interstellar atoms are driven into the heliosphere by the relative Sun-cloud motion of 26.2 km s^{-1}, where they become ionized and processed into pickup ions and anomalous cosmic rays. The question arises: 'Is the interstellar cloud feeding gas and dust into the heliosphere a typical interstellar cloud?' We know more about the circumheliospheric interstellar material (CHISM) than other clouds, because theoretical models combine *in situ* data with observations of nearby stars to model the cloud opacity profile. In this paper, the kinematics, temperature, ionization, composition and density of the CHISM are compared to low density ISM seen in the solar neighborhood. The answer to the title question is 'yes'.

P.C. Frisch (✉)
Dept. of Astronomy and Astrophysics, University of Chicago, Chicago, USA
e-mail: frisch@oddjob.uchicago.edu

DOI: 10.1007/978-1-4419-0247-4_15

The properties of the heliosphere are governed by the Local Bubble (LB) void and the Local Interstellar Cloud (LIC). The LB is transparent to radiation, exposing the LIC to a diffuse interstellar radiation field that includes both a soft X-ray background that anticorrelates with the H^0 column density (Bowyer et al. 1968), and radiation from distant hot and white dwarf stars (Gondhalekar et al. 1980; Vallerga 1998). The LIC and ISM surrounding the Sun are part of the shell of a superbubble expanding into the low density interior of the LB. The LIC ionization caused by the radiation field of the hot stars bordering the LB then dominates the heliosphere boundary conditions, while the ram pressure of gas in the expanding superbubble configures the heliosphere into the familiar raindrop shape.

2 Galactic Environment of the Sun

By now it is well known that the Sun resides in a region of space with very low average interstellar densities, the Local Bubble, formed by ISM associated with the ring of young stars bounding this void and known as Gould's Belt (Frisch 1995). The missing part of our understanding has been the origin of the Gould's Belt stars, or the Orion spur containing these stars. The Orion spur, which is located on the leading edge or convex side of the Sagittarius arm, is not included in models of the Milky Way spiral arms. However recent advances in our understanding of the formation of spurs, or 'feathers', on spiral arms naturally explains the origin of the Gould's Belt stars and the Orion spur. The interaction between the gaseous disk and the gravitational potential of spiral arms in the presence of a magnetic field induces self-gravitating perturbations that develop into two-dimensional flows that become unstable and fragment, driving spurs, or 'feathers', of star-forming material into low-density interarm regions (e.g. Shetty and Ostriker 2006). The Orion spur can be seen extending between galactic longitude $\ell \sim 60°$ distance ~ 1.2 kpc, and $\ell \sim 170°$ distance ~ 0.7 kpc, in Fig. 10 in Lucke (1978), together with two other spurs. These three spurs have pitch angles with respect to a circle around the galactic center of ~ 40–$55°$, and are separated by 550–1500 pc, values that are normal for MHD models of spur formation (Shetty and Ostriker 2006). Star formation occurs in spurs, and the Gould's Belt stars can thus naturally be explained by the Orion spur of which they are part. The Local Bubble is at the inner edge of the Orion spur. Based on MHD models of spur formation, we must conclude that the Gould's Belt environment of the Sun is normal, answering the question posed above with a 'yes'.

3 Heliospheric ISM

The CHISM ionization balance is governed by photoionization and recombination, so that neutral atoms and dust in the heliosphere trace the cloud physics, as well as the composition, temperature, density, and origin (Slavin and Frisch 2008). *In situ* gas and dust data, combined with radiative transfer models of CHISM ionization, test the "missing-mass" premise that assumes the combined interstellar atoms in gas and dust provide an invariant tracer of the chemical composition of the ISM (Savage and Sembach 1996; Frisch et al. 1999; Slavin and Frisch 2008). This test is potentially interesting because Gruen and Landgraf (2000) have shown that large and small dust grains couple to interstellar gas over different spatial scales, so that in the presence of active or recent grain shattering by interstellar shocks, local and global values for the gas-to-dust mass ratio may differ.

Interstellar particles with gyroradii larger than the distance between the particle and heliopause typically penetrate the heliosphere. If thermal and magnetic pressures are equal in the CHISM, then the magnetic field strength is ~ 2.7 μG (Slavin and Frisch 2008). Depending on the strength of the radiation field responsible for grain charging through photoejection of electrons, interstellar dust grains entering the heliosphere have radii larger than ~ 0.06–0.2 μm (Krueger and Gruen 2008; Frisch et al. 1999; Czechowski and Mann 2003). Grains with radii ~ 0.01–0.09 μm traverse the bow shock region, but are deflected around the heliopause with other charged populations. When STARDUST observations of interstellar grains become available, it will be possible to verify the missing-mass premise that the composition of the CHISM consists of the sum of elements in the gas and dust phases, and check whether solar abundances apply to the CHISM. If the missing-mass assumption is wrong it would explain the $\sim 50\%$ difference between the gas-to-dust mass ratios found from *in situ* observations of interstellar grains, and missing mass arguments utilizing radiative transfer models (Slavin and Frisch 2008).

Only the most abundant interstellar elements with first ionization potentials $\gtrsim$13.6 eV are observed in the heliosphere in detectable quantities, including H, He, N, O, Ne, and Ar. Each of these elements is observed in at least two forms, pickup ions (PUI) and anomalous cosmic rays (ACR). ACRs are accelerated PUIs. Pickup ions are formed when interstellar neutrals become ionized through either charge-exchange with the solar wind (H, N, O, Ne, Ar), photoionization (He, H), or electron impact ionization (He, N, Ar). Helium data yield the best temperature, He density, and velocity data, since the He charge-exchange cross-section with the solar wind is low and He penetrates to within ~ 0.5 AU of the Sun before ionization by photons and electron impact become significant (Möbius et al. 2004). The He data indicate for the CHISM: $T = 6300 \pm 400$ K, $n(\mathrm{He}^0) = 0.015 \pm 0.002$ cm^{-3}, and $V = -26.2 \pm 0.5$ km s^{-1}, and an upwind direction of $\lambda = 255.0^\circ \pm 0.6^\circ$, $\beta = 5.2 \pm 0.3^\circ$ (corrected for J2000 coordinates Möbius et al. 2004; Witte 2004). Early December each year the Earth passes through a cone of gravitationally focused He, extending over 5 AU downwind of the Sun.

Hydrogen is the most abundant ISM observed in the heliosphere, however the initial thermal interstellar velocity distribution of H^0 is modified and deformed as H^0 enters and propagates through the heliosphere. Interpretation of the Lyα fluorescence and PUI data require corrections for the weak coupling between H^0 and the interstellar magnetic field outside of the heliosphere due to H^0–H^+ interactions, strong filtration through charge-exchange between interstellar protons and H^0 in the heliosheath regions, deformation of the H^0 velocity distribution as H atoms enter and propagate through the heliosphere, and the solar-cycle dependent variation in the ratio of radiation pressure and gravitational forces. These effects are discussed elsewhere in this issue (e.g. Bzowski 2008; Quemerais 2008; Wood 2008; Opher 2008b; Pogorelov 2008). These observations and models of PUI H inside of the heliosphere are consistent with an H density at the termination shock of ~ 0.11 cm^{-3}, and when combined with filtration values yield an interstellar density of $n(\mathrm{H}^0) \sim 0.195 \pm 0.02$ cm^{-3} for the CHISM (Bzowski et al. 2007). The H filtration factor is based on the Moscow Monte Carlo model, which also yields a CHISM plasma density of $n(\mathrm{H}^+) = 0.04 \pm 0.02$ cm^{-3}. These results are in excellent agreement with the completely independent radiative transfer results that conclude $n(\mathrm{H}^0) = 0.19$–0.20 cm^{-3}, and $n(\mathrm{H}^+) = 0.07 \pm 0.02$ cm^{-3} for the CHISM (Slavin and Frisch 2008).

Comparisons between abundances of neutrals in the CHISM as predicted by radiative transfer studies, with interstellar neutral abundances based on PUI and ACR densities corrected to values at the termination shock, require that the filtration of neutrals crossing outer heliosheath regions is understood (Cummings et al. 2002; Müller and Zank 2004; Izmodenov et al. 1999). In the heliosheath regions, charge-exchange with interstellar protons increases filtration of O, and reverse charge-exchange potentially allows interstellar O^+

into the heliosphere. Electron impact ionization contributes to filtration of N and Ar. Reverse charge-exchange between interstellar ions and protons in the outer heliosheath is insignificant for all elements except possibly O and H. The range of filtration factors found for H, He, N, O, Ne, and Ar are listed in Slavin and Frisch (2008). Correcting PUI densities given by Gloeckler and Fisk (2007) for the termination shock with calculated filtration factors yields interstellar densities for these elements that are consistent, to within uncertainties, with the range of neutral densities predicted by the CHISM radiative transfer models for these elements (Table 4 in Slavin and Frisch 2008). In the radiative transfer models, abundances of H, He, N and O are variables that are required to match the data, but Ne and Ar abundances are assumed at 123 ppm and 2.82 ppm, respectively.

The only measurements of Ne in the local ISM are the *in situ* PUI and ACR data; Ne is a sensitive tracer of the ionization conditions of the CHISM because three ionization states are present in significant quantities, Ne^0:Ne^+:Ne^{++} $\sim$1.0:3.3:0.8 (Slavin and Frisch 2008). The radiative transfer model results are also consistent with the Ne abundance of $\sim$ 100 ppm found in the Orion nebula (Simpson et al. 2004), and within the range of uncertainties for the solar Ne abundance.

Neutral Ar traces the equilibrium status of the CHISM because Ar^0 and H^0 are the end products of processes with similar recombination rates, but have different photoionization rates (Slavin 2008). The radiative transfer models (Slavin and Frisch 2008) together with the PUI Ar data indicate the CHISM is in ionization equilibrium. The ratio $Ar^\circ/H^0 \sim 1.0 \times 10^{-6}$ in the CHISM found from PUI data and radiative transfer models is comparable to interstellar values towards nearby stars based on the FUV data, $Ar^0/H^0 \sim 1.2 \times 10^{-6}$ (Jenkins et al. 2000). Agreement with the FUSE data can be achieved by a small increase in the assumed Ar abundances in Slavin and Frisch (2008).

Isotopes in PUIs, ACRs, and He indicate that the CHISM is formed from similar material as the Sun. The ratios $^{22}Ne/^{20}Ne \sim 0.073$ and $^{18}O/^{16}O \sim 0.002$ are close to isotopic ratios in the solar wind (Cummings and Stone 2007; Leske 2000). He data gives $^3He/^4He \sim 1.7 \times 10^{-4}$, which is similar to meteoritic and HII region values (Salerno et al. 2003; Gloeckler and Fisk 2007). Evidently the expected ^3He enrichment of the ISM by nucleosynthesis in low-mass stars has not affected the CHISM. The ^{22}Ne isotope indicates that the CHISM is not significantly mixed with ejecta from Wolf Rayet stars common to OB associations, where ^{22}Ne would be enriched by He-burning. The CHISM gas therefore appears isotopically similar to solar system material, and ^3He values are consistent with isotopic ratios in HII regions.

Summarizing, observations of interstellar products inside of the heliosphere yield densities and abundances for H, He, N, O, Ne, and Ar that are in agreement with radiative transfer models of LIC absorption components in the star ε CMa. Argon has similar abundances, Ar^0/H^0, in the CHISM and towards near white dwarf stars. Isotopic ratios suggests that the CHISM has a solar composition. *In situ* observations of interstellar dust grains yield a gas-to-dust mass ratio that varies by 50% or more from values predicted by radiative transfer models, indicating that the either the abundances of elements depleted onto dust grains or the true metallicity of the CHISM is not understood. The CHISM abundances determined from *in situ* data are consistent with abundances typical of low density ISM, so that based on *in situ* observations of ISM we conclude that the answer posed above is 'yes'.

4 Kinematics and Temperatures of Very Local ISM versus Warm Interstellar Gas

Using *Copernicus*, IUE, and optical data inside of the heliosphere and towards nearby stars such as α Oph at 14 pc, Frisch (1981) showed that the ISM inside and close to

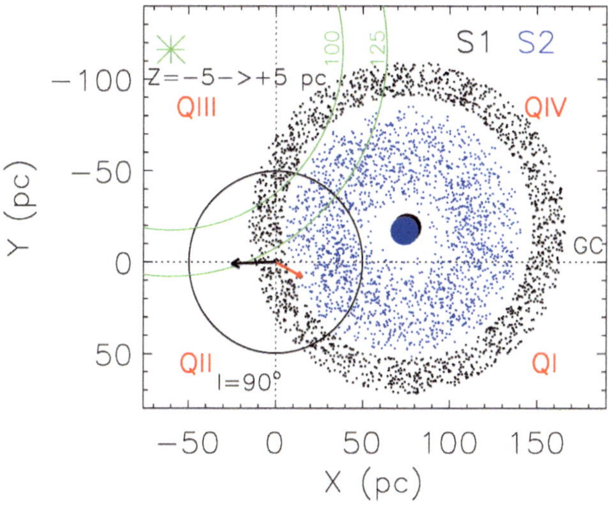

Fig. 1 A slice of the S1 (*black dots*) and S2 (*blue dots*) shells within 5 pc of the galactic plane ($|Z| < 5$ pc) are plotted in x, y coordinates, where the x-axis is directed towards the galactic center (from Frisch 2008). *Red and black arrows* show the LSR motions of the Sun and CLIC, respectively. The *black circle* is 50 pc from the Sun. The two superimposed *blue and black large dots* show the centers of the two shells. The notation Q1–QIV indicate galactic quadrants. The *green star* shows the x, y position of the brightest extreme ultraviolet (EUV) source in the sky, ε CMa, which is located 15 pc below the galactic plane. The *two concentric green circles* show regions 100 pc and 125 pc equidistant from ε CMa. The S1 and S2 shells are constructed using the shell parameters in Wolleben (2007)

the heliosphere has the kinematic and abundance properties expected for an origin related to the Loop I superbubble. The first spectrum of Lyα fluorescence from interstellar H^0 inside of the heliosphere, acquired by *Copernicus* during 1975 (Adams and Frisch 1977), yielded the H^0 velocity in the upwind direction of ~ -24.7 km s^{-1} (neglecting heliospheric acceleration and converting to the current upstream direction (Witte 2004; Frisch 2008)). This H^0 velocity projects to ~ -21.1 km s^{-1} in the α Oph direction, and differs somewhat from the dominant cloud velocities known for that direction of $\sim -24 \pm 1$ km s^{-1} (Marschall and Hobbs 1972). It is now known that the Lyα line backscattered emission has a significant contribution from secondary H^0 atoms, and also that the LIC velocity observed inside of the heliosphere differs by ~ 1 km s^{-1} from the gas velocity towards the nearest star α Cen, $\sim 50°$ from the heliosphere nose (Linsky and Wood 1996), and ~ 3 km s^{-1} from velocities of nearby gas in the upwind direction towards 36 Oph (Wood et al. 2000). A more complete picture of the kinematics and temperature structure of the LISM is now available. The Sun is embedded in an ISM flow, the complex of local interstellar clouds (CLIC), which has an upwind direction in the Local Standard of Rest (LSR) directed towards the center of the S1 subshell of the Loop I superbubble shell around the Scorpius–Centaurus Association (Frisch and York 1986; Frisch 1995; Frisch et al. 2002; Wolleben 2007; Redfield and Linsky 2008). Figure 1 shows the S1 shell of Wolleben, found by fitting 1.4 GHz and 23 GHz polarization data, the solar apex motion, and the bulk motion of the CLIC through the local standard of rest (LSR). The CLIC LSR upwind direction, $(\ell, b) \sim (358°, -5°)$ (Frisch et al. 2002; Frisch and Slavin 2006) is $\sim 10°$ from the center of the S1 shell at $(\ell, b) \sim (346 \pm 5°, 3 \pm 5°)$. The CLIC kinematics is thus naturally explained by the expansion of the S1 shell to the solar location (Frisch 1981). The expansion of Loop I has been modeled by Frisch (1995, 1996), and corresponds to an origin during a star formation epoch $\sim$ 4–5 Myrs ago.

Morphologically prominent shells such as the S1 shell are common features in the ISM, often found between spiral arms where spurs are seen. Shell properties have been surveyed in the H^0 21-cm hyperfine transition, revealing filamentary structures consisting partly of warm neutral material (WNM). Column densities for WNM are typically $N(H^0) > 10^{19}$ cm^{-2}. WNM with column densities comparable to the CLIC, $N(H^0) < 10^{18.5}$ cm^{-2}, or LIC, $N(H^0) \sim 10^{17.6}$ cm^{-2}, are not yet observed. Zeeman data show that shell structures are associated with magnetic fields of $\sim$ 6.4 μG or less. Unfortunately, Zeeman-splitting data show that flux-freezing does not occur in low density ISM, $n < 10^3$ cm^{-3} (Crutcher et al. 2003), so the magnetic field strength at the solar location can not be inferred from the magnetic field in more distant portions of the S1 shell.

Turning back to the question "Is the Sun embedded in a typical interstellar cloud". I use the Arecibo Millennium Survey of the H^0 21 cm line to define the meaning of "typical". The Arecibo survey is a complete and unbiased survey of warm and cold interstellar clouds, as seen from the tropical Arecibo sky (Heiles and Troland 2003). The systematic fitting of Gaussian components to the emission profiles revealed that 60% of the ISM mass is contained in warm neutral material (WNM), with median cloud column densities of 1.3×10^{20} cm^{-2}, compared to the lower median column density of the cold neutral medium (CNM) of 5×10^{19} cm^{-2}.

The kinematics of the Arecibo clouds can be used as a benchmark for answering the above question as it applies to cloud kinematics. In Fig. 2, left, the kinematics of the CLIC cloudlets (LSR velocities) are compared to the kinematics of the WNM and CNM. For the CLIC velocities I use Ca^+ and UV absorption line data such as D^0 (Frisch et al. 2002; Redfield and Linsky 2004b; Wood et al. 2005), and for plotting purposes the ratios $Ca^+/H^0 = 10^{-8}$ and $D^0/H^0 = 10^{-4.82}$. It is immediately apparent that the kinematics of CLIC clouds are comparable to the global kinematics of both WNM and CNM clouds in the Arecibo survey.

A second test using the Arecibo is also made. The range of temperatures for the WNM are shown in Fig. 2, right. CLIC temperatures from Redfield and Linsky (2004a) are also plotted, although there are still poorly understood aspects of these temperatures (Sect. 6). The Arecibo temperatures shown for WNM include both spin temperatures (red arrows) and the kinetic temperatures (red triangles) based on the FWHM of the fitted components. For the WNM, the kinetic and spin temperatures are upper and lower limits on the thermal temperature, respectively, because turbulence is not removed, and the true spin temperature is a function of a limit on the cloud opacity (Heiles and Troland 2003). The median kinetic temperatures for the WNM for clouds for latitudes $b < 30°$ versus $b > 30°$ are, respectively, 5,962 K, and 5,182 K. The same ratio of low-to-high latitude WNM temperatures is found for spin temperatures (1.15). Low and high latitude WNM median column densities are, respectively, $N(H^0) = 10^{20.68}$ cm^{-2} and $N(H^0) = 10^{19.98}$ cm^{-2}. Since the CHISM temperatures is 6,300 K, it is within the WNM temperature range. Figure 2, right, shows that the temperatures of the CLIC clouds (green dots) fall consistently between the upper and lower limits on the WNM temperatures. In addition, the CHISM temperature is close to the median kinetic temperature of WNM components with $b < 30°$. The typical CLIC column densities, $N(H^0) < 10^{18.5}$ cm^{-2}, are below the range of detected WNM column densities. Since photoionization dominates the heating of the CHISM (Slavin and Frisch 2008), the higher low-latitude WNM temperatures also suggest that radiation heating of the ISM is stronger at low latitudes than high latitudes.

Based on the kinematical and temperature information in the Arecibo Millennium Survey, the answer to the question posed above is again 'yes'.

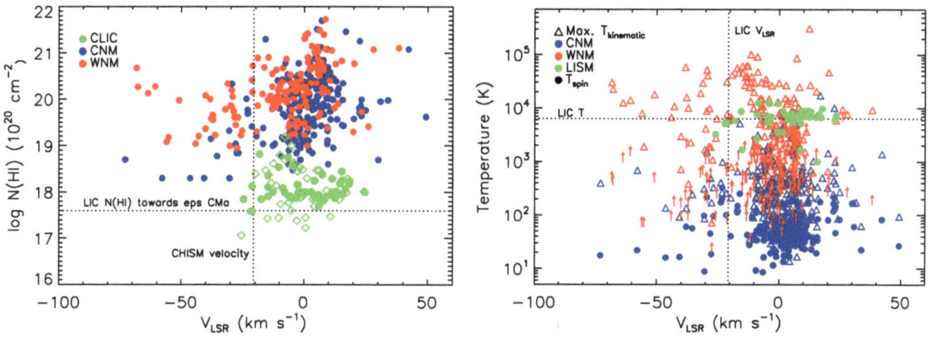

Fig. 2 The column densities $N(H^0)$, *left*, and H^0 temperatures, *right*, are plotted as a function of velocity for the WNM (*red*) and CNM (*blue*) components in the Arecibo survey (Heiles and Troland 2003) and for the CLIC. The spin temperatures (*red arrows*) obtained for WNM by Heiles and Troland (2003) are lower limits. The CLIC components (*green dots*) are based on Ca^+ (*diamonds*, e.g. Frisch et al. 2002; Frisch 2008), H^0, and D^0 (*dots*, Redfield and Linsky 2004b; Wood et al. 2005). The CHISM LSR velocity is indicated, based on the Standard solar apex motion

5 Partially Ionized Gas and the Interstellar Radiation Field

Two coupled attributes dominate the CHISM: it is partially ionized, and it is low column density. The first attribute follows from the second in the presence of photons with energies > 13.6 eV able to ionize H^0. The earliest observations of H^0 and He^0 inside the solar system found ratios of $H^0/He^0 \sim 6$ (Ajello et al. 1987; Weller and Meier 1981), in contrast to EUV observations of five white dwarf stars with average distances of 57 pc and $H^0/He^0 \sim 14$ (Frisch 1995). Cheng and Bruhweiler (1990) found that hot star radiation dominates H ionization of the LISM but soft X-rays produced He ionization, and therefore yielding higher ionization levels for He compared to H. More recent studies show that He ionization is produced by EUV emission from a conductive interface between the LIC and LB plasma, white dwarf stars, and the low energy tail of the soft X-ray background (Slavin and Frisch 2008; Slavin 2008; Shelton 2008, this issue). The low H/He ratio found inside of the heliosphere thus becomes evidence of the strong H filtration in heliosheath regions.

Ionized gas is a major component of the solar neighborhood. FUSE observations of ISM towards white dwarf stars within 70 pc find up to $\sim 70\%$ ionization levels, and electron densities in the range 0.025–0.25 cm^{-3} for stars with $N(H^0) = 10^{18.8}$–$10^{19.6}$ cm^{-2} (Lehner et al. 2003). Hydrogen is $\sim 23\%$ ionized at the heliosphere, which is within the ionization range obtained by FUSE. Radiative transfer models (Slavin and Frisch 2008) that predict the heliosphere boundary conditions show that the CHISM electron densities of ~ 0.05–0.09 cm^{-3} are similar to electron densities found by FUSE, and also in the diffuse ionized gas sampled by pulsar dispersion measures and Hα recombination lines.

The distribution of ionized gas near the LB is dominated by classic HII regions around hot stars; the Wisconsin H-α Mapper (WHAM) survey of the red Hα line shows these regions beautifully (Haffner et al. 2003). However, recombination emission from low density ionized gas carries more subtle information about partially ionized regions such as the LIC. Ionized gas in the solar vicinity fills $\sim 20\%$ of the disk and is contained in warm diffuse low density regions with $n \sim 0.1$ cm^{-3} and $T \sim 10^4$ K. Ionization of this gas is powered by O-stars, and requires transparent voids through which the O-star radiation can propagate; the Local Bubble is such a void. A detailed comparison of H^0 and Hα in a ~ 120 square-degree of sky showed that at least 30% of the Hα emission is both spatially and kinematically associated with warm H^0 21-cm features, many of which are filamentary (Reynolds et al. 1995). Some of this H^+ is in regions physically distinct from the H^0 gas. Ionization levels reach 40% for these low density, ~ 0.2–$0.3\ cm^{-3}$, clouds. The temperature of diffuse ionized gas varies between 6,000 K and 9,000 K, with higher temperatures at higher latitudes (Haffner et al. 1999). This result follows from the temperature dependence of the Hα intensity of $\sim T^{-0.9}$, and [N^+] and [S^+] data. The CHISM temperature of 6300 K is within the range for the Hα clouds.

Is the diffuse Hα emission formed in partially ionized gas similar to the LIC? The answer to this is 'probably', however whether or not the LIC radiation field is typical of diffuse gas is an open question. Observations of the He^0 5876 A recombination line in the diffuse ionized gas yield low levels of ionized He compared to H, although the dominant O-star ionization source would predict higher levels of He ionization. Reynolds (2004) compared LIC radiative transfer model results with the partially ionized LIC gas for four sightlines through diffuse ionized gas where the forbidden 6300 A O° line is measured. These sightlines indicated H ionization fractions of $> 70\%$, compared to the LIC value of $\sim 23\%$. In addition, for these diffuse gas data, the ionization fraction of He is 30–60% of that of H, but the absolute He ionization level is similar to the LIC. Together these results suggest that the radiation field at the LIC is harder than the diffuse radiation field that maintains the warm ionized medium. From the relative H and He recombination lines, one thus might conclude that the LIC is not typical of diffuse ionized gas. However, the LIC emission measure is $EM \sim 0.003\ cm^{-6}$ pc, which is below the WHAM sensitivity. Radiative transfer models show that very low column density clouds such as the LIC are transparent to H-ionizing radiation, and such clouds may be invisible to WHAM. The question as to whether the relative ionizations of H and He in the LIC is typical of ionized gas thus remains an open question, but the low LIC column density probably explains the hardness of the local radiation field compared to more distant regions.

Other properties of the interstellar radiation field that are important for LIC ionization include the EUV and soft X-ray fluxes (Slavin and Frisch 2008). Some doubt has been cast on the absolute flux level of interstellar photons with energies < 0.25 keV because of contamination of the soft X-ray background (SXRB) by heliospheric emissions at energies > 0.4 keV from charge-exchange between interstellar neutrals and the solar wind (Shelton 2008; Koutroumpa 2008). At 0.1 eV, LB emission has been modeled as contributing $\sim 50\%$ of the flux (Henley et al. 2007). Clumping in the ISM may change this picture, however, since a typical value of $N(H^0) \sim 10^{21}\ cm^{-2}$ may include tiny cold clouds such as the $N(H^0) \sim 10^{18}\ cm^{-2}$ structures that are completely opaque at low energies (Stanimirovic and Heiles 2005). If the X-ray emitting plasma contains embedded clumps of ISM with significant opacity at ~ 0.1–0.2 keV, the energy dependence of the ISM opacity will be significantly altered from that of a homogeneously distributed ISM (Kahn and Jakobsen 1988). This effect will be significant for Loop I X-ray emission, where embedded molecular clouds are found. The physical properties of the Local Bubble plasma need to be revisited by including not only foreground emission from charge-exchange between the solar wind and interstellar H^0, but also clumping in the ISM as noted by Kahn and Jakobsen (1988).

There is one point that is not yet appreciated. Since the hot star ε CMa dominates the 13.6 eV radiation field in the solar vicinity, and therefore the flux of H-ionizing photons, sightlines through the third and fourth galactic quadrants (QIII, QIV), $\ell = 180°$ to $\ell = 360°$, will sample ISM with higher ionization levels than sightlines through the first two galactic quadrants (Frisch 2008). This occurs because ISM associated with the S1 shell structure is closer to ε CMa in QIII and QIV, than in the first two galactic quadrants. The relative locations of the S1 shell and ε CMa are shown in Fig. 1.

6 Chemical Composition of the ISM at the Sun

The outstanding feature of warm low density interstellar clouds is that the abundances of refractory elements such as Fe, Ti, and Ca, are enhanced by an order of magnitude when compared to abundances in cold clouds at ~ 50 K. The enhanced abundances were originally discovered for the Ca^+ line seen in high-velocity clouds (Routly et al. 1952), although the importance of the ionization balance between Ca^{++} and Ca^+, which favors Ca^{++} in warm ionized gas, was not fully appreciated at that time. Enhanced abundances are particularly strong for Ti, which is one of the first elements to condense onto dust grains with $T_{\mathrm{condensation}} \sim 1500$ K. Column densities of Ti^+ can be directly compared with H^0 abundances without ionization corrections because Ti^+ and H^0 have similar ionization potentials. Enhanced refractory element abundances in warm gas at higher velocities has been modeled as due to the destruction in shock fronts of refractory-laden interstellar dust grains composed of silicates and/or carbonaceous material (e.g. Slavin et al. 2004). The CLIC gas shows such abundances, requiring the CLIC grains to have been processed through shocks of $\sim 80\ \mathrm{km\,s^{-1}}$ (Slavin 2008; Frisch et al. 1999). Refractory elements such as Mg, Si, Fe, and Ca are predominantly singly ionized in the LIC, so that ionization corrections are required to obtain accurate abundance information. Ionization corrections are generally not available for determining abundances of distant warm gas; however the range of uncertainty in elemental abundances is large enough that with or without ionization corrections, the CLIC gas has typical abundances for low density clouds (e.g. Welty et al. 1999). The radiative transfer models provide accurate CHISM abundances that are discussed by Slavin (2008), and except for one sightline CHISM abundances are typical for low density ISM.

There is only one sightline through the CLIC that shows a poorly understood abundance pattern, and this is the sightline of α Oph that led to my original conclusion that the Loop I superbubble shell has expanded to the solar location (Frisch 1981). The strongest observed Ca^+ line in the CLIC is towards α Oph, where strong Ti^+ is also seen. The star α Oph is 14 pc from the Sun in the direction of the North Polar Spur, and the interstellar gas in this sightline may be in the region where the S1 and S2 shells are in collision, so that shock destruction of the grains is underway.

Two caveats must be attached to most determinations of elemental abundances: (1) Common refractory elements tend to have FIP's < 13.6 eV, so that ionization corrections are required to obtain accurate abundances. (2) Accurate H^0 column densities are also required so that abundances per H-atom can be calculated. The first requirement is seldom met, because cloud ionization data at best typically return electron densities calculated based on either Mg^0/Mg^+ or C^{+*}/C^{++} ratios. Total H^+ column densities are not directly measured and must be inferred. Optimally, radiative transfer models of each cloud could provide the same quality of results now available for the LIC. The second requirement is notoriously difficult to achieve for H^0 values relying on the heavily saturated Lyα line.

There are difficulties with extracting reliable column densities from lines where thermal line-broadening dominates turbulent line-broadening. The Voigt profile used to determine the parameters of absorption lines invokes the Doppler b-value ($b_{\mathrm{D}} \sim FWHM/1.7$):

$$b_{\mathrm{D}}(T,m)^2 = b_{\mathrm{thermal}}(T,m)^2 + b_{\mathrm{turbulent}}^2, \quad (1)$$

where $b_{\mathrm{turbulent}}$ has no mass (m) or temperature (T) dependence. This assumption appears to break down for stars within 10 pc of the Sun, as is shown in Fig. 3 (using data from Redfield and Linsky 2004b) by the correlation between $N(\mathrm{D}^0)$ and temperature T, and the anticorrelation between $b_{\mathrm{thermal}}(T,2)$ and turbulence $\xi = b_{\mathrm{turbulent}}$. No known cloud physics

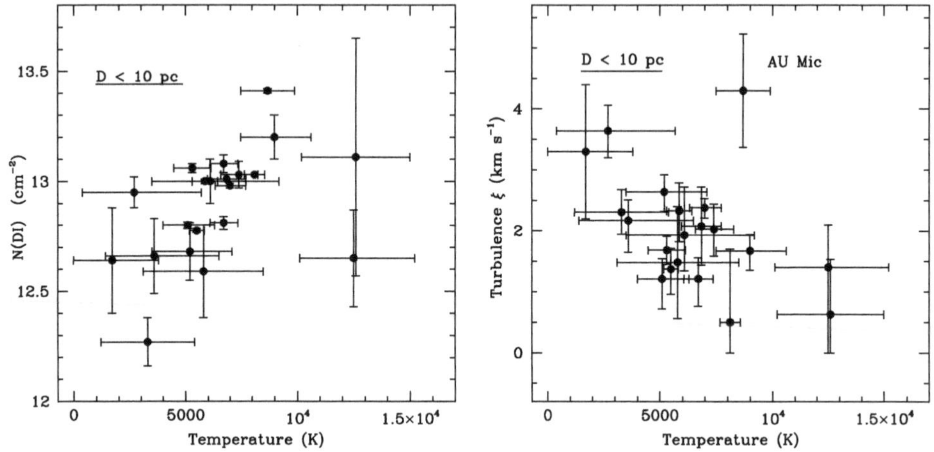

Fig. 3 Temperature versus column density N(DI) (*left*) and temperature versus turbulence ξ (*right*) for interstellar absorption components seen towards stars within 10 pc of the Sun (data from Redfield and Linsky 2004b)

explains a correlation between D^0 and T that is accompanied by an anticorrelation between turbulent and thermal broadening. One explanation for this effect is that the assumption of isotropic Maxwellian gas velocities and mass-independent turbulence breaks down in a partially ionized low density ISM due to the coupling between ions and magnetic fields.

The summary conclusion of this section is that CLIC and CHISM abundances are similar to abundances in partially ionized gas. Because of the uncertainties, this statement holds true when elemental abundances are correctly compared to $H^0 + H^+$, or H^0 alone. The one caveat on this statement is that D^0 column densities for stars within 10 pc show evidence of correlations that indicate the line-broadening parameter is incorrectly defined. The one sightline that is not typical is α Oph, which may hold hidden clues about colliding superbubbles near the Sun.

7 Interstellar Magnetic Field at the Solar Location

The orientation, but not the polarity, of the interstellar magnetic field (ISMF) at or near the heliosphere can be derived from optical polarization vectors for nearby stars, < 30 pc. This orientation can then be compared with the local magnetic field direction derived from the S1 shell low frequency radio continuum polarization (1.4 GHz, Wolleben 2007; Frisch 2008). The strongest optical polarizations are seen for stars located along the ecliptic plane and with a peak in the polarization that is offset by $\lambda \sim 40°$ from the direction of the heliosphere nose (Frisch 2008). The orientation of the S1 shell magnetic field in the heliosphere nose region agrees with the values obtained from the optical polarization direction, to within the uncertainties, for the Wolleben (2007) angle parameter $B_\phi = -42°$. The magnetic field at the position of the polarized stars forms an angle of $\sim 65 \pm 9°$ with respect to the ecliptic plane, and $\sim 55 \pm 9°$ with respect to the galactic plane. At the position of the He^0 inflow direction, the S1 shell configuration consistent with the polarization data gives a magnetic field inclination of $\sim 73 \pm 10°$ with respect to the ecliptic plane, and $\sim 44 \pm 10°$ with respect to the galactic plane. When the uncertainties on the upwind directions of interstellar H^0 and He^0 flowing into the heliosphere are considered, then the offset angle between these two inflow directions is $4.8 \pm 0.6°$, and these two upwind directions define an angle of $55 \pm 20°$ with respect to an ecliptic parallel (Frisch 2008; Möbius et al. 2004; Lallement et al. 2005). When the uncertainties are considered the H^0–He^0 offset angle, and the S1 shell direction that is consistent with the optical polarization data, yield consistent ISMF orientations.

A non-zero angle between the ISMF direction and the inflowing ISM velocity vector causes an asymmetric heliosphere, including a possible tilt of $\sim 12°$ between the heliosphere nose, as defined by the maximum outer heliosheath plasma density, and the ISM velocity (e.g. Ratkiewicz et al. 1998; Linde et al. 1998; Ratkiewicz et al. 2008). Is the S1 shell field orientation at the heliosphere consistent with models of the known asymmetries of the heliosphere? Pogorelov et al. (2008) argue that large angles between the upwind ISM and magnetic field directions are required to reproduce the heliospheric asymmetry seen by the Voyager 1 and Voyager 2, which encountered different termination shock distances at ~ 94 AU and ~ 84 AU (Stone 2008). A magnetic field direction tilted by $\sim 60°$ with respect to the ecliptic plane reproduces the offset angle between H^0–He^0, but not the heliospheric asymmetry seen by the Voyager satellites (Pogorelov et al. 2008). Opher (2008a) reports that an interstellar field direction inclined by ~ 60–$90°$ with respect to the galactic plane reproduces the Voyager results, including particle streaming in the outer heliosheath. Ratkiewicz et al. (2008) find that an ISMF directed towards galactic coordinates $\ell = 217 \pm 14°$, $b = -50 \pm 9°$ explains the position of the Lyα maximum observed by the Voyager spacecraft in the outer heliosphere. These models, the interstellar polarization data, and the S1 shell predictions of the ISMF direction at the heliosphere agree to within the large uncertainties remaining in this problem.

If magnetic and thermal pressures in the CHISM are approximately equal, then the CHISM field strength is ~ 2.7 μG (Slavin and Frisch 2008). The polarity of the CHISM field is a more difficult question, and can presently only be inferred from the polarity of the nearby global magnetic field. The global magnetic field direction in the solar vicinity is directed towards $\ell \sim 80°$ (Han 2006). For a classical expanding superbubble model (MacLow and McCray 1988), where a shock front sweeps up interstellar material and compresses magnetic field lines in the expanding shell, the S1 shell expansion would have preserved the global field polarity so that the S1 shell field direction at the Sun is directed from the south to north. With this polarity, the direction of the interstellar magnetic field at the heliosphere nose is shown in Fig. 4 (see Frisch 2008 for the field direction in galactic coordinates). This model neglects possible additional rotation of the field direction, such as may arise from coupling between the ISMF and $\sim 18°$ tilt of the plane of Gould's Belt with respect to the galactic plane.

Based on the similar magnetic field directions obtained from the S1 shell magnetic field at the heliosphere and the polarization of light for stars close to the Sun, the magnetic field in the CLIC and CHISM is typical. The field strength inferred from pressure equilibrium in the CHISM, ~ 2.7 μG, is typical of field strengths found from Zeeman splitting of the 21-cm line for WNM in the Arecibo survey, and $\sim 50\%$ larger than the large-scale ordered magnetic component inferred from pulsar data (Rand and Kulkarni 1989). Given that there is no evidence that the interstellar magnetic field at the heliosphere is anomalous, again the answer to the posed question is 'yes'.

8 Conclusions

By all of the standard measures of interstellar clouds, such as temperature, velocity, composition, ionization, and magnetic field, the interstellar gas inside of the heliosphere and in

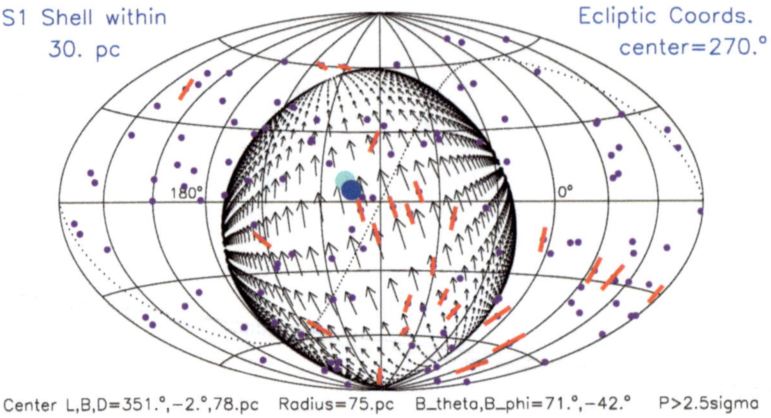

Fig. 4 The magnetic field associated with the parts of the S1 shell within 30 pc is plotted in ecliptic coordinates for an aitoff projection. The parameters for the S1 shell given in Wolleben (2007) have been varied within the range of allowed uncertainties to yield the best match to polarization data towards stars in the heliosphere nose region (indicated by the *dark blue dots*). The *dark* and *light blue dots* show the inflow directions of interstellar He^0 and H^0 into the heliosphere. The S1 subshell parameters used in the above figures correspond to a shell center at $(\ell, b) = (351°, -2°)$ and 78 pc away, shell radius of 75 pc, and magnetic field angles $B_\theta = 71°$ and $B_\phi = -42°$. The *dots* show stars within 50 pc with polarization data, and the *red bars* show polarization vectors for stars where polarizations are larger than 2.5σ (Tinbergen 1982; Piirola 1977; Frisch 2007)

the LIC are typical of warm partially ionized gas seen elsewhere in the neighborhood of the Sun. Unfortunately clouds with low LIC-like column densities are not yet observable in either H^0 21-cm or $H\alpha$ recombination lines, so that clouds with hard radiation fields similar to the LIC can only be identified through ultraviolet absorption lines.

The association of the LISM and LIC gas with the expanding S1 superbubble shell, and possibly the S2 shell, naturally explains the kinematics of ISM within ~ 30 pc. Furthermore, the S1 shell structure leads to specific predictions about the relative ionizations of different parts of the shell due to proximity to ε CMa and other nearby hot stars (Frisch et al. 2008). The S1 shell also predicts a direction of the interstellar magnetic field at the heliosphere that is consistent with observations of optical polarizations towards nearby stars.

It is evident that the answer to the question posed by the title of this paper is 'yes', so this paper will close with a more difficult question posed years ago by Eugene Parker: "What is an interstellar cloud". Originally clouds like the LIC and other LISM clouds were named "intercloud medium". The LIC column density towards Sirius, α CMa, suggests the Sun has entered the LIC within the past few thousand years (Frisch 1994), while the velocity discrepancy between interstellar He^0 inside of the heliosphere and ISM in the upwind direction towards α Cen and 36 Oph suggests the Sun is at the edge of the LIC (Lallement et al. 1995; Wood et al. 2000). Are there two separate clouds adjacent to the heliosphere? Or instead are we crossing a pocket of microturbulence with scale sizes of ~ 0.02 pc? What is a cloud anyway? The LIC is $\sim 10^{21}$ orders-of-magnitude less dense than the terrestrial atmosphere.

Acknowledgements The author would like to thank NASA for support through grants NAG5-13107, NNG06GE33G, and through the IBEX mission. My collaborator Jon Slavin has made important contributions to much of this work. The author would also like to thank the International Space Sciences Institute for sponsoring the workshop "From the Outer Heliosphere to the Local Bubble: Comparison of New Observations with Theory".

References

T.F. Adams, P.C. Frisch, Astrophys. J. **212**, 300–308 (1977)
J.M. Ajello, A.I. Stewart, G.E. Thomas, A. Graps, Astrophys. J. **317**, 964–986 (1987)
C.S. Bowyer, G.B. Field, J.F. Mack, Nature **217**, 32 (1968)
M. Bzowski, Space Sci. Rev. (2008, this issue)
M. Bzowski, E. Moebius, S. Tarnopolski, V. Izmodenov, G. Gloeckler, Density of neutral interstellar hydrogen at the termination shock from Ulysses pickup ion observations. ArXiv e-prints, 710 (2007)
K. Cheng, F.C. Bruhweiler, Astrophys. J. **364**, 573–581 (1990)
R. Crutcher, C. Heiles, T. Troland, Lect. Notes Phys. Berlin Springer **614**, 155–181 (2003)
A.C. Cummings, E.C. Stone, Space Sci. Rev. **130**, 389–399 (2007)
A.C. Cummings, E.C. Stone, C.D. Steenberg, Astrophys. J. **578**, 194–210 (2002)
A. Czechowski, I. Mann, Astron. Astrophys. **410**, 165–173 (2003)
B. Flynn, J. Vallerga, F. Dalaudier, G.R. Gladstone, J. Geophys. Res. **103**, 6483 (1998)
P. Frisch, D.G. York, Interstellar clouds near the Sun, in *The Galaxy and the Solar System* (University of Arizona Press, Arizona, 1986), pp. 83–100
P.C. Frisch, Nature **293**, 377–379 (1981)
P.C. Frisch, Science **265**, 1423 (1994)
P.C. Frisch, Space Sci. Rev. **72**, 499–592 (1995)
P.C. Frisch, Space Sci. Rev. **78**, 213–222 (1996)
P.C. Frisch, Astrophys. J. (2007). arXiv:0707.2970
P.C. Frisch, The S1 Shell and interstellar magnetic field and gas near the heliosphere. Astrophys. J. (2008, submitted)
P.C. Frisch, J.D. Slavin, Short term variations in the galactic environment of the Sun, in *Solar Journey: The Significance of Our Galactic Environment for the Heliosphere and Earth*, ed. by P.C. Frisch (2006)
P.C. Frisch, A. Choi, D.G. York, L.M. Hobbs, Observations of CaII and NaI in the local interstellar medium (2008, in preparation)
P.C. Frisch, J.M. Dorschner, J. Geiss, J.M. Greenberg, E. Grün, M. Landgraf, P. Hoppe, A.P. Jones, W. Krätschmer, T.J. Linde, G.E. Morfill, W. Reach, J.D. Slavin, J. Svestka, A.N. Witt, G.P. Zank, Astrophys. J. **525**, 492–516 (1999)
P.C. Frisch, L. Grodnicki, D.E. Welty, Astrophys. J. **574**, 834–846 (2002)
G. Gloeckler, L. Fisk, Johannes Geiss' investigations of solar, heliospheric and interstellar matter. "Composition of Matter", Space Sciences Series of ISSI, publ. Springer, vol. 27 (2007)
P.M. Gondhalekar, A.P. Phillips, R. Wilson, Astron. Astrophys. **85**, 272 (1980)
E. Gruen, M. Landgraf, J. Geophys. Res. **105**, 10291–10298 (2000)
L.M. Haffner, R.J. Reynolds, S.L. Tufte, Astrophys. J. **523**, 223–233 (1999)
L.M. Haffner, R.J. Reynolds, S.L. Tufte, G.J. Madsen, K.P. Jaehnig, J.W. Percival, Astrophys. J. Suppl. **149**, 405–422 (2003)
J.L. Han, Chin. J. Astron. Astrophys. Suppl. **6**, 211–217 (2006)
C. Heiles, T.H. Troland, Astrophys. J. Suppl. **145**, 329–354 (2003)
D.B. Henley, R.L. Shelton, K.D. Kuntz, An XMM-Newton observation of the local bubble using a shadowing filament in the southern galactic hemisphere. ArXiv Astrophysics e-prints (2007)
V.V. Izmodenov, J. Geiss, R. Lallement, G. Gloeckler, V.B. Baranov, Y.G. Malama, J. Geophys. Res. **104**, 4731–4742 (1999)
E.B. Jenkins, W.R. Oegerle, C. Gry, J. Vallerga, K.R. Sembach, R.L. Shelton, R. Ferlet, A. Vidal-Madjar, D.G. York, J.L. Linsky, K.C. Roth, A.K. Dupree, J. Edelstein, Astrophys. J. Lett. **538**, L81–L85 (2000)
S.M. Kahn, P. Jakobsen, Astrophys. J. **329**, 406–409 (1988)
D. Koutroumpa, Space Sci. Rev. (2008, this issue)
H. Krueger, E. Gruen, Interstellar dust inside and outside the heliosphere. ArXiv e-prints, 802 (2008)
R. Lallement, R. Ferlet, A.M. Lagrange, M. Lemoine, A. Vidal-Madjar, Astron. Astrophys. **304**, 461–474 (1995)
R. Lallement, E. Quémerais, J.L. Bertaux, S. Ferron, D. Koutroumpa, R. Pellinen, Science **307**, 1447–1449 (2005)
N. Lehner, E. Jenkins, C. Gry, H. Moos, P. Chayer, S. Lacour, Astrophys. J. **595**, 858–879 (2003)
R.A. Leske, Anomalous cosmic ray composition from ACE, in *AIP Conf. Proc. 516: 26th International Cosmic Ray Conference, ICRC XXVI*, ed. by B.L. Dingus, D.B. Kieda, M.H. Salamon (2000), p. 274
T.J. Linde, T.I. Gombosi, P.L. Roe, K.G. Powell, D.L. DeZeeuw, J. Geophys. Res. **103**, 1889–1904 (1998)
J.L. Linsky, B.E. Wood, Astrophys. J. **463**, 254–270 (1996)
P.B. Lucke, Astron. Astrophys. **64**, 367–377 (1978)
E. Möbius, M. Bzowski, S. Chalov, H.-J. Fahr, G. Gloeckler, V. Izmodenov, R. Kallenbach, R. Lallement, D. McMullin, H. Noda, M. Oka, A. Pauluhn, J. Raymond, D. Ruciński, R. Skoug, T. Terasawa, W. Thompson, J. Vallerga, R. von Steiger, M. Witte, Astron. Astrophys. **426**, 897–907 (2004)

H.-R. Müller, G.P. Zank, J. Geophys. Res. 7104–7116 (2004)
M. MacLow, R. McCray, Astrophys. J. **324**, 776–785 (1988)
L.A. Marschall, L.M. Hobbs, Astrophys. J. **173**, 43–62 (1972)
M. Opher, Role of the interstellar magnetic field in the flows in the heliosheath. AGU-Spring Meeting, Florida, pages Paper SH24A–08 (2008a)
M. Opher, Space Sci. Rev. (2008b, this issue)
V. Piirola, Astron. Astrophys. Suppl. Ser. **30**, 213 (1977)
N.V. Pogorelov, Space Sci. Rev. (2008, this issue)
N.V. Pogorelov, G.P. Zank, T. Ogino, Adv. Space Res. **41**, 306–317 (2008)
E. Quemerais, Space Sci. Rev. (2008, this issue)
R.J. Rand, S.R. Kulkarni, Astrophys. J. **343**, 760–772 (1989)
R. Ratkiewicz, A. Barnes, G.A. Molvik, J.R. Spreiter, S.S. Stahara, M. Vinokur, S. Venkateswaran, Astron. Astrophys. **335**, 363–369 (1998)
R. Ratkiewicz, L. Ben-Jaffel, J. Grygorczuk, What do we know about the orientation of the local interstellar magnetic field? in *Numerical Modeling of Space Plasma Flows*, vol. 385, ed. by N.V. Pogorelov, E. Audit, G.P. Zank. Astronomical Society of the Pacific Conference Series (2008), p. 189
S. Redfield, J.L. Linsky, Astrophys. J. **602**, 776–802 (2004a)
S. Redfield, J.L. Linsky, Astrophys. J. **613**, 1004–1022 (2004b)
S. Redfield, J.L. Linsky, Astrophys. J. **673**, 283–314 (2008)
R.J. Reynolds, Adv. Space Res. **34**, 27–34 (2004)
R.J. Reynolds, S.L. Tufte, D.T. Kung, P.R. McCullough, C. Heiles, Astrophys. J. **448**, 715–726 (1995)
P. Routly, Jr., L. Spitzer, Astrophys. J. **115**, 227 (1952)
E. Salerno, F. Bühler, P. Bochsler, H. Busemann, M.L. Bassi, G.N. Zastenker, Y.N. Agafonov, N.A. Eismont, Astrophys. J. **585**, 840–849 (2003)
B.D. Savage, K.R. Sembach, Annu. Rev. Astron. Astrophys. **34**, 279–330 (1996)
R.L. Shelton, Space Sci. Rev. (2008, this issue)
R. Shetty, E.C. Ostriker, Astrophys. J. **647**, 997–1017 (2006)
J.P. Simpson, R.H. Rubin, S.W.J. Colgan, E.F. Erickson, M.R. Haas, Astrophys. J. **611**, 338–352 (2004)
J.D. Slavin, Space Sci. Rev. (2008, this issue)
J.D. Slavin, P.C. Frisch, Astron. Astrophys. (2008, in press)
J.D. Slavin, A.P. Jones, A.G.G.M. Tielens, Astrophys. J. **614**, 796–806 (2004)
S. Stanimirovic, C. Heiles, The thinnest cold HI clouds in the diffuse interstellar medium? ArXiv Astrophysics e-prints, May 2005
E. Stone, Voyager 2. Observations of the solar wind termination shock and heliosheath. *AGU-Spring Meeting, Florida*, pages Paper SH24A–01 (2008)
J. Tinbergen, Astron. Astrophys. **105**, 53–64 (1982)
J. Vallerga, Astrophys. J. **497**, 921–927 (1998)
C.S. Weller, R.R. Meier, Astrophys. J. **246**, 386–393 (1981)
D.E. Welty, L.M. Hobbs, J.T. Lauroesch, D.C. Morton, L. Spitzer, D.G. York, Astrophys. J. Suppl. **124**, 465–501 (1999)
M. Witte, Astron. Astrophys. **426**, 835–844 (2004)
M. Wolleben, Astrophys. J. **664**, 349–356 (2007)
B.E. Wood, J.L. Linsky, G.P. Zank, Astrophys. J. **537**, 304–311 (2000)
B.E. Wood, Space Sci. Rev. (2008, this issue)
B.E. Wood, S. Redfield, J.L. Linsky, H.-R. Müller, G.P. Zank, Astrophys. J. Suppl. **159**, 118–140 (2005)

Section IV: What Are the Energy and Pressure Balances in the Local Bubble?

Pressure and Ionization Balances in the Circum-Heliospheric Interstellar Medium and the Local Bubble

Edward B. Jenkins

Originally published in the journal Space Science Reviews, Volume 143, Nos 1–4, 205–216.
DOI: 10.1007/s11214-008-9352-1

Abstract A disconcerting mismatch of thermal pressures for two media in contact with each other, (1) the warm, Circum-Heliospheric Interstellar Medium (CHISM) and (2) the very hot material within a much larger region called the Local Bubble (LB), has troubled astronomers for over two decades. A possible resolution of this problem, at least in part, now seems possible. We now understand that earlier estimates for the average electron density in the very hot LB plasma were inflated by an unrecognized foreground contamination to the low energy diffuse X-ray background measurements. This foreground illumination arises from photons emitted by charge exchange reactions between solar wind ions and neutral atoms from the interstellar medium that enter into the heliosphere. However, with the resolution of this problem comes a new one. The high ionization fraction of helium in the CHISM, relative to that of hydrogen, could be understood in terms of the effects from a strong flux of EUV and X-ray radiation coming from both the Local Bubble and a conductive interface around the CHISM. A revision of this interpretation may now be in order, now that the photoionization rate from radiation emitted by hot gas the Local Bubble is lower than previously assumed.

Keywords Galaxy: solar neighborhood · ISM: bubbles · ISM: clouds · X-rays: ISM

*"When a thing ceases to be a subject of controversy,
it ceases to be a subject of interest"*—William Hazlitt (1778–1830)

1 Introduction: The Pressure Problem

The history of science has highlighted many instances where the need to resolve an incongruity in our perception of the natural world goaded us into abandoning an entrenched idea—a process that almost always has led to an important new threshold for progress. A ma-

E.B. Jenkins (✉)
Princeton University Observatory, Princeton, NJ 08544-1001, USA
e-mail: ebj@astro.princeton.edu

jor theme we will address here is one such disparity, one which takes the form of an apparent mismatch in thermal pressures[1] between two phases of interstellar material that are in contact with each other. One such medium is the warm gas in the Circum-Heliospheric Interstellar Medium (CHISM)[2], while the other is a much lower density medium that surrounds the CHISM, which is understood to assume the form of a hot plasma that resides within a large volume known as the Local Bubble (LB). There are several possible ways to confront this issue: (1) acknowledge the presence of other forms of pressure (turbulent or magnetic) whose differences can conceivably compensate for the thermal pressure mismatch, (2) accept the possibility that a dynamical response is underway and that this response will eventually resolve the difference or (3) overcome the clash by revising our notions about the physical properties of one or both regions. We will touch upon each of these three themes later in this article.

2 Development of the Conventional View of the Local Bubble (LB)

The diffuse soft X-ray background detected by instruments on sounding rocket flights in the 1970's was recognized by Williamson et al. (1974) to originate from hot plasmas in spaces within and beyond the Milky Way (see also Burstein et al. 1977). Soon afterward, it became clear that much of this radiation at the lowest energies must come from an irregularly shaped volume surrounding the Sun, since the X-ray energy distribution showed little evidence for foreground absorption by cool, foreground material, and in directions away from the Galactic plane the intensities were anticorrelated with 21-cm emission by H I, probably as a result of the fact that the hot and cold gases displace one another (Sanders et al. 1977; Hayakawa et al. 1978; Marshall and Clark 1984).

From early evidence that a very limited amount of H I was present out to a radius of about 100 pc from the Sun (determined from various investigations, summarized later by Paresce 1984), Fried et al. (1980) estimated that the thermal pressure of the X-ray emitting gas was of order $p/k = 10^4$ cm^{-3} K, a figure that was consistent with more refined estimates that came later (Marshall and Clark 1984; Cox and Reynolds 1987; Snowden et al. 1990, 1998). Determinations of the average electron densities that were perhaps the most straightforward to interpret came from measurements toward dense clouds whose distances were known and which were expected to block the radiation. A study of such cases by Snowden et al. (1993), Kuntz et al. (1997) and Burrows and Guo (1998) indicated that $n(e) \approx 0.005$ cm^{-3}; for an assumed temperature in the range 1–2 $\times 10^6$ K, this pointed toward a pressure $10^4 < p/k < 2 \times 10^4$ cm^{-3} K. Similar cloud shadowing measurements conducted by Bowyer et al. (1995) and Berghöfer et al. (1998), but using EUV radiation instead of diffuse X-rays, indicated that $p/k = 19{,}500$ and 16,500 cm^{-3} K, respectively. While these values for the thermal pressure inside the LB are higher than those generally seen elsewhere in the ISM (Jenkins and Shaya

[1]Throughout this paper, pressures will be stated in terms of real pressures divided by the Boltzmann constant, i.e., $p/k = nT$ for thermal pressures, because they are easier to relate to particle densities and temperatures.

[2]Many interpretations in the recent literature depict the heliosphere as being located close to the edge (but still inside) a cloud of warm gas called the Local Interstellar Cloud (LIC) that has coherent kinematics and is seen over much of the sky. However a recent investigation by Redfield and Linsky (2008) indicates that our location is in a transition region between this cloud and another cloud that is situated in the general direction of the Galactic center called the G Cloud. In recognition of this more complicated picture, we avoid calling the material that surrounds us as the Local Interstellar Cloud, since it now has a more restricted meaning, and replace it with the term Circum-Heliospheric Interstellar Medium (CHISM) that comprises both the LIC and G Cloud.

1979; Jenkins et al. 1983; Jenkins and Tripp 2001), they are not far below an estimate of 2.8×10^4 cm^{-3} K for the expected *total* pressure in the Galactic plane (Boulares and Cox 1990) created by the weight of interstellar gases in the Galaxy's gravitational potential.

A means for outlining the shape of the Local Bubble was developed by Welsh et al. (1999) and Sfeir et al. (1999), who measured either the sudden attenuation of EUV radiation or onset of strong Na I absorption features for stars at different distances and in different directions, as sight lines started to penetrate the much denser gas beyond the transparent hot medium. As of now, the most refined picture of the Local Bubble has been presented by Lallement et al. (2003), who were able to draw upon a larger number of stars than those available earlier.

3 The Circum-Heliospheric Interstellar Medium (CHISM)

Well before our awareness of the LB, early investigations of an anisotropic backscatter of solar Lα from outside the geocorona (Chambers et al. 1970; Bertaux and Blamont 1971; Thomas and Krassa 1971) indicated that the solar system is moving through a low-density medium containing some neutral hydrogen, consistent with a description made shortly beforehand by Blum and Fahr (1969). With more refined observations and a theory of this backscattered radiation, coupled with measurements of the absorption of EUV radiation from nearby stars (Vallerga 1996), there emerged a picture where the solar system is moving through a partially ionized medium with a total particle density $n(\mathrm{H}+\mathrm{He}+\mathrm{e}) \approx 0.35$ cm^{-3} and a temperature $T = 7000$ K (Bertaux et al. 1985; Frisch 1995, 2004; Slavin and Frisch 2002), which leads to a thermal pressure $p/k = 2500$ cm^{-3} K. This material is similar in nature to that found in other gas complexes in the local vicinity (Lallement 1996; Redfield and Linsky 2004b, 2008). As indicated in the introduction (Sect. 1), the local thermal pressure is in conflict with the much higher value for the surrounding hot medium in the LB; this disparity has been a nagging problem for many years and must be reconciled in some way.

4 Attempted Solutions that Seem to Fail

An imbalance is acceptable At one time, when the age of the LB and our location with respect to the outer boundary of the CHISM were not very certain, one could imagine[3] that perhaps the imbalance is real and, as a consequence, a shock front that started at the perimeter of the CHISM is now approaching us from all directions. This shock should have a Mach number about equal to the square-root of the relative pressure ratio, i.e., $v_{\mathrm{shock}} \approx$ 20 km s^{-1}, and it has not yet reached us. However, no evidence has been found for such a front (i.e., from UV absorption lines, which are very sensitive to small amounts of gas), and moreover we now know that we are no more than a few pc away from the edge of the CHISM and that the age of the LB is probably much longer than 10^5 yr.

Magnetic support Since the local medium is partly ionized, it should be strongly coupled to any magnetic field that could be present. For some time, the value and direction of the magnetic field in the CHISM was unknown, so one could hold out some hope that magnetic pressure could keep the CHISM from collapsing under the external pressure from the LB.

---

[3] As the author of this article conjectured during a panel discussion at a conference on the local interstellar medium, IAU Colloquium 81 in 1984.

Now, however, 3 independent studies using different observational methods indicate that the magnetic field is inclined relative to the upwind direction and creates an asymmetry in the shape of the heliosphere, but it has a strength that ranges somewhere between 1.8 and 2.5 μG (Lallement et al. 2005; Opher et al. 2007; Wood et al. 2007). If the strength were as high as 3 μG, $B^2/(8k\pi) = 1700$ cm^{-3} K. This much pressure supplementing the thermal pressure still falls well short of the apparent pressure of the LB.

Turbulent support If the LIC had a high level of turbulence, it is conceivable that a random dynamical pressure $\rho v_{\rm rms}^2$ could prevent it from collapsing under the external pressure. If we draw upon determinations based on other local clouds, this seems unlikely. Redfield and Linsky (2004b) found that on average the thermal pressure was about 26 times larger than the pressure from turbulence.

5 Early Indications of Trouble with the X-ray Background Measurements

By the time that the soft X-ray background started to be observed by satellite missions, as opposed to the very brief sounding rocket flights, it became apparent that enhancements in the background count rates occurred during periods of increased solar activity (Singh et al. 1987; Garmire et al. 1992; Snowden et al. 1994). At the time, the real cause of these increased event rates was not known; the presumption was that they originated from the direct sensing of charged particles associated with the solar events. The investigators applied corrections to the data to eliminate the time-variable components, but any rate increases not identifiable with increased solar activity would have gone unnoticed and would have contaminated the maps of X-ray background intensity.

An important clue to the origin of the background arose from an observation of EUV and X-ray emission (0.1–2.0 keV) from Comet C/Hyakutake 1996 B2 reported by Lisse et al. (1996), which was interpreted by Cravens (1997) to arise from the charge-transfer of solar wind heavy ions as they interacted with the atmosphere of the comet. At a conference on the Local Bubble held in 1997, Freyberg (1998) and Cox (1998) took note of the possibility that the solar wind charge exchange (SWCX) mechanism could operate not only locally, producing the X-ray enhancements clearly correlated in time with solar activity, but also over the very large distances to the edge of the heliosphere, where an interaction would occur with the incoming hydrogen from the CHISM. As a result, the signal would be so badly smeared out in time that the contamination of the X-ray intensity maps would not be recognized. This contamination would have the consequence of misleading us into thinking that the overall LB X-ray intensities, and thus the emission measures[4] that generally ranged from about 0.0018 to 0.0058 cm^{-6} pc (Snowden 1998), are stronger than reality. Later, Cravens (2000) stated, "A simple model demonstrates that heliospheric X-ray emission can account for about 25–50% of the observed soft X-ray background intensities."

While the need to understand the strength of the SWCX contamination is crucial for a proper reassessment of the LB emission measures in various directions, it is a complex problem to solve. We have some understanding of the types of processes that create energetic photons when SWCX occurs (Kharchenko and Dalgarno 2000; Pepino et al. 2004), but there is nevertheless a multitude of factors that enters into any attempt to correct the

[4]Emission measures (EM) denote the integral of the square of the electron density along a line of sight for temperatures that are relevant to the emission mechanism.

X-ray maps for this contamination. One must account for differences between the interactions with the fast and slow solar winds (along with the occasional coronal mass ejections), the differing geometrical distributions for different charge states as the highly charged ions undergo successive charge exchanges, the changes in the density of incoming H and He within the heliosphere, and so forth. Lallement (2004) constructed a model for the emissivity of X-rays as a function of location in the sky when the ROSAT[5] survey was created, but initially with an uncertain global scaling factor for the intensities. She then derived this factor empirically by determining the best fit of a polar plot of the corrected intensities to the shape of the bubble's perimeter (Lallement et al. 2003) and concluded that a subtraction of the SWCX contribution could lower the inferred thermal pressure by a factor 2.5, i.e., to as low as 4000–6000 cm^{-3} K. A similar study by Bellm and Vaillancourt (2005) arrived at a pressure reduction factor of 1.7. These downward revisions approached, but still did not equal, the pressure that seems evident for the CHISM (Sect. 3).

In the above paragraphs, we have followed a trail of reasoning that has led to the conclusion that a simple interpretation of the apparent intensities of the soft X-ray background is biased toward a value for the emission measure in the Local Bubble that is too high. This, in turn, suggests that the average electron density (and hence the thermal pressure) has been overestimated by a substantial factor. One way to overcome this systematic error is to measure the difference in brightness between an opaque cloud at some known distance within the LB and that toward a more distant screen in nearly the same direction, such as the wall of the LB. Here, an overall elevation of the flux by a foreground emission source would have no effect. This was the principle behind the determinations of $n(e)$ using differential emission measures of EUV radiation made by Berghöfer et al. (1998). Unfortunately, their investigation did not incorporate the knowledge (obtained later) that the telescope they used on EUVE had some response to X-rays at energies above 0.28 keV (B. Welsh, private communication). Since their fiducial foreground blocking clouds (with $E_{b-y} = 0.02$) had a transmission of about 0.7 to such X-rays, it is possible that some of the differential signal they sensed came from the X-rays, which could have added to the decrement seen for the EUV radiation.

6 How Else to Measure the Pressure in the LB?

6.1 The Fine-Structure Excitation of Neutral Carbon

There are alternate means for sensing the pressure inside the LB, although they are less direct than determining the emission measures in either the EUV or soft X-ray bands. One is to examine conditions inside clouds other than the CHISM that are inside the LB. If such clouds are situated in front of bright, early-type stars and have a sufficient internal density to show absorption features from C I[6], we can determine allowed combinations of internal densities and temperature (or upper limits thereof) by studying the relative strengths of absorption out of different fine-structure levels of the ground electronic state. The populations of the

[5]The mapping of the soft X-ray background that came from the German *Röntgensatellit* (ROSAT) launched in 1990 is the most comprehensive picture of the soft X-ray sky at the present time.

[6]The dominant ionization stage of carbon is the singly ionized form. The fraction of these atoms that appear in neutral form, usually quite small, is governed by the balance between photoionization and recombinations with free electrons or negatively charged dust grains, hence the absolute abundance of neutral carbon scales roughly in proportion to the square of the local density.

upper levels are governed by an equilibrium between collisional excitations (which scale with density) and spontaneous radiative decays.

Jenkins (2002) studied the C I absorption features toward 4 stars that were either inside or near the edge of the LB. From a combination of fine-structure population ratios and temperature constraints arising from either line widths or thermal and ionization equilibria, he concluded that $10^3 < p/k < 10^4\ \mathrm{cm}^{-3}$ K for the clouds under study.

6.2 EUV Diffuse Background

Hurwitz et al. (2005) reported upper limits obtained from observations by the *CHIPS* spectrometer[7] for the emission of radiation from highly ionized iron atoms in collisional ionization equilibrium within the LB. If the abundance of iron conformed to the solar abundance ratio to hydrogen, their upper limit for this emission at moderate Galactic latitudes indicated emission measures EM $< 0.0004\ \mathrm{cm}^{-6}$ pc (at a 95% confidence level) for plasmas in the temperature range $10^{5.55} < T < 10^{6.4}$ K. This limit is about an order of magnitude below the typical EM values seen in the soft X-ray background without corrections for the SWCX emission, and it is consistent with an upper limit for EUV iron-line emission detected by *ALEXIS*[8] (Bloch et al. 2002). Evidently, the relative SWCX contamination for the Fe emission lines is smaller than for the soft X-ray emission.

A troubling issue with the Fe-line measurements is the possibility that the iron atoms that should accompany the gas may be depleted onto dust grains, thus causing the abundance of these ions relative to protons in the plasma to be below the solar abundance ratio. While one might conclude that the constraint on the EM determined by *CHIPS* could be weakened considerably by this depletion, it appears unlikely that this is happening: observations of X-ray absorption features from highly ionized Fe and Ne elsewhere within the Galaxy by Yao et al. (2006) indicate a normal abundance ratio in the hot ISM.

6.3 Cloud Lifetimes

Warm clouds immersed in a hot medium should have their outer layers heated by conduction. If radiative losses are small, this heating will cause an evaporation of atoms on the cloud's surface into the surrounding hot medium. (If the radiative losses are large, then the reverse process of condensation will take place instead.) In the absence of a magnetic field that could inhibit conduction by some modest factor, the characteristic time scale for the shrinkage of a cloud due to a steady erosion by evaporation is given by

$$M/\dot{M} = 0.65 n p_4^{-5/6} r_{\mathrm{pc}}^{7/6}\ \mathrm{Myr} \qquad (1)$$

(Slavin 2007), where n is the cloud's internal gas particle density, p_4 is the thermal pressure in units of $10^4\ \mathrm{cm}^{-3}$ K, and r_{pc} is the cloud's radius in pc. If $p_4 = 0.2$, a cloud with $n = 0.2\ \mathrm{cm}^{-3}$ and $r_{\mathrm{pc}} = 2$ (i.e., the CHISM or similar clouds in the local vicinity) should have a shrinkage lifetime of about 1 Myr, but if $p_4 = 1.5$, this characteristic time decreases to only 0.2 Myr, which becomes inordinately short compared to the estimated age of 10–15 Myr for the LB (Maíz-Apellániz 2001; Berghöfer and Breitschwerdt 2002; Fuchs et al. 2006).

[7]The *Cosmic Hot Interstellar Plasma Spectrometer* (*CHIPS*) was launched in early 2003 to measure the strengths of emission lines from the LB hot plasma over the wavelength range 90–265 Å.

[8]The *Array of Low Energy X-ray Imaging Sensors* (*ALEXIS*) satellite experiment was launched in 1993 and observed radiation in narrow energy bands centered on 186, 172, and 130 Å.

Of course, it is possible that the clouds we see at the present time arose from a relatively recent ejection or tearing away of neutral material from the wall of the LB (Breitschwerdt et al. 2000; Cox and Helenius 2003). Also, there is evidence that to some extent the clouds mutually protect one another from evaporation, as we will consider in more detail later (Sect. 8.3).

7 Pressure Balance Between the LB and Galactic Disk

If we now accept the proposition that the internal pressure in the LB is well below 10^4 cm^{-3} K, and that this thermal pressure dominates over other forms of pressure (which may not be true), we must then contemplate a different imbalance, namely, the fact that the total expected pressure in the Galactic disk that immediately surrounds the LB should be two to three times larger (Boulares and Cox 1990). The LB has extensions of very low density gas above and below the Galactic plane that seem limitless; these are called "chimneys" (Welsh et al. 1999, 2002)—see Figs. 5, 7 and 8 of Lallement et al. (2003). Thus the LB is not directly exposed to the usual amount of weight of a vertical column of gas, but still we must address the issue of what prevents the LB from being strangled by the higher pressure gas that surrounds it in the Galactic plane.

Perhaps in this instance, unlike what we found for the CHISM/LB interface in Sect. 4, we can draw upon magnetic fields to provide the necessary support for the LB wall. If a supernova remnant develops in an initially weakly magnetized medium, the field lines should be compressed as the remnant expands, creating a magnetic wall at the perimeter that has a field strength that is strong enough to ward off any encroachment of the outside material as the remnant cools radiatively and its thermal pressure drops (Slavin and Cox 1992; Balsara et al. 2001). Instabilities at the wall that would start to allow localized intrusions of the outside disk gas would carry the magnetic field with them and, as a result, create convex field configurations whose tensions toward the outside could inhibit further inward motion.

Is there any evidence of an enhanced magnetic field at the boundary of the LB? The answer is yes: Andersson and Potter (2006) measured the dispersion of polarization angles of starlight emitted by stars just beyond the LB wall and combined them with measurements of the velocity dispersion of the gas and determinations of n(H) using C I fine-structure level populations. They then used the Chandrasekhar-Fermi method to estimate that the field strength is 8^{+5}_{-3} μG, which leads to a pressure $B^2/(8\pi k) = 1.8^{+3}_{-1} \times 10^4$ cm^{-3} K.

8 Ionization State of the CHISM

8.1 Initial Findings

One of the early surprises about the ionization of the local gas was that observations of neutral H and He absorptions to local white dwarfs indicated that the fractional ionization of helium was about equal to that of hydrogen (Kimble et al. 1993a, 1993b; Vennes et al. 1993) or even greater (Green et al. 1990; Dupuis et al. 1995), in spite of the fact that the recombination coefficient of He with free electrons is greater than that of H over temperatures of interest. There are many sources of UV radiation that can partially photoionize these two elements in the CHISM. A dominant source of such photons is the star ϵ CMa (Vallerga and Welsh 1995), by far the brightest star in the sky in the EUV spectral region. However this flux, supplemented by radiation from nearby white dwarfs, only amounts to photoionization rates $\Gamma(\mathrm{H}) = 1.56 \times 10^{-15}\ \mathrm{s}^{-1}$ and $\Gamma(\mathrm{He}) = 8.25 \times 10^{-16}\ \mathrm{s}^{-1}$ according to Vallerga (1998). (Vallerga calculated that emissions from unobserved late-type stars could raise the He ionization rate somewhat, but only by about 14%.) These values are not sufficient to maintain the observed levels of ionization.

8.2 Incomplete Recombination?

One might question the validity of the assumption that the ionization of the CHISM has reached an equilibrium, as did Lyu and Bruhweiler (1996). They suggested that a shock created by a nearby supernova about 1 to a few Myr ago might have strongly heated and ionized the gas, leading a condition where the material is now overionized (for its current temperature) because the e-folding time for He recombination $[\alpha(\mathrm{He})n(e)]^{-1} = 1.8$ Myr, whereas the gas can cool much more rapidly. The shock seemed to be the only mechanism that could work: both Lyu and Bruhweiler (1996) and Frisch and Slavin (1996) concluded that the UV flash from any very recent, nearby supernova would not be able to appreciably ionize the local gas (unless it were very much closer than the closest known supernova in the constellation of Vela).

Sofia and Jenkins (1998) proposed that a way to test the proposal of incomplete recombination would be to examine the abundance of Ar I relative to H I in the local gas. Argon is not likely to be depleted in the ISM, so measurements of the ratio of neutral forms of these two elements indicates their relative levels of ionization. The recombination rates of Ar and H are about equal, whereas the photoionization cross section of Ar is much higher than that of H. Thus, as Jenkins and Sofia pointed out, a condition of steady-state ionization would yield Ar I/H I less than the underlying abundance ratio of these elements (assumed to be solar), and conversely an incomplete recombination from a much more highly ionized state would yield a ratio more in accord with the real value of Ar/H. Later measurements of N(Ar I) relative to N(O I)[9] toward nearby white dwarf stars revealed that Ar I is indeed deficient (Lehner et al. 2003), supporting the notion that the ionization of the local medium is maintained by a steady-state photoionization.

8.3 Diffuse Hard Photoionization Flux

If we now accept the idea that the ionization is in equilibrium, a possible solution to the apparent excess helium ionization is the ionizing effect provided by the additional, much harder radiation emitted by the conductive interface around the CHISM (Slavin 1989), supplemented by the soft X-rays from the LB. Slavin and Frisch (2002) have calculated that these two sources of hard radiation, working in conjunction with the fluxes from stars, could explain the observed level of ionization of H and He in the CHISM.

Slavin and Frisch (2002) made reasonable allowances for a number of possible effects that could alter the conduction rate, and hence the intensity of the emitted radiation, from the local cloud's conductive interface. They considered the possible influence of a randomly oriented (but not tangled) magnetic field that would slightly suppress the conduction, the effect of conduction saturation (Cowie and McKee 1977; Dalton and Balbus 1993), and the modification of the cooling rate due to the partial ionization of H and He. From the discussion in

[9] O I is a good surrogate for H I because oxygen atoms are not heavily depleted onto dust grains and their relative ionization is strongly coupled to that of H through a strong charge-exchange reaction (Field and Steigman 1971). Thus, taking N(O I) and dividing by the solar abundance ratio of O to H should yield a good estimate for N(H I).

Sect. 6.3, it is clear that the conduction and evaporation rates depend on the pressure of the external hot gas. Thus one might suppose that if this pressure were to be revised downwards, the inferred ionizing flux from the interface should be lowered as well. Evidently this is not the case for the flux calculated by Slavin and Frisch (2002), since their integration of the outward evaporation flow and its changing state (including the emission rate) started with a boundary condition that corresponded to the pressure inside the cloud, instead of imposing the requirement that there be a match to what now seems to be an unreasonably high pressure for the surrounding hot medium (J. Slavin, private communication). While this may be so, there is still a need to lower the estimate for the portion of the ionizing flux that comes from the hot gases within the LB itself.

It is quite possible that the CHISM does not have a well established conduction and evaporation front because it is shielded from the hot plasma by other clouds in the local vicinity. Balbus (1985) derived solutions for the conduction flows for complex geometries. One outcome of his study was that individual clouds well inside an ensemble of neighboring clouds should have significantly reduced flows because of obstruction by neighboring clouds, in a manner analogous to electrostatics where a conducting body is shielded from external potential differences by (grounded) nearby bodies that surround it (as in the extreme case of a Faraday cage). He expressed a criterion for the maximum volume filling factor f of the clouds having radii a within a spherical volume of radius R,

$$f \ll \frac{2}{3}\frac{a^2}{R^2} \tag{2}$$

that is required for a cloud in the center to be unaffected by the shielding.

In order to make use of (2), we need to know something about the cloud environment in the local vicinity. From the appearance of UV absorption lines for stars out to various distances (Redfield and Linsky 2004a) and the detections of astrospheres[10] around stars in different locations (Wood et al. 2005), it seems that the CHISM is near the center of an ensemble of similar clouds that fills a volume out to a radius of about 10 pc from the Sun. Beyond that distance, the systems that contain some neutral hydrogen are much more sparsely distributed. From the frequency of appearance of astrospheres, Wood et al. (2005) estimated that $f = 0.59$. Redfield and Linsky (2008) arrived at a different estimate of $0.055 < f < 0.19$, based on the coverage of UV absorption systems that exhibited coherent velocities over patches of the sky. In either event, for characteristic cloud sizes of 2 pc or smaller[11] within the volume with radius $R = 10\,\text{pc}$, the inequality shown in (2) is not satisfied. Thus, shielding appears to be a significant effect that can reduce or virtually eliminate the conduction front around the CHISM, as well as its closest neighboring clouds.

Two different observational outcomes appear to support an absence of multiple conduction fronts within the ensemble of local clouds, but still allow for a single front at the edge of the complex. In a collisionally ionized medium, five-times ionized oxy-

[10] Astrospheres are structures created by the interaction of a star's stellar wind with an ambient medium, much like the heliosphere around the Sun. They are detected by unique Lα absorption features that can be seen on top of the stellar Lα emission line, as long as we view the star from a direction that is generally near the upwind direction for the star's motion through its surroundings.

[11] In some instances, what may seem to be multiple clouds may in fact be velocity bunching within a single, large cloud. While this may be so, it does not alter the basic conclusion that significant shielding may still occur.

gen appears mostly at temperatures of around 3×10^5 K,[12] a condition that should arise in the outflow of gas from the surface of an evaporating cloud. For the first measurement, we note that for a single direction in the sky, Shelton (2003) found a 2σ upper limit of 800 photons $cm^{-2}\,s^{-1}\,str^{-1}$ for the brightness of the 1032, 1038 Å doublet of O VI, and this limit is about consistent with the predicted intensity from a conduction front of 600–1000 photons $cm^{-2}\,s^{-1}\,str^{-1}$, depending on the age of the front (Borkowski et al. 1990). This limit seems to rule out multiple fronts (but,again, for only one particular direction). The second type of evidence arises from a survey of O VI absorption features in the spectra of white dwarf stars within the LB (Oegerle et al. 2005; Savage and Lehner 2006)[13]. Generally, when detected, the O VI features exhibited kinematic properties that were consistent with our looking outwards through a single evaporation front: the lines had the correct width, and they had a tendency to show a positive velocity shift relative to absorptions from C II that arise from the cloud material.

Since the cloud ensemble ultimately has an outer boundary that meets the hot plasma of the LB, the gas within the CHISM will still be exposed to ionizing radiation from all directions—the only difference here is that this diffuse illumination will come from a much greater distance ($\approx$ 10 pc) than the CHISM boundary. For this reason, the radiation will be absorbed by gaseous material in the clouds with a total column density of hydrogen equivalent to about 10^{18} cm^{-2} [but for a few directions, much more than this value, according to Wood et al. (2005)]. A column this large will attenuate photons at energies just above the He I ionization limit by about a factor of 4, but this absorption will diminish for the higher energies.[14]

In short, the rate of ionization arising from both the X-ray background and the conductive interfaces could be lower than the estimate furnished by Slavin and Frisch (2002). While their investigation was a major step in the right direction, perhaps some revised thinking on this important topic is justified by the more recent findings.

Acknowledgements The author thanks B. Welsh for alerting him about the X-ray leak problem with EUVE. J. Slavin provided useful advice on conduction fronts.

References

B.-G. Andersson, S.B. Potter, Astrophys. J. (Lett.) **640**, L51–L54 (2006)
S.A. Balbus, Astrophys. J. **291**, 518–522 (1985)
D. Balsara, R.A. Benjamin, D.P. Cox, Astrophys. J. **563**, 800–805 (2001)
E.C. Bellm, J.E. Vaillancourt, Astrophys. J. **622**, 959–964 (2005)
T.W. Berghöfer, D. Breitschwerdt, Astron. Astrophys. **390**, 299–306 (2002)

[12] This value is for an equilibrium condition. In evaporation and condensation flows, the preferred temperature can be somewhat lower or higher because the temperature changes more rapidly than the time needed to establish an equilibrium.

[13] The reality of the interstellar O VI absorptions has recently been questioned by M. Barstow and colleagues. At the time of writing of this article, the evidence that supports this conclusion has not yet appeared in the literature, but the article by B. Welsh in this volume presents a brief report on the principal claims by this group.

[14] According to Hurwitz et al. (2005), for a solar abundance plasma at $T = 10^6$ K, 90% of the helium-ionizing photons come from an EUV complex of Fe lines near 180 Å (i.e., the same lines that they were looking for—see Sect. 6.2)—the calculation of this percentage includes a weighting by the change of the helium ionization cross section with energy. At this wavelength, the attenuation is much less than that near the He ionization edge. If Fe is depleted, however, these lines should no longer be as dominant.

T.W. Berghöfer, S. Bowyer, R. Lieu, J. Knude, Astrophys. J. **500**, 838 (1998)
J.L. Bertaux, J.E. Blamont, Astron. Astrophys. **11**, 200–217 (1971)
J.L. Bertaux, R. Lallement, V.G. Kurt, E.N. Mironova, Astron. Astrophys. **150**, 1–20 (1985)
J.J. Bloch, D. Roussel-Dupré, J. Theiler, E. Johnson, in *Continuing the Challenge of EUV Astronomy: Current Analysis and Prospects for the Future*. ASP Conf. Ser., ed. by S. Howell, J. Dupuis, D. Golombek, F. Walter, J. Cullison (Astr. Soc. Pacific, San Francisco, 2002), pp. 243–251
P.W. Blum, H.J. Fahr, Nature **223**, 936–937 (1969)
K.J. Borkowski, S.A. Balbus, C.C. Fristrom, Astrophys. J. **355**, 501–517 (1990)
A. Boulares, D.P. Cox, Astrophys. J. **365**, 544–558 (1990)
S. Bowyer, R. Lieu, S.D. Sidher, M. Lampton, J. Knude, Nature **375**, 212–214 (1995)
D. Breitschwerdt, M.J. Freyberg, R. Egger, Astron. Astrophys. **361**, 303–320 (2000)
D.N. Burrows, Z. Guo, in *The Local Bubble and Beyond*, ed. by D. Breitschwerdt, M.J. Freyberg, J. Trümper (Springer, Berlin, 1998), pp. 279–282
P. Burstein, R.J. Borken, W.L. Kraushaar, W.T. Sanders, Astrophys. J. **213**, 405–420 (1977)
W.H. Chambers, P.E. Fehlau, J.C. Fuller, W.E. Kunz, Nature **225**, 713–714 (1970)
L.L. Cowie, C.F. McKee, Astrophys. J. **211**, 135–146 (1977)
D.P. Cox, in *The Local Bubble and Beyond*, ed. by D. Breitschwerdt, M.J. Freyberg, J. Trümper (Springer, Berlin, 1998), pp. 121–131
D.P. Cox, L. Helenius, Astrophys. J. **583**, 205–228 (2003)
D.P. Cox, R.J. Reynolds, Annu. Rev. Astron. Astrophys. **25**, 303–344 (1987)
T.E. Cravens, Geophys. Res. Lett. **24**, 105–108 (1997)
T.E. Cravens, Astrophys. J. (Lett.) **532**, L153–L156 (2000)
W.W. Dalton, S.A. Balbus, Astrophys. J. **404**, 625–635 (1993)
J. Dupuis, S. Vennes, S. Bowyer, A.K. Pradhan, P. Thejll, Astrophys. J. **455**, 574–589 (1995)
G.B. Field, G. Steigman, Astrophys. J. **166**, 59–64 (1971)
M.J. Freyberg, in *The Local Bubble and Beyond*, ed. by D. Breitschwerdt, M.J. Freyberg, J. Trümper (Springer, Berlin, 1998), pp. 113–116
P.M. Fried, J.A. Nousek, W.T. Sanders, W.L. Kraushaar, Astrpophys. J. **242**, 987–1004 (1980)
P.C. Frisch, Space Sci. Rev. **72**, 499–592 (1995)
P.C. Frisch, Adv. Space Res. **34**, 20–26 (2004)
P.C. Frisch, J.D. Slavin, Space Sci. Rev. **78**, 223–228 (1996)
B. Fuchs, D. Breitschwerdt, M.A. de Avillez, C. Dettbarn, C. Flynn, Mon. Not. R. Astron. Soc. **373**, 993–1003 (2006)
G.P. Garmire et al., Astrophys. J. **399**, 694 (1992)
J. Green, P. Jelinsky, S. Bowyer, Astrophys. J. **359**, 499–505 (1990)
S. Hayakawa, T. Kato, F. Nagase, K. Yamashita, Y. Tanaka, Astron. Astrophys. **62**, 21–28 (1978)
M. Hurwitz, T.P. Sasseen, M.M. Sirk, Astrophys. J. **623**, 911–916 (2005)
E.B. Jenkins, Astrophys. J. **580**, 938–949 (2002)
E.B. Jenkins, M. Jura, M. Loewenstein, Astrophys. J. **270**, 88–104 (1983)
E.B. Jenkins, E.J. Shaya, Astrophys. J. **231**, 55–72 (1979)
E.B. Jenkins, T.M. Tripp, Astrophys. J. (Suppl.) **137**, 297–340 (2001)
V. Kharchenko, A. Dalgarno, JGR Space Phys. **105**, 18351–18359 (2000)
R.A. Kimble et al., Astrophys. J. **404**, 663–672 (1993a)
R.A. Kimble, A.F. Davidsen, K.S. Long, P.D. Feldman, Astrophys. J. (Lett.) **408**, L41–L44 (1993b)
K.D. Kuntz, S.L. Snowden, F. Verter, Astrophys. J. **484**, 245 (1997)
R. Lallement, Space Sci. Rev. **78**, 361–374 (1996)
R. Lallement, Astron. Astrophys. **418**, 143–150 (2004)
R. Lallement, B.Y. Welsh, J.L. Vergely, F. Crifo, D. Sfeir, Astron. Astrophys. **411**, 447–464 (2003)
R. Lallement, E. Quémerais, J.L. Bertaux, S. Ferron, D. Koutroumpa, R. Pellinen, Science **307**, 1447–1449 (2005)
N. Lehner, E.B. Jenkins, C. Gry, H.W. Moos, P. Chayer, S. Lacour, Astrophys. J. **595**, 858–879 (2003)
C.M. Lisse et al., Science **274**, 205–289 (1996)
C.H. Lyu, F.C. Bruhweiler, Astrophys. J. **459**, 216–225 (1996)
J. Maíz-Apellániz, Astrophys. J. **560**, L83–L86 (2001)
F.J. Marshall, G.W. Clark, Astrophys. J. **287**, 633–652 (1984)
W.R. Oegerle, E.B. Jenkins, R. Shelton, D.V. Bowen, P. Chayer, Astrophys. J. **622**, 377–389 (2005)
M. Opher, E.C. Stone, T.I. Gombosi, Science **316**, 875–878 (2007)
F. Paresce, Astron. J. **89**, 1022–1037 (1984)
R. Pepino, V. Kharchenko, A. Dalgarno, R. Lallement, Astrophys. J. **617**, 1347–1352 (2004)
S. Redfield, J.L. Linsky, Astrophys. J. **602**, 776–802 (2004a)
S. Redfield, J.L. Linsky, Astrophys. J. **613**, 1004–1022 (2004b)

S. Redfield, J.L. Linsky, Astrophys. J. **673**, 283–314 (2008)
W.T. Sanders, W.L. Kraushaar, J.A. Nousek, P.M. Fried, Astrpophys. J. (Lett.) **217**, L87–L91 (1977)
B.D. Savage, N. Lehner, Astrophys. J. (Suppl.) **162**, 134–160 (2006)
D.M. Sfeir, R. Lallement, F. Crifo, B.Y. Welsh, Astron. Astrophys. **346**, 785–797 (1999)
R.L. Shelton, Astrophys. J. **589**, 261–269 (2003)
K.P. Singh, J.A. Nousek, D.N. Burrows, G.P. Garmire, Astrophys. J. **313**, 185–191 (1987)
J.D. Slavin, Astrophys. J. **346**, 718–727 (1989)
J.D. Slavin, in *SINS – Small Ionized and Neutral Structures in the Diffuse Interstellar Medium*, ed. by M. Halverkorn, W.M. Goss. ASP Conf. Ser., vol. 365 (Astr. Soc. Pacific, San Francisco, 2007), pp. 113–120
J.D. Slavin, D.P. Cox, Astrophys. J. **392**, 131–144 (1992)
J.D. Slavin, P.C. Frisch, Astrophys. J. **565**, 364–379 (2002)
S.L. Snowden, in *The Local Bubble and Beyond*, ed. by D. Breitschwerdt, M.J. Freyberg, J. Trümper (Springer, Berlin, 1998), pp. 103–112
S.L. Snowden, D. McCammon, F. Verter, Astrophys. J. (Lett.) **409**, L21–L24 (1993)
S.L. Snowden, D.P. Cox, D. McCammon, W.T. Sanders, Astrophys. J. **354**, 211–219 (1990)
S.L. Snowden, D. McCammon, D.N. Burrows, J.A. Mendenhall, Astrophys. J. **424**, 714–728 (1994)
S.L. Snowden, R. Egger, D.P. Finkbeiner, M.J. Freyberg, P.P. Plucinsky, Astrophys. J. **493**, 715–729 (1998)
U.J. Sofia, E.B. Jenkins, Astrophys. J. **499**, 951–965 (1998)
G.E. Thomas, R.F. Krassa, Astron. Astrophys. **11**, 218–233 (1971)
J. Vallerga, Space Sci. Rev. **78**, 277–288 (1996)
J. Vallerga, Astrophys. J. **497**, 921 (1998)
J.V. Vallerga, B.Y. Welsh, Astrophys. J. **444**, 702–707 (1995)
S. Vennes et al., Astrophys. J. (Lett.) **410**, L119–L122 (1993)
B.Y. Welsh, D.M. Sfeir, M.M. Sirk, R. Lallement, Astron. Astrophys. **352**, 308–316 (1999)
B.Y. Welsh, S. Sallmen, D. Sfeir, R.L. Shelton, R. Lallement, Astron. Astrophys. **394**, 691–699 (2002)
F.O. Williamson, W.T. Sanders, W.L. Kraushaar, D. McCammon, R. Borken, A.N. Bunner, Astrophys. J. (Lett.) **193**, L133–L137 (1974)
B.E. Wood, S. Redfield, J.L. Linsky, H.-R. Müller, G.P. Zank, Astrophys. J. (Suppl.) **159**, 118–140 (2005)
B.E. Wood, V.V. Izmodenov, J.L. Linsky, D. Alexashov, Astrophys. J. **659**, 1784–1791 (2007)
Y. Yao, N. Schulz, Q.D. Wang, M. Nowak, Astrophys. J. (Lett.) **653**, L121–L124 (2006)

The Solar Wind Charge-eXchange Contribution to the Local Soft X-ray Background

Model to Data Comparison in the 0.1–1.0 keV Band

Dimitra Koutroumpa · Rosine Lallement · Vasili Kharchenko · Alex Dalgarno

Originally published in the journal Space Science Reviews, Volume 143, Nos 1–4, 217–230.
DOI: 10.1007/s11214-008-9381-9

Abstract The major sources of the Soft X-ray Background (SXRB), besides distinct structures as supernovae and superbubbles (e.g. Loop I), are: (i) an absorbed extragalactic emission following a power law, (ii) an absorbed thermal component ($\sim 2 \times 10^6$ K) from the galactic disk and halo, (iii) an unabsorbed thermal component, supposedly at 10^6 K, attributed to the Local Bubble and (iv) the very recently identified unabsorbed Solar Wind Charge-eXchange (SWCX) emission from the heliosphere and the geocorona.

We study the SWCX heliospheric component and its contribution to observed data. In a first part, we apply a SWCX heliospheric simulation to model the oxygen lines (3/4 keV) local intensities during shadowing observations of the MBM 12 molecular cloud and a dense filament in the south galactic hemisphere with Chandra, XMM-Newton, and Suzaku telescopes. In a second part, we present a preliminary comparison of SWCX model results with ROSAT and Wisconsin surveys data in the 1/4 keV band.

We conclude that, in the 3/4 keV band, the total local intensity is entirely heliospheric, while in the 1/4 keV band, the heliospheric component seems to contribute significantly to the local SXRB intensity and has potentially a strong influence on the interpretation of the ROSAT and Wisconsin surveys data in terms of Local Bubble hot gas temperature.

Keywords SWCX · Heliosphere · SXRB · ISM · Local bubble

D. Koutroumpa (✉)
UMR 7620, IPSL/Service d'Aéronomie, CNRS, Université Pierre et Marie Curie, Université Versailles-Saint-Quentin, Verrières-le-Buisson, France
e-mail: dimitra.koutroumpa@aerov.jussieu.fr

R. Lallement
UMR 7620, IPSL/Service d'Aéronomie, Verrières-le-Buisson, France

V. Kharchenko · A. Dalgarno
Harvard-Smithsonian Center for Astrophysics, Cambridge, MA, USA

DOI: 10.1007/978-1-4419-0247-4_17

1 Introduction

The surprising discovery of X-ray emission from comets (Lisse et al. 1996) led to establishing Charge-eXchange (CX) collisions between highly charged Solar Wind ions and solar system neutrals as a very efficient non-thermal mechanism of soft X-ray emission (Cravens 1997). Cox (1998) suggested that X-ray emission induced in Solar Wind Charge-eXchange (SWCX) collisions with interstellar (IS) neutrals flowing across the Heliosphere would be an additional component of the diffuse soft X-ray background (SXRB).

Until that discovery, the SXRB was generally assumed to consist of an absorbed extragalactic emission following a power law, an absorbed thermal component ($\sim 2 \times 10^6$ K) associated with the galactic disk and halo, and an unabsorbed thermal component, supposedly at 10^6 K, attributed to the Local Bubble (McCammon and Sanders 1990; Kuntz and Snowden 2000, and references within). The Local Bubble (or Local Cavity) is defined as the region within $\sim$100 pc of the solar system particularly devoid of dense (and cold) gas as suggested by absorption lines in nearby star spectra (Lallement et al. 2003).

The $1/4$ keV background generally presents a negative correlation with the column density of Galactic neutral hydrogen (dominated by a Galactic plane to-pole variation) that strongly suggests an "absorption" model, in which the spatial structure of the SXRB is produced by absorption of a distant X-ray component e.g., Bowyer et al. 1968; Marshall and Clark 1984). However, the strong absorption predicted in such models was not confirmed by observation, neither was the large energy dependence of the absorption, so different solutions were needed.

At the time, the model most consistent with the data was the so-called displacement model (Sanders et al. 1977; Snowden et al. 1990a). In this model, the hot gas is in the foreground (in the Local Hot Bubble) with respect to the absorbing cool-gas regions, and higher X-ray intensities are produced in directions where the cavity has a greater extent (towards the LB chimneys at high galactic latitudes) and thus, a greater emission measure of the plasma in it.

Since the discovery of SWCX emission, it became clear very quickly that it should contaminate at an unknown level all soft X-ray observations and it should depend strongly on SW flux and abundance variations as confirmed by the Long Term Enhancements detected during the ROSAT all-sky survey (Snowden et al. 1995) and recent X-ray observations (Snowden et al. 2004; Smith et al. 2005; Henley and Shelton 2007a). The debate concerned the amount of contamination, and how the SWCX emission level would interfere with the Local Bubble thermal emission, since the two are most probably the major components of the unabsorbed fraction of the SXRB. Post-CX geocoronal emission can also contribute, but only during very intense SW enhancements and for short and easily identified intervals (Cravens et al. 2001).

The two components, although due to completely different mechanisms (collisional excitation for thermal plasma emission and electron capture for CX emission), produce spectra in the same energy range ($E \leq 1.5$ keV) that are frequently confused. Cravens (2000) estimated that SWCX emission could be as much as the LB emission. Line ratios, though, are very sensitive to gas temperature (and abundance) for thermal emission. For example, a 10^6 keV plasma might be responsible for all IS diffuse emission at 0.25 keV, while at 2–4 $\times$ 10^6 K it might significantly contribute to the $3/4$ keV diffuse emission as well (Raymond 1988; McCammon and Sanders 1990, and references therein). CX line ratios mainly depend on heavy SW ion relative abundances and collision energy (Beiersdorfer et al. 2001; Kharchenko and Dalgarno 2001; Wargelin et al. 2008). Detailed spectral comparison of

SWCX and thermal emission needs a spectral resolution that should allow to resolve characteristic features as the O VII triplet at 0.57 keV and in general individual lines from the continuum. Such a comparison is beyond the scope of this paper.

In this paper we mainly focus on the intensity contribution of the SWCX emission to the observed (foreground) data in the 0.1–1.0 keV range. In Sect. 2 we briefly present the basic simulation model that calculates the SWCX emission.

In Sect. 3 we analyse the simulation results for the 3/4 keV range and in particular the oxygen lines (O VII at 0.57 keV keV and O VIII at 0.65 keV). We present typical stationary maps of oxygen lines in Sect. 3.1 and in Sect. 3.2 we present the dynamic model we developed to account for short-scale temporal variations of SWCX emission, with an application on the MBM 12 Suzaku observation of February, 2006. Finally, we conclude Sect. 3 with a summary of shadowing observations on which we applied our SWCX dynamic model and discuss the results in the 3/4 keV range (Sect. 3.3).

In Sect. 4 we present a preliminary analysis of SWCX simulation results in the 1/4 keV range that we compare to observation data from the Wisconsin sounding rocket and ROSAT satellite surveys.

Finally, in Sect. 5 we close our paper with a general discussion on our results.

2 SWCX Model Description

In Koutroumpa et al. (2006) we have presented the basic stationary model calculating the SWCX emission in the inner heliosphere. We calculate self-consistently the neutral H and He density distributions in the inner heliosphere (up to ∼100 AU), in response to solar gravity, radiation pressure and anisotropic ionization processes for the two neutral species. Ionization is mainly due to charge-exchange with SW protons for H atoms and to solar EUV photons for He atoms. We also consider the impact of CX on the solar wind ions distributions. This interaction is described in the following reaction:

$$X^{Q+} + [\mathrm{H}, \mathrm{He}] \rightarrow X^{*(Q-1)+} + [\mathrm{H}^+, \mathrm{He}^+]. \tag{1}$$

The collision rate per volume unit $R_{X^{Q+}}$ (cm^{-3} s^{-1}) of X^{Q+} ions with the neutral heliospheric atoms is given by the equation:

$$R_{X^{Q+}}(r) = N_{X^{Q+}}(r)\, v_r\, (\sigma_{(\mathrm{H},X^{Q+})}\, n_{\mathrm{H}}(r) + \sigma_{(\mathrm{He},X^{Q+})}\, n_{\mathrm{He}}(r)) = R_{(X^{Q+},\mathrm{H})}(r) + R_{(X^{Q+},\mathrm{He})}(r) \tag{2}$$

where $\sigma_{(\mathrm{H},X^{Q+})}$ and $\sigma_{(\mathrm{He},X^{Q+})}$ are the hydrogen and helium CX cross-sections, $n_{\mathrm{H}}(r)$ and $n_{\mathrm{He}}(r)$ are the hydrogen and helium density distributions respectively, $\bar{v}_r = \bar{V}_{\mathrm{SW}} - \bar{v}_n \approx \bar{V}_{\mathrm{SW}}$ the relative velocity between SW ions and IS neutrals in the inner heliosphere, and $N_{X^{Q+}}(r)$ is the self-consistent solution to the differential equation:

$$\begin{aligned} \frac{dN_{X^{Q+}}}{dx} = & - N_{X^{Q+}}\, (\sigma_{(\mathrm{H},X^{Q+})}\, n_{\mathrm{H}}(x) + \sigma_{(\mathrm{He},X^{Q+})}\, n_{He}(x)) \\ & + N_{X^{(Q+1)+}}\, (\sigma_{(\mathrm{H},X^{(Q+1)+})}\, n_{\mathrm{H}}(x) + \sigma_{(\mathrm{He},X^{(Q+1)+})}\, n_{\mathrm{He}}(x)) \end{aligned} \tag{3}$$

expressing the evolution of the density distribution of ion X^{Q+} along SW streamlines due to production (from CX reactions of ion $X^{(Q+1)+}$) and loss terms.

Cross-section uncertainties are mainly due to instrumental systematic errors and most important to collision energy dependance of cross-sections. Detailed uncertainties for individual ions are not given in literature, but average uncertainties of ∼30% at most are reported (Wargelin et al. 2008).

Then, we establish emissivity grids in units of (photons cm^{-3} s^{-1}):

$$\varepsilon_i(r) = R_{(\mathrm{X}^{Q+},\mathrm{H})}(r)\,Y_{(E_i,\mathrm{H})} + R_{(\mathrm{X}^{Q+},\mathrm{He})}(r)\,Y_{(E_i,\mathrm{He})} \tag{4}$$

where $Y_{(E_i,M)}$ is the photon emission probability of spectral line E$_i$ following CX with the corresponding neutral species M (H or He individually). For any line of sight (LOS) and observation date, this spectral line is given by:

$$I_{E_i}\,(\mathrm{LU}) = \frac{1}{4\pi}\int_0^{\sim 100\ \mathrm{AU}} \varepsilon_i(s)\,ds \tag{5}$$

which defines the average level emission of the spectral line for the particular date and LOS, as well as the solar cycle phase (minimum or maximum) corresponding at this date.

3 SWCX in the 3/4 keV Band

An example of calculated SWCX emission spectra in the 3/4 keV (0.5–1.0 keV) range is presented in Fig. 1. The major emission lines contributing in this energy range are:

(i) the He-like O VII multiplet, consisting of the following transitions:
  - the 2^3S_1 triplet at 560.9 eV, called forbidden (O6f)
  - the 2^3P_1 triplet at 568.5 eV, called intercombination line (O6i), and
  - the singlet state 2^1P_1 at 574 eV, usually referred to as resonance line (O6r), and

(ii) the H-like O VIII Lyman-α line at 653.1 eV.

In this list we can also add the He-like Ne IX multiplet (905.1 eV, 914.7 eV and 922.1 eV), but actual observing instruments do not always allow the detection of this line.

3.1 Stationary Model Maps

In Fig. 2 we present an example of calculated full-sky maps of heliospheric emission for line O VIII (0.65 keV) in ecliptic coordinates, on December, 5 (when the Earth is located downwind) of a typical solar maximum (upper panel) and solar minimum (lower panel) year. Color scales represent O VIII line intensity given in units of 10^{-9} erg cm^{-2} sr^{-1} s^{-1}, the red

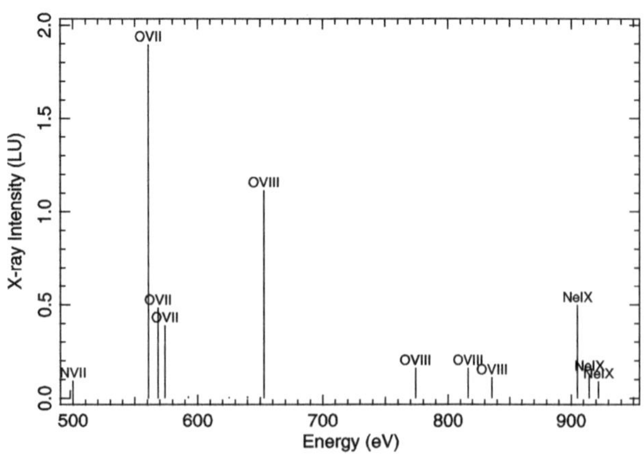

Fig. 1 SWCX model spectra in Line Units (photons cm^{-2} s^{-1} sr^{-1}) for the 3/4 keV band. The emitting ion is noted above each line

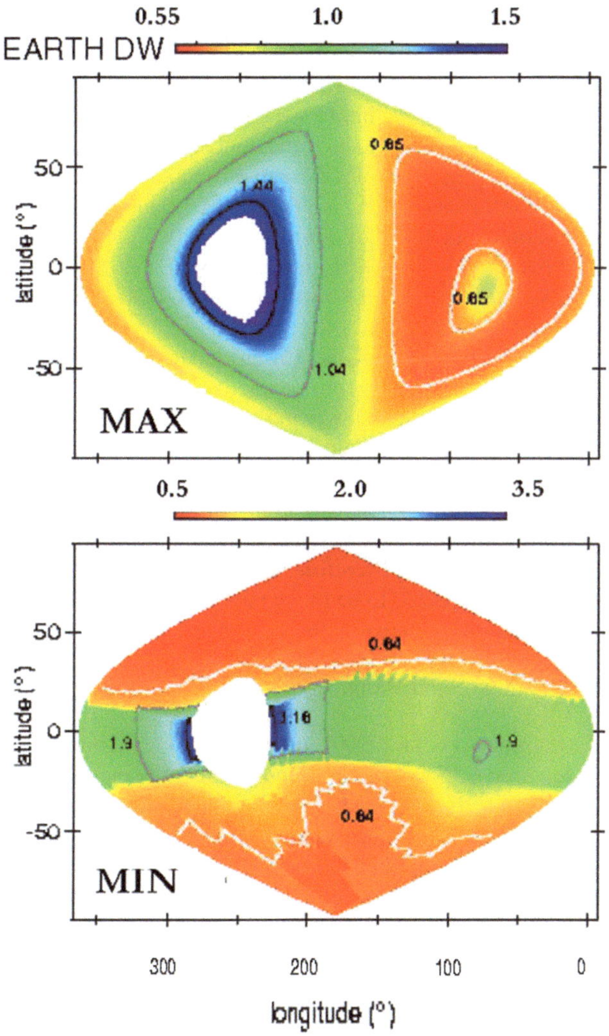

Fig. 2 Calculated full sky maps in ecliptic coordinates of the 0.65 keV O VIII line for solar maximum (*upper panel*) and solar minimum (*lower panel*) conditions. The observer is situated downwind (ecliptic longitude 75°). Color scales are in units of 10^{-9} erg cm^{-2} s^{-1} sr^{-1}

color corresponding to minimum and the blue to maximum values. We have removed from every map a data portion of $20^\circ \times 20^\circ$ around the solar disk where no instrument can observe.

What we need to point out here are the striking differences between maximum and minimum solar conditions. At low latitudes, near the solar equator where slow SW dominates, X-ray emission is more intense for solar minimum than for solar maximum because the neutral atom content is higher during solar minimum. During solar minimum, photo-ionization is less efficient and H and He are less readily destroyed by photo-ionization. Since gravitational pressure exceeds radiation pressure, neutral H atoms have incoming trajectories that fill the ionization cavity.

On the other hand, at high ecliptic latitudes, LOS's are mostly affected by fast wind during solar minimum, and the X-ray emission is dominated by differences of ion relative abundances with respect to maximum cycle phase. Indeed, O^{8+} is completely absent from the fast SW and the O VIII line is expected to be much fainter at solar minimum than solar maximum, as it is seen in the maps in Fig. 2.

O^{7+} has the same trend in relative abundance variations as O^{8+} between solar minimum and solar maximum. Nevertheless, instead of being completely absent in the fast wind, it is only strongly depleted ($[O^{7+}/O] = 0.03$) with respect to slow wind ($[O^{7+}/O] = 0.2$). Therefore, we expect that O VII line (0.57 keV) variations from solar maximum to solar minimum will also have the same trend as O VIII shown in the maps.

Two excellent examples of such differences between solar minimum and maximum conditions for the same LOS, are the MBM 12 and South Galactic Filament (SGF) shadowing observations, analysed by Smith et al. (2005, 2007) and by Henley et al. (2007), Henley and Shelton (2007b, 2007a) respectively. These fields were thoroughly analysed for their SWCX contamination in Koutroumpa et al. (2007) and the main outline and conclusions are presented in Sect. 3.3.

The MBM 12 LOS points toward galactic coordinates (159.2°, −34.47°), which translates to (47°, 3°) in helioecliptic coordinates, very close to the ecliptic plane and inside the limits of the equatorial slow wind zone. Therefore, during solar minimum the MBM 12 exposures (Suzaku) will yield higher SWCX contamination than solar maximum exposures (Chandra) in the 3/4 keV range.

In the case of SGF, at very high southern ecliptic latitude (353°, −73°) the effect is the opposite compared to MBM 12, because it is dominated by differences in the solar ions relative abundances. Therefore, the SWCX oxygen emission in the 3/4 keV range for the SGF field is expected to be much fainter at solar minimum (Suzaku observations) than solar maximum (XMM-Newton observations).

3.2 Dynamic Simulations

In Koutroumpa et al. (2007) we presented a dynamic variant of the basic stationary SWCX model, used to simulate the heliospheric X-ray modulation due to SW proton flux and abundance variations. For each field simulated we use real-time flux and abundance measurements from *in situ* solar wind instruments (Wind, ACE/SWICS) and we model SW enhancements as simplified Corotating Interaction Regions (CIR) of a step function form. The total width of the CIR is defined such that the total duration of the step function is the same as the measured enhancement in SW instruments. The step function amplitude is defined so that its total area is equal to the integral of the measured flux during the same period of time.

At each instant we define the form of the CIR (as a Parker spiral) and its propagation in the interplanetary space, taking into account solid solar rotation (27-day period), the radial propagation speed, the "ignition" time on the solar disk towards each radial direction, and the total width of the CIR. Only in the cases of CMEs, we neglect the solar rotation, since radial propagation is dominating the CME structure. Depending on the CIR's width and propagation instant, we calculate the fragment of the LOS affected by the CIR and the local emissivity $\varepsilon_i'(r)$ modified by the SW proton flux and heavy ion abundance variations. We can then reproduce the temporal variation of the X-ray intensity levels during the periods of observation in simulated light curves for each of our targets. Abundance variations, can be correlated or anticorrelated with proton fluxes, so they can either emphasize or compensate for the influence of the SW proton flux enhancements.

3.2.1 The Case of the MBM 12 Suzaku Observation of February 2006

The observation of MBM 12 with Suzaku on February 3–8, 2006 was performed in two consecutive exposures, ON-CLOUD (3–6/02/2006, for a total of 231 ks) and immediately after that, OFF-CLOUD (6–8/02/2006, for a total of 168 ks) (Smith et al. 2007). A detailed description of the simulation we performed on this field is given in Koutroumpa et al. (2007).

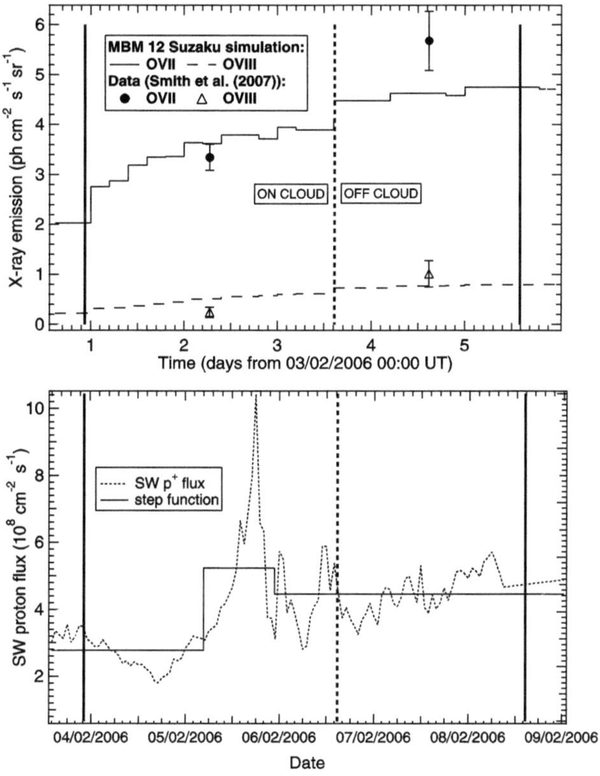

Fig. 3 *Top*: O VII and O VIII line simulated light curves (in LU) for the MBM12 ON and OFF observations with Suzaku during the period 03–08/02/2006. Plain line stands for the O VII emission and *dashed line* stands for the O VIII emission. *Dots* represent the measured ON and OFF O VII line flux, and *triangles* represent the measured ON and OFF O VIII line flux (Smith et al. 2007). *Bottom*: Solar wind proton flux (*dotted line*) in units of 10^8 cm^{-2} s^{-1} for the same period. The step function simulating the SW enhancement is presented by the *plain black line*. In both panels, the *vertical plain lines* represent the start and end of the observation period, while the *dashed vertical line* is the separation between the ON and OFF exposures

As we already explained, the MBM 12 LOS is located at (47°, 3°) of helioecliptic coordinates. In the February Suzaku observation this geometry is pointing directly toward the He cone, which is only ~2 AU away from the observer, where the emissivity on the LOS is maximum. The observation geometry was presented in figure 1 of Koutroumpa et al. (2007). The cone, with a denser distribution of helium atoms, is acting as an amplifier to any SW flux variation.

The MBM 12 Suzaku observation was influenced by a short SW perturbation recorded at the end of the ON-CLOUD pointing. We model this enhancement as a step function in particle flux of 5.2×10^8 cm^{-2} s^{-1} propagating at a low speed ($V_{SW} = 350$ km s^{-1}) for 0.75 d. The SW proton flux remained at a high, but stable, level equal to 4.42×10^8 cm^{-2} s^{-1} after the spike. In the lower panel of Fig. 3 we present the measured SW flux curve with the dotted line and the modeled step function with the plain line.

The resulting simulated light curves for O VII and O VIII are shown in the upper panel of Fig. 3 in plain and dashed lines respectively. The plain vertical lines mark the limits of the total observing period and the dashed vertical line shows the limit between the ON and OFF cloud exposures. The dynamic model, taking into account SW measurements, is predicting a smooth but gradual rise of the oxygen line intensities in the ON and OFF exposures. The O VII and O VIII line fluxes averaged for the ON-CLOUD exposure yield 3.56 LU and 0.5 LU respectively, while averages on the OFF-CLOUD exposure yield 4.62 LU and 0.77 LU for O VII and O VIII respectively. The equivalent averaged Suzaku data for O VII and O VIII are shown in the figure with plain dots and hollow triangles respectively.

The model-data comparison for the ON-cloud exposure shows that LB emission is negligible, since the SWCX model accounts for all of the observed intensity. The X-ray flux increase (~30% and ~55% for O VII and O VIII respectively) in the OFF-CLOUD exposure would change the data interpretation in terms of hot halo gas temperature, since Smith et al. (2007) assumed a constant foreground emission.

3.3 Summary of Shadowing Observations

In Table 1 we resume the MBM 12 and South Galactic Filament shadowing observations analysed in detail in Koutroumpa et al. (2007). All four observations were simulated with the SWCX model with SW conditions as close as possible to measurements in solar instruments, to account for the SWCX contamination of the oxygen line emission in each field.

Besides the basic differences due to solar maximum or solar minimum conditions during the observing period for each field, that we reported in Sect. 3.1, three of the four observations (MBM 12 Chandra, Suzaku and SGF XMM observations respectively) were found to be highly contaminated by short-term SWCX emission variations. In particular, in our simulations we confirmed the contamination of the MBM 12 Chandra observation by a strong CME as suggested by Smith et al. (2005). We suggested that the SGF XMM observation was also highly contaminated, most probably by a CME, which was confirmed by Henley and Shelton (2007a) when they re-analysed and compared their Suzaku and XMM data.

The observation data represent unabsorbed (attributed to the LB) O VII and O VIII line intensities as derived from the authors (Smith et al., 2005, 2007; Henley et al. 2007; Henley and Shelton, 2007a, 2007b). In particular, for the SGF field we give two sets of data: from the initial analysis published by Henley et al. (2007) and Henley and Shelton (2007b) (values we used in Koutroumpa et al. 2007) and the revised values published by Henley and Shelton (2007a).

The shadows block more or less efficiently the Galactic Halo oxygen emission, and in their analysis the authors derive the unabsorbed foreground emission of the oxygen lines, that we use to compare to the SWCX heliospheric emission. The only exception is in the Suzaku/MBM 12 observation (Smith et al. 2007) where the O VII and O VIII halo emission is added to the foreground emission for the OFF-CLOUD exposure. Contrary to the constant foreground assumption in the Smith et al. (2007) analysis, we demonstrated that there is a ~30% and ~55% increase in the OFF-CLOUD simulated SWCX O VII and O VII line

Table 1 Comparison of shadowing observation foreground data to SWCX model for O VII and O VIII line intensities

Target	Observation[a]	O VII (LU)			O VIII (LU)		
		Data	Data[c]	SWCX	Data	Data[c]	SWCX
MBM 12	Aug. '00, C	1.79 ± 0.55		1.49	2.34 ± 0.36		2.13
MBM 12	Feb. '06, S-on	3.34 ± 0.26		3.56	0.24 ± 0.10		0.50
MBM 12	Feb. '06, S-off[b]	5.68 ± 0.59		4.62	1.01 ± 0.26		0.77
SGF	May, '02, X	3.40	$6.2^{+2.8}_{-2.9}$	3.16	1.00	...	1.02
SGF	Mar. '06, S	0.13	$1.1^{+1.1}_{-1.4}$	0.34	N.C.	1.0 ± 1.1	0.02

[a]Observation date and observing instrument: C for Chandra, X for XMM-Newton and S for Suzaku

[b]MBM 12 Suzaku-off exposure is included in the analysis, as foreground emission was not constant (see Sect. 3.2.1 for details)

[c]Revised values for SGF foreground line intensities, from Henley and Shelton (2007a)

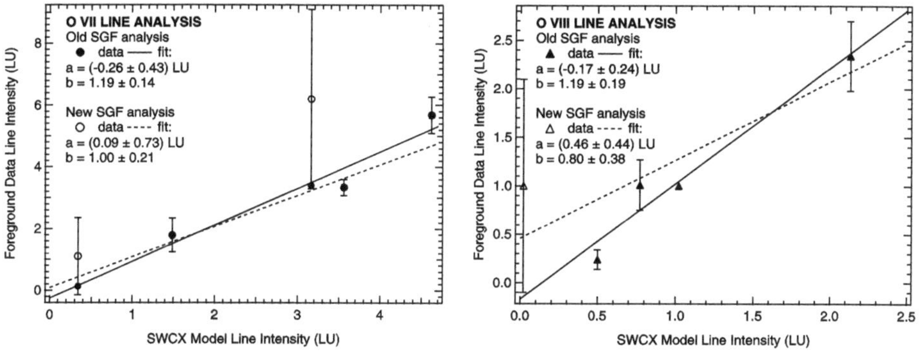

Fig. 4 Linear fits of oxygen line intensities. Foreground emission data from shadowing observations in literature (MBM 12: Smith et al. 2005, 2007) and (SGF: Henley et al. 2007; Henley and Shelton 2007b) versus SWCX model results for the O VII (*left panel*) and O VIII (*right panel*) line intensities. All linear fits are calculated with no error-bar weighting except for the O VII New SGF analysis fit (*left panel, dashed line*). See Sect. 3.3 for more details

intensities respectively, due to the brief SW enhancement at the end of the ON-CLOUD exposure (see Sect. 3.2.1). This increase was erroneously attributed to the Galactic Halo emission, which should be revised, and thus we include these values in the analysis as well.

In Fig. 4 we linearly fit the data over the SWCX model results for the O VII (left panel) and O VIII (right panel) lines for the initial (plain lines) and revised (dashed lines) data sets. We did not use data errorbars as standard deviation to weight the fit, because not all data errorbars were communicated (Henley et al. 2007; Henley and Shelton 2007b). The only exception is the O VII line fit, since Henley and Shelton (2007a) provided errorbars for both XMM and Suzaku O VII detections. The old and new fit coefficients ($y(LU) = (a + bx)(LU)$) for the O VII and O VIII lines are also noted on the figure.

The first linear fits (in plain lines in Fig. 4) we performed (Koutroumpa et al. 2007) had a slope of 1.19 (± 0.14; 0.19 at 1σ for O VII and O VIII resp.) meaning that the SWCX model reproduced very well the local foreground emission measured in shadows. Moreover, the fit also gave an estimate of the residual foreground emission to be attributed to the LB. The foreground was found to be (-0.26 ± 0.43 at 1σ) LU for O VII and (-0.17 ± 0.24 at 1σ) LU for O VIII, which means that with a high probability the LB O VII and O VIII emission is negligible compared to the heliospheric emission.

The new fits (in dashed lines, Fig. 4), where we substituted the SGF observed foreground data with the revised values of Henley and Shelton (2007a) give slightly different results, although the correlation to the simulated values remains very satisfying (the correlation is even improved for the O VII line intensities) and the residual LB emission is still consistent to zero.

4 SWCX in the 1/4 keV Band

In this section we present a preliminary study of the SWCX background in the 1/4 keV (0.1–0.3 keV) band that we compare with the ROSAT and Wisconsin surveys data. In Sect. 4.1 we describe briefly the SWCX simulation and in Sect. 4.2 we compare the model results with the survey data sets.

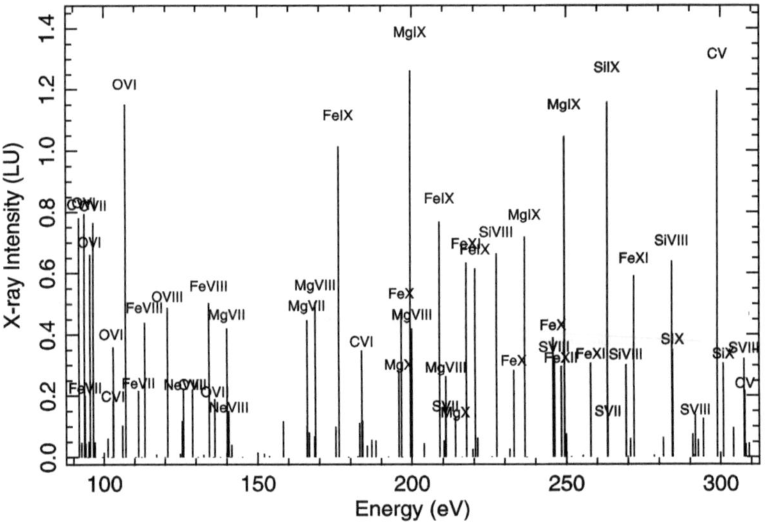

Fig. 5 Calculated SWCX spectra in Line Units for the 1/4 keV band. Emitting ions are marked above the *most intense lines*

4.1 Simulations

For the study on the 1/4 keV band, we updated our atomic database to include Fe, Si, S, Mg ions that emit intense lines in the 0.1–0.3 keV range. Individual radiative transition probabilities (photon yields) for these ions were calculated assuming that the ions are hydrogenic. Moreover, photon yields were calculated using a unique neutral species, which means that no distinction between H and He was made. The hydrogenic ion assumption and unique neutral species choice are only approximations, but the results give a good estimate of the order of magnitude of the X-ray intensity for this energy range. An example of calculated spectra in the 0.1–0.3 keV range is presented in Fig. 5, with the emitting ion identifying the most intense lines.

We have calculated spectra for two LOS, upwind (UW) and downwind (DW), for an observer assumed to be at 1 AU and in a crosswind position with respect to the LOS. This observing geometry was chosen to match the ROSAT satellite observing constraints. In the 1/4 keV (0.1–0.3 keV) range the model yields a total of 28.35 LU for the UW LOS and a total of 20.27 LU for the DW LOS. The intensity is probably underestimated because of the SW ion propagation in the heliosheath up to the heliopause, and in the heliotail up to ~3 000 AU, where all ions are used up. The outer heliospheric region is neglected (for the moment) in our model, but preliminary estimates yield a maximum additional ~20% contribution in the DW direction. In general, because of parallax effects as well, we can assume that within a 50% error bar, these values give an good estimate of average values of the heliospheric SWCX emission in the 1/4 keV range.

4.2 Model-Data Comparison

The sounding rocket survey of the University of Wisconsin and the ROSAT satellite survey have dominated the diffuse SXRB observations from 1972 to 1991. The Wisconsin survey allowed a full-sky mapping of the IS SXRB with a spatial resolution of ~7° (McCammon

Table 2 Summary of main band fluxes for the SWCX model in the upwind (UW) and downwind (DW) directions and measured data. For ROSAT data only the unabsorbed fraction I_o is given

Band	SWCX model		Data range
	UW	DW	
B (cts s^{-1})	11.99	8.04	[20–100]
C (cts s^{-1})	73.32	56.24	[50–250]
R1 (RU[a])	141.14	104.38	
R2 (RU)	192.07	147.27	
R12 (RU)	333.21	251.65	[250–820]

[a]RU = 10^{-6} cts s^{-1} $arcmin^{-1}$

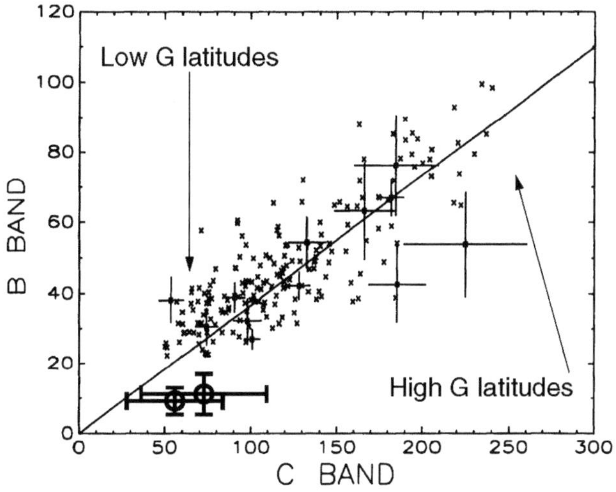

Fig. 6 Scatter plot of the total intensity calculated in the SWCX model (*large circles*) compared to total measured (*crosses*) values in B and C bands in the Wisconsin survey. The model reproduces the observed C band fluxes in the galactic plane, but not the B band fluxes. Figure extracted and adapted for our study from Snowden et al. (1990b)

et al. 1983). The ROSAT survey improved the spatial resolution to ∼12′ (Snowden et al. 1997). However, the instruments did not have any spectral resolution so they provided only total fluxes for each band.

In order to compare the modeled SWCX emission level to the measured SXRB intensities at 1/4 keV, we convolve the calculated spectra for the UW and DW LOS with each band response in this energy range of the instruments. The bands concerning our study are B (0.13 – 0.188 keV), C (0.16 – 0.284 keV) for the Wisconsin survey and R1 (0.08 – 0.284 keV), R2 (0.14 – 0.284 keV) for the ROSAT survey. The corresponding band responses are presented in figures: 1(a) of McCammon et al. (1983) for Wisconsin and 1 of Snowden et al. (2000) for ROSAT. In Table 2 we summarize the total band fluxes derived from the SWCX spectra convolution with the band responses, and the measured data range for each band (evaluated from Figs. 6 and 7).

4.2.1 Wisconsin Sounding Rocket Survey

The SXRB survey of the University of Wisconsin was performed with a series of 10 sounding rocket flights between 1972 and 1979. The only spectral information in the 1/4 keV range is extracted by the boron (B) and carbon (C) filters incorporated into the detector windows that separate the two bands B and C presented in McCammon et al. (1983), Fig. 1(a).

Detailed maps derived from the complete survey were presented in McCammon et al. (1983). The maps present a clear negative correlation with the column density of galactic

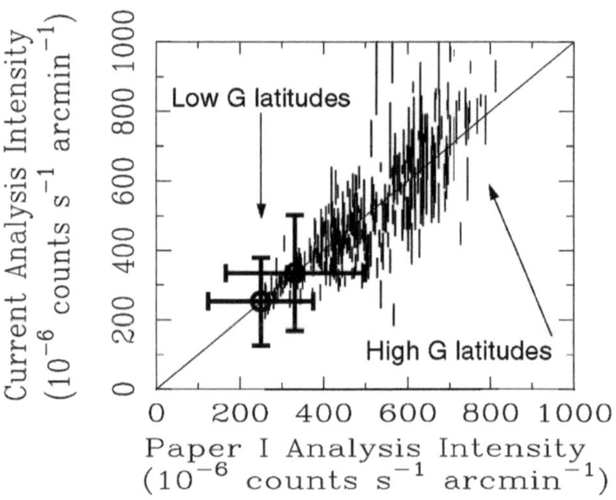

Fig. 7 Same as Fig. 6 but for the unabsorbed (I_o) emission of ROSAT shadowing observations in band R12 (R1 + R2). The model reproduces most of the R12 band flux in the galactic plane. Figure extracted and adapted for our study from Snowden et al. (2000)

neutral hydrogen N_{HI}. The X-ray intensity is lowest toward the galactic plane and highest at high galactic latitudes. Snowden et al. (1990b) investigated the band intensity ratio B/C (low to high energies) and found a variation between 0.25 to 0.46 in a dipole-like spatial correlation, aligned with a roughly galactic center-anticenter direction, that they attributed to an anisotropy of the Local Hot Bubble temperature ranging between $10^{6.2}$ and $10^{5.9}$ K respectively.

From Snowden et al. (1990b) we extracted a scatter plot (Fig. 6) comparing the B and C band measurements in the Wisconsin survey. The data range is summarized in Table 2 and gives ∼[20–100] cts s^{-1} for the B band and ∼[50–250] cts s^{-1} for the C band. We superposed, in this figure, the two points (wide circles) corresponding to our simulated spectra convolved with the B and C band responses (see Table 2). Error bars are given only for information at 50% of the SWCX band flux, in order to give an estimate of the heliospheric flux range in these bands.

The SWCX model in the C band predicts fluxes well within the observed values in the galactic plane (lower limits), but simulated B band flux is about two times lower than the lower observed limits. The band ratio B/C for the SWCX flux is accordingly inconsistent with measured values, and fairly constant from UW to DW directions, with a mean value of B/C ∼ 0.15.

4.2.2 The ROSAT Survey

The ROSAT all-sky survey was performed in a scan mode with imaging telescopes in the soft X-ray band of 0.1–2 keV, during the first year of the mission. The observing geometry constraints imposed to scan the sky in a big circle, during one orbit, perpendicular to the spin axis that coincided with the Sun-Earth direction. Therefore a complete scan of the sky was performed in a six-month period.

The ROSAT all-sky maps in the 1/4 keV range (Snowden et al. 1995; Snowden et al. 1997) confirmed the Wisconsin results and the negative correlation with the H I column density. The improved spatial resolution of ROSAT, with respect to previous surveys, allowed a better use of shadows as tools to separate the foreground from distant components of the SXRB. An extensive catalog of shadows in the keV1/4 band, with fitted foreground and distant emission values was presented in Snowden et al. (2000).

In Fig. 7 we present a scatter plot of foreground intensity I_o measured in band R12, which is the sum of intensities in bands R1 and R2. Units are in 10^{-6} cts s^{-1} $arcmin^{-1}$, better known as ROSAT Units (RU). The figure was extracted from Snowden et al. (2000) and the authors compare the data set analysed with two different methods. The exact comparison of the two methods is beyond the scope of this paper, since what we need is just the data intensity range (250–820 RU). In the figure $y = x$, thus the fitted slope is one. Simulated spectra convolved with R1 and R2 band responses and summed for the total result in R12 band yield 333.21 and 251.65 RU for the UW and DW direction respectively. These values, represented by the wide circles in Fig. 7 with error bars at 50% (to give the SWCX flux range), are within the lower limits of observational data, that correspond to the low galactic latitude regions.

In the same paper (Snowden et al. 2000) confirm the dipole-like spatial correlation of the R2/R1 (high to low energies) flux ratio, although the variation is less pronounced than the one derived from the Wisconsin analysis, 1.25 to 1.04 from UW to DW. Our model results yield a constant R2/R1 ratio at an average value of 1.38.

5 Discussion

We have presented simulation results on the SWCX soft X-ray emission in the heliosphere in the 0.1–1.0 keV energy range, that we compared to data from past and present observing instruments. We separated our analysis into two energy ranges: the 3/4 keV, dominated by the oxygen (O VII and O VIII) lines and the 1/4 keV, where heavier ion lines as Mg, Si, S and Fe dominate the SWCX spectra. This analysis yields estimates of the SWCX heliospheric component within the diffuse Soft X-ray Background and confirms the large contamination of X-ray data by the heliospheric emission that needs to be taken into account in future determinations of the LB temperature and pressure.

The simulation results on shadowing clouds observed with Chandra, XMM-Newton and Suzaku suggest that the local 3/4 keV emission detected in front of shadows is entirely explained by the heliospheric SWCX emission (see Table 1 and Fig. 4) and no emission from the LB is needed at these energies. Thus, combining our results in the 3/4 keV range and previous results published on the 3/4 and 1/4 keV ROSAT band ratios, the strict upper limit of the LB temperature is 10^6 K. Indeed, if the LB plasma is in collisional equilibrium at 10^6 K, then it mainly emits in the 1/4 keV range, and only very little emission is produced in the oxygen lines at 3/4 keV.

But, then, how does SWCX affect the data interpretation in the 1/4 keV range? We have compared SWCX simulation results with data from the ROSAT and Wisconsin surveys. Both sets of measured data were originally interpreted in terms of hot gas emission in the LB at a temperature of $\sim 10^6$ K. Variations in the observed X-ray intensity were attributed to variations in the extent of the emission volume and therefore the emission measure of the plasma.

SWCX emission flux calculated for the R12 and C bands (which practically cover the same energy range) yielded average values that explain most of the observed emission in the galactic plane, previously attributed to the restricted regions of the LB. Indeed, according to Snowden et al. (1998), in the galactic center and anticenter directions (roughly UW and DW respectively), the LHB is found to have an extent of $R_{LHB} \sim 50$ pc (see upper left panel of figure 10 in Snowden et al. 1998), which corresponds to an unabsorbed 1/4 keV emission of $I_o = R_{LHB}/0.155 \approx 322$ RU measured in the R12 band, while our model predicts ~[250–333] RU.

On the other hand, the SXRB intensity is typically brighter at higher Galactic latitudes, and SWCX cannot account for all the emission. This may be explained by the fact that at high latitudes the LOS's point through the chimneys allowing for non-absorbed halo emission to reach the observer. Moreover in the B band, which is the least overlapping in energy range with respect to the other 1/4 keV bands, the SWCX simulations predict about half of the observed lower limits. Besides, the SWCX model fails to reproduce the observed R2/R1 ratio by 10–33% and the B/C ratio by 40–67%.

The answer, then, to the local unabsorbed SXRB puzzle might lie in the proper mix of SWCX emission and warm (rather than hot) LB that will fill-in the gap in the B band and reproduce the observed band ratios (Koutroumpa et al., in preparation).

Acknowledgements DK and RL acknowledge The Institute for Theoretical Atomic, Molecular and Optical Physics (ITAMP) for travel and living expenses support during a visit to ITAMP facilities. The authors would like to thank Dan McCammon for providing accurate values of the effective area functions of the Wisconsin and ROSAT bands. We are also thankful to Steve Snowden for useful (and lively) discussions. Finally, we are really greatfull to our referee Jeffrey Linsky for the thorough report and constructive remarks, as well as for the organization, along with the organizing committee of the ISSI Local Bubble Workshop, of this most interesting meeting.

References

P. Beiersdorfer, C.M. Lisse, R.E. Olson, G.V. Brown, H. Chen, Astrophys. J. Lett. **549**, L147–L150 (2001)
C.S. Bowyer, G.B. Field, J.F. Mack, Nature **217**, 32 (1968)
D.P. Cox, in *IAU Colloq. 166: The Local Bubble and Beyond*, ed. by D. Breitschwerdt, M.J. Freyberg, J. Truemper. Lecture Notes in Physics, vol. 506 (Berlin, Springer, 1998), pp. 121–131
T.E. Cravens, Geophys. Res. Lett. **24**, 105–108 (1997)
T.E. Cravens, Astrophys. J. Lett. **532**, L153–L156 (2000)
T.E. Cravens, I.P. Robertson, S.L. Snowden, J. Geophys. Res. **106**, 24,883–24,892 (2001)
D.B. Henley, R.L. Shelton, ArXiv e-prints **712** (2007a)
D.B. Henley, R.L. Shelton, ArXiv Astrophysics e-prints (2007b)
D.B. Henley, R.L. Shelton, K.D. Kuntz, Astrophys. J. **661**, 304–319 (2007)
V. Kharchenko, A. Dalgarno, Astrophys. J. Lett. **554**, L99–L102 (2001)
D. Koutroumpa, R. Lallement, V. Kharchenko, A. Dalgarno, R. Pepino, V. Izmodenov, E. Quémerais, Astron. Astrophys. **460**, 289–300 (2006)
D. Koutroumpa, F. Acero, R. Lallement, J. Ballet, V. Kharchenko, Astron. Astrophys. **475**, 901–914 (2007)
K.D. Kuntz, S.L. Snowden, Astrophys. J. **543**, 195–215 (2000)
R. Lallement, B.Y. Welsh, J.L. Vergely, F. Crifo, D. Sfeir, Astron. Astrophys. **411**, 447–464 (2003)
C.M. Lisse, K. Dennerl, J. Englhauser, M. Harden, F.E. Marshall, M.J. Mumma, R. Petre, J.P. Pye, M.J. Ricketts, J. Schmitt, J. Trumper, R.G. West, Science **274**, 205–209 (1996)
F.J. Marshall, G.W. Clark, Astrophys. J. **287**, 633–652 (1984)
D. McCammon, D.N. Burrows, W.T. Sanders, W.L. Kraushaar, Astrophys. J. **269**, 107–135 (1983)
D. McCammon, W.T. Sanders, Ann. Rev. Astron. Astrophys. **28**, 657–688 (1990)
J.C. Raymond, in *NATO ASIC Proc. 249: Hot Thin Plasmas in Astrophysics*, ed. by R. Pallavicini (1988), p. 3
W.T. Sanders, W.L. Kraushaar, J.A. Nousek, P.M. Fried, Astrophys. J. Lett. **217**, L87–L91 (1977)
R.K. Smith, R.J. Edgar, P.P. Plucinsky, B.J. Wargelin, P.E. Freeman, B.A. Biller, Astrophys. J. **623**, 225–234 (2005)
R.K. Smith, M.W. Bautz, R.J. Edgar, R. Fujimoto, K. Hamaguchi, J.P. Hughes, M. Ishida, R. Kelley, C.A. Kilbourne, K.D. Kuntz, D. McCammon, E. Miller, K. Mitsuda, K. Mukai, P.P. Plucinsky, F.S. Porter, S.L. Snowden, Y. Takei, Y. Terada, Y. Tsuboi, N.Y. Yamasaki, Publ. Astron. Soc. Jpn. **59**, 141–150 (2007)
S.L. Snowden, D.P. Cox, D. McCammon, W.T. Sanders, Astrophys. J. **354**, 211–219 (1990a)
S.L. Snowden, J.H.M.M. Schmitt, B.C. Edwards, Astrophys. J. **364**, 118–122 (1990b)
S.L. Snowden, M.J. Freyberg, P.P. Plucinsky, J.H.M.M. Schmitt, J. Truemper, W. Voges, R.J. Edgar, D. McCammon, W.T. Sanders, Astrophys. J. **454**, 643 (1995)
S.L. Snowden, R. Egger, M.J. Freyberg, D. McCammon, P.P. Plucinsky, W.T. Sanders, J.H.M.M. Schmitt, J. Truemper, W. Voges, Astrophys. J. **485**, 125 (1997)
S.L. Snowden, R. Egger, D.P. Finkbeiner, M.J. Freyberg, P.P. Plucinsky, Astrophys. J. **493**, 715 (1998)
S.L. Snowden, M.J. Freyberg, K.D. Kuntz, W.T. Sanders, Astrophys. J. Supp. **128**, 171–212 (2000)
S.L. Snowden, M.R. Collier, K.D. Kuntz, Astrophys. J. **610**, 1182–1190 (2004)
B.J. Wargelin, P. Beiersdorfer, G.V. Brown, Can. J. Phys. **86**, 151–169 (2008)

Revising the Local Bubble Model due to Solar Wind Charge Exchange X-ray Emission

Robin L. Shelton

Originally published in the journal Space Science Reviews, Volume 143, Nos 1–4, 231–239.
DOI: 10.1007/s11214-008-9358-8

Abstract The hot Local Bubble surrounding the solar neighborhood has been primarily studied through observations of its soft X-ray emission. The measurements were obtained by attributing all of the observed local soft X-rays to the bubble. However, mounting evidence shows that the heliosphere also produces diffuse X-rays. The source is solar wind ions that have received an electron from another atom. The presence of this alternate explanation for locally produced diffuse X-rays calls into question the existence and character of the Local Bubble. This article addresses these questions. It reviews the literature on solar wind charge exchange (SWCX) X-ray production, finding that SWCX accounts for roughly half of the observed local 1/4 keV X-rays found at low latitudes. This article also makes predictions for the heliospheric O VI column density and intensity, finding them to be smaller than the observational error bars. Evidence for the continued belief that the Local Bubble contains hot gas includes the remaining local 1/4 keV intensity, the observed local O VI column density, and the need to fill the local region with some sort of plasma. If the true Local Bubble is half as bright as previously thought, then its electron density and thermal pressure are $1/\sqrt{2}$ as great as previously thought, and its energy requirements and emission measure are 1/2 as great as previously thought. These adjustments can be accommodated easily, and, in fact, bring the Local Bubble's pressure more in line with that of the adjacent material. Suggestions for future work are made.

Keywords Local Bubble · Interstellar medium · Solar wind charge exchange · Solar wind · SWCX · Diffuse X-rays · O VII

1 Introduction

The Local Bubble (LB) is traditionally thought of as a large ($\sim$ 60 pc in radius) bubble filled with hot ($\sim 10^6$ K) plasma surrounding the Solar neighborhood. Several clouds of

R.L. Shelton (✉)
Department of Physics and Astronomy, the University of Georgia, Athens, GA 30602, USA
e-mail: rls@physast.uga.edu

warm ($\sim 10^4$ K) gas reside within the Local Bubble. One of these clouds, the Local Cloud, envelops the Solar System.

The Local Bubble was discovered in the 1970's through 1/4 keV X-ray observations. Early X-ray instruments detected soft X-rays from every direction and with B and C band (bandpasses: ~70–188 eV and ~160–284 eV, respectively) intensities that anti-correlate with absorbing column density but correlate well with each other. These characteristics suggested that the Earth resides within an X-ray emissive bubble which has displaced neutral material in the disk (McCammon and Sanders 1990). That some of the observed X-rays were produced locally and not, for example, in the Galactic halo or bulge, has been confirmed by shadowing studies in which absorbing interstellar clouds are used to block X-rays from more distant sources (Burrows and Mendenhall 1991; Snowden et al. 1991). The X-ray intensity observed in the direction of the cloud, minus the intensity of distant photons that have leaked through the cloud, indicates the brightness of the local region. Technically, these studies indicate only that X-rays were produced "locally", i.e. somewhere between the satellite and the absorbing cloud; these studies do not reveal the precise location of the X-ray emitting gas. Although astronomers have long recognized the possibility that the Solar System contributed to the observed local X-ray intensity, the agreement between 1/4 keV observations taken more than a year apart (Bunner et al. 1973) required a constant emission source and so favored the interstellar interpretation. This changed in 1996 when *ROSAT* observed X-rays from the coma of Comet Hyakutake (Lisse et al. 1996). These observations unequivocally demonstrated that the heliosphere contains X-ray sources aside from the Sun and therefore called into question the assumption that all of the locally produced diffuse X-ray flux was created by the Local Bubble. The existence and nature of the Local Bubble are now being reevaluated. This article addresses these issues. After further describing the historical Local Bubble model and diffuse Solar System X-ray emission, the paper proceeds to a discussion of whether or not the Solar System can entirely replace the Local Bubble as the explanation for the "local" emission (it cannot) and to a new picture of the Local Bubble which takes into account the presence of Solar System X-rays.

2 The Historical Local Bubble Model

As mentioned in the introduction, the historical Local Bubble's temperature is about a million degrees Kelvin. This temperature was found by comparing observed X-ray band ratios with models for optically thin hot gas in collisional ionizational equilibrium. Even the data from the earliest 1/4 keV X-ray telescopes, in which the Local Bubble emission was not distinguished from that of the Galactic halo, matched that of a 0.9 to 1.2×10^6 K plasma (McCammon and Sanders 1990). Later, *ROSAT* All Sky Survey data, for which it was possible to separate the Local Bubble and Galactic halo contributions, yielded a similar Local Bubble temperature ($T_{LB} = 1.3 \times 10^6$ K, with variation from $T_{LB} = 1.1$ to 1.9×10^6 K depending upon direction (Kuntz and Snowden 2000)).

Warm clouds are buried deep within the bubble. These have been observed via their absorption of light from nearby stars and by analyses of the material that abuts the heliosphere (see other articles in this volume for further details). In addition to these temperature extremes, we expect to find intermediate temperature gas in the transition zones between the clouds and the hot bubble plasma. Thus, O VI, which traces 3×10^5 K gas was sought. It was found, both in a statistical analysis of dozens of *Copernicus* column density measurements (Shelton and Cox 1994) and, also in analyses of *FUSE* data for stars within about a hundred parsecs of the Sun (Oegerle et al. 2005; Savage and Lehner 2006). This distance range is thought to be similar to the Local Bubble's radius, which was determined from data on a molecular cloud residing just inside the bubble boundary (Snowden and McCammon 1993). Once the radius was known for one direction, the radii for other directions were estimated from the observed intensities in those directions and the assumption that the temperature and density are the same in all directions. By this method, the distance from the Sun to the Local Bubble's periphery was found to vary from about 40 pc to slightly more than 100 pc (Snowden et al. 1998). Measurements of Na I column densities toward nearby stars indicate that a larger cavity, called the Local Cavity, also surrounds the solar neighborhood. This region of exceptionally low volume density varies in radius from ~ 40 to ~ 200 pc (Lallement et al. 2003). The Local Bubble fits easily within the Local Cavity.

Several models have been proposed to explain the Local Bubble, including a single supernova explosion (Cox and Anderson 1982; Edgar 1986), multiple supernova explosions (de Avillez and Breitschwerdt 2003), and a bubble that has broken out of its nascent molecular cloud (Breitschwerdt and Schmutzler 1994). Multiple supernova models are currently favored because of the large input energy requirement to blow such a large and energetic bubble and because the observed 1/4 keV X-ray to O VI and C III ratios disallow the blow-out model (Welsh et al. 2002; Shelton 2003; Oegerle et al. 2005).

It is noteworthy that time variability was seen in the *ROSAT* data. The *ROSAT* All Sky Survey scans overlapped each other. By comparing data taken during one orbit with that taken during the next, it was possible to see that the flux level varied with time in addition to position on the sky. The excess fluxes, called "long term enhancements" were nominally removed from the *ROSAT* All Sky Survey data before the data were used to measure the Local Bubble's temperature and brightness. The long term enhancements were unexplained and troubling. But, after X-rays were seen from Comet Hyakutake, Cravens (2000) suggested that solar wind X-rays could have caused the long term enhancements in the *ROSAT* survey. A more troubling notion is the possibility that solar wind X-rays may cause additional contamination.

3 Charge Exchange in the Solar System

3.1 Solar Wind Charge Exchange X-rays

The physical mechanism by which Comet Hyakutake emitted X-rays also acts throughout the Solar System. In this mechanism, highly charged ions in the solar wind collide with and receive electrons from other atoms. The electrons transfer into high n levels in the solar wind ions, then they radiatively decay by emitting one or more photons. For example, $O^{7+} + H^0 \rightarrow O^{6+*} + H^+ \rightarrow O^{6+} + H^+ +$ photons. During the radiative decay, the electron may pass through several excitation levels. The last transition, the $n \geq 2$ to $n = 1$ transition, will produce an X-ray photon. For example, in O^{6+}, each $n = 2$ to $n = 1$ transition produces a ~ 570 eV X-ray photon.

The charge transfer interactions, called Solar Wind Charge Exchange (SWCX), are expected to occur between solar wind ions and neutral interstellar atoms that have drifted into the heliosphere (called heliospheric SWCX), and between solar wind ions and material in the Earth's upper atmosphere (called geocoronal SWCX). SWCX interactions with material flowing into the heliosphere were found to outnumber SWCX interactions with material in the Earth's atmosphere (Robertson and Cravens 2003). The solar wind is non-isotropic and time varying on long and short time scales. The longest timescale is ~ 11 years and is associated with the Solar activity cycle. The shortest timescale is on the order of an hour and is caused by Coronal Mass Ejections.

Multiple research groups have modeled the SWCX emission of soft X-ray photons. Concentrating on the 1/4 keV X-ray band, and accounting for the state of the Solar Cycle and the sight line geometries during the *ROSAT* All Sky Survey Robertson and Cravens (2003) calculated the heliospheric and geocoronal SWCX contributions to the 1/4 keV flux observed by *ROSAT*. Their SWCX emission map (Figs. 9 and 10 in their paper) is markedly non-isotropic, with two dim regions centered on $\ell \sim 220°$, $b \sim -45°$ and $\ell \sim 130°$, $b \sim 0°$ and with an extended bright region whose intensity peaks near $\ell = 240°$, $b = 45°$. Their conservatively estimated intensity of 1/4 keV emission accounts for about half of the diffuse 1/4 keV X-ray intensity seen in the Galactic plane and thus accounts for about half of the intensity previously attributed to the Local Bubble in the Galactic plane. The comparison at higher latitudes is more complicated due to the Galactic halo contribution. Subsequently, Koutroumpa et al. (2006) modeled the heliospheric O VII SWCX emission at ~ 570 eV and estimated that as many as 0.8 to 3.4 photons cm^{-2} s^{-1} sr^{-1} could result from SWCX in the slowly varying solar wind, depending on the stage of the Solar cycle and the viewing angle. This range is of the order of what is expected from the Local Bubble. Note that the O VII triplet at $\sim$570 eV lies outside *ROSAT*'s 1/4 keV band, but within *ROSAT*'s 3/4 keV band, as well as *XMM*, *Chandra*, and *Suzaku*'s bandpasses. The Local Bubble is much dimmer in the 3/4 keV band than the 1/4 keV band. Therefore the LB's O VII spectral features are not expected to be bright and the SWCX photons may account for a relatively larger fraction of the observed intensity.

In addition to the heliospheric and geocoronal SWCX components, there seems to be an even brighter component which may be associated with Coronal Mass Ejections (CMEs). Large CMEs discharge $\sim 10^{11}$ to $\sim 10^{13}$ kg of ionized material non-isotropically into the heliosphere over periods of hours. As this material flows outwards through the heliosphere, it is subject to charge exchange that leads to X-ray emission. This is the conclusion drawn by Koutroumpa et al. (2007) and used to explain the relatively high O VII intensities seen in *Chandra* and *XMM* shadowing observations. In one example, *XMM* observed a pair of shadowing sight lines in 2002, during the solar maximum when CMEs are most common. *Suzaku* observed the sightlines in 2006, during Solar minimum when CMEs are least common. Between the 2002 and 2006 observations, the "locally produced" O VII intensity dropped from $6.1^{+2.8}_{-3.0}$ to 0.4 ± 1.1 photons cm^{-2} s^{-1} sr^{-1} (Henley and Shelton 2008). Not only does a CME provide enough material to explain the large intensity difference, but, as Henley and Shelton (2008) point out, it also provides an explanation for why there was no anomaly in the solar wind proton data recorded by *ACE* SWEPAM near the time of the *XMM* observation. *ACE* monitors the solar wind ~ 0.01 AU upstream from the Earth, and it would not have noticed a CME that did not pass through the satellite.

3.2 Solar Wind Charge Exchange O VI

Although SWCX is an important contaminant to the observations of diffuse X-rays, it is not a significant contaminant to the O VI observations. This can be verified by the following estimates. In the first estimate, the O VI column density, N_{OVI}, is found from $\int n_{\mathrm{OVI}}\,dl$, where the volume density of O VI ions, n_{OVI}, can be estimated from the volume density of solar wind protons, n_p, the oxygen-to-hydrogen ratio, [O/H], and the fraction of solar wind oxygen atoms in the O^{5+} ionization state. The solar wind proton density decreases with distance from the Sun, r, as $n_p = n_{p1}(r_1/r)^2$, where r_1 is 1 AU and n_{p1} is the proton density at r_1. The proton density at r_1 can be found from the relationship between the proton density, proton flux, and solar wind velocity (i.e. $n = F/v$), using Schwadron and Cravens (2000)'s estimates for the solar wind proton flux at r_1, ($F_1 = 2 \times 10^8$ cm^{-2} s^{-1}), and solar wind velocity (fast wind: 810 km s^{-1}, slow wind: 442 km s^{-1}, simple average: 626 km s^{-1}). Thus $n_{p1} = 3.2$ cm^{-3}. Schwadron and Cravens (2000)'s estimate for the oxygen-to-hydrogen ratio in the solar wind is also used. The ratio varies from $1/1780$ for the slow wind to $1/1550$ for the fast wind. The simple average of these numbers yields [O/H] $\sim 6.0 \times 10^{-4}$ oxygens per hydrogen. All of the solar wind oxygen atoms are ionized at the O^{6+} level or above when they leave the Sun (Schwadron and Cravens 2000). Some of these ions will charge exchange to the O^{5+} level while in the Solar System, so the upper limit on the O^{5+} to oxygen ratio along a given sightline is 1.0. As a result, the upper limit on $N_{\rm O\,VI}$ is 1.9×10^{-3} cm^{-3} $\int (r_1/r)^2\, dl$. The integral is equal to 0.99 AU, for the simplest path, which begins at the Earth and proceeds directly away from the Sun to the heliopause located about 100 AU downstream. Thus, the estimated $N_{\rm O\,VI} \lesssim 2.9 \times 10^{10}$ cm^2. This value is much smaller than the error bars on some of the smallest column densities observed by *FUSE*, such as $N_{\rm O\,VI} = 2.4 \pm 1.4 \times 10^{12}$ cm^{-2} (Savage and Lehner 2006), and $N_{\rm O\,VI} = 0.3 \pm 2.3 \times 10^{12}$ cm^{-2} (Oegerle et al. 2005).

The *intensity* of SWCX-induced O VI resonance line photons (1032, 1038 Å) has not been estimated by Robertson and Cravens (2003), Koutroumpa et al. (2006, 2007). However, an estimate can be made from the information in Koutroumpa et al. (2006). They predicted the intensity of heliospheric SWCX 72 and 82 eV photons, which are produced by O^{5+} ions undergoing transitions from the $n = 3$ shell to the $n = 2$ shell. Note that some of these transitions will place the electron into the p subshell, while others will place it into the s subshell. Those left in the p subshell will then undergo the $2p\ ^2P_{3/2}$ to $2s\ ^2S_{1/2}$ or the $2p\ ^2P_{1/2}$ to $2s\ ^2S_{1/2}$ transitions that yield 1032 and 1038 Å O VI resonance line photons. Thus, the upper limit on the O VI resonance line intensity due to heliospheric SWCX is set by Koutroumpa et al. (2006)'s predictions for the 72 and 82 eV photons. The greatest intensity of these photons shown in their Fig. 1 maps is 3.3×10^{-9} ergs s^{-1} cm^{-2} sr^{-1}. Thus, the O VI resonance line intensity can be estimated as < 26 photons s^{-1} cm^{-2} sr^{-1}. Even if this upper limit were to be elevated by a factor of several due to a CME along the line of sight, it would be much smaller than the 1σ error bars on the *FUSE* observations of the local O VI resonance line intensity (~ 200 photons s^{-1} cm^{-2} sr^{-1} for each of the resonance lines (Shelton 2003)), showing that SWCX is not an important factor when evaluating O VI emission observations.

4 Is There a Hot Local Bubble After All?

There is a hot Local Bubble. Robertson et al. (2003) found that maps of the $1/4$ keV X-ray sky are still bright, even after the SWCX intensity is subtracted. Especially noteworthy is that there are regions in the Galactic plane with non-zero net intensity (countrate $\sim 200 \times 10^6$ counts s^{-1} arcmin^{-2}). Due to the high opacity of the Galactic disk, only soft X-rays that are made locally and a small fraction of the X-rays that come from very bright non-local sources will be observed at Earth. Thus, low latitude X-rays are almost entirely attributable to the local region and the mapped intensity of low latitude X-rays remaining after SWCX X-rays have been subtracted is almost entirely attributable to the Local Bubble. The story becomes more complicated when one considers the $3/4$ keV band, or one of its most prominent features, the O VII triplet. According to Koutroumpa et al. (2007), it is possible that SWCX might account for as much as 100% of the measured local O VII intensity. However, given the inherent uncertainty in the SWCX O VII estimates, there is room for both the

Local Bubble and Solar Wind Charge Exchange. Also, even if the majority of local O VII photons resulted from SWCX events, the Local Bubble would not be doomed. The Local Bubble is not thought to be especially bright in 3/4 keV X-rays (the local 3/4 keV countrate is only $\sim 1/15$ as bright as the local 1/4 keV countrate (Snowden and McCammon 1993)), which makes it easier for a contaminant such as SWCX to produce a large fraction of the observed countrate.

One of the arguments for the continued belief in the Local Bubble is that the O VI column density found between the Earth and nearby stars (Jenkins 1978a; Oegerle et al. 2005; Savage and Lehner 2006) implies the presence of hot gas and cannot be explained by SWCX. The O VI ion is a tracer of $\sim 3 \times 10^5$ K gas. Gas of this temperature is thermally unstable; it does not remain at this temperature for long periods unless hotter gas is present to resupply it. Therefore, it is primarily located in transition zones between hotter and cooler gas. Its existence within the local region has been found by *Copernicus*, which observed measureable quantities of O VI ions on 2 sight lines terminating within 90 pc of the Sun and several more sight lines terminating within 200 pc of the Sun (Jenkins 1978a), and later by *FUSE*, which observed O VI ions on dozens of sight lines with similar distance ranges (Oegerle et al. 2005; Savage and Lehner 2006). It should be noted that the original *FUSE* work has been questioned by Barstow et al. (2008), who have reprocessed the original data using a more recent version of the *FUSE* data processing pipeline. They found that some of the previously observed O VI absorption features may be artifacts of the older data processing, while other features might be circumstellar features due to the white dwarf targets. Nonetheless, Barstow et al. (2008) did not contradict all of the earlier *FUSE* detections. They found interstellar O VI on several close sight lines, including WD0004+330, WD0354−368, WD2000−561, and WD2331−475. Because they found far less O VI within the LB's radius than had Oegerle et al. (2005), Savage and Lehner (2006), they concluded that the interstellar O VI only resides near the LB's periphery, and not within the LB as Oegerle et al. (2005) had concluded. The Barstow et al. (2008) reanalysis does not extend to the *Copernicus* data. None of the *Copernicus* targets were white dwarfs; all of the nearby targets were B stars. Furthermore, the original work by Jenkins (1978a, 1978b) ruled out circumstellar origins for the O VI found in the *Copernicus* data set. As shown in Sect. 3.2, the observed O VI ions cannot be attributed to the solar wind ions or SWCX, and so the only choice is that they are associated with hot interstellar gas in the local region, i.e. in the Local Bubble.

Another argument for the Local Bubble's existence is that the Local Cavity must be occupied. If there is no Local Bubble to balance the thermal and magnetic pressure of the material outside the Local Cavity, then the material outside the cavity would expand inwards at the sound speed. Similarly, the Local Cloud that surrounds the Solar System and other clouds that are also embedded in the Local Bubble's volume would also expand at the sound speed. Such expansion, with predicted velocities of about 15 km s^{-1}, has not been observed (Slavin 2007, personal communication). It might be argued that it is the total pressure (including magnetic and cosmic ray pressures) rather than the thermal pressure, alone, which governs the dynamics. Therefore, it may be possible for the total pressure in the Local Cavity to balance that in the embedded clouds and the surrounding material if the non-thermal pressure in the Local Cavity is unusually large. In response, one could point out that some researchers find low magnetic field strengths ($B = 0.7$ μG; Spangler (2007)) although other researchers find high magnetic field strengths ($B = 8$ μG; Andersson and Potter (2006)) on lines of sight terminating within 200 pc of the Sun.

Also note that during the conference, some participants added additional reasons to this list. Steve Snowden pointed out that there is good agreement between the Wisconsin, HEAO 1, and *ROSAT* 1/4 keV band data (Snowden et al. 1995), yet these surveys were made at different times and with different angles between the target and the Sun. Such agreement is not possible if most of the X-rays derive from a strongly time and look-angle dependent source such as the solar system. Furthermore, the anticorrelation between neutral hydrogen column density and soft X-ray intensity cannot be explained by SWCX. For a debate of the arguments for and against the existence of the Local Bubble, the reader is referred to the session report in this volume (Shelton 2008).

5 Revising the Local Bubble

If, as Robertson and Cravens (2003) conservatively estimated, half of the local 1/4 keV X-rays are due to SWCX and half are due to the hot Local Bubble, then the Local Bubble must be considerably weaker than previously thought. Its electron density and thermal pressure would be reduced by a factor of $1/\sqrt{2}$ from their previous values to $n_e \sim 0.005$ cm^{-3} and $P_{th}/k \sim 11{,}000$ K cm^{-3}, respectively, if the gas temperature and path-length remain unchanged. These reductions are comforting, because the previous pressure estimate ($P_{th}/k \sim 15{,}000$ K cm^{-3}; Snowden et al. (1998)) was far greater than that of the embedded warm clouds ($P_{th}/k \sim 2000$ to 3000 K cm^{-3}).

The reductions may also help to explain the dearth of Fe IX, Fe X, and Fe XI ~ 72 eV photons observed by the X-ray calorimeter (McCammon et al. 2002) and *CHIPS* (Hurwitz et al. 2005). This is a muddy issue, because *CHIPS* observed a significantly lower intensity than did the calorimeter and because preferential depletion of gas phase iron relative to other elements can partly explain the dearth of photons observed by *CHIPS*. Nonetheless, if SWCX caused half of the observed local 1/4 keV emission but did not contribute Fe IX, Fe X, or Fe XI photons, then the LB's emission measure would be half that previously expected (thus half of 0.0038 pc^{-1} cm^{-6} expected for the *CHIPS* observations assuming a temperature of $10^{6.1}$ K; Hurwitz et al. (2005)) bringing it well within the 95% upper limit contour found from their observations (0.0036 pc^{-1} cm^{-6}).

A weaker Local Bubble requires $1/\sqrt{2}$ less initial input energy. Rather than a few supernova explosions, perhaps only one or two would be sufficient for evacuating and heating the bubble. Another consequence of considering a significant SWCX contribution is the possibility that it has skewed the LB temperature measurements. The present observational evidence tells us the shape of the combined LB+SWCX spectrum. The SWCX spectrum must be subtracted in order to find the true shape of the LB spectrum. Therefore, the Local Bubble could be hotter or cooler than previously believed. There may already be observational hints of this effect. For example, *XMM* and *Suzaku* observed the same cloud shadow 4 years apart. The *XMM* data was later found to be contaminated by a SWCX CME. Before that was known, the *XMM* data, in combination with *ROSAT* 1/4 keV data, was used to find the temperature of the local plasma (i.e. the temperature of the SWCX and LB spectra, combined; $\log(T) = 6.06^{+0.02}_{-0.04}$ K (Henley et al. 2007)). But, the *Suzaku* data, which is not known to be contaminated by a CME, in combination with the same *ROSAT* data, yielded $\log(T) = 5.96^{+0.05}_{-0.04}$ K (Henley and Shelton 2008).

Lastly, as a result of SWCX contamination, some LB models that had previously been found to be in conflict with the observations may need to be reevaluated. For example, the break-out model (in which the Local Bubble gas was heated and then expanded so fast that the ions adiabatically cooled faster than they recombined (Breitschwerdt and Schmutzler 1994)) had been eliminated on the grounds that its ratios of 1/4 keV to O VI emission, C III emission, and O VI absorption column density were higher than the observations indicated. But, if the LB's soft X-ray intensity were to be revised downwards significantly, then the break-out model would require reconsideration.

6 How We Can Make Progress

Continued work on SWCX estimates and the uncertainties on those estimates will enable continued progress in determining how best to apportion the observed local X-ray intensity between the SWCX and LB components. Continued work on estimating the SWCX 1/4 to 3/4 keV spectrum and its subtraction from the observationally derived local spectrum will enable revised estimates of the LB's spectrum and hence the plasma's temperature. But, such a statement must be qualified, because 1) some estimates already exist (Cravens 2000) although the spectral subtraction has yet to be done; and 2) if the SWXC contribution is large, then the uncertainty in the LB spectrum will be very large.

Clever strategies are needed in order to set observational constraints on the SWCX intensity. Perhaps these strategies will use the temporal and spatial variations in the model SWCX emission. Or perhaps, they will use a shadowing strategy to observe the SWCX spectrum without "contamination" by Galactic X-rays. Observations in the 1/4 keV band would be very useful, although they may not be attainable with current instruments. Concentrating on this band is necessary because the observed local X-ray spectrum peaks in the 1/4 keV band, and as a result, observations in higher bands alone are not sufficient for constraining fits to spectral models. In the future, very high spectral resolution observations will be useful. Few to several-eV resolution is necessary for unambiguous measurements of atomic transitions in the cascade from high n levels. Such measurements will include the determination of the ratio of intensities in the O VII "triplet" at 561, 569, and 574 eV and measurements of the weak O VIII Lyδ line at 836 eV and the weak C VI Lyγ line at 459 eV lines predicted by Wargelin et al. (2008).

Acknowledgements I would like to thank all of the members of my splinter group for their lively and informative contributions to the discussions of the Local Bubble and to thank the conference organizers for enabling these exchanges. I would also like to acknowledge helpful comments on the manuscript from Jeff Linsky, informative conversations with Dave Henley, an excellent explanation of solar wind charge exchange by Tom Cravens, and funding through the NASA LTSA grant NNG04GD78G.

References

B.-G. Andersson, S.B. Potter, Astrophys. J. **640**, L51–L54 (2006)
M.A. Barstow, D.D. Boyce, J.K. Barstow, A.E. Forbes, B.Y. Welsh, R. Lallement, Astrophys. Space Sci. (2008, in press)
D. Breitschwerdt, T. Schmutzler, Nature **371**, 774–777 (1994)
A.N. Bunner, P.L. Coleman, W.L. Kraushaar, D. McCammon, F.O. Williamson, Astrophys. J. **179**, 781–788 (1973)
D.N. Burrows, J.A. Mendenhall, Nature **351**, 629–631 (1991)
D.P. Cox, P.R. Anderson, Astrophys. J. **253**, 268–289 (1982)
T.E. Cravens, Astrophys. J. **532**, L153–L156 (2000)
M.A. de Avillez, D. Breitschwerdt, Rev. Mex. AA Ser. Conf. **15**, 299–299 (2003)
R.J. Edgar, Astrophys. J. **308**, 389–400 (1986)
D.B. Henley, R.L. Shelton, Astrophys. J. **676**, 335–350 (2008)
D.B. Henley, R.L. Shelton, K.D. Kuntz, Astrophys. J. **661**, 304–319 (2007)
M. Hurwitz, T.P. Sasseen, M.M. Sirk, Astrophys. J. **623**, 911–916 (2005)
E.B. Jenkins, Astrophys. J. **219**, 845–860 (1978a)
E.B. Jenkins, Astrophys. J. **220**, 107–123 (1978b)
D. Koutroumpa, R. Lallement, V. Kharchenko, A. Dalgarno, R. Pepino, V. Izmodenov, E. Quémerais, Astron. Astrophys. **460**, 289–300 (2006)
D. Koutroumpa, F. Acero, R. Lallement, J. Ballet, V. Kharchenko, Astron. Astrophys. **475**, 901–914 (2007)
K.D. Kuntz, S.L. Snowden, Astrophys. J. **543**, 195–215 (2000)
R. Lallement, B.Y. Welsh, J.L. Vergely, F. Crifo, D. Sfeir, Astron. Astrophys. **411**, 447–464 (2003)

C.M. Lisse, K. Dennerl, J. Englhauser, M. Harden, F.E. Marshall, M.J. Mumma, R. Petre, J.P. Pye, M.J. Ricketts, J. Schmitt, J. Trümper, R.G. West, Science **274**, 205–209 (1996)
D. McCammon, W.T. Sanders, Annu. Rev. Astron. Astrophys. **28**, 657–658 (1990)
D. McCammon et al., Astrophys. J. **576**, 188–203 (2002)
W.R. Oegerle, E.B. Jenkins, R.L. Shelton, D.V. Bowen, P. Chayer, Astrophys. J. **622**, 377–389 (2005)
I.P. Robertson, T.E. Cravens, J. Geophys. Res. **108** (A10) 8031, LIS6-1-6-10 (2003)
I.P. Robertson, T.E. Cravens, S.L. Snowden, American Geophysical Union, 2003, SH11C1126 (2003)
B.D. Savage, N. Lehner, Astrophys. J. Suppl. Ser. **162**, 134–160 (2006)
N.A. Schwadron, T.E. Cravens, Astrophys. J. **544**, 558 (2000)
R.L. Shelton, Astrophys. J. **589**, 261–269 (2003)
R.L. Shelton, Space Sci. Rev. (2008, this issue)
R.L. Shelton, D.P. Cox, Astrophys. J. **434**, 599–613 (1994)
S.L. Snowden, U. Mebold, W. Hirth, U. Herbstmeier, J.H.M. Schmitt, Science **252**, 1529–1532 (1991)
S.L. Snowden, D. McCammon, F. Verter, Astrophys. J. **409**, L21–L24 (1993)
S.L. Snowden, M.J. Freyberg, P.P. Plucinsky, J.H.M.M. Schmitt, J. Trümper, W. Voges, R.J. Edgar, D. McCammon, W.T. Sanders, Astrophys. J. **454**, 643–653 (1995)
S.L. Snowden, R. Egger, D.P. Finkbeiner, M.J. Freyberg, P.P. Plucinsky, Astrophys. J. **493**, 715–729 (1998)
S. Spangler, personal communication (2007)
B.J. Wargelin, P. Beiersdorfer, G.V. Brown, Can. J. Phys. **86**, 151–169 (2008)
B.Y. Welsh, S. Sallmen, D. Sfeir, R.L. Shelton, R. Lallement, Astron. Astrophys. **394**, 691–699 (2002)

Warm and Hot Gas in the Local ISM

Barry Y. Welsh

Originally published in the journal Space Science Reviews, Volume 143, Nos 1–4, 241–252.
DOI: 10.1007/s11214-008-9351-2

Abstract I present a short review of the physical properties and relationships between warm partially ionized and hot and highly ionized gas in the local interstellar-medium (ISM). The majority of the measurements reviewed have been obtained through ultraviolet absorption spectroscopy recorded towards stellar targets located within $\sim$ 150 pc of the Sun. Many questions still remain to be answered concerning the location, temperature, density and source of ionization of both of these phases of the diffuse interstellar gas. In particular, we focus on absorption observations of the ions of C IV λ 1550 Å and Si IV λ 1394 Å and discuss if highly ionized gas (as traced by the O VI λ 1032 Å ion) has in fact actually been detected within the Local Cavity.

Keywords Local ISM · Radiation environment

1 Introduction

Although it has long been established that the Local Cavity (LC), the volume of the interstellar medium that surrounds the Sun to a radius of $\sim$ 80 pc in most galactic directions, is a rarefied region of unusually low neutral gas density ($n_{\rm HI} < 0.1$ cm^{-3}), the physical characteristics of the plasma that fills this void are still highly debated. The existence of cold ($T < 1000$ K), warm (3000 K $< T <$ 20,000 K) and hot (50,000 K $< T <$ 300,000 K) diffuse gas located within the LC would apparently seem well-established (Meyer et al. 2006; Redfield and Linsky 2008; Savage and Lehner 2006). However, new results have raised important questions regarding the previous interpretation of some of these data with regard to the actual presence of hot ($T >$ 150,000 K) gas within 100 pc (Barstow et al. 2008; Koutroumpa et al. 2007). Additionally, it is still presently unclear if these three phases of the ISM are, or can be, physically co-located within the LC. An added point of controversy is their relationship to the 10^6 K soft X-ray emitting local gas, that some believe to be ubiquitously present within the LC (Snowden et al. 1998). In addition, there is still no consensus

B.Y. Welsh (✉)
Experimental Astrophysics Group, Space Sciences Laboratory, UC Berkeley, CA 94720, USA
e-mail: bwelsh@ssl.berkeley.edu

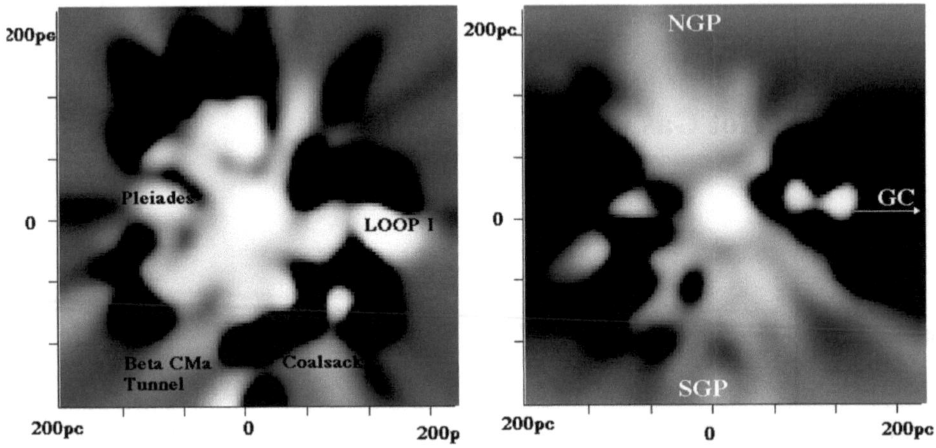

Fig. 1 Views of the distribution of cold and neutral gas within 250 pc of the Sun as traced by Na I absorption. The *Left-hand view* is a view along the galactic plane, the *right-hand view* is in the meridian plane (the Local Chimney). *Dark regions* represent regions of high volumic density of cold gas, derived from inversion of the measured column densities along each sight-line direction (Lallement et al. 2003)

on what are the mechanisms that heat and ionize the diffuse gas clouds that are known to be present within the LC, and are there density and ionization state variations present within this cavity? Finally, we may ask what is the origin of the warm and hot and highly ionized gas in the LC and is this linked to the origin of the LC itself?

In this short article, we shall mainly concentrate on reviewing observations of the warm and hot ionized gas components of the LC, performed using high resolution ultraviolet absorption measurements towards stellar targets located within $\sim$ 150 pc of the Sun. These, and Na I and Ca II absorption observations carried out at visible wavelengths, have revealed the presence of many diffuse cloudlets located within the rarefied cavity (Lallement et al. 2003; Vallerga et al. 1993). In some directions several warm and partially ionized gas clouds are detected over very short path lengths, while in other directions the sight-line structure is quite simple (Redfield and Linsky 2008). Thus, some 'clouds' may be organized structures and others could be more chaotic and filamentary in nature. These gas 'cloudlets' typically have hydrogen column densities, log N(H) $\sim 10^{17-18}$ cm^{-2}, and the most studied and nearest ($d < 5$ pc) assemblage of this gas is the Local Interstellar Cloud (Lallement and Bertin 1992), or LIC, within which the Sun is thought to reside. Since observations along more distant sight-lines mostly pass through this dynamical complex of clouds which possess bulk motions differing by only 5–10 km s^{-1}, then understanding the physical properties of this foreground warm and partially ionized gas is of paramount importance.

As a guide to the region of space currently under discussion, in Fig. 1 we show the spatial distribution of cold and neutral interstellar gas, as traced by high resolution spectroscopic observations of Na I absorption towards $\sim$ 1500 stars within 250 pc of the Sun (Lallement et al. 2003). Note that the central region (white) is essentially devoid of major condensations of cold and dense gas, and this absence of neutral gas essentially defines the extent of the rarefied LC. Other low-density interstellar regions lie adjacent to the LC (i.e. Loop I, the β CMa tunnel), which suggests that the structure of the surrounding ISM is typified by many cavities that may be linked to each other. Finally, we note that none of the local clouds are visible on either of these plots, since their volumic neutral density is at the lowest plotting level of the figures.

2 Local Warm Gas

2.1 Density and Temperature

The electron density (n_e) of the local warm gas clouds (T < 20,000 K) has generally been determined in two different ways: (i) use of the Mg II/Mg I or C II/C II* column density ratios (Lallement et al. 1994), or (ii) EUV measurement of the fractional ionizations of H and He in sight-lines towards nearby hot white dwarf stars (Wolff et al. 1999). A general consensus of all of these data suggests that $n_e \sim 0.1$ cm^{-3} for several local clouds within 15 pc of the Sun (Slavin and Frisch 2002).

The temperature of these local gas clouds is usually determined from the doppler widths of UV absorption lines recorded towards nearby stars. Using the HI Ly-α line, and taking into account turbulence, Linsky et al. (1995) found $T = 7000 \pm 900$ K for gas within 13 pc. This value agrees well with that of 6300 $\pm$340 K derived from in-situ solar system measurements of He ions by the Ulysses spacecraft (Witte 2004). In a survey of 26 sight-lines within 100 pc Redfield and Linsky (2004a, 2004b) found an average temperature of 6680 ± 1500 K and a turbulence value of 2.25 km s^{-1}. From all of these measurements, it seems that many of the local gas clouds have similar physical properties, and also that this gas is similar in nature to the Warm Ionized Medium (WIM) which represents $\sim$ 20% of the volume of the general ISM (Reynolds 2004). However, as we shall see in the next section, the level of ionization within the local clouds is considerably lower than that generally found for the WIM. In addition, if the majority of clouds within the LC possess temperatures < 10^4 K and they are also immersed within a million degree hot medium, as some authors infer from measurements of the soft X-ray background flux (Snowden et al. 1998), then a sharp temperature gradient should exist at the interface between the two media. This would cause heat to flow into the clouds and would also create an evaporative outflow that eventually would destroy the clouds themselves. We point out that the survival of the local cloud(s) in an environment of a putative 10^6 K gas is therefore problematical. If the clouds predate the occurrence of a nearby supernova event whose blast wave may have created the LC, they should not have survived the outburst. Even within their current interstellar environment, the local clouds must have a short lifetime because of evaporation by the surrounding hot LC gas (1). Thus it may be that these clouds are fairly young, in contrast to the age of the rarefied LC which is currently estimated to be $\sim$ 14 Myr (Fuchs et al. 2006).

2.2 Local Ionization

The local clouds contain neutral, partially ionized and fully ionized regions. Ionization fractions of $X(\mathrm{H}) = 0.45$ and $X(\mathrm{He}) = 0.57$ have been determined for the LIC (Wood and Linsky 1997) and for other local gas clouds (Gry and Jenkins 2001) using the previously derived values of electron density. The EUV spectra of local hot white dwarfs can also be used (via modeling) to determine the H I, He I and He II column densities from which ionization fractions can be inferred. All these data suggest that T, n_e and $X(\mathrm{H})$ all vary over distance scales of a few parsecs, as demonstrated for 17 local sight-lines within 60 pc by Wolff et al. (1999). These data also suggest that hydrogen is preferentially ionized in the galactic direction towards the stars ε and β CMa, which are known to dominate the local EUV radiation field (Vallerga 1998).

One unexpected result from EUV observations of the absorption edge at 228 Å seen in nearby hot white dwarf spectra is the resulting high ionization fraction of 0.3–0.4 for He II (Barstow et al. 1997). Such a value implies that He is slightly more ionized than H in the local ISM. The *FUSE* satellite has found a deficiency of Ar I absorption for nearby white dwarf stars, implying that the local clouds are strongly photo-ionized thus supporting the notion of an over-ionization of He (Jenkins et al. 2000). As mentioned previously, the local radiation field is thought to be dominated by contributions from two nearby B stars (ε and β CMa), local hot white dwarfs and the emission that originates in the hot million degree gas that is thought to fill the LC (Slavin and Frisch 2002). However, finding a potential source with sufficient ionization to explain the observed high level of He ionization found in the local clouds still remains a significant problem (Vallerga 1998)). One explanation is that the LC may not be in ionization equilibrium but rather a recombining plasma (Lyu and Bruhweiler 1996). Another possible explanation could be that the amount of He II absorption has been over-estimated, since recent high resolution EUV spectra of the white dwarf G191-B2B by Cruddace et al. (2002) have revealed the flux distribution in the wavelength region around the 228 Å edge to be far from simplistic. This is due to the presence of many other (white dwarf) stellar lines that would have appeared as a spectral blend with the interstellar HeII edge in the lower resolution spectra recorded with the *EUVE* satellite spectrometer.

2.3 Velocity Structure

High resolution ($R < 5$ km s^{-1}) Ca II observations towards numerous B and A-type stars within 30 pc of the Sun have revealed a very complex absorption velocity structure (Lallement et al. 1986; Crawford et al. 1997). Not only is the absorption structure complicated due to the number of absorbing gas clouds, but in some sight-line directions gas with very different temperatures is observed over quite short distance scales. Redfield and Linsky (2002) list 7 sight-lines within 100 pc that possess both high resolution ground-based Ca II-K and UV (Fe II and Mg II) absorption line data, from which an absorption comparison can be made. They find that most, but not all, sight-line components are seen in both Ca II and the UV lines to within a velocity error of ± 1.5 km^{-1}, supporting the notion that the Ca II-K line is a good indicator of the absorption structure that one might expect to detect from the warm and ionized gas sampled by the UV lines. The good agreement between these absorption velocities also suggests a physical co-location for the Ca II and UV absorbing ions within the same warm and diffuse gas clouds. However, it should be noted that the optical depths probed by the Ca II and Fe II lines differ by at least an order of magnitude due to abundance differences, such that for intrinsically weak column densities (especially over short sight-lines), Ca II provides little information.

In Fig. 2 we show a similar absorption velocity comparison between the Fe II and Ca II lines and the C IV line, which samples hotter and more highly ionized interstellar regions. These data represent local gas sampled along two sight-lines in the direction of the Loop I superbubble for which both C IV and Si IV (and N V) absorption has been unambiguously detected (Welsh and Lallement 2005). From these data it appears that neither the absorption profiles of Fe II nor Ca II *alone* can trace all of the observed C IV absorption features. However, when taken together the combined Fe II and Ca II profiles track the majority of the various absorption components revealed by the more highly ionized and/or hotter C IV absorbing regions. Although such velocity agreement could be a chance occurrence for these two sight-lines, these data do suggest a spatial relationship between the partially ionized warm and highly ionized (hot) local gas. Since the lower ions cannot simultaneously exist with C IV under the same ionization environment, it therefore seems likely that the highly ionized gas may be located in an evaporating shell surrounding (and shielding) a denser gas cloud (moving with the same interstellar velocity) that contains the lower ionization components. Unfortunately, we are unable to test such a hypothesis further for the LC, since

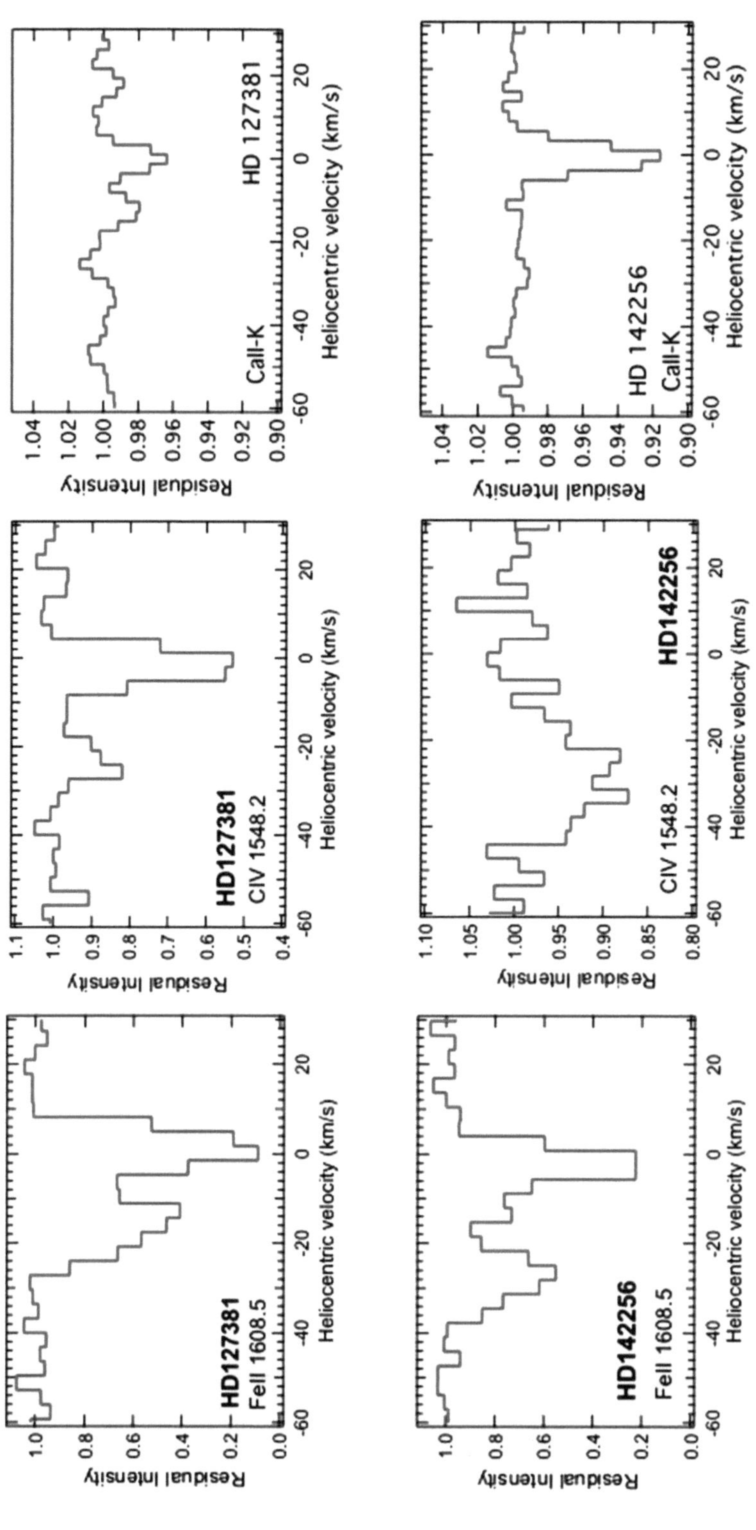

Fig. 2 Comparison of the extent of absorption velocities for the lines of Fe II, C IV and Ca II observed towards HD 127381 ($d = 176$ pc) and HD 142256 ($d = 186$ pc). Note that when combined, the Ca II + Fe II profiles track all of the C IV absorption velocities

no other detections of the three ion of C IV, Si IV and Ca II currently exist for any other local sight-line.

2.4 Element Abundances

Element abundances for the local gas have been determined for many sight-lines < 100 pc by Lehner et al. (2003). Most metals are depleted by at least a factor of 10 in the local ISM, while C, N and O are essentially near-solar. Such a depletion pattern shows no strong evidence for interstellar grain destruction by shocks, but instead favors an origin for the LIC as an expelled fragment of HI from the LC-Loop I interaction zone (Kimura et al. 2003). We note that very little is known about the element depletion patterns in other nearby clouds.

Much recent scientific attention has been focussed on measurement of the local D/H ratio, which can place important constraints on the chemical evolution of the Galaxy's ISM. For sight-lines within the LC a value of D/H $\sim$ 15 ppm is obtained, whereas for more distant sight-lines this ratio can reach values $\sim 60\%$ higher (Linsky et al. 2006). The most favored current explanation for the observed variation in the D/H ratio is the depletion of gaseous elements onto interstellar dust grains.

3 The Detectability of Hot Gas in the Local Cavity

3.1 Local C IV and Si IV Absorption

Although the high ions of Si IV λ 1394 Å, C IV λ 1550 Å, N V λ 1238 Å and O VI λ 1032 Å have been widely observed in absorption throughout most of the Galaxy, the halo and within other galaxies, their detection within 150 pc has proven extremely difficult. Although theoretical models predict column density ratios for Mg II/Mg I and C II/C II* that are close to those observed in the LIC (Slavin and Frisch 2002), these same models also predict that the LIC should be evaporating via thermal conduction. However, observational evidence is contrary to this assumption, since absorption line observations of high ions (C IV and N V)) towards nearby stars using *HST* (Bertin et al. 1995; Holberg et al. 1999) give smaller column densities than those generally predicted by cloud evaporation models (Bohringer and Hartquist 1987; Borkowski et al. 1990). A further complication arises in the choice of stellar background source for such observations (i.e. hot white dwarfs), since Bannister et al. (2003) argue that many of the C IV and Si IV absorption lines seen in such UV spectra are of a photospheric or circumstellar, and not interstellar, origin. This view is further supported by the detection of photospheric N V in the UV spectrum of a white dwarf with a temperature of only 38,500 K, thus indicating that one is unable to assume that high ionization species cannot be present in the photospheres of cool (i.e. $T < 40{,}000$ K) white dwarfs (Holberg et al. 1997).

An alternative choice of UV source continuum is that of nearby B-type stars. Although they are not numerous within 150 pc (and they are often too bright to observe due to their very high UV fluxes), they do have the distinct advantage of possessing well-known (Hipparcos) distances and their stellar atmospheres are well understood with regard to their ability to produce high ionization stellar lines that could be mistaken for those of an interstellar origin. Since the β CMa interstellar tunnel is thought to provide a direct conduit into the LC for EUV ionizing radiation of the local clouds, then it would seem an excellent place to search for absorption from the high ions of C IV and Si IV. Weak C IV absorption has been detected towards both of the B-type stars, β CMa ($d = 153$ pc) and ε CMa ($d = 132$ pc), by

Dupin and Gry (1998) and Gry and Jenkins (2001). Although the absorption velocity of both C IV detections is consistent with that of the LIC, no highly ionized gas has been detected towards the very nearby star α CMa ($d = 2.6$ pc), whose sight-line lies in the same galactic direction. This has prompted several authors to question whether this highly ionized gas is not of local origin, but arises in a far more distant interstellar cloud whose velocity is similar to that of the LIC.

Appreciable Si IV, C IV and N V absorption has been detected towards two B-type stars ($d < 150$ pc) along the sight-line towards the Loop I superbubble region by Welsh and Lallement (2005). It was found that the observed absorption profiles were all characterized by quite narrow doppler line-widths, indicative of gas temperatures < 20,000 K. The authors favor a strong local source of photo-ionization, or a highly non-equilibrium process, to produce both the observed strength and shape of these high ionization absorption lines. Collisional ionization processes are seen to be incompatible with these presently observed narrow line widths. Furthermore, in a *FUSE* absorption survey of 30 stars within 200 pc by Lehner et al. (2003), it was determined that photo-ionization was the major influence on the physical condition of the local gas and that photo-ionization conditions may not be constant throughout the LC region.

Finally, we note that several authors have reported detecting unusually high amounts of interstellar absorption from the Si III λ 1206 Å line along several local sight-lines (Holberg et al. 1999; Dupin and Gry 1998; Gry and Jenkins 2001). Its detection at a similar velocity to that of C IV has prompted Gry and Jenkins (2001) to suggest that the Si III may be located in an ionized interface that surrounds many of the local clouds. Unfortunately observations of this line can normally only be made towards very low column density sight-lines, since it lies very close to the Si III stellar line and saturation effects can easily contaminate the observed data.

3.2 Local O VI Absorption

The high ionization line of O VI λ 1032 Å is an important probe of any hot gas that exists within the LC and, as such, its observability was one of the prime scientific objectives of the *FUSE* mission. Previous observations of local O VI by the *Copernicus* satellite suggested a foreground (i.e. LC) component of $\log N(\text{O VI}) = 13.2$ cm^{-2} to all more distant sight-line measurements (Shelton and Cox 1994). However, many of these local detections have now been placed beyond the neutral boundary to the LC (due to improved sight-line distance determinations by Hipparcos), such that this high foreground column of O VI may now be reconciled with observations if it is mostly confined to a thin 'shell' conduction interface at the edge of the LC, as predicted by many theories.

Extensive observations of local hot white dwarfs by the *FUSE* satellite have resulted in only 7 detections of > 3-σ significance for targets with distances < 70 pc, all typically with low column densities, $\log N(\text{OVI}) \sim 12.9$ cm^{-2} (Savage and Lehner 2006; Oegerle et al. 2005). These new data have been interpreted as being consistent with thermally Doppler broadened gas absorption near 2.8×10^5 K originating in either interfaces or hot bubbles and/or the evaporative outflow of a conductive interface between warm and hot gas. The spatial distribution of local O VI absorption can best be described as being 'patchy'. There are some sight-lines < 70 pc that have no detectable O VI and others that have appreciable detections. Similarly, there are sight-lines beyond the neutral boundary of the LC that have appreciable O VI absorption and others that have null detections. In order to explain this apparent 'randomness' in O VI detectability, Oegerle et al. (2005) have suggested that O VI may be formed in evaporative interfaces between cool clouds and the hot, million degree gas within the LC. The presence of tangled magnetic fields within the gas clouds will tend to quench thermal conduction over much of the cloud's surface, such that O VI can only form in small regions where conduction is allowed to operate (Cox and Helenius 2003). We note however, that such a formation mechanism requires a LC filled with million degree gas and that recent observations of the soft X-ray background emission suggest a very local origin (i.e within the heliosphere) for a large percentage of this background signal (Koutroumpa et al. 2007). We also note that hot white dwarf distances are generally not known to within $\pm 25\%$ and thus their positions with respect to the neutral boundary of the LC is quite uncertain for distances in the 60–120 pc range.

The apparent randomness in local O VI detectability has led Barstow et al. (2008) to re-examine and re-assess the *FUSE* hot white dwarf data. Using the latest (and improved) version of the *FUSE* data reduction pipeline, many of the $< 3\sigma$ O VI detections claimed by Savage and Lehner (2006) are now revealed as instrumental background noise features that mimic weak absorption dips near the wavelength of the expected position of the interstellar O VI line. In addition, in Fig. 3 we show several of these re-processed white dwarf absorption spectra that clearly show that the positions of most of the ($> 3\sigma$) O VI absorption features are more consistent with that of the white dwarf stellar velocity than that of an interstellar origin. The authors argue that O VI can be formed in both the photospheres and in the circumstellar environment around white dwarfs, such that the *FUSE* data is equally consistent with a non-interstellar origin for these weak O VI absorption features. They claim that if any interstellar O VI exists within the LC, then it has a column density of $\log N(\text{O VI}) < 12.5\ \text{cm}^{-2}$, which is a level below the amounts generally predicted by ionization models of the local ISM (Slavin 1989; Bohringer and Hartquist 1987)

An alternative approach to measurement of local O VI absorption is to use nearby B-stars as probes of the highly ionized gas in the LC. Although such stars do not have the problems of stellar O VI contamination or large uncertainties in their distance estimates, due to their high FUV brightness (which can cause detector saturation effects) very few (< 25) have been observed by the *FUSE* satellite. However, in Fig. 4 we show firm detections of interstellar O VI absorption towards the 2 B-type stars, HD 149730 ($d = 80$ pc) and HD 149630 ($d = 93$ pc), with respective column densities of $\log N(\text{O VI}) = 13.16$ and $13.32\ \text{cm}^{-2}$ (Welsh et al. 2008). Both these stars lie just beyond (or even within) the neutral boundary of the LC and the doppler widths of their O VI profiles are equivalent to a temperature of $\sim 280{,}000$ K, suggesting that they are formed in a thermally conductive interface (presumably at the neutral boundary of the LC). Our *FUSE* observations of 14 other B-stars within 190 pc have resulted in placing an upper limit of $\log N(\text{O VI}) < 12.5\ \text{cm}^{-2}$ for sight-lines < 75 pc, which is consistent with the recent determination of Barstow et al. (2008). This number is rather 'inconvenient', since it represents the typical instrumental detection limit for most *FUSE* observations along any sight-line, and thus unfortunately does not confirm or deny the existence of highly ionized gas within the LC.

Finally, it should also be mentioned that the O VI ion can also be observed in *emission*. Using cloud-shadowing techniques, Shelton (2003) has determined surprisingly low upper limits to any local O VI emission. These observations have placed severe constraints on several models of the LC, in particular those that invoke a hot (10^6 K) plasma in collisional ionizational equilibrium. Additionally, the *SPEAR* satellite has mapped the all-sky O VI emission and determined that its spatial distribution is *not* correlated with that of the diffuse soft X-ray background (Kregenow et al. 2008). It should also be noted that both the NASA *EUVE* and *CHIPS* missions have failed to detect the EUV line emission which is to be expected from a diffuse hot LC plasma emitting with a temperature of $\sim 10^6$ K (Jelinsky et

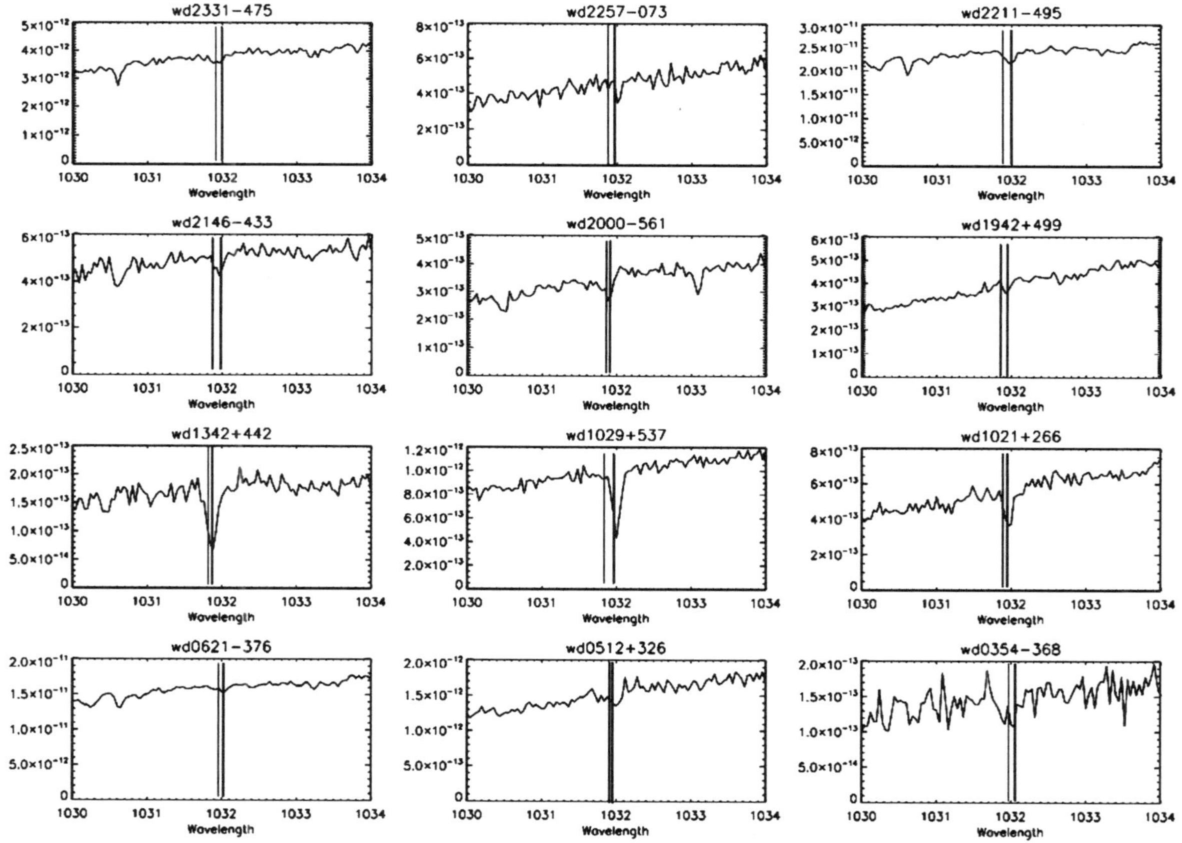

Fig. 3 Re-processed *FUSE* absorption spectra for 12 local hot white dwarfs (Barstow et al. 2008). The *thick line* represents the position of the radial velocity of the star, whereas the *thin line* represents the position of the interstellar velocity derived from nearby O I and C II lines. The weak 'absorption' features (near λ 1032 Å) are more consistent with that of the stellar radial velocity than that of the expected interstellar velocity, which in all cases appears blue-ward of the stellar velocity

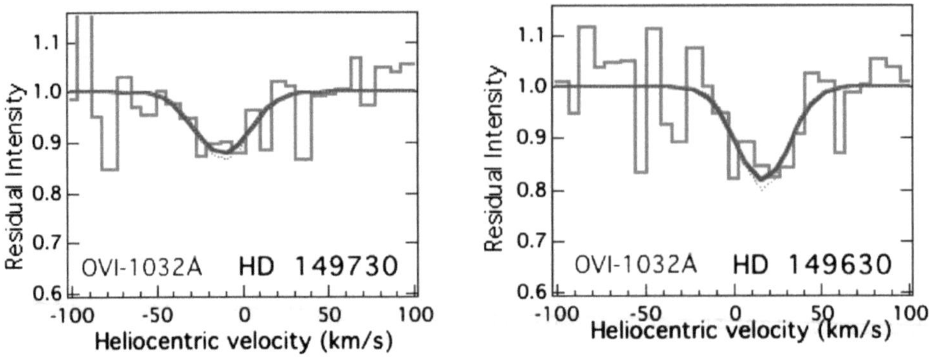

Fig. 4 *FUSE* OVI absorption spectra for the two nearby B-type stars, HD 149730 ($d = 80$ pc) and HD 149630 ($d = 93$ pc). Both stars lie just beyond the neutral boundary of the LC and these profiles may represent the formation of O VI in an evaporating conductive interface at that interstellar boundary

al. 1995; Hurwitz et al. 2005). The non-detection of this EUV background emission signal has been interpreted as (a) the local hot gas having a very low iron abundance and a far lower gas density than previously thought, or (b) being due to the absence of appreciable levels of million degree hot gas within the LC.

Thus, the present situation concerning the existence of highly ionized gas within the LC is still an open topic. The *FUSE* (O VI) absorption spectra recorded towards hot white dwarfs can be interpreted as being photospheric/circumstellar features, instrumental background noise or real (but weak) interstellar lines. If O VI does exist within the LC then, according to currently favored production models, there must also be million degree gas within it. If there is no (detectable) O VI within the LC then the O VI detected over more distant sight-lines most probably contains a contribution from the conduction interface between the neutral gas of the LC boundary and hot gas bubbles lying *beyond* that boundary. Our observations shown in Fig. 4 may be the best indicators of what that particular conductive interface is physically like.

4 The Local Chimney Connection

The LC has no dense and neutral boundary at high galactic latitudes (see Fig. 1) and this 'Local Chimney' 'opens' directly into the upper and lower galactic halo. Na I and Ca II interstellar absorption measurements towards halo stars with sight-lines < 1 kpc along the two axes of the Local Chimney reveal significant numbers of negative velocity components, indicative of gas that is falling towards the LC at velocities in the -20 to -60 km s^{-1} range (Crawford et al. 2002). Gas with similar velocities (in the northern galactic hemisphere) has been associated with the Intermediate Velocity Arch, which is an extensive intermediate velocity cloud (IVC) complex that may have been ejected into the lower halo by the break-out of a stellar wind-driven bubble through a galactic fountain. It may be possible that this gaseous infall from the inner halo could appear as the diffuse gas clouds located within the LC, in addition to the outflowing gas from the interface region between the LC and the adjacent Loop I superbubble. It is intriguing that the elemental abundances of the infalling (IVC) gas clouds and the diffuse clouds within the LC are very similar (Welsh et al. 2004), and that they could account for the seeming mis-match between the age of the LC and the presence (and survival) of warm diffuse gas clouds within such a supposed hot and highly ionized region.

5 Future Observations

It is clear that much still needs to be explained concerning the presence (or even existence) of warm and hot and highly ionized gas within 100 pc of the Sun. Unfortunately with the recent demise of the *FUSE* satellite and the on-orbit failure of the HST-*STIS* instrument, future access to high resolution FUV and UV absorption spectra of the ions that probe the warm and hot gas will be severely compromised. NASA intends to place the Cosmic Origins Spectrograph on HST in mid-2008, and although this instrument has a high UV sensitivity its spectral resolution ($R \sim 20{,}000$) is not ideal for local interstellar medium work since even over short sight-lines in the LC the absorption structure often contains many closely spaced velocity components. It is hoped that with the possible future repair of the HST-*STIS* instrument, we shall at least be able to probe the local warm gas through many more sight-lines within 100 pc using high resolution spectroscopy of the Si IV, C IV and N V lines. From such future observations we may well be able to determine the still controversial physical relationships between the local hot and local warm phases of the interstellar gas.

Acknowledgements BYW acknowledges funding from NASA grant NNX07AE33G and also many fruitful discussions with John Vallerga, Rosine Lallement, Steve Snowden and Martin Barstow.

References

N. Bannister, M. Barstow, J. Holberg et al., Mon. Not. R. Astron. Soc. **341**, 477 (2003)
M. Barstow, P. Dobbie, J. Holberg et al., Mon. Not. R. Astron. Soc. **286**, 58 (1997)
M. Barstow, D. Boyce, J. Barstow et al., Astrophys. Space Sci. (2008, in press)
P. Bertin, A. Vidal-Madjar, R. Lallement et al., Astron. Astrophys. **302**, 889 (1995)
H. Bohringer, T. Hartquist, Mon. Not. R. Astron. Soc. **228**, 915 (1987)
K. Borkowski, S. Balbus, C. Fristrom, Astrophys. J. **355**, 501 (1990)
D. Cox, L. Helenius, Astrophys. J. **583**, 205 (2003)
I. Crawford, N. Craig, B.Y. Welsh, Astron. Astrophys. **317**, 889 (1997)
I. Crawford, R. Lallement, R. Price et al., Mon. Not. R. Astron. Soc. **337**, 720 (2002)
R. Cruddace, M. Kowalski, D. Yentis et al., Astrophys. J. **565**, 47 (2002)
O. Dupin, C. Gry, Astron. Astrophys. **335**, 661 (1998)
B. Fuchs, D. Breitschwerdt, M. de Avillez et al., Mon. Not. R. Astron. Soc. **373**, 993 (2006)
C. Gry, E. Jenkins, Astron. Astrophys. **367**, 617 (2001)
J. Holberg, M. Barstow, T. Lanz, I. Hubeny, Astrophys. J. **484**, 871 (1997)
J. Holberg, F. Bruhweiler, M. Barstow et al., Astrophys. J. **517**, 841 (1999)
M. Hurwitz, T. Sasseen, M. Sirk, Astrophys. J. **623**, 911 (2005)
P. Jelinsky, J. Vallerga, J. Edelstein, Astrophys. J. **442**, 653 (1995)
Jenkins, Space Sci. Rev. (2008, this issue)
E. Jenkins, W. Oegerle, C. Gry et al., Astrophys. J. **538**, L81 (2000)
H. Kimura, I. Mann, E. Jesseberger, Astrophys. J. **582**, 846 (2003)
D. Koutroumpa, F. Acero, R. Lallement et al., Astron. Astrophys. **475**, 901 (2007)
J. Kregenow et al., Astrophys. J. (2008, in preparation)
R. Lallement, P. Bertin, Astron. Astrophys. **266**, 479 (1992)
R. Lallement, A. Vidal-Madjar, R. Ferlet, Astron. Astrophys. **168**, 225 (1986)
R. Lallement, P. Bertin, R. Ferlet et al., Astron. Astrophys. **286**, 898 (1994)
R. Lallement, B.Y. Welsh, J.L. Vergely et al., Astron. Astrophys. **411**, 447 (2003)
N. Lehner, E. Jenkins, C. Gry et al., Astrophys. J. **595**, 858 (2003)
J. Linsky, A. Diplas, B. Wood et al., Astrophys. J. **451**, 335 (1995)
J. Linsky, B. Draine, H. Moos et al., Astrophys. J. **647**, 1106 (2006)
C.-H. Lyu, F. Bruhweiler, Astrophys. J. **459**, 216 (1996)
D. Meyer, J. Lauroesch, C. Heiles et al., Astrophys. J. **650**, L67 (2006)
W. Oegerle, E. Jenkins, R. Shelton et al., Astrophys. J. **622**, 377 (2005)
S. Redfield, J. Linsky, Astrophys. J. Suppl. Ser. **139**, 439 (2002)
S. Redfield, J. Linsky, Astrophys. J. **602**, 776 (2004a)

S. Redfield, J. Linsky, Astrophys. J. **613**, 1004 (2004b)
S. Redfield, J. Linsky, Astrophys. J. **673**, 283 (2008)
R. Reynolds, Adv. Space Res. **34**, 27 (2004)
B. Savage, N. Lehner, Astrophys. J. Suppl. Ser. **162**, 134 (2006)
R. Shelton, Astrophys. J. **589**, 261 (2003)
R. Shelton, D. Cox, Astrophys. J. **638**, 206 (1994)
J. Slavin, Astrophys. J. **346**, 718 (1989)
J. Slavin, P. Frisch, Astrophys. J. **565**, 364 (2002)
S. Snowden, R. Egger, D. Finkbeiner et al., Astrophys. J. **493**, 715 (1998)
J. Vallerga, Astrophys. J. **497**, 921 (1998)
J. Vallerga, P. Vedder, N. Craig, B.Y. Welsh, Astrophys. J. **411**, 729 (1993)
B.Y. Welsh, R. Lallement, Astron. Astrophys. **436**, 615 (2005)
B.Y. Welsh, S. Sallmen, R. Lallement, Astron. Astrophys. **414**, 261 (2004)
B.Y. Welsh, R. Lallement, Astron. Astrophys. (2008, in preparation)
M. Witte, Astron. Astrophys. **426**, 835 (2004)
B. Wolff, D. Koester, R. Lallement, Astron. Astrophys. **346**, 969 (1999)
B. Wood, J. Linsky, Astrophys. J. **474**, 304 (1997)

What Can Be Learned from X-ray Spectroscopy Concerning Hot Gas in the Local Bubble and Charge Exchange Processes?

S.L. Snowden

Originally published in the journal Space Science Reviews, Volume 143, Nos 1–4, 253–262.
DOI: 10.1007/s11214-008-9343-2

Abstract Both solar wind charge exchange emission and diffuse thermal emission from the Local Bubble are strongly dominated in the soft X-ray band by lines from highly ionized elements. While both processes share many of the same lines, the spectra should differ significantly due to the different production mechanisms, abundances, and ionization states. Despite their distinct spectral signatures, current and past observatories have lacked the spectral resolution to adequately distinguish between the two sources. High-resolution X-ray spectroscopy instrumentation proposed for future missions has the potential to answer fundamental questions such as whether there is any hot plasma in the Local Hot Bubble, and if so what are the abundances of the emitting plasma and whether the plasma is in equilibrium. Such instrumentation will provide dynamic information about the solar wind including data on ion species which are currently difficult to track. It will also make possible remote sensing of the solar wind.

Keywords X-rays · Solar system · Interstellar medium

1 Introduction

At lower Galactic latitudes the emission regions capable of producing the observable diffuse X-ray background are strongly limited by interstellar absorption to be within the nearest couple hundred parsecs. Even at higher Galactic latitudes the minimum optical depth for $\frac{1}{4}$ keV X-rays originating beyond the Galactic disk is $\sim$0.5. In the late 1980s Cox and Snowden (1986) (see also Snowden et al. 1990), building on work from the late 1970s (Sanders et al. 1977; Tanaka and Bleeker 1977), used the existence of the finite flux in the Galactic plane and the angular structure of the $\frac{1}{4}$ keV background to attribute the observed emission to a $\sim 10^6$ K plasma filling a cavity (Knapp 1975) in the local interstellar medium in which

This work was supported by a NASA ADP grant.

S.L. Snowden (✉)
Code 662, NASA/GSFC, Greenbelt, MD 20771, USA
e-mail: steven.l.snowden@nasa.gov

the Sun is located, the Local Hot Bubble (LHB). This cavity is well mapped by interstellar absorption line measurements (e.g., Frisch and York 1983; Paresce 1984; Sfeir et al. 1999; Lallement et al. 2003) and extends from a few tens of parsecs in some directions in the Galactic plane to a couple hundred parsecs at higher Galactic latitudes. For general historical reviews of our understanding of the Galactic diffuse X-ray background see Snowden (2001) and McCammon and Sanders (1990).

In the late 1990s it was found that charge exchange emission from the solar wind interacting with cometary neutrals could produce an easily observable flux of X-rays for these objects (Lisse et al. 1996; Cravens 1997). Subsequent to that discovery it was realized that the solar wind could also charge exchange with interstellar neutrals passing through the solar system along with exospheric material in Earth's magnetosheath. These interactions are responsible for a variable and sometimes intense (relative to the typical cosmic background at $\frac{1}{4}$ keV) contamination component of X-ray observations (Cox 1998).

In the early 2000s it was suggested that the quiescent SWCX flux could also be responsible for some if not all of the emission which had previously been attributed to the LHB (Lallement 2004). It has also been suggested that SWCX may be responsible for all of the observed general diffuse X-ray background at $\frac{3}{4}$ keV (primarily emission from O VII and O VIII) not associated with the extragalactic background of unresolved objects at cosmological distances (Koutroumpa et al. 2007; Wargelin et al. 2004). Whether the emission comes from the nearest 100 AU or from the Galactic halo/intergalactic medium has obvious and considerable impacts on our understanding of the energy balance of the Milky Way, its halo, and the more global environment.

Currently it is not possible to distinguish observationally between the two emission mechanisms except for times of significantly enhanced SWCX production where both the light curve of the diffuse emission varies and certain emission lines (e.g., O VIII and Mg XI) can be strongly enhanced. The $\frac{1}{4}$ keV band, where most of the LHB emission occurs and the SWCX emission can be the most intense, is for the most part out of the useful instrumental band passes. Current missions typically have CCDs as their imaging detectors, and CCDs have energy resolutions of ~50 eV where resolutions of order a few eV are necessary to resolve individual emission lines. Further, when the CCDs do have response at $\frac{1}{4}$ keV the energy resolution is degraded significantly.

2 A Little History—Instrumentation

The first measurements of the soft X-ray diffuse background and the following all-sky surveys were done with proportional counters which have spectral resolutions ($E/\Delta E$) of about 1 at $\frac{1}{4}$ keV. The data were separated into broad bands by the use of filters, with the $\frac{1}{4}$ keV band isolated by the carbon in the plastic entrance windows and typically a carbon coating on the windows which is necessary for electrical conduction. Proportional counters were the norm from the late 1960s to the *ROSAT* (Trümper 1992) era. Figure 1a shows *ROSAT* proportional counter spectra of the diffuse background during a low-background time period and a period of intense SWCX emission. Both spectra are from the same direction on the sky and were taken during the same 24-hour period. Note that because of the low spectral resolution the two spectra have no significantly different characteristics besides their different intensities at lower energies. Above ~1 keV they are quite similar.

As noted above, current observatories (e.g., *XMM-Newton*, *Suzaku*, and *Chandra*) use CCDs as their X-ray imagers. While the energy resolution is considerably improved over proportional counters, $E/\Delta E \sim 10$ in the $\frac{3}{4}$ keV energy range, spectral emission lines still

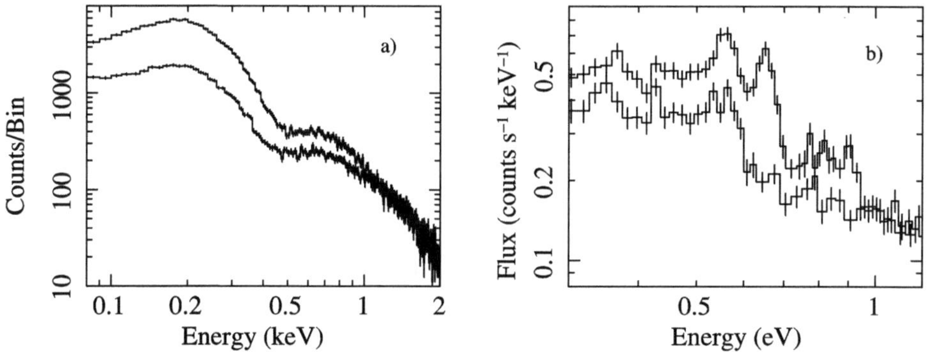

Fig. 1 (**a**) *ROSAT* spectra of the diffuse X-ray background. The two spectra are from the same observation with the upper spectrum from times of strongly enhanced SWCX emission. (**b**) *XMM-Newton* spectra of the diffuse X-ray background. The two spectra are from different observations of the same direction on the sky. The upper spectrum has a strong contribution from SWCX while the lower spectrum does not

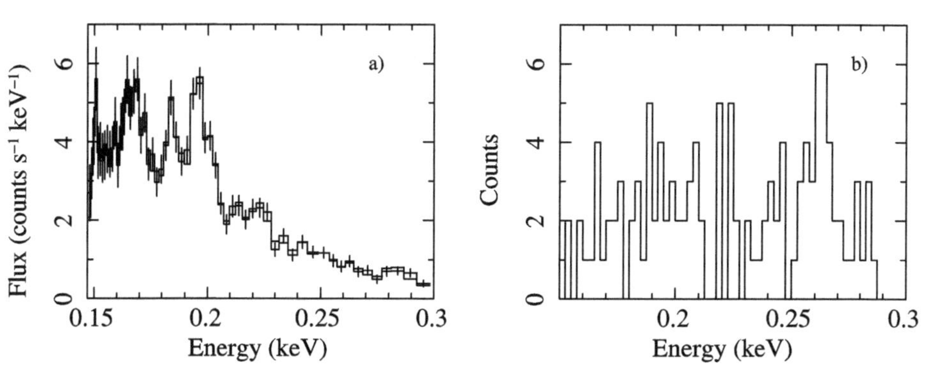

Fig. 2 (**a**) *DXS* spectrum (Sanders et al. 2001) of the diffuse X-ray background showing the data (*points with error bars*) and best-fit model spectrum of 23 Gaussians (*stepped curve*). (**b**) Calorimeter spectrum (McCammon et al. 2002) of the diffuse X-ray background

appear as relatively broad features, plus both their resolution and sensitivity are significantly reduced at lower energies. Figure 1b shows *XMM-Newton* spectra of the diffuse X-ray background from same direction on the sky, the Hubble Deep Field North, during periods of intense SWCX emission and negligible SWCX emission (Snowden et al. 2004). The SWCX emission contributes strongly below ~1 keV with a particularly strong enhancement in the O VIII line at 0.65 keV. The O VII triplet of lines at 0.56 keV had a strong contribution and lines of C VI, Ne IX, and Mg XI were also detected. While O VII is often seen in observations of the diffuse X-ray background, the other lines, even the O VIII line, often are only marginally observed if at all with current instrumentation.

Two additional detector types have been flown for limited durations. The Diffuse X-ray Spectrometer (DXS, Sanders et al. 2001) was a space shuttle mission which used Bragg crystals and proportional counters to achieve a resolution of $E/\Delta E \sim 30$ to observe the diffuse background at $\frac{1}{4}$ keV. Figure 2a shows the DXS $\frac{1}{4}$ keV spectrum of a ~0.25 sr region of the Galactic plane away from bright objects and extended structure. Because of the short mean free path of $\frac{1}{4}$ keV X-rays in the interstellar medium, this is a reliable spectrum of nearby emission (Local Hot Bubble and/or SWCX).

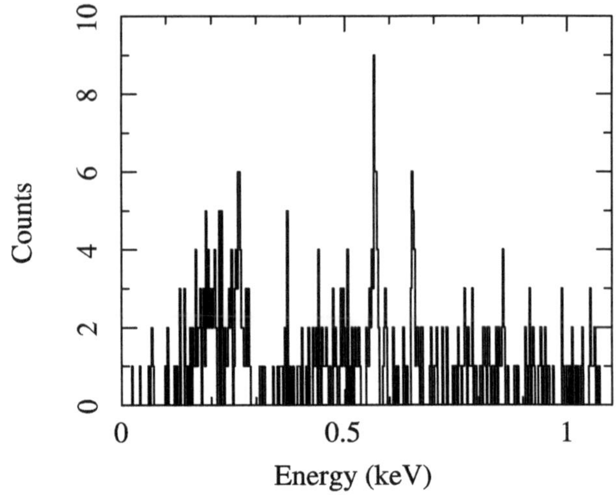

Fig. 3 Calorimeter spectrum of the diffuse X-ray background in the 0.05–1.1 keV band. This plot clearly shows the O VII line at 0.56 keV and the O VIII line at 0.65 keV. The greater width of the O VII line when compared to that of the O VIII line is due to the triplet nature of the O VII line

The other innovative detector is the X-ray Quantum Calorimeter (XQC, McCammon et al. 2002). The XQC was flown on a sounding rocket (with more flights to follow) and achieved an energy resolution of ~6 eV, for $E/\Delta E \sim 40$ at $\frac{1}{4}$ keV. Unfortunately, sounding rocket flights provide only ~6 minutes of data so the statistics are rather limited. Figure 2b shows the XQC spectrum of the diffuse background from a ~1 sr field of view in a high Galactic latitude direction. Because of the different directions on the sky (high and low Galactic latitudes), the DXS and XQC spectra are not directly comparable as the XQC spectrum has a significant contribution from emission originating in the Galactic halo. Part of the difference between the two spectra also lies in the different functional forms of the effective areas of the two detectors with the DXS response being relatively flat while the XQC response sharply increases from lower to higher energy. However, both spectra were very reassuring in that they indicate the presence of strong emission lines in the $\frac{1}{4}$ keV band. The dominance by line emission has long been a staple in models for the $\frac{1}{4}$ keV background, but they had not been observed previously because of instrumental limitations. Figure 3 shows the XQC spectrum now including the $\frac{3}{4}$ keV band which clearly shows the O VII triplet at 0.56 keV and the O VIII line at 0.65 keV.

3 Spectrum Röntgen Gamma

With calorimeter detectors, like those used on XQC, the non-dispersive spectrometer technology now exists to provide moderate-angular-resolution, high-spectral-resolution, high-precision measurements of the diffuse X-ray background. What is necessary is an orbital platform which will provide long exposures. Fortunately, a calorimeter is on track for inclusion in the Spectrum Röntgen Gamma mission to be launched in 2011.

The projected spectral resolution for the calorimeter is $E/\Delta E \sim 100$ (a few eV at $\frac{1}{4}$ keV). It will be matched to a mirror with a ~10′ field of view so the 6 × 6 pixel array will have ~1.6′ angular resolution, a reasonably good match to the resolution of the mirrors. It will also have a reasonable response at $\frac{1}{4}$ keV which is critical for studying the diffuse X-ray emission originating either as thermal emission in the LHB or SWCX emission in the solar system, and therefore distinguishing their relative contributions to the observed flux.

The Spectrum Röntgen Gamma mission will combine a limited period of pointed observations followed by an extended survey phase lasting several years. The survey phase will consist of scans covering approximate great circles on the sky which precess with Earth's orbit around the Sun. Over the three-year survey period each point on the sky will be covered at half-year intervals for two or three orbits at the ecliptic plane, to many times near the ecliptic poles. Each sample will last only a few seconds but the data can be binned into larger solid-angle bins to achieve the desired statistics. This is the same strategy used in the analysis of the *ROSAT* All-Sky Survey, which also had limited statistics relative to the angular resolution of the instrument.

4 What Can Be Learned?

With the good low-energy response and high spectral resolution of the calorimeter, and the long effective exposures provided by binning the survey data, it will be possible to decipher the physical conditions, temperatures, abundances and abundance variations, and origin (solar system or LHB) of the local diffuse X-ray background. The two primary emission mechanisms involved, SWCX for the solar system and thermal emission from the LHB, have radically different spectral signatures. SWCX emission starts with an electron capture into an excited state in a highly ionized atom. After capture the electron will cascade to lower energy states producing various emission lines following the branching probabilities. Advances in the study of the expected emission are being made both experimentally (e.g., measuring interaction cross sections with EBIT,[1] Otranto et al. 2006) and theoretically (e.g., Kharchenko 2004). The SWCX X-ray emission comes from the recipient ion in the solar wind, not the donor atom so the observed flux is a direct measure of the composition of the wind.

SWCX emission should likely be considered as two separate components, one which is varying slowly if at all and one which is episodic with strong enhancements lasting on the order of a day. The latter can often be detected in an observation light curve and are responsible for the excess SWCX emission seen in Fig. 1. These enhancements are the ones referred to as "long-term enhancements" (LTEs) in Snowden et al. (1997) and likely originate in Earth's magnetosheath or at least within a couple AU of the Sun. The less variable, or quiescent SWCX component originates from the solar wind interacting with interstellar neutrals throughout the entire heliosphere and perhaps in the heliopause. Because the solar wind travels roughly a quarter of an AU per day the quiescent heliospheric emission is an integral of the solar wind conditions over the previous year (or longer if the heliopause contributes significant emission). Since the variable part of the SWCX emission was fairly effectively removed from the *ROSAT* All-Sky Survey data, it is the quiescent component which is in competition with the Local Hot Bubble for producing the observed $\frac{1}{4}$ keV background which is known to be local (originating within the nearest few hundred parsecs).

Thermal emission in the soft X-ray band is produced when bound electrons in the plasma are collisionally excited to higher energy levels and then radiatively decay, or when the ions capture an electron in an excited state and again radiatively decay. While the abundance and ionization ratios in the solar wind are determined by processes on the Sun and are essentially frozen in at that point, in a thin plasma (such as postulated for the Local Hot Bubble)

[1]The Electron Beam Ion Trap (EBIT) Facility of the National Institute of Standards and Technology makes possible the measurement of the interaction cross sections of highly charged ions, http://physics.nist.gov/MajResFac/EBIT/ebit.html.

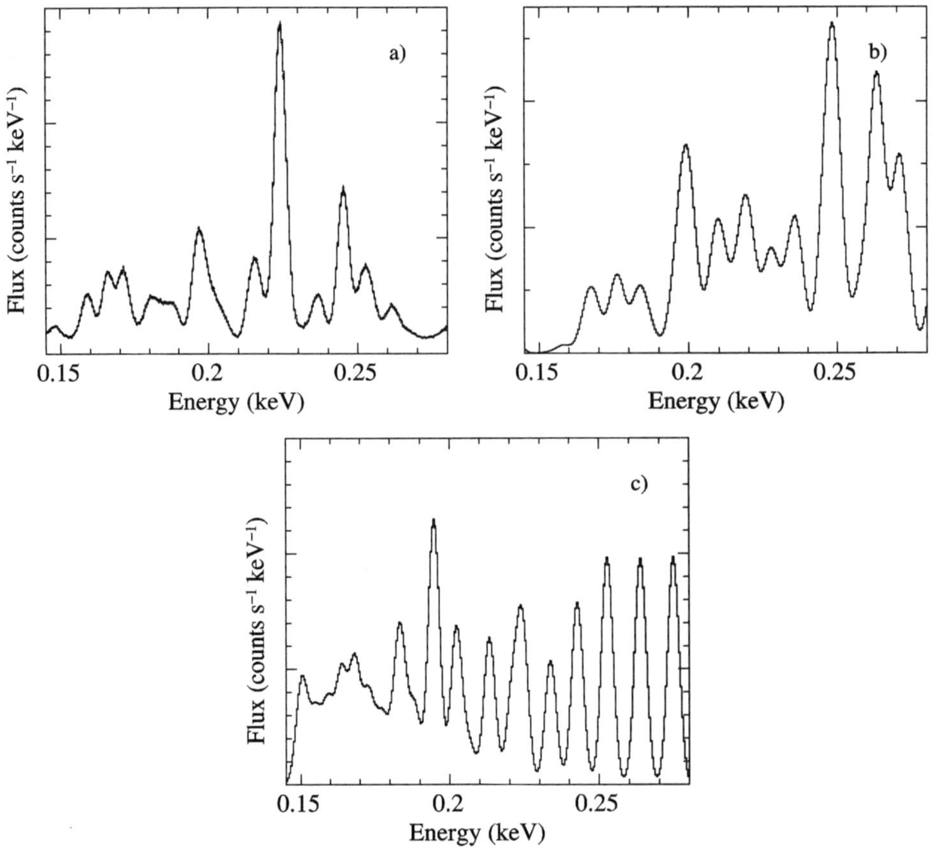

Fig. 4 Model spectra folded through the Spectrum Röntgen Gamma calorimeter response. (**a**) $\sim 10^{6}$ K thermal bremsstrahlung spectrum typically used to model emission from the Local Hot Bubble. (**b**) Model solar wind charge exchange spectrum of Koutroumpa and Karchenko (private communication). (**c**) The 23 Gaussians used to fit the *DXS* spectrum of the diffuse X-ray background in the $\frac{1}{4}$ keV band

the ratios are determined by the properties of the plasma (the abundances, temperature, density, and equilibrium status). In addition, the spectra from the temporally variable part of the SWCX emission should vary significantly as the composition of the solar wind varies. The quiescent SWCX component, because of the long integration interval for solar wind conditions, should be relatively constant while LHB spectra (for a given direction) should be entirely constant with time. Thermal spectra are relatively well understood with standard models going back to the late 1970s (e.g., Raymond and Smith 1977). Considerable effort has been expended over the intervening years (e.g., Smith et al. 2001) and the codes are being improved, particularly at lower energies ($\sim\frac{1}{4}$ keV) where emission from L- and M-shells of ions is important. In addition, models for both equilibrium and non-equilibrium conditions have been developed.

Figure 4 shows three different model $\frac{1}{4}$ keV spectra folded through the preliminary Spectrum Röntgen Gamma calorimeter response. Panel (a) shows an APEC 0.1 keV thermal emission model with solar abundances, panel (b) shows a model SWCX spectrum (Koutroumpa and Kharchenko, private communication), and panel (c) shows the best-fit

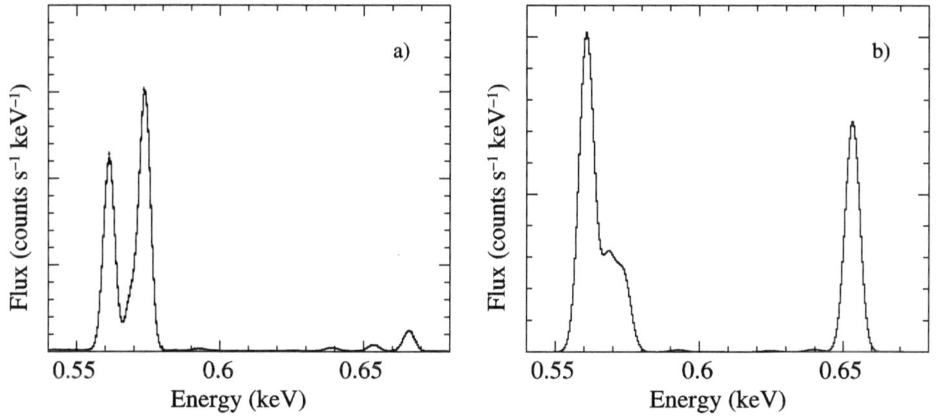

Fig. 5 (**a**) Model thermal bremsstrahlung and (**b**) model solar wind charge exchange spectra of the diffuse X-ray background O VII and O VIII lines folded through the Spectrum Röntgen Gamma calorimeter response

DXS spectrum (23 Gaussians, Sanders et al. 2001). What is overwhelmingly clear from these plots is that nobody has it right yet. The two model spectra are very distinctly different, and yet neither comes close to matching the observed DXS spectrum. This is also the conclusion which Sanders et al. (2001) came to, although without the SWCX spectrum to judge. No current thermal equilibrium emission model with standard abundances can fit the data. Extreme abundance depletions and non-equilibrium states may come closer to a reasonable fit, but the results are unsatisfying. Refined SWCX models fit to high resolution data may better fit the observed quiescent diffuse background, but they certainly will be very productive in probing the solar wind. During time periods of strong SWCX emission the spectra will offer a wealth of information on the abundances and ionization states of solar-wind material.

Figure 5 shows the model O VII and O VIII lines from thermal and SWCX emission folded through the Spectrum Röntgen Gamma calorimeter response. Again, the spectra are radically different providing a simple diagnostic to determine whether SWCX or thermal emission is the dominant source of the local (SWCX or LHB) soft X-ray background. Note that the strong contribution of the SWCX model in the O VIII line is qualitatively consistent with the SWCX enhancement in the *XMM-Newton* spectrum of the Hubble Deep Field North (Fig. 1b).

5 Random Thoughts

A critical issue that must be considered when studying the contribution of solar wind charge exchange emission to the diffuse X-ray background is that the spectrum is extremely variable, at least for those components most strongly associated with emission from the magnetosheath where the charge exchange is with exospheric material. Figure 6 presents a scatter plot of the measured solar wind charge exchange emission at $\frac{3}{4}$ keV versus that at $\frac{1}{4}$ keV. (Data are from the analysis of Snowden et al. 1997, where the SWCX emission was called, for the lack of a better term, long-term enhancements, and was treated empirically as a contamination component and subtracted.) There is very little in the way of detailed correlation between the two bands indicating the strong spectral variation of the SWCX emission.

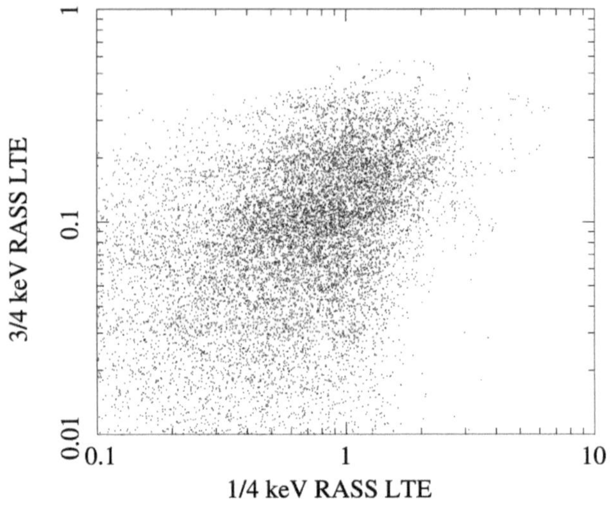

Fig. 6 Scatter plot of the $\frac{3}{4}$ keV band long-term enhancement value (SWCX emission) versus that for the $\frac{1}{4}$ keV band from the processing of the *ROSAT* All-Sky Survey (RASS). The units are counts s^{-1} over the *ROSAT* field of view. These excesses attributed to SWCX were determined by analyzing the scan-to-scan count-rate variations of the nearly same regions on the sky provided by the multiple coverage of the survey. The *ROSAT* field of view was nearly 2° and the maximum displacement of consecutive scans was 4′

This also suggests that the SWCX emission at $\frac{3}{4}$ keV measured by *Chandra*, *XMM-Newton*, and *Suzaku* is, empirically, a different quantity from the SWCX emission at $\frac{1}{4}$ keV contributing to the discussion as to whether there actually is hot plasma in the LHB. The SWCX emission spectrum associated with a strong enhancement in the solar wind in Earth's neighborhood, which is likely the source of the SWCX emission reported in recent observations (e.g., Wargelin et al. 2004; Snowden et al. 2004; Fujimoto et al. 2007), is unlikely to have the same spectrum as the quiescent SWCX emission integrated through the heliosphere. The variability of the SWCX spectrum is not surprising as the contributions arise from different elements and ionization states of those elements, which all vary significantly in the solar wind. The SWCX $\frac{3}{4}$ keV emission is dominated by O VII and O VIII with some help from C VI, depending on the softness of the instrumental response, as well as some emission from Ne IX. SWCX emission at $\frac{1}{4}$ keV is not produced by those elements and ions but by M- and L-shell emission from higher atomic-number metals. There is no monolithic or standard SWCX spectrum, at least for the variable part, and so one must be careful about what energy band and what solar wind conditions are being considered for the conclusions being drawn. One must also not confuse the SWCX emission spectrum from solar wind enhancements which likely produce the reported spectra with the quiescent SWCX emission spectrum which represents the solar wind averaged over the previous year.

To illustrate the differences in the relative intensities of the $\frac{3}{4}$ keV and $\frac{1}{4}$ keV band emission in the temporally variable SWCX component, Fig. 7 displays $\frac{1}{4}$ keV, $\frac{3}{4}$ keV, and 1.5 keV maps from the *ROSAT* All-Sky Survey (Snowden et al. 1997) before cleaning of the long-term enhancements (SWCX). The variable SWCX emission clearly has a much greater significance in the $\frac{1}{4}$ keV band. That need not necessarily translate to a greater quiescent SWCX emission contribution to the all-sky map.

6 Summary

High resolution soft X-ray spectroscopy of diffuse soft X-ray emission is the Rosetta stone for understanding the relative contributions of thermal emission from the Local Hot Bubble and solar wind charge exchange emission from the heliosphere. If the local cavity in the H I

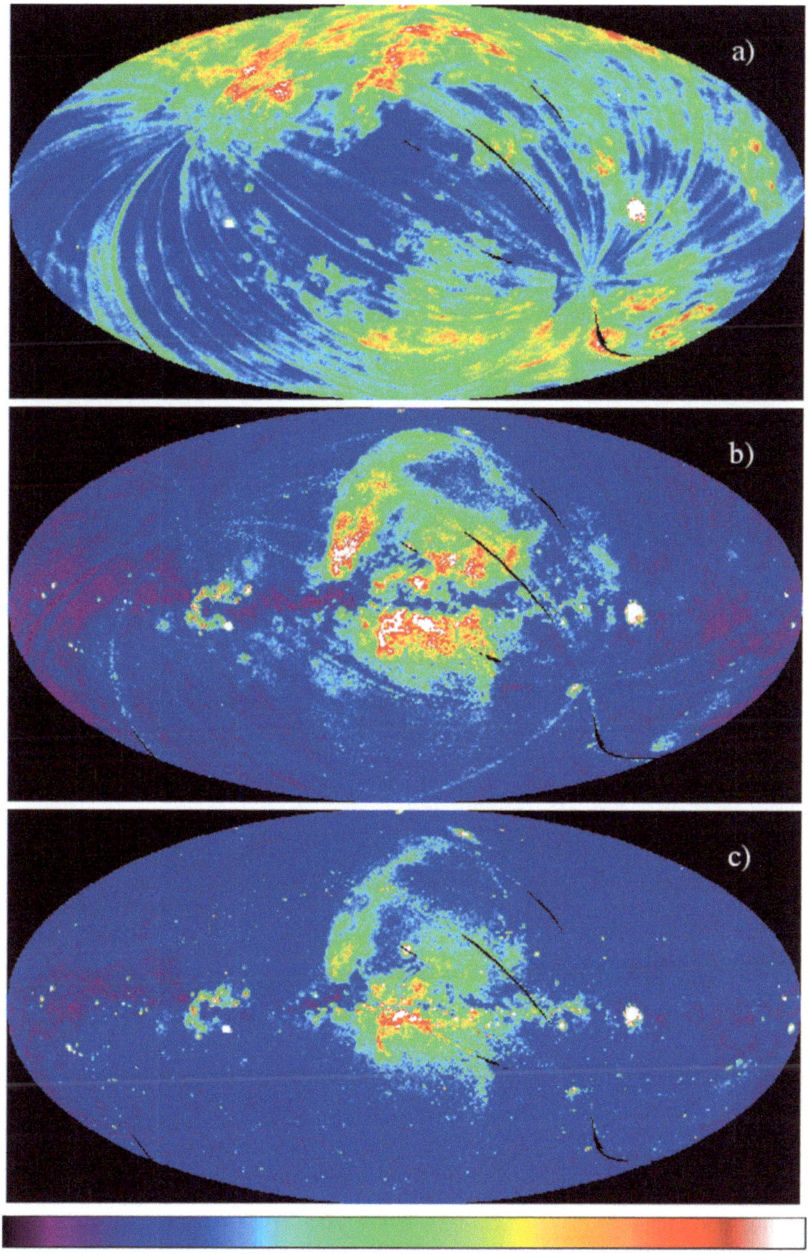

Fig. 7 *ROSAT* All-Sky Survey maps for the (**a**) $\frac{1}{4}$ keV, (**b**) $\frac{3}{4}$ keV, and (**c**) 1.5 keV bands in an Aitoff-Hammer equal-area projection in Galactic coordinates with the Galactic center at the center and with longitude increasing to the left. The striping, which is along great circles with constant ecliptic longitude, are the so-called long-term enhancements which are now understood as SWCX emission. The color bar is linear with a range of 0–0.002 counts s^{-1} $arcmin^{-2}$ for the $\frac{1}{4}$ keV band and a range of 0–0.0006 counts s^{-1} $arcmin^{-2}$ for the $\frac{3}{4}$ keV and 1.5 keV bands

of the Galactic disk is filled with hot plasma then the high-resolution spectra will provide information on the ionization states and abundances of the plasma and whether the plasma is in equilibrium, or more likely in what way the plasma is out of equilibrium.

With the long temporal coverage provided by satellite instrumentation, it will be possible to determine the nominal quiescent level of SWCX emission and separate the observed SWCX flux into heliospheric (interstellar neutrals passing through the solar system providing the target atoms) or geospheric (emission from the magnetosheath with exospheric material providing the target atoms). By the nature of the observations, this will provide a remote sensing of the solar wind.

Acknowledgements The author thanks members of the GSFC calorimeter team for providing details on their prospective contribution to the Spectrum Röntgen Gamma project, Wilt Sanders for providing the DXS data, Dan McCammon for providing the XQC data, and Dimitra Koutroumpa and Vasili Karchenko for providing the model SWCX emission spectrum. The author would also like to thank K.D. Kuntz, Randall Smith, and Jeffrey Linsky for their helpful comments.

References

D.P. Cox, in *Proceedings of the IAU Colluquium #166*, ed. by D. Breitschwerdt, M.J. Freyberg, J. Truemper. Lecture Notes in Physics, vol. 506 (Springer, Berlin, 1998), pp. 121–131
D.P. Cox, S.L. Snowden, Adv. Space Res. **6**(2), 97–107 (1986)
T.E. Cravens, Geophys. Res. Lett. **24**(1), 105–108 (1997)
P.C. Frisch, D.G. York, Astrophys. J. Lett. **271**, L59–L63 (1983)
R. Fujimoto, K. Mitsuda, D. McCammon, Y. Takei, M. Bauer, Y. Ishisaki, S.F. Porter, H. Yamaguchi, K. Hayashida, N.Y. Yamasaki, Publ. Astron. Soc. Jpn. **59**, 133–140 (2007)
V. Kharchenko, in *Physics of the Outer Heliosphere*, ed. by V. Florinski, N.V. Pogorelov, G.P. Zank. American Institute of Physics Conference Series, vol. 719 (Am. Inst. Phys., New York, 2004), pp. 416–421
G.R. Knapp, Astron. J. **80**, 111–116 (1975)
D. Koutroumpa, F. Acero, R. Lallement, J. Ballet, V. Kharchenko, Astron. Astrophys. **475**, 901–914 (2007)
R. Lallement, Astron. Astrophys. **418**, 143–150 (2004)
R. Lallement, B.Y. Welsh, J.L. Vergely, F. Crifo, D. Sfier, Astron. Astrophys. **411**, 447–464 (2003)
C.M. Lisse, K. Dennerl, J. Englhauser, M. Harden, F.E. Marshall, M.J. Mumma, R. Petre, J.P. Pye, M.J. Ricketts, J. Schmitt, J. Trumper, R.G. West, Science **274**, 205–209 (1996)
D. McCammon, R. Almy, E. Apodaca, W. Bergmann Tiest, W. Cui, S. Deiker, M. Galeazzi, M. Juda, A. Lesser, T. Mihara, J.P. Morgenthaler, W.T. Sanders, J. Zhang, Astrophys. J. **576**, 188–203 (2002)
D. McCammon, W.T. Sanders, Annu. Rev. Astron. Astrophys. **28**, 657–688 (1990)
S. Otranto, R.E. Olson, P. Beiersdorfer, Phys. Rev. A **73**, 022723 (2006)
F. Paresce, Astron. J. **73**, 1022–1037 (1984)
J.C. Raymond, B.W. Smith, Astrophys. J. Suppl. Ser. **35**, 419–439 (1977)
W.T. Sanders, W.L. Kraushaar, J.A. Nousek, P.M. Fried, Astrophys. J. Lett. **217**, L87–L91 (1977)
W.T. Sanders, R.J. Edgar, W.L. Kraushaar, D. McCammon, J.P. Morgenthaler, Astrophys. J. **554**, 694–709 (2001)
D.M. Sfeir, R. Lallement, F. Crifo, B.Y. Welsh, Astron. Astrophys. **346**, 785–797 (1999)
R.K. Smith, N.S. Brickhouse, D.A. Liedahl, J.C. Raymond, Astrophys. J. **556**, L91–L95 (2001)
S.L. Snowden, in *The Century of Space Science*, ed. by J.A.M. Bleeker, J. Geiss, M.C.E. Huber (Kluwer Academic, Dordrecht, 2001), pp. 581–606
S.L. Snowden, M.R. Collier, K.D. Kuntz, Astrophys. J. **610**, 1182–1190 (2004)
S.L. Snowden, D.P. Cox, D. McCammon, W.T. Sanders, Astrophys. J. **354**, 211–219 (1990)
S.L. Snowden, R. Egger, M.J. Freyberg, D. McCammon, P.P. Plucinsky, W.T. Sanders, J.H.M.M. Schmitt, J. Trümper, W. Voges, Astrophys. J. **485**, 125–133 (1997)
Y. Tanaka, J.A.M. Bleeker, Space Sci. Rev. **20**, 815–888 (1977)
J. Trümper, Q. J. R. Astron. Soc. **33**, 165–174 (1992)
B.J. Wargelin, M. Markevitch, M. Juda, V. Kharchenko, R. Edgar, A. Dalgarno, Astrophys. J. **607**, 596–610 (2004)

Section V: What Physical Processes Drive the Multiphase Interstellar Medium in the Local Bubble?

What Physical Processes Drive the Interstellar Medium in the Local Bubble?

D. Breitschwerdt · M.A. de Avillez · B. Fuchs · C. Dettbarn

Originally published in the journal Space Science Reviews, Volume 143, Nos 1–4, 263–276.
DOI: 10.1007/s11214-008-9425-1

Abstract Recent 3D high-resolution simulations of the interstellar medium in a star forming galaxy like the Milky Way show that supernova explosions are the main driver of the structure and evolution of the gas. Its physical state is largely controlled by turbulence due to the high Reynolds numbers of the average flows. For a constant supernova rate a dynamical equilibrium is established within 200 Myr of simulation as a consequence of the setup of a galactic fountain. The resulting interstellar medium reveals a typical density/pressure pattern, i.e. distribution of so-called gas phases, on scales of 500–700 pc, with interstellar bubbles being a common phenomenon just like the Local Bubble and the Loop I superbubble, which are assumed to be interacting. However, modeling the Local Bubble is special, because it is driven by a moving group, passing through its volume, as it is inferred from the analysis of Hipparcos data. A detailed analysis reveals that between 14 and 19 supernovae have exploded during the last 15 Myr. The age of the Local Bubble is derived from comparison with HI and UV absorption line data to be $14.5\pm^{0.7}_{0.4}$ Myr. We further predict the merging of the two bubbles in about 3 Myr from now, when the interaction shell starts to fragment. The Local Cloud and its companion HI clouds are the consequence of a dynamical instability in the interaction shell between the Local and the Loop I bubble.

Keywords Interstellar matter · Superbubbles · Local bubble · Turbulence

D. Breitschwerdt (✉)
Institut für Astronomie, University of Vienna, Türkenschanzstraße 17, 1180 Vienna, Austria
e-mail: breitschwerdt@astro.univie.ac.at

M.A. de Avillez
Department of Mathematics, University of Évora, R. Romão Ramalho 59, 7000 Évora, Portugal

B. Fuchs · C. Dettbarn
Astronomisches Rechen-Institut, ZAH Universität Heidelberg, Mönchhofstraße 12-14,
69120 Heidelberg, Germany

1 Introduction

The interstellar medium (ISM) in galaxies like our Milky Way is continuously evolving by forming stars in dense and cool clouds, and by enriching the gas with heavy elements from stellar explosions thus closing a galactic matter cycle. It has been known for a long time that such feedback mechanisms regulate star formation. More debatable, and hence less accepted, was the proposed outflow of hot gas from the disk as a galactic fountain (Shapiro and Field 1976; Bregman 1980; Kahn 1981; de Avillez 2000; Breitschwerdt and Komossa 2000) or even as a wind (e.g. Mathews and Baker 1971; Breitschwerdt et al. 1991; Everett et al. 2008) driven by supernova heated plasma and cosmic rays. One reason for skepticism lay in the lack of direct observational evidence. Absorption line studies of the galactic halo (e.g. Savage and de Boer 1981) were overwhelmingly pointing to negative velocities and hence infalling gas. However, the existence of hot gaseous X-ray halos observed with ROSAT, and now with Chandra and XMM-Newton, argue beyond doubt for outflow in star forming galaxies like NGC 891 (Tüllmann et al. 2006) sometimes labeled as a twin of our galaxy.

Why is the outflow from the disk into the halo so important for the evolution of the disk ISM? The simplest answer is: it acts as a pressure release valve. Without it, most of the disk volume, perhaps 50–70% as envisaged by McKee and Ostriker (1977), would be occupied by the hot medium. The evolution of a superbubble like our Local Bubble (LB), in which the solar system is embedded, would be quite different in a high pressure, low density environment. However, the story to tell about the ISM in general and the local ISM (LISM) in particular is not quite as simple and more subtle. In the ISM the estimated Reynolds' numbers are quite high, about 10^5–10^7 (Elmegreen and Scalo 2004), and the gas is highly compressible. Therefore its dynamical and thermal state is largely determined by turbulence, as was first pointed out by von Weizsäcker (1951), which can be both supersonic and super-alfvénic. The implications for the evolution of the ISM, as we shall show in the following sections, are profound. Turbulence does not add only extra pressure to the medium, as it is often simplistically treated, but also transfers energy from larger to smaller scales, or vice versa (inverse cascade), and promotes efficient mixing of chemically enriched material allowing for the existence of gas in classical thermally unstable density-temperature regimes.

From all we know the LISM on scales of a few hundred parsecs is not vastly different from the general ISM. Star formation seems to be ongoing as can be inferred from the existence of several nearby superbubbles. The Loop I superbubble, with the North Polar Spur being its shock illuminated outer shell and also the largest coherent X-ray structure in the sky, is excited by the Sco Cen association (or one of its distinct subgroups). On the other hand, there is no young stellar cluster known inside the LB. However, a detailed investigation of Hipparcos and radial velocity data of the local stellar population revealed (Berghöfer and Breitschwerdt 2002; Fuchs et al. 2006) the existence of a moving group passing through the volume occupied by the LB during the last 15 million years and generating the LB by the successive explosions of about 20 SNe.

This review paper is organized as follows: In Sect. 2 the modeling of the general ISM and the LB are discussed, while in Sect. 3 the results of our models are compared with observations, notably in HI and UV. In Sect. 4 we summarize our results and present some conclusions.

2 Modeling the ISM and the Local Bubble

The limitations of analytical and semianalytical models of the state and evolution of the ISM are obvious. One is forced to make assumptions, e.g. of global pressure equilibrium, mass balance, etc. like in the so-called three phase model. Transitions from one phase to another are instantaneous and regulated by the temperature perturbations at constant pressure of the net heat loss function $\mathcal{L} = n^2 \Lambda(T) - n \Gamma$, where n, T, Λ, Γ are the number density, temperature, interstellar metallicity dependent cooling and heating functions, respectively. According to Field's (1965) criterion $(\frac{\partial \mathcal{L}}{\partial T})_P > 0$ guarantees stability of a gas phase. However, as we shall see, this is only true if the cooling time is less than the dynamical time scale. In particular turbulence can have a stabilizing effect on gas in the transition regimes.

Turbulent flows are fundamentally different in nature from laminar flows. Generally, turbulence is fed by the mean flow field via Reynolds stresses, which arise from the randomly fluctuating velocity components in a fluid. Much insight can be gained by looking at the evolution and distribution of vorticity ($\omega = \nabla \times \mathbf{u}$). In turbulence, the velocities are correlated over a characteristic distance, which can be described by so-called "eddies". The largest eddies contain the bulk of the kinetic energy and their characteristics are often depending directly on the nature of the source. Since vorticity can neither be created nor destroyed in the interior of a flow, but rather in boundary layers, its evolution is determined by diffusion and stretching of turbulent blobs. The latter leads to the break-up of eddies into smaller structures and a subsequent cascade of energy from the largest to the smallest scales, where viscous dissipation sets in. A simple manifestation of a turbulent flow is shear flow. In the ISM a lot of shear is introduced by interacting flows or the interaction of pressure driven flows with the ambient medium. Therefore a multitude of sources for turbulence exists: stellar (e.g. from HII regions, stellar winds, SNe), differential galactic rotation, fluid instabilities (Kelvin-Helmholtz, Parker, magnetorotational instability, etc.), self-gravity, etc. An estimate of their relevance in the ISM has been carried out with the result that SNe dominate by at least a factor of a few (MacLow and Klessen 2004). Therefore it seems justified to start numerical investigations of the evolution of the ISM with SNe (and stellar winds) as the driving sources of mass, momentum and energy input. We note, however, that our simulations like those of other groups, suffer from a fundamental shortcoming: we cannot resolve the viscous scale, where energy dissipation occurs. This problem is well known in turbulence studies, and there exist basically two approaches. In direct numerical simulations (DNS), the maximum Reynolds number is limited to about $Re \sim 1000$, since in incompressible turbulence $Re \sim (L/l_{\mathrm{box}})^{4/3} N_x^{4/9}$, where L, l_{box}, and N_x are the outer scale at which turbulence is fed in, the box size, and the number of data points required at any instant, respectively. For ISM simulations this is far below realistic Reynolds numbers. In connection with laboratory studies, large eddy simulations (LES) are carried out, in which only the largest scales are resolved, since they only require a tiny fraction of the computing time, and yet carry the most important properties of turbulence, like e.g. the energy flux cascading down to smaller scales. To cut off the flow at some wavenumber k_{max} can only be done at the price of supplying an empirically motivated closure model (from comparison to laboratory studies), which accounts for the hopefully correct energy drainage out of the turbulent cascade. In case of the ISM this is far from trivial and is the subject of ongoing research.

In the following we give a brief overview on modeling a SN driven ISM on large and small scales.

2.1 A Supernova Driven ISM Model

Before going into details, a few general remarks are appropriate. Modeling a whole galaxy with sufficiently high resolution is beyond the current limit of computing power. We have therefore chosen a representative patch of the ISM, large enough to cover all, even the largest, structures, and small enough to still retain high resolution locally. A common feature in our models is the inclusion of the complete large scale disk–halo–disk circulation flow, and, as a prerequisite of this, a sufficiently long evolution time of typically 400 Myr. Calculations have shown that simulations with dynamical time scales shorter than 200 Myr still bear memory effects of the initial conditions, as the fountain return flow time is of about the same order. Another feature of the models to be described is that they contain the most important physical processes, but are not exhaustive in the sense, that we need to understand their physical implications first, before we add new physics to the model. In this "bottom-up" manner we have included so far: gravitational field of stars, adopted from Kuijken and Gilmore (1989) (but not self-gravity, which is currently tested), detailed interstellar heating and cooling processes assuming an optically thin gas in collisional ionization equilibrium (CIE, i.e. ionization rate by collisions between electrons and ions is balanced by radiative recombination rate) using the cooling functions of Sutherland and Dopita (1993) for $T \geq 10^4$ K and Dalgarno and McCray (1972) for $T < 10^4$ K, with a temperature cut-off at 10 K, and uniform heating due to starlight varying with height z (Wolfire et al. 1995).

The computational box was chosen to be $\Delta x \times \Delta y \times \Delta z = (1 \times 1 \times \pm 10)$ (in units of kpc) on a Cartesian grid, centred on the solar circle, and evolved on parallel processors with adaptive mesh refinement (AMR). The hydrodynamics was based on a Riemann solver scheme using the piecewise parabolic method with Lagrangian remap (PPMLR). For further details see de Avillez and Breitschwerdt (2004), and for the setup of the MHD runs de Avillez and Breitschwerdt (2005a). We have included SNe Types Ia, b, c and Type II with their appropriate scale heights taken from observation. After terminating their main sequence life time, half of these explosions occur in a clustered fashion, the others randomly in the field. As a criterion for stars to form we have applied a density/temperature threshold criterion of $n > 10\,\mathrm{cm}^{-3}$, $T < 100$ K if a sufficient amount of mass is available to distribute the stars according to their initial mass function (IMF). Stars drift away from their parent association with a random velocity of ~ 5 km/s.

2.2 Results on the General ISM

After having reached global dynamical equilibrium after about 200 Myr, the initial conditions for the setup have been completely erased. In Fig. 1 it can be seen that both the density and temperature distributions in the Galactic midplane span more than 5 orders of magnitude in values and that the morphology of the gas is very filamentary. It is interesting to note that the probability distribution functions (PDFs) in Fig. 7 of de Avillez and Breitschwerdt (2005a, 2005b) for density of the warm HI and the diffuse ionized gas (DIG) show variations of about 3–4 orders of magnitude at the level of 0.01, and that the PDFs represent a lognormal distribution. This is more or less exactly what has been found by Berkhuijsen and Fletcher (2008) in recent observations of a sample of pulsars by analyzing the dispersion and emission measures.

Textbook superbubbles are found nowhere, except if they are caught during their very early stages of evolution. Interestingly the pressure distribution (middle panel of Fig. 1) shows also large variations, not only locally, but also on larger scales. This means that the pressure in the ISM is certainly not uniform, as one of the key assumptions in the model of McKee and Ostriker (1977) requires. This may again be regarded as an effect of turbulence, since ram pressure dominates the flow between 10^2–10^6 K, as can be inferred from Fig. 2 (bottom). Finally, as a consequence of the existence of a large scale galactic fountain, the volume filling factors are also vastly different from equilibrium models. In our simulations (both HD and MHD) the filling factor of the hot gas is always less than 20% (see Fig. 2 top). Another surprising result is that the gas is not distributed into distinct phases like in the classical model, but also exists in intermediate and thermally unstable temperature regimes. This is most certainly due to the stabilizing effect of turbulence on local condensation modes.

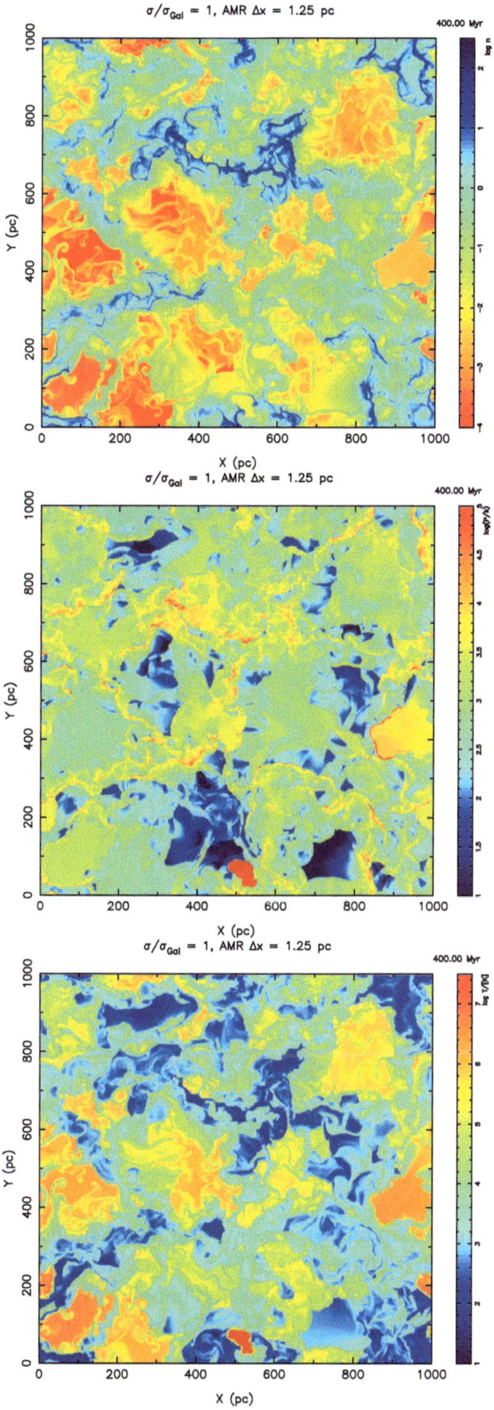

Fig. 1 Two-dimensional cuts through the 3D data cube: number density n (*top*), thermal pressure in units of Boltzmann's constant (cgs), P/k (*middle*), and temperature, T in K (*bottom*), distribution in the Galactic midplane plane for a Galactic supernova rate ($\sigma/\sigma_{Gal} = 1$)

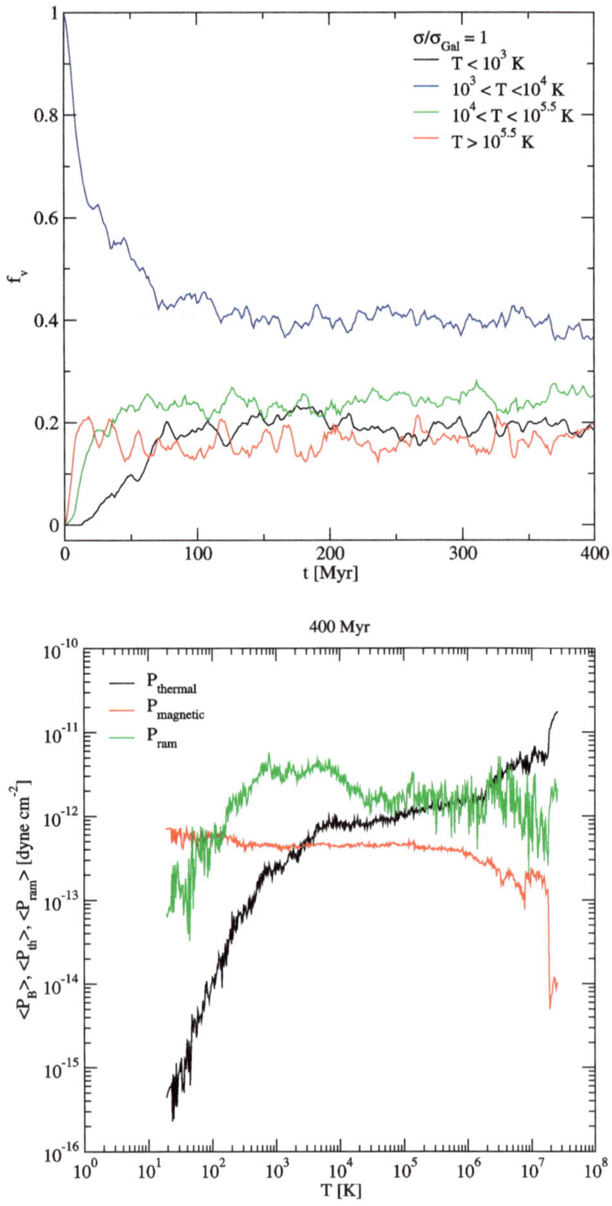

Fig. 2 *Top*: Time evolution of the volume filling factors of the cold ($T \leq 10^3$ K, *black curve*), cool ($10^3 < T \leq 10^4$ K, *blue curve*), warm ($10^4 < T \leq 10^{5.5}$ K, *green curve*), and hot ($T > 10^{5.5}$ K, *red curve*) gas phases for a Galactic supernova rate $\sigma/\sigma_{Gal} = 1$. The finest AMR level resolution is 1.25 pc. *Bottom*: Comparison of the average pressures $\langle P_{\rm ram} \rangle$ (ram pressure, *green*), $\langle P_{\rm th} \rangle$ (thermal pressure, *black*) and $\langle P_{\rm mag} \rangle$ (magnetic pressure, *red*) as a function of temperature in the Galactic disk ($|z| \leq 250$ pc) averaged over temperature bins of $\Delta \log T = 0.1$ K

In Fig. 3 (top panel) we show the magnetic field structure perpendicular to the Galactic plane, which is characterized by a tangled topology, with a significant vertical field component, as well as showing loops and bubbles expanding away from the disk. The temperature field in the midplane (Fig. 3, bottom panel), clearly visualizes the elongated bubble structures due to magnetic tension forces of an initially disk parallel field geometry (for details see de Avillez and Breitschwerdt 2005a).

In order to proceed modeling the Local Bubble, we will now use these global ISM simulations as a typical background medium, in which a SN driven bubble evolves. Therefore

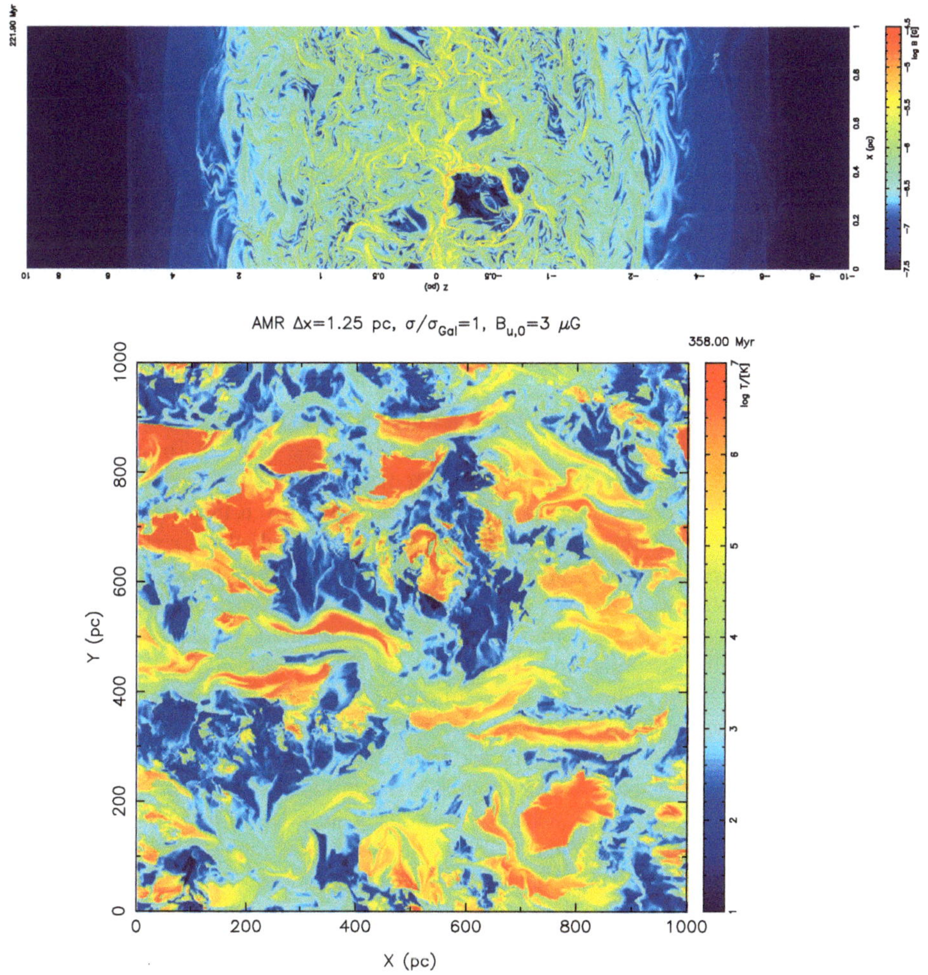

Fig. 3 2D slices of a magnetized ISM as inferred from 3D AMR MHD simulations (cf. de Avillez and Breitschwerdt 2005a). The colour coding refers to *red* equals high temperature/high magnetic field, and *blue* low temperature/low field. *Top*: Magnetic field distribution perpendicular to the Galactic disk at an evolution time of $t = 221.9$ Myr. The average initial field in the galactic plane was $B = 3.1$ μG and parallel to the disk. The *scale above and below* $z = \pm 0.5$ kpc is shrunk in size to fit the picture. Magnetized loops far above the disk are clearly visible, as well as small scale turbulent structures. *Bottom*: Temperature distribution in the Galactic disk after 358 Myr of evolution. Interstellar bubbles, visible as *red blotches*, are preferentially elongated due to the presence of the magnetic field

our next task will be to determine the stellar cluster which is responsible for the generation of the LB.

2.3 Modeling the Local Bubble

Modeling the LB has been so far a considerable challenge. The reasons are the following: firstly we are located inside, and therefore have some difficulty in determining the bubble extension observationally. Considering the soft X-ray flux, this can only be done by making some restrictive assumptions about the emission measure, e.g. by calibrating the flux in some direction (cf. Snowden et al. 1998). In contrast, the extension of the local HI cavity seems to be known to higher precision, since NaI absorption line studies towards more than 1000 background stars (Lallement et al. 2003) yield a better spatial resolution, at least at low galactic latitudes. Secondly, we have no information about an exciting star cluster, and hence lack the most important initial condition: the energy input rate as a function of time. This is in noteworthy contrast to the neighbouring Loop I superbubble, which still harbours young massive stars that feed energy into the bubble by winds and SN explosions. The LB however seems to be extinct. Therefore in the past modeling its evolution has been based on leaving the SN rate as a free parameter, and assuming some ambient density distribution to begin with. Smith and Cox (2001) have argued that 3 SN explosions, with the last one occurring about 5 Myr ago would be sufficient to reproduce the soft X-ray data. However, as we shall see below, the assumption of a constant ambient density is already critical for an evolved superbubble, because the older and the larger the bubble becomes, the more its shape and expansion will be controlled by the ambient density and pressure gradients. Thus initially, the bubble will be roughly spherical, but later on will become more and more asymmetric according to the inhomogeneity of the surrounding medium. We have therefore ventured to provide a "realistic" background medium, by evolving a SN driven ISM (see Sect. 2.2) for 200 Myr and then let the bubble expand into such a pre-structured inhomogeneous medium.

We have solved the problem of the missing OB star cluster, responsible for the creation of the LB, in the following way. We were searching the Hipparcos data base for moving groups with the idea that stars have traveled in the past through the region now occupied by the LB. In a first analysis (Berghöfer and Breitschwerdt 2002), we used the data provided by Asiain et al. (1999a, 1999b) who searched for moving groups in the solar neighbourhood and identified the Pleiades subgroup B1 as one of its youngest members. We then calculated the trajectory of this group backwards in time and found that it crossed the LB region, entering it about 15 Myr ago and leaving it just about 1 Myr from now. The remaining 27 stars of B1 are of spectral type A or later, and, if their masses are fitted to an IMF (e.g., Massey et al. 1995), yields the cumulative mass of stars more massive than 8 solar masses. If we bin this, somewhat arbitrarily, into stars with integer solar masses with bin size of 1 solar mass, we obtain 21 SN candidates, two of which are now in the neighbouring Loop I bubble. Hence we concluded that 19 SN explosions must have occurred, with the most massive star having had 20 solar masses. The main sequence life times were calculated using a formula by Stothers (1972). This analysis, although possessing some ad hoc elements, reveals a conclusive picture of the origin of the LB. The fact that Loop I is co-evolving is in our view rather a support than a counter argument for this model. Taking into account the existence of other nearby bubbles, like e.g. the Eridanus superbubble, seems to be consistent with the idea that the nearby ISM is a region of active star formation. Nevertheless, we tried to remove the assumption that the LB was the result of a *particular* moving group star cluster, and searched ab initio the whole region accessible to Hipparcos (supplemented by radial velocity data of the ARIVEL catalogue), which turned out to be a volume of 400 pc in diameter centered on the solar circle. This analysis is described in detail by Fuchs et al. (2006, 2008). It proved that our idea of a moving group was basically correct with no other groups or interlopers being present. Hence the SN explosions inferred from the Pleiades subgroup B1 and the ones inferred from the analysis of Fuchs et al. (2006) must have been the same (see also Fuchs et al. 2008).

ROSAT PSPC observations have shown (Egger and Aschenbach 1995) that soft X-rays are absorbed by a nearby neutral shell, which is most likely the result of an interaction between the LB and the Loop I superbubble. Therefore the expansion of the LB has to be simulated along with that of Loop I, which is driven by explosions of the Sco Cen association members. Again, using the stellar content from observations and applying an IMF, we were able to follow the expansion of the Loop I bubble in space and time. The explosion site was chosen such as to give the correct distance from the solar circle.

The simulations were carried out by using the 3D parallelized adaptive mesh refinement (AMR) hydrocode EVAF (Évora-Vienna Astrophysical Fluid) code to track small scale structures down to 1.25 pc, where necessary, and to follow the LB and Loop I evolution in an ISM, driven by SNe types Ia, Ib+c and II (with canonical explosion energies of 10^{51} erg) occurring at the Galactic rate (taken from Cappellaro et al. 1999), on a Cartesian grid of $0 \leq (x, y) \leq 1$ kpc size in the Galactic plane, and $-10 \leq z \leq 10$ kpc perpendicular to it (see also Sect. 2.1).

2.4 Results on the Local Bubble

In order to provide as a realistic as possible background medium for the LB to expand into, we chose an ISM simulation, for which global dynamical equilibrium has been established, i.e. after 200 Myr of Galactic evolution time. Moreover, we ensured that a sufficient amount of cold gas, to be consumed by star formation, was present in the region where we placed the origin of the LB (and Loop I as well). Indeed, the Sco Cen cluster, was put (arbitrarily) at $(375, 400)$ pc (cf. Fig. 4). We then followed the trajectory of the Pleiades moving group B1, whose SNe in the LB went off along a path crossing the location at $(175, 400)$ pc, chosen to match the present observed distances from the Sun and the Sco Cen. Due to small peculiar motions of the local gas, the LB will be tied to the local standard of rest (LSR; for details of the moving group kinematics see Berghöfer and Breitschwerdt, 2002). Periodic boundary conditions are applied along the four vertical boundary faces, while outflow boundary conditions are imposed at the top ($z = 10$ kpc) and bottom ($z = -10$ kpc) boundaries. The simulation time of the LB and Loop I evolution run was an additional 30 Myr, which is somewhat larger than the age of the exciting star clusters.

The SN explosions along the cluster trajectory were generating a coherent LB structure (for details see Breitschwerdt and de Avillez 2006), which was meticulously followed in time. The bubbles, which, as it happens, are generated almost coevally, are smooth and spherical in the beginning, but develop internal structure after 8 Myr of evolution time. This is due to the inhomogeneity of the ambient ISM and becomes progressively more important with time. The present time is fixed by comparing the LB extension with observations, which amounts to $t_{\rm LB} \simeq 14$–15 Myr after the first explosion. As we shall see below, the average column density of O VI absorption lines toward background stars (inside and beyond the LB) as measured by the FUSE and Copernicus satellites can be nicely reproduced for $t_{\rm LB} = 14.5$ Myr. Thus the last SN explosion just occurred 0.5 Myr before the present time, and the LB is becoming extinct, while Loop I with its larger stellar content is still active (see Fig. 4 top).

Both bubbles are bounded by shells whose interaction generates Rayleigh-Taylor instabilities due to a larger pressure in Loop I with respect to the LB, in agreement with a linear stability analysis of Breitschwerdt et al. (2000). After becoming nonlinear this leads to the formation of cloudlets, which will travel towards the solar system from the Sco Cen region, in agreement with our simulations and observations of local clouds with measured velocities of about 26 km/s. On the basis of our simulations we further predict that the interaction shell will break up in 3 Myr from now, allowing mass transfer of hot gas from Loop I to the LB, and in ~ 15 Myr from now the bubbles will have merged (see Fig. 4 bottom).

In CIE, O VI is most abundant at $T \sim 3 \times 10^5$ K, whereas soft X-Ray emission of a thermal plasma requires a temperature of $\sim 10^6$ K. It turns out that between 14.5 and 14.7 Myr,

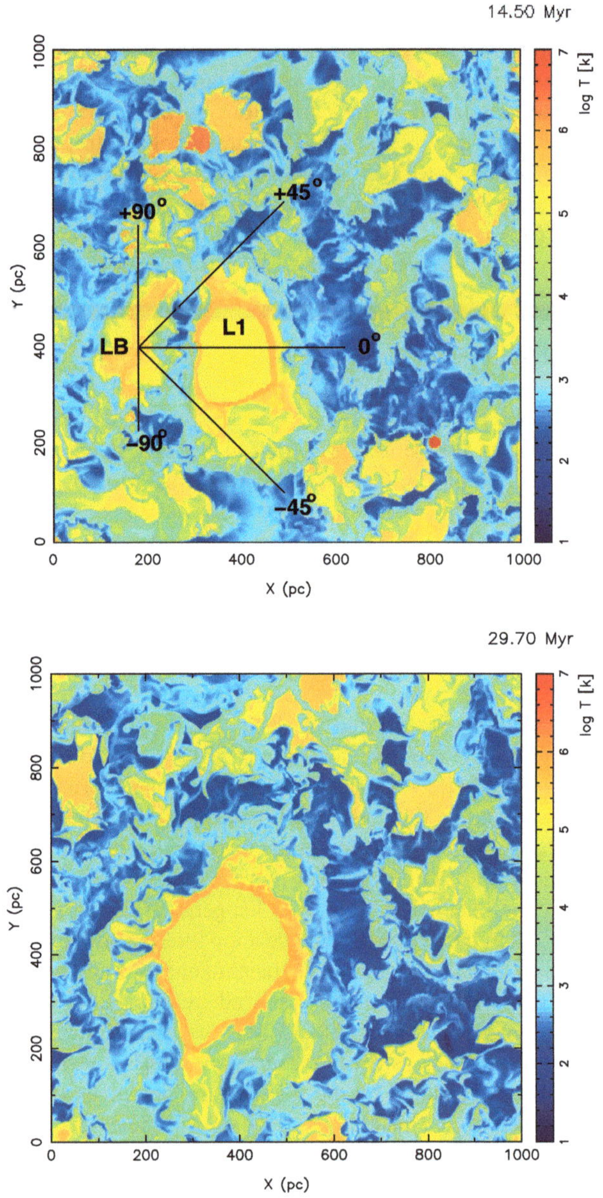

Fig. 4 *Top*: Colour-coded temperature map in the range $10 \leq T \leq 10^7$ K for a slice through the data cube (representing the Galactic midplane) of a 3D LB high-resolution simulation representing the present time (i.e. 14.5 Myr after the first explosion) with the LB centred at (175, 400) pc and Loop I shifted 200 pc to the right. *Bottom:* Same, showing the "future" of the LB and Loop I at $t = 29.7$ Myr. Note that part of the LB has merged with the ISM, part has been engulfed by Loop I (for details, see Breitschwerdt and de Avillez 2006)

cooler (O VI absorbing) and hotter (X-ray emitting) gas *co-exist*. This is a direct result of the turbulent and inhomogeneous density and temperature structure in realistically evolved superbubbles. Turbulent mixing generates gas in the whole thermally unstable regime between 10^5–10^6 K. While the amount of O VI in the LB has been measured fairly accurately (see Sect. 3), the source of the soft X-ray emission is more elusive. First of all, it is more difficult to localize, as there are most likely no dense clouds inside the LB, which can be used for shadowing experiments, thus allowing us to separate back- and foreground emission. Secondly, an as yet uncertain amount of soft X-rays will be produced locally by solar wind charge exchange reactions with heliospheric plasma (Cravens 2000; see also Koutroumpa 2008; Shelton 2008). And thirdly, the LB plasma need not be in CIE. It has been shown that if a plasma undergoes fast adiabatic expansion its kinetic temperature will drop, but recombination can not follow fast enough, thus mimicking an "overionised" plasma (similar to the solar wind), where recombination is delayed. In essence, the gas could be at a few times 10^5 K, but still emit copiously in soft X-rays (Breitschwerdt and Schmutzler 1994). We are currently investigating this possibility.

3 Comparison with Observations

We have determined the amount of O VI along the lines of sight (LOS) for an ISM plasma in CIE with solar abundances (Anders and Grevesse 1989; for details see de Avillez and Breitschwerdt 2005b). To measure the O VI column density distribution inside LB and Loop I, we took 91 LOS extending from the Sun and crossing Loop I (the hot pressured region 200 pc to the right of the LB) from an angle of $-45°$ to $+45°$ as marked in the top panel of Fig. 4. Figure 5 (top) shows the history (for $14.3 \leq t \leq 14.7$ Myr) of the variation of the LOS averaged O VI column densities, $\langle$N(O VI)$\rangle$, with distance along these LOS. They decrease steeply with time after the last SN in the LB has exploded, from 4×10^{13} to 3×10^{11} cm^{-2}, because no further heating is taking place. For LOS sampling gas from outside the LB (i.e., $l_{LOS} > 100$ pc) we have $\langle$N(O VI)$\rangle > 2 \times 10^{13}$ cm^{-2}. The histograms of column densities obtained in the 91 LOS for $t = 14.5$ and 14.6 Myr (Fig. 5, bottom) show that for $t = 14.6$ Myr all the LOS have column densities smaller than $10^{12.9}$ cm^{-2}, while for $t = 14.5$ Myr 67% of the lines have column densities smaller than 10^{13} cm^{-2} and about 50% of the

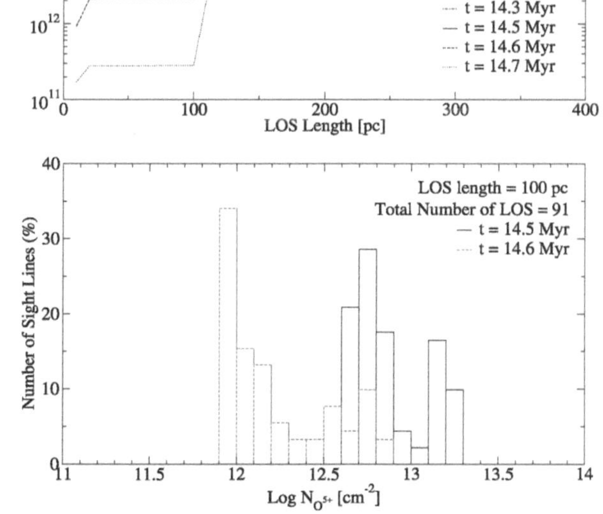

Fig. 5 *Top*: Averaged O VI column densities (over 91 LOS between ±45° as marked in Fig. 4 top) as a function of LOS path length at $14.3 \leq t \leq 14.7$ Myr of Local and Loop I bubbles evolution. *Bottom*: Histogram of the percentage of LOS as a function of N(O VI) within the LB at $t = 14.5$ and 14.6 Myr

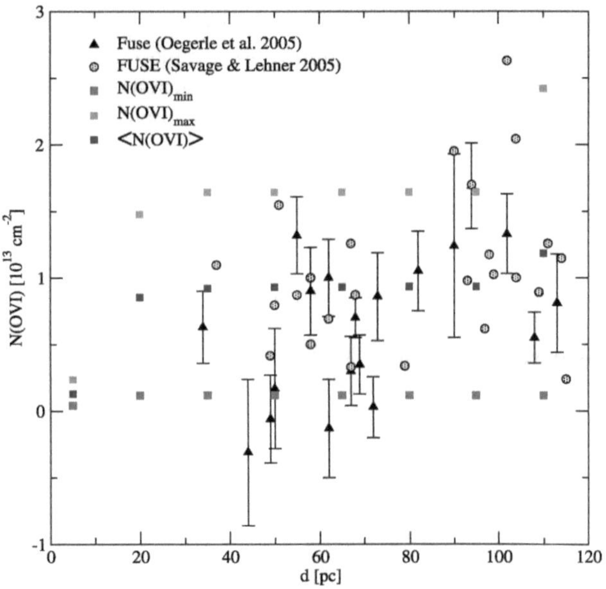

Fig. 6 Comparison between FUSE (*triangles:* Oegerle et al. 2005; *circles*: Savage and Lehner 2006) O VI column densities with the calculated minimum, maximum and average column densities along the same 91 lines of sight (at $t = 14.5$ Myr) pointing towards Loop I as shown in Fig. 1 (top). Negative values on the ordinate derive from essentially zero equivalent absorption line widths, i.e. non-detections, due to the noise in the spectrum (cf. Oegerle et al. 2005)

lines have N(O VI) $\leq 7.9 \times 10^{12}$ cm^{-2}. The maximum average column density is 1.78×10^{13} cm^{-2} in excellent agreement with the observations discussed in Savage and Lehner (2006), which is 1.1 times the average value of 1.6×10^{13} cm^{-2} inferred by Shelton and Cox (1994) from a reanalysis of COPERNICUS absorption line data (cf. Jenkins 1978).

In Fig. 6 we compare FUSE data by Oegerle et al. (2005; triangles) and Savage and Lehner (2006; circles) with simulated minimum (red squares), maximum (green squares) and averaged (blue squares) column density of O VI measurements along the 91 LOS outlined in Fig. 4 (top). It can be clearly seen in Fig. 6 that for $t = 14.5$ Myr the calculated N(O VI) distribution in the LB is similar to that observed with FUSE. These results allow us to constrain the age of the LB to be $14.5\pm^{0.7}_{0.4}$ Myr,[1] and capture at the same time all the relevant LB properties, such as size, age, dynamical instabilities in the interaction shell, amongst others.

Successive blast waves advance very fast through the low-density cavity, but slow down considerably later, as they run into the dense shell. As a consequence an asymmetric reverse shock leads to shear flow and turbulence inside the cavity, with a largest eddy size of a fraction of the bubble diameter, typically $l \leq 75$ pc (de Avillez and Breitschwerdt 2007). Thus the turnover time scale is about $\tau_m \sim l/c_s \leq 3.7 \times 10^5$ yr, for an averaged sound speed of $c_s \approx 200$ km/s after a SN explosion. As the last SN occurred about 0.5 Myr ago, we expect that SN ejected oxygen inside the LB at present time should have a rather uniform O/H distribution, a result that is also confirmed by FUSE observations of the O/H ratio (Moos et al. 2002).

[1] We note that some uncertainty in the LB age is due to the fact that the O VI column density can be matched both during temperature increase ("ionization phase") as well as during temperature decrease ("recomnination phase") after the last explosion (see de Avillez et al. 2008).

4 Summary and Conclusions

The solar system seems to be embedded in a region of ongoing star formation, which is testified by the existence of several nearby bubbles, like the Eridanus bubble or the Gum Nebula, and possibly radio loops II and III, just to name a few. It is therefore suggestive to assume that also the Local Bubble is the result of recent supernova explosions. The fact that the trajectory of the moving group, which harboured the progenitor stars, is not going through the geometrical centre of the LB is immaterial, since the expansion always proceeds in the direction of the largest pressure gradient. The expansion of the neighbouring Loop I superbubble, which was generated almost at the same time, corroborates the argument, because its high pressure will force the LB to expand preferentially away from it, which is fully confirmed by our numerical simulations. The analysis of the stellar content of the moving group reveals that substantially more than 3 SNe occurred during the last 15 Myr. Again, the expansion proceeds most rapidly in the direction of least resistance, i.e. perpendicular to the Galactic disk, once a disk scale has been reached. Therefore, it is not surprising that a number of at least 14 SNe (cf. Fuchs et al. 2006) will create a local chimney in agreement with the observations of Lallement et al. (2003).

The Local Cloud and its companions occur naturally in our model as a consequence of Rayleigh-Taylor instabilities, ensuing from the interaction between the LB and Loop I. It is probably not fortuitous that its velocity vector points away from the interaction shell of the two bubbles.

High-resolution 3D hydrodynamical simulations of the LB and Loop I show that UV data from FUSE and Copernicus—both for lines of sight within and outside the LB—can be nicely reproduced. It should be emphasized that our simulations were performed *before* the FUSE data were available; hence there is no "tuning" involved here. It can also be concluded from these calculations that within the computational box of 1 kpc $\times$ 1 kpc in the Galactic plane several bubble regions are present. Hence the physical processes that drive the general ISM structures are also responsible for structuring the LISM in general and the LB in particular. A key element of the physics governing the ISM is (compressible) turbulence. Turbulent diffusion is an efficient transport process and is capable of inhibiting local thermal instabilities, thus keeping a lot of interstellar gas in thermally unstable regimes. For example, about half of the warm neutral medium is between 500 and 5000 K, in agreement with observations by Heiles and Troland (2003).

Since the Lyman-Spitzer Colloquium "The Local Bubble and Beyond" in 1997, it is fair to state that modeling the LB has become less of a guess work, and a coherent picture seems to emerge, although of course lots of details still need to be explained.

Acknowledgements DB thanks ISSI, Ruedi von Steiger and S.F. Saliba for their organizational help with the presentation of the talk.

References

E. Anders, N. Grevesse, Geochim. Cosmochim. Acta **53**, 197 (1989)
R. Asiain, F. Figueras, J. Torra, B. Chen, Astron. Astrophys. **341**, 427 (1999a)
R. Asiain, F. Figueras, J. Torra, Astron. Astrophys. **350**, 434 (1999b)
T. Berghöfer, D. Breitschwerdt, Astron. Astrophys. **390**, 299 (2002)
E.B. Berkhuijsen, A. Fletcher (2008). arXiv:0806.4316
J.N. Bregman, Astrophys. J. **236**, 577 (1980)
D. Breitschwerdt, M.A. de Avillez, Astron. Astrophys. **452**, L1 (2006)
D. Breitschwerdt, S. Komossa, Astrophys. Space Sci. **272**, 3 (2000)

D. Breitschwerdt, T. Schmutzler, Nature **371**, 774 (1994)
D. Breitschwerdt, J.F. McKenzie, H.J. Völk, Astron. Astrophys. **245**, 79 (1991)
D. Breitschwerdt, R. Egger, M.J. Freyberg, Astron. Astrophys. **361**, 303 (2000)
M.A. de Avillez, Mon. Not. R. Astron. Soc. **315**, 479 (2000)
M.A. de Avillez, D. Breitschwerdt, Astron. Astrophys. **425**, 899 (2004)
M.A. de Avillez, D. Breitschwerdt, Astron. Astrophys. **436**, 585 (2005a)
M.A. de Avillez, D. Breitschwerdt, Astrophys. J. **634**, L65 (2005b)
M.A. de Avillez, D. Breitschwerdt, Astrophys. J. **665**, L35 (2007)
M.A. de Avillez, et al., Space Sci. Rev. (2008, this issue)
E. Cappellaro, R. Evans, M. Turatto, Astron. Astrophys. **351**, 459 (1999)
D.P. Cox, R.K. Smith, Astrophys. J. Suppl. Ser. **134**, 283 (2001)
T.E. Cravens, Astrophys. J. **532**, L153 (2000)
A. Dalgarno, R.A. McCray, Annu. Rev. Astron. Astrophys. **10**, 375 (1972)
R. Egger, B. Aschenbach, Astron. Astrophys. **294**, L25 (1995)
B.G. Elmegreen, J. Scalo, Annu. Rev. Astron. Astrophys. **42**, 211 (2004)
J.E. Everett, E.G. Zweibel, R.A. Benjamin, D. McCammon et al., Astrophys. J. **674**, 258 (2008)
G.B. Field, Astrophys. J. **142**, 531 (1965)
B. Fuchs, D. Breitschwerdt, M.A. de Avillez, C. Dettbarn, C. Flynn, Mon. Not. R. Astron. Soc. **373**, 993 (2006)
Fuchs et al., Space Sci. Rev. (2008, this issue)
C. Heiles, T.H. Troland, Astrophys. J. **586**, 1067 (2003)
E.B. Jenkins, Astrophys. J. **219**, 845 (1978)
F.D. Kahn, in *Investigating the Universe*, ed. by F.D. Kahn (Reidel, Dordrecht, 1981)
Koutroumpa, Space Sci. Rev. (2008, this issue)
K. Kuijken, G. Gilmore, Mon. Not. R. Astron. Soc. **239**, 651 (1989)
R. Lallement, B.Y. Welsh, J.L. Vergely, F. Crifo, D. Sfeir, Astron. Astrophys. **411**, 447 (2003)
M.-M. MacLow, R.S. Klessen, Rev. Mod. Phys. **76**, 125 (2004)
P. Massey, K.E. Johnson, K. Degioia-Eastwood, Astrophys. J. **454**, 151 (1995)
W.G. Mathews, J.C. Baker, Astrophys. J. **170**, 241 (1971)
C.F. McKee, J.P. Ostriker, Astrophys. J. **218**, 148 (1977)
H.W. Moos, K.R. Sembach, A. Vidal-Madjar et al., Astrophys. J. Suppl. Ser. **140**, 3 (2002)
W.R. Oegerle, E.B. Jenkins, R.L. Shelton, D.V. Bowen, P. Chayer, Astrophys. J. **622**, 377 (2005)
B.D. Savage, K.S. de Boer, Astrophys. J. **243**, 460 (1981)
B.D. Savage, N. Lehner, Astrophys. J. Suppl. Ser. **162**, 134 (2006)
P.M. Shapiro, G.B. Field, Astrophys. J. **205**, 762 (1976)
R. Shelton, Space Sci. Rev. (2008, this issue)
R. Shelton, D.P. Cox, Astrophys. J. **434**, 599 (1994)
S.L. Snowden, R. Egger, D.P. Finkbeiner, D. McCammon et al., Astrophys. J. **493**, 715 (1998)
R. Stothers, Astrophys. J. **175**, 431 (1972)
R.S. Sutherland, M.A. Dopita, Astrophys. J. Suppl. Ser. **88**, 253 (1993)
Tüllmann, W. Pietsch, J. Rossa, D. Breitschwerdt, R.-J. Dettmar, Astron. Astrophys. **448**, 43 (2006)
C.F. von Weizsäcker, Astrophys. J. **114**, 165 (1951)
M.G. Wolfire, C.F. McKee, D. Hollenbach, A.G.G.M. Tielens, E.L.O. Bakes, Astrophys. J. **443**, 152 (1995)

Plasma Turbulence in the Local Bubble

Steven R. Spangler

Originally published in the journal Space Science Reviews, Volume 143, Nos 1–4, 277–290.
DOI: 10.1007/s11214-008-9391-7

Abstract Turbulence in the Local Bubble could play an important role in the thermodynamics of the gas that is there. This turbulence could also determine the transport of cosmic rays and perhaps heat flow through this phase of the interstellar medium. The best astronomical technique for measuring turbulence in astrophysical plasmas is radio scintillation. Scintillation measurements yield information on the intensity and spectral characteristics of plasma turbulence between the source of the radio waves and the observer. Measurements of the level of scattering to the nearby pulsar B0950+08 by Philips and Clegg in 1992 showed a markedly lower value for the line-of-sight averaged turbulent intensity parameter $\langle C_N^2 \rangle$ than is observed for other pulsars, qualitatively consistent with radio wave propagation through a highly rarefied plasma. In this paper, we discuss the observational progress that has been made since that time. The main development has been improved measurements of pulsar parallaxes with the Very Long Baseline Array. This provides better knowledge of the media along the lines of sight. At present, there are four pulsars (B0950+08, B1133+16, J0437-4715, and B0809+74) whose lines of sight seem to lie mainly within the local bubble. The mean densities and line of sight components of the interstellar magnetic field along these lines of sight are smaller than nominal values for pulsars, but not by as large a factor as might be expected. Three of the four pulsars also have measurements of interstellar scintillation. The value of the parameter $\langle C_N^2 \rangle$ is smaller than normal for two of them, but is completely nominal for the third. This inconclusive status of affairs could be improved by measurements and analysis of "arcs" in "secondary spectra" of pulsars, which contain information on the location and intensity of localized screens of turbulence along the lines of sight. Similar data could be obtained from observations of highly compact extragalactic radio sources which show the "intraday variability" phenomenon.

Keywords ISM: scintillation and scattering

PACS 98.38.Am

S.R. Spangler (✉)
Department of Physics and Astronomy, University of Iowa, Iowa City, IA 52242, USA
e-mail: steven-spangler@uiowa.edu

J.L. Linsky et al. (eds.), *From the Outer Heliosphere to the Local Bubble.*
DOI: 10.1007/978-1-4419-0247-4_22

1 Introduction

The Local Bubble is a medium of low density and high temperature, and thus is a plasma medium. The interstellar magnetic field doubtlessly threads through this region, making magnetized plasma phenomena important. We would like to diagnose the usual plasma parameters in this medium as well as the properties of turbulence, if it exists. These properties would include the "wave mode" which most naturally describes the excitations, the energy density of turbulence, and its spatial power spectrum.

Such knowledge could be important in better understanding the dynamics of the Local Bubble, and the role it plays in the general galactic interstellar medium. Turbulence has the potential of dissipating on fast timescales; if the energy density of the turbulence is comparable to other energy densities, turbulent dissipation can play an important role in the heating of this medium, as it seems to do in the solar corona as well as other parts of the interstellar medium.

Turbulence in the Local Bubble could also play an important role in the galactic transport of cosmic rays. At times in the past it has been thought that "tunnels" like the Local Bubble could fill a large fraction of the volume of the interstellar medium. If the mean free path of cosmic rays in these bubbles or tunnels is very long, longer than in other parts of the interstellar medium, these bubbles might play a major role in the transport of cosmic rays throughout the Galaxy. In a collisionless plasma, turbulent fluctuations will determine parameters such as the pitch angle and spatial diffusion coefficients, and effectively the mean free path of particles.

2 Radio Astronomy and Diagnostics of Interstellar Plasmas

The refractive index of radio waves in a plasma depends on the density of the plasma and, to a much lesser but nonetheless observable degree, the magnetic field. As such, radioastronomical propagation measurements can yield information on plasma characteristics such as the plasma density and magnetic field. Path-integrated measurements of the plasma density can be obtained from measurement of the dispersion measure (DM). Observationally, the dispersion measure describes the degree to which radio waves of different frequencies travel at different speeds. Physically, it is the path integral of the electron density along the sight. It is only measurable for pulsars. The rotation measure (RM) is also a path integral along the line of sight. Observationally, it is a measure of how the polarization position angle changes with wavelength, and it can be measured for pulsars or extragalactic radio sources. The integrand in the RM is the product of the electron density and the line-of-sight component of the magnetic field.

Of equal interest, radio waves propagating through a turbulent medium with random density fluctuations undergo a number of modifications in amplitude and phase which are collectively described as interstellar scintillations (ISS). This topic was last extensively reviewed by Rickett (1990). The major ISS phenomena which have been observed and analysed are shown in cartoon form in Fig. 1. The different scintillation observations illustrated in Fig. 1 yield different information about the properties of interstellar turbulence.

In general, ISS observations yield information on the path integral of density turbulence properties, with different observational techniques featuring different kernels in the path integral. However, as will be discussed below, ISS observations sometimes contain additional information such as the distribution of turbulent plasma along the line of sight.

Some of the phenomena illustrated in Fig. 1 can be observed for either pulsars or extragalactic radio sources. Others are observable only for pulsars. In this paper, I concentrate on

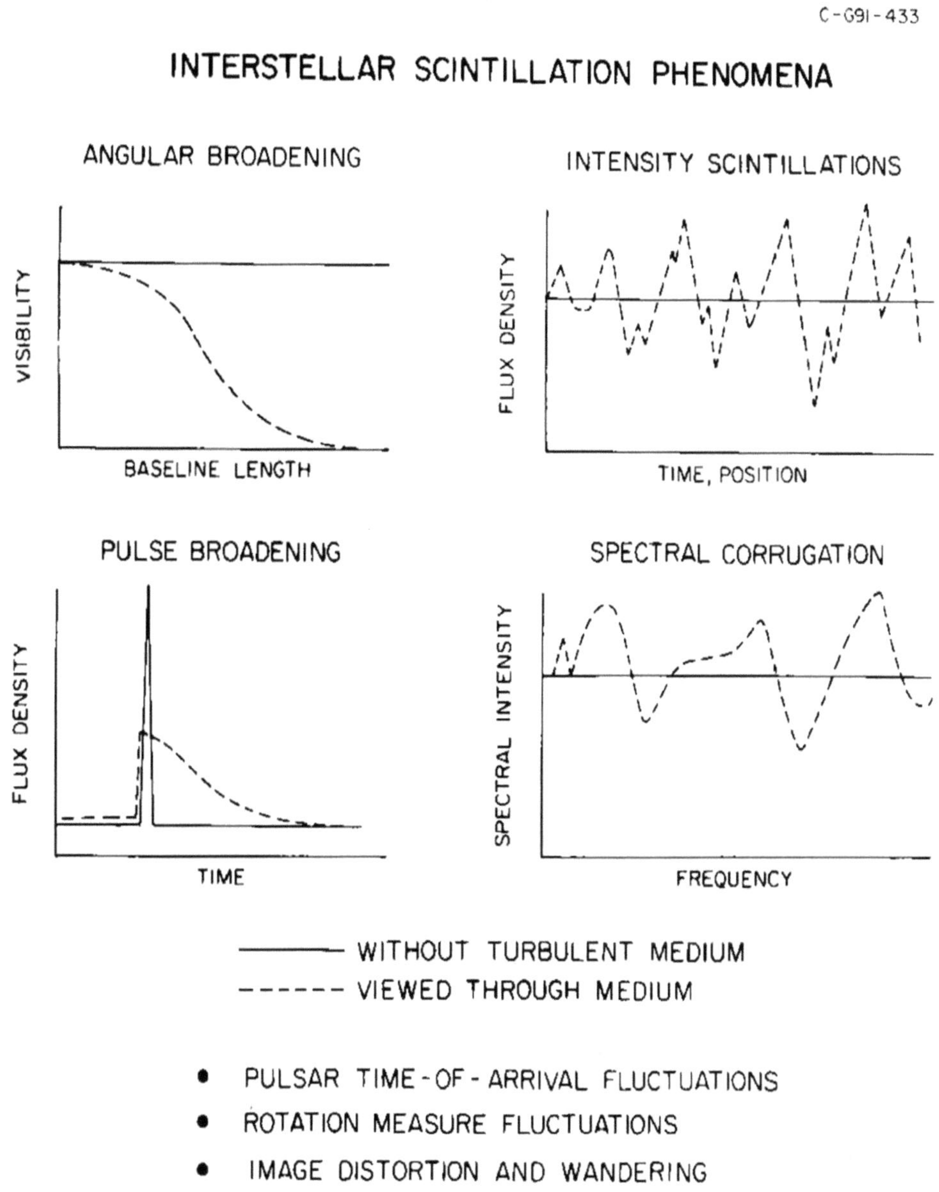

Fig. 1 The variety of ISS phenomena. The *solid lines* indicate what is measured if the source is viewed without an intervening turbulent screen. The *dashed lines* indicate what is observed in the presence of turbulence. For the representation of angular broadening we show the measurement of a radio interferometer. The magnitude of the interferometer visibility (the properly normalized correlation coefficient of the wave electric field at two antennas) is plotted versus the baseline length between the interferometer elements. The term "spectral corrugation" refers to a stochastic variation in the flux density as a function of radio frequency, induced by wave passage through a turbulent medium. Figure taken from Spangler (2007)

pulsars because the lines of sight are confined within the Galaxy, and in the most interesting case would lie completely within the Local Bubble.

It is typical in interstellar scintillation observations to model the density spatial power spectrum by

$$P_{\delta n}(q) = \frac{C_N^2}{[q_z^2 + \Lambda^2(q_x^2 + q_y^2)]^{\alpha/2}} \tag{1}$$

where q_x, q_y, and q_z are three Cartesian components of the spatial wavenumber. The parameter Λ describes the anisotropy of the irregularities. The power spectrum is rotationally symmetric about the z axis. In many analyses, an isotropic spectrum has been assumed in which $\Lambda = 1$, and such will be assumed for most of this paper. However, ISS observations have presented compelling evidence that the interstellar density spectrum is anisotropic; a recent summary of the evidence for this anisotropy is given in Rickett (2007). Anisotropy of the turbulence is expected theoretically on the basis of recent theories of magnetohydrodynamic (MHD) turbulence (Spangler 1999; Chandran and Backer 2002).

The parameter C_N^2 can be related to more intuitive properties of plasma turbulence via the following formula, which relates C_N^2 to the mean electron density $\bar{n_e}$, the "modulation index" of the turbulence $\epsilon \equiv \delta n/\bar{n_e}$, and its outer scale l_0 (Spangler and Cordes 1998)

$$C_N^2 = f(\alpha)\left[\frac{\epsilon^2}{(1+\epsilon^2)l_0^{2/3}}\right]\bar{n_e}^2 \tag{2}$$

In (2), $f(\alpha)$ is a dimensionless number which depends on α; for the Kolmogorov value $\alpha = 11/3$, $f(\alpha) = 0.181$. Analyses of interstellar scintillation data attempt to retrieve C_N^2, α, and Λ. Most ISS observations yield an estimate of C_N^2 which is averaged over the line of sight. This observational quantity is referred to as $\langle C_N^2 \rangle$. The discussion in this paper will concentrate on the value of $\langle C_N^2 \rangle$ which can be retrieved from published observations of lines of sight which have passed through the Local Bubble. I am particularly interested in comparing $\langle C_N^2 \rangle$ for lines of sight which are largely within the Local Bubble to those which pass through the normal ISM in the vicinity of the Sun.

3 Why the Hot Phase of the Interstellar Medium is of Interest

What can selected pulsar observations tell us about plasma turbulence in the local bubble? What makes this part of the interstellar medium (ISM) of particular interest? One response is that it has plasma parameters which may differ in important ways from the plasma in the more extensively diagnosed Diffuse Ionized Gas (DIG; Cox and Reynolds 1987). One of the more important such parameters is the plasma beta, which may be defined as

$$\beta \equiv \frac{c_s^2}{V_A^2} \tag{3}$$

where $c_s \equiv \sqrt{\frac{5/3 k_B(T_e+T_i)}{m_p}}$ is the ion acoustic wave speed, and V_A is the Alfvén speed. In the expression for the ion acoustic speed, T_e and T_i are the electron and ion temperatures, and m_p is the proton mass. Choosing nominal Local Bubble parameters of $n_e \simeq 5 \times 10^{-3}$ cm^{-3}, $T_e = T_i \simeq 10^6$ K, and a standard ISM magnetic field strength of $B = 4\,\mu$G, we have $\beta = 2.2$. This is larger than the corresponding value for the DIG of $\beta \simeq 0.4$. It should be noted that the magnetic field strength in the Local Bubble may be lower than we have assumed here. Opher et al. (2006) estimate $|B| \simeq 2$ μG immediately outside the heliosphere. Such a value for the magnetic field in the Local Bubble would increase our estimate to $\beta \simeq 9$. Finally, it is possible that the magnetic field strength is lower still in the truly low density Local Bubble, as opposed to the immediate vicinity of the Sun. This would result in an even higher estimate for β in the Local Bubble.

The plasma β plays a very important role in plasma wave and turbulence dynamics. Damping mechanisms such as Landau damping can depend strongly on the plasma β (e.g., Barnes 1966). Nonlinear processes such as parametric decay instabilities also depend on the beta, being allowed for certain values and forbidden for others. As a result, the dynamics of waves and turbulence could be quite different in hot phase plasmas like the Local Bubble relative to the DIG and colder phases. The differences in the properties and dynamics of waves and turbulence in a high β plasma could result in similar differences in the interaction between turbulence and charged particles in the Local Bubble.

4 Probing the Local Bubble: The Case of Pulsar B0950+08

A difficulty in interpreting radio scintillation observations is that a measurement has contributions from gas all along the line of sight. Since most parts of the interstellar medium are much denser than the Local Bubble (particularly the DIG, which appears to be correlated with scintillations), the Local Bubble will typically make a small and inseparable contribution to the measured signal. It has been clear for a long time that the key to remote sensing of the Local Bubble is to find a radio source which scintillates (in one of the observables shown in Fig. 1) and whose line of sight is completely or mainly within the local bubble.

It was clear 20 years ago that the pulsar B0950+08 was the best, and possibly only suitable object for this type of measurement. Based on a then-current parallax distance of 127 parsecs, the line of sight should have been entirely within the local bubble.

Several years later, Phillips and Clegg (1992) published observations of B0950+08 showing interstellar scintillation, and deduced a value of $\langle C_N^2 \rangle = 2.82 \times 10^{-5}$ $\mathrm{m}^{-20/3}$. This was the lowest value for $\langle C_N^2 \rangle$ that had been reported at that time, and was qualitatively consistent with expectations for a line of sight that was mainly within a low density bubble. By way of context for this measurement, the extensive survey of Cordes et al. (1985) gives a typical value of $\langle C_N^2 \rangle$ in the diffuse, or "Type A" scattering medium of about $\langle C_N^2 \rangle = 3.16 \times 10^{-4}$ $\mathrm{m}^{-20/3}$. In the remainder of this paper, I will typically quote the base-10 logarithm of $\langle C_N^2 \rangle$, which for B0950+08 and the Type A medium are -4.55 and -3.5, respectively.

Nonetheless, the value reported by Phillips and Clegg (1992), while about a factor of 10 lower than the average value of $\langle C_N^2 \rangle$ for the interstellar medium, was larger than might have been expected. The basis of this statement is the formula for C_N^2 in (2). The electron density in the local bubble is at least a factor of 5 less than that in the DIG (line-of-sight average of 5×10^{-3} cm^{-3} rather than 2–3 $\times 10^{-2}$ cm^{-3} for the DIG), and the outer scale l_0 is probably larger than that in the DIG, so $\langle C_N^2 \rangle$ should arguably be even smaller than that reported by Phillips and Clegg (1992). The fact that it is not might indicate that the Local Bubble, despite being of a lower electron density than the general interstellar medium, is more turbulently "agitated". A more mundane possibility is that the measurements of B0950+08 are "contaminated" by denser plasma along the line of sight (see Sect. 8).

5 Relevant Recent Progress in Pulsar Astronomy

It has now been sixteen years since the paper by Phillips and Clegg (1992), and we should have a clearer idea of the properties of turbulence in the Local Bubble. There are several developments that are particularly relevant in this respect.

1. The most important development has been the completion of the Very Long Baseline Array (VLBA) of the National Radio Astronomy Observatory[1], and the full achievement of its astrometric capability (Brisken et al. 2000, 2002). This has permitted precise parallaxes to be measured for many pulsars. The distance to a pulsar obviously allows us to determine what regions of the interstellar medium lie along the line of sight.
2. The number of pulsars known has increased from 558 (Taylor et al. 1993) at the time of Phillips and Clegg (1992) to 1775 at the writing of this paper. A catalog of the properties of all known pulsars is maintained by the pulsar group at the Australia Telescope National Facility (ATNF) under Richard Manchester in a published form (Manchester et al. 2005) as well as an updated Internet site (http://www.atnf.csiro.au/research/pulsar/psrcat/). This catalog is an important research aid in its own right, and provided data for this paper.
3. Lallement et al. (2003) have published a three-dimensional map of the Local Bubble, using NaD absorption data (a diagnostic of the neutral phase) for 1005 lines of sight to relatively nearby stars. This study has provided an accurate representation of the structure of the bubble.
4. During this period there has been steady progress in our understanding of pulsars in the galactic environment, including the information they contain about the interstellar medium. A prime example of a relevant recent discovery which has particular promise is discussed in Sect. 9 below.

6 A New Set of Pulsars to Probe the Local Bubble

We used the ATNF catalog to select a set of pulsars which can be used to probe the properties of the plasma in the Local Bubble. The information extracted from the catalog included, most importantly, the parallaxes measured by Brisken et al. (2000, 2002). The criteria used in selecting these pulsars were as follows.

1. The distance to the pulsar had to be less than 500 parsecs.
2. The part of the sky in which the pulsar is observed was required to be in the direction of the Local Bubble.

The purpose of these selection criteria was to select pulsars whose lines of sight are, to the greatest possible extent, within the Local Bubble. The desired goal was to find a pulsar whose line of sight lies entirely within the Local Bubble, so that its radio propagation measurements would diagnose the plasma in that region.

The pulsars so selected are given in Table 1. Column 1 gives the pulsar name and Column 2 gives its distance. The parameters in the remaining columns are described in the next section. In addition to distance and direction information, measurements of dispersion measure (DM) and Faraday rotation measure (RM) were taken from the ATNF catalog. Data on ISS parameters were taken from elsewhere in the literature. The final row gives parameters for the Type A scattering medium, to which the properties of the Local Bubble lines of sight are compared.

[1]The Very Long Baseline Array is an instrument of The National Radio Astronomy. NRAO is a facility of the National Science Foundation, operated under cooperative agreement with Associated Universities Inc.

7 Results on Plasma and Plasma Turbulence in the Local Bubble

Using the new, VLBA-provided data on pulsar distances in Table 1, we can use published measurements of propagation and scattering effects to obtain measurements of path-averaged plasma properties such as the plasma density (from the pulsar dispersion measure), line-of-sight component of the magnetic field (from Faraday rotation measure), and $\langle C_N^2 \rangle$ (from ISS phenomena such as spectral corrugation or pulse broadening).

The dispersion measure (DM) is defined as

$$DM \equiv \int_{LOS} n_e \, ds = \bar{n_e} L. \tag{4}$$

If n_e is expressed in cm^{-3} and ds in parsecs, DM has the customary, non-cgs units of pc-cm^{-3}. The variable L is the effective path length (parsecs) through the medium. The Faraday rotation measure (RM) can be expressed as (Minter and Spangler 1996)

$$RM \equiv 0.81 \int_{LOS} n_e \mathbf{B} \cdot \mathbf{ds} \tag{5}$$

with n_e and ds again in cm^{-3} and parsecs, and $|B|$ in microgauss. The odd coefficient 0.81 is the transformed version of the conventional set of fundamental constants which appears in Faraday rotation formulas, when this set of units is chosen. The units of RM are rad/m^2.

An estimate of the path-averaged, line-of-sight component of the interstellar magnetic field is

$$\bar{B}_{\parallel} = \frac{\int_{LOS} n_e \mathbf{B} \cdot \mathbf{ds}}{\int_{LOS} n_e \, ds} = 1.24 \frac{RM}{DM}. \tag{6}$$

In (6), $\bar{B}_{\parallel}$ is in microgauss. Formulas (4) and (6) are used to obtain $\bar{n_e}$ and $\bar{B}_{\parallel}$, which are given in columns 3 and 4 of Table 1. Measurements of scintillation parameters (such as the angular width of a broadened image) are proportional to a quantity called the scattering measure (SM) which is defined in a manner similar to dispersion measure and rotation measure,

$$SM \equiv \int_{LOS} C_N^2 \, ds = \langle C_N^2 \rangle L. \tag{7}$$

In the simplest case, $\langle C_N^2 \rangle$ is obtained by dividing the measured SM by the path length through the medium. The process is usually more complicated in the case of pulsars, which are embedded in the ISM.

The values for $\langle C_N^2 \rangle$ were taken from three different sources for the three pulsars with ISS measurements. These are Phillips and Clegg (1992) for B0950+08, Bhat et al. (1998)

Table 1 Pulsar probes of the local bubble

Pulsar	Distance (pc)	$\bar{n_e}$ (cm^{-3})	$\bar{B}_{\parallel}$ (μG)	log ($\langle C_N^2 \rangle$)	Corr.	log ($\langle C_N^2 \rangle_A$)
B0950+08	262	0.011	0.57	−4.55	A+B	−5.42
B1133+16	350	0.014	1.00	−3.32	B	−3.82
J0437-4715	170	0.016	0.70	...	...	...
B0809+74	433	0.014	2.37	−5.16	none	−5.16
Type A medium	...	0.025	2.0	−3.50	A+B	−4.38

for B1133+16, and Rickett et al. (2000) for B0809+74. The published values taken from these papers are given in column 5 of Table 1.

These values cannot be directly compared with each other, or with the Type A background value of $\langle C_N^2 \rangle$, as I now discuss. The measurements of B0950+08 by Phillips and Clegg (1992) were made in the same way, and analysed with the same formulas as Cordes et al. (1985); the factor of 10 difference in $\langle C_N^2 \rangle$ between B0950+08 and the general Type A background appears to be a sound result.

Bhat et al. (1998) made repeated ISS observations of all pulsars in their sample, with each pulsar being observed from 10–90 times. They found that the fundamental ISS observable, the decorrelation bandwidth ($\Delta\nu_d$), varied markedly from one observing session to another. The reasons for this variability are not entirely clear, but appear to be due to distortion of diffractive ISS processes by refractive ones. To minimize this variability, Bhat et al. (1998) averaged the two dimensional autocorrelation functions from the dynamic spectra to obtain a Global Average Correlation Function (GACF) prior to fitting for the decorrelation bandwidth. With the resulting value of $\Delta\nu_d$, they used the same formula as Cordes et al. (1985) to obtain a value of $\langle C_N^2 \rangle$.

As discussed by Bhat et al. (1998), and as is clearly visible in Table 2 of their paper, there is a systematic bias in the values of $\langle C_N^2 \rangle$ relative to those of Cordes et al. (1985), in the sense that the values of $\langle C_N^2 \rangle$ from the GACF correlation functions are lower. This bias is presumably due to the aforementioned, poorly-understood refractive bias. We calculated the difference in $\log\langle C_N^2 \rangle$ between Bhat et al. (1998) and Cordes et al. (1985) for the 14 pulsars in common, and obtained a mean value of 0.37, with a standard deviation of 0.35. If $\langle C_N^2 \rangle$ data from Phillips and Clegg (1992) and Cordes et al. (1985) are to compared with Bhat et al. (1998), this correction for refractive bias (hereafter indicated by the unimaginative name "Correction A") must be applied to the former data sets.

The value of $\langle C_N^2 \rangle$ for B0809+74 was taken from Rickett et al. (2000). Scintillation parameters were obtained from measurements of intensity scintillations at a frequency of 933 MHz, rather than the autocorrelation function of a dynamic spectrum. Rickett et al. (2000) obtain a smaller value for the scattering measure, and thus $\langle C_N^2 \rangle$, from these data than had previous observations. Rickett et al. (2000) argue that previous investigations, including Phillips and Clegg (1992), Cordes et al. (1985), and Bhat et al. (1998) had used an inappropriate value for the multiplicative constant in a formula which converted from $\Delta\nu_d$ to $\langle C_N^2 \rangle$ (equation (6) of Cordes et al. (1985)). To correct $\langle C_N^2 \rangle$ for this effect, Rickett et al. (2000) suggest multiplying the published values of $\langle C_N^2 \rangle$ by a factor of 0.315, or adding -0.50 to $\log\langle C_N^2 \rangle$. I refer to this as "Correction B".

These corrections are reported in columns 6 and 7 of Table 1. Column 6 indicates which (if any) correction should be applied to the published value of $\langle C_N^2 \rangle$ to obtain a consistent value. Column 7 gives the corrected value of $\langle C_N^2 \rangle$. The adjusted values are noted by $\langle C_N^2 \rangle_A$. The values in this column are, one hopes, directly comparable.

The following conclusions may be drawn from Table 1.

1. The crucial pulsar B0950+08 is now known to be twice as far as indicated by earlier parallax measurements. This result comes from the improved parallax measurements of Brisken et al. (2000). This means that much of the line of sight lies beyond the probable back wall of the Local Bubble, and renders the implications of the B0950+08 data for Local Bubble plasma physics much less certain.
2. The tremendous increase in the number of known pulsars over the last 15 years did not assist the present endeavor. Most of the new pulsars were found in the direction of the galactic center, and had evaded earlier detection because heavy interstellar scattering broadened the pulses sufficiently to make the pulsars undetectable. The Local Bubble, on

the other hand, lies in the direction of the galactic anticenter, where scattering is much less pronounced and earlier surveys had found the pulsars which can be detected.

3. Turning to the plasma parameters in Table 1, the mean electron densities $\bar{n_e}$ are lower than the standard value for long lines of sight through the interstellar medium ($\bar{n_e} \simeq 0.025$), but they are substantially larger than expected for the hot gas in the Local Bubble, which would be $n_e \simeq 0.005$. This probably results from the lines of sight being "contaminated" by denser plasma outside the Local Bubble.
4. Intriguingly, the inferred values for the line of sight component of the magnetic field are only slight less, or completely consistent with, lines of sight through the general interstellar medium.
5. The scattering for two of the pulsars (B0950+08 and B0809+74) is less than on typical lines of sight through the ISM. However, $\langle C_N^2 \rangle$ is not depressed by as much as one would expect on the basis of the lower density in the bubble. Furthermore, one of the three pulsars (B1133+16) has a completely "nominal" scattering measure and associated value of $\langle C_N^2 \rangle$. In considering the scattering measurements, we are again confronted with the question of how much of the measured propagation effects are due to plasmas outside the Local Bubble, and how much is due to plasma within the Local Bubble.

8 Estimates of Other Media Along Pulsar Lines of Sight

From the summarizing points of the previous section, we conclude that either the Local Bubble has higher density, turbulent intensity, etc., than expected, or the measurements are being contaminated by other media along the lines of sight. In this section we briefly discuss what we know of these other media, and estimate the magnitude of the contamination.

The first medium to consider is the denser ISM behind the Local Bubble. The lengths of the lines of sight given in Table 1 are larger than estimates of chord lengths in the Local Bubble. To estimate the degree of contamination by the ISM beyond the Local Bubble, we used the meridian slices through the Local Bubble and its vicinity published in Lallement et al. (2003). The slices shown in Figs. 6–8 of Lallement et al. (2003) were used to plot the length and orientation of each pulsar's line of sight. In no case did the line of sight remain entirely within the Local Bubble. It appears to be the case that no known pulsar is sufficiently close, and in the right direction, to be entirely within the Local Bubble. Typically, about half of the line of sight appears to be in the denser (presumably typical) ISM behind the far wall of the Local Bubble. These rough estimates appear consistent with the result from Table 1 that the line-of-sight averaged densities are about half of what is typically observed for pulsars on longer pathlengths. They are also consistent with the normal, or nearly normal, values of $\bar{B}_{\|}$ for the pulsars in Table 1.

The values of $\langle C_N^2 \rangle_A$ given in Table 1 for B0950+08 and B0809+74 appear to be lower, relative to the adopted mean for the Type A medium, by more than a factor of 0.50 (0.30 in the logarithm). The portion of the line of sight from the far wall to the pulsar is at distances above the galactic plane of order 100 parsecs and greater for both pulsars. The intensity of interstellar turbulence at these altitudes may be less than in the galactic plane. Furthermore, the portion of the line of sight in the denser ISM behind the wall is between the pulsar and the midpoint of the line of sight. The properties of intensity scintillations are dependent on the location of the turbulence along the line of sight; turbulent plasma close to the pulsar has less of an effect than plasma at the midpoint.

The other medium known to be on the lines of sight to these pulsars consists of clouds relatively close to the Sun and embedded in the Local Bubble, such as the LIC cloud and G cloud. These have been discussed recently by Frisch (2007) and Redfield and Linsky (2008). To estimate the contributions of these clouds to DM, RM, and SM for the pulsars, we used the mean cloud properties given in Sect. 4.2 of Redfield and Linsky (2008). We assume that these clouds have an electron density $n_e = 0.12$ cm^{-3} (Redfield and Falcon 2008), radius $R_c = 1.5$ parsecs, and that they fill space out to a distance of $L_c = 15$ parsecs from the Sun with a filling factor $0.055 \leq f \leq 0.19$. Rotation measures were calculated with the same assumed parameters, and an adopted value of $B_{\|} = 4$ μG. Since the magnitude of the interstellar magnetic field in the DIG is estimated to be 4 μG, the calculated RMs will be upper limits.

The turbulence parameter C_N^2 was calculated via (2), using the same estimate for n_e as used for DM and RM. The outer scale to the turbulence l_0 in the local clouds is totally unknown, but it is not unreasonable to assume that it is a fraction of the radius of the clouds, so we adopt $l_0 \simeq R_c/3 \simeq 0.5$ pc. Equation (2) shows that the value of C_N^2 is not strongly dependent on the assumed value of l_0. The modulation index ϵ is also an "imponderable". In the absence of other information, we adopt values measured for the density turbulence in the solar wind. Spangler (2002) and Spangler and Spitler (2004) obtained estimates for this quantity in the slow solar wind, utilizing radio remote sensing observations for the solar wind in the heliocentric distance range 16–26 $R_\odot$, and spacecraft measurements at 1 au. Based on these results, we adopt an estimate of $0.050 \leq \epsilon \leq 0.15$.

With these assumed parameters, our estimates for the contribution of the local clouds to the pulsar propagation parameters are as follows. Calculated dispersion measures are in the range $0.10 \leq DM \leq 0.34$ pc cm^{-3}, depending on the assumed value of the filling factor. The observed dispersion measures for the pulsars in Table 1 range from 2.6 to 6.1 pc cm^{-3} (Manchester et al. 2005). The local clouds might make a small contribution to the observed dispersion measure, but most of DM must come from the local bubble or the ISM beyond the Local Bubble.

Estimated local cloud values for RM are in the range 0.32–1.11 rad/m^2, again dependent on the assumed filling factor. The observed RMs have absolute values between 1.35 and 11.7 rad/m^2, with three of four pulsars having $|RM| < 4$ rad/m^2 (Manchester et al. 2005). These RM estimates should be considered upper limits; the true values will be reduced by the unknown but generally nonzero angle between the line of sight and the magnetic field in the clouds. Nonetheless, it is possible that the local clouds could make a significant contribution to the rotation measures for these pulsars.

Our estimates of the turbulence parameter C_N^2 in the clouds are $-4.0 \leq \log C_N^2 \leq -3.0$. These values are mostly larger than those given in column 7 of Table 1, although a direct comparison is misleading. The values of $\langle C_N^2 \rangle_A$ given in Table 1 are averages over paths of several hundred parsecs. Since the effective path lengths through the local clouds are very small compared to the typical pulsar distance, the effective value of $\langle C_N^2 \rangle_A$ due only to these clouds would be much smaller than the local value of C_N^2. To roughly estimate this correction, we multiplied the above, calculated values of C_N^2 by a factor of $f L_c/D$. This is equivalent to taking the SM due to the clouds and dividing by the distance to the pulsar. These estimates of the effective cloud values of $\langle C_N^2 \rangle_A$ show a very wide range, reflecting uncertainty in both the cloud filling factor as well as the modulation index ϵ. The maximum possible value is comparable to the observed $\langle C_N^2 \rangle_A$ for both B0950+08 and B0809+74. The observed scattering to B1133+16 is larger than can be plausibly accounted for by the local clouds.

This exercise indicates that the observations of ISS for two of the pulsars in Table 1 (B0950+08 and B0809+74) are not incompatible with a significant role for turbulence in the set of local clouds described by Frisch (2007) and Redfield and Linsky (2008). It is worth noting that such a role is also strongly indicated by the interpretation of the intraday variability of three compact quasars by Linsky et al. (2008).

This section may be summarized as follows. The four pulsars in Table 1 furnish our best chance of diagnosing the plasma and plasma turbulence in the Local Bubble by radio propagation effects. However, the lines of sight to these pulsars pass not only through the Local Bubble, but also the system of small clouds close to the Sun and the general ISM beyond the far wall of the Local Bubble. The estimates presented in this section indicate that the contributions of these two media to DM, RM, and SM, separately as well as together, are at least plausibly consistent with the measured values. While further investigation is certainly warranted, it is difficult to reject the null hypothesis that all of the radio propagation effects occur in media outside the Local Bubble.

9 New Diagnostic Measurements for Local Interstellar Plasmas

The previous section remains inconclusive because we cannot be sure how much of an observed propagation effect is due to plasma within the Local Bubble, and how much is due to (arguably less interesting) plasma outside the bubble. In the last few years, new observational advances have been made which hold the prospect of progress. A particularly promising new observational development is the discovery of "arcs" in pulsar "secondary spectra". Secondary spectra are defined in the next paragraph. This intriguing arc phenomenon was discovered by Daniel Stinebring and his students at Oberlin College. Recent expositions and physical discussions of the phenomenon are given by Walker et al. (2004) and Cordes et al. (2006).

For many years, pulsar astronomers have studied pulsar dynamic spectra, which are two dimensional plots of the pulsar intensity as a function of frequency and time, $I(\nu, t)$. For at least 20 years, it has been realized that these plots often showed variations which are not entirely random, but instead consisted of patterns which suggested constructive and destructive interference between a few rays.

Something remarkable was noted when Stinebring and colleagues examined the "secondary spectrum", which consists of Fourier transforming the dynamic spectrum. The secondary spectrum may be considered a plot of the intensity as a function of the conjugate variables to frequency and time, f_ν and f_t respectively. What was found for many pulsars is most of energy is concentrated on parabolic arcs described by the simple relation

$$f_\nu = a f_t^2. \tag{8}$$

In general, pulsars have more than one such parabolic arc in the secondary spectrum. Clear examples are shown, for example, in Walker et al. (2004) and Cordes et al. (2006). These papers also demonstrate that such behavior is easily understood as interference between a non-scattered central ray from the pulsar to an observer, and a scattered secondary ray. Each pair of rays produces its own parabolic arc.

For these arcs to be present, it is necessary for there to be relative motion between the medium containing the turbulent irregularities, the pulsar, and the observer. It also appears to be necessary, rather unexpectedly, for the turbulent density fluctuations to be concentrated in relatively narrow sheets or screens along the line of sight. For many years the "thin screen approximation" has been employed in the theory of wave propagation in a random medium, in which the turbulent medium is idealized as being concentrated in a small fraction of the line of sight. It now seems to be the case that this approximation is physically accurate for the interstellar medium.

The coefficient of the parabolic arc contains information on the location of the screen responsible for the pulsar arcs. The coefficient a in (2) may be shown to be (Cordes et al. 2006)

$$a = \frac{Ds(1-s)\lambda^2}{2cV_\perp^2} \tag{9}$$

where D is the distance to the pulsar, s is the fractional distance of the screen along the line of sight (i.e. sD is the distance from the pulsar to the screen), c is the speed of light, and $V_\perp$ is the speed of the observer relative to the scintillation pattern, in a direction perpendicular to the direction of the pulsar. The wavelength of observation is λ.

The discovery of pulsar arcs opens the possibility of determining what fraction, if any, of the scattering of the pulsars in Table 1 is due to turbulence in the Local Bubble. One of the pulsars, B1133+16, has prominent arcs in its secondary spectrum. These data and their analysis are discussed by Trang and Rickett (2007). The curvature a of the primary arc gives a screen distance of 140 parsecs from Earth, which may place it within the Local Bubble. The charts of Lallement et al. (2003) are not entirely clear as to whether a point on the line of sight to B1133+16 140 parsecs from the Sun is within the Local Bubble, or in the denser surrounding medium. If it can be determined that this point is within the Local Bubble, the ISS observations of Trang and Rickett (2007) are of great interest to the topic of this paper.

Observations of ISS for B0950+08 and B0809+74 have been published, but the data do not seem to have had secondary spectra calculated, and analyses of the sort of interest here have apparently not been made. The archived data could be reanalysed with this in mind, or new observations could be made. In view of the suggestive results for B1133+16, it would be highly desirable to search for pulsar arcs for all the pulsars in Table 1. These data could indicate what part of the scattering is due to plasmas in the Local Bubble, and what part is due to more remote plasmas.

Another technique with considerable promise is that of intraday variability of highly compact extragalactic radio sources (Rickett 2007). A disadvantage of extragalactic radio sources with respect to pulsars is that the lines of sight pass all the way out of the Galaxy, so there is no prospect of having a line of sight begin and end within the Local Bubble. A compensating advantage of extragalactic sources is their high surface density in the direction of the galactic anticenter and the Local Bubble, where very few pulsars are to be found. Furthermore, a model-dependent estimate of the distance to the scattering plasma can be made from these intensity scintillation measurements. Intraday variability observations of three radio quasars were used by Linsky et al. (2008) to indicate that the turbulence responsible for radio scattering comes from the ionized edges of a number of local clouds within the Local Bubble. A unique aspect of the Linsky et al. (2008) study was the ability to identify the three dimensional velocities of the various clouds with the observed "pattern speed" of the scintillations with respect to the radio telescope.

It is worth noting that the successful modeling of the quasar intraday variability by Linsky et al. (2008) indicates that the density fluctuations responsible for this component of ISS are confined to sheets or screens in the local clouds or boundaries between local clouds. This conclusion is similar to that drawn for the pulsar arcs discussed above, and may reveal an important and general property of astrophysical turbulence. In view of the success of the intraday variability technique, further observations should be carried out. Once again, it would be interesting to determine how much of the scattering measure can be attributed to these clouds, and how much might be contributed by other plasmas in the Local Bubble and beyond.

10 Summary and Conclusions

1. There are at least four pulsars which are potential probes of the plasma in the Local Bubble. These pulsars are listed in Table 1.
2. Even these pulsars are probably too far away to have their lines of sight entirely "fit" within the Local Bubble. Unfortunately, it seems to be the case that there are no radio pulsars which are close enough to have the entire line of sight within the bubble.
3. Two of the three pulsars with ISS observations have lower values of scattering measure, and thus the scattering parameter $\langle C_N^2 \rangle$, than is typical for pulsars without special lines of sight.
4. A significant, or even dominant portion of the observed dispersion measure, rotation measure, and scattering measure values to these pulsars may arise in the denser ISM beyond the far wall of the Local Bubble, or in the local clouds in the vicinity of the Sun. The local clouds may contribute a significant portion of the observed ISS.
5. The one pulsar which has had its scintillation data analysed for the presence of "pulsar arcs" (B1133+16) has a prominent arc that might be caused by a turbulent screen within the Local Bubble.
6. If most of the scattering of B1133+16 is due to plasma within the Local Bubble, it would present a challenge to our understanding of interstellar turbulence in two ways. First, the intensity of the turbulence, parameterized by the coefficient $\langle C_N^2 \rangle$ is unexpectedly high. Second, it is hard to understand how or why plasma turbulence in the hot, high β plasma of the Local Bubble would be concentrated in a thin sheet within the bubble.
7. The remaining three pulsars in Table 1 should be examined for arcs in their secondary spectra, either by reanalysing archival data, or by making new observations.
8. Additional, targeted observations of highly compact extragalactic sources for intraday variability could provide similar information on the location of turbulent plasma which might be located within the Local Bubble.

Acknowledgements This work was supported by the National Science Foundation through grant ATM03-54782 to the University of Iowa.

References

A. Barnes, Phys. Fl. **9**, 1483–1495 (1966)
N.D.R. Bhat, Y. Gupta, A.P. Rao, Astrophys. J. **500**, 262–279 (1998)
W.F. Brisken, J.M. Benson, A.J. Beasley, E.B. Fomalont, W.M. Goss, S.E. Thorsett, Astrophys. J. **541**, 959–962 (2000)
W.F. Brisken, J.M. Benson, W.M. Goss, S.E. Thorsett, Astrophys. J. **571**, 906–917 (2002)
B.D.G. Chandran, D.C. Backer, Astrophys. J. **576**, 176–187 (2002)
J.M. Cordes, J.W. Weisberg, V. Boriakoff, Astrophys. J. **288**, 221–247 (1985)
J.M. Cordes, B.J. Rickett, D.R. Stinebring, W.A. Coles, Astrophys. J. **637**, 346–365 (2006)
D.P. Cox, R.J. Reynolds, Ann. Rev. Astron. Astrophys. **25**, 303–341 (1987)
P.C. Frisch, Space Sci. Rev. **130**, 355–365 (2007)
R. Lallement, B.Y. Welsh, J.L. Vergely, F. Crifo, D. Sfeir, Astron. Astrophys. **411**, 447–464 (2003)
J.L. Linsky, B.J. Rickett, S. Redfield, Astrophys. J. **675**, 413–419 (2008)
R.N. Manchester, G.B. Hobbs, A. Teoh, M. Hobbs, Astron. J. **129**, 1993–2006 (2005)
A.H. Minter, S.R. Spangler, Astrophys. J. **458**, 194–214 (1996)
M. Opher, E.C. Stone, P.C. Liewer, Astrophys. J. **640**, L71–L74 (2006)
J.A. Phillips, A.W. Clegg, Nature **360**, 137–139 (1992)
S. Redfield, J.L. Linsky, Astrophys. J. **673**, 283–314 (2008)
S. Redfield, R.E. Falcon, Astrophys. J. (2008, in press). ArXiv:0804.1802
B.J. Rickett, Ann. Rev. Astron. Astrophys. **28**, 561–605 (1990)

B.J. Rickett, W.A. Coles, J. Markkanen, Astrophys. J. **533**, 304–319 (2000)
B.J. Rickett, in *SINS: Small Ionized and Neutral Structures in the Interstellar Medium*, ed. by M. Haverkorn, W.M. Goss. Astronomical Society of the Pacific Conferences Proceedings Series, vol. 365 (2007), pp. 207–218
S.R. Spangler, J.M. Cordes, Astrophys. J. **505**, 766–783 (1998)
S.R. Spangler, Astrophys. J. **522**, 879–896 (1999)
S.R. Spangler, Astrophys. J. **576**, 997–1004 (2002)
S.R. Spangler, L.G. Spitler, Phys. Plasm. **11**, 1969–1977 (2004)
S.R. Spangler, in *Turbulence and Nonlinear Processes in Astrophysical Plasmas*, ed. by D. Shaikh, G.P. Zank. American Institute of Physics Conference Proceedings Series, vol. 932 (2007), pp. 85–95
J.H. Taylor, R.N. Manchester, A.G. Lyne, Astrophys. J. Suppl. **88**, 529–568 (1993)
F.S. Trang, B.J. Rickett, Astrophys. J. **661**, 1064–1072 (2007)
M.A. Walker, D.B. Melrose, D.R. Stinebring, C.M. Zhang, Mon. Not. R. Astron. Soc. **354**, 43–53 (2004)

Exotic Clouds in the Local Interstellar Medium

Snežana Stanimirović

Originally published in the journal Space Science Reviews, Volume 143, Nos 1–4, 291–301.
DOI: 10.1007/s11214-008-9363-y

Abstract The neutral interstellar medium (ISM) inside the Local Bubble (LB) has been known to have properties typical of the warm neutral medium (WNM). However, several recent neutral hydrogen (HI) absorption experiments show evidence for the existence of at least several cold diffuse clouds inside or at the boundary of the LB, with properties highly unusual relative to the traditional cold neutral medium. These cold clouds have a low HI column density, and AU-scale sizes. As the kinematics of cold and warm gas inside the LB are similar, this suggests a possibility of all these different flavors of the local ISM belonging to the same interstellar flow. The co-existence of warm and cold phases inside the LB is exciting as it can be used to probe the thermal pressure inside the LB. In addition to cold clouds, several discrete screens of ionized scattering material are clearly located inside the LB.

The cold exotic clouds inside the LB are most likely long-lived, and we expect many more clouds with similar properties to be discovered in the future with more sensitive radio observations. While physical mechanisms responsible for the production of such clouds are still poorly understood, dynamical triggering of phase conversion and/or interstellar turbulence are likely to play an important role.

Keywords Interstellar medium: Physical properties · Interstellar medium: Solar neighborhood

1 Introduction

The diffuse interstellar medium (ISM) in the Galaxy contains structure over a wide range of spatial scales. While on scales $\gtrsim 1$ pc we can observationally trace the entire hierarchy of structures, the extremely small-scale end of this spectrum, on scales from < 1 pc to tens of AUs, is still largely unexplored. In terms of the diversity of physical properties, the diffuse neutral ISM is traditionally viewed as two types of "clouds", referred to as the Cold Neutral

S. Stanimirović (✉)
Department of Astronomy, University of Wisconsin, 475 N. Charter Street, Madison, WI 53706, USA
e-mail: sstanimi@astro.wisc.edu

Medium (CNM) and the Warm Neutral Medium (WNM). The CNM and the WNM have very different temperature and volume densities, however they co-exist spatially and are expected to be (from a theoretical point of view), at least locally, in pressure equilibrium. The CNM has a temperature of $T \sim 70$ K and a hydrogen volume density $n \sim 40$ cm^{-3}, while the WNM has $T \sim 5000$ K and $n \sim 0.5$ cm^{-3} (Heiles and Troland 2003). In terms of HI column density, the CNM has typically 5×10^{19} cm^{-2} (mean value), and the WNM has 1.3×10^{20} cm^{-2} (Heiles and Troland 2003). The CNM clouds were traditionally assumed to be spherically symmetric with a size of 1–2 pc. Interestingly, recent observations are finding much larger aspect ratios.

In this paper, we focus on the structure in the diffuse ISM on spatial scales of a few tens to a few thousands of AUs. We show that these features have observationally inferred properties not in accord with the traditional ISM picture, and therefore sample flavors of the most exotic diffuse ISM. In particular, we discuss the tiny scale atomic and ionized structures, and CNM clouds with very low HI column densities. Several of these exotic features are known to be very close by, inside or at the boundary of the Local Bubble (LB). The most intriguing question is whether these exotic clouds exist because they are related to the LB, or the LB has nothing to do with their existence and physical properties. We also explore how the exotic ISM compare with warm clouds within the LB, and look briefly into mechanisms that could play an important role for the formation and survival of clouds within the Local Cavity.

2 Tiny Scale Atomic Structure (TSAS)

The phenomenon of Tiny Scale Atomic Structure (TSAS) was discovered in the late 1970s when interferometric observations revealed an atomic cloud with a size of 70 AUs (Dieter et al. 1976). When compared with traditional CNM clouds, TSAS has very exotic properties: a typical observed neutral hydrogen (HI) column density of 10^{18-19} cm^{-2} and a (typical) size of about 30 AU, imply a volume density $n = 10^{3-4}$ cm^{-3} and a thermal pressure of $P/k = nT = 10^{4-6}$ cm^{-3} K. This is at least two orders of magnitude higher that what is expected for the traditional CNM. TSAS has caused a lot of controversy since its discovery as it obviously has properties grossly outside of the range expected for the ISM in pressure equilibrium. Yet, TSAS continues to be frequently encountered observationally.

How do we observe TSAS? There are several different observational methods, one being temporal variability of HI absorption spectra against pulsars. Pulsars have high transverse velocities, and typically travel over 5–50 AU per year. By comparing HI absorption spectra obtained at two different epochs we can probe density inhomogeneities on AU-scales in the CNM clouds that happen to lie between us and pulsars. This is a powerful technique, however the difference in absorption spectra we are after are tiny and therefore high resolution and sensitivity observations are essential, as well as careful instrumental calibration and data reduction.

Results from the latest study of TSAS using pulsars are presented in Stanimirović et al. (2008) and Weisberg and Stanimirović (2007). These studies show that one pulsar in particular, B1929 + 10 (Fig. 1), exhibits significant changes in HI absorption spectra at several observing epochs. The significant fluctuations, with an estimated likelihood for being real of 90–100% (for the full analysis please see Stanimirović et al. 2008), correspond to TSAS with a typical velocity FWHM of 1 km s^{-1}, $\Delta N(\mathrm{HI}) = 10^{18}$ cm^{-2}, and a size $L = 30$–45 AU. If TSAS is assumed to be spherically symmetric then, $n =$ a few $\times 10^3$ cm^{-3} and thermal

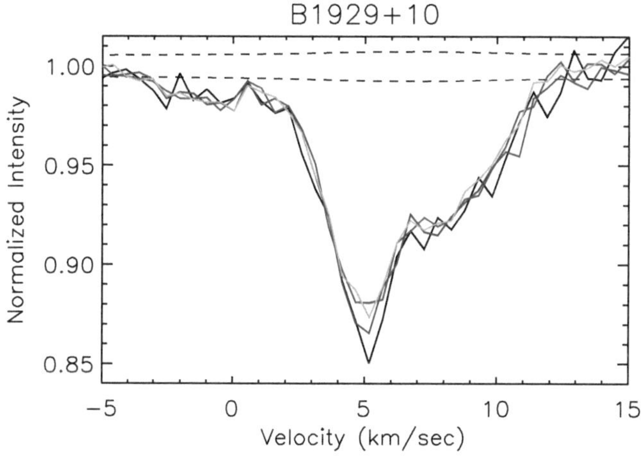

Fig. 1 HI absorption spectra in the direction of B1929 + 10 ($l = 47°$, $b = -3.9°$) obtained with the Arecibo telescope at four observing epochs. *Dashed lines* show a typical, $\pm 1 - \sigma$ noise level in the absorption spectra. The absorption feature at 5 $\mathrm{km\,s^{-1}}$ has changed significantly (2–3-σ level) with time. For full analysis please see Stanimirović et al. (2008)

pressure is $P/k =$ a few $\times 10^4\ \mathrm{cm^{-3}}$ K. However, a geometrical elongation along the line-of-sight, by a factor of $\sim$10, would be able to bring the volume density and thermal pressure to what is typically expected for the CNM.

B1929 + 10 is the closest pulsar in our sample, at a distance of 361 (+8, −10) pc (Chatterjee et al. 2004). Interestingly, the line-of-sight toward B1929 + 10 runs along the elongated finger of dense neutral gas bounding the LB, based on Na I observations by Lallement et al. (2003). Several studies, including Phillips and Clegg (1992), reported enhanced turbulence and scattering at the boundaries of the LB. As half of the line-of-sight to B1929 + 10 is either inside or along the wall of the Local Bubble, and also because this is the only pulsar in our sample that shows consistently significant fluctuations, the change in HI absorption spectra probably originates in either the presence of TSAS inside the LB or the enhanced turbulent fluctuations along the wall of the LB.

3 Tiny Scale Ionized Structure (TSIS)

Analogues of TSAS exist in the ionized medium, and are called Tiny Scale Ionized Structures or TSIS. This phenomenon is observed by measuring the scattering of pulsar signals as they propagate through the intervening ionized medium. For excellent reviews of this phenomenon see Rickett (2007) and Stinebring (2007). In order to explain the observed scattering properties, two phenomena are needed in the intervening ionized medium: (i) a smooth density distribution, with random fluctuations described with a Kolmogorov spectrum; and (ii) discrete structures located along the line-of-sight, with a thickness of a few tens of parsecs (observed as "scintillation arcs" which were discovered by Stinebring et al. 2001). The discrete structures are often referred to as "thin scattering screens" to describe their localized and discrete nature. Furthermore, sometimes scattering screens have well-defined sub-structure ("arclets"), in the form of isolated (ionized) density fluctuations, with a size of about 1 AU and a volume density of about 100 $\mathrm{cm^{-3}}$ (Hill et al. 2005). The volume density of arclets is more than 10^3 times larger than the average electron density in the ISM and implies a high ionization fraction, with $T \gtrsim 10^4$ K and $P \sim 10^6$ cm^{-3} (Heiles and Stinebring 2007).

Scintillation properties have been measured in directions to many pulsars and thin scattering screens are frequently found. If the distance to the pulsar and its proper motion are known, the location of screens along the line-of-sight can be determined. Stinebring (2007) summarizes the locations of currently known screens, and it is interesting to note that at least 6–7 screens are located inside the LB. In particular, pulsars B1929 + 10 and B0823 + 26 each have 3 screens inside the LB, while B1133 + 16 has a single screen at a distance of 140 pc (Trang and Rickett 2007).

4 Low Column Density Clouds

Braun and Kanekar (2005) and Stanimirović and Heiles (2005) recently revealed a new population of CNM clouds which have HI column densities about 50–100 times lower than typical CNM clouds. We call these clouds the "thinnest" or low-N(HI) clouds. Interestingly, in terms of volume density and size, low-N(HI) clouds occupy a regime between the traditional CNM clouds and the extreme TSAS. The work that motivated the discovery of a large sample of low-N(HI) clouds is the Millennium survey (Heiles and Troland 2003), which observed 79 continuum sources to establish temperature and column densities of the CNM and the WNM. Heiles and Troland (2003) noticed that 20% of their sight lines indicated either no or a very low column density of CNM. Stanimirović and Heiles (2005) and Stanimirović et al. (2007) used the Arecibo telescope to observe about 20 continuum sources which either had no previously detected CNM or very weak CNM features. In total, 23 new CNM clouds were detected, with a typical integration time of the order of a few hours per source. Essentially, with long integration times weak HI absorption lines emerge easily. As an example, Fig. 2 shows one of the sources, 3C264.0, where two new CNM clouds were found with a peak optical depth of 5×10^{-3} and 3×10^{-3}, or an HI column density of 9×10^{18} and 5×10^{17} cm^{-2}, respectively. Clearly, these HI column densities are among the lowest ever detected for the CNM. Another, not so extreme example is shown in Fig. 3, where a single new CNM cloud is detected at a LSR velocity of -0.8 km s^{-1}.

There are several important points emerging from recent studies. (i) The detection rate of the Stanimirović et al. (2007) experiment is very high, suggesting that weak HI absorption features are common in the ISM. Essentially, we just need to integrate long enough. (ii) The preliminary addition of the new population of CNM clouds to the "traditional" CNM population by Heiles and Troland (2003) resulted in the combined column density histogram that can be easily fitted with a single function, $\propto N(\mathrm{HI})^{-1}$. This suggests that low-N(HI) clouds most likely do not represent a separate population, but could simply be a low column density extension of the traditional CNM. (iii) However, the low-N(HI) clouds are probably smaller than typical CNM clouds. If we assume an equilibrium pressure of 3000 cm^{-3} K and a volume density of about 40 cm^{-3}, we can estimate a cloud line-of-sight size of about 800–4000 AU. If over-pressured, clouds would have even smaller line-of-sight length. (iv) The low-N(HI) clouds have an extremely low ratio of the CNM to total HI column density, <5%. This suggests that cold low-N(HI) clouds are most likely surrounded by a large amount of warmer gas. This may be important for cloud survival by providing a blanket against evaporation.

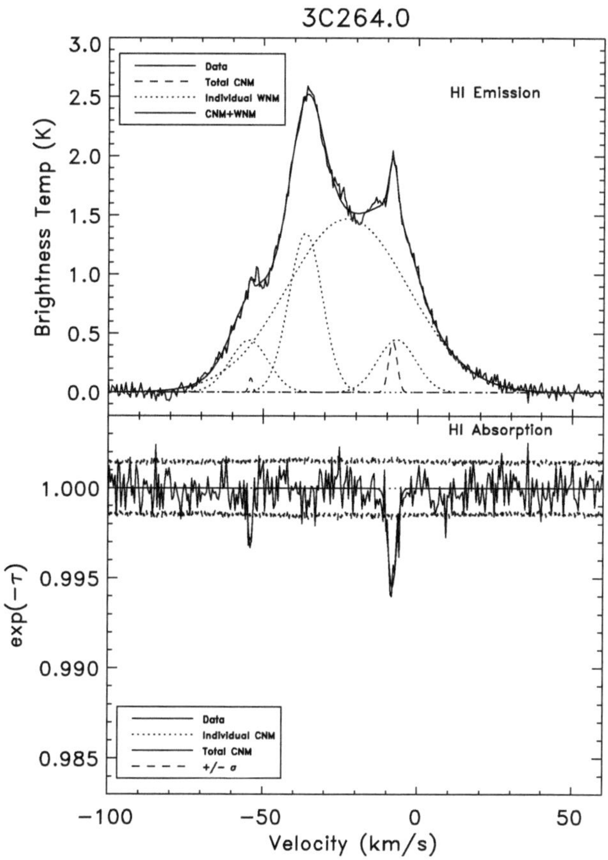

Fig. 2 HI emission and absorption spectra in the direction of 3C264.0, obtained with the Arecibo telescope (Stanimirović et al. 2008; De Back et al. 2008). Two new CNM components were detected. The *upper panel* shows the HI emission profile with separate contributions from the CNM and WNM shown as *dashed* and *dot-dashed lines*, respectively, while the final (simultaneous) fit is shown with the *thick solid line*. The *lower panel* shows the HI absorption spectrum with two fitted Gaussian components, with *dotted lines* representing $\pm 1 - \sigma$ noise level. The decomposition of HI emission and absorption spectra into Gaussian components was performed to estimate the HI spin temperature, employing the method of Heiles and Troland (2003)

5 Cold Clouds Inside the Local Bubble

For one of the low-N(HI) clouds (3C270, $l = 184.8°$, $b = 5.8°$) we have obtained recently Na I absorption spectra in directions of two close by stars with known distances (Meyer and Stanimirović 2008). As shown in Fig. 4, in the direction of star HD 107258, which is 12 arcmin to the south-east from 3C270 and at a distance of 108 (+11, −9) pc, an absorption feature was detected at a LSR velocity of 0.55 km s^{-1}and with a velocity FWHM of 3–4 km s^{-1}. The HI absorption spectrum in Fig. 3 shows a feature at a velocity of −0.8 km s^{-1}, with a velocity FWHM of 3.2 km s^{-1}. The two absorption features most likely trace the same interstellar gas. This suggests that the local HI absorbing gas, seen in Fig. 3, is likely to be located within 100–120 pc from the Sun. In this direction, Na I observations by Lallement et al. (2003) suggest that the boundary of the LB, with $N(\mathrm{HI}) > 10^{20}$ cm^{-2}, is at a distance of ∼130 pc. Therefore, the HI cloud in the direction of 3C270 could easily be lying within the Local cavity. This is a cold HI cloud, with the velocity linewidth of only 3.2 km s^{-1}, the estimated spin temperature of 115 K, and the peak HI column density of 10^{19} cm^{-2}. As the 0.55 km s^{-1}absorption feature is absent in the Na I spectrum of HD 107238, a star which is 14 arcmin to the north-east from 3C270 and at a distance of 192 (+40, −28) pc, the spatial extent of the HI absorbing cloud maybe only ∼0.3–0.4 pc. Figure 4 also shows a strong Na I absorption feature at a velocity of −2.5 km s^{-1}seen only in

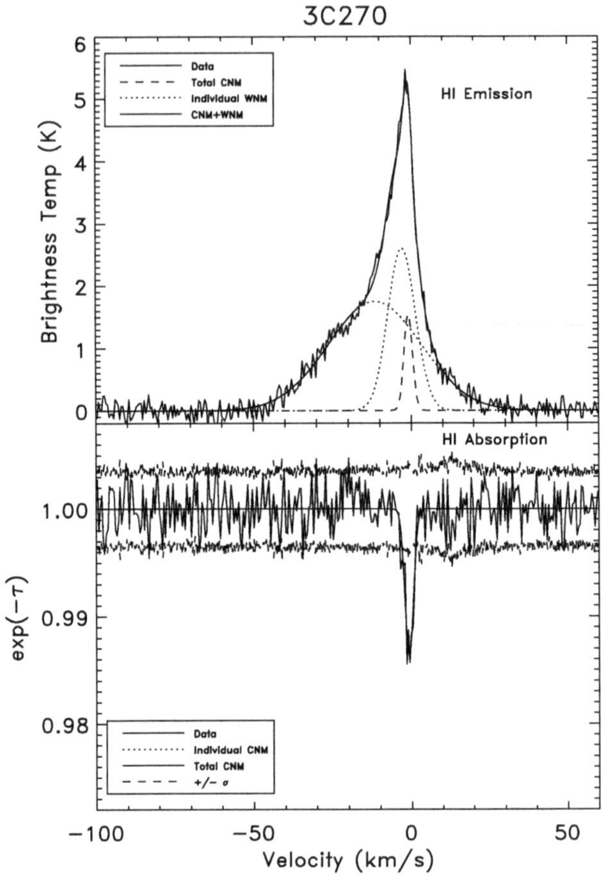

Fig. 3 HI emission and absorption spectra in the direction of 3C270, obtained with the Arecibo telescope. One CNM component was detected. Plotted quantities are the same as in Fig. 2

the spectrum against the more distant star and without a corresponding HI absorption. This (warmer) cloud is likely to be at a distance of 110–190 pc. Followup observations of several stars around 3C270 are underway with the Kitt Peak 3.5 m telescope to constrain better cloud distances.

Another especially clear example of a cold HI cloud found inside the LB was studied by Meyer et al. (2006). This Leo cloud holds a double record, being the coldest diffuse cloud with a spin temperature of only 20 K, and at the same time the closest diffuse cloud with a firm upper distance limit of 45 pc (constrained using 33 stars with known distances). If in pressure equilibrium with $P = 2300$ K cm^{-3}, it has a size of $1.4 \times 4.9 \times 0.07$ pc, being more ribbon-like than spherically symmetric. It is worth noting that a thermal pressure close to the standard ISM pressure was assumed and this resulted in the pc-scale size of the cold cloud (a higher thermal pressure would decrease the cloud size).

The Riegel-Crutcher cloud is another cold, highly filamentary cloud, that appears to be located right at the edge of the LB (Crutcher and Lien 1984; McClure-Griffiths et al. 2006). This cloud extends over ~ 40 degrees of Galactic longitude ($l = 5°$ to $355°$) and ~ 10 degrees of latitude ($b = -5°$ to $5°$). It has a distance of 125 ± 25 pc, a size of $80 \times 20 \times$ (1–5) pc, and a HI spin temperature of 40 K. If filaments are cylindrical with the thickness similar to their plane-of-sky width of 0.1 pc, then they have an average volume density of 450 cm^{-3} and a thermal pressure of 2×10^4 K cm^{-3}. However, an edge-on ribbon geometry

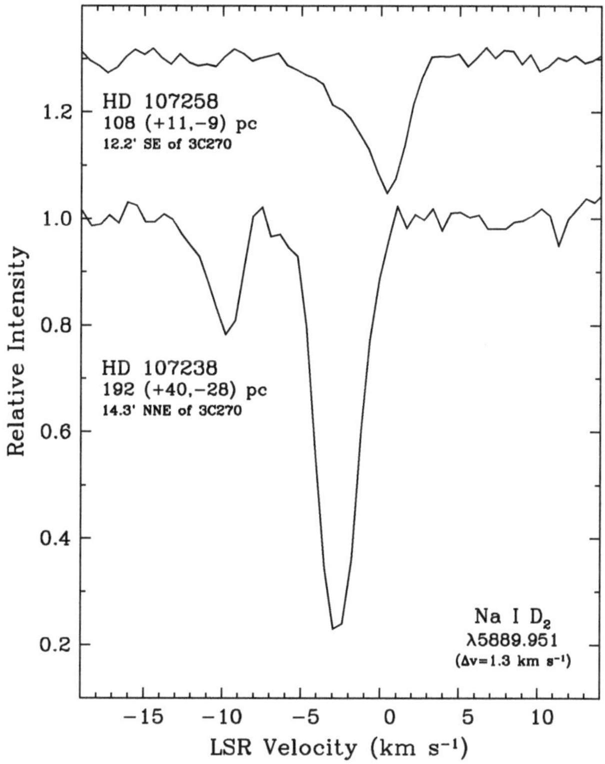

Fig. 4 Na I absorption spectra in the direction of 3C270. HD 107258 is located 12 arcmin to the south-east from 3C270 and is at a distance of 108 pc, while HD 107238 is 14 arcmin to the north-east from 3C270 and at a distance of 190 pc. Spectra were obtained with the 0.9 m coudé feed telescope and spectrograph at Kitt Peak National Observatory

Table 1 Approximate properties of clouds inside the Local Bubble

Type	Temperature (K)	Density cm^{-3}	Size (AU)	Ionization state
CNM	100	40	10^5	Neutral
TSAS	$\lesssim 100$	10^{3-4}	10–100	Neutral
TSIS	$\gtrsim 10^4$	10^2	1	Ionized (arclets)
Low-N(HI)	$\lesssim 100$	10^2	800–4000	Neutral
CLIC	6800	0.2	$\sim 10^5$	partially ionized

would imply a line-of-sight thickness of 1 pc, an average volume density of 46 cm^{-3}, and a thermal pressure of 2×10^3 K cm^{-3}.

6 Comparison with the Cluster of Local Interstellar Clouds

The cluster of local interstellar clouds (CLIC) is a clumpy flow of warm, low-density material inside the LB, and within 35 pc from the Sun. Table 1 provides a summary of basic properties of CLIC and clouds discussed in previous sections. Further from the Sun, within 350 pc, the ISM includes low-density, hot gas inside the LB, but also a range of clouds with cooler temperatures. For example, Welty et al. (1999) found four neutral, cold

Table 2 Summary of exotic clouds inside or just outside of the Local Bubble

Object	Distance (pc)	Type	Comment
B1929 + 10 cloud(s)	<360	TSAS, TSIS	Cold clouds inside/edge of LB, several scintillation screens
B1133 + 16 screen	140	TSIS	Scintillation screens (Trang and Rickett 2007), no TSAS
3C270 cold cloud	$\lesssim$ 100–120	Low-N(HI)	Most likely inside LB
Leo cloud	$\lesssim$ 45	CNM	Clearly inside LB
R-C cloud	125	CNM	At the edge of LB

($T \sim 100$ K) and dense (N(HI) $\sim 10^{20.7}$ cm^{-2} and $n \sim 10$–15 cm^{-3}) clouds, and many warm ($T \sim 3000$ K), dense ($n \sim 15$–20 cm^{-3}) clouds.

While CLIC clouds have column densities similar to those of TSAS and low-N(HI) clouds, they are significantly warmer and more ionized. In terms of fractional ionization, CLIC is intermediate between the WNM and the warm ionized medium (Dickey 2004). Müller et al. (2006) compared the HI column density and the LSR velocity of CNM and WNM clouds at high Galactic latitudes (from Heiles and Troland 2003) with those of CLIC clouds within 50 pc of the Sun. While CLIC clouds have typically a low HI column density (N(HI) $< 10^{19}$ cm^{-2}), their kinematic properties appear similar to those of WNM clouds. The central velocity of the majority of low-N(HI) clouds is in the range -17 to 10 km s^{-1}, and the column density range is 3×10^{17} to 4×10^{19} cm^{-2}. In terms of column density and kinematics, low-N(HI) clouds (as well as the cold Leo cloud and the TSAS clouds in the direction of B1929 + 10) and CLIC occupy a very similar parameter space. This may suggest that all these clouds belong to the same interstellar flow.

Table 2 provides a summary of exotic clouds inside or at the boundary of the LB. Clearly, Leo cold clouds are undoubtly inside the LB. The low-N(HI) cloud in the direction of 3C270 and the TSAS cloud(s) in the direction of B1929 + 10 are very likely to be located inside the LB. In addition, several TSIS scintillation screens are clearly located inside the LB. The key question is whether warm and cold clouds inside the LB are physically related. For example, could CLIC clouds represent warm envelopes of colder TSAS and low-N(HI) clouds? And, could scintillation screens represent ionized fronts of CLIC? As hinted in Sect. 4, there is evidence for the existence of large warm envelopes around low-N(HI) clouds. The existence of cold clouds inside the LB with kinematic properties similar to those of CLIC rises an exciting possibility that at least some cold clouds are physically co-located with CLIC. Based on heating and cooling considerations, the co-existance of warm and cold phases requires a well-defined range of thermal pressures. Hence, future distance measurements of low-N(HI) clouds may allow us to constrain better the thermal pressure inside the LB.

7 Formation and Survival of Cold Clouds

Is it surprising to find cold clouds inside the LB? At first glance, we would expect that cold "thin" clouds would be subjected to the heat from the surrounding hot medium and be eaten away by evaporation very quickly. However, careful consideration of conductive heat transfer, especially in the small-scale regime, suggests that such clouds are surprisingly long-lived although are most likely undergoing evaporation. Their typical evaporation timescale is of order of 10^6 years (see Stanimirović and Heiles 2005 for details; also Slavin 2007). If surrounded by large WNM envelopes (as suggested in Sect. 4), and/or magnetic field, the clouds would be even longer lived, and therefore could be common in the local ISM.

However, physical mechanisms responsible for the production of such clouds in the ISM in general are not well understood. There are several potentially interesting theoretical avenues that await a closer comparison with observations.

Condensation of WNM into CNM triggered by the collision of turbulent flows is capable of producing a large number of small CNM clouds with low column densities, as seen in simulations by Audit and Hennebelle (2005) and Hennebelle and Audit (2007). The CNM clouds produced in simulations are thermally stable and embedded in large, unstable WNM filaments, their typical properties are: $n \sim 50$ cm^{-3}, $T \sim 80$ K, $R \sim 0.1$ pc. The number of cold clouds, as well as their properties, depend heavily on the properties of the underlying turbulent flows. Within this scenario, exotic cold clouds inside the LB may be remnants of collisions between warmer CLIC clouds.

General ISM turbulence envisions interstellar clouds as dynamic entities that are constantly changing in response to the turbulent "weather", a picture very different from the traditional approach. Numerical simulations of turbulent ISM, for example Vazquez-Semadeni et al. (1997) show many clouds with very small sizes and low column densities ($<10^{19}$ cm^{-2}). These clouds are out of dynamical equilibrium and probably very transient. In this scenario, the exotic cold clouds may be a tail-end of the turbulent spectrum of general ISM clouds.

CNM destruction by shocks in simulations by Nakamura et al. (2006) can also produce a spray of small HI 'shreds' that could be related to low-N(HI) clouds. These shreds are expected to have large aspect ratios, up to 2000. As several successive supernova explosions have been postulated to be responsible for the formation of the LB (Breitschwerdt and de Avillez 2006), it would not be hard to imagine shock interactions with clouds inside the LB and the production of exotic cloudlets. Similarly, Breitschwerdt and de Avillez (2006) suggest that the interaction between shells that bound bubbles caused by supernova explosions can drive Rayleigh-Taylor instability and the formation of cloudlets.

8 Conclusions

We have summarized evidence for the existence of at least several exotic clouds inside or just outside the LB. The importance of these clouds is twofold. Firstly, they provide extreme examples of the interstellar clouds that the Sun may encounter over a timescale of several million years. As shown by Müller et al. (2006), highly dense and cold clouds would have a significant effect on the size of the heliosphere. Secondly, studies of different phases of the very local ISM can shed some light on the properties and the formation mechanism of the LB.

The cold exotic clouds (TSAS and low-N(HI) clouds) have AU-scale sizes and low HI column densities. While TSAS has high HI volume density and thermal pressure, low-N(HI) clouds are in this regard similar to traditional CNM clouds. While significantly colder and less ionized than CLIC, the exotic (cold) clouds agree well with CLIC clouds in terms of HI column density and kinematics. This may suggest the physical co-existence of different populations, however distances of most of low-N(HI) clouds are unknown. If physically co-located, these clouds would imply a well-defined range of thermal pressures inside the LB. Future distance constrains of cold clouds are essential to explore this avenue.

While most likely undergoing thermal evaporation, cold exotic clouds inside the LB could last for a long time. Therefore, it is not surprising to encounter them frequently. However, their production mechanisms are still not understood. Potential avenues highlight the importance of dynamical triggering of phase conversion (through collision of warm clouds or interaction of shocks with warm clouds), and interstellar turbulence.

Acknowledgements It is a great pleasure to acknowledge my collaborators Joel Weisberg, Carl Heiles, Dave Meyer, and Evan De Back, without whom this work would not have been so much fun. I would also like to thank the organizers for inviting me to participate in this highly stimulating conference. Finally, this article has greatly benefited from the editors' insightful comments.

References

E. Audit, P. Hennebelle, Astron. Astrophys. **433**, 1 (2005)
R. Braun, N. Kanekar, Astron. Astrophys. **436**, 53 (2005)
D. Breitschwerdt, M.A. de Avillez, Astron. Astrophys. **452**, L1 (2006)
S. Chatterjee, J.M. Cordes, W.H.T. Vlemmings, Z. Arzoumanian, W.M. Goss, T.J.W. Lazio, Astrophys. J. **604**, 339–345 (2004)
R.M. Crutcher, D.J. Lien, in *IAU Colloq.: Local Interstellar Medium*, vol. 81, ed. by Y. Kondo, F.C. Bruhweiler, B.D. Savage (NASA, Washington, 1984), p. 117
De Back et al. (2008, in preparation)
J.M. Dickey, Adv. Space Res. **34**, 14 (2004)
N.H. Dieter, W.J. Welch, J.D. Romney, Astrophys. J. **206**, L113 (1976)
C. Heiles, T.H. Troland, Astrophys. J. **586**, 1067–1093 (2003)
C. Heiles, D. Stinebring, in *SINS—Small Ionized and Neutral Structures in the Diffuse Interstellar Medium*, ed. by M. Haverkorn, W.M. Goss. Astronomical Society of the Pacific Conference Series, vol. 365 (Astronomical Society of the Pacific, San Francisco, 2007), p. 331
P. Hennebelle, E. Audit, Astron. Astrophys. **465**, 431 (2007)
A.S. Hill, D.R. Stinebring, C.T. Asplund, D.E. Berwick, W.B. Everett, N.R. Hinkel, Astrophys. J. **619**, L171 (2005)
R. Lallement, B.Y. Welsh, J.L. Vergely, F. Crifo, D. Sfeir, Astron. Astrophys. **411**, 447–464 (2003)
N.M. McClure-Griffiths, J.M. Dickey, B.M. Gaensler, A.J. Green, M. Haverkorn, Astrophys. J. **652**, 1339 (2006)
D.M. Meyer, J.T. Lauroesch, C. Heiles, J.E.G. Peek, K. Engelhorn, Astrophys. J. **650**, L67–L70 (2006)
D.M. Meyer, S. Stanimirović (2008, in preparation)
H.-R. Müller, P.C. Frisch, V. Florinski, G.P. Zank, Astrophys. J. **647**, 1491 (2006)
F. Nakamura, C.F. McKee, R.I. Klein, R.T. Fisher, Astrophys. J. **164**, 477 (2006)
J.A. Phillips, A.W. Clegg, Nature **360**, 137–139 (1992)
B.J. Rickett, in *SINS—Small Ionized and Neutral Structures in the Diffuse Interstellar Medium*, ed. by M. Haverkorn, W.M. Goss. Astronomical Society of the Pacific Conference Series, vol. 365 (Astronomical Society of the Pacific, San Francisco, 2007), p. 207
J.D. Slavin, in *SINS—Small Ionized and Neutral Structures in the Diffuse Interstellar Medium*, ed. by M. Haverkorn, W.M. Goss. Astronomical Society of the Pacific Conference Series, vol. 365 (Astronomical Society of the Pacific, San Francisco, 2007), p. 113
S. Stanimirović, C. Heiles, Astrophys. J. **631**, 371 (2005)
S. Stanimirović, C. Heiles, N. Kanekar, in *SINS—Small Ionized and Neutral Structures in the Diffuse Interstellar Medium*, ed. by M. Haverkorn, W.M. Goss. Astronomical Society of the Pacific Conference Series, vol. 365 (Astronomical Society of the Pacific, San Francisco, 2007), p. 22
S. Stanimirović et al. (2008, in preparation)
D. Stinebring, in *SINS—Small Ionized and Neutral Structures in the Diffuse Interstellar Medium*, ed. by M. Haverkorn, W.M. Goss. Astronomical Society of the Pacific Conference Series, vol. 365 (Astronomical Society of the Pacific, San Francisco, 2007), p. 254
D.R. Stinebring, M.A. McLaughlin, J.M. Cordes, K.M. Becker, J.E.E. Goodman, M.A. Kramer, J.L. Sheckard, C.T. Smith, Astrophys. J. **549**, L97–L100 (2001)
F.S. Trang, B.J. Rickett, Astrophys. J. **661**, 1064 (2007)

E. Vazquez-Semadeni, J. Ballesteros-Paredes, L.F. Rodriguez, Astrophys. J. **474**, 292 (1997)
J.M. Weisberg, S. Stanimirović, in *SINS—Small Ionized and Neutral Structures in the Diffuse Interstellar Medium*, ed. by M. Haverkorn, W.M. Goss. Astronomical Society of the Pacific Conference Series, vol. 365 (Astronomical Society of the Pacific, San Francisco, 2007), p. 28
D.E. Welty, L.M. Hobbs, J.T. Lauroesch, D.C. Morton, L. Spitzer, D.G. York, Astrophys. J. **124**, 465 (1999)

The Local Bubble Debate

Report from Sessions 1 and 3

Robin L. Shelton · Members of Sessions 1 and 3

Originally published in the journal Space Science Reviews, Volume 143, Nos 1–4, 303–309.
DOI: 10.1007/s11214-008-9359-7

Abstract This report summarizes the discussions in the Session 1 and Session 3 groups which met to discuss the questions: "What Physical Processes Drive the Multiphase Interstellar Medium in the Local Bubble?", and "What are the Energy and Pressure Balances in the Local Bubble?" Most of our understanding of the Local Bubble has come from soft X-ray observations, but recent appreciation of the importance of solar wind charge exchange (SWCX) reactions has shown that the heliosphere produces some fraction of the soft X-rays that were previously ascribed to the Local Bubble. Some astronomers suggest that the SWCX X-rays rather than Local Bubble emission could explain most of the locally produced X-rays. Our discussions, therefore, also included a debate concerning the Local Bubble's existence.

Keywords Galaxy: Local Bubble · Solar System: Solar Wind · Galaxy: ISM · Observations: Diffuse X-rays · Ultraviolet: O VI

1 Introduction

The workshop conveners created small groups which were asked to discuss issues related to the heliosphere and the Local Bubble (LB). The members of Sessions 1 and 3 combined to discuss the physical processes within and pressure and energy balances of the LB. While these topics assume that the LB actually exists, this assumption is debatable because the traditional evidence for a hot LB (soft X-rays that originate within a few hundred parsecs of the Earth and are seen in all directions) could possibly be due to another, recently discovered source. That source, solar wind charge exchange (SWCX), is outlined in Sect. 2. The debate over the Local Bubble's existence is presented in Sect. 3. In Sect. 4, we summarize the discussion of the pressure and energy balances within the Local Bubble. The discussion of the pressure balance allows for the possibility that the LB gas pressure is smaller than previously assumed due to the SWCX component, but the later discussions assume the

R.L. Shelton (✉) · Members of Sessions 1 and 3
Department of Physics and Astronomy, The University of Georgia, Athens, GA 30602, USA
e-mail: rls@physast.uga.edu

traditional conception of the LB as containing hot gas. In Sect. 5, we summarize the discussion of clouds within the Local Bubble, and in Sect. 6, we discuss other important physical processes. This report follows the arguments presented at the session and later presented to the workshop participants for their comments.

2 Solar Wind Charge Exchange (SWCX)

1. The solar wind is highly charged. When solar wind ions encounter neutral gas (in the coma of comets, in planetary atmospheres, and dispersed within the heliosphere), they have a high probability of capturing an electron from the neutral atom.
2. An example using solar wind element "X": $X^{(n+1)+} + H$ or $He \Rightarrow X^{n+*} + H^+$ or He^+.
3. Since the electron is transferred into a highly excited orbital (denoted by $*$), the charge transfer is soon followed by de-excitation and radiation of at least one X-ray photon. The resulting X-rays are colloquially referred to as SWCX X-rays.
4. Cravens (2000) and Robertson and Cravens (2003) proposed that ~50% of the observed 1/4 keV flux previously attributed to the LB is due to SWCX in the heliosphere.
5. In the 3/4 keV band, Koutroumpa et al. (2007) proposed that SWCX can explain ~100% of the observed O VII X-ray flux previously attributed to the LB.
6. Charge exchange and the subsequent release of X-ray photons are being studied currently with more refined reaction rates expected in the near future.
7. The SWCX intensity depends on the Sun's activity level, which varies with time.

3 After We Consider the Effects of SWCX, Is There Reason to Believe that the LB Actually Exists?—the Local Bubble vs. Solar Wind Charge Exchange Debate

Below, we present the arguments for (left column) and against (right column) the existence of the Local Bubble:

Pro Local Bubble

(1) *X-Ray Argument: SWCX explains only a fraction of the observed "locally produced" X-rays, the remainder must come from the Local Bubble*:

A Cravens (2000) and Robertson and Cravens (2003) predicted that ~1/2 of the diffuse 1/4 keV X-rays observed in the Galactic plane by *ROSAT* are from SWCX. The remainder must be from the LB.

- The Robertson and Cravens predictions were made for the same lines of sight through the heliosphere as were used in the *ROSAT* survey.

Anti Local Bubble

(1) *X-Ray Argument: SWCX generates all of the "locally produced" 3/4 keV X-rays, so there is no need to assume the existence of a Local Bubble*:

A SWCX predictions for 2 sample LB observations explain all of the observed local O VII and O VIII emission at 0.57 and 0.65 keV (Koutroumpa et al. 2007).

- These calculations were done for the same lines of sight through the heliosphere as the *Chandra*, *XMM*, and *Suzaku* observations.
- These calculations included the SWCX enhancements associated with coronal mass ejections.

Pro Local Bubble, continued

B The distribution of SWCX 1/4 keV photons (see map in Robertson and Cravens 2003) differs significantly from the observed distribution (see map in Snowden et al. 1998). The difference must be due to the hot LB contribution.

C SWCX intensities vary with time and direction, but several soft X-ray surveys made during different years and using different angles between the target and Sun found similar soft X-ray background fluxes.

- The fluxes observed by the *ROSAT*, SAS 3, and Wisconsin survey are shown to be correlated in Fig. 7 of Snowden et al. (1995).

D The observed 1/4 keV X-ray flux is bright where the H I column density is dim and *vice versa*. This anticorrelation implies that the X-ray emitting gas occupies a cavity in the H I distribution. In general, the effect is not due to absorption, which would simultaneously harden the observed spectrum (see McCammon and Sanders 1990).

(2) O VI *Argument*: O VI, *which traces* $T = 3 \times 10^5$ K *gas at interfaces between hotter and cooler gas, has been observed, so there must be hot gas in the LB*:

A Absorption by O VI ions has been seen on short sight lines that are within or extend somewhat beyond the radius of the LB (Jenkins 1978; Oegerle et al. 2005; Savage and Lehner 2006.) The existence of O VI implies the existence of a hot LB.

- Response: Surely, not all of the O VI observations could be wrong: Some of the *FUSE* observations used cool ($T < 40{,}000$ K) white dwarf stars as background sources and the *Copernicus* observations used B stars. These stars should not have photospheric O VI.

(2) O VI *Counter-Argument*: *Most of the observed* O VI *absorption features can be explained without a LB*:

A Many of the observed O VI absorption features can be explained in other ways

- Savage and Lehner used nearby white dwarfs. Some of their white dwarf spectra were contaminated by stellar or circumstellar O VI absorption, which Savage and Lehner incorrectly attributed to the local ISM (Barstow et al. 2008; Welch 2008).
- Savage and Lehner processed their *FUSE* data using an old version (v2.5) of the FUSE pipeline; when a newer version was used, some of the O VI features disappeared (Barstow et al. 2008).

(3) *EUV Counter-Argument*: *The CHIPS and EUVE measurements are ambiguous*

A Responses:

- The models assumed CIE and solar abundances of gas-phase iron. Either assumption could be incorrect.
- The *EUVE* observations are thought to have been contaminated by a very soft X-ray leak.
- The calorimeter (McCammon et al. 2002) saw a much larger intensity in the 170 Å region than did *CHIPS* or *EUVE*.

(3) *EUV Argument*: *LB models predict Extreme UV emission that was not seen by EUVE or CHIPS*

A *EUVE* and *CHIPS* did not find as much Fe VIII to Fe XI emission around 170 Å as expected from models of the hot LB (Jelinsky et al. 1995; Vallerga and Slavin 1998; Hurwitz et al. 2005).

(4) *Local Cavity and Local Cloud Argument* (*also see Sect. 4*):

A The Local Cavity with little H I (Lallement et al. 2003) must be filled with ionized gas (such as the LB), otherwise it would collapse. Similarly, the Local Interstellar Cloud (see other articles in this issue) would expand if it were not compressed by the Local Bubble's pressure.

- Response: Invoking a large magnetic pressure in the Local Cavity requires an unreasonably strong magnetic field.
- Response: If the gas is not in pressure balance, then the Local Cloud would be expanding at 10 $\mathrm{km\,s^{-1}}$ which is inconsistent with observations.

B Stars within the Local Bubble region have outflows that terminate where they interact with the ISM. Therefore, for termination shocks to exist here, there must be material in the LB (Wood 2007, personal communication).

(4) *Local Cavity and Local Cloud Counter-Argument*:

A There is an unexplained gap between the LB and the Local Cavity wall, so the Milky Way does contain regions with apparent imbalances in the thermal pressure.

- Magnetic pressure may compensate for the imbalance in thermal pressure.
- Total pressures need not be balanced if the system is dynamic.

3.1 What Measurements Are Needed to Determine How Much Hot Gas May Exist in the Local Bubble?

1. More X-ray spectroscopic observations are needed to test the SWCX predictions.
2. X-ray observations with much higher spectral resolution are needed to compare the spectral details of the SWCX models with observed spectra.

3. The high spectral resolution observations must have sufficient spatial resolution for shadowing experiments, in which opaque clouds are used to block the X-rays from more distant diffuse sources.
4. Such high spectral and spatial resolution observations are needed in the 1/4 keV band, which carries much of the flux of the presumably hot ($T \sim 10^6$ K) Local Bubble.

4 Pressure and Energy Balances in the Local Bubble

4.1 Pressure Balance

1. Previously, the LB's thermal pressure had been calculated from the flux of 1/4 keV photons produced between the Sun and a shadowing cloud located ~90 pc from the Sun. The resulting thermal pressure was $P_{th}/k \sim 15{,}000\ \mathrm{K\,cm^{-3}}$ (Snowden et al. 1998).
   (a) The LB pressure quoted above is far larger than the thermal pressure of the local warm clouds and other cool gas in the Galactic disk, $P_{th}/k \sim 2200\ \mathrm{K\,cm^{-3}}$ (Jenkins and Tripp 2001). The pressure discrepancy between the LB and local warm clouds could be alleviated somewhat if half of the soft X-ray flux previously attributed to the LB is actually due to SWCX ions. Since the flux is proportional to n_e^2, the thermal pressure is proportional to the square root of the flux. If the LB flux were reduced by a factor of 2, as suggested by Cravens (2000) and Robertson and Cravens (2003), the LB pressure would decrease by a factor of $\sqrt{2}$, but this is not sufficient to explain the large pressure discrepancy between the LB and the local warm cloud pressures.
   (b) On the other hand, we do not need to invoke SWCX X-rays in order to lower the LB thermal pressure which could be smaller than the estimated $15{,}000\ \mathrm{K\,cm^{-3}}$ if the gas is far from CIE. For example, Breitschwerdt (2001) explained the LB's soft X-ray flux as emission from a recombining plasma whose P_{th}/k was only $\sim 2000\ \mathrm{K\,cm^{-3}}$.
   (c) Alternatively, a high thermal pressure in the Local Bubble may well be correct; it may be needed to balance the pressure of the material above the LB ($P_{th}/k \sim 20{,}000\ \mathrm{K\,cm^{-3}}$, Bourares and Cox 1990).
   (d) Lastly, it is possible that there is not sufficient pressure support to balance the weight of overlying disk and halo material. Some observations show material moving downwards at $\sim 20\ \mathrm{km\ s^{-1}}$.
2. If one concludes that the LB does not exist, then one must find some other material to fill the space within the Local Cavity and provide pressure to balance the warm clouds and the overlying material. Suggestions for solving this problem include:
   (a) Diffuse material having a high magnetic field could provide sufficient magnetic pressure to support the overlaying material. Andersson and Potter (2006) found that the magnetic field strength in the plane of the sky is 8^{+5}_{-3} μG, creating a magnetic pressure of $P_B/k \sim 18{,}000\ \mathrm{K\,cm^{-3}}$ on sightlines terminating within about 200 pc of the Sun. However, in his conference presentation, Steve Spangler reported a LB sight line (toward pulsar J0437-4715, located 170 pc from the Sun), for which the interstellar magnetic field strength is 0.7 μG. The magnetic pressure along that sight line would be far too small to balance the pressure of the adjacent gas.
   (b) Photoionized gas with a temperature of $T \sim 20{,}000$ K could fill the space, but such gas may be disallowed by observations.
   (c) We also considered turbulent pressure and ram pressure. We abandoned turbulent pressure because it does not provide enough support (Redfield and Linsky 2004), but ram pressure could provide significant support (de Avillez 2007, personal communication).

3. Is the Local Bubble expanding? Probably not. If anything, the Loop I Bubble is expanding into the Local Bubble.

4.2 Energy Balance

1. The hot gas in the LB is cooling, but it's energy loss is not balanced by energy gain.
2. The embedded warm clouds are also losing energy by radiation, but they may also gain energy via photoionization by the extreme ultraviolet and soft X-ray photons from the LB (see Slavin and Frisch 2002).
3. We considered the possibility that heat may be conducted into the clouds from the surrounding hot LB plasma (see Slavin 1989).
   (a) However, cloud interfaces may be complicated and the conditions on either side of the interface are not well known.
   (b) If the cloud is surrounded by a tangled magnetic field, then the magnetic field will quench thermal conduction.
   (c) Some lines of sight through embedded clouds do not contain observable numbers of O VI ions (a tracer of evaporative interfaces), but this may be result from the suppression of thermal conduction by magnetic fields and, therefore, cannot be taken as evidence against the existence of hot gas in the Local Bubble.

5 Small Scale Structure in the LB

1. Warm ($T \sim 8000$ K), parsec-scale clouds exist within the LB (Redfield and Linsky 2008).
2. One explanation for the presence of such clouds within the hostile environment of the Local Bubble is the Cox and Helenius (2003) model, in which the magnetic field drags material from the LB wall into the interior of the bubble.
3. In addition to the population of parsec-scale clouds, there may also be a population of smaller clouds. Scintillation observed in QSO signals implies that very small structures exist in very local ISM (Linsky et al. 2008).
4. The very small scale clouds (diameter = 800 to 4000 AU) that exist in the general interstellar medium may also exist within the Local Bubble (see Snezana Stanimirovic's paper in this issue). They could be neutral or ionized and may be fragments from the LB wall, as in the Cox and Helenius model or they may result from collisions of the parsec-scale clouds (Redfield 2007, personal communication).
5. However, evaporation should destroy clouds on Myr timescales unless the clouds are well protected by the magnetic field.

6 Other Important Physical Processes?

1. The standard assumption used when analyzing the X-ray data has been that the gas is in CIE but this assumption should be re-examined.
2. Breitschwerdt and Schmutzler (1994) and Breitschwerdt (2001) computed drastically underionized models of the LB which explained the observed quantity of soft X-ray photons.
3. The Breitschwerdt and Schmutzler scenario is disallowed because the ratios of the observed X-rays to O VI and C III ions and intensity are higher than expected from the model (Welsh et al. 2002; Shelton 2003; Oegerle et al. 2005).
4. In order for the Breitschwerdt and Schmutzler scenario to be valid, the LB's soft X-ray intensity would need to be reduced by a large factor.

Acknowledgements I would like to thank the conference conveners (Dieter Breitschwerdt, Priscilla Frisch, Vlad Izmodenov, Jeffrey Linsky, and Eberhard Möbius) for organizing an extremely productive and enjoyable conference. Special thanks go to Jeffrey Linsky, who ably facilitated the discussions in Sessions 1 and 3, asked me to write up this session report for the conference proceedings, and improved the manuscript. It is a pleasure to credit my fellow members of Sessions 1 and 3 (Miguel de Avillez, Dieter Breitschwerdt, Berkhard Fuchs, Ed Jenkins, Dimitra Koutroumpa, Rosine Lallement, Steve Snowden, Steve Spangler, Snezana Stanimirovic, and Barry Welsh) for providing the discussions recorded in this report. In the question and answer periods during and after the session report was orally delivered, the larger group of workshop participants refined and supplemented some of our thinking. Brian Wood and Seth Redfield added ideas that have been incorporated into this manuscript. In addition, I would like to acknowledge the generosity of the International Space Science Institute (ISSI) and NASA through LTSA grant NNG04GD78G for funding my trip to the conference.

References

B.-G. Andersson, S.B. Potter, Astrophys. J. **640**, L51–L54 (2006)
M. Barstow, D.D. Boyce, J.K. Barstow, A.E. Forbes, B.Y. Welsh, R. Lallement, Astrophys. Space Sci. (2008, in press)
A. Bourares, D.P. Cox, Astrophys. J. **365**, 544–558 (1990)
D. Breitschwerdt, Astrophys. Space Sci. **276**, 163–176 (2001)
D. Breitschwerdt, T. Schmutzler, Nature **371**, 774–776 (1994)
D.P. Cox, L. Helenius, Astrophys. J. **583**, 205–228 (2003)
T.E. Cravens, Astrophys. J. **532**, L153–156 (2000)
M. Hurwitz, T.P. Sasseen, M.M. Sirk, Astrophys. J. **623**, 911–916 (2005)
P. Jelinsky, J.V. Vallerga, J. Edelstein, Astrophys. J. **442**, 653–661 (1995)
E.B. Jenkins, Astrophys. J. **219**, 845–860 (1978)
E.B. Jenkins, T.M. Tripp, Astrophys. J. Suppl. Ser. **137**, 297–340 (2001)
D. Koutroumpa, F. Acero, R. Lallement, J. Ballet, V. Kharchenko, Astron. Astrophys. **475**, 901–914 (2007)
R. Lallement, B.Y. Welsh, J.L. Vergely, F. Crifo, D. Sfeir, Astron. Astrophys. **411**, 447–464 (2003)
J.L. Linsky, B.J. Rickett, S. Redfield, Astrophys. J. **675**, 413 (2008)
D. McCammon et al., Astrophys. J. **576**, 188–203 (2002)
D. McCammon, W.T. Sanders, Annu. Rev. Astron. Astrophys. **28**, 657–688 (1990)
W.R. Oegerle, E.B. Jenkins, R.L. Shelton, D.V. Bowen, P. Chayer, Astrophys. J. **622**, 377–389 (2005)
S. Redfield, J.L. Linsky, Astrophys. J. **613**, 1004–1022 (2004)
S. Redfield, J.L. Linsky, Astrophys. J. **673**, 283 (2008)
I.P. Robertson, T.E. Cravens, J. Geophys. Res. **108**(A10), 6-1–6-10 (2003)
B.D. Savage, N. Lehner, Astrophys. J. Suppl. Ser. **162**, 134–160 (2006)
R.L. Shelton, Astrophys. J. **589**, 261–269 (2003)
J.D. Slavin, Astrophys. J. **346**, 718–727 (1989)
J.D. Slavin, P.C. Frisch, Astrophys. J. **565**, 364–379 (2002)
S.L. Snowden, R. Egger, D.P. Finkbeiner, M.J. Freyberg, P.P. Plucinsky, Astrophys. J. **493**, 715–729 (1998)
S.L. Snowden, M.J. Freyberg, P.P. Plucinsky, J.H.M.M. Schmitt, J. Trümper, W. Voges, R.J. Edgar, D. McCammon, W.T. Sanders, Astrophys. J. **454**, 643–653 (1995)
J. Vallerga, J. Slavin, Lect. Notes Phys. **506**, 79–82 (1998)
B.Y. Welch, Space Sci. Rev. (2008, this issue)
B.Y. Welsh, S. Sallmen, D. Sfeir, R.L. Shelton, R. Lallement, Astron. Astrophys. **394**, 691–699 (2002)

Section VI: What Are the Origin and Physical Properties of the Clump of Very Local ISM?

The Origins and Physical Properties of the Complex of Local Interstellar Clouds

Jonathan D. Slavin

Originally published in the journal Space Science Reviews, Volume 143, Nos 1–4, 311–322.
DOI: 10.1007/s11214-008-9342-3

Abstract The Complex of Local Interstellar Clouds (CLIC) is a relatively tight grouping of low density, warm, partially ionized clouds within about 30 pc of the Solar System. The Local Interstellar Cloud (LIC) is the cloud observed on most lines of sight and may be the cloud that immediately surrounds our Solar System, the properties of which set the outer boundary conditions of the heliosphere. Using absorption line data toward nearby stars, *in situ* observations of inflowing interstellar gas from spacecraft in the Solar System, and theoretical modeling of the interstellar radiation field and radiative transfer, we can deduce many characteristics of the LIC. We find that the LIC is partially ionized with modest electron density, $n_e \approx 0.07\ \text{cm}^{-3}$. The combination of its temperature and ionization favor photoionization/thermal equilibrium over a non-equilibrium cooling cloud picture. The abundances in the LIC suggest moderate dust destruction for silicate dust but complete destruction of carbonaceous grains. An origin for the LIC as a density enhancement in the ambient medium that has been overrun by a shock seems likely, while its velocity away from the Sco-Cen association points to a possible connection to that region and the Loop I bubble.

Keywords Interstellar medium: Physical properties · Interstellar medium: Solar neighborhood · Interstellar medium: Atomic processes

1 Introduction

The interstellar medium (ISM) that surrounds the Solar System impinges on the outflowing Solar wind creating the heliosphere. The particular characteristics of that circumheliospheric interstellar medium (CHISM), such as its density, ionization, temperature and magnetic field, determine in detail the interaction of the gas with the solar wind. Thus understanding the nature of the heliosphere requires knowledge of the state of the CHISM while at the same time interpretation of some of the *in situ* observations relevant to discerning the nature of the CHISM demands an understanding of the heliosphere.

J.D. Slavin (✉)
Harvard-Smithsonian Center for Astrophysics, 60 Garden Street, MS 83, Cambridge, MA 02138, USA
e-mail: jslavin@cfa.harvard.edu

It has long been assumed, because it appears on so many lines of sight, that the velocity component (or cloud) identified as the Local Interstellar Cloud (LIC) surrounds the heliosphere. Recently Redfield and Linsky (2008) have analyzed many lines of sight toward nearby stars and identified 15 distinct velocity vectors, which are attributed to coherent clouds in the local ISM. The LIC component, found in by far the most lines of sight, was found to have an associated temperature of $T = 7500 \pm 1300$ K. Redfield and Linsky (2008) claim that the heliosphere is located in a transition zone between the LIC and G cloud based on the fact that the temperature and velocity of the gas are intermediate between those found for the LIC and G cloud components. We believe that this is somewhat speculative as yet since the temperature of the inflowing gas is within the error bars on the LIC while falling outside of the error bars for the G cloud value. The inflowing gas velocity does appear to be somewhat discrepant (2–3 $\mathrm{km\,s^{-1}}$) with their LIC vector, but that may be due to some weak disturbance in the cloud. Much of what we discuss does not depend specifically on whether the heliosphere is within the LIC as long as their properties are not very different, which appears to be the case. The primary assumption that we make related to this is that the portion of the line of sight between the Solar System and the star ε CMa that traverses the LIC can be treated as passing through a cloud that is in thermal and photoionization equilibrium.

2 The Physical Characteristics of the Complex of Local Interstellar Clouds

2.1 Primary Observational Evidence

The complex of local interstellar clouds, of which the Local Interstellar Cloud (LIC) that surrounds the Solar System is one member, are nearby, low density and partially ionized patches of gas that are surrounded by much lower density and (probably) hot ISM, the Local Bubble. We have the most information about the LIC since most sight lines to nearby stars pass through it, but the LIC is not extremely different from other clouds within the CLIC (see Redfield and Linsky 2008, also Redfield 2008, this issue).

Among the information on the LIC that we need to understand its origins are:

- ionization—is the LIC gas in photoionization equilibrium or not?
- abundances—important for evaluating ionization corrections, radiative cooling rate and as evidence for dust destruction or enrichment,
- temperature/density/pressure—where does the LIC fit into typical ISM phase picture? Is the LIC in thermal equilibrium?
- velocity—can provide hints as to the dynamics, history and past environment of the LIC,
- magnetic field—both the strength and orientation affect the size and shape of the heliosphere. The field also provides pressure support for the cloud and can play an important role in determining the nature of the cloud boundary.

Some of this information is relatively directly measurable, e.g. the temperature of the cloud at the Solar System via He^0 temperature observations (Witte 2004), but most require some modeling to infer their values.

Evidence on the nature of the LIC comes from a wide variety of sources. There is a large database of ion absorption lines toward nearby stars wherein particular velocity components have been identified as being due to the LIC (see Redfield and Linsky 2008, and references therein). This has been accomplished by finding a consistent velocity vector for the cloud given the observed projections for many lines of sight.

Among the important ions for which column densities have been observed for the LIC are: Mg II, Mg I, C II*, S II, and Fe II. The importance of these ions lies in the constraints they place on the physical state of the cloud. As we discuss below, with Mg II and Mg I we can determine the electron density. With the addition of C II*, Si II and Fe II we gain information on the elemental abundances of dust components. These along with S II provide necessary information on the total cooling within the cloud, which is primarily due to forbidden line emission. In order to understand the ionization and thermal properties of the cloud it is necessary to have data on these column densities and data on more ions will generally help to constrain the models better.

The most direct evidence on the state of the CHISM comes from observations of neutral He. Because of its small cross section for charge transfer reactions, He^0 is believed to sail through the bowshock, heliopause and termination shock essentially unaffected. As a result the measurements of He^0 density and temperature, $n(He^0) = 0.015 \pm 0.003$ cm^{-3} and $T(He^0) = 6300 \pm 340$ K (Witte 2004), are the best constraints we have on the CHISM. Other *in situ* data including backscattered H Lyman α, anomalous cosmic rays and pickup ions provide further constraints, but each comes with a more model-dependent interpretation. Direct observations of interstellar dust flowing into the Solar System (Baguhl et al. 1995; Landgraf et al. 2000) is very interesting for what it tells us about the chemical composition of the LIC and clues about its history, but does not directly inform us about the gas phase abundances that govern its thermal balance. Dust can be an important heat source in the ISM via photoelectron ejection, but we find the dust heating is small relative to photoionization heating for the conditions of the LIC.

An additional set of input data needed to understand the physical conditions in the LIC is the ionizing radiation field. Both the current ionization and the sources of photoionization are needed if we are to make sense of the present state of the cloud. Portions of the field have been directly observed, namely the far UV and extreme UV from stellar sources. There have also been direct observations of diffuse soft X-rays, though the source of those remain controversial (see Koutroumpa et al. 2008, this issue). As of now it appears that most of the softest X-rays (Wisconsin B and C band, $E = 70$–280 eV) do come from hot gas within the Local Bubble and these are the most important X-rays for the ionization of the cloud. Unfortunately the potentially even more important diffuse EUV background has yet to be observed and instrumental limitations make it unlikely that such an observation can be made for some time to come.

Another important observational fact is that the LIC, and indeed the CLIC, exist in a large, extremely low density cavity. The location of these clouds within the Local Bubble is clearly important to understanding their origins and evolution.

2.2 Electron Density and Temperature of the LIC

As noted above, the observations of Mg II and Mg I place constraints on the electron density of the LIC. Balance of FUV ionization, radiative and dielectronic recombination and charge transfer gives us

$$n_e = \frac{\Gamma(\mathrm{Mg}^0)}{\alpha(\mathrm{Mg}^+) + C^{CT}(n(\mathrm{H}^+)/n_e)} \frac{N(\mathrm{Mg\,I})}{N(\mathrm{Mg\,II})} \quad (1)$$

where $\Gamma(\mathrm{Mg}^0)$ is photoionization rate, $\alpha(\mathrm{Mg}^+)$ is (total) recombination rate and C^{CT} is charge transfer rate. We derive the photoionization rate using the observed and modeled FUV background from Gondhalekar et al. (1980). If we have $N(\mathrm{C\,II})$ and $N(\mathrm{C\,II}^*)$ we can also get T:

$$\frac{N(\mathrm{C\,II}^*)}{N(\mathrm{C\,II})} = \frac{\gamma_{12}(T)\,n_e}{A_{21} + \gamma_{21}(T)\,n_e} \quad (2)$$

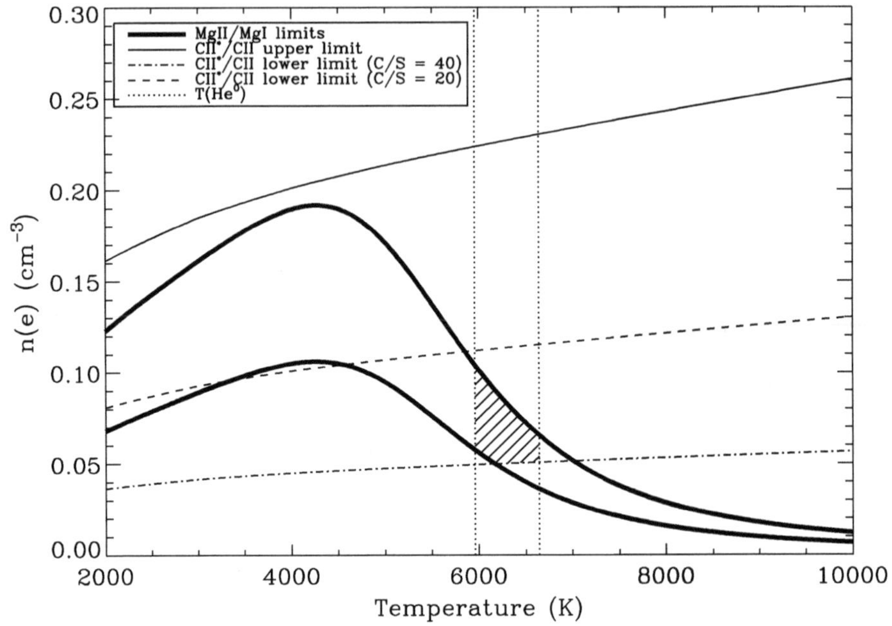

Fig. 1 Constraints on the electron density and temperature in the LIC derived from Mg II, Mg I, C II*, C II and *in situ* He^0 temperature observations

Unfortunately, the most easily observable C II line at 1334.5Å is nearly always saturated, making the column density difficult to derive with any certainty.

For the ε CMa line of sight, n_e and T have been derived by Gry and Jenkins (2001) by making assumptions for the abundance of C to S and using the well observed S II line results to derive a upper limit for $N(\mathrm{C\,II})$. If we use the constraints on T from *in situ* observations and up-to-date values for the Mg^+ recombination coefficients, we find that we need an abundance ratio of $\mathrm{C/S} > 20$ to find a viable solution for n_e and T. The combination of the upper limit on $N(\mathrm{C\,II})$, the observed limits on Mg II/Mg I and the observed limits on T lead to tight limits, $n_e = 0.05$–$0.104\ \mathrm{cm}^{-3}$, as illustrated in Fig. 1. We discuss these limits and the implications for the C abundance in more detail in Slavin and Frisch (2006). When we carry out more detailed modeling including models for the interstellar radiation field, thermal balance and radiative transfer (Slavin and Frisch 2008), we derive even tighter limits on n_e, $n_e = 0.07 \pm 0.01\ \mathrm{cm}^{-3}$.

2.3 Radiative Transfer Models

To go from ion column densities to abundances requires ionization corrections, especially for $N(\mathrm{H})$. In general elements with first ionization potential $E_0 > 13.6$ eV will require an ionization correction to derive the total element column density and these include He, N, O, Ne, and Ar. Oxygen is a special case, however, because its ionization is tightly coupled to H ionization by charge transfer. Nitrogen is also similarly coupled to H ionization but much less tightly. These corrections are particularly important because the H I column is not very well determined in many cases (e.g. ε CMa). Data from the Extreme Ultraviolet Explorer (EUVE) can give us the total H I column, but does not give us the fraction of the total attributable to each velocity component if there are multiple velocity components along the line of sight.

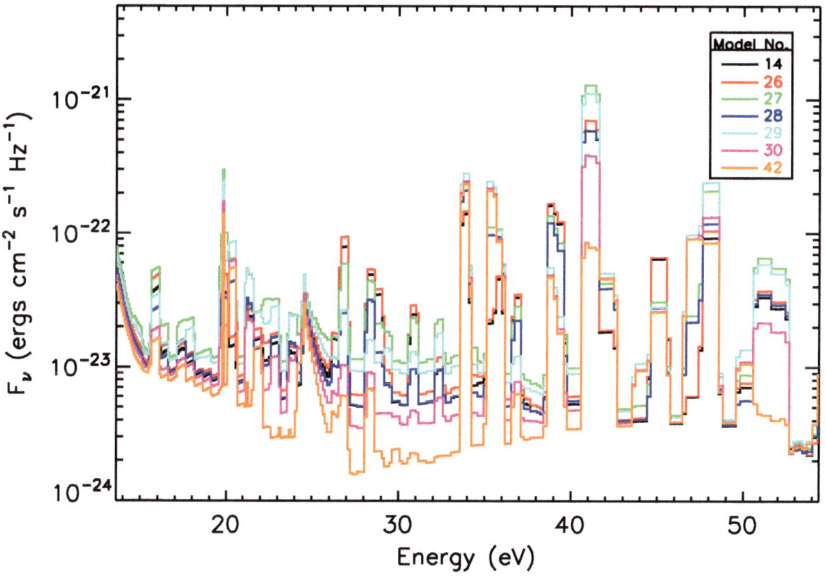

Fig. 2 Model ionizing radiation fields for several different models from Slavin and Frisch (2008), all of which match the observational constraints

In order to derive the ionization corrections one must have a model for the ionization of the cloud and in general this demands a model for the ionizing radiation field, as well as the radiative transfer within the cloud. To model the field we use the sum of emission from stellar sources (FUV and EUV) and diffuse sources including the hot gas that we assume fills the Local Bubble and an evaporative boundary between the warm cloud and the hot gas.

The nature of the boundary between the warm cloud gas and the surrounding hot gas is still quite uncertain as there have been no definitive observations establishing the existence of an evaporative boundary. It is known that the magnetic field will suppress thermal conduction across field lines, yet a completely magnetically isolated cloud would seem unlikely. In our models we assume a magnetic suppression of conductivity of a factor of 2, appropriate to the case in which the field is at a 45° angle with the cloud surface. This part of the radiation field is clearly quite uncertain and we intend to explore possible variants in future work.

We find that the details of the field do not seem to strongly influence our results for the ionization of the cloud. We show in Fig. 2 the H ionizing radiation field at the position of the Sun in several models that are consistent with the observations. The requirements we impose on the models to match $n(\mathrm{He}^0)$, $T(\mathrm{He}^0)$ and the ion column densities act to fix the H ionization, $X(\mathrm{H})$ and n_e.

We find successful models for range of values for $N(\mathrm{H\,I})$ towards εCMa and T(hot gas). Our results for the Solar location are:

- $n(\mathrm{H}^0) = 0.19$–0.20 cm^{-3},
- $n_e = 0.07 \pm 0.01$ cm^{-3},
- $n(\mathrm{H}) = 0.23$–0.27 cm^{-3},
- $B = 2.1, 2.5$ μG for two best models (the magnetic field affects emission intensity from cloud boundary).

3 Dust and Elemental Abundances in the LIC

Our approach to modeling the abundances in the LIC is to force our models to match the column densities observed by adjusting the elemental abundances. Thus the abundances are an output of the modeling rather than in input. Figure 3 illustrates the abundance results for C, N, O, S, Si, Mg and Fe relative to a particular assumed solar abundance of the elements (Asplund et al. 2005). The x-axis is the condensation temperature for the element, a quantity often presumed to correlate with the amount of depletion from the gas phase. In Table 1 we list the derived abundances in our models that are consistent with observations.

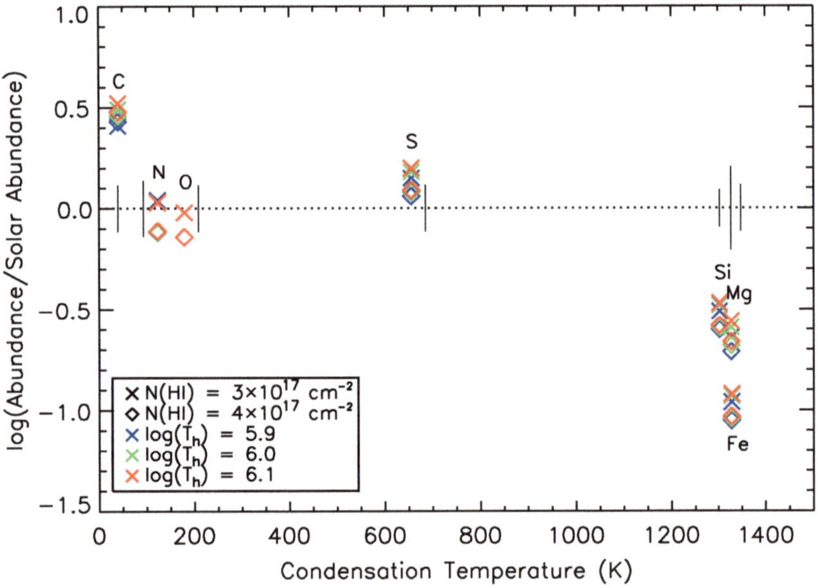

Fig. 3 Abundances relative to Asplund et al. (2005) solar abundances vs. condensation temperature. Results from models of the ionization for the ε CMa line of sight including radiative transfer (Slavin and Frisch 2008). The *symbol shape* indicates the assumed H I column density for the model, while the *color* indicates the assumed temperature for the hot gas of the Local Bubble. The abundances are fixed so as to match the observed ion column densities

Table 1 Elemental gas phase abundances (ppm)

Model no.	Element						
	C	N	O	Mg	Si	S	Fe
14	589	40.7	295	5.89	7.24	14.1	2.24
25	631	66.1	437	7.76	10.0	19.5	3.09
26	661	46.8	331	6.61	8.13	15.8	2.51
27	759	64.6	437	8.71	10.7	20.9	3.31
28	708	45.7	331	7.08	8.32	16.6	2.57
29	813	64.6	437	9.33	11.0	21.9	3.39
30	741	46.8	331	7.41	8.51	17.0	2.63
42	724	39.8	295	6.76	7.76	15.1	2.34

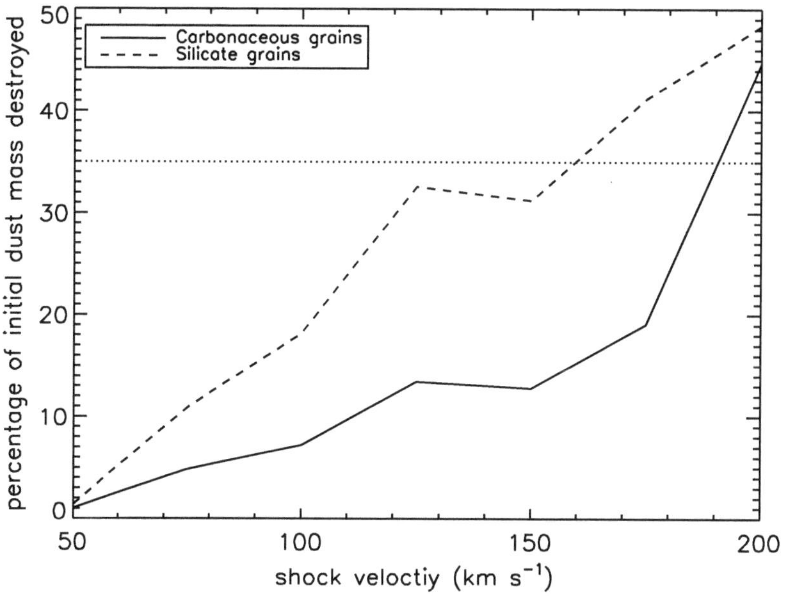

Fig. 4 Percentage of initial total grain mass lost because of dust destruction in the shock vs. shock speed. To get the Si gas phase abundance determined for the LIC we need $\lesssim$35% destruction of silicate dust which is consistent with $v_{\rm shock} \lesssim 150$ km s^{-1}. Such shocks should destroy less than 15% of the cabonaceous dust, however, in contradiction with the derived large gas phase abundance of C

The derived abundances indicate modest depletion of the constituents of silicate dust, Si, Fe, Mg, and O, implying that at least some destruction of this type of dust. Depending on ones assumptions about the initial depletion of Si (Savage and Sembach 1996, quote values of 70–95%), the total Solar abundance of Si (recent determinations range from about 30 to 43 ppm) and its current gas phase abundance (we find ~ 7.2–11.5 ppm) the fraction of the silicate grain mass destroyed (i.e. returned to the gas phase) ranges from 0 to 35%. The high C abundance on the other hand seems to indicate that all the carbonaceous dust has been destroyed. Radiative shocks destroy dust via various processes: sputtering, vaporization, shattering. Detailed calculations by Jones et al. (1996) find that silicate dust should be *more* destroyed than carbonaceous dust by shocks, as illustrated in Fig. 4. Therefore either some other process is strongly influencing the gas phase abundances in the LIC or the models of shock processing of grains need revision.

4 LIC Ionization and Thermal Balance

Observations with the Extreme Ultraviolet Explorer (EUVE) toward nearby stars found unexpected results for the ratio of H I to He I column density (Dupuis et al. 1995). Instead of the expected ratio of ≤ 10 that one would get if the cosmic He abundance is 0.1 and H is more ionized than He, it was found that $N(\text{H I})/N(\text{He I}) \sim 14$, indicating that He is more ionized than H. This unusual ionization of the local ISM has long been considered puzzling and has led to the suggestion that the LIC is out of ionization equilibrium, being overionized for its temperature because of an earlier ionizing event (e.g. a shock) (see, e.g., Lyu and Bruhweiler 1996). The long timescale for recombination, particularly of H, it was reasoned,

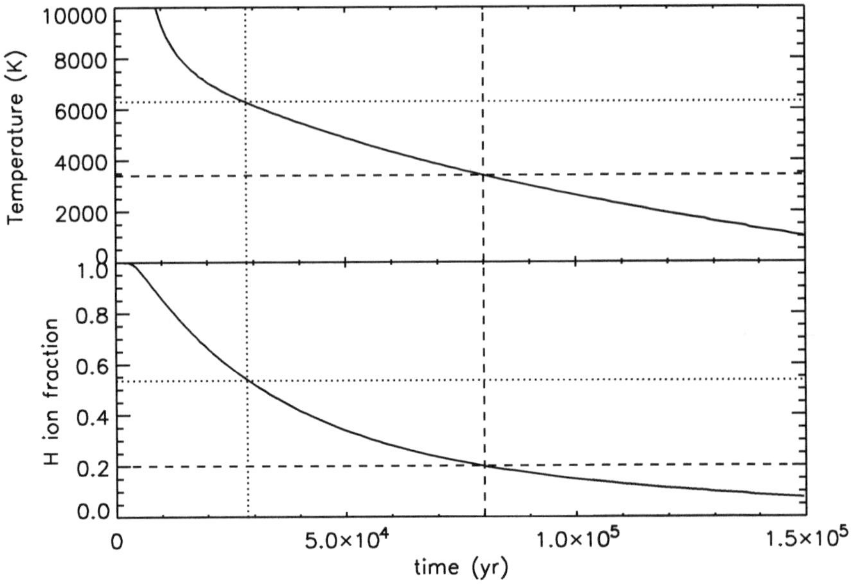

Fig. 5 Time evolution of temperature and H ion fraction behind a 100 $\mathrm{km\,s^{-1}}$ shock (Raymond, private communication). The *dotted lines* show the time, temperature and H ion fraction when the gas cools to the LIC temperature, while the *dashed lines* show the same for the LIC H ion fraction. Isobaric cooling shows a similar lag of H recombination. The gas cools too fast to allow the relatively low degree of ionization in the LIC at temperatures of ~ 6300 K without some substantial heating source

makes it likely that the LIC is out of ionization equilibrium. However, the cooling rate of the gas also has to be considered in such a model. Doing this one finds that in fact the cooling time for the gas in any likely scenario is considerably less than the recombination time. As a result, if the LIC were cooling from a hotter and more ionized state, it should still be quite highly ionized by the time that it has cooled to the observed temperature of $T \approx 6300$ K. In Fig. 5 we illustrate this by showing the temperature and ionization evolution behind a 100 $\mathrm{km\,s^{-1}}$ shock. Even greater disparity between the cooling time and recombination time is found for a simple isobaric cooling model. Therefore the fact that the LIC is in fact mostly neutral, $X(\mathrm{H}^+) \sim 0.2$, implies that the cloud has had time to recombine while being maintained at a warm temperature. This requires a heat source to balance the cooling. While alternative sources have been proposed, such as turbulent dissipation (Minter and Spangler 1997), the most likely heat source appears to be photoionization heating. Such heating is also accompanied by ionization, suggesting that the cloud is at least close to thermal and photoionization equilibrium.

Another argument in favor of the cloud being in ionization equilibrium was first suggested by Jenkins et al. (2000) based on Ar I and O I data. Since O ionization is tied to H ionization by charge exchange, if we assume an O abundance we can then compare the ionization of Ar and H. In recombining gas, it is found that Ar and H have roughly equal ionization fractions because H^+ and Ar^+ have similar recombination coefficients. However, for gas in photoionization equilibrium, Ar I is deficient relative to H I (or O I) because the photoionization cross section for Ar^0 is 5–30 times larger than that for H^0. Observations for the LIC (Jenkins et al. 2000) find $X(\mathrm{Ar}^0)/X(\mathrm{H}^0) \sim 0.4$ toward nearby white dwarfs. Detailed NEI calculations for cooling gas show that $X(\mathrm{Ar}^0)/X(\mathrm{H}^0)$ remains ~ 1 until gas nears equilibrium *and* is photoionized.

Despite this evidence that the LIC is currently close to ionization and thermal equilibrium, there are reasons to believe that it has not always been so. The LIC is clearly many times denser than its surrounding gas in the Local Bubble as can be deduced from the low absorption by neutral gas within the bubble and lack of observable optical emission from possible warm ionized gas that could conceivably fill the cavity. That leaves only highly ionized and very low density gas as the primary volume filling gas in the bubble. Therefore it appears highly likely that the gas that presently makes up the LIC and other nearby clouds was at one time substantially overdense compared with the surrounding medium before becoming incorporated into the Local Bubble. The most likely scenario is that cold neutral medium gas, with $n \sim 100$ cm^{-3}, $T \sim 100$ K, that was embedded in warm gas was hit by a shock. However, it is important to note that any shock *no matter what speed* hitting such a dense cloud will go radiative in the cloud. Thus one needs to find a means to heat the cloud to warm neutral medium temperatures. An origin in a fragmented shell implies a similar radiative shock and heating requirements. The means to heat the shocked warm clouds seems to require their expansion to lower density as the pressure of the bubble drops at the same time as ionizing flux from the hot gas and possibly from the cloud boundary regions provides heating.

For diffuse ISM conditions, calculated heating and cooling rates typically lead to the possibility of thermal balance with two stable thermal phases within a limited range of thermal pressures, with a cold neutral phase and a warm neutral or (perhaps partially) ionized phase. Figure 6 shows a density vs. pressure plot or phase diagram showing two different phase equilibrium curves. The one for "Low Ionization" comes from the work of Wolfire et al. (2003) and assumes low ionizing flux whereas the "LIC ionization" one is calculated using one of our model ionizing radiation fields for the LIC. We note that the thermal pressure will generally not dominate the total dynamical pressure because other pressure forms including magnetic, cosmic ray and turbulent, are typically estimated to be of the same order of magnitude as the thermal pressure. This does not affect the phase curves, however, since it is the components of the thermal pressure (i.e. density and temperature) that directly affect the heating-cooling balance. In the diffuse ISM cosmic ray heating is small compared to dust and photoionization heating. It may be that turbulent dissipation, particularly in concert with MHD turbulence, provides significant heating, however the rate for that remains quite uncertain and is neglected in the figure.

The arrows on the plot indicate how gas parcels will evolve under the influence of shocks, adiabatic cooling (cooling via expansion) and evaporation via thermal conduction. The shock arrow indicates a relatively small increase in pressure, which would require only a mach 2.5 (relative to the cold gas) shock. A shock that could heat typical warm (ionized or neutral) medium gas to about 10^6 K would need to be much faster, $v_s \approx 270$ km s^{-1}, or mach 27 in the warm medium. The pressure would thus be increased to $P/k_B \sim 2 \times 10^6$ cm^{-3} K. In order for the local clouds to become warm would require the pressure to drop by more than two orders of magnitude after the shock passed over them. This could be achieved after sufficient expansion, e.g. a factor of ~ 4 in radius assuming adiabatic expansion of a spherical bubble. This requires that the clouds were close enough to the center of the superbubble that the shock had not gone radiative yet and that the clouds (or at least a fraction of them) could survive long enough to persist until our current state in which the surrounding bubble has a relatively low pressure.

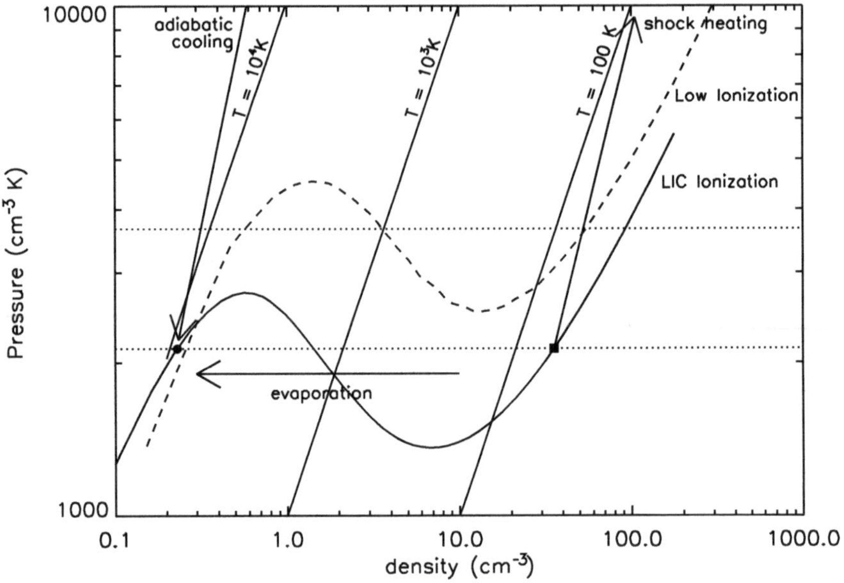

Fig. 6 Phase diagram for the diffuse interstellar medium for either low ionizing flux conditions (*dashed curve*, using rates from Wolfire et al. 2003) or the moderate ionizing flux as modeled for the LIC (*solid curve* using our modeled radiation field). *Points* on the curve are conditions of thermal equilibrium, below the curve heating exceeds cooling and above the curve cooling exceeds heating. The *diagonal lines* are curves of constant temperature. A shock will tend to move a gas parcel up and to the right in the diagram, while adiabatic cooling moves points down and to the left as illustrated by the arrows. Evaporation via thermal conduction moves points to left. Note that the pressure in the plot is only thermal pressure and thus neglects the dynamically important magnetic, cosmic ray and turbulent pressures in the ISM

5 The Origin the Complex of Local Interstellar Clouds

The above discussion lays out some of the challenges facing any model for the origin of the complex of local interstellar clouds. In summary we would like a theory to explain these facts:

- The density, temperature and ionization of the clouds are in sharp contrast to the surrounding Local Bubble gas (though we don't know all the properties of that gas),
- the CLIC has a significant velocity relative to the LSR and direction roughly away from Galactic center,
- the ionization of the LIC is unusual with He apparently more ionized than H,
- the abundances in the gas seem to imply that carbonaceous dust has been destroyed, and yet interstellar dust observed in the Solar System implies a relatively low gas-to-dust ratio.

A number of theories have been put forward to explain the CLIC. The clouds have been variously proposed to be: (1) pieces of the Sco-Cen bubble from an earlier epoch of star formation (Frisch 1981), (2) a fragment from Sco-Cen/LB interaction (Breitschwerdt et al. 2000), and (3) a flux tube/filament that has broken away from the bubble wall (Cox and Helenius 2003). We would add to this list, (4) a dense cloud in the ambient medium overrun by an expanding bubble shock, a model that we have discussed briefly above but that has yet to be fully explored.

Each of these models has its problems. The velocity and relative positions of the CLIC and Loop I bubble strongly suggest a connection between them but detailed modeling of how these clouds could have come from that bubble is lacking. Breitschwerdt et al. (2000) propose that the Local Bubble and the Loop I bubble are interacting and that the CLIC is associated with the wall that separates the bubbles. In their model the clouds are created by instabilities generated in the interaction region. The LIC is currently about 70 pc from that neutral wall and moving about 20 km s^{-1} away from the Sco-Cen association that is believed to be responsible for creating the Loop I bubble. It is unclear how cloudlets like the CLIC could have been traveling for 3.5 million years away from this interaction zone and yet the wall is apparently intact between the two bubbles. Frisch (1981) suggests that the clouds as well as the Local Bubble are associated with a previous epoch of star formation of Sco-Cen. This requires that somehow a cold neutral wall was reformed within the bubble between these epochs of star formation. The mechanism for doing that is left unexplained. The flux tube theory of the origins of the CLIC by Cox and Helenius (2003) requires that a flux tube sprang from the wall of the Local Bubble pulling warm gas along with it into the bubble interior. The magnetohydrodynamics of this explanation seem questionable however, in particular that one flux tube can spring from the bubble wall while the rest of the bubble is not collapsing. Finally, our idea that the clouds originated as cold clouds in a warm intercloud medium seems reasonable but does not explain why the velocity of the CLIC is directed away from the Sco-Cen association and towards the center of the Local Bubble rather than away from it. We must appeal to a random velocity of the gas prior to being overrun by the expanding Local Bubble to explain this.

6 Summary

The wide range of data that we have on the LIC has lead to a fairly complete picture of the cloud. We find that it is:

- partially ionized, $X(\mathrm{H}^+) \sim 0.2$–0.3, $X(\mathrm{He}^+) \sim 0.3$–$0.4$, $n_e \approx 0.07\ \mathrm{cm}^{-3}$,
- has experienced mixed dust destruction – moderate for silicate dust, complete for carbonaceous,
- at or close to ionization equilibrium.

An origin as a cloud embedded in a lower density medium that was shocked seems likely, and some association with the Loop I bubble and Sco-Cen OB association remains a possibility. Many mysteries remain about its abundances and origins within the local ISM.

Acknowledgements I would like to thank the organizers of the "Outer Heliosphere to the Local Bubble" conference for inviting me to give this talk and Priscilla Frisch, my collaborator in much of the work I presented. This research was supported by NASA Solar and Heliospheric Program grants NNG05GD36G and NNG06GE33G to the University of Chicago.

References

M. Asplund, N. Grevesse, A.J. Sauval, in *Cosmic Abundances as Records of Stellar Evolution and Nucleosynthesis*. ASP Conf. Ser., vol. 336 (ASP, San Francisco, 2005), pp. 25–37
M. Baguhl, E. Grün, D.P. Hamilton, G. Linkert, R. Riemann, P. Staubach, Space Sci. Rev. **72**, 471 (1995)
D. Breitschwerdt, M.J. Freyberg, R. Egger, Astron. Astrophys. **361**, 303–320 (2000)
D.P. Cox, L. Helenius, Astrophys. J. **583**, 205–228 (2003)
J. Dupuis, S. Vennes, S. Bowyer, A.K. Pradhan, P. Thejll, Astrophys. J. **455**, 574 (1995)

P.C. Frisch, Nature **293**, 377–379 (1981)
P.M. Gondhalekar, A.P. Phillips, R. Wilson, Astron. Astrophys. **85**, 272 (1980)
C. Gry, E.B. Jenkins, Astron. Astrophys. **367**, 617–628 (2001)
E.B. Jenkins, W.R. Oegerle, C. Gry, J. Vallerga, K.R. Sembach, R.L. Shelton, R. Ferlet, A. Vidal-Madjar, D.G. York, J.L. Linsky, K.C. Roth, A.K. Dupree, J. Edelstein, Astrophys. J. Lett. **538**, L81–L85 (2000)
A.P. Jones, A.G.G.M. Tielens, D.J. Hollenbach, Astrophys. J. **469**, 740 (1996)
D. Koutroumpa, R. Lallement, V. Kharchenko, Space Sci. Rev. (2008, this issue)
M. Landgraf, W.J. Baggaley, E. Grün, H. Krüger, G. Linkert, J. Geophys. Res. **105**, 10343–10352 (2000)
C.-H. Lyu, F.C. Bruhweiler, Astrophys. J. **459**, 216 (1996)
A.H. Minter, S.R. Spangler, Astrophys. J. **485**, 182 (1997)
S. Redfield, Space Sci. Rev. (2008, this issue)
S. Redfield, J.L. Linsky, Astrophys. J. **673**, 283–314 (2008)
B.D. Savage, K.R. Sembach, Astrophys. J. **470**, 893 (1996)
J.D. Slavin, P.C. Frisch, Astrophys. J. Lett. **651**, L37–L40 (2006)
J.D. Slavin, P.C. Frisch, Astron. Astrophys. (2008, submitted)
M. Witte, Astron. Astrophys. **426**, 835–844 (2004)
M.G. Wolfire, C.F. McKee, D. Hollenbach, A.G.G.M. Tielens, Astrophys. J. **587**, 278–311 (2003)

Physical Properties of the Local Interstellar Medium

Seth Redfield

Originally published in the journal Space Science Reviews, Volume 143, Nos 1–4, 323–331.
DOI: 10.1007/s11214-008-9422-4

Abstract The observed properties of the local interstellar medium (LISM) have been facilitated by a growing ultraviolet and optical database of high spectral resolution observations of interstellar absorption toward nearby stars. Such observations provide insight into the physical properties (e.g., temperature, turbulent velocity, and depletion onto dust grains) of the population of warm clouds (e.g., 7000 K) that reside within the Local Bubble. In particular, I will focus on the dynamical properties of clouds within ~15 pc of the Sun. This simple dynamical model addresses a wide range of issues, including, the location of the Sun as it pertains to the relationship between local interstellar clouds and the circumheliospheric interstellar medium (CHISM), the creation of small cold clouds inside the Local Bubble, and the association of interacting warm clouds and small-scale density fluctuations that cause interstellar scintillation. Local interstellar clouds that can be easily distinguished based on their dynamical properties also differentiate themselves by other physical properties. For example, the two nearest local clouds, the Local Interstellar Cloud (LIC) and the Galactic (G) Cloud, show distinct properties in temperature and depletion of iron and magnesium. The availability of large-scale observational surveys allows for studies of the global characteristics of our local interstellar environment, which will ultimately be necessary to address fundamental questions regarding the origins and evolution of the local interstellar medium.

Keywords Local interstellar medium · High resolution spectroscopy · Dynamics · Abundances

1 Introduction

The interstellar medium (ISM) in our local neighborhood (within ~100 pc) is a diverse collection of material that informs us about general ISM phenomena that are occurring elsewhere in our galaxy and other galaxies, as well as, about the properties of our own solar system, namely the structure of the heliosphere. It is the momentum balance between the ever-changing circumheliospheric interstellar medium (CHISM) and the solar wind that defines

S. Redfield (✉)
Department of Astronomy, University of Texas, 1 University Station, C1400, Austin, TX 78712, USA
e-mail: sredfield@astro.as.utexas.edu

DOI: 10.1007/978-1-4419-0247-4_26

the location of the heliosphere, which today is located at approximately 100 AU. Therefore, if we are to study the general ISM or the heliosphere, it is important to understand the properties of the LISM. For thorough reviews of recent work on the local interstellar medium (LISM), see Redfield (2006), Frisch (1995, 2004), and Ferlet (1999).

Multiple phases of interstellar media have been identified within the LISM. Hot (i.e., a few $\times 10^5$ to 10^6 K) gas, that presumably pervades the Local Bubble cavity has been measured in soft X-rays (Snowden et al. 1998), O VI absorption lines (Oegerle et al. 2005), O VII and O VIII emission lines (Smith et al. 2005), extreme ultraviolet (EUV) emission lines (Hurwitz et al. 2005), and by the general lack of cold Na I gas (Lallement et al. 2003). Cold (i.e., $\sim$20 K) gas, though rare, has been detected well within the Local Bubble, using H I 21 cm and Na I absorption (Meyer et al. 2006; Heiles and Troland 2003), and CO emission (Magnani et al. 1996; Chol Minh et al. 2003). Warm (i.e., $\sim$7000 K) gas, is quite prevalent within the Local Bubble, and is actually the material that currently surrounds our solar system. This most local of interstellar material can be studied using Lyman-α backscatter measurements (Quémerais et al. 2000), *in situ* He I observations (Witte et al. 1996), and ultraviolet (UV) and optical absorption line spectroscopy (Redfield and Linsky 2002, 2004a; Welty et al. 1996; Crawford 2001; Frisch et al. 2002). Absorption line spectroscopy, and the resulting measurements of fundamental physical characteristics of the warm LISM, will be the focus of this particular paper.

Due to the relatively low column densities and short path lengths, measuring the properties of the LISM has been remarkably difficult. Ironically, from an observational perspective, it is much easier to detect the signature of more distant ISM structures with much higher column densities, than the nearest ISM clouds just beyond the solar system. In order to effectively measure the properties of LISM clouds, very sensitive atomic transitions are required in order to be sensitive to such low column densities. Only a few dozen such lines exist, and are predominately located in the ultraviolet (UV), see Fig. 2 in Redfield (2006). Although the LISM is more difficult to observe, it is actually much simpler to investigate physical characteristics once the observations have been made, because only a handful of absorbers are located along the line of sight, as opposed to possibly dozens of blended components along distant sight lines. For stars within 100 pc, on average, only 1.7 absorbers are observed along the line of sight in high resolution ($R \equiv \lambda/\Delta\lambda \sim 100{,}000$) UV spectra of multiple ions (Redfield and Linsky 2004a). Observations at even higher spectral resolution could possibly uncover additional absorption components Welty et al. (1996).

2 LISM Observational Database

The deployment of high spectral resolution UV instruments on space-based telescopes (e.g., the Goddard High Resolution Spectrograph (GHRS) and the Space Telescope Imaging Spectrograph (STIS) onboard the *Hubble Space Telescope* (*HST*)) has enabled the acquisition of high quality absorption spectra of the LISM. Because all observations of objects beyond are solar system necessarily traverse the LISM, a growing database of measurements have been accumulating, even though LISM absorption measurements did not typically motivate the observations. The current high resolution spectral database of LISM observations are summarized in Fig. 1. Observations using both *HST* high resolution spectrographs, GHRS and STIS, are shown as circles. Sight lines that also have far-ultraviolet (FUV) measurements from the *Far Ultraviolet Spectroscopic Explorer* (*FUSE*) are shown as triangles. High resolution optical observations of Ca II are indicated by squares. The current database of observations includes $\sim$160 stars within 100 pc, showing $\sim$270 LISM absorption components, where $\sim$60% of the observations were taken for purposes other than to study the

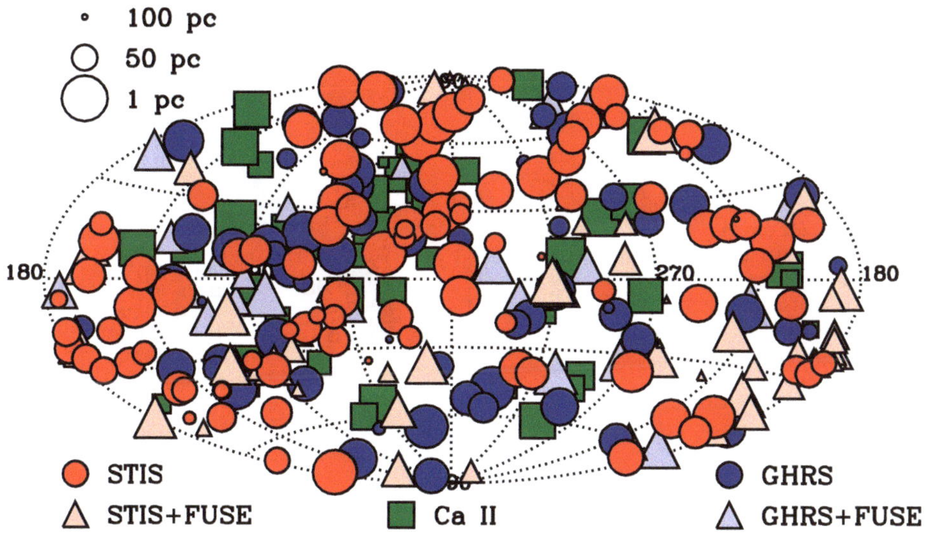

Fig. 1 Distribution of current LISM observational database. More than 270 individual velocity components are detected toward ~160 stars within 100 pc. The high resolution UV spectrographs on *HST* have provided a rich database with which to study the LISM

LISM. Clearly, a substantial database of LISM observations have quickly accumulated, and provide an excellent opportunity to study the physical properties of our local interstellar environment.

The scientific value of multiple observations over many sight lines is immediately clear from Fig. 1, in that a dense sampling of observations provides a detailed mapping of both large- and small-scale structures in the LISM. Scientific value is also dramatically increased by another multiplicity: observations of multiple absorption ions. A single ion observed along a single sight line provides two fundamental physical measurements of the absorbing material, its radial velocity and column density. The line width, or Doppler width, is also measured, but alone, provides little insight into any property of the absorption gas. If multiple ions are observed, even along just a single sight line, a large number of physical properties become accessible, including temperature and turbulence (Redfield and Linsky 2004b), electron density (Redfield and Falcon 2008), depletion (Linsky et al. 2006), volume density (Redfield and Linsky 2000), and ionization fraction (Jenkins et al. 2000). If ultimately, multiple ions are observed along multiple sight lines, fundamental issues, such as the origin and evolution of the LISM and the interaction of the different phases of material in the LISM, can be addressed by studying the global morphology (Redfield and Linsky 2008), global dynamics (Frisch et al. 2002), and small-scale structure (Redfield and Linsky 2001).

3 Physical Properties of Individual Structures in the LISM

3.1 Dynamics

One of the most fundamental measurements of LISM absorption is the radial velocity of the absorbing medium along the line of sight. Due to the proximity of LISM clouds, a single

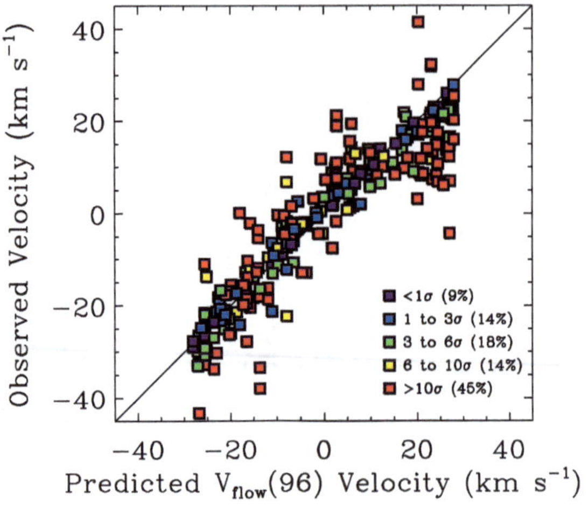

Fig. 2 Comparison of observed LISM radial velocities and the $V_{\mathrm{flow}}(96)$ global LISM velocity vector derived from Frisch et al. (2002). All LISM absorption velocities can be *roughly* described by a single velocity vector, although there is clear evidence for dynamical variations from this global flow. In particular, a deceleration is detected at the leading edge of the LISM cloud complex (Redfield and Linsky 2001), and the majority of measurements deviate $>3\sigma$ from that predicted from the global flow vector

structure can extend over much of the sky. The Local Interstellar Cloud (LIC) and Galactic (G) Cloud each occupy large areas of the sky, 45% and 20%, respectively (Redfield and Linsky 2008). Because multiple radial velocity measurements of the same collection of gas can be made along various directions, a three-dimensional velocity vector can be fit to characterize the bulk flow of the LISM cloud. Frisch et al. (2002) presented a global model of the entire LISM which led to a single velocity vector, $V_{\mathrm{flow}}(96)$, based on 96 velocity components. The collection of sight lines extend to distance of 132 pc, although the majority (62%) are within 30 pc. Figure 2 shows the observed radial velocity for all 270 velocity components (limited to sight lines within 100 pc) shown in Fig. 1 compared to that predicted by the $V_{\mathrm{flow}}(96)$ velocity vector. Although both samples utilize sight lines out to $\sim$100 pc, the absorbing material is likely located within $\sim$30 pc, as indicated by the lack of an increase of absorbing components for sight lines extending from 30–100 pc (Redfield and Linsky 2004a). Figure 2 demonstrates that as a whole, the entire LISM can be roughly characterized by a single velocity vector, indicating a common history or common dynamical driver for all LISM clouds. However, highly significant deviations from this velocity vector are also clear. First, the leading edge of the LISM, where the greatest positive velocities are measured, show considerable deceleration, as if the gas is colliding with other material and being compressed and slowed down. This deceleration was identified by Redfield and Linsky (2001) using a sample of LISM observations toward stars in the Hyades, which fortuitously is located very closely to the downwind direction of the LISM flow. Second, the vast majority of measurements (77%) are $>3\sigma$ from the general flow velocity, and almost half (45%) are $>10\sigma$.

Redfield and Linsky (2008) developed a large-scale empirical dynamical model of 15 LISM clouds within 15 pc. This work builds on previous research on LISM dynamics by several groups (e.g., Crutcher 1982; Lallement et al. 1986; Bzowski 1988; Lallement and Bertin 1992; Frisch et al. 2002). Crutcher (1982) calculated the average LISM velocity flow from 7 Ti II observations of nearby stars, Lallement and Bertin (1992) identified two independent velocity vectors for the LIC and G Cloud based on 16 optical and UV observations, and Frisch et al. (2002) measured a LISM average velocity vector from 96 velocity components toward 60 stars, as well as identified vectors for several individual clouds. The Redfield and Linsky (2008) dynamical model is based on 270 velocity components toward $\sim$160 stars. Velocity vectors for 15 independent cloud structures are determined. All velocity vectors are roughly parallel in flow direction, but vary in velocity magnitude from 0–60 km s^{-1} relative to the motion of the Sun.

3.2 Solar Location Relative to LISM Clouds

Interplanetary *in situ* measurements of warm gas streaming into our solar system indicates that we are currently surrounded by warm gas that is similar to the LISM material we observe via absorption line spectroscopy (Möbius et al. 2004). Early estimates of the velocity vector of the LIC, based on approximately a dozen lines of sight, were within the errors of the *in situ* measurements, so it was claimed that the solar system was surrounded by the LIC. However, LIC material is not observed in all directions, which therefore indicates that if we are indeed surrounded by the LIC, that it extends only a very short distance in some directions. Indeed, based on the motion of the Sun toward the G Cloud, and the roughly antiparallel relative motion of the LIC, we might expect to travel outside the boundaries of the LIC, toward the G Cloud, in $\lesssim$7000 years (Redfield and Linsky 2000).

The new velocity vectors for the LIC and G clouds by Redfield and Linsky (2008) now include 79 and 21 lines of sight, respectively. This substantial improvement has led to a refinement of the velocity vectors. Now the *in situ* measurements in velocity, direction, and temperature, fall intermediate to the LIC and G Cloud values. The discrepancy in velocity is $\sim 3\sigma$ for both the LIC and G Cloud. This argues that the Sun may actually be located in the transition zone between the LIC and G Cloud.

3.3 Turbulence and Depletion

Physical characteristics of specific LISM clouds, other than their dynamical properties, such as, abundances, depletion onto dust grains, temperature, and turbulence, can also be analyzed in a global sense. Redfield and Linsky (2008) found that cloud-averaged measurements of the depletion of iron and magnesium are correlated with cloud-averaged turbulent velocity measurements. The linear correlation coefficients are 0.69 and 0.73, respectively, with $\sim$1.2–1.7% likelihoods that the distributions could have been drawn from an uncorrelated parent sample. This correlation may arise if turbulent shocks can destroy dust grains and return ions to the gas phase and thus decrease the magnitude of the depletion. Weak shock grain destruction has also been invoked to explain depleted deuterium (Linsky et al. 2006). It is important to keep in mind that the absorption line measurements of both abundance and turbulent velocity are line-of-sight-averages, and extreme environments, such as where dust destruction may be occurring, would be difficult to identify.

3.4 Inter-cloud Variation of Physical Properties

Figure 3 indicates the distribution of physical measurements made for LISM absorbers. Physical properties of the LIC and G Cloud, the two nearest LISM clouds which have the most independent measurements, are also shown. The LIC and G Cloud appear to have distinct properties, in terms of, temperature and depletion of iron and magnesium. The LIC is significantly warmer, with an average temperature of 7500 $\pm$ 1300 K, while the average temperature of G Cloud gas is 5500 $\pm$ 400 K. The LIC is also significantly more depleted in both magnesium and iron, with average depletions of –0.97 $\pm$ 0.23 and –1.12 $\pm$ 0.10, respectively, while the G Cloud depletions are –0.36 $\pm$ 0.35 and –0.54 $\pm$ 0.11. Note that the depletions do not take into account partial ionization of hydrogen, which is likely important although unknown for almost all sight lines, or neutral or doubly ionized magnesium and

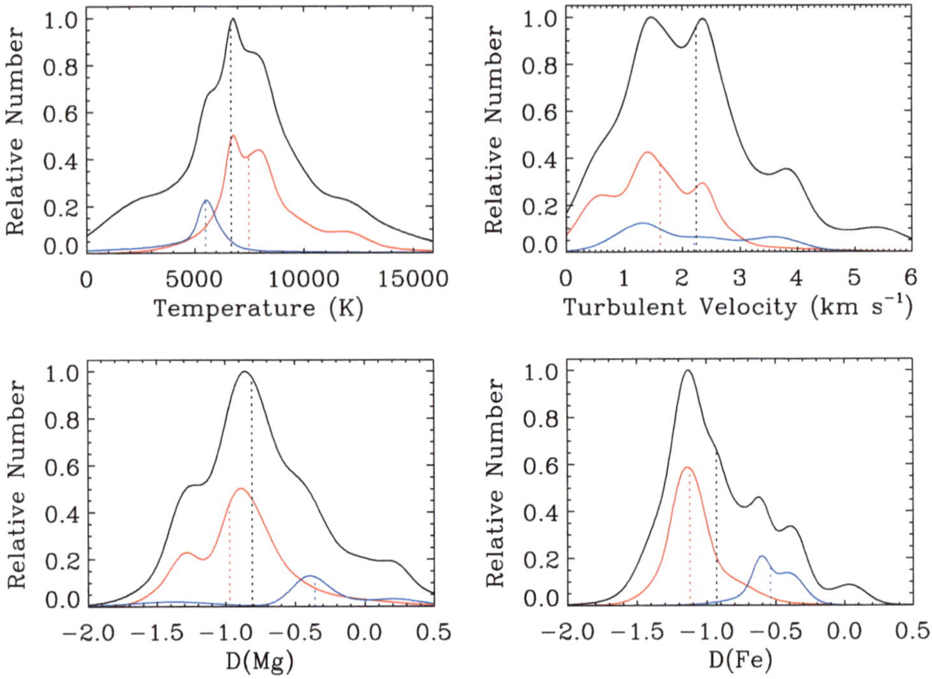

Fig. 3 Distribution of full LISM (black), LIC (red), and G Cloud (blue) samples. The distributions are created using Gaussian profiles for each measurement consistent with their particular errors. For temperature and depletion, the LIC and G Cloud are significantly different

iron, which are likely much less important since they are not expected to become a dominant ionization stage of either element (e.g., Slavin and Frisch 2002; Lehner et al. 2003). Redfield and Linsky (2008) searched for a gradient in properties as a function of flow direction, and did not find any for temperature. Although differences are not dramatic, Fig. 3 argues that the individual cloud structures, such as the LIC and G Cloud, have distinct properties, with modest internal variation.

3.5 Dynamical Cloud Interactions

Although the flow directions of LISM clouds are similar, they are not necessarily identical, and the flow velocity of adjacent clouds can very by 10's of km s^{-1}. For these reasons, neighboring clouds with different velocity vectors could produce regions of dynamical disturbance where the clouds are interacting. Indeed, the dynamical studies of LISM clouds relate to two recent discoveries of structures in our cosmic neighborhood: (1) the existence of a cold (e.g., ~20 K) cloud, well inside the Local Bubble, <40 pc away (Meyer et al. 2006), and (2) distance measurements to turbulent electron density fields that cause interstellar scintillation of high redshift radio sources, which place these screens at ~10 pc (Rickett et al. 2002; Bignall et al. 2006).

Cold Clouds The Local Bubble has typically been assumed to be a region devoid of any cold neutral gas. Lallement et al. (2003) took advantage of this fact to define the Local Bubble boundaries using Na I absorption line observations of nearby stars. Stars within ~100 pc typically show no Na I absorption, indicating that no cold gas exists locally, whereas more distant stars, beyond the Local Bubble boundary, show very strong Na I absorption lines, as the sight lines traverse the nearby star-forming clouds, such as, Taurus, Ophiuchus, and Chamaeleon. The rapid onset of Na I absorption is used to define the shape of the Local Bubble. However, a few small cold clouds have been identified well within the Local Bubble (e.g., Meyer et al. 2006; Lallement et al. 2003). The origins and evolution of these objects is unknown, though the dynamical properties of nearby warm LISM clouds may provide an answer. In simulations of turbulent interstellar clouds, collisions of warm gas clouds can produce small sheetlike clouds of cold gas (Vázquez-Semadeni et al. 2006; Audit and Hennebelle 2005). The three-dimensional velocity vectors of warm LISM clouds can be used to search for compressive flows between neighboring clouds that may create local cold clouds. In the direction of the cold Leo cloud studied by Meyer et al. (2006), differential velocities in the radial direction reach magnitudes of $\sim$15 km s^{-1} (Redfield and Linsky 2008), which may act to produce the small, sheetlike cold cloud that has been observed in 21 cm H I absorption (Heiles and Troland 2003). Using the dynamical properties of the LISM clouds, we can identify regions that are conducive to cold cloud creation, and search for further evidence of small cold clouds within the Local Bubble.

Interstellar Radio Scintillation Some high redshift, compact quasars show variability on timescales of hours at radio (3–8 GHz) frequencies (e.g., Quirrenbach et al. 1992). The characteristic timescale of variability itself varies precisely over the course of a year. This behavior has been attributed to interstellar scintillation, caused by small-scale density fluctuations in the galactic ISM. Recent observations have placed the location of the scintillation screen very nearby, on the order of $\sim$10 pc (Dennett-Thorpe and de Bruyn 2003; Macquart and de Bruyn 2006). Observations of scintillation toward nearby pulsars also show evidence for a local source (Putney and Stinebring 2006). Since the scintillation screens and the warm LISM clouds share the same volume, the question arises, whether these two phenomena are physically related. Linsky et al. (2008) used the dynamical properties of both structures in order to argue that indeed the two structures are physically associated. The yearly variation of the characteristic timescale of scintillation provides an estimate of the transverse flow of the scintillation screen, while the velocity vectors of the warm LISM clouds allows for the calculation of the transverse flow of the warm gas. For three independent scintillating sight lines, the warm LISM clouds located along the line of sight provide much better fits to the transverse flow of the scintillating screen than the Local Standard of Rest, indicating a physical relationship between the warm LISM clouds and the scintillation source (Linsky et al. 2008). Interactions at the boundaries of neighboring clouds could produce small-scale density fluctuations, and thereby facilitate scintillation. Additional distance measurements of warm LISM clouds and scintillating screens are needed in order to further confirm the relationship between these two phenomena.

4 Conclusions and Future Work

The dynamical properties of the warm LISM clouds provide tremendous insight into the morphological and physical structure of our most immediate interstellar environment, as well as possible explanations for associated phenomena, such as the creation of small cold clouds and density fluctuations that cause interstellar scintillation. This work was made possible by a large-scale survey of interstellar absorption toward nearby stars (UV observations of $\sim$160 stars). Such a dense sampling was required to deconvolve the subtle velocity difference of LISM clouds and to provide distance constraints on the absorbing medium. Future studies will be enabled by even larger surveys. One such survey, the Ca II LISM Optical Survey of our Environment (CLOSE), will provide absorption measurements toward ~400 nearby stars (Redfield et al. 2006). The observations utilize the two highest resolution spectrographs in the world, the Coúde Spectrograph (maximum resolving power: $R \equiv \lambda/\Delta\lambda \sim$500,000) on the 2.7 m Harlan J. Smith Telescope at McDonald Observatory in the northern hemisphere, and the Ultra High Resolution Facility (UHRF; maximum resolving power: $R \sim$1,000,000) on the 3.9 m Anglo-Australian Telescope at the Anglo-Australian Observatory in the southern hemisphere. Although the Ca II line is not as sensitive as many lines in the UV, access to ultra high resolution ground-based spectrographs allows for a large-scale dedicated survey of individual warm LISM absorbers. With the anticipated repair of the only high resolution UV spectrograph in space, STIS on *HST*, the UV LISM database will hopefully continue to grow. Such high spectral resolution observations will continue to provide a rich database in order to probe the detailed physical conditions of our most local interstellar environment.

Acknowledgements I would like to thank my collaborators on various projects that were mentioned in passing in this review, they include Jeff Linsky, Barney Rickett, Brad Gibson, Chris Thom, Annie Hughes, Naomi McClure-Griffiths, and Povilas Palunas. I thank the organizers of this ISSI workshop for inviting me to participate. The reviewer of this manuscript provided many careful and helpful comments, which were much appreciated. I acknowledge support provided by NASA through Hubble Fellowship grant HST-HF-01190.01 awarded by the Space Telescope Science Institute, which is operated by the Association of Universities for Research in Astronomy, Inc., for NASA, under contract NAS 5-26555.

References

E. Audit, P. Hennebelle, Astrophys. J. **433**, 1–13 (2005)
H.E. Bignall, J.-P. Macquart, D.L. Jauncey, J.E.J. Lovell, A.K. Tzioumis, L. Kedziora-Chudczer, Astrophys. J. **652**, 1050–1058 (2006)
M. Bzowski, Acta Astron. **38**, 443–453 (1988)
Y.C.Y. Chol Minh et al., New Astron. **8**, 795–803 (2003)
I.A. Crawford, Mon. Not. R. Astron. Soc. **327**, 841–848 (2001)
R.M. Crutcher, Astrophys. J. **254**, 82–87 (1982)
J. Dennett-Thorpe, A.G. de Bruyn, Astron. Astrophys. **404**, 113–132 (2003)
R. Ferlet, Astron. Astrophys. Rev. **9**, 153–169 (1999)
P.C. Frisch, Space Sci. Rev. **72**, 499–592 (1995)
P.C. Frisch, Adv. Space Res. **34**, 20–26 (2004)
P.C. Frisch, L. Grodnicki, D.E. Welty, Astrophys. J. **574**, 834–846 (2002)
C. Heiles, T.H. Troland, Astrophys. J. Suppl. Ser. **145**, 329–354 (2003)
M. Hurwitz, T.P. Sasseen, M.M. Sirk, Astrophys. J. **623**, 911–916 (2005)
E.B. Jenkins et al., Astrophys. J. Lett. **538**, L81–L85 (2000)
R. Lallement, P. Bertin, Astron. Astrophys. **266**, 479–485 (1992)
R. Lallement, A. Vidal-Madjar, R. Ferlet, Astron. Astrophys. **168**, 225–236 (1986)
R. Lallement, B.Y. Welsh, J.L. Vergely, F. Crifo, D. Sfeir, Astron. Astrophys. **411**, 447–464 (2003)
N. Lehner, E.B. Jenkins, C. Gry, H.W. Moos, P. Chayer, S. Lacour, Astrophys. J. **595**, 858–879 (2003)
J.L. Linsky et al., Astrophys. J. **647**, 1106–1124 (2006)
J.L. Linsky, B.J. Rickett, S. Redfield, Astrophys. J. **675**, 413–419 (2008)
J.-P. Macquart, A.G. de Bruyn, Astron. Astrophys. **446**, 185–200 (2006)
L. Magnani, D. Hartmann, B.G. Speck, Astrophys. J. Suppl. Ser. **106**, 447 (1996)
D.M. Meyer, J.T. Lauroesch, C. Heiles, J.E.G. Peek, K. Engelhorn, Astrophys. J. **650**, L67–L70 (2006)
E. Möbius et al., Astron. Astrophys. **426**, 897–907 (2004)
W. Oegerle, E. Jenkins, R. Shelton, D. Bowen, P. Chayer, Astrophys. J. **622**, 377–389 (2005)
M.L. Putney, D.R. Stinebring, Chin. J. Astron. Astrophys. Suppl. **6**, 233–236 (2006)
E. Quémerais, J.-L. Bertaux, R. Lallement, M. Berthé, E. Kyrölä, W. Schmidt, Adv. Space Res. **26**, 815–818 (2000)

A. Quirrenbach, A. Witzel, T.P. Kirchbaum, C.A. Hummel, R. Wegner, C.J. Schalinski, M. Ott, A. Alberdi, M. Rioja, Astron. Astrophys. **258**, 279–284 (1992)
S. Redfield, in *ASP Conf. Ser. 352, New Horizons in Astronomy, Frank N. Bash Symposium 2005*, ed. by S.J. Kannappan, S. Redfield, J.E. Kessler-Silacci, M. Landriau, N. Drory (ASP, San Francisco, 2006), p. 79
S. Redfield, R.E. Falcon, Astrophys. J. **683**, 207 (2008)
S. Redfield, J.L. Linsky, Astrophys. J. **534**, 825–837 (2000)
S. Redfield, J.L. Linsky, Astrophys. J. **551**, 413–428 (2001)
S. Redfield, J.L. Linsky, Astrophys. J. Suppl. Ser. **139**, 439–465 (2002)
S. Redfield, J.L. Linsky, Astrophys. J. **602**, 776–802 (2004a)
S. Redfield, J.L. Linsky, Astrophys. J. **613**, 1004–1022 (2004b)
S. Redfield, J.L. Linsky, Astrophys. J. **673**, 283–314 (2008)
S. Redfield, B. Gibson, C. Thom, A. Hughes, N. McClure-Griffiths, P. Palunas, Anglo-Aust. Obs. Epp. Newslett. **110**, 9 (2006)
B.J. Rickett, L. Kedziora-Chudczer, D.L. Jauncey, Astrophys. J. **581**, 103–126 (2002)
J.D. Slavin, P.C. Frisch, Astrophys. J. **565**, 364–379 (2002)
R.K. Smith, R.J. Edgar, P.P. Plucinsky, B.J. Wargelin, P.E. Freeman, B.A. Biller, Astrophys. J. **623**, 225–234 (2005)
S.L. Snowden, R. Egger, D.P. Finkbeiner, M.J. Freyberg, P.P. Plucinsky, Astrophys. J. **493**, 715 (1998)
E. Vázquez-Semadeni, D. Ryu, T. Passot, R.F. González, A. Gazol, Astrophys. J. **643**, 245–259 (2006)
D.E. Welty, D.C. Morton, L.M. Hobbs, Astrophys. J. Suppl. Ser. **106**, 533 (1996)
M. Witte, M. Banaszkiewicz, H. Rosenbauer, Space Sci. Rev. **78**, 289–296 (1996)

Perspectives on Interstellar Dust Inside and Outside of the Heliosphere

B.T. Draine

Originally published in the journal Space Science Reviews, Volume 143, Nos 1–4, 333–345.
DOI: 10.1007/s11214-008-9411-7

Abstract Measurements by dust detectors on interplanetary spacecraft appear to indicate a substantial flux of interstellar particles with masses $> 10^{-12}$ g. The reported abundance of these massive grains cannot be typical of interstellar gas: it is incompatible with both interstellar elemental abundances *and* the observed extinction properties of the interstellar dust population. We discuss the likelihood that the Solar System is by chance located near an unusual concentration of massive grains and conclude that this is unlikely, unless dynamical processes in the ISM are responsible for such concentrations. Radiation pressure might conceivably drive large grains into "magnetic valleys". If the influx direction of interstellar gas and dust is varying on a ~ 10 yr timescale, as suggested by some observations, this would have dramatic implications for the small-scale structure of the interstellar medium.

Keywords Dust · Interstellar dust · Heliosphere · Interstellar matter

1 Introduction

The interstellar medium (ISM) consists of a partially-ionized, magnetized gas mixed with solid particles of dust. The ionization state and molecular fraction of the gas depend primarily on the gas density and the local intensity of ultraviolet radiation that can photodissociate molecules and photoionize molecules and atoms. The dust content is determined by the prior history of the gas, including injection of newly-formed dust in stellar winds and supernova explosions, grain destruction in violent events such as supernova blast waves, and grain growth in the interstellar medium by both vapor deposition and coagulation in dense regions.

While we do not know the properties of interstellar dust with precision, they are strongly-constrained by a variety of observations. The observed wavelength dependence of interstellar extinction—the so-called "reddening curve" (reviewed in Sect. 2)—provides strong constraints on both the composition and size distribution of interstellar dust. In the local regions

B.T. Draine (✉)
Princeton University Observatory, Peyton Hall, Princeton University, Princeton, NJ 08544-1001, USA
e-mail: draine@astro.princeton.edu

of the Milky Way, interstellar dust is abundant, containing a large fraction of the elements (such as Mg, Si, and Fe) that can be incorporated into refractory solids. As discussed in Sect. 3, interstellar abundances therefore provide a strong constraint on grain models. The size distribution of interstellar grains can be inferred from the observed average reddening curve together with interstellar abundance constraints.

Microparticle impacts on detectors on Ulysses and Galileo have been interpreted as showing a flux of solid particles entering the Solar system from the local interstellar medium (Grun et al. 1993). In Sect. 3 we show that the population of large grains inferred from dust impact detectors on Ulysses and Galileo (Landgraf et al. 2000; Krüger et al. 2007; Krueger and Gruen 2008) is incompatible with average elemental abundances in the ISM. In Sect. 4, we show that such a large grain population would result in wavelength-dependent extinction very different from what is observed.

The Ulysses and Galileo data, if correctly interpreted, imply that the Solar System is, by chance, located in a very atypical spot in the ISM, with an overabundance of very large grains. The likelihood of such a scenario is discussed in Sect. 5. In Sect. 6 we comment on suggestions that the interstellar dust inflow vector might have changed appreciably over only $\sim$5 yrs. Our conclusions are summarized in Sect. 7.

2 Dust in the Diffuse Interstellar Medium

In the Milky Way and many other galaxies, a substantial fraction of the "refractory elements" in the ISM are in solid materials, in submicron dust particles. At least on large scales, the dust and gas are well-mixed, with the density of dust tending to be proportional to the density of gas.

The properties of the dust—size distribution, shapes, composition—are inferred from a wide range of observations (for a review, see Draine 2003) including: wavelength-dependent extinction and polarization of starlight, light scattering in the visible and ultraviolet, small-angle scattering of X-rays, thermal emission from infrared to submm wavelengths, and microwave radiation from spinning dust. Studies of the strength and wavelength-dependence of interstellar extinction $A_\lambda \equiv (2.5/\ln 10)\tau_{\rm ext}(\lambda) = 1.086\tau_{\rm ext}(\lambda)$ provide strong constraints on the size distribution and composition of interstellar dust. Figure 1 shows an empirical parameterization of the extinction by dust in "diffuse clouds" (Cardelli et al. 1989; Fitzpatrick 1999). A "diffuse cloud" is simply a region with visual extinction $A_{\rm V} \lesssim 1$ mag; most of the interstellar H I is in such regions. The interstellar material surrounding the heliosphere consists of diffuse H I, and it was natural to expect that the interstellar dust outside the heliosphere would be typical "diffuse cloud" dust—typical in both its size distribution and its abundance relative to the gas.

The wavelength-dependence of A_λ is known to vary from one sightline to another. Extinction curves are often characterized by $R_V \equiv A_V/(A_B - A_V)$. Average diffuse clouds have $R_V \approx 3.1$ but R_V can be as small as ~ 2.2 in some diffuse clouds (e.g., $R_V = 2.22 \pm 0.14$ toward HD 210121: Fitzpatrick 1999) and can reach values as large as ~ 5.8 in dense regions (e.g., $R_V = 5.8 \pm 0.6$ toward HD 36982: Fitzpatrick 1999). The extinction law shown in Fig. 1 is intended to be an average curve for diffuse clouds, with $R_V \approx 3.1$. The most notable characteristic of the extinction curve in Fig. 1 is the continuing rise into the vacuum ultraviolet; this requires that the size distribution be such that the total surface area of the dust is dominated by very small grains with radii $a \lesssim 200$ Å. The second notable characteristic is the prominent "bump" in the extinction at $\lambda \approx 2175$ Å. While this feature has not yet been identified with complete certainty (see, e.g., Draine 1989), it is thought

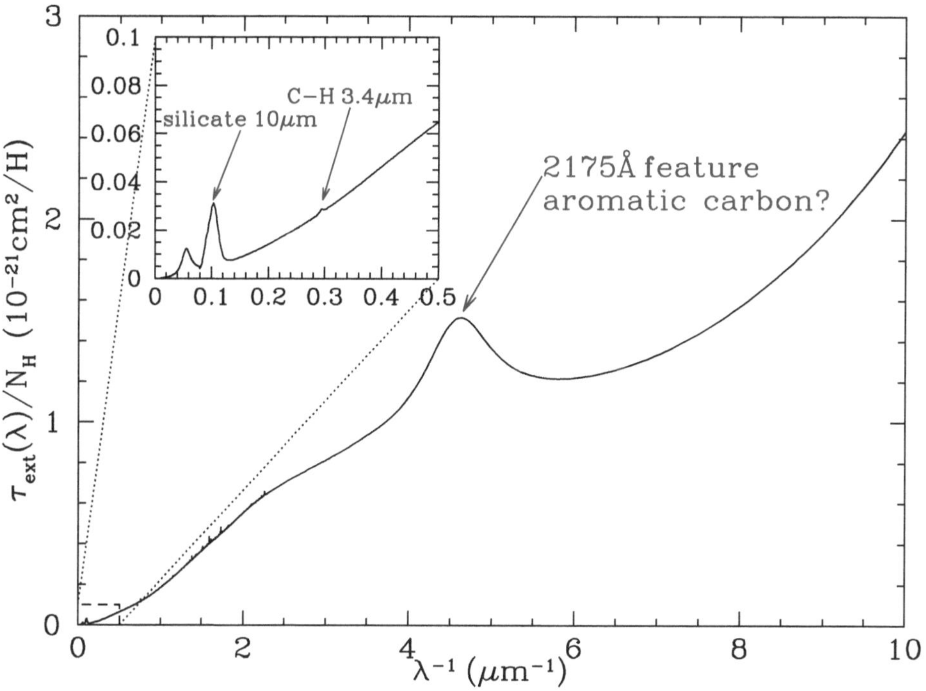

Fig. 1 The average observed extinction per H nucleon, as a function of inverse wavelength $1/\lambda$, in diffuse regions of the Milky Way. The prominent "bump" at $\lambda \approx 2175$ Å is probably due to $\pi \rightarrow \pi^*$ electronic transitions in sp^2-bonded (aromatic) carbon. The strong infrared extinction features (see *inset*) are produced by the Si–O stretching mode (9.8 μm) and the O–Si–O bending mode (18 μm). There is also a weak feature at 3.4 μm due to the C–H stretch in aliphatic (chainlike) hydrocarbons

to be produced by $\pi \rightarrow \pi^*$ electronic transitions in aromatic carbon, such as the carbon in graphite or in polycyclic aromatic hydrocarbons (PAHs). The 2175 Å bump traces only the aromatic carbon in particles with masses $m \lesssim 10^{-16}$ g: the feature is suppressed in larger grains. The 2175 Å feature therefore gives only a lower bound on the carbon content of the dust population: $\gtrsim 15\%$ of interstellar carbon is in aromatic structures.

There are also two spectroscopic features in the infrared—strong absorption features peaking near 9.7 μm and 18 μm. These features are characteristic of Si–O stretching and O–Si–O bending modes in amorphous silicates. The strength of the features requires that most interstellar Si atoms be incorporated into these silicates, together with corresponding amounts of Mg, Fe, and O. Amorphous silicates and carbonaceous materials are together thought to account for the bulk of the mass of interstellar dust in diffuse clouds. In dense and dark clouds, ices are also present, but the heliosphere is not located near a dark cloud, hence ices are not expected to be present in the dust entering the solar system from the ISM.

As discussed in Sect. 3, observations of the elements that are "depleted" from the gas phase in interstellar clouds provide an indication of what elements are in grains—the bulk of the mass of interstellar dust is contributed by the elements C, O, Mg, Si, and Fe. Based on the spectroscopic evidence in Fig. 1, it can be concluded that the dominant materials are some form of amorphous silicate (with composition $\sim$ $MgFeSiO_4$) and some mixture of carbonaceous materials—PAHs, amorphous carbon, graphite, and perhaps even diamond.

Table 1 Dust Mass per H from Milky Way Abundances. $(N_X/N_H)_\odot$ and $(N_X/N_H)_{gas}$ are the abundances of element X, by number, relative to H in the Sun and in the gas phase of a "standard" interstellar cloud (see text). $M_{X,dust}/M_H$ is the mass of element X in dust relative to the total mass of H

X	$(N_X/N_H)_\odot$ (ppm)[a]	$(N_X/N_H)_{gas}/(N_X/N_H)_\odot$[a]	$M_{X,dust}/M_H$
C	247	0.57	0.0013
N	85	0.72	0.0003
O	490	0.73	0.0021
Mg	38	0.08	0.0008
Al	3[b]	$\lesssim$0.1[c]	0.0001
Si	32	0.05	0.0009
Ca	2[b]	0.0002[d]	0.0001
Fe	29	0.007	0.0016
Ni	2	0.004	0.0001
total			0.0073

[a]Jenkins (2004) except as noted

[b]$(N_X/N_H)_\odot$ from Grevesse and Sauval (1998)

[c] assumed

[d]Savage and Sembach (1996)

3 Models for Interstellar Dust: Extinction vs. Elemental Abundances

Observations of the spectra of recently-formed stars, together with absorption lines produced by interstellar gas, have led to estimates of elemental abundances in the local interstellar material. Abundances of many elements relative to hydrogen in the ISM can also be deduced from emission lines from H II regions. Although the Sun was formed out of the ISM 4.5 Gyr ago, the elemental abundances in the ISM today appear to be close to those in the solar photosphere, and "solar abundances" are generally considered to be a good guide to interstellar abundances, although "solar abundances" are themselves uncertain: e.g., recent estimates of O/H in the solar photosphere range from (457 ± 56) ppm (Asplund et al. 2004) to (730 ± 100) ppm (Centeno and Socas-Navarro 2008).

The second column of Table 1 lists the solar abundances of the elements that are sufficiently abundant to contribute 1% or more of the mass of interstellar dust. Elements such as Ti do not appear in Table 1 because they are too rare: the abundance of Ti, by mass, is only about 0.3% of the abundance of Fe. Therefore, even though most interstellar Ti is in fact locked up in grains, Ti is not a major grain constituent.

The third column gives observed gas-phase abundances, relative to solar, in "standard" interstellar diffuse clouds, such as the well-studied cloud on the line-of-sight to the bright star ζ Oph. The gas-phase abundance of C appears to be only ~57% of the total C abundance, implying that ~43% of the carbon is sequestered in grains. For elements such as Mg, Si, or Fe the "depletions" are more severe, with 90% or more of the material locked up in grains.

Based on these observations alone, we can estimate the mass of interstellar dust: ~0.73% of the mass of the hydrogen in a "standard" cloud. It is important to recognize that "solar" abundances of elements such as C, Mg, Si, and Fe remain uncertain, and interstellar abundances might be a bit higher than solar abundances, but it is difficult to imagine that the

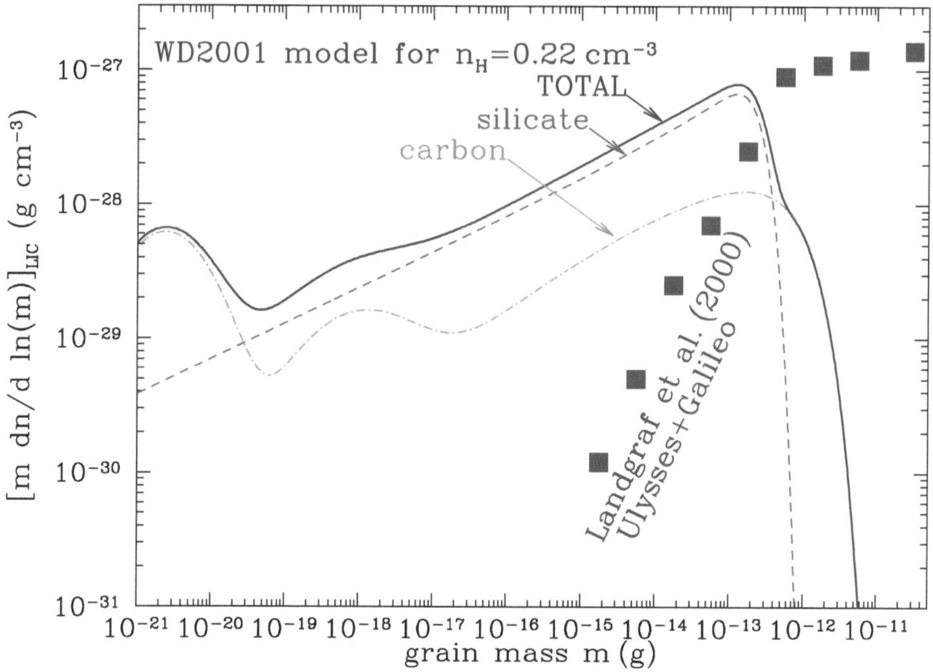

Fig. 2 The mass distribution from Weingartner and Draine (2001a) scaled to the density $n_H \approx 0.22\ cm^{-3}$ of the local interstellar cloud. The peak near $\sim 3 \times 10^{-21}$ g consists of PAHs. Also shown is the mass distribution estimated from impacts on Ulysses and Galileo (Landgraf et al. 2000). No correction for "filtration" by the heliospheric magnetic field has been applied. For $5 \times 10^{-13} < m < 3 \times 10^{-11}$ g the mass flux observed by Ulysses and Galileo is far above that expected for interstellar dust (see text)

total mass of dust in "standard" diffuse clouds could be much more than ~1.0% of the total hydrogen mass.

Various authors have obtained dust grain size distributions that reproduce the observed extinction per H as shown in Fig. 1, subject to the constraint that the mass of the dust in the model should be consistent with the "observed" mass given in Table 1 (e.g., Mathis et al. 1977; Draine and Lee 1984; Weingartner and Draine 2001a; Zubko et al. 2004). This turns out not to be an easy task: models that reproduce the observed extinction—even when trying to also minimize the total grain mass—tend to consume 100% or more of the "available" material. Modest discrepancies between the mass in the dust model and the "observed" dust mass in Table 1 would not be unexpected, given uncertainties in the observations, and given that the theoretical models make simplifying assumptions, e.g., typically assuming spherical grains. Overall, one draws the conclusion that the bulk of the interstellar grain mass is in dust grains with masses $\lesssim 5 \times 10^{-13}$ g—these grains are *needed* to produce the observed extinction, and there isn't much dust mass "left over" once the observed extinction has been reproduced.

Assuming that the interstellar grain population consists of two distinct compositions—amorphous silicate grains and carbonaceous grains—Weingartner and Draine (2001a, hereafter WD01) found size distributions for these two components that would produce extinction close to the observed extinction curve in Fig. 1, and which would incorporate amounts of C, Mg, Si, and Fe approximately consistent with current estimates of elemental abundances in the ISM. The same dust model, heated by starlight, is consistent with observa-

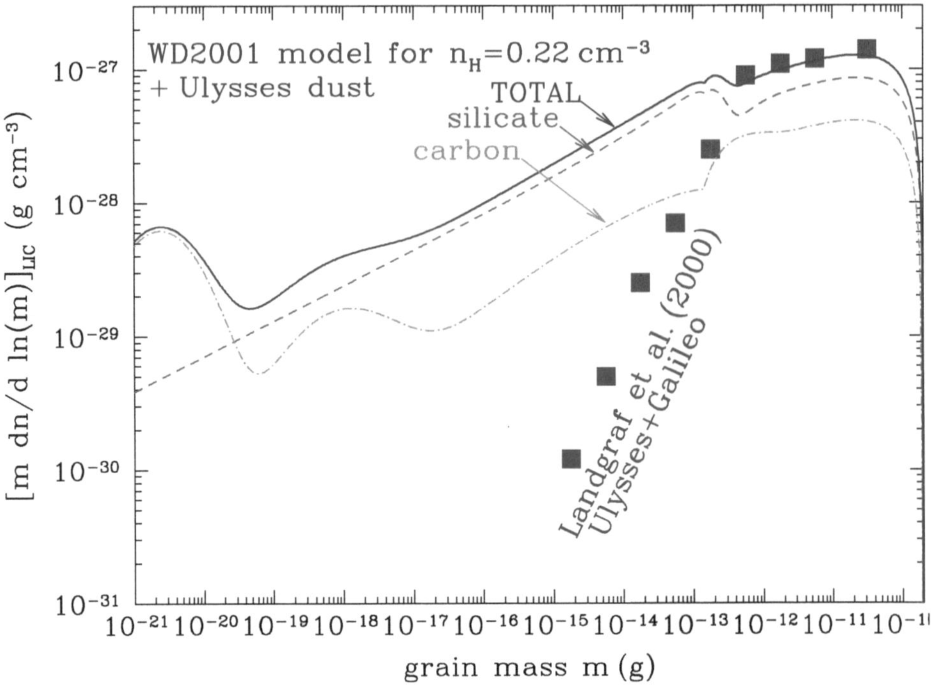

Fig. 3 The mass distribution from Weingartner and Draine (2001a) plus the "big grain" component from the Ulysses and Galileo measurements. This model has a dust/H mass ratio of 0.028, much larger than the value 0.010 of the WD01 size distribution in Fig. 2

tions of infrared emission from the Milky Way and similar galaxies (Draine and Li 2007; Draine et al. 2007). The resulting mass distributions are shown in Fig. 2, for an H nucleon density $n_{\mathrm{H}} = 0.22\ \mathrm{cm}^{-3}$, the value currently estimated for the very local ISM based on observations of inflowing He^0 (Lallement et al. 2004) and photoionization models for the nearby ISM (Slavin and Frisch 2007).

Also shown in Fig. 2 is the mass distribution of particles entering the heliosphere from the local ISM as estimated by Landgraf et al. (2000) from the dust impact detectors on Ulysses and Galileo. Because the magnetic field of the heliosphere is expected to substantially deflect incoming particles with masses $m \lesssim 3 \times 10^{-13}$ g, the fact that the Landgraf et al. (2000) results fall well below the WD2001 model for $m < 10^{-13}$ g is not surprising. However, the reported flux of $m \gtrsim 3 \times 10^{-13}$ g particles is quite unexpected if the local ISM has a dust/gas ratio typical of diffuse regions in our Galaxy.

First of all, there is the question of overall mass: as seen from Table 1, current estimates for solar and interstellar abundances would allow $M_{\mathrm{dust}}/M_{\mathrm{H}}$ of only 0.0073. Given uncertainties in both measured abundances and grain modeling, it can be argued that the WD2001 dust model ($M_{\mathrm{dust}}/M_{\mathrm{H}} \approx 0.010$—a factor 1.4 greater than the total in Table 1) is within tolerances. However, extending the size distribution to include the Ulysses results, as in Fig. 3, raises $M_{\mathrm{dust}}/M_{\mathrm{H}}$ to 0.028–3.9 times higher than the estimated total in Table 1. This is incompatible with our current understanding of elemental abundances in the general ISM.

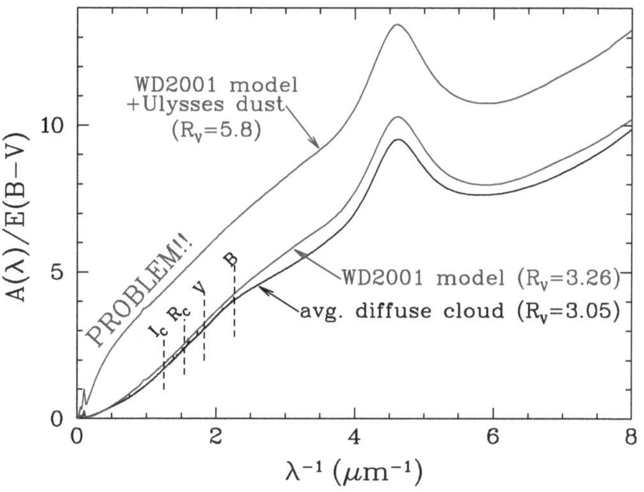

Fig. 4 Reddening law calculated for the mass distribution of Fig. 3 (note: $E(B-V) \equiv A(B) - A(V)$, where $A(\lambda)$ is the extinction at wavelength λ). The large grains contribute substantial amounts of additional extinction, and the resulting reddening curve differs strongly from observed reddening: $A(\lambda)/E(B-V)$ exceeds observed values by factors 1.5 – 2 for the commonly-observed B, V, R, and I bands

4 Contribution of Massive Grains to Extinction

If the massive grains detected by Landgraf et al. (2000) were part of the general interstellar grain population, they would have conspicuous effects on the interstellar extinction. To see this, we have taken the WD01 size distribution, and added to it an additional population of carbonaceous and silicate particles so as to approximately reproduce the Landgraf et al. (2000) size distribution at $m > 3 \times 10^{-13}$ g. We arbitrarily assume that 2/3 of the added mass is contributed by amorphous silicates and 1/3 by graphite. The adopted size distribution is shown in Fig. 3. Approximating the particles as spheres, the extinction as a function of wavelength has been calculated for the extended size distribution of Figure 3. The resulting "reddening curve" $A(\lambda)/E(B-V)$ is shown in Fig. 4.

On suitable sightlines, $A(\lambda)/E(B-V)$ can be determined observationally to accuracies of $\sim 10\%$ for $0.5 \lesssim (\lambda/\mu\text{m})^{-1} \lesssim 3$. The reddening law shown in Fig. 4 is well outside the range of what is observed (see, e.g. Mathis 1990). The synthetic curve in Fig. 4 has $R_V \approx 5.8$—such large values of R_V are not seen in diffuse clouds, being found only in regions with $A_V \gtrsim 2$.

It does not seem possible for the dust in the general ISM to have the size distribution for $m \gtrsim 3 \times 10^{-13}$ g *reported by* Landgraf et al. *(*2000*):* (1) as shown in Sect. 3, there are simply not enough atoms of C, Mg, Si, and Fe to constitute such a large mass in dust, and, (2) as seen here, if such dust were pervasive, the wavelength-dependence of interstellar extinction would be totally unlike what is actually observed.

5 Could the Dust in the Local ISM Be Atypical?

We have shown above that the large-grain population reported by Landgraf et al. (2000) cannot be pervasive. However, it is important to realize that the dust detectors on Ulysses and Galileo have only probed a tiny portion of the ISM: a cylindrical volume with diameter ~10 AU, and length increasing by ~5 AU/yr due to the solar-system's motion of 26.2 km/s relative to the local ISM (Möbius et al. 2004). We have therefore probed only a "microscopic" sample of the ISM—how representative do we expect this sample to be?

5.1 Turbulent Mixing in the ISM

MHD turbulence appears to be pervasive in the ISM. Although not understood in detail, the turbulence appears to be the result of "driving" by energetic phenomena on large scales $L_{\rm max}$—e.g., H II regions, stellar winds, supernova explosions. The turbulent cascade to smaller scales appears to approximately follow the "Kolmogorov" power-law scaling, with the velocity differences on length-scale L varying as

$$v_L \approx v_{\rm max}(L/L_{\rm max})^{1/3} \quad \text{for } L_{\rm diss} < L < L_{\rm max}, \tag{1}$$

where $L_{\rm diss}$ is the length scale below which dissipation is dominant. Because of the magnetic field, the turbulence is anisotropic, but the scaling law (1) approximately applies to turbulent motions perpendicular to the magnetic field.

Observations of turbulence within ~ 100 pc are more-or-less consistent with $v_{\rm max} \approx 10$ km s^{-1} and $L_{\rm max} \approx 100$ pc. Nonuniformities on a scale L will be erased on timescales

$$\tau_{\rm diff} \approx \frac{L}{v_L} \approx \frac{L^{2/3} L_{\rm max}^{1/3}}{v_{\rm max}} \approx 0.5 \left(\frac{L}{\rm pc}\right)^{2/3} \text{Myr}, \tag{2}$$

where we have adopted $v_{\rm max} \approx 10\,{\rm km\,s^{-1}}$ and $L_{\rm max} \approx 100$ pc. A mixing timescale of $\lesssim$Myr is short relative to Galactic timescales. Therefore we do not expect to find small-scale abundance inhomogeneities unless they were very recently injected, or unless some specific mechanism sustains them. What injection mechanisms might produce local enhancements in the population of large dust particles?

5.2 Enrichment by Supernova Explosions?

One possible source of inhomogeneity is Type II supernova explosions following core collapse in massive stars. Each such explosion enriches the nearby ISM with $\sim 5\,M_\odot$ of heavy elements, a fraction of which may be in grains. Hydrodynamic instabilities in the supernova remnant will mix these heavy elements with a mass $M_{\rm mix}$ of the ISM. The normal ISM has a heavy-element mass fraction $Z \approx 0.02$; this will be enhanced by $\Delta Z \approx 0.05(10^2\,M_\odot/M_{\rm mix})$. For the average density $\langle n_{\rm H}\rangle \approx 1$ cm^{-3} of the ISM in the solar neighborhood, this corresponds to a lengthscale $L_{\rm mix} \approx 14$ pc$(M_{\rm mix}/10^2 M_\odot)^{1/3}$ and from (2), we would expect inhomogeneities on this length scale to be erased in a time

$$\tau_{\rm diff} \approx 2.9(M_{\rm mix}/10^2 M_\odot)^{2/9} \text{ Myr} \tag{3}$$

The supernova rate/volume in the Galactic disk is $S \approx 10^{-13}$ pc^{-3} yr^{-1}. The probability that a SN exploded within a distance $L_{\rm mix}$ within a time $\tau_{\rm diff}$ is only $\sim L_{\rm mix}^3 S \tau_{\rm diff} \approx 10^{-3}$. It is therefore very improbable that the local interstellar cloud has been heavily enriched by a recent SN explosion. The Local Bubble is believed to have been caused by one or more SN explosions over the past 10–15 Myr, but these were located at a distance of ~ 100 pc (Fuchs et al. 2006). The strongest argument against enrichment by SN ejecta is the fact that the gas-phase abundances of Mg and Fe appear to show normal depletions relative to solar abundances (Redfield and Linsky 2008).

5.3 Wake of an Evolved Star?

Cool AGB stars have dusty winds that may pollute the ISM with fresh grain material—for example, the wind from Mira = o Ceti (Martin et al. 2007). What is the likelihood that a recent passage by an AGB star left behind a concentration of large grains that might account for the excess of large particles seen by Ulysses?

Consider a star moving at speed $v_\star$ relative to the ISM, losing mass at a rate $\dot{M}$. It will leave behind a wake, with radius R_w, filled with gas with density n_w and temperature T_w. Mass conservation and balance with the interstellar pressure $p_{\rm ISM}$ give

$$n_w 1.4 m_{\rm H} \pi R_w^2 v_\star = \dot{M}, \tag{4}$$

$$n_w k T_w = p_{\rm ISM}. \tag{5}$$

These two equations can be solved for the wake radius

$$R_w = \left[\frac{\dot{M}}{1.4 m_{\rm H} \pi v_\star} \frac{T_w}{p_{\rm ISM}/k}\right]^{1/2} = 0.13\ {\rm pc} \left[\frac{\dot{M}_{-6}}{v_{\star,6}} \frac{T_{w2}}{(p_{\rm ISM}/k)_{5000}}\right]^{1/2},$$

$$\dot{M}_{-6} \equiv \frac{\dot{M}}{10^{-6} M_\odot\ {\rm yr}^{-1}}, \quad v_{\star,6} \equiv \frac{v_\star}{10\ {\rm km\ s}^{-1}}, \quad T_{w2} \equiv \frac{T}{100\ {\rm K}},$$

$$\left(\frac{p_{\rm ISM}}{k}\right)_{5000} \equiv \frac{p_{\rm ISM}/k}{5000\ {\rm cm}^{-3}\ {\rm K}}. \tag{6}$$

The trailing wake will be mixed with the ISM by turbulent diffusion on a time given by (3):

$$\tau_{\rm diff} \approx 0.5 (R_w/{\rm pc})^{2/3}\ {\rm Myr} \approx 0.13 \left[\frac{\dot{M}_{-6}}{v_{\star,6}} \frac{T_{w2}}{(p_{\rm ISM}/k)_{5000}}\right]^{1/3}\ {\rm Myr}. \tag{7}$$

If the duration of the mass loss phase is longer than $\tau_{\rm diff}$, the wake volume will be

$$V_w \approx \pi R_w^2 v_\star \tau_{\rm diff} \approx 0.07 \left[\frac{\dot{M}_{-6} T_{w,2}}{(p_{\rm ISM}/k)_{5000}}\right]^{4/3} v_{\star,6}^{-1/3}\ {\rm pc}^3. \tag{8}$$

The *total* rate of stellar mass loss in the Milky Way is $\sim 1\ M_\odot\ {\rm yr}^{-1}$, e.g., 10^6 stars, each with $\dot{M}_{-6} \approx 1$. If $\sim 10^6$ stars are randomly-distributed in a disk of full-thickness ~ 200 pc and radius 12 kpc, then the stellar density is $n_\star \approx 1 \times 10^{-5}\ {\rm pc}^{-3}$, and the nearest-neighbor distance is $n_\star^{-1/3} \approx 50$ pc. [The distance to Mira, $D = 107$ pc, is in rough agreement with our estimate for $n_\star^{-1/3}$.] With $n_\star \approx 10^{-5}\ {\rm pc}^{-3}$, the fraction of the volume occupied by "wakes" is very small:

$$n_\star V_w \approx 7 \times 10^{-7} \left[\frac{\dot{M}_{-6} T_{w,2}}{(p_{\rm ISM}/k)_{5000}}\right]^{4/3} v_{\star,6}^{-1/3}\ . \tag{9}$$

It is therefore *extremely* unlikely that the Solar System would by chance be located today within such a stellar wake; this conclusion will not be changed for any plausible variation of uncertain parameters such as $\dot{M}_{-6}$, $T_{w,2}$, or $v_{\star,6}$.

5.4 Dynamical Concentration of Massive Grains?

We have seen above that it is highly unlikely that the Solar System is by chance passing through gas that was recently enriched in very large grains from either a supernova explosion

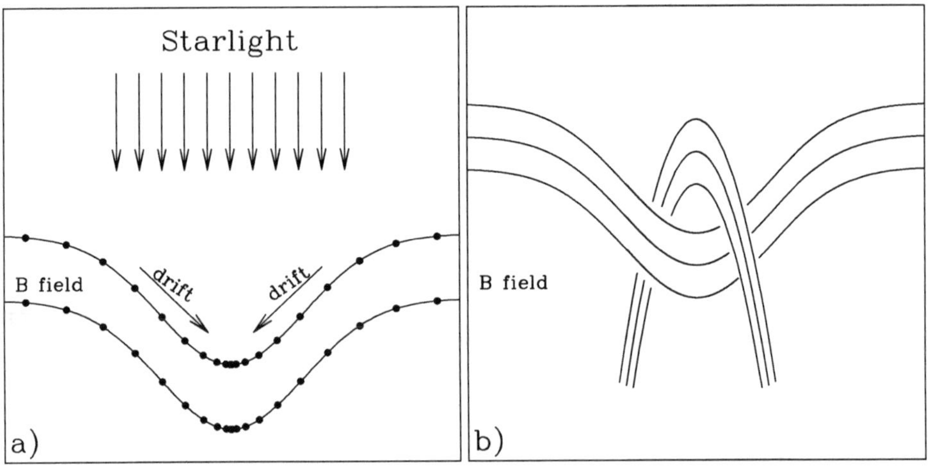

Fig. 5 (**a**) Radiation pressure-driven drift of dust grains along magnetic field lines could concentrate grains in magnetic "valleys" (see text). (**b**) Tangled magnetic fields could produce magnetic valleys

or an evolved star. What other process might produce the anomalous concentration of dust grains that appears to be present in the portion of the ISM we are now passing through?

Dust grains and gas atoms are subject to different forces, and in general the dust grains will *drift* relative to the gas. Charged dust grains are coupled to the magnetic field, which inhibits drift across magnetic field lines, but the grains are free to drift *along* field lines. Drift velocities resulting from radiation pressure and other effects of anisotropic starlight have been discussed by Weingartner and Draine (2001b). The drift velocities are not large, but can attain $\sim 0.5\,\mathrm{km\,s^{-1}}$ in the "warm neutral medium" conditions characteristic of the region the Solar System is now moving through (see Figs. 17, 18 of Weingartner and Draine 2001b). If sustained for long enough, these drifts might result in variations in the dust/gas ratio.

One possible scenario for concentrating dust is illustrated in Fig. 5a. If field lines are bent, radiation pressure could push grains into magnetic "valleys", as shown. If the width and depth of the "valley" are both of size L_B, then dust might accumulate on a time scale

$$t_{\mathrm{accum}} \approx \frac{L_B}{v_{\mathrm{drift}}} \approx 10^3\,\mathrm{yr}\left(\frac{L_B}{100\ \mathrm{AU}}\right)\left(\frac{0.5\,\mathrm{km\,s^{-1}}}{v_{\mathrm{drift}}}\right). \tag{10}$$

Magnetic stresses will act to try to straighten the field lines. Radiation pressure acting on the grains, if strong enough, could keep the field deformed, and could even cause the field deformation to grow, in a manner akin to the Parker instability, except with radiation pressure on dust playing the role of gravity on gas. However, this would require balancing the magnetic force per volume $\sim \nabla(B^2/8\pi) \approx B^2/8\pi L_B$ with the radiation pressure force per volume $\kappa\rho J_{\mathrm{rad}}/c$, where κ is the dust opacity, ρ is the dust mass density, and J_{rad} is the net flux of starlight. With the magnetic force/volume scaling as $1/L_B$, and parameters appropriate to the Milky Way, it does not appear that radiation pressure on dust could deform the magnetic field on length scales $L_B \lesssim 10$ pc. However, local field curvature might be maintained by magnetic stresses if the magnetic field is tangled, as shown in Fig. 5b.

If radiation-pressure-driven drift is responsible for concentrating very large grains, it should also have acted to concentrate the $m \approx 10^{-13}$ g grains that are thought to dominate the grain size distribution in the average interstellar medium (recall Fig. 2), as their drift velocities will be similar to those of larger grains. It is not clear that the "filtration" effects in the heliosphere will be able to suppress the flux of these particles to the values observed by Ulysses and Galileo.

6 Structure of the Very Local ISM

Recent analyses of microparticle impacts on the Ulysses spacecraft appear to indicate that the impacting dust velocity vector in heliocentric coordinates has shifted by 30° over the 15 years of observation (Krüger et al. 2007). The interstellar dust mass flux at 4–5 AU also appears to have varied by a factor $\sim$3 over 1992–2006. These variations might be due to solar-cycle-related changes in the interplanetary B field at $\gtrsim 5$ AU (Landgraf et al. 2003; Krüger et al. 2007), but variations in such electromagnetic "filtration" would be expected to result in variations in velocity vector and flux as a function of grain size, with electromagnetic deflection expected to be minimal for $m \gtrsim 2 \times 10^{-12}$ g. Surprisingly, size-dependence of the velocity vector is not evident in the data (Krüger et al. 2007), so we must consider the possibility that the grain mass flux impinging on the heliosphere is variable. Since 15 years of observation corresponds to a spatial scale of only $\sim$83 AU, variations in the grain flux incident on the heliosphere would require substantial variations in both grain density and velocity over length scales of only tens of AU. Such small-scale variations in the dust density in the local interstellar medium, if present, would appear to require an active mechanism, such as described in Sect. 5.4, to maintain it. While slow dust drift relative to the gas might account for density variations, one would not expect large velocity variations in a quiescent medium (Weingartner and Draine 2001b, estimated $v_{\rm drift} \lesssim 0.5$ km s^{-1}).

It is interesting to note that the velocity of the inflowing He^0 does not coincide with the velocity of two closest interstellar clouds: the "Local Interstellar Cloud" (LIC) and "Cloud G": the velocity of the local He^0 is close to the *average* of the LIC and G cloud velocity vectors (Redfield and Linsky 2008). In view of this, it is natural to consider the possibility that the heliosphere might, by chance, be located in the narrow shock transition where the two clouds interact: the timc-dependence of the mass flux and velocity of inflowing atoms and dust grains may be revealing structure in a multifluid shock transition layer.

7 Summary

The size distribution of interstellar grains entering the heliosphere, as inferred from observations by Ulysses and Galileo (Landgraf et al. 2000; Krüger et al. 2007) cannot be typical of the general interstellar medium, as can be demonstrated by two independent arguments:

1. The required abundance of elements in grains would substantially exceed what is available in the interstellar medium.
2. If such a size distribution were generally present, it would produce an interstellar "reddening law" very different from what is observed.

Therefore, if the size distribution of local interstellar dust does have the large grain population reported by Landgraf et al. (2000), the dust grain/gas ratio in the interstellar medium must be quite nonuniform. The length scale characterizing these nonuniformities is not known. If the velocity vector of the incoming dust flow is actually changing over time scales of only years—one possible explanation for the variations in the directions of impacting particles reported by Krüger et al. (2007)—this would require that the dust velocity vary over lengthscales of only tens of AU. Such small scale structure was not expected.

Mechanisms that might account for such nonuniformity are considered. It seems extremely unlikely that the Sun is passing through a region that has recently been enriched with dust from a stellar source. The least unlikely scenario may involve concentration of dust in certain regions, and removal of dust from other regions, by dynamical processes. One possible mechanism involving anisotropic starlight driving dust grains along deformed magnetic field lines is outlined. Whether this can compete with the diffusive effects of turbulent mixing is far from clear, however.

It is important to carry out additional observations to confirm the enhanced grain size distribution, and to confirm the time-dependence of the density and velocity vector of the inflowing dust and gas. If the reported density of large grains, and the time-dependence of the inflow, are confirmed, this may require revision of our understanding of the small-scale structure of the ISM. Absorption line studies seem to suggest that, by coincidence, the heliosphere is just now passing through the transition zone—possibly a shock transition—between the "Local Interstellar Cloud" and "Cloud G". If so, the flow into the heliosphere offers the opportunity to study the small-scale structure in this transition zone. The Ulysses observations indicate that this region is heavily enriched with large dust particles, although why this should be so remains unclear.

Acknowledgements I thank the organizers for inviting me to participate in the Workshop. I am grateful to R.H. Lupton for availability of the SM graphics program, used in preparation of the figures in this paper. I thank E. Grün, H. Krüger, J. Linsky, and an anonymous referee for helpful discussions and comments. This research was supported in part by NSF grant AST-0406883.

References

M. Asplund, N. Grevesse, A.J. Sauval, C. Allende Prieto, D. Kiselman, Astron. Astrophys. **417**, 751–768 (2004)
J.A. Cardelli, G.C. Clayton, J.S. Mathis, Astrophys. J. **345**, 245–256 (1989)
R. Centeno, H. Socas-Navarro, ArXiv:0803.0990v3 (2008)
B.T. Draine, in *IAU Symp. 135: Interstellar Dust* (1989), pp. 313–327
B.T. Draine, Annu. Rev. Astron. Astrophys. **41**, 241–289 (2003)
B.T. Draine, D.A. Dale, G. Bendo, K.D. Gordon, J.D.T. Smith, L. Armus, C.W. Engelbracht, G. Helou, R.C. Kennicutt, A. Li, H. Roussel, F. Walter, D. Calzetti, J. Moustakas, E.J. Murphy, G.H. Rieke, C. Bot, D.J. Hollenbach, K. Sheth, H.I. Teplitz, Astrophys. J. **663**, 866–894 (2007)
B.T. Draine, H.M. Lee, Astrophys. J. **285**, 89–108 (1984)
B.T. Draine, A. Li, Astrophys. J. **657**, 810–837 (2007)
E.L. Fitzpatrick, Publ. Astron. Soc. Pac. **111**, 63–75 (1999)
B. Fuchs, D. Breitschwerdt, M.A. de Avillez, C. Dettbarn, C. Flynn, Mon. Not. R. Astron. Soc. **373**, 993–1003 (2006)
N. Grevesse, A.J. Sauval, Space Sci. Rev. **85**, 161–174 (1998)
E. Grun, H.A. Zook, M. Baguhl, A. Balogh, S.J. Bame, H. Fechtig, R. Forsyth, M.S. Hanner, M. Horanyi, J. Kissel, B.-A. Lindblad, D. Linkert, G. Linkert, I. Mann, J.A.M. McDonnell, G.E. Morfill, J.L. Phillips, C. Polanskey, G. Schwehm, N. Siddique, P. Staubach, J. Svestka, A. Taylor, Nature **362**, 428–430 (1993)
E.B. Jenkins, in *Origin and Evolution of the Elements*, ed. by A. McWilliam, M. Rauch (2004), pp. 336–353
H. Krueger, E. Gruen, ArXiv e-prints **802** (2008)
H. Krüger, M. Landgraf, N. Altobelli, E. Grün, Space Sci. Rev. **130**, 401–408 (2007)
R. Lallement, J.C. Raymond, J. Vallerga, M. Lemoine, F. Dalaudier, J.L. Bertaux, Astron. Astrophys. **426**, 875–884 (2004)
M. Landgraf, W.J. Baggaley, E. Grün, H. Krüger, G. Linkert, J. Geophys. Res. **105**, 10,343–10,352 (2000)
M. Landgraf, H. Krüger, N. Altobelli, E. Grün, J. Geophys. Res. (Space Phys.) **108**, 8030 (2003)

D.C. Martin, M. Seibert, J.D. Neill, D. Schiminovich, K. Forster, R.M. Rich, B.Y. Welsh, B.F. Madore, J.M. Wheatley, P. Morrissey, T.A. Barlow, Nature **448**, 780–783 (2007)
J.S. Mathis, Annu. Rev. Astron. Astrophys. **28**, 37–70 (1990)
J.S. Mathis, W. Rumpl, K.H. Nordsieck, Astrophys. J. **217**, 425–433 (1977)
E. Möbius, M. Bzowski, S. Chalov, H.-J. Fahr, G. Gloeckler, V. Izmodenov, R. Kallenbach, R. Lallement, D. McMullin, H. Noda, M. Oka, A. Pauluhn, J. Raymond, D. Ruciński, R. Skoug, T. Terasawa, W. Thompson, J. Vallerga, R. von Steiger, M. Witte, Astron. Astrophys. **426**, 897–907 (2004)
S. Redfield, J.L. Linsky, Astrophys. J. **673**, 283–314 (2008)
B.D. Savage, K.R. Sembach, Annu. Rev. Astron. Astrophys. **34**, 279–330 (1996)
J.D. Slavin, P.C. Frisch, Space Sci. Rev. **130**, 409–414 (2007)
J.C. Weingartner, B.T. Draine, Astrophys. J. **548**, 296–309 (2001a)
J.C. Weingartner, B.T. Draine, Astrophys. J. **553**, 581–594 (2001b)
V. Zubko, E. Dwek, R.G. Arendt, Astrophys. J. Suppl. Ser. **152**, 211–249 (2004)

Interstellar Dust Inside and Outside the Heliosphere

Harald Krüger · Eberhard Grün

Originally published in the journal Space Science Reviews, Volume 143, Nos 1–4, 347–356.
DOI: 10.1007/s11214-008-9431-3

Abstract In the early 1990s, after its Jupiter flyby, the Ulysses spacecraft identified interstellar dust in the solar system. Since then the in-situ dust detector on board Ulysses continuously monitored interstellar grains with masses up to 10^{-13} kg, penetrating deep into the solar system. While Ulysses measured the interstellar dust stream at high ecliptic latitudes between 3 and 5 AU, interstellar impactors were also measured with the in-situ dust detectors on board Cassini, Galileo and Helios, covering a heliocentric distance range between 0.3 and 3 AU in the ecliptic plane. The interstellar dust stream in the inner solar system is altered by the solar radiation pressure force, gravitational focussing and interaction of charged grains with the time varying interplanetary magnetic field. The grains act as tracers of the physical conditions in the local interstellar cloud (LIC). Our in-situ measurements imply the existence of a population of 'big' interstellar grains (up to 10^{-13} kg) and a gas-to-dust-mass ratio in the LIC which is a factor of 1.5–2 larger than the one derived from astronomical observations, indicating a concentration of interstellar dust in the very local interstellar medium. Until 2004, the interstellar dust flow direction measured by Ulysses was close to the mean apex of the Sun's motion through the LIC, while in 2005, the data showed a 30° shift, the reason of which is presently unknown. We review the results from spacecraft-based in-situ interstellar dust measurements in the solar system and their implications for the physical and chemical state of the LIC.

Keywords Dust · Interstellar dust · Heliosphere · Interstellar matter

H. Krüger (✉)
Max-Planck-Institut für Sonnensystemforschung, Max-Planck-Str. 2, 37191 Katlenburg-Lindau, Germany
e-mail: krueger@linmpi.mpg.de

H. Krüger · E. Grün
Max-Planck-Institut für Kernphysik, Saupfercheckweg 1, 69117 Heidelberg, Germany

E. Grün
Laboratory for Atmospheric and Space Physics, University of Colorado, Boulder, CO, 80303-7814, USA

1 Introduction

Interstellar dust (ISD) became a topic of astrophysical research in the early 1930s when the existence of extinction, weakening, and scattering of starlight in the interstellar medium (ISM) was realised. At that time, astronomical observations provided the only information about the properties of the dust in the ISM. With the advent of dust detectors onboard spacecraft, it became possible to investigate dust particles in-situ. About 30 years ago, analysis of the data obtained with the dust instruments flown on a couple of spacecraft suggested that ISD grains can cross the heliospheric boundary and penetrate deeply into the heliosphere (Bertaux and Blamont 1976; Wolf et al. 1976). In the 1990s, this was undoubtedly demonstrated with the dust instrument carried by the Ulysses spacecraft: the Ulysses dust detector, which measured mass, speed and approach direction of the impacting grains, identified ISD grains with radius above 0.1 μm sweeping through the heliosphere (Grün et al. 1993, 1994, 1995).

The galactic setting of the Sun is dominated by the Local Bubble void in the interstellar medium which is a low-density, presumably hot region of irregular shape that extends to approximately 100 pc from the Sun. The Local Bubble was excavated by supernova explosions and winds of massive stars in the neighbouring starforming regions of the Scorpius-Centaurus and Orion Associations. The hot gas in the Local Bubble is usually called Local Interstellar Medium (LISM). Figure 1 shows the projection in galactic coordinates of 15 interstellar clouds identified within 15 pc of the Sun. According to recent analyses (Redfield and Linsky 2008) the Sun appears to be located in the transition zone between two of the closest clouds, the Local Interstellar Cloud (LIC) and the G cloud. The only direct observation of ISD close to the Sun is weak polarization observed along the sightline towards 36 Oph (distance about 6 pc) which is due to magnetically aligned dust grains (Tinbergen 1982). Therefore, in-situ sampling of dust from the LIC can greatly improve our understand-

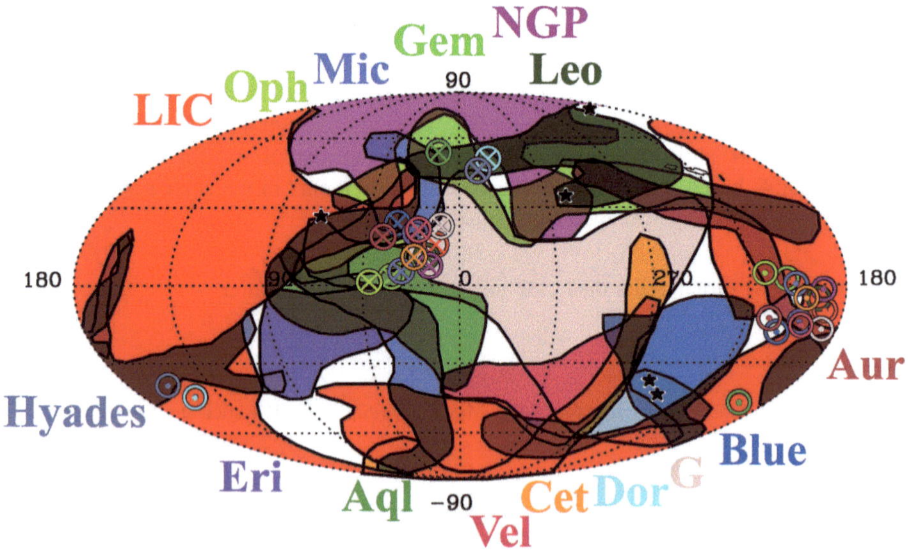

Fig. 1 Locations in galactic coordinates of 15 interstellar clouds located within 15 pc of the Sun (from Redfield and Linsky 2008). The upwind heliocentric direction of the velocity vector for each cloud is indicated by the *circled cross*, while the downwind heliocentric direction is indicated by a *circled dot*

ing of the nature and processing of dust in various galactic environments and can cast new light on the chemical composition and homogeneity of the interstellar medium.

In this paper we review the results from in-situ ISD measurements obtained with the Ulysses and other space-borne dust detectors. We review our current knowledge about ISD inside the heliosphere and in our local interstellar environment.

2 Interstellar Dust Inside the Heliosphere

The Ulysses in-situ dust measurements showed that the grain motion through the solar system is parallel to the flow of neutral interstellar hydrogen and helium gas, both gas and dust travelling with a speed of 26 km s^{-1} (Grün et al. 1994; Baguhl et al. 1995; Witte et al. 1996; Witte 2004). The upstream direction of the dust flow lies at 259° ecliptic longitude and 8° latitude (Landgraf 1998). The interstellar dust flow persists at high ecliptic latitudes above and below the ecliptic plane and even over the poles of the Sun, whereas interplanetary dust is strongly depleted at high latitudes (Grün et al. 1997). The interstellar dust flux measured at a distance of about 3 AU from the Sun is time-dependent, and the mean mass of the grains is about 3×10^{-16} kg (Landgraf et al. 2000), corresponding to a grain radius of approximately 0.3 μm. Measurements with the identical dust instrument onboard Galileo performed in the ecliptic plane showed that beyond about 3 AU the interstellar dust flux even exceeds the flux of micron-sized interplanetary grains.

Results from the first decade of the Ulysses mission showed that the radii of clearly identified interstellar grains range from 0.05 μm to above 1 μm. The data show distance-dependent alteration of the interstellar dust stream caused by gravitational focussing by the Sun, solar radiation pressure, and electromagnetic interaction with the time-varying interplanetary magnetic field (IMF; Altobelli et al. 2003, 2005a, 2005b; Mann and Kimura 2000; Landgraf 2000; Czechowski and Mann 2003). Radiation pressure and electromagnetic forces strongly depend on grain size, leading to a strong modification of the size distribution and fluxes of grains measured inside the heliosphere (Landgraf et al. 1999, 2003). Strong filtration of small grains due to electromagnetic forces also occurs at the heliospheric boundary (Linde and Gombosi 2000). The overall size distribution measured in-situ with Ulysses within 5 AU from the Sun shows a deficiency of small grains below 0.3 μm compared to astronomically observed ISD (Mathis 2000; Draine 2003; Frisch et al. 1999). In addition, solar radiation pressure deflects grains with sizes of about 0.4 μm and was found to be effective at solar distances below 4 AU (Landgraf et al. 1999).

Significant differences in the particle sizes were also recorded at different heliocentric distances. In addition to the Ulysses measurements which revealed a lack of small 0.3 μm ISD grains within 3 AU heliocentric distance, measurements by Cassini and Galileo between 0.7 and 3 AU showed that the detected interstellar particles were bigger than 0.5 μm, with grain masses increasing closer to the Sun (Altobelli et al. 2003, 2005b; Altobelli 2004). The flux of these bigger particles did not exhibit temporal variations due to the solar-wind magnetic field like the flux of smaller particles observed by Ulysses. The trend of increasing particle masses continues even closer to the Sun, as demonstrated by Helios which recorded particles of about 1 μm down to 0.3 AU (Altobelli et al. 2005a, 2006). These facts support the idea that the ISD stream is strongly filtered by solar radiation pressure. Interstellar particles with optical properties of astronomical silicates or organic refractory materials are consistent with the observed radiation pressure effects (Landgraf et al. 1999).

In addition to studies of the distribution of grain masses, the Ulysses dust instrument monitored the flux of the interstellar particles in the heliosphere (Fig. 2). In mid 1996, we

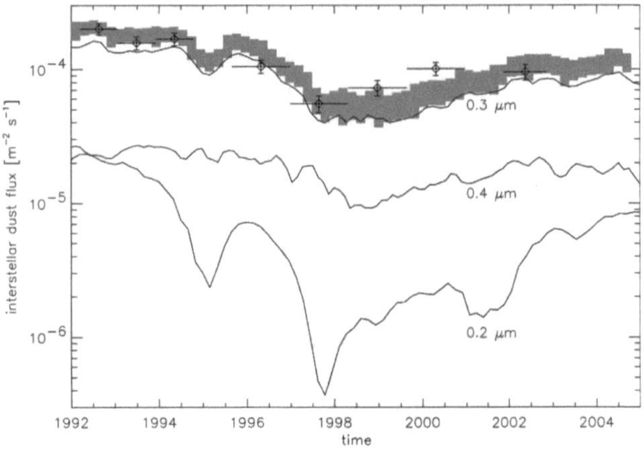

Fig. 2 Fit of simulated to measured flux (Landgraf et al. 2003). The fit parameters are the relative contributions of grains of sizes between 0.1 and 0.4 μm (the 0.1 μm curve is not shown, because it did not contribute to the fit). The *solid lines* show the flux profiles of the simulated grains of various sizes, scaled with their best-fit relative contributions. The *shaded region* indicates the best-fit total predicted flux, with its vertical extent giving the 1σ uncertainty

observed a decrease of the interstellar dust flux by a factor of 3 from an initial value of 1.5×10^{-4} m^{-2} s^{-1} down to 0.5×10^{-4} m^{-2} s^{-1}. This drop was attributed to increased filtering of small grains by the solar wind driven IMF during solar minimum conditions (Landgraf 1998, 2000; Landgraf et al. 2000). Since early 2000, Ulysses has again detected interstellar dust flux levels above 10^{-4} m^{-2} s^{-1} (Landgraf et al. 2003; Krüger et al. 2007). Monte-Carlo simulations of the grain dynamics in the heliosphere showed that the dominant contribution to the dust flux comes from grains with a charge-to-mass ratio of $q/m = 0.59$ C kg^{-1} and a radiation pressure efficiency of $\beta = 1.1$, corresponding to grain radii of 0.3 μm (Landgraf et al. 2003).

Particles even bigger (40 μm) than the grains measured in-situ with the spacecraft detectors were reliably identified by meteor radar observations (Taylor et al. 1996; Baggaley 2000; Baggaley and Neslušan 2002; Meisel et al. 2002). The grains were identified by their hyperbolic speeds, and their flow direction varied over a significantly wider angular range than that of the much smaller grains observed by spacecraft. Baggaley (2000) identified a general background influx of extra-solar system particles from southern ecliptic latitudes with enhanced fluxes from discrete sources. More sensitive meteor observations with the Arecibo radar found micron-sized interstellar meteor particles radiating from the direction of the Geminga pulsar (Meisel et al. 2002). This is particularly interesting because the supernova that formed the Geminga pulsar is a potential candidate which may have created the Local Bubble.

Ulysses has monitored the interstellar dust flow through the solar system for more than 15 years now. This time period covers more than two and a half revolutions of the spacecraft about the Sun through more than 2/3 of a complete 22-year solar cycle. Thus, Ulysses measured interstellar dust during solar minimum and solar maximum conditions of the interplanetary magnetic field (IMF). The interstellar dust flux modulation due to grain interaction with the magnetic field during solar minimum could be well explained (Landgraf 1998, 2000; Landgraf et al. 2003). By taking into account the sensor side wall in the instrument field of view we could recently improve the flux determination (Altobelli et al. 2004).

Until early 2005 the approach direction of the interstellar grains was in agreement with the interstellar helium flow direction as determined by the Ulysses/GAS measurements (Witte 2004; Witte et al. 2004; Landgraf and Grün 1998; Frisch et al. 1999; Krüger et al. 2006). An example is shown in the left panel of Fig. 3 which shows the impact direction of the interstellar impactors in the period from 1996 to 2000. Six years

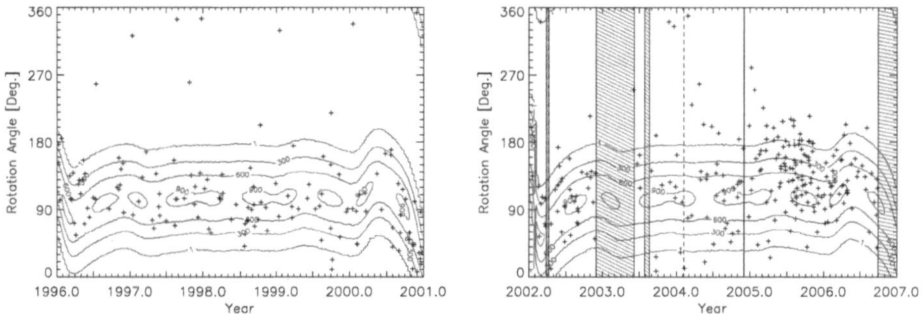

Fig. 3 Impact direction (i.e. spacecraft rotation angle at dust particle impact) of interstellar grains measured with Ulysses in two time intervals (from Krüger et al. 2007). Each cross indicates an individual dust impact. *Left:* 1 January 1996 to 31 December 2000; *right:* 1 January 2002 to 31 December 2006. In the time intervals shown here, the spacecraft rotation axis was pointing towards Earth. The dust detector was mounted almost perpendicular to the spacecraft–Earth line so that it scanned through a plane perpendicular to the spacecraft–Earth line (Grün et al. 1992, 1997; Krüger et al. 2001, 2006). Ecliptic north is close to 0°. Contour lines show the effective sensor area for particles approaching from the upstream direction of interstellar helium (Witte 2004; Witte et al. 2004). In the right panel, a vertical dashed line shows Jupiter closest approach on 5 February 2004, *five shaded areas* indicate periods when the dust instrument was switched off

later, when Ulysses was travelling through almost the same spatial region and had an almost identical detection geometry for interstellar grains, the situation was vastly different: first, the range in approach directions of the grains was somewhat wider (best seen in 2004; right panel of Fig. 3; see also Krüger et al. 2007); second, and more noticeable, in 2005/06 the approach direction of the majority of grains was shifted away from the helium flow direction. Our preliminary analysis indicates that this shift is about 30° away from the ecliptic plane towards southern ecliptic latitudes (Krüger et al. 2007). At the moment, we do not know whether it is a temporary shift limited to the time period stated above or whether it continues to the present time. Furthermore, the reason for this shift remains mysterious. Whether it is connected to a secondary stream of interstellar neutral atoms shifted from the main neutral gas flow (Collier et al. 2004; Wurz et al. 2004; Nakagawa et al. 2006) is presently unclear. Given, however, that the neutral gas stream is shifted along the ecliptic plane while the shift in the dust flow is offset from the ecliptic, a connection between both phenomena seems unlikely.

Even though Ulysses' position in the heliosphere and the dust detection conditions were very similar during both time intervals considered in Fig. 3, the configurations of the solar wind driven interplanetary magnetic field (IMF), which strongly affects the dynamics of the smallest grains, were completely different. We have to consider that the interstellar grains need approximately twenty years to travel from the heliospheric boundary to the inner solar system where they are detected by Ulysses. Thus, the effect of the IMF on the grain dynamics is the accumulated effect caused by the interaction with the IMF over several years: In the earlier time interval (1997–1999) the grains had a recent dynamic history dominated by solar minimum conditions (Landgraf 2000), while the grains detected during the second interval (2002–2005) had a recent history dominated by the much more disturbed solar maximum conditions of the IMF. During the solar maximum conditions the overall magnetic dipole field changed polarity. Morfill and Grün (1979) predicted that due to this effect in a 22-year cycle, small interstellar grains experience either focussing or defocusing conditions. During these times they are systematically deflected by the solar wind magnetic field either towards or away from the solar magnetic equator plane (close to the ecliptic plane). This latter configuration likely has a strong influence on the dust dynamics and the total interstellar flux in the inner heliosphere but it is not modelled in detail in the presently existing models. An explanation of the grain interaction with the IMF at the recent solar maximum conditions is still pending.

The fact that the models fit the flux variation by assuming a constant dust concentration in the Local Interstellar Cloud (LIC) implies that the dust phase of the LIC must be homogeneously distributed over length scales of 50 AU, which is the distance inside the LIC travelled by the Sun during the measurement period of Ulysses from end 1992 to end 2002 (Landgraf et al. 2003). This conclusion is supported by the more recent Ulysses data until the end of 2004 (Krüger et al. 2006). The 2005/06 data, on the other hand, put a question mark onto this conclusion because if the observed shift in impact direction turns out to be intrinsic, it would imply that this homogeneity breaks down on larger length scales.

3 Interstellar Dust in the Local Interstellar Cloud

ISD grains carry information about their past dynamics outside the heliosphere and are thus of strong interest to understand the dynamical processes in the Local Interstellar Cloud (LIC). They provide the main reservoir and transport mechanism of heavy elements in the interstellar medium (Li and Greenberg 1997). The dynamics of the grains is crucial for understanding of nucleation, growth and collisional destruction processes (Draine 2003) which strongly depend on the relative velocities of the grains. The most important phenomena responsible for the spread of velocities in the LIC are gas drag, interaction with the local interstellar magnetic field, radiation pressure and photoelectric emission (Frisch et al. 1999). The relative strengths of the different forces strongly depend on the size and the charge of the grains, together with local conditions of the interstellar medium (ISM), like gas or magnetic field turbulences.

Observations of interstellar material (ISM) towards nearby stars and inside of the solar system, combined with radiative transfer models, constrain the conditions in the LIC (Frisch 1998, 1999; Frisch and Slavin 2003; Slavin and Frisch 2006, 2007, 2008). To our present knowledge, the main characteristics of the LIC are: atomic neutral hydrogen density $n(H^0) = 0.19–0.20\,cm^{-3}$, electron and ion density $n(e^-) = 0.05–0.08\,cm^{-3}$, temperature $\sim$6300 K, and a relative Sun-cloud velocity $\sim 26\,km\,s^{-1}$. The physical conditions in the LIC are those of the intercloud medium—warm, low density, partially ionized gas ($X(H^+) \sim 0.2$, $X(He^+) \sim 0.4$). An enhancement of refractory elements (such as Fe, Mg, Mn) in LIC gas, compared to cool interstellar clouds, points to the destruction of interstellar dust grains by interstellar shocks (velocity 100–200 $km\,s^{-1}$; Frisch et al. 1999).

At least 5 distinct cloudlets are found within 5 pc of the Sun, with differing compositions and physical properties (Redfield and Linsky 2002, 2004a, 2004b, 2008). Temperatures range from 5,400 K (towards α Cen) to 10,000 K (Blue Cloud towards ε CMa) and total densities from $>0.04\,cm^{-3}$ (Blue Cloud towards ε CMa) to possibly $>5\,cm^{-3}$ (G-cloud towards α Cen, Frisch 2003; Gry and Jenkins 2001). The gas-phase abundance of Fe, with respect to undepleted S, varies by $\sim$50% within 3 pc of the Sun, evidently due to grain destruction processes (Frisch and Slavin 2003).

If the ISM is chemically homogeneous, elements absent from the gas phase must be depleted onto dust grains. This argument can be used to evaluate the gas-to-dust mass ratio $R_{g/d}$ over the integrated LIC column, and $R_{g/d}$ can be compared with that of other nearby interstellar clouds. However, the required knowledge of the total chemical composition of the ISM is an elusive quantity that has not been reliably determined. A 40–50% variation in Fe^+/S^+ and Si^+/S^+ for the two clouds towards ε CMa indicates different grain histories for two similar clouds within 3 pc of each other. If atoms not observed in the gas are concentrated in the dust, $R_{g/d}$ can be calculated from observations of interstellar absorption lines towards nearby stars. When evidence for 60–70% subsolar abundances is included, $R_{g/d} \sim 600$ integrated over the diameter of the LIC (Frisch et al. 1999; Frisch and Slavin 2003). Gas-to-dust mass ratios calculated from more recent models with improved solar abundances are in the range $R_{g/d} \sim 149$–217, again depending on solar abundances (Slavin and Frisch 2008). Interestingly, $R_{g/d}$ determined from comparisons of the Ulysses in-situ measurements inside of the solar system, compared to gas densities from these models, yield $R_{g/d} = 116$–127 (Landgraf et al. 2000; Altobelli et al. 2004). It should be emphasised that the $R_{g/d}$ obtained from the in-situ measurements is an upper limit, since the smallest interstellar dust grains (radii 0.1 μm) are prevented from entering the heliosphere.

The analysis of the interstellar dust mass distribution by Landgraf (1998) was based on only a few years of dust measurements from the early Ulysses mission. We are presently re-evaluating the mass distribution with the data set until 2007. Our preliminary analysis confirms the earlier results. In particular, the existence and abundance of big interstellar grains as discussed above are confirmed by the more recent data. Furthermore, after 2002 the overall dust flux has levelled off to the values before 1996 (Krüger et al. 2007).

Overall, the in-situ value is a factor of 1.5–2 larger than the one derived from astronomical observations, indicating a relative concentration of interstellar dust in the ISM close to the Sun compared to the ~0.5 pc LIC cloud length towards ε CMa. The gas-to-dust mass ratio also varies by more than 30% over the nearest 3 pc. If ISM abundances are solar, the in-situ and astronomical methods of determining $R_{g/d}$ are—generally—in better agreement, but interstellar absorption line data towards weakly reddened stars remain unexplained. These differences are not yet understood. The chemical composition of interstellar dust grains observed within the solar system thus provides a window on the chemical composition and homogeneity of the ISM.

The combination of absorption line data toward ε CMa and the modelled photoionization also lead to the conclusion that the LIC has a very interesting pattern of gas phase elemental abundances (Slavin and Frisch 2007, 2008): C appears to be substantially supersolar while Fe, Mg and Si are subsolar. O and N are close to solar. This indicates that carbonaceous grains have been destroyed in the LIC while silicate grains have survived. The extra C in the gas has not been explained but may be evidence for a local enhancement of carbonaceous dust followed by grain destruction in a shock.

The masses of interstellar grains measured in-situ with the spacecraft detectors range from 10^{-18} kg to above 10^{-13} kg. If we compare the mass distribution of these interstellar impactors with the dust mass distribution derived from astronomical observations, we find that the in-situ measurements overlap only with the largest masses observed astronomically. This is further supported by the radar measurements which revealed even bigger grains. These measurements imply that the intrinsic size distribution of interstellar grains in the LIC extends to sizes much larger than those grains which are detectable by astronomical observations (Frisch et al. 1999; Frisch and Slavin 2003; Landgraf et al. 2000; Grün and Landgraf 2000).

There are no direct observations of interstellar dust within 5 pc and outside of the solar system. The observations of very weak starlight polarization towards nearby stars (<40 pc) may originate from magnetically aligned dust grains close to the solar system. The observed polarization strength is consistent with the average interstellar density of ~0.1 cm^{-3} over tens of parsecs in the upwind direction (Frisch 1990). The in-situ grains have a size distribution consistent with these classical dust grains.

Interstellar gas and dust couple through collisional processes, and through coupling of ions and charged grains to the interstellar magnetic field. Over distance scales of 100–500 pc, gas-dust coupling is demonstrated through the correlation of starlight-reddening dust grains (measured as color excess E(B-V)) and interstellar hydrogen ($N(H^0) + 2N(H_2)$) (Bohlin et al. 1978). For the multicloud structure observed within 5 pc, gas-dust coupling is not proven. The collisional lifetimes for classical dust grains (radii $\sim$0.2 μm) in the LIC are $\sim 0.3 \times 10^6$ years, during which time the LIC will move $\sim$5 pc through local space. The gyroradius is $\sim$0.1 pc in a field of 3 μG (Grün and Landgraf 2000). The result will be magnetically captured dust grains that are collisionally destroyed over the lifetime of the cloud. Grün and Landgraf (2000) suggested that the small "classical" grains are replenished by the collisional destruction of larger dust grains. Alternatively, silicate grain destruction peaks near shock column densities of $N(H) \sim 6 \times 10^{17}$ cm^{-2} (Jones et al. 1994), allowing the breakdown of gas-dust coupling locally over $\sim$1 pc length scales.

There are important consequences from the existence of the big particle population in the LIC. While particles observed by spacecraft couple to the interstellar medium on length scales of less than 1 pc via electromagnetic interactions, more massive grains couple to the gas over much longer scales of 100 to 1000 pc (Grün and Landgraf 2000). Therefore, big interstellar meteor particles travel unaffected over much longer distances and may come directly from their source region.

4 Outlook

In 2007 and early 2008 Ulysses was in the inner solar system where interstellar grains cannot reliably be separated from interplanetary impactors but in mid-2008 the dust instrument could begin to monitor the interstellar grains again. Due to a serious failure in the power system on board, however, the spacecraft is not fully operational at the moment and the dust detector is switched off. It is presently unclear whether Ulysses can be put back into normal operation. Before the onboard failure occurred, a mission extension until 2011 was considered and would have provided interstellar dust data from the outer heliosphere again. With this extension the Ulysses measurements from the entire mission would have covered an almost entire 22-year solar cycle. But even now the Ulysses data is a unique set of dust measurements from interplanetary space for decades to come. Together with detailed modelling of the grain interaction with the IMF during the highly disordered solar maximum conditions we will hopefully be able to reveal the origin of the observed 30° shift. If the shift turned out to be intrinsic, being potentially connected with a secondary population of interstellar grains, it would put strong constraints on the small-scale structure of the LIC. This would also be highly relevant for the interpretation of results from the Stardust mission which recently brought a sample of collected interstellar grains to Earth (A. Westphal 2006, priv. comm.), and for future dust astronomy space missions aiming at the in-situ analysis and sample return of interstellar dust (CosmicDUNE, SARIM; Grün et al. 2005; Srama et al. 2008).

Acknowledgements We thank the Ulysses project at ESA and NASA/JPL for effective and successful mission operations. This work has been supported by the Deutsches Zentrum für Luft- und Raumfahrt e.V. (DLR) under grants 50 0N 9107 and 50 QJ 9503. Support by Max-Planck-Institut für Kernphysik and Max-Planck-Institut für Sonnensystemforschung is also gratefully acknowledged.

References

N. Altobelli, Monitoring of the interstellar dust stream in the inner solar system using data of different spacecraft. Ph.D. Thesis, Ruprecht-Karls-Universität Heidelberg (2004)

N. Altobelli, S. Kempf, M. Landgraf, R. Srama, V. Dikarev, H. Krüger, G. Moragas-Klostermeyer, E. Grün, J. Geophys. Res. **108**(A10), 7–1 (2003)
N. Altobelli, R. Moissl, H. Krüger, M. Landgraf, E. Grün, Planet. Space Sci. **52**, 1287–1295 (2004)
N. Altobelli, S. Kempf, H. Krüger, M. Landgraf, R. Srama, E. Grün, in *AIP Conf. Proc. 761: The Spectral Energy Distributions of Gas-Rich Galaxies: Confronting Models with Data* (2005a), pp. 149–152
N. Altobelli, S. Kempf, H. Krüger, M. Landgraf, M. Roy, E. Grün, J. Geophys. Res. **110**, 7102–7115 (2005b)
N. Altobelli, E. Grün, M. Landgraf, Astron. Astrophys. **448**, 243–252 (2006)
W.J. Baggaley, J. Geophys. Res. **105**(A5), 10,353–10,361 (2000)
W.J. Baggaley, L. Neslušan, Astron. Astrophys. **382**, 1118–1124 (2002)
M. Baguhl, E. Grün, D.P. Hamilton, G. Linkert, R. Riemann, P. Staubach, Space Sci. Rev. **72**, 471–476 (1995)
J.L. Bertaux, J.F. Blamont, Nature **262**, 263–266 (1976)
R.C. Bohlin, B.D. Savage, J.F. Drake, Astrophys. J. **224**, 132–142 (1978)
M.R. Collier, T.E. Moore, D. Simpson, A. Roberts, A. Szabo, S. Fuselier, P. Wurz, M.A. Lee, B.T. Tsurutani, Adv. Space Res. **34**, 166–171 (2004)
A. Czechowski, I. Mann, J. Geophys. Res. **108**(A10), 8038 (2003). doi:10.1029/2003JA009.917
B.T. Draine, Ann. Rev. Astron. Astrophys. **41**, 241–289 (2003)
P.C. Frisch, in *Physics of the Outer Heliosphere*, ed. by S. Grzedzielski, D.E. Page, 1st COSPAR Colloquium, Warsaw (1990), pp. 19–22
P.C. Frisch, Space Sci. Rev. **86**, 107 (1998)
P.C. Frisch, in *The Galactic Environments of Nearby Cool Stars*, ed. by L.M. Celnikier, J.T.T. Van (Editions Frontieres, 1999), pp. 3–10
P.C. Frisch, Astrophys. J. **593**, 868–873 (2003)
P.C. Frisch, J.D. Slavin, Astrophys. J. **594**, 844–858 (2003)
P.C. Frisch, J. Dorschner, J. Geiß, J.M. Greenberg, E. Grün, M. Landgraf, P. Hoppe, A.P. Jones, W. Krätschmer, T.J. Linde, G.E. Morfill, W.T. Reach, J. Slavin, J. Svestka, A. Witt, G.P. Zank, Astrophys. J. **525**, 492–516 (1999)
E. Grün, M. Landgraf, J. Geophys. Res. **105**(A5), 10,291–10,298 (2000)
E. Grün, H. Fechtig, J. Kissel, D. Linkert, D. Maas, J.A.M. McDonnell, G.E. Morfill, G.H. Schwehm, H.A. Zook, R.H. Giese, Astron. Astrophys. Suppl. **92**, 411–423 (1992)
E. Grün, H.A. Zook, M. Baguhl, A. Balogh, S.J. Bame, H. Fechtig, R. Forsyth, M.S. Hanner, M. Horányi, J. Kissel, B.A. Lindblad, D. Linkert, G. Linkert, I. Mann, J.A.M. McDonnell, G.E. Morfill, J.L. Phillips, C. Polanskey, G.H. Schwehm, N. Siddique, P. Staubach, J. Svestka, A. Taylor, Nature **362**, 428–430 (1993)
E. Grün, B.E. Gustafson, I. Mann, M. Baguhl, G.E. Morfill, P. Staubach, A. Taylor, H.A. Zook, Astron. Astrophys. **286**, 915–924 (1994)
E. Grün, M. Baguhl, D.P. Hamilton, J. Kissel, D. Linkert, G. Linkert, R. Riemann, Planet. Space Sci. **43**, 941–951 (1995). Paper I
E. Grün, P. Staubach, M. Baguhl, D.P. Hamilton, H.A. Zook, S.F. Dermott, B.A. Gustafson, H. Fechtig, J. Kissel, D. Linkert, G. Linkert, R. Srama, M.S. Hanner, C. Polanskey, M. Horányi, B.A. Lindblad, I. Mann, J.A.M. McDonnell, G.E. Morfill, G.H. Schwehm, Icarus **129**, 270–288 (1997)
E. Grün, R. Srama, H. Krüger, S. Kempf, V. Dikarev, S. Helfert, G. Moragas-Klostermeyer, Icarus **174**, 1–14 (2005)
C. Gry, E.B. Jenkins, Astron. Astrophys. **367**, 617–628 (2001)
A.P. Jones, A.G.G.M. Tielens, D.J. Hollenbach, C.F. McKee, Astrophys. J. **433**, 797–810 (1994)
H. Krüger, E. Grün, M. Landgraf, S.F. Dermott, H. Fechtig, B.A. Gustafson, D.P. Hamilton, M.S. Hanner, M. Horányi, J. Kissel, B. Lindblad, D. Linkert, G. Linkert, I. Mann, J.A.M. McDonnell, G.E. Morfill, C. Polanskey, G.H. Schwehm, R. Srama, H.A. Zook, Planet. Space Sci. **49**, 1303–1324 (2001). Paper VII
H. Krüger, N. Altobelli, B. Anweiler, S.F. Dermott, V. Dikarev, A.L. Graps, E. Grün, B.A. Gustafson, D.P. Hamilton, M.S. Hanner, M. Horányi, J. Kissel, M. Landgraf, B. Lindblad, D. Linkert, G. Linkert, I. Mann, J.A.M. McDonnell, G.E. Morfill, C. Polanskey, G.H. Schwehm, R. Srama, H.A. Zook, Planet. Space Sci. **54**, 932–956 (2006). Paper IX
H. Krüger, M. Landgraf, N. Altobelli, E. Grün, Space Sci. Rev. **130**, 401–408 (2007)
M. Landgraf, Modellierung der Dynamik und Interpretation der In-situ-Messung interstellaren Staubs in der lokalen Umgebung des Sonnensystems. Ph.D. Thesis, Ruprecht-Karls-Universität Heidelberg (1998)
M. Landgraf, J. Geophys. Res. **105**(A5), 10,303–10,316 (2000)
M. Landgraf, E. Grün, in *Proceedings of the IAU Colloquium No. 166 on the Local Bubble and Beyond*, ed. by D. Breitschwerdt, M.J. Freyberg, J. Trümper. Lecture Notes in Physics, vol. 506 (Springer, Heidelberg, 1998), pp. 381–384
M. Landgraf, K. Augustsson, E. Grün, B.A.S. Gustafson, Science **286**, 2,319–2,322 (1999)
M. Landgraf, W.J. Baggaley, E. Grün, H. Krüger, G. Linker, J. Geophys. Res. **105**(A5), 10,343–10,352 (2000)

M. Landgraf, H. Krüger, N. Altobelli, E. Grün, J. Geophys. Res. **108**, 5–1 (2003)
A. Li, J.M. Greenberg, Astron. Astrophys. **323**, 566–584 (1997)
T.J. Linde, T.I. Gombosi, J. Geophys. Res. **105**, 10,411–10,418 (2000)
I. Mann, H. Kimura, J. Geophys. Res. **105**(A5), 10,317–10,328 (2000)
J.S. Mathis, J. Geophys. Res. **105**, 10,269–10,278 (2000)
D.D. Meisel, D. Janches, J.D. Mathews, Astrophys. J. **567**, 323–341 (2002)
G.E. Morfill, E. Grün, Planet. Space Sci. **27**, 1283–1292 (1979)
H. Nakagawa, M. Bzowski, A. Yamazaki, H. Fukunishi, S. Watanabe, Y. Takahashi, M. Taguchi, in *36th COSPAR Scientific Assembly*, vol. 36. COSPAR, Plenary Meeting (2006), pp. 1170–+
S. Redfield, J.L. Linsky, Astrophys. J. Suppl. **139**, 439–465 (2002)
S. Redfield, J.L. Linsky, Astrophys. J. **602**, 776–802 (2004a)
S. Redfield, J.L. Linsky, Astrophys. J. **613**, 1004–1022 (2004b)
S. Redfield, J.L. Linsky, Astrophys. J. **673**, 283–314 (2008)
J.D. Slavin, P.C. Frisch, Astrophys. J. Lett. **651**, L37–L40 (2006)
J.D. Slavin, P. Frisch, Space Sci. Rev. **130**, 409–414 (2007)
J.D. Slavin, P.C. Frisch, Astron. Astrophys. (2008, submitted). http://arxiv.org/abs/0704.0657
R. Srama, T. Stephan, E. Grün, N. Pailer, A. Kearsley, A.L. Graps, R. Laufer, P. Ehrenfreund, N. Altobelli, K. Altwegg, S. Auer, J. Baggaley, M. Burchell, J. Carpenter, L. Colangeli, F. Esposito, S. Green, H. Henkel, M. Horányi, A. Jäckel, S. Kempf, N. McBride, G. Moragas-Klostermeyer, H. Krüger, A. Srowig, M. Trieloff, P. Tsou, Z. Sternovsky, O. Zeile, H.-P. Röser, Exp. Astron. (2008, submitted)
A.D. Taylor, W.J. Baggaley, D.I. Steel, Nature **380**, 323–325 (1996)
J. Tinbergen, Astron. Astrophys. **105**, 53–64 (1982)
M. Witte, Astron. Astrophys. **426**, 835–844 (2004)
M. Witte, H. Banaszkiewicz, H. Rosenbauer, Space Sci. Rev. **78**(1/2), 289–296 (1996)
M. Witte, M. Banaszkiewicz, H. Rosenbauer, D. McMullin, Adv. Space Res. **34**, 61–65 (2004)
H. Wolf, J. Rhee, O.E. Berg, in *Interplanetary Dust and Zodiacal Light*, ed. by H. Elsaesser, H. Fechtig. Lecture Notes in Physics, vol. 48 (Springer, Berlin, 1976), pp. 165–169
P. Wurz, M.R. Collier, T.E. Moore, D. Simpson, S. Fuselier, W. Lennartson, in *Physics of the Outer Heliosphere*, ed. by V. Florinski, N.V. Pogorelov, G.P. Zank. *American Institute of Physics Conference Series*, vol. 719 (2004), pp. 195–200

Section VII: What Are the Roles That Magnetic Fields Play in the Local Bubble and Its Constituents?

Obtaining Spectra of Turbulent Velocity from Observations

A. Lazarian

Originally published in the journal Space Science Reviews, Volume 143, Nos 1–4, 357–385.
DOI: 10.1007/s11214-008-9460-y

Abstract We discuss a long-standing problem of how turbulence can be studied using observations of Doppler broadened emission and absorption lines. The focus of the present review is on two new techniques, the Velocity–Channel Analysis (VCA), which makes use of the channel maps, and the Velocity Coordinate Spectrum (VCS), which utilizes the fluctuations measured along the velocity axis of the Position–Position Velocity (PPV) data cubes. Both techniques have solid theoretical foundations based on analytical calculations as well as on numerical testings. Among the two the VCS, which has been developed quite recently, has two unique features. First of all, it is applicable to turbulent volumes that are not spatially resolved. Second, it can be used with absorption lines that do not provide good spatial sampling of different lags over the image of the turbulent object. In fact, recent studies show that measurements of absorption line along less than 10 absorption directions are sufficient for a reliable recovering of the underlying spectrum of the turbulence. Moreover, both weak absorption lines and absorption lines in saturated regime can be used, which extends the applicability of the technique. Our comparison of the VCA and the VCS with a more traditional technique of Velocity Centroids shows that the former two techniques recover reliably the spectra of supersonic turbulence, while the Velocity Centroids may be used only for studying subsonic turbulence. We discuss spectra of astrophysical turbulence obtained with the VCA and the VCS techniques.

Keywords Turbulence · MHD · Interstellar medium

PACS 95.30.Qd · 52.30.Cv · 96.50.Tf · 98.38.Gt

1 Introduction

The ISM is turbulent on scales ranging from AUs to kpc (see Armstrong et al. 1995; Elmegreen and Scalo 2004), with an embedded magnetic field that influences almost all

A. Lazarian (✉)
Department of Astronomy, University of Wisconsin-Madison, Madison, USA
e-mail: lazarian@astro.wisc.edu

of its properties. MHD turbulence is accepted to be of key importance for fundamental astrophysical processes, e.g. star formation, propagation and acceleration of cosmic rays.

How to study astrophysical turbulence? This review stresses the importance of observational studies of the turbulence spectra. We feel that the progress of numerical modeling of astrophysical turbulence shifted somewhat the attention of the astrophysical community from observational studies. Therefore we believe that stressing of the synergy of the observational and numerical studies is due.

It is generally accepted that numerical simulations have tremendously influenced our understanding of the physical conditions of turbulence (see Vazquez-Semadeni et al. 2000; Mac Low and Klessen 2004; Ballesteros-Paredes et al. 2006; McKee and Ostriker 2007 and ref. therein). Present codes can produce simulations that resemble observations, but because of their limited numerical resolution, there are serious concerns about how well they reproduce reality, especially astrophysical turbulence (see e.g. McKee 1999; Shu et al. 2004). In this respect, observational studies of turbulence can test to what extent the numerical approach represents the actual physical processes.

The turbulence spectrum, which is a statistical measure of turbulence, can be used to compare observations with both numerical simulations and theoretical predictions. Note that statistical descriptions are nearly indispensable strategy when dealing with turbulence. The big advantage of statistical techniques is that they extract underlying regularities of the flow and reject incidental details. The Kolmogorov description of unmagnetized incompressible turbulence is a statistical one. For instance it predicts that the difference in velocities at different points in turbulent fluid increases on average with the separation between points as a cubic root of the separation, i.e. $|\delta v| \sim l^{1/3}$. In terms of direction-averaged energy spectrum this gives the famous Kolmogorov scaling $E(k) \sim 4\pi k^2 P(\mathbf{k}) \sim k^{5/3}$, where $P(\mathbf{k})$ is a *3D* energy spectrum defined as the Fourier transform of the correlation function of velocity fluctuations $\xi(\mathbf{r}) = \langle \delta v(\mathbf{x}) \delta v(\mathbf{x}+\mathbf{r}) \rangle$. Note that in this paper we use $\langle \ldots \rangle$ to denote averaging procedure.

However, the importance of obtaining turbulence spectrum from observations goes well beyond testing the accuracy of numerical modeling. For instance, the energy spectrum $E(k)dk$ characterizes how much energy resides at the interval of scales $k, k+dk$. At large scales l which correspond to small wavenumbers k (i.e. $l \sim 1/k$) one expects to observe features reflecting energy injection. At small scales one should see the scales corresponding to sinks of energy. In general, the shape of the spectrum is determined by a complex process of non-linear energy transfer and dissipation. Thus, observational studies of the turbulence spectrum can determine sinks and sources of astrophysical turbulence.

In view of the above it is not surprising that attempts to obtain spectra of interstellar turbulence have been numerous since 1950s (see Munch 1958). However, various directions of research achieved various degree of success (see Armstrong et al. 1995). For instance, studies of turbulence statistics of ionized media accompanied by theoretical advancements in understanding scattering and scintillations of radio waves in ionized medium (see Goodman and Narayan 1985) were rather successful (see Spangler and Gwinn 1990). They provided the information of the statistics of plasma density at scales 10^8–10^{15} cm. However, these sort of measurements provide only the density statistics, which is an indirect measure of turbulence.

Velocity statistics is a much more coveted turbulence measure. Although it is clear that Doppler broadened lines are affected by turbulence, recovering of velocity statistics is extremely challenging without an adequate theoretical insight. Indeed, both z-component velocity and density contribute to fluctuations of the energy density $\rho_s(\mathbf{X}, V_z)$ in the Position–Position–Velocity (PPV) space.

Traditionally, the information on turbulence spectra is obtained with the measure of Doppler shift termed Velocity Centroids, $\sim \int V_z \rho_s d V_z$, where the integration is taking place over the range of the velocities relevant to the object under study. In this situation it is easy to see that the Velocity Centroids are also proportional to $\int V_z \rho ds$, where ρ is an actual three dimensional density and the integration is performed along the line of sight (see Lazarian and Esquivel 2003). While usually the Velocity Centroids are normalized over the integrated intensity over the line of sight (see Stenholm 1990), Esquivel and Lazarian (2005) showed that this normalization does not change the statistical properties of the measure. The numerical and analytical analysis in (Esquivel and Lazarian 2005) and Esquivel et al. (2007) showed that the Velocity Centroids fail for studying supersonic turbulence. This provides extremely bad news for the studies of velocity statistics in molecular clouds and diffuse cold ISM (see Dickman and Kleiner 1985; Miesch et al. 1999; Miville-Deschênes et al. 2003). The studies for HII regions (O'dell and Castaneda 1987) are less affected, as in most cases, the turbulence there is subsonic.

There were attempts to analyze PPV data cubes in other ways, for instance, Crovisier and Dickey (1983), Green (1993) and Stanimirovic et al. (1999) analyzed power spectra of velocity channels of HI data. The spatial spectrum of fluctuations of these velocity slices of PPV revealed power-law dependences, but the physical meaning of these dependences was absolutely unclear.

The analytical study of the statistical properties of the PPV energy density ρ_s has been initiated in Lazarian and Pogosyan (2000). There the observed statistics of ρ_s was related to the underlying 3D spectra of velocity and density in the astrophysical turbulent volume. Initially, the volume was considered transparent, but later the treatment was generalized for the volume with self-absorption and for studies of turbulence using absorption lines. In what follows, we discuss how the observable Doppler-shifted lines can be used to recover the spectrum of turbulent velocity using two new techniques that, unlike other mostly empirical techniques, have solid theoretical foundations. How to obtain using spectroscopic observations other characteristics of turbulence, e.g. higher order statistics, anisotropies has been reviewed earlier (see Lazarian 2004).

The new techniques of studying astrophysical turbulence are intended to make use of extensive spectroscopic surveys. In the review, we mostly discuss HI and CO data. However, many more lines, not necessarily radio lines, can be used. For instance, absorption optical and UV lines can also be used within the framework of the new techniques.

In general, this review attempts to provide an intuitive approach to the understanding of the new techniques of data analysis. We start with a general discussion of how the Doppler-broaderned spectral lines can be analyzed in Sect. 2, introduce the basics of the analytical description of the statistics of the spectral lines in Sect. 3. In Sect. 4 for power-law spectra of velocities and densities we present the spectra of one, two and three dimensional fluctuations observable spectral line fluctuations and discuss the effects of additional factors, e.g. finite resolution of a telescope, presence of self-absorption. We describe the studies of turbulence with saturated absorption lines in Sect. 5, compare the two major techniques for studying turbulence with spectral lines in Sect. 6, and present the results of numerical testing of the techniques in Sect. 7. In Sect. 8 we discuss the ways of extending the techniques, e.g. to allow for a) studies of turbulence that does not obey a simple power-law, b) studies of turbulence in the local ISM, and c) studies of turbulence with lines the strength of which is proportional to density squared. A short summary of the observational studies of interstellar turbulence using the new techniques is presented in Sect. 9. We analyze the alternative approaches to studying velocity turbulence in Sect. 10. In Sect.11 we discuss the synergy that can be obtained by combining different techniques of observational turbulence studies and outline the prospects for the field.

2 Ways to Analyze Position–Position–Velocity Data Cubes

The most detailed information on turbulent velocity observationally available from an emitting or absorbing turbulent volume is a Position–Position–Velocity (PPV) data cube. Evidently, the resolution in the P-P plane is provided by the spatial resolution of the telescope, while the V-resolution requires spectroscopic resolution of the Doppler shifted lines at different locations over the image of the emitting volume.

As we mentioned above, frequently, astronomers analyze the distribution of PPV intensities within a particular velocity range or a channel map. This velocity range may be the minimal interval corresponding to the maximum of spectral resolution (see Green 1993) or chosen to be wider, e.g. to decrease the noise (Stanimirovic et al. 1999). *Spatial* spectra obtained by taking Fourier transform of channel maps had been used to study HI before we conducted our theoretical study of what those spectra mean in Lazarian and Pogosyan, (2000, henceforth LP00). These studies were providing power spectra of channel map intensity distributions which varied from study to study. The relation of the power spectra to the underlying velocity fluctuations was a subject of speculation with claims that the spectral index of channel maps coincides with the spectral index of velocity fluctuation in 2D slices of the actual turbulent volume (see Green 1989 and references therein). We showed in LP00 that the latter naive claim, as well as other speculations related to the interpretation of the spatial intensity variations within channel maps, are not correct. This was done through establishing the explicit relation between the statistics of the intensity fluctuations within the channel maps and the statistics of the underlying velocity and density.

To understand the essence the Velocity Channel Analysis (VCA) formulated in LP00, consider a PPV cube arising from measuring Doppler shifted spectra from a turbulent volume (see Fig. 1). The channel maps in Fig. 1 correspond to the velocity slices of the PPV cube. One may ask a question whether fluctuations depend on the thickness of the velocity slice. It is intuitively clear that if the medium is optically thin and the velocity is integrated over the entire spectral line, the fluctuations can depend only on density inhomogeneities. It is also suggestive that the contribution of the velocity fluctuations may depend on whether the images of the eddies under study fit within a velocity slice or if their velocity extent is larger than the slice thickness (see Fig. 1, left). In the former case the slice is "thick" for

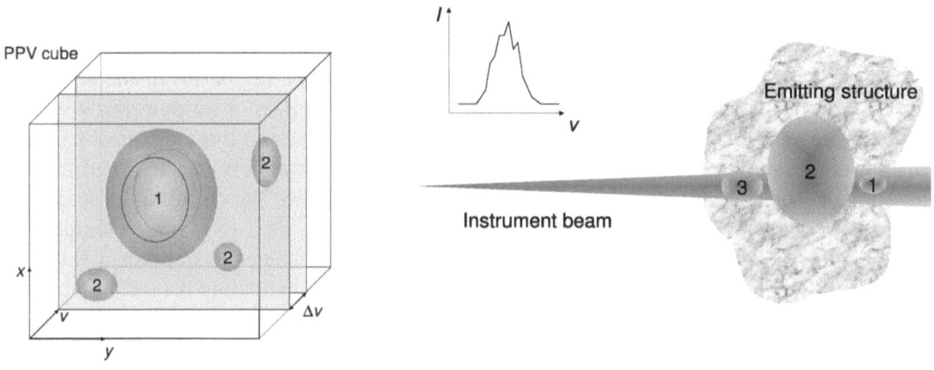

Fig. 1 *Left panel*: PPV data cube. Illustration of the concepts of the thick and thin velocity slices. The slices are thin for the PPV images of the large eddies, but thick for the images of small eddies. *Right panel*: Illustration of the VCS technique. For a given instrument resolution large eddies are in the high resolution limit, while small eddies are in the low resolution limit. The velocity profile shown illustrates the turbulence-brodened profiles of lines that are used for the VCS studies

eddies and in the latter case it is "thin". According to LP00 the spectra of fluctuations that correspond to "thin" and "thick" slices are different and varying the thickness (i.e. effective channel width) of slices it is possible to disentangle the statistics of underlying velocities and densities in the turbulent volume.

Note, that the questions of whether the spectra of intensity fluctuations within channel maps may depend on the channel map thickness were not posed by the research prior to LP00 study. This resulted in comparing of spectra obtained with the channel map spectra of different thickness, which, as we understand now, is incorrect. LP00 showed that some of the differences of the earlier reported spectral indexes were due the differences in the thickness of the velocity slices analyzed (see Stanimirovic and Lazarian 2001). By now the VCA technique has been applied to a number of different sets of spectral line data to determine the underlying velocity and density of turbulent fluctuations (see Sect. 9).

LP00 dealt with optically thin data. Optically thick CO data was traditionally analyzed differently from HI (see Falgarone and Puget 1995). Namely, spectra of total intensities were studied. The origin of such a spectrum and its relation to the underlying velocity and density fluctuations was established in Lazarian and Pogosyan (2004, henceforth LP04). This work also clarified the effects of absorption that were reported for HI data. In terms of the techniques of turbulence study, LP04 extends the study of the VCA technique in LP00 to the case of studying turbulence within an emitting turbulent volume in the presence of absorption. Intuitively, results of LP04 are easy to understand. In turbulence there is a unique statistical relation between the physical scales and the turbulent velocities. LP04 proved that if the thickness of a velocity slice is larger than the dispersion of velocities of the eddies which get optically thick, the effects of self-absorption should be taken into account (see Sect. 4 for more details).

A radically different way of analyzing spectroscopic data is presented in Lazarian and Pogosyan (2006, henceforth LP06). There the spectra of intensity fluctuations along the V-axis of the PPV cube are studied (see Fig. 1, right). The mathematical foundations of the technique can be traced to LP00, but there the high potential of the Velocity Coordinate Spectrum (VCS), as the technique was later termed in Lazarian (2004), required further studies. Indeed, it took some time to understand the advantages that the VCS provides for the practical handling of the observational data. Numerical testing of the technique (see Chepurnov and Lazarian 2008) was also important.

LP06 deals with studies of turbulence using emission lines and absorption lines in the limit of weak absorption. VCS allows us to analyze turbulence when spatial information is either not available or sparse. For instance, as we discuss in Sect. 6 one can potentially study turbulence and obtain the turbulence spectrum in a spatially unresolved turbulent volume.[1] One can also study turbulence having just a couple of absorption lines, which corresponds to sampling of the PPV volume along a few directions only (see Fig. 8). Naturally, this stems from the fact that the fluctuations along the V-axis are studied by the VCS. This makes the VCS technique really unique for the velocity turbulence studies. We illustrate the technique discussing its first applications in Sect. 9.

A more recent study by Lazarian and Pogosyan (2008, henceforth LP08) deals with the studies of turbulence using saturated absorption lines. There we showed that the saturation of the line acts as a sort of window function in the velocity space. In the presence of this window function one can still use the unsaturated wings of the line to get a high frequency input on turbulence.

[1] Note, that poor resolution in terms of the PPV cube is equivalent to averaging over P-P dimensions.

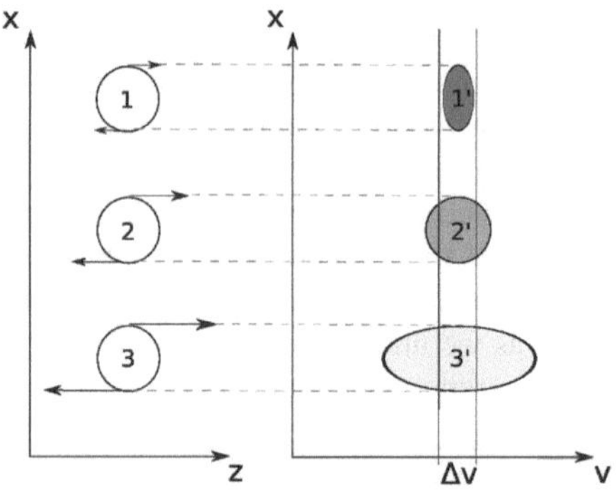

Fig. 2 An illustration of the mapping from the real space to the PPV space. In the real space 3 eddies above have the same spatial size, but different velocities. They are being mapped to the PPV space and there they have the same PP dimensions, but a different V-size. The larger is the velocity of eddies, the larger the V-extent of the eddies, the less density of atoms over the image of the eddy. Therefore, in terms of the intensity of fluctuations in the velocity channel Δv, the largest contribution is coming from the least energetic eddy, i.e. eddy 1, while the most energetic eddy, i.e. eddy 3, contributes the least

The non-trivial nature of the statistics of the eddies in the PPV space is illustrated in Fig. 2. The figure illustrates the fact that from 3 equal size and equal density eddies, the one with the smallest velocity provides the largest contribution to the PPV intensity. This explains the asymptotical scalings of power spectra in Table 3, which indicates that a spectrum of eddies that corresponds to most of turbulent energy at *large* scales corresponds to the spectrum of thin channel map intensity fluctuations having most of the energy at *small* scales. It is also clear that if the channel map or velocity slice of PPV data gets thicker than the velocity extent of the eddy 3, all the eddies contribute to the intensity fluctuations the same way, i.e. in proportion to the total number of atoms within the eddies. Similarly, in terms of the spectrum of fluctuations along the V-axis, the weak velocity eddy 1 provides the most singular small-scale contribution, which is important in terms of the VCS analysis.

3 Basics of the Formalism

While the mathematical foundations that relate the statistics of the turbulent Doppler-shifted lines with the underlying turbulent spectra are provided in mathematically-intensive papers, e.g., LP00, LP04, LP06, LP08. Below we provide a brief introduction to this formalism and introduce the mathematical foundations of the VCA and the VCS techniques (see more in LP00, LP04 and LP06). Our goal is to relate the statistics that can be obtained through spectral line observations, for instance, the structure function of the intensity of emission $I_{\mathbf{X}}(v)$

$$\mathcal{D}(\mathbf{X}, v_1, v_2) \equiv \left\langle [I_{\mathbf{X}}(v_1) - I_{\mathbf{X}}(v_2)]^2 \right\rangle, \quad (1)$$

where the z-axis component velocity v is measured in the direction defined by the two dimensional vector[2] $\mathbf{X}$, to the underlying properties of the turbulent cascade.

[2]Henceforth we denote by the capital bold letters the two dimensional position–position vectors that specify the line of sight. Small bold letters are reserved to describe the vectors of three dimensional spatial position. The z-axis is chosen to be along the line of sight.

Consider, for simplicity the case of no absorption. In this case one can easily see that the density of energy in the PPV space

$$\rho_s(\mathbf{X}, v) \sim \int_{-S}^{S} dz \rho(\mathbf{X}, z) \phi_{vz}(\mathbf{X}, z) \tag{2}$$

where the emission is coming from the cloud of the $2S$ and the intensity of emission is assumed to be proportional to the density ρ. The distribution function of the z-component of velocity is given by ϕ_{vz}.

What is the distribution function $\phi_v z$? It is a Maxwellian shifted by the z-components of the turbulent u velocity

$$\phi_{vz}(\mathbf{x})\mathrm{d}v = \frac{1}{(2\pi\beta)^{1/2}} \exp\left[-\frac{(v - u(\mathbf{x}))^2}{2\beta}\right] \mathrm{d}v, \tag{3}$$

where $\beta = \kappa_B T/m$, m being the mass of atoms. For $T \to 0$ the function ϕ_v tends to a delta-function that depends on regular gas flow and the turbulent velocity u.

Several things are clear from (2). First of all, both densities and velocities contribute to the energy density in the PPV space. Velocity and density enter the expression in different ways. Thus one can expect that the expressions for the statistics of ρ_s will depend differently on the statistics of ρ and u.

What are the statistics that we are dealing with? We discuss the correlation and structure functions and their Fourier components, which are the spectra. For the z-component of the turbulent velocity field (i.e. u), we use the structure function

$$D_z(\mathbf{r}) = \langle (u(\mathbf{x}+\mathbf{r}) - u(\mathbf{x}))^2 \rangle, \tag{4}$$

which for a self-similar power-law turbulent motions provide

$$D_z \sim D(L)(r/L)^m, \tag{5}$$

where L is the turbulent injection scale, $D_z(L)$ is the variance of velocity at this scale, m is the scaling exponent, which is $m = 1/3$ for the Kolmogorov turbulence. These velocity correlations together with the correlations of over-density

$$\xi(r) = \xi(\mathbf{r}) = \langle \rho(\mathbf{x}) \rho(\mathbf{x}+\mathbf{r}) \rangle, \tag{6}$$

enter the correlation function that can be constructed from the PPV densities ρ_s, which are available through spectroscopic observations. If the gas is confined in an isolated cloud of size S, the zero-temperature correlation function is (see LP06)

$$\xi_s(R, v) \equiv \langle \rho_s(\mathbf{X_1}, v_1) \rho(\mathbf{X_2}, v_2) \rangle$$
$$\propto \int_{-S}^{S} \mathrm{d}z \left(1 - \frac{|z|}{S}\right) \frac{\xi(r)}{D_z^{1/2}(\mathbf{r})} \exp\left[-\frac{v^2}{2D_z(\mathbf{r})}\right], \tag{7}$$

where the correlation function ξ_s is defined in the PPV space, where R is the spatial separation between points in the plane-of-sky and v is the separation along the V-axis.

The correlation function of over-density given by (6) has a constant part that depends on the mean density only and a part that changes with r. For instance, for the power-law

Table 1 The short-wave asymptotical behavior of power spectra P_s in PPV space. Results are presented for one-dimensional spectrum of fluctuations along the velocity coordinate $P_s(k_v)$, two-dimensional spectrum of fluctuations $P_s(K)$, three-dimensional anisotropic spectrum of PPV fluctuations $P_s(K, k_v)$

	1D: $P_s(k_v)$ $k_v D_z^{1/2}(S) \gg 1$	2D: $P_s(K)$ $KS \gg 1$	3D: $P_s(K, k_v)$ $k_v^2 D_z(S) \gg (kS)^m$
P_ρ :	$(r_0/S)^\gamma [k_v D_z^{1/2}(S)]^{2(\gamma-1)/m}$	$(r_0/S)^\gamma [KS]^{\gamma+m/2-3}$	$(r_0/S)^\gamma [k_v D_z^{1/2}(S)]^{-2(3-\gamma)/m}$
P_v :	$[k_v D_z^{1/2}(S)]^{-2/m}$	$[KS]^{m/2-3}$	$[k_v D_z^{1/2}(S)]^{-6/m}$

We use the convention that capital letters denote 2D vectors in PP-plane. The adopted convention is that the variables related to spatial coordinates are denoted with capital letters. The component of P_s arising from pure density fluctuations is P_ρ, while the component affected both by the density and velocity fluctuations is P_v. The power-law underlying statistics of density and velocity are assumed: γ is the spectral index of the density correlation function, m is the spectral index of the velocity correlation function. The size of the turbulent cloud is S. From LP06

density spectrum the correlation functions of over-density take the form (see LP06 for the discussion of cases of $\gamma < 0$ and $\gamma > 0$):

$$\xi(r) = \langle\rho\rangle^2 \left(1 + \left[\frac{r_0}{r}\right]^\gamma\right), \tag{8}$$

where r_0 has the physical meaning of the scale at which fluctuations are of the order of the mean density (see more in LP06). Substituting (8) in (7) it is easy to see that the PPV correlation function ξ_s can be presented as a sum of two terms, one of which does depend on the fluctuations of density, the other does not. Taking Fourier transform of ξ_s one gets the PPV spectrum P_s, which is also a sum of two terms P_ρ and P_v, namely:

$$P_s = P_\rho + P_v, \tag{9}$$

where the asymptotics for P_ρ and P_v in one dimension (along the velocity axis), two dimensions (in the velocity slice) and three dimensions (the entire PPV space) are presented in Table 1.

It is worth noting that while in the expressions for density and velocity correlations, i.e. (8) and (5), the spectral indexes of the γ and m enter the same way. However, they enter the expression for P_v, which is the part of the spectrum affected both by density and velocity, in very different ways. The origin of this difference can be seen in the expression for the correlation function of PPV intensity given by (8). This is a mathematical consequence of the effect illustrated by Fig. 2. Depending on the values of γ and m either P_ρ or P_v dominates in the asymptotical regime.

Expression (8) and its generalizations may be used directly to solve the inverse problem to find the properties of the underlying astrophysical turbulence for an arbitrary spectrum. However, most attention so far was given to the astrophysically important case of power-law turbulence. Note, that using spectra rather than the correlation function has advantages. For instance, the correlations along the V-axis of the PPV cube may be dominated by large-scale gradients, while spectra provide correct result[3] (see explanation in LP04 and LP06). The results for 1, 2 and 3 dimensional spectra are presented in Table 1.

[3]Modified correlation functions suggested in LP08 take care of the large-scale gradients and also provide correct results for the VCS studies.

Table 2 Conditions for the impact of density inhomogeneities to the PPV statistics exceeds the velocity contribution

$m \geq \max[\frac{2}{3}, \frac{2}{3}(1-\gamma)]$	$v^2 < D_z(S)(r_0/S)^m$
$\frac{2}{3}(1-\gamma) < m < \frac{2}{3}$	$v^2 < D_z(S)(r_0/S)^{\frac{2/3\gamma m}{m-2/3(1-\gamma)}}$
$m \leq \min[\frac{2}{3}, \frac{2}{3}(1-\gamma)]$	$r_0/S > 1$

Spectral index of density fluctuations γ must be larger than 0, i.e. the intensity of fluctuations increases with the decrease of scale. For $\gamma < 0$ the velocity fluctuations always dominate in creating small-scale ripples in the PPV space. From LP06

Whether P_ρ or P_v dominates depends on the statistical properties of density given by the spectral index γ. It was shown in LP07 that for $\gamma < 0$ the P_ρ contribution is always subdominant. For $\gamma > 0$ Table 2 presents special cases when the contribution of P_ρ exceeds that of P_v. For instance, two first lines of the Table 2 define for a range of values γ the minimal velocity for which $P_\rho > P_v$.

Above we did not consider the galactic rotational velocities, which can be large. The justification for this can be most easily understood if one deals with a spectral representation of the statistics. Taking Fourier transforms we deal with velocity gradients, which are larger for turbulent motions than for large-scale sheer. For instance, the latter for the Galactic rotation is given by the Oort's constant, which is 14 km s^{-1} kpc^{-1}. In comparison, the shear due to typical Kolmogorov-type turbulent motions in the Galaxy with the injection of energy at 10 km s^{-1} at the scale of $L \sim 30$ pc is ~ 300 km s$^{-1}(L/l)^{2/3}$ kpc^{-1}. Thus, in spite of the fact, that regular large-scale galactic shear velocities may be much larger than the turbulent velocities, they can be neglected for our analysis (LP00 and a numerical study in Esquivel et al. 2003).

In addition, the simplified discussion above ignored the self-absorption of the radiation. The intensities $I_{\mathbf{X}}(v)$, in general, are affected by both turbulence and absorption. To quantify these effects one can consider the standard equation of radiative transfer

$$dI_\nu = -g_\nu I_\nu ds + j_\nu ds, \tag{10}$$

where, for absorption and emissivity coefficients α and $\tilde{\varepsilon}$, $g_\nu = \alpha(\mathbf{x})\rho(\mathbf{x})\phi_v(\mathbf{x})$, $j_\nu = \tilde{\varepsilon}\rho(\mathbf{x})\phi_v(\mathbf{x})$, $\mathbf{x}$ is a three dimensional position vector $(\mathbf{X}, z)$, $\rho(\mathbf{x})$ is the density and $\phi_v(\mathbf{x})$ is the velocity distribution of the atoms. Equation (10) is the starting point of a detailed discussion in LP04, LP06, LP08.

Other complications include the geometry of lines of sight and the dependences of the emissivities on the density squared. Indeed, Eq. (8) assumes that the emission is collected along the parallel lines of sight, which is an approximation valid for studying turbulence in a distant cloud, but not valid for studying turbulence in gas at high galactic latitudes or in nearby regions of space. The corresponding generalization is provided in (Chepurnov and Lazarian 2008). There also the expressions in Table 1 are modified for the case of emissivities $\sim \rho^2$, as discussed in Sect. 8. There we also show that the practical data handling may provide better accuracy if one uses fitting of the PPV observational data with turbulence models, where the energy injection scale and the temperature of gas are used as additional fitting parameters. General expressions obtained in the aforementioned theoretical studies should be used for this purpose.

Table 3 The VCA asymptotics

Slice thickness	Shallow 3-D density $P_n \propto k^{-3+\gamma}, \gamma > 0$	Steep 3-D density $P_n \propto k^{-3+\gamma}, \gamma < 0$
2-D intensity spectrum for thin slice	$\propto K^{-3+\gamma+m/2}$	$\propto K^{-3+m/2}$
2-D intensity spectrum for thick slice	$\propto K^{-3+\gamma}$	$\propto K^{-3-m/2}$
2-D intensity spectrum for very thick slice	$\propto K^{-3+\gamma}$	$\propto K^{-3+\gamma}$

Thin means that the channel width < velocity dispersion at the scale under study; *thick* means that the channel width > velocity dispersion at the scale under study; *very thick* means that a substantial part of the velocity profile is integrated over

4 Illustration of VCS and VCA Techniques for Power-Law Velocities and Densities

Turbulence at its inertial range exhibits power-law spectra of velocity and density (see Biskamp 2003 and ref. therein). Therefore, it is natural to study the relation between the PPV statistics and power-law underlying statistics of velocity and density.

4.1 Velocity Channel Analysis

The interpretation of the channel maps is the domain of the VCA. Table 3 shows how the power spectrum of the intensity fluctuations depends on the thickness of the velocity channel. Below we provide quantitative discussion of the VCA. It is easy to see that both for steep and shallow underlying density the power law index *steepens* with the increase of velocity slice thickness. In the thickest velocity slices the velocity information is averaged out and we get the density spectral index $-3+\gamma$. The velocity fluctuations dominate in thin slices, and the index m that characterizes the velocity fluctuation can be obtained using thin velocity slices (see Table 1). As we mentioned earlier, the notion of thin and thick slices depends on the turbulence scale under study and the same slice can be thick for small scale turbulent fluctuations and thin for large scale ones (see Fig. 1).

One may notice that the spectrum of intensity in a thin slice gets shallower as the underlying velocity get steeper. To understand this effect let us consider turbulence in an incompressible optically thin medium. The intensity of emission in a velocity slice is proportional to the number of atoms per velocity interval given by the thickness of the slice. Thin slice means that the velocity dispersion at the scale of study is larger than the thickness of a slice. The increase of the velocity dispersion at a particular scales means that less and less energy is being emitted within the velocity interval that defines the slice (see Fig. 2). Mathematically this effect results in the dependences in Table 3.

If density variations are also present they modify this result. However, for small-scale asymptotics of thin slices this happens only when the density spectrum is shallow (i.e. $\gamma > 0$), i.e. dominated by fluctuations at small scales (see (8)).

4.2 Velocity Coordinate Spectrum

The VCS is a brand new technique, which, unlike the VCA, was not motivated by the interpretation of the existing observations. In the case of the VCS it was theoretical advances that induced the subsequent data analysis.

Unlike the standard spatial spectra, that are functions of angular wavenumber, the VCS is a function of the wave number $k_v \sim 1/v$, which means that large k_v correspond to small velocity differences, while small k_v correspond to large velocity differences.

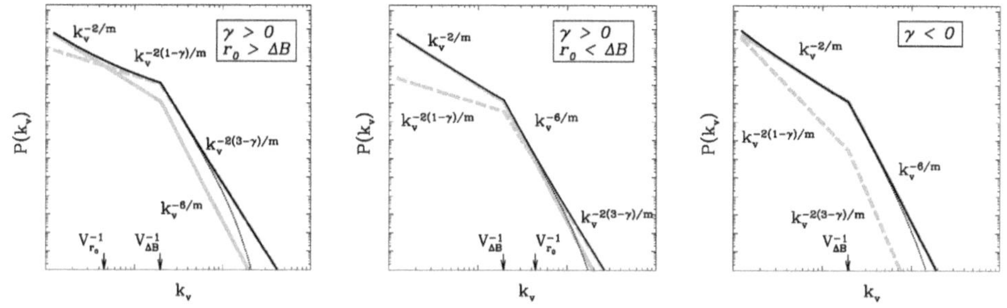

Fig. 3 Particular cases of studying turbulence with the VCS technique. In every panel *light lines* show contributions from the ρ-term (density modified by velocity, *dashed line*) and v-term (pure velocity effect, *solid line*) separately, while the *dark solid line* shows the combined total VCS power spectrum. Thermal suppression of fluctuations is shown by the *dotted line*. *Labels below the dark solid lines* mark the scaling of the subdominant contributions. For the *left* and *middle panels* the density power spectrum is taken to be shallow, i.e., $\gamma > 0$. The *left panel* corresponds to high amplitude of the density correlations, $r_0 > \Delta B$, i.e. when density effects become dominant at relatively long wavelengths for which the beam is narrow. In the *middle panel*, the amplitude of density correlations is low $r_0 < \Delta B$ and they dominate only the smallest scales which results in the intermediate steepening of the VCS scaling. The *right panel* corresponds to the steep density spectrum. In this case the density contribution is always subdominant. In this example the thermal scale is five times shorter than the resolution scale $V_{\Delta B}$. From Lazarian and Pogosyan (2008)

Table 4 Scalings of VCS for shallow and steep densities for measurements taken with the telescope with a finite beam size. From LP06

Spectral term	$\Delta B < S[k_v^2 D_z(S)]^{-\frac{1}{m}}$	$\Delta B > S[k_v^2 D_z(S)]^{-\frac{1}{m}}$
$P_\rho(k_v)$	$\propto (k_v D_z^{1/2}(S))^{-2(1-\gamma)/m}$	$\propto (k_v D_z^{1/2}(S))^{-2(3-\gamma)/m}$
$P_v(k_v)$	$\propto (k_v D_z^{1/2}(S))^{-2/m}$	$\propto (k_v D_z^{1/2}(S))^{-6/m}$

Assume that the maximal resolution of a telescope corresponds to resolving the scale ΔB of a cloud at distance $L_{\rm cloud} \gg S$. At this scale the turbulent velocity is $V_{\Delta B} \equiv \sqrt{D(S)(\Delta B/S)^m}$. It is not difficult to see that when $k_v^{-1} > V_{\Delta B}$ the beam is narrow, while for observations of smaller scales its width is important. The spectrum of fluctuations along the V-coordinate at the scale k_v depends on whether the instrument resolves the correspondent spatial scale $[k_v^2 D_z(S)]^{-1/m} S$. If this scale is resolved then $P_v(k_v) \propto k_v^{-2/m}$ and $P_\rho(k_v) \propto k_v^{-2(1-\gamma)/m}$. If the scale is not resolved then $P_v(k_v) \propto k_v^{-6/m}$ and $P_\rho(k_v) \propto k_v^{-2(3-\gamma)/m}$. These results are presented in a compact form in Table 4.

The transition from the low to the high resolution regimes happens as the velocity scale under study gets comparable to the turbulent velocity at the minimal spatially resolved scale. As the change of slope is a velocity-induced effect, it is not surprising that the difference in spectral indexes in the low and high resolution limit is $4/m$ for both the P_v and P_ρ terms, i.e. it does not depend on the density.[4] This allows for separation of the velocity and density contributions. For instance, Fig. 3 illustrates that in the case of shallow density both the density and velocity spectra can be obtained. Potentially, procedures for extracting information on 3D turbulent density can be developed for the steep density case as well. However, this

[4] In the situation where the available telescope resolution is not sufficient, i.e. in the case of extragalactic turbulence research, the high spatial resolution VCS can be obtained via studies of the absorption lines from point sources.

requires careful accounting for errors as the contribution from density is subdominant in this case.[5]

4.3 Effects of Self-Absorption

The issues of absorption were worrisome for researchers from the very start (see Munch 1958). Unfortunately, erroneous statements about the effects of absorption on the observed turbulence statistics are widely spread (see discussion in LP04).

Using transitions that are less affected by absorption, e.g. HI, may allow us to avoid the problem. However, it is regretful not to use the wealth of spectroscopic data only because absorption may be present. A study of absorption effects in is presented in LP04 and LP06. For the VCA it was found that for sufficiently thin[6] slices the scalings obtained in the absence of absorption still hold provided that the absorption on the scales under study is negligible.

When dealing with self-absorption, one should start with (10) (see LP04). The criterion for the absorption to be important is $\alpha^2\langle(\rho_s(\mathbf{X}, v_1) - \rho_s(\mathbf{X}, v_2))^2\rangle \sim 1$, which for $\gamma < 0$ results in the critical size of the slice thickness V_c given by (LP06)

$$\begin{aligned} V_c/D_z(S)^{1/2} &\approx (\alpha\bar{\rho}_s)^{\frac{2m}{m-2}}, \quad m > 2/3 \\ V_c/D_z(S)^{1/2} &\approx (\alpha\bar{\rho}_s)^{-1}, \quad m < 2/3, \end{aligned} \tag{11}$$

where $\bar{\rho}_s$ is the mean PPV density. The absorption is dominant for the slices thicker than V_c. The difference with the case of $\gamma > 0$ is that, in the latter case, one should also consider the possibility that the density contribution can be important (see Table 2). The criterion above coincides with one for the VCS, if we identify the critical k_v with $1/V_c$. If the resolution of the telescope is low, another limitation applies. The resolved scale should be less than the critical spatial scale that arises from the condition $\alpha^2\langle(\rho_s(\mathbf{X_1}, v) - \rho_s(\mathbf{X_2}, v))^2\rangle \sim 1$ which for $\gamma < 1$ results in $R_c/S \approx (\alpha\bar{\rho}_s)^{\frac{2}{m-2}}$ (LV06). If only scales larger than R_c are resolved, the information on turbulence is lost.

If integrated intensity of spectral lines is studied in the presence of absorption non-trivial effects emerge. Indeed, for optically thin medium the spectral line integration results in PPV intensity fluctuations that reflect the density statistics. LP04 showed that this may not be any more true for lines affected by absorption. When velocity is dominant a very interesting regime for which intensity fluctuations show universal behavior, i.e. the power spectrum $P(K) \sim K^{-3}$ emerges. When density is dominant (see Table 2), the spectral index of intensity fluctuations in those two situations is the same as in the case of an optically thin cloud integrated through its volume. This means that for $\gamma > 0$, i.e. for steep spectrum of density, in the range of parameter space defined by Table 2 the measurements of intensity fluctuations of the integrated spectral lines reflect the *actual* underlying density spectrum in spite of the absorption effects.

5 Studies of Turbulence with Absorption Lines

The analysis of weak absorption lines is analogous to the analysis of the weak emission lines. For instance, the weak absorption data from extended sources, e.g. of atomic gas

[5]Needless to say, when the turbulent object is resolved, the easiest way to obtain the density spectral index is to study the integrated intensity maps, provided that the absorption is negligible (see criteria for this in LP04).

[6]The thermal broadening limits to what extent the slice can be thin. This means that in some cases that the actual turbulent velocity spectrum may not be recoverable.

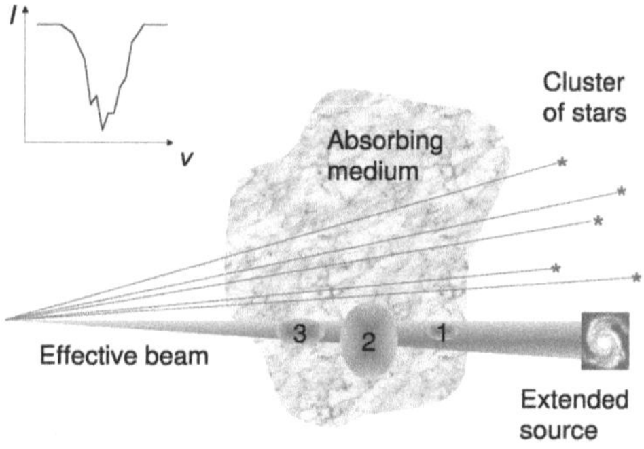

Fig. 4 Illustration of VCS absorption studies of turbulence. *Left panel*: Schematic of measuring turbulence with absorption lines from point sources, e.g. stars, and extended source, e.g. a galaxy. From Chepurnov and Lazarian (2008)

towards Cassiopeia A and Cygnus A in Deshpande et al. (2000), can be interpreted with the traditional VCA technique. VCS can handle both the weak absorption data from both point sources and extended sources (see Fig. 4).

New effects arise when strong absorption lines, which are in a saturated regime, are studied. This problem was addressed in Lazarian and Pogosyan (2008, henceforth LP08), who proposed to analyze optical depth fluctuations. These correspond to the logarithms of the measured intensities. The analysis was made for the VCS technique, as this is the technique that requires sampling of a turbulent volume just over a few lines of sight. Indeed, numerical simulations in Chepurnov and Lazarian (2008) proved that the sampling along from 5 to 10 directions to absorbing sources, e.g. stars, is enough to recover properly the underlying spectrum of turbulent velocity (see Fig. 8).

At what optical depth the is the recovery of the turbulence spectra feasible? It is well known that for optical depth τ larger than 10^5 the wings are totally dominated by Lorentz factor (see Spitzer 1978). For the range of optical depth less than 10^3, the line width is determined by Doppler shifts rather than the atomic constants, which simplifies the study. While formally the entire line profile provides information about the turbulence, in reality, the flat saturated part of the profile will contain only noise and will not be useful for any statistical study. Thus, it is the wings of saturated absorption lines that can be used for turbulence studies.

If however, the absorption lines are studied using an extended emission source, then the VCA analysis is also possible with the logarithms of the intensities in the velocity channels. For both the VCA and VCS techniques, the effect of saturation of the absorption line results in limiting the range of k_v at which the information on turbulence is available. If one approximates the effect of saturation with the help of a Gaussian mask of width Δ, centered in the middle of the wing, Δ measures the fraction of the line that is available for studies of turbulence. Our studies (see Fig. 5) in LP08 show that the recovery of the turbulent spectrum with the VCS is possible for $k_v > 3\Delta^{-1}$, where all the quantities are normalized over the total turbulent velocity dispersion. In terms of the VCA this translates into the requirement that the thickness of the velocity channel over which it is feasible to analyze the spectrum of the *logarithm* of the intensity fluctuations is $< 1/3\Delta$. Thermal effects provide additional limitations for the range of scales available for observations, i.e. for finite $\beta' = \beta/D_z(S)$ (see (3)), where $D_z(S)$ is the velocity dispersion of the line the range of k_v for studies of asymptotic power-law solutions is limited to $3\Delta^{-1} < k_v < 1/(3\beta)$. For k_v beyond this range, the recovery of the turbulence spectrum is still possible, but fitting of the integral expressions, rather than the use of the asymptotical solutions is necessary (see Sect. 8).

When several absorption lines from different species are available along the same line of sight, one can improve the recovery of the turbulence spectrum by combining them together. We believe that piecewise analyses of the wings belonging to different absorption

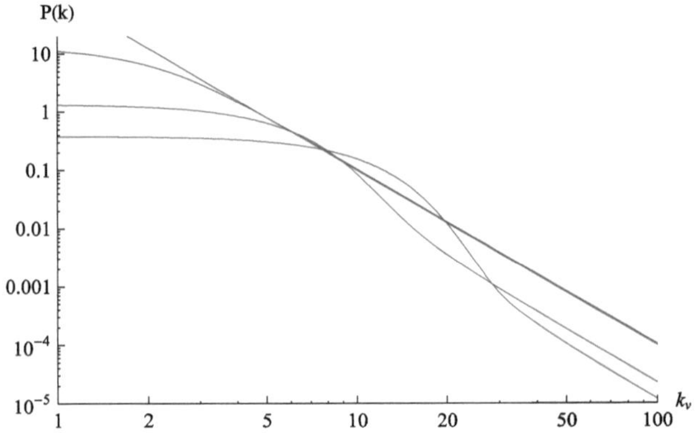

Fig. 5 Power spectrum of the optical depth fluctuations from line wings. All parameters, k_v, are dimensionless, in the units of the variance of the turbulent velocity at the scale of the cloud. Only the effect of the turbulent motions and not spatial inhomogeneity of the absorbers is taken into account. The underlying scaling of the turbulent velocities is Kolmogorov, $m = 2/3$. The *left panel* illustrates the power aliasing due to finite width of the window. The power spectrum is plotted, from top to bottom, for $\Delta = 1, 0.2, 0.1$, i.e. the widths of the wing ranges from the complete line to one-tenth of the line width. Effects of thermal broadening that distort the spectrum at large k_v are not shown. From Lazarian and Pogosyan (2008)

lines is advantageous. Optical and UV absorption lines are the primary targets for such an analysis. Formally, for lines with weak absorption, i.e. $\tau_0 < 1$, there is no need for other measurements. However, in the presence of inevitable noise, the situation may be far from trivial. Naturally, noise of a constant level, e.g. instrumental noise, will affect more weak absorption lines. The strong absorption lines, in terms of VCS, sample turbulence only for sufficiently large k_v. This limits the range of turbulent scales that can be sampled with the technique. However, the contrast that is obtained with the strong absorption lines is higher, which provides an opportunity of increasing signal to noise ratio for the range of k_v that is sampled by the absorption lines.

6 Comparison of VCA and VCS

Traditionally the techniques to study velocity turbulence, e.g. velocity centroids or VCA, require observations to spatially resolve the scale of the turbulence under study.[7] This constrains the variety of astrophysical objects where turbulence can be studied. In this way, the VCS, is a unique tool that allows studies of astrophysical turbulence even when the instrument does not resolve spatially the turbulent fluctuations.

Can the VCS technique recover the turbulence spectrum while dealing with spatially unresolved astrophysical objects? From a pure theoretical standpoint, this should be feasible. Indeed, if we deal with fluctuations at very small scales we can identify different parts of the spectral line with different statistical realizations of the small-scale stochastic process. This enables us to perform the averaging using those parts. From a practical point of view, thermal broadening limits the range of the scales that can be resolved spectroscopically.

[7] As it was discussed in LP00, the VCA can be applied directly to the raw interferometric data, rather than to images that require good coverage of all spatial frequencies. However, even with interferometers, the application of the VCA to extragalactic objects is restricted.

Fig. 6 *Top panel*: Studies of turbulence using a single absorption spectral line. *Bottom panel*: Velocity Coordinate Spectrum obtained using sampling of a turbulent volume along 10 lines of sight. The *solid line* corresponds to the theoretical expectations. From Chepurnov and Lazarian (2008)

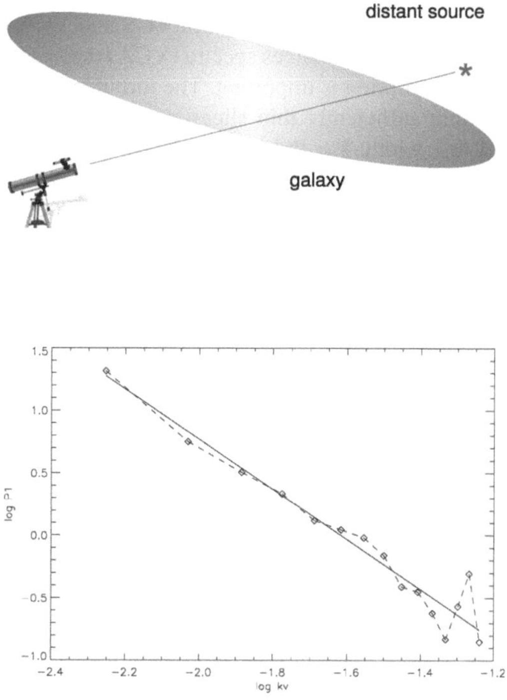

Thus, reliable studies of turbulence may require spatial averaging. Chepurnov and Lazarian (2008) established with numerical simulations that 5–10 measurements of the spectrum can be sufficient for performing an adequate spatial averaging. This means that for a turbulent volume with marginal spatial resolution but good spectral resolution, studies of turbulence with the VCS are possible using emission lines. In addition, if the turbulent volume rotates, the resulting Doppler broadened lines with width much larger than the turbulent width can be chopped into spectral pieces that can be used to perform statistical averaging. This makes feasible the recovery of the spectrum with a single spectral measurement (see Fig. 6).

Our study of the effect of finite temperatures for the technique reveals that, unlike the VCA, the temperature broadening does not prevent the turbulence spectrum from being recovered from observations. Indeed, in VCA, gas temperature acts in the same way as the width of a channel. Within the VCS the term with temperature gets factorized and it influences the amplitude of fluctuations (LP06). One can correct for this term,[8] which also allows for a new way of estimating the interstellar gas temperature.

Another advantage of the VCS compared to the VCA is that it reveals the spectrum of turbulence directly, while within the VCA the slope of the spectrum should be inferred from varying the thickness of the channel. As the thermal line width acts in a similar way as the channel thickness, additional care (see LP04) should be exercised not to confuse the channel that is still thick due to thermal velocity broadening with the channel that shows the thin slice asymptotics. A simultaneous use of the VCA and the VCS makes the turbulence spectrum identification more reliable.

[8] To do this, one may attempt to fit for the temperature that would remove the exponential fall off in the spectrum of fluctuations along the velocity coordinate (Chepurnov and Lazarian 2006).

The introduction of absorption in VCS and VCA brings about different results. Within the analysis of velocity slices spectra (VCA) the absorption results in new scalings for slices for which absorption is important. The turbulence spectral indexes can be recovered for the VCA within sufficiently thin slices, provided that the thickness of the slices exceeds the thermal line width. For the VCS at large k_v for which absorption becomes important the spectra get exponentially damped.

Both VCA and VCS are applicable to studies of not only emission, but also absorption lines. However, the necessity of using extended emission sources limits the extent of possible VCA studies of turbulence. This is not an issue for the VCS, for which absorption lines from *point sources* can be used (see Fig. 4). Interestingly enough, in this case the asymptotics for the high resolution limit for the VCS technique should be used irrespectively of the actual beam size of the instrument.

7 Numerical Testing

VCA predictions were tested in Lazarian et al. (2001), Esquivel et al. (2003), Padoan et al. (2006) and in Chepurnov and Lazarian (2008) using synthetic maps obtained with synthetic power-law data as well as with numerical compressible MHD simulations. Simulated data cubes allowed both density and velocity statistics to be measured directly. Then these data cubes were used to produce synthetic spectra which were analyzed using the VCA. As the result, the velocity and density statistics were successfully recovered.

The most extensive high resolution testing of VCA and the first numerical testing of VCS was performed in Chepurnov and Lazarian (2008). The results for the velocity spectrum corresponding to shocks, i.e. for the $E(k) \sim k^{-2}$, are shown in Fig. 7. The spectra for more shallow velocities, e.g. the Kolmogorov spectrum[9] $E(k) \sim k^{-5/3}$ show more noise, which increases at small scales. This noise originates at velocities corresponding to the velocities of the adjacent numerical points and is caused by the discrete nature of the data set involved in

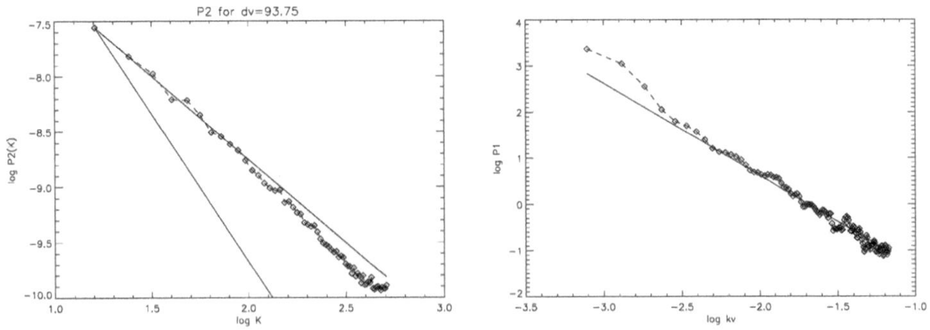

Fig. 7 Testing of the predictions for the VCA and VCS techniques with synthetic observations. The underlying spectral index of velocity fluctuations is −4. *Left panel*: VCA spatial spectrum for different velocity slice thickness. Thin slice: shallower solid *line shows* the expected velocity-dominated spectrum, Thick slice: steeper *solid line* shows density-dominated spectrum. *Right panel*: Spectrum measured from the synthetic data with the VCS techniques versus the theoretical expectations

[9]Note, that in terms of three dimensional spectrum, Kolmogorov spectrum corresponds to $k^{-11/3}$. This difference arises from the averaging over directions in k-space.

Fig. 8 Spectrum of turbulence recovered from synthetic observations when the signal is sampled along 10 absorption lines. The solid line corresponds to the underlying spectrum and points correspond to the recovered spectrum. From Chepurnov and Lazarian (2008)

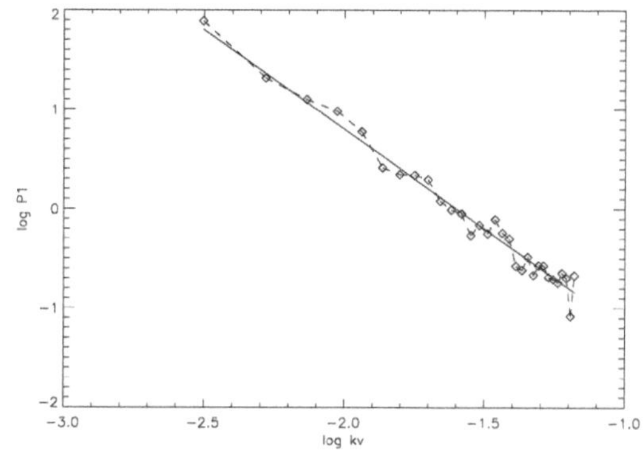

the simulations. Therefore, this noise is not expected for the real-world smooth distribution of turbulent fluid. This noise is not observed while the actual astrophysical data is handled either. However, this is an important effect for *numerical* testing.[10]

Apart from testing of the VCA and VCS techniques, Chepurnov and Lazarian (2008) tested the effect of limited data samples on the noise in the spectra obtained. In particular, applying the VCS technique to synthetic observations of absorption lines, Chepurnov and Lazarian (2008) showed that having just several spectral absorption measurements is sufficient for recovering the underlying turbulent velocity spectrum (see Fig. 8).

8 Extending VCA and VCS

8.1 Studying Non-power-Law Turbulence

Historically, the VCA technique was initiated to explain the power-law dependences of emission in velocity channels reported by observers. These power-law dependencies are only approximate, as it follows from the theory (see LP00). Indeed, a velocity slice may contain the images of both small eddies, for which the slice is thick, and large eddies, for which the slice is thin (see Fig. 1). Similarly, thermal broadening distorts the VCS output even for an underlying power-law turbulence. Needless to say, studies of the interstellar turbulence spectrum with injection and dissipation of energy at various scales are most exciting. This underlying spectrum is not a power-law and therefore the VCA and VCS data handling are not going to deliver power-law dependences either.[11]

While the analytical dependences that we presented in the tables are very useful for understanding the nature the fluctuations observed and the initial analysis of the data, we believe that real progress in quantitative studies of turbulence can be achieved by using the integral expressions obtained in LP00, LP04, LP06, LP08 to fit the data with the turbulence models. The first work in this direction has been done with the VCS (see Sect. 9.2). We expect similar work to be done with the VCA.

[10]Low resolution testing of VCA in Miville-Deschênes et al. (2003) had inadequate resolution and therefore brought erroneous results.

[11]In general, studying turbulence dissipation is an exciting avenue for getting insight into the intimate details of the turbulence cascade. For instance, it has been shown numerically that MHD turbulence creates separate cascades of Alfvenic, fast and slow modes (Cho and Lazarian 2002, 2003), with these modes having dissipation scales. This should result in the velocity spectrum deviating from the power law at small scales.

8.2 Studies of Nearby Gas

The notion of "nearby gas" in terms of the turbulence studies is related to situations when the lines of sight coming from the telescope cannot be considered parallel. For instance, studies of gas in the Local Bubble may require taking into account the convergence of the lines of sight. Naturally, if the sampling happens along an infinitely thin line, there is no difference for the nearby gas and the distant gas in terms of the VCS technique. This is the case, for instance, for turbulence studies using absorption from a star. The spectral index for P_v is given by Table 4, i.e. $P_v \sim k_v^{-2/m}$.

If, on the contrary, the resolution is low, the difference between the distant object and nearby gas result in the a very different solutions for P_v, in particular, in the asymptotic of $P_v \sim k_v^{-4/m}$ (Chepurnov and Lazarian 2008). Note, that this asymptotical spectrum is not as steep as the spectrum arising in the case of low resolution and parallel lines of sight (cf. Table 4). This simplifies recovering of the velocity spectrum from observations in the presence of noise.

8.3 Studies of Emission Lines with Emissivity $\sim \rho^2$

The quantities we deal with in spectral line observations are the velocity and the gas emissivity. Lazarian & Pogosyan (see LP00, LP04, LP06, LP08) treated the emissivities proportional to the density to the first power. Therefore, in terms of scalings, the emissivities and densities were interchangeable. This is not true, however, when the emissivities are proportional to ρ^2, as is the case of the recombination lines in plasma. The latter regime modifies the analysis. In particular, for the shallow spectrum of density, Chepurnov and Lazarian (2008) showed that the spectral index of the correlation function of emissivity $\gamma_\varepsilon = 2\gamma$, where γ is the index of the correlation function of density, which we discussed all the way above in the paper.

The first consequence of this is that if the density is shallow the emissivity is shallow. The second consequence is that one can use the asymptotics the second column of Table 4, but substituting there $\gamma_\varepsilon = 2\gamma$ instead of γ. The same relation, i.e. $\gamma_\varepsilon = 2\gamma$ is applicable for shallow densities when the VCS is considered.

For steep density spectra the spectral index of emissivity γ_ε should be instead of γ when the emissivities are proportional to ρ^2. This is true for both the VCA and VCS techniques, which once again stresses their close connection.

9 Observational Studies

9.1 Applying VCA

Table 5 illustrates the results obtained with VCA by different groups analyzing observations. It displays the variety of objects to which VCA has been applied. The studies of spectra in channel maps predated VCA (see lines 1, 8, 9 in Table 5 as well as Crovisier and Dickey 1983; Stanimirovic et al. 1999), but researchers were choosing Δv arbitrarily, making any sensible comparisons impossible. For these cases, we performed the VCA analysis using the published data. The results presented are based on the detailed analytical treatment of different cases in Lazarian and Pogosyan (2000, 2004). For instance, Lazarian and Pogosyan

Table 5 Selected VCA results

N	Data	Object	P_{PPV}^{thin}	P_{PPV}^{thick}	Depth	E_v	E_ρ	Ref. obs.	Ref. theor.
1	HI	Anticenter^g	$K^{-2.7}$	N/A	Thin	$k^{-1.7}$	N/A	[1]	[2]
2	HI	→CygA	$K^{-(2.7)}$	$K^{-(2.8)}$	Thin	N/A	$k^{-(0.8)}$	[3]	[3]
3	HI	SMC^e	$K^{-2.7}$	$K^{-3.4}$	Thin	$k^{-1.7}$	$k^{-1.4}$	[4]	[4]
4	HI	Center^g	K^{-3}	K^{-3}	Thick	N/A	N/A	[5]	[6]
5	HI	B. Mag.^{g}	$K^{-2.6}$	$K^{3.4}$	Thin	$k^{-1.8}$	$k^{-1.2}$	[7]	[7]
6	HI	Arm^g	K^{-3}	K^{-3}	Thick	N/A	N/A	[8]	[9]
7	HI	DDO 210^e	K^{-3}	K^{-3}	Thick	N/A	N/A	[10]	[10] [9]
8	^{12}CO	L1512	N/A	$K^{-2.8}$	Thick	N/A	$k^{-0.8}$	[11]	[5]
9	^{13}CO	L1512	N/A	$K^{-2.8}$	Thick	N/A	$k^{-0.8}$	[11]	[10]
10	^{13}CO	Perseus	$K^{-(2.7)}$	K^{-3}	Thick	$k^{-(1.7)}$	N/A	[12]	[12]
11	^{13}CO	Perseus	$K^{-2.6}$	K^{-3}	Thick	$k^{-1.8}$	N/A	[13]	[13]
12	C^{18}O	L1551	$K^{-2.7}$	$K^{-2.8}$	Thin	$k^{-1.7}$	$k^{-0.8}$	[14]	[14]

Superscript "^{g}" denotes galactic objects, "^{e}"—extragalactic. P_{PPV}^{thin} and P_{PPV}^{thick} are the power law spectrum in thin and thick PPV slices, respectively. "Ref. obs." and "Ref. theor." correspond to papers where the measurement were done and interpreted using VCA, respectively. Indexes in round brackets correspond to substantial observational errors correspond to consistency only. → CygA is used to denote material towards Cygnus A. [1] is Green (1993), [2] is Lazarian and Pogosyan (2006), [3] is Deshpande et al. (2000), [4] is Stanimirovic and Lazarian (2001), [5] is Dickey et al. (2001), [6] is Lazarian and Pogosyan (2004), [7] is Muller et al. (2004), [8] is Khalil et al. (2006), [9] is Lazarian (2006), [10] is Begum et al. (2006), [11] is Stutzki et al. (1998), [12] is Sun et al. (2006), [13] is Padoan et al. (2006), [14] is Swift (2006) and Swift and Welch (2008)

(2004) predict that absorption can induce a universal[12] spectrum $\sim K^{-3}$, where K is the observational analog, in the $2D$ plane of the sky, of the wavenumber k. If this spectrum dominates in both thin and thick slices, the only conclusion that can be made is that the density spectrum $E_\rho \sim k^{-\alpha}$ corresponds to $\alpha > 1$ (see lines 3, 6, 7), while the details of the spectrum are not available. When the spectrum of intensities in a thick slice is different from K^{-3} in the presence of absorption, the information about the underlying densities is available (see lines 8, 9). It is encouraging that the observed spectral indexes correspond to those in simulations (see Beresnyak et al. 2005; Kowal et al. 2007), which show a tendency of having the spectrum of density getting flatter as the spectrum of velocity gets steeper. A stronger statement about the good quantitative correspondence between the VCA analysis of Perseus data and their numerical simulations is made in Padoan et al. (2006) and also in Kritsuk et al. (2007).

One of the first applications was in Stanimirovic and Lazarian (2001), where the technique was applied to the Small Magellanic Cloud (SMC) data. The analysis revealed spectra of 3D velocity fluctuations roughly consistent with the Kolmogorov scaling (a bit more shallow). Esquivel et al. (2003) used simulations of MHD turbulent flows to show that, in spite of the presence of anisotropy caused by magnetic field, the expected scaling of fluctuations is indeed Kolmogorov. Studies by Cho and Lazarian (2002, 2003) revealed that the Kolmogorov-type scaling is also expected in the compressible MHD flows. This also supports the conclusion in LP00 that the data in Green (1993) is consistent with MHD turbulence scaling.

[12] This assumes that the velocity spectrum $E_v \sim k^{-\beta}$ has $\beta > 1$. However, this is true for all turbulence spectra that we know.

Studies of turbulence are more complicated for the inner parts of the Galaxy, where (a) two distinct regions at different distances from the observer contribute to the emissivity for a given velocity and (b) effects of the absorption are important. However, the analysis in Dickey et al. (2001) showed that some progress may be made even in those unfavorable circumstances. Dickey et al. (2001) found the steepening of the spectral index with the increase of the velocity slice thickness. They also observed the spectral index for strongly absorbing direction approached -3 in accordance with the conclusions in LP04. Note, that the effects of optical depths may explain some other case when the spectral index stayed the same, e.g. -3, while the thickness of the slice was varying (see Khalil et al. 2006). Incidently, this situation can be confused with the situation when the fluctuations arise from density only (see Begum et al. 2006).

21-cm absorption provides another way of probing turbulence on small scales. The absorption depends on the density to temperature ratio ρ/T, rather than to ρ as in the case of emission.[13] However, in terms of the VCA this change is not important and we still expect to see the emissivity index steepening as velocity slice thickness increases, provided that velocity effects are present. In view of this, results of Deshpande et al. (2000), who did not see such steepening, can be interpreted as the evidence of the viscous suppression of turbulence on the scales less than 1 pc. The fluctuations in this case should be due to density and their shallow spectrum $\sim k^{-2.8}$ maybe related to damped magnetic structures below the viscous cutoff (see Lazarian et al. 2004). This may be also a consequence of the shallow density spectrum in compressible MHD (see Beresnyak et al. 2005).

Historically, the CO data was analyzed after integration over the entire emission line. Stutzki et al. (1998) presented the power spectra of ^{12}CO and ^{13}CO fluctuations obtained via integrating the intensity over the entire emission line for L1512 molecular cloud. Counter-intuitively, Stutzki et al. (1998) found for both isotopes the power spectrum with a similar spectral index. According to LP04 this may correspond to optically thick asymptotics (i.e. the integration range of velocities is larger than V_c (see (11)). If the velocity fluctuations dominate, the expected index is universal and equal to -3 (meaning K^{-3}), if the density fluctuations dominate (see Table 2) the expected index is $-3+\gamma$ (meaning $K^{-3+\gamma}$). The index measured in Stutzki et al. (1998) is ~ 2.8, which may either correspond to -3 within the experimental errors, or more likely indicate that $\gamma \approx 0.2$, i.e. the density spectrum is shallow. The latter possibility is indirectly supported by ^{18}CO measurement for L1551 cloud in Swift (2006), used the VCA (observing the changes of the channel map spectral index while changing the velocity slice thickness) and obtained the shallow density spectrum with $\gamma \approx 0.2$, while his measured velocity spectrum was approximately Kolmogorov (the index is -3.72). Padoan et al. (2006) both successfully tested the VCA with high resolution numerical simulations that included radiative transfer and applied the technique to Five College Radioastronomy Observatory (FCRAO) survey of the Perseus molecular cloud complex. He obtained the velocity index around -3.81.

The practical applications of the VCA have only just began. Nevertheless, it has already provided some intriguing results and proved to be a promising way of using the wealth of spectroscopic surveys for studies of astrophysical turbulence.

9.2 Applying VCS

Figure 9 (left) shows the results of our VCS-analysis of galactic high latitude data. Rather than first correcting for the gas thermal broadening, then fitting the power law into the VCS

[13] In the case of an isobaric medium the product of density and temperature are constant and the problem is similar to studies of transitions for which the emissivity is proportional to ρ^2 that we discussed earlier.

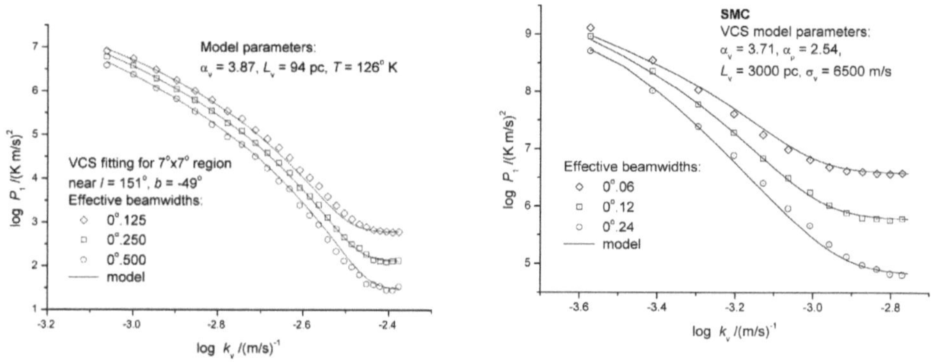

Fig. 9 Fitting of turbulent models and observational data for different resolutions. *Left panel*: Application of the VCS high latitude HI Arecibo data. The spatial resolution of the maps was decreased to illustrate the VCS in both high and low resolution regimes. ε here is the power spectrum index, which for the Kolmogorov turbulence is $11/3$. The measured ε is in the range [3.52; 3.57]. The energy injection scale is 94 pc. *Right panel*: Application of the VCS to Small Magellanic Cloud HI data. The spectral index for velocity is a bit more shallow. The model corresponds to the energy injection at 3 kpc. From Chepurnov and Lazarian (2008)

spectrum as discussed in Lazarian and Pogosyan (2006), we used our analytical expressions to find the model that fits the data set corresponding to VCS for data at different spatial resolutions. The resolutions play for the VCS a similar[14] role as the thickness of slices Δv for the VCA and therefore PPV data cubes at different resolutions are non-trivially related, as far as the VCS analysis is concerned.

Fitting data to the models opens ways of studying non-power law turbulence, e.g. turbulence at the injection or dissipation scales. It allows also studies of turbulence when thermal broadening is important. Our results in Fig. 9 (left) show that the model of turbulence with spectrum steeper than Kolmogorov, i.e. with $E_v \sim k^{-1.9}$, the temperature[15] of gas around 130K and a single injection scale of 100 pc.

Different lines correspond to studies of maps with different resolution. The fitting was done with a solenoidal component of turbulence only, which results in variations of the fitting precision as the resolution changes.[16] This is the first example of the analysis in which we obtained more than just the turbulence spectral index from the spectroscopic data. Our first applications of a similar approach to the VCA have also produced encouraging results. In particular, we improved the precision of determining turbulence spectrum in HI of Small Magellanic Cloud (SMC) (see line 3 in Table 5) and obtained the injection scale of around 4 kpc, and which corresponds to the ideas of exciting turbulence in SMC during its encounter with its neighbor, Large Magellanic Cloud. Figure 9 (right) illustrates the application of the VCS to the same set of SMC data. The correspondence in the spectral indexes of velocity turbulence which were obtained by independent application of the VCA and VCS techniques is encouraging.

---

[14]The important difference is that with VCS, we can restore the velocity spectrum for any resolution, while for the VCA, the velocity spectrum is available only for the thinnest slices.

[15]The analysis in LP06 shows that the contribution of turbulence in warm gas to fluctuations in PPV is exponentially suppressed compared to that in cold gas.

[16]In general, both the solenoidal and potential components of turbulence contribute to the observed signal. The relative contribution of the two components varies with the resolution. With higher precision data, one can potentially disentangle the two components by varying the resolution.

9.3 Expected Developments

Our discussion above shows that fitting of the observed spectrum of fluctuations with turbulence models is more informative than just using asymptotic solutions. Although the theoretical work so far was aimed at obtaining the expected asymptotical analytical relations, the formalism developed in this process contains integral equations that can be used for the detailed fitting similar to the one described in Sect. 9.2. We expect that both the VCA and VCS techniques will be used this way in future modeling.

We also expect that VCA and VCS are to be used simultaneously, whenever possible. In addition, we expect that combining many emission and absorption lines will provide insight into astrophysical turbulence in its complexity. Indeed, as we discussed earlier, the techniques are most sensitive to the colder component of gas. Therefore, for an idealized situation of two phase HI, the turbulence is being sampled mostly in cold phase. To sample turbulence in the warm phase or hot phase as, for instance, is the case for the clouds in the Local Bubble, one may use other species as tracers. Note, that the properties of turbulence may differ from one phase to another, which makes these type of studies very important.

10 Alternative Approaches

10.1 Velocity Centroids

As we mentioned in the introduction, a more traditional approach to turbulence studies includes velocity centroids, i.e. $S(\mathbf{X}) = \int v_z \; \rho_s(\mathbf{X}, v_z) \; dv_z$, where ρ_s is the density of emitters in the PPV space. Analytical expressions for structure functions[17] of centroids, i.e. $\langle [S(\mathbf{X_1}) - S(\mathbf{X_2})]^2 \rangle$ were derived in Lazarian and Esquivel (2003). In that paper a necessary criterion for centroids to reflect the statistics of velocity was established. Esquivel and Lazarian (2005) confirmed the utility of the criterion and revealed that for MHD turbulence simulations it holds for subsonic or slightly supersonic turbulence (see also Ossenkopf et al. 2006). A subsequent study by Esquivel et al. (2007) showed that there are fundamental problems with using centroids for supersonic turbulence. This is in contrast to the VCA and the VCS that provide reliable ways to study supersonic turbulence. At the same time, velocity centroids can be successfully used to study anisotropies arising from the existence of the mean field (see Sect. 11.1).

10.2 Wavelets and Principal Component Analysis

The use of different *wavelets* for the analysis of data is frequently treated in the literature as different statistical techniques of turbulence studies (Gill and Henriksen 1990; Stutzki et al. 1998; Cambresy 1999; Khalil et al. 2006), which creates an illusion of an excessive wealth of tools and approaches. In reality, while Fourier transforms use harmonics of $e^{i\mathbf{kr}}$, wavelets use more sophisticated basis functions, which may be more appropriate for problems at hand. In our studies we also use wavelets both to analyze the results of computations (see Kowal and Lazarian 2006) and synthetic maps (Ossenkopf et al. 2006; Esquivel et al. 2007), along with or instead of Fourier transforms or correlation functions.

[17] Expressions for the correlation functions are straightforwardly related to those of structure functions. The statistics of centroids using correlation functions was studied in a follow-up paper by Levier (2004).

Wavelets may reduce the noise arising from inhomogeneity of data, but we found in the situations when correlation functions of centroids that we studied were failing, a popular wavelet (Δ-variance) was also failing (cp. Esquivel and Lazarian 2005; Ossenkopf et al. 2006; Esquivel et al. 2007). While in wavelets the basis functions are fixed, a more sophisticated technique, Principal Component Analysis (PCA), chooses basis functions that are, in some sense, the most descriptive. Nevertheless, the empirical relations obtained with PCA for extracting velocity statistics provide, according to Padoan et al. (2006), results for low order statistics, $(\delta v)^p$, where $p < 0.5$ (see also Brunt et al. 2003), while the spectrum corresponds to $p = 2$. In addition, while the PI's research shows that for density spectra $E_\rho \sim k^{-\alpha}$, for $\alpha < 1$ both velocity and density fluctuations influence the statistics of PPV cubes.

It is also worrisome that no dependencies of PPV statistics on density have been reported so far in PCA studies. We do know from the analysis in LP00, that, for shallow density the fluctuations, PPV statistics should depend both on velocities and densities. Therefore no detection of the density spectrum may reflect the problem of finding the underlying relations empirically with data cubes of limited resolution. The latter may induce a special kind of shot noise (Lazarian et al. 2001; Esquivel et al. 2003; Chepurnov and Lazarian 2006, 2008). One way or another, this seem to illustrate the difficulty of empirical establishing of the relations between the PPV statistics and the underlying velocities and densities.

We feel that the analytical insight which was obtained in the process of VCA and VCS development should be used to get the corresponding calibration of the techniques which use wavelet or PCA decomposition instead of the Fourier one. Such a calibration can be done in many cases analytically.

10.3 Spectral Correlation Function

Spectral Correlation Function (SCF) was developed by Alyssa Goodman's group at the same time we developed the VCA technique (see Rosolowsky et al. 1999 and Lazarian 1999, respectively). Further development of the SCF technique in Padoan et al. (2001) removed the adjustable parameters from the original expression for the SCF and made the technique rather similar to VCA in terms of the observational data analysis. Indeed, both SCF and VCA measure correlations of intensity in PPV "slices" (channel maps with a given velocity window Δv) (see Fig. 1), but if SCF treats the outcome empirically, the analytical relations in Lazarian and Pogosyan (2000) relate the VCA measures to the underlying velocity and density statistics. In fact, we predicted several physically-motivated regimes for VCA studies. For instance, slices are "thick" for eddies with velocity ranges less than Δv and "thin" otherwise. VCA relates the spectral index of intensity fluctuations within channel maps to the thickness of the velocity channel and to the underlying velocity and density in the emitting turbulent volume. We showed that much of the earlier confusion stemmed from different observational groups having used velocity channels of different thicknesses (compare, e.g., Green 1993 and Stanimirovic et al. 1999). In the VCA these variations of indexes with the thickness of PPV "slice" are used to disentangle velocity and density contributions.

We suspect that similar "thick" and "thin" slice regimes should be present in the SCF analysis of data, but they have not been reported yet. While the VCA can be used for all the purposes the SCF is used (e.g. for an empirical comparisons of simulations and observations), the opposite is not true. In fact, Padoan et al. (2006) stressed that VCA eliminates errors inevitable for empirical attempts to calibrate PPV fluctuations in terms of underlying 3D velocity spectrum. With the VCA one can relate both observations and simulations to *turbulence theory*. Using explicit expressions for the VCA one can also study non-power law velocity fluctuations.

SCF was successfully used in Padoan et al. (2003) to relate the results of numerical simulations and observations. The advantage of the VCA and VCS techniques is that they can also relate the simulations and observations to the turbulence theory. Therefore, it looks advantageous to repeat the analysis in Padoan et al. (2003) using both the VCA and VCS techniques.

10.4 Identifying Objects in PPV Space

There exist a number of algorithms that identify shells and clumps using PPV data. The most widely-used codes, namely "Gaussclumps" (Stutzki and Guesten 1990) and "CLUMPFIND" (Williams et al. 1994) have been used to identify clumps in many observational data sets.

In view of our discussion of the complexity of the PPV space, one can potentially encounter problems arising from the caustics created in the velocity space. Indeed, we know both from the analytical theory and numerical testing that the ripples along the velocity axis are strongly affected by the turbulent velocities. This calls for more studies aimed at understanding at what conditions the features identified in the PPV space are real and in what cases they are caused by velocity crowding. Future research can identify the characteristic features of the latter effect.

11 Synergy and Future Work

11.1 Synergetic Statistical Studies

We would like to stress, however, that both VCS and VCA are the ways of obtaining of the spectrum of turbulent velocities.The velocity spectrum is not the only characteristics that the astronomers would like to know about the turbulence. Density spectrum is another characteristics that is available from the techniques above. The spectra of magnetic fields are potentially available through measurements of Faraday rotation and synchrotron emission (see Cho and Lazarian 2002; Haverkorn et al. 2008). However, even having 3 different spectra above, one does not have a full description of turbulence. The problem is that the spectrum, as a measure, does not preserve the information about the phases of turbulent motions. Below we list techniques that could, potentially, provide a complementary insight.

Genus analysis is a tool to study topology used in Cosmology (Gott et al. 1990). The 2D "genus", $G(\nu)$, is the difference between the number of regions with a projected density higher than ν and those with densities lower than ν. Unlike visual inspection, the genus quantifies the topology and allows us to compare numerical results with observations. In Kowal et al. (2007) we performed a systematic study of genus statistics for synthetic maps obtained via MHD turbulence simulations and studied the variations of genus with sonic and Alfven Mach numbers. We determined the cases when these variations were more prominent than the corresponding variations in spectral index. Genus was applied to observational data in Lazarian et al. (2002), Kim and Park (2007), Chepurnov et al. (2008). Note that genus at large scales (obtained with smoothing of the high resolution maps) is expected to be different for models of ISM where hot hydrogen forms tunnels (see Cox 2005) and in the classical McKee and Ostriker (1977) model.

Bispectrum is another tool which was first used in Cosmology (see Scoccimarro (1997, 2000) for a review). Its prospects for ISM studies were discussed in Lazarian (1999) and Lazarian et al. (2008). The standard power spectrum is obtained by multiplying the Fourier transform of a given **k** by its complex conjugate. In this process, the complex component is eliminated along with all information about *wave phases*. In the Bispectrum technique, the correlation of the Fourier transforms of different **k**'s is studied, and the wave phases are kept, allowing the turbulent fields corresponding to the same spectrum to be distinguished by their bispectrum. This method is ideal for the study of scale-correlations in multidimensional problems. Recently we applied bispectrum to the synthetic column density maps obtained from numerical simulation cubes. We showed that MHD turbulence presents a larger band of non-linearly interacting waves than non-magnetized turbulence does (Burkhart et al. 2008).

Studies of Anisotropy. A technique for studies of magnetic field direction, which makes use of preferred correlation of turbulent motions along the field, was proposed in Lazarian et al. (2002). The technique was originally discussed in terms of channel maps and velocity centroids (see also Vestuto et al. 2003) The application of it to synthetic maps in Esquivel and Lazarian (2005) showed that the anisotropies in the statistics of velocity centroids reflect well the projected direction of the magnetic field, even for high Mach number turbulence. This makes the technique very appealing to studies of magnetic fields in molecular clouds. A study of the anisotropy using the PCA decomposition was successfully performed by (Heyer et al. 2008).

11.2 Prospects of Quantiative Studies of PPV Data

VCA and VCS are two new techniques that emerged as an attempt to disentangle velocity and density contributions in studies of turbulence with spectroscopic data. They evolved from modest attempts to explain the existing velocity channel data in LP00 to formulating new ways of data analysis as in LP06, LP08 and Chepurnov and Lazarian (2008).

In LP00, LP04, LP06 we used HI as an example of species to which the technique is to be applied. Using heavier species that have lower thermal Doppler width of spectral lines allows one to study turbulence up to smaller scales. In addition, we would like to stress that the VCS technique can be used at different wavelength. For instance, the X-ray spectrometers with high spatial resolution can be used to study of turbulence in hot plasma. In particular, the potential of VCS is high for studies of turbulence in clusters of galaxies (cf. Sunyaev et al. 2003 and references therein). A simulated example of such a study with the future mission Constellation X is provided in Fig. 10.

Studies of turbulence in objects which are poorly resolved spatially is a natural avenue for the VCS applications. Interestingly enough, in this case one can combine the absorption line

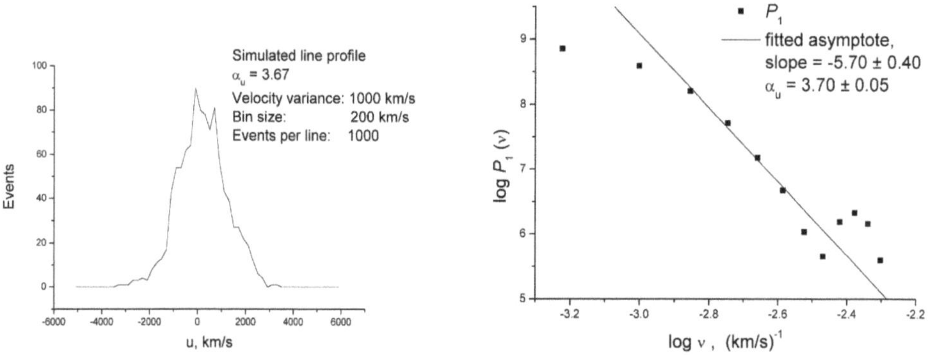

Fig. 10 Synthetic spectral line (*left*) and the velocity spectrum (*right*) that can be obtained with Constellation X for turbulence in clusters of galaxies. Calculations are done assuming one hour of observations

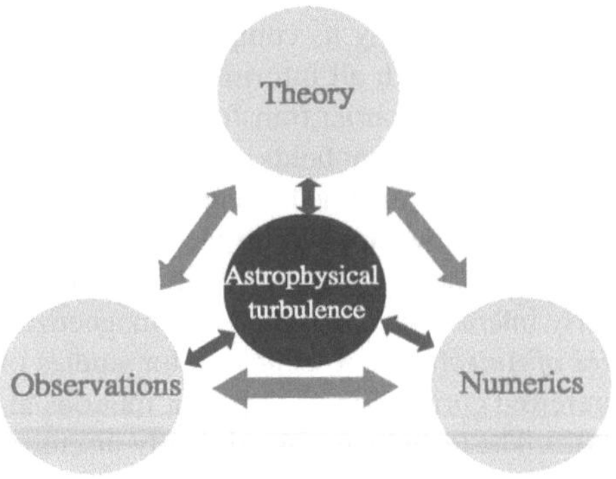

Fig. 11 Big Picture: interrelation between the approaches to studying astrophysical turbulence

studies, which would provide the VCS for the pencil beam, i.e. for the high resolution, with the emission studies that would provide the VCS in the poor resolution limit. Potentially, both velocity and density spectra can be obtained this way.

The importance of this work goes beyond the actual recovery of the particular power-law indexes. First of all, as we mentioned earlier, the techniques can be generalized to solve the inverse problem to recover non-power law turbulence spectra. This may be important for studying turbulence at scales at which either injection or dissipation of energy happens. Such studies are important for identifying astrophysical sources and sinks of turbulent energy. Second, studies of the VCS transition from low resolution to high resolution regimes (see Fig. 9) allows one to separate thermal and non-thermal contributions to the line-widths as it is discussed in LP06. This could both test the thermal correction that can be applied to extend the power-law into sub-thermal velocity range (see also Chepurnov and Lazarian 2006) and enable studies of temperature distribution of the gas in atomic clouds (cf. Heiles and Troland 2003).

In Sect. 6 we mentioned several advantages of using VCA and VCS simultaneously. One may mention an additional one. In compressible fluid the power spectrum of velocity can be decomposed into the spectra of solenoidal (incompressible) and potential (compressible) motions. Interestingly enough, the VCA and VCS depend on these two components differently. This opens prospects of studying the effects of compressibility by combining the two techniques.

We feel that one should not feel constrained by the framework of the VCA and VCS techniques. The most important, in terms of theory, was the development of the general description of the correlations within PPV cubes for different situations, including self-absorbing data and saturated absorption lines. This opens avenues to developing completely new ways of analyzing observational data. As we discussed in Sect. 10.2 the use of wavelets instead of Fourier transforms may be a straightforward generalization of the techniques. More sophisticated work may be required to get higher order correlations and bispectrum from the PPV data.

11.3 Outlook onto Big Picture

Several decades ago measurements of spectral profiles were rather challenging. The situation is radically different these days where the challenge is to use productively the enormous wealth of spectroscopic surveys. Therefore the development of the theoretical description of the PPV statistics, which resulted in the birth of the VCA and VCS techniques is a very timely achievement.

Astrophysical turbulence is a very tough subject and dealing with it requires a cohesive use of analytical, numerical and observational approaches (see Fig. 11). Analytical studies most often predict power spectra for turbulence in different regimes (see Galtier et al. 2002; Goldreich and Sridhar 1995; Lazarian et al. 2004; Boldyrev 2006; Lithwick et al. 2007; Beresnyak and Lazarian 2008; Chandran 2008). At relatively high effective viscosities and resistivities, which are the only ones that are available with the present-day computational facilities, these predictions can be tested (see Cho and Vishniac 2000; Maron and Goldreich 2001; Cho et al. 2002a, 2002b, 2003; Cho and Lazarian 2002, 2003; Beresnyak and Lazarian 2006, 2008; Galtier 2008). We expect the VCS and VCA techniques to test the turbulence spectra in astrophysical circumstances.

We expect that as the turbulence theory matures, it starts posing more sophisticated questions, e.g. on the scaling of the higher order velocity correlations (She and Leveque 1994; Dubrulle 1994, see Biskamp 2003). In view of this the statistical description of the PPV space obtained so far can serve as a starting ground for developing new techniques.

This is not the only role of the quantitative description of the PPV statistics, however. Astrophysical turbulence happens in complex environments and has multiple sources and sinks.[18] Many of numerical simulations, for instance, numerical simulations of interstellar medium, try to simulate the astrophysical settings in all its complexity. Theory in these situations provides rough guidance and insight, while through observational studies of actual shape of the power spectra can clarify many of the relevant issues.

For all the applications it is important that the developed techniques can be used for studying turbulence in different phases of the astrophysical media. Indeed, the VCA and VCS techniques can employ various spectral lines: emission and absorption, optically thin and thick, saturated and not saturated.

Acknowledgements I thank Alexey Chepurnov, Dmitry Pogosyan for their input. The research is supported by the NSF Center for Magnetic Self Organization in Laboratory and Astrophysical Plasmas and NSF grant AST 0808118. Helpful comments by the referee were appreciated. I thank Blakesley Burkhard for reading the manuscript and her suggestions.

References

J.W. Armstrong, B.J. Rickett, S.R. Spangler, Astrophys. J. **443**, 209 (1995)
J. Ballesteros-Paredes, R. Klessen, M. Mac Low, E. Vasquez-Semadeni, in *Protostars and Planets V*, ed. by D. Reipurth, X. Jewitt, K. Keil (University of Arizona Press, Tucson, 2006), pp. 63–80
A. Begum, J. Chengalur, S. Bharadwaj, Mon. Not. R. Astron. Soc. **372**, L33–L37 (2006)
A. Beresnyak, A. Lazarian, Astrophys. J. Lett. **640**, L175 (2006)
A. Beresnyak, A. Lazarian, Astrophys. J. **682**, 1070 (2008)
A. Beresnyak, A. Lazarian, J. Cho, Astrophys. J. Lett. **624**, 93 (2005)
D. Biskamp, Magnetohydrodynamic Turbulence (CUP, 2003)
S. Boldyrev, Phys. Rev. Lett. **96**, 115002 (2006)
C.M. Brunt, M.H. Heyer, E. Vázquez-Semadeni, B. Pichardo, Astrophys. J. **595**, 824 (2003)
B. Burkhart, D. Falceta-Goncalves, G. Kowal, A. Lazarian, Astrophys. J. **692** (2008, in press)
L. Cambresy, Astron. Astrophys. **345**, 965–976 (1999)
B.D.G. Chandran, Astrophys. J. **685**, 1413 (2008)
A. Chepurnov, A. Lazarian (2006). astro-ph/0611463
A. Chepurnov, A. Lazarian, Astrophys. J. (2008, in press)
A. Chepurnov, A. Lazarian, J. Gordon, S. Stanimirovic, Astrophys. J. (2008, in press). arXiv:0711.2530

[18]The energy is being injected by supernovae explosions, stellar winds, magnetorotational instability etc. In addition, cosmic rays may redistribute the turbulent energy absorbing it at the large scales and reinjecting at small spatial scales (Lazarian and Beresnyak 2006).

J. Cho, A. Lazarian, Astrophys. J. **575**, 63 (2002)
J. Cho, E. Vishniac, Astrophys. J. **539**, 273 (2000)
J. Cho, A. Lazarian, Mon. Not. R. Astron. Soc. **345**, 325 (2003)
J. Cho, A. Lazarian, E.T. Vishniac, Astrophys. J. **564**, 291 (2002a)
J. Cho, A. Lazarian, E. Vishniac, Astrophys. J. **566**, L49 (2002b)
J. Cho, A. Lazarian, E. Vishniac, Astrophys. J. **595**, 812 (2003)
D.P. Cox, Ann. Rev. Astron. Astrophys. **43**, 337 (2005)
J. Crovisier, J.M. Dickey, Astron. Astrophys. **122**, 282 (1983)
A.A. Deshpande, K.S. Dwarakanath, W.M. Goss, Astrophys. J. **543**, 227 (2000)
J. Dickey, N. McClure-Griffiths, S. Stanimirovic, B. Gaensler, A. Green, Astrophys. J. **561**, 264 (2001)
R.L. Dickman, S.C. Kleiner, Astrophys. J. **295**, 479 (1985)
B. Dubrulle, Phys. Rev. Lett. **73**, 959–962 (1994)
E. Falgarone, J.-L. Puget, Astron. Astrophys. **293**, 840 (1995)
S. Galtier, J. Geophys. Res. (Space Phys.) **113**, 1102 (2008)
S. Galtier, S. Nazarenko, A. Newell, A. Pouquet, Astrophys. J. **564**, L49 (2002)
A. Gill, R.N. Henriksen, Astrophys. J. **365**, L27 (1990)
D.A. Green, Astron. Astrophys. Suppl. **78**, 277 (1989)
D.A. Green, Mon. Not. R. Astron. Soc. **262**, 328 (1993)
P. Goldreich, S. Sridhar, Astrophys. J. **438**, 763 (1995)
J. Goodman, R. Narayan, Mon. Not. R. Astron. Soc. **214**, 519 (1985)
J.R. Gott, C. Park, R. Juskiewicz, W. Bies, F. Bouchet, A. Stebbins, Astrophys. J. **352**, 1 (1990)
M. Haverkorn, J.C. Brown, B.M. Gaensler, N.M. McClure-Griffiths, Astrophys. J. **680**, 362 (2008)
C. Heiles, T. Troland, Astrophys. J. Suppl. **151**, 271 (2003)
M. Heyer, H. Gong, E. Ostriker, C. Brunt, Astrophys. J. **680**, 420 (2008)
B. Elmegreen, J. Scalo, Ann. Rev. Astron. Astrophys. **42**, 211 (2004)
A. Esquivel, A. Lazarian, Astrophys. J. **784**, 320 (2005)
A. Esquivel, A. Lazarian, S. Horibe, V. Ossenkopf, J. Stutzki, J. Cho, Mon. Not. R. Astron. Soc. **381**, 1733 (2007)
A. Esquivel, A. Lazarian, D. Pogosyan, J. Cho, Mon. Not. R. Astron. Soc. **342**, 325 (2003)
S. Kim, C. Park, Astrophys. J. **663**, 244 (2007)
G. Kowal, A. Lazarian (2006). astro-ph/0611396
G. Kowal, A. Lazarian, A. Beresnyak, Astrophys. J. **658**, 423 (2007)
A. Khalil, G. Joncas, F. Nekka, P. Kestener, A. Arneodo, Astrophys. J. Suppl. **165**, 512 (2006)
A.G. Kritsuk, M.L. Norman, P. Padoan, R. Wagner, Astrophys. J. **665**, 416 (2007)
A. Lazarian, in *Plasma Turbulence and Energetic Particles in Astrophysics*, ed. by M. Ostrowski, R. Schlickeiser (Cracow, 1999), p. 28
A. Lazarian, J. Korean Astron. Soc. **37**, 563 (2004)
A. Lazarian, Spectral Line Shapes: XVIII, AIP Conf. Proc. **874**, 301 (2006)
A. Lazarian, D. Pogosyan, A. Esquivel, Astronomical Society of the Pacific Conference Series, vol. 276, 182 (2002)
A. Lazarian, G. Kowal, A. Beresnyak, Astronomical Society of the Pacific Conference Series, vol. 385, 3 (2008)
A. Lazarian, A. Beresnyak, Mon. Not. R. Astron. Soc. **373**, 1195–1202 (2006)
A. Lazarian, A. Esquivel, Astrophys. J. Lett. **592**, L37 (2003)
A. Lazarian, D. Pogosyan, Astrophys. J. **537**, 720 (2000)
A. Lazarian, D. Pogosyan, Astrophys. J. **616**, 943 (2004)
A. Lazarian, D. Pogosyan, Astrophys. J. **652**, 1348 (2006)
A. Lazarian, D. Pogosyan, Astrophys. J. **686**, 650 (2008)
A. Lazarian, D. Pogosyan, E. Vazquez-Semadeni, B. Pichardo, Astrophys. J. **555**, 130 (2001)
A. Lazarian, E. Vishniac, J. Cho, Astrophys. J. **603**, 180 (2004)
F. Levier, Astron. Astrophys. **421**, 387 (2004)
Y. Lithwick, P. Goldreich, S. Sridhar, Astrophys. J. **655**, 269 (2007)
M. Mac Low, R. Klessen, Rev. Mod. Phys. **76**, 125 (2004)
J. Maron, P. Goldreich, Astrophys. J. **554**, 1175–1196 (2001)
C. McKee, in *The Origin of Stars and Planetary Systems*, ed. by Ch.L. Lada, N.D. Kylafis (Kluwer, Dordrecht, 1999), p. 29
C. McKee, J. Ostriker, Astrophys. J. **218**, 148–169 (1977)
C. McKee, E. Ostriker, Ann. Rev. Astron. Astrophys. **45**, 565–687 (2007)
M.S. Miesch, J. Scalo, J. Bally, Astrophys. J. **524**, 895 (1999)
M. Miville-Deschênes, F. Levrier, E. Falgarone, Astron. Astrophys. **593**, 831 (2003)
E. Muller, S. Stanimirovic, E. Rosolowsky, L. Staveley-Smith, Astrophys. J., **616**, 845

G. Munch, Rev. Mod. Phys. **30**, 1035 (1958)
V. Ossenkopf, A. Esquivel, A. Lazarian, J. Stutzki, Astron. Astrophys. **452**, 223 (2006)
E.W. Rosolowsky, A.A. Goodman, D.J. Wilner, J.P. Williams, Astrophys. J. **524**, 887–894 (1999)
P. Padoan, E. Rosolowsky, A. Goodman, Astrophys. J. **547**, 862 (2001)
P. Padoan, M. Juvla, A. Kritsuk, M. Norman, Astrophys. J. **653**, 125 (2006)
P. Padoan, A. Goodman, M. Juvela, Astrophys. J. **588**, 881 (2003)
C.R. O'dell, H.O. Castaneda, Astrophys. J. **317**, 686 (1987)
R. Scoccimarro, Astrophys. J. **487**, 1 (1997)
R. Scoccimarro, Astrophys. J. **544**, 597 (2000)
Z.-S. She, E. Leveque, Phys. Rev. Lett. **72**, 336 (1994)
F. Shu, Z.-Y. Li, A. Allen, Astrophys. J. **601**, 930 (2004)
S. Spangler, C. Gwinn, Astrophys. J. **353**, L29 (1990)
L. Spitzer, *Physical Processes in the Interstellar Medium* (Wiley, New York, 1978)
L.G. Stenholm, Astron. Astrophys. **232**, 495 (1990)
S. Stanimirovic, A. Lazarian, Astrophys. J. **551**, L53 (2001)
S. Stanimirovic, L. Staveley-Smith, J.M. Dickey, R.J. Sault, S.L. Snowden, Mon. Not. R. Astron. Soc. **302**, 417 (1999)
J. Stutzki, R. Guesten, Astrophys. J. **356**, 513 (1990)
J. Stutzki, F. Bensch, A. Heithausen, V. Ossenkopf, M. Zielinsky, Astron. Astrophys. **336**, 697 (1998)
R.A. Sunyaev, M.L. Norman, G. Bryan, Astron. Lett. **29**, 783 (2003)
K. Sun, C. Kramer, V. Ossenkopf, F. Bensch, J. Stutzki, M. Miller, Astron. Astrophys. **451**, 539–549 (2006)
J. Swift, PhD Thesis, Univ. of California-Berkeley (2006)
J.J. Swift, W.J. Welch, Astrophys. J. Suppl. **174**, 202 (2008)
E. Vazquez-Semadeni, E.C. Ostriker, T. Passot, C.F. Gammie, J.M. Stone, in *Protostars and Planets IV* (University of Arisona Press, Tucson, 2000), p. 3
J. Vestuto, E. Ostriker, J. Stone, Astrophys. J. **590**, 858 (2003)
J. Williams, E. de Geus, L. Blitz, Astrophys. J. **428**, 693 (1994)

Properties and Selected Implications of Magnetic Turbulence for Interstellar Medium, Local Bubble and Solar Wind

A. Lazarian · A. Beresnyak · H. Yan · M. Opher · Y. Liu

Originally published in the journal Space Science Reviews, Volume 143, Nos 1–4, 387–413.
DOI: 10.1007/s11214-008-9452-y

Abstract Astrophysical fluids, including interstellar and interplanetary medium, are magnetized and turbulent. Their appearance, evolution, and overall properties are determined by the magnetic turbulence that stirs it. We argue that examining magnetic turbulence at a fundamental level is vital to understanding many processes. A point that frequently escapes the attention of researchers is that magnetic turbulence cannot be confidently understood only using "brute force" numerical approaches. In this review we illustrate this point on a number of examples, including intermittent heating of plasma by turbulence, interactions of turbulence with cosmic rays and effects of turbulence on the rate of magnetic reconnection. We show that the properties of magnetic turbulence may vary considerably in various environments, e.g. imbalanced (or cross-helical) turbulence in solar wind differs from balanced turbulence and both of these differ from turbulence in partially ionized gas. Appealing for the necessity of more observational data on magnetic fields, we discuss a possibility of studying interplanetary turbulence using alignment of Sodium atoms in the tail of comets.

Keywords Turbulence · MHD · Interstellar medium

PACS 95.30.Qd · 52.30.Cv · 96.50.Tf

A. Lazarian · A. Beresnyak (✉)
Department of Astronomy, University of Wisconsin-Madison, Madison, USA
e-mail: beresnyak@wisc.edu

H. Yan
CITA, University of Toronto, Toronto, Canada

M. Opher
Department of Physics and Astronomy, George Mason University, Fairfax, USA

Y. Liu
University of New Hampshire, Durham, USA

1 Introduction

It is well known that astrophysical fluids are magnetized and turbulent (see Armstrong and Woo 1981; Armstrong et al. 1995; Verdini and Velli 2007 and ref. therein). Turbulence, a highly stochastic fluid flow, is known to affect most of the properties of fluids, e.g. thermal conductivity, propagation of waves and energetic particles, magnetic field generation etc. (see Kazantsev 1968; Moffatt 1978; Schlickeiser 2002; Vishniac et al. 2003; Cranmer and van Ballegooijen 2005 and references therein). Local Bubble, solar wind and interstellar medium are not exceptions: the fluids there are turbulent. A substantial progress in understanding of the media above has been achieved using numerical simulations. It is very encouraging that present codes can produce simulations that resemble observations. However, one may wonder to what extent numerical results reflect reality. The answer to this question is not as simple as it may look.

It is easy to justify a cautious approach to the interpretation of the numerical simulations. Indeed, a meaningful numerical representation of the turbulent fluid requires some basic non-dimensional combinations of the physical parameters of the simulation to be similar to those of the real ISM. One such is the "Reynolds number", *Re*, the ratio of the eddy turnover time of a parcel of gas to the time required for viscous forces to slow it appreciably. A similar parameter, the "magnetic Reynolds number", *Rm*, is the ratio of the magnetic field decay time to the eddy turnover time. The properties of flows on all scales depend on *Re* and *Rm*. It is not realistic to expect that we can in any foreseeable future simulate turbulent flows with $Re > 10^8$ and $Rm > 10^{16}$. Note that 3D simulations for 512 cubes can have *Re* and *Rm* up to $\sim$2000 and are limited by their grid sizes. It should be kept in mind that while low-resolution observations show true large-scale features, low-resolution numerics may potentially produce an incorrect physical picture.

How feasible is it, then, to strive to understand the complex microphysics of astrophysical MHD turbulence? Substantial progress in this direction is possible by means of "*scaling laws*", or analytical relations between non-dimensional combinations of physical quantities that allow a prediction of the motions over a wide range of *Re* and *Rm*. Even with its limited resolution, numerics is a great tool to *test* scaling laws.

On a basis of several selected examples we show that a "brute force" numerical approach has limited applicability while dealing with interstellar medium, Local Bubble and solar wind. Instead we claim that to create an appropriate "tool box" that can be incorporated into simulations one must understand better underlying fundamental physical processes. Moreover, a better understanding of these processes would provide insight concerning the circumstances when one should or should not rely on present-day numerical simulations. For instance, if magnetic reconnection is slow in collisional astrophysical environments, numerical simulations *cannot* represent fluids there. We argue below that in turbulent fluids reconnection is fast and therefore numerical simulations are not very different from the reality in this respect. We believe that the solid progress in the studies of complex processes in magnetized astrophysical fluids must be based on verification of theoretical constructions with observations. For this purpose we discuss a new technique of studying interplanetary magnetic fields using a subtle phenomenon of atomic alignment.

We realize that within a relatively short review, one cannot cover in depth the properties and numerous implications of astrophysical turbulence. Therefore while covering a few relevant subjects within the area of our expertise, we provide, wherever appropriate, references to extensive reviews and monographs. Although our review is intended for the special volume of the proceedings on the heliosphere and the Local Bubble, a number of examples in the review are from the area of interstellar medium. In fact, we feel that the existence of the gap between the interstellar and the heliospheric/interplanetary communities is not healthy, as the fundamental underlying physics that researchers deal with in these two cases is very similar. Moreover, we believe that with the Voyager spacecrafts approaching the interstellar medium (see Stone et al. 2008, http://voyager.jpl.nasa.gov/), the cross-pollination between the fields should be enhanced.

In what follows we discuss the spectrum and anisotropy of magnetic turbulence in different regimes, including the regime of balanced turbulence, imbalanced turbulence and viscosity-damped turbulence (Sect. 2). We consider turbulence intermittency in Sect. 3 and the modification of the magnetic turbulence in the presence of cosmic rays in Sect. 4. Section 5 is devoted to the selected implications of turbulence in interstellar and interplanetary medium, including scattering of cosmic rays, turbulent reconnection and perpendicular diffusion of cosmic rays and heat. We present an example of interplanetary turbulence studies using observations of the sodium tail of a comet in Sect. 6. Our main results are summarized in Sect. 7.

2 Spectrum of Turbulence and Its Anisotropy

2.1 Anisotropic MHD Turbulence

Magnetized turbulence is a tough and complex problem with many excellent monographs and reviews devoted to different aspect of it (see Biskamp 2003 and references therein). A broad outlook on the astrophysical implications of the turbulence can be found in a review by Elmegreen and Scalo (2004), while the effects of turbulence on molecular clouds and star formation are reviewed in McKee and Ostriker (2007) and Ballesteros-Paredes et al. (2006). However, the issues of turbulence spectrum and its anisotropies, we feel, are frequently given less attention than they deserve.

While turbulence is an extremely complex chaotic non-linear phenomenon, it allows for a remarkably simple statistical description (see Biskamp 2003). If the injections and sinks of the energy are correctly identified, we can describe turbulence for *arbitrary Re* and *Rm*. The simplest description of the complex spatial variations of any physical variable, $X(\mathbf{r})$, is related to the amount of change of X between points separated by a chosen displacement $\mathbf{l}$, averaged over the entire volume of interest. Usually the result is given in terms of the Fourier transform of this average, with the displacement $\mathbf{l}$ being replaced by the wavenumber $\mathbf{k}$ parallel to $\mathbf{l}$ and $|\mathbf{k}| = 1/|\mathbf{l}|$. For example, for isotropic turbulence the kinetic energy spectrum, $E(k)dk$, characterizes how much energy resides at the interval $k, k+dk$. At some large scale L (i.e., small k), one expects to observe features reflecting energy injection. At small scales, energy dissipation should be seen. Between these two scales we expect to see a self-similar power-law scaling reflecting the process of non-linear energy transfer.

Thus, in spite of its complexity, the turbulent flow is similar on small and large scales when one goes into the frame of the fluid. The physical variables are proportional to simple powers of the eddy sizes over a large range of sizes, leading to scaling laws expressing the dependence of certain non-dimensional combinations of physical variables on the eddy size. Robust scaling relations can predict turbulent properties on the whole range of scales, including those that no large-scale numerical simulation can hope to resolve. These scaling relations are extremely important for obtaining an insight of processes on the small scales.

The presence of a magnetic field makes MHD turbulence anisotropic (Montgomery and Turner 1981; Matthaeus et al. 1983; Shebalin et al. 1983; Higdon 1984; Goldreich and Sridhar 1995 see Ougthton et al. 2003 for a review). The relative importance of hydrodynamic and magnetic forces changes with scale, so the anisotropy of MHD turbulence does too. Many astrophysical results, e.g. the dynamics of dust, scattering and acceleration of energetic particles, thermal conduction, can be obtained if the turbulence spectrum and its anisotropy are known. As we discuss below, additional important insight can be obtained if we know turbulence intermittency.

Estimates of turbulence anisotropy obtained in relation to the observations of magnetic fluctuation of the outer heliosphere and solar wind (see Zank and Matthaeus 1998 and references therein) provided, for an extended period of time, the only guidance for theoretical advances. This resulted in a picture of MHD turbulence consisting of 2D "reduced MHD" perturbations carrying approximately 80% of energy and "slab" Alfvenic waves (having wavevector parallel to the magnetic field) carrying the remaining 20% of energy (see Matthaeus et al. 2003 and references therein). In other words, in the suggested picture the MHD turbulence was presented by two anisotropic components, one having wave vectors mostly perpendicular to magnetic field (the 2D one), the other having them mostly parallel to magnetic field (the slab one). This model became a default one for many calculations of the propagation of cosmic rays (see Bieber et al. 1988, 1994). On the contrary, guided mostly by compressible MHD numerical simulations, the interstellar community adopted a model of the MHD turbulence where the basic MHD modes, i.e. slow, fast and Alfven are well coupled together and efficiently dissipate energy in shocks (Stone et al. 2008; Mac Low 1999). Little cross-talk between the two communities did not stimulate the interdisciplinary debates on the nature of MHD turbulence, which was regretful, as the magnetospheric community has the advantage of the in-situ spacecraft measurements.

In spite of the intrinsic limitations of the "brute force" approach, we feel that reliable results can be obtained numerically if the studies are focused on a particular property of turbulence in order to get a clear picture of the underlying physics occurring on small scales ("microphysics") that cannot be resolved in "global" interstellar simulations.[1]

We feel that it were the "focused" numerical simulations that allowed the validation of the Goldreich and Shridhar (1995, henceforth GS95) model of MHD turbulence. Indeed, GS95 made predictions regarding relative motions parallel and perpendicular to **B** for Alfvenic turbulence. The model did not predict the generation of any "slab" modes and, instead of pure 2D Alfvenic modes, predicted that most of the Alfvenic energy is concentrated in the modes with a so-called "critical balance" between the parallel and perpendicular motions. The latter can be understood in an intuitive picture where eddies mixing magnetic field perpendicular to its local[2] direction induce Alfvenic waves with the period equal to the period of the eddy rotation. This results in the scale-dependent anisotropy of velocity and magnetic perturbations, with the anisotropy being larger for smaller eddies.

The relations predicted in GS95 were confirmed numerically for incompressible[3] (Cho and Vishniac 2000; Maron and Goldreich 2001; Cho et al. 2002a) and compressible MHD turbulence[4] (Cho and Lazarian 2002, 2003, henceforth CL02 and CL03). (See Cho et al.

---

[1]By contrast, numerical simulations that deal with many physical conditions simultaneously cannot distinguish between the effects of different processes. Moreover, they inevitably have a more restricted interval of scales on which energy is injected by numerics, initial conditions, or boundary conditions. Their results are, therefore, difficult to interpret in physical terms.

[2]The notion of the direction being local is critical. Small eddies are affected by magnetic field in their vicinity, rather than a global field. No universal scalings are possible to obtain in the frame of the mean magnetic field.

[3]As in any developing field, ongoing controversies and competing ideas exist on how to improve the GS95 scalings (see Boldyrev 2006; Beresnyak and Lazarian 2006; Gogoberidze 2007).

[4]Some studies of MHD compressible turbulence, e.g. Vestuto et al. (2003) did not perform a decomposition of MHD perturbations into Alfven, slow and fast modes as it is done in CL02 and CL03. They did not use

2003 for a review.) They are in good agreement with observed and inferred astrophysical spectra. A remarkable fact revealed in Cho et al. (2002a) is that fluid motions perpendicular to **B** are identical to hydrodynamic motions. This provides an essential physical insight and explains why in some respects MHD turbulence and hydrodynamic turbulence are similar, while in other respects they are different.

Any MHD perturbation can be decomposed into four fundamental modes: entropy, slow, fast and Alfven mode (see, e.g., Biskamp 2003). GS95 provided theoretical arguments in favor of low coupling between fast and Alfven modes, and low impact of slow modes to Alfven modes (see also Lithwick and Goldreich 2001). This challenged the paradigm accepted by the interstellar community. While the decomposition of MHD perturbations into fundamental MHD waves was widely used in the literature (see Dobrowolny et al. 1980) it was usually assumed that the Alfvenic waves exist and interact with other waves for many periods (see a discussion in Zweibel et al. 2003). This is not the case of the GS95 model of turbulence, where the Alfven modes non-linearly decay within one wave period. This reduces the time of interaction and therefore the coupling. Interestingly enough, in GS95 model, the Alfvenic modes can affect slow modes, but the opposite is not true. These results were successfully tested in CL02 and CL03.

Insight into fundamental properties of interstellar turbulence has already paid dividends. For instance, theories of cosmic ray propagation and dynamics of interstellar grains have been revised in view of better understanding of MHD turbulence[5] (see Cho and Lazarian 2005 and references therein). Similarly, the issue of turbulent support of molecular clouds requires revisions. Indeed, a widely accepted view that the rapid decay of turbulence is caused by the coupling of compressible and incompressible motions was disproved with numerical simulations in CL02. In fact, it was shown there that the rate of turbulent decay depends on the turbulence driving (Cho et al. 2003a). Moreover, contrary to common beliefs, MHD motions do not die away at the scale at which ambipolar diffusion (when magnetic field "slips" through the fluid containing large percentage of neutrals) damps hydrodynamic motions. As we discuss in Sect. 2.1 a new regime of viscosity-damped MHD turbulence emerges (Cho et al. 2002b).

Some of the relevant results are illustrated in Fig. 1. Contrary to the common expectation, the modes exhibited nice scaling laws that allow further analytical and numerical applications. For instance, numerical studies in CL02 and in CL03 reveal that the Goldreich-Shridhar scalings are valid for the Alfvenic part of the turbulence cascade even in the highly compressible regime (see also Beresnyak and Lazarian 2006).

The density spectrum of MHD turbulence was studied in Beresnyak et al. (2006) and Kowal et al. (2007). The spectrum was shown to be approximately Kolmogorov for low Mach numbers, and get flatter as the Mach number increases. The compression of gas by shocks was identified as the reason for the spectrum flattening. At the same time, it was shown that the logarithms of density have a Kolmogorov spectrum ($-5/3$ spectral slope) and GS95 anisotropies.

The statistical decomposition of MHD turbulence into Alfven, slow and fast modes[6] suggested in CL02 was successfully tested for slow modes of magnetically dominated plasmas

---

local system of reference for which the GS95 scaling is formulated. Therefore a direct comparison of their results with the GS95 predictions is difficult.

[5] Incidentally, the concept of the scale-dependent anisotropy in respect to the *local* direction of magnetic field is very important. This local magnetic field is what actually affects energetic particles.

[6] While Alfven mode can be thought of as a transverse oscillation of a magnetic field line, slow and fast modes are the generalization of sound wave to magnetized fluid and rely on gas and magnetic pressure as restoring force.

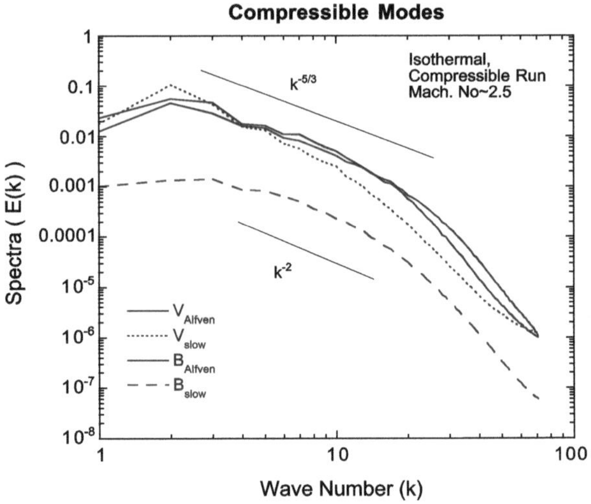

(a)

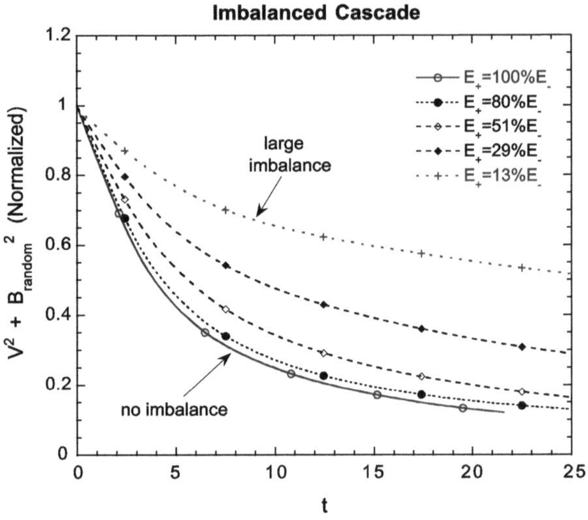

(b)

Fig. 1 Statistics of MHD Turbulence: (**a**) Scaling of compressible motions for plasma with magnetic pressure ten times the gas pressure. This regime is important for molecular clouds (from CL02). The velocity (*solid line*) and **B** (*dashed line*) spectra are plotted against k ($\equiv$ 1/(eddy size)). They show well defined statistical properties that allow further fruitful applications. (**b**) The simulations of decaying turbulence show that the rate of the decay of the total energy is a strong function of the imbalance of the energy contained in waves initially moving in opposite directions. In the lowest curve, the waves have the same amplitude. The energy decays more slowly with imbalance, affecting how long turbulence can support molecular clouds and whether transfer of energy between clouds is efficient (Cho et al. 2002b). E_+ and E_- are the initial energies of opposite waves. (**c**) The evolution of kinetic energy in 3-D compressible turbulence, initially started with $\nabla \cdot v = 0$ as if the gas were incompressible. The dashed curve shows the evolution of the $\nabla \cdot v = 0$ motions. We see that Alfvenic turbulence creates only a marginal amount of compressible motions, suggesting that Alfvenic modes should evolve independently of the compressible cascade (CL02). (**d**) Magnetic fluctuations persist beyond the turbulent damping scale at large k, while hydrodynamic fluctuations damp out in partially ionized gas (Lazarian et al. 2004). This viscosity-dominated regime of turbulence may dominate small scale structure of partially ionized gas. (**e–f**) Isocontours of equal correlation for Alfven and fast modes (CL02). (**e**) The Alfvenic motions are much more correlated along **B** than perpendicular to it. (**f**) In contrast, fast magnetosonic fluctuations show essentially circular (isotropic) isocontours of correlation

Fig. 1 (*Continued*)

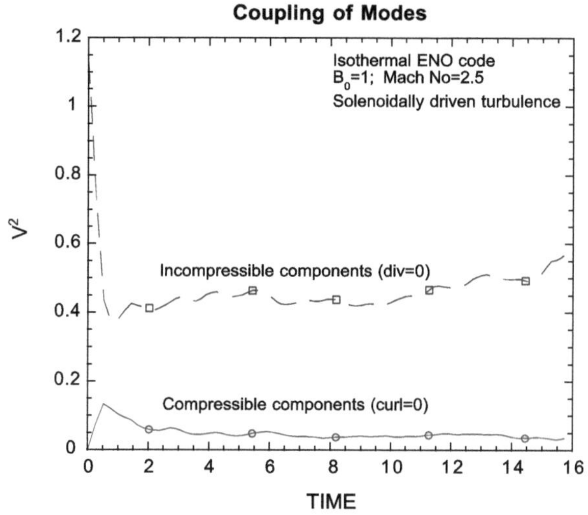

(c)

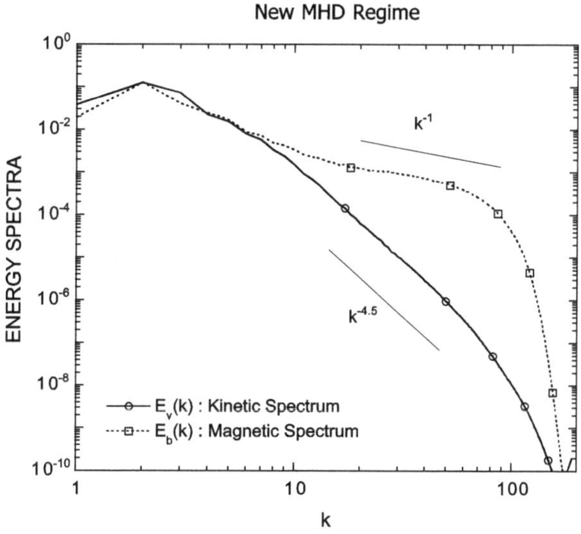

(d)

in CL03. The statistical procedure was further improved in Kowal and Lazarian (2006), where wavelets rather than the Fourier transforms were used. There the study of spectra and anisotropy of velocity, magnetic field and density of compressible MHD turbulence was performed and the advantages of the wavelets in comparison with the Fourier technique were revealed for studying turbulence with weak mean field. For turbulence with $B_{\mathrm{mean}} \sim \delta B$ the obtained results are consistent with the CL02 and CL03 studies. Nevertheless, it would be wrong to say that we have a complete understanding of the scaling of MHD modes and their interactions. First of all, one should distinguish weak and strong Alfvenic turbulence. The weak turbulence is essentially 2D, with the turbulent cascade creating more structure perpendicular to magnetic field as the turbulence cascades (see Galtier et al. 2000). Such a cascade emerges when the driving of turbulence at the outer scale is weak, i.e. the in-

Fig. 1 (*Continued*)

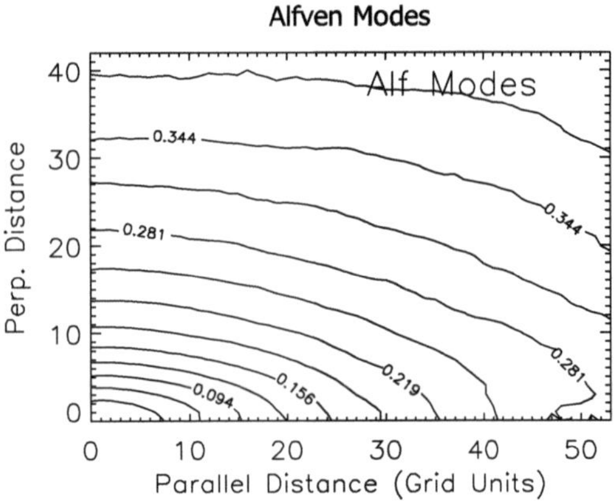

(e)

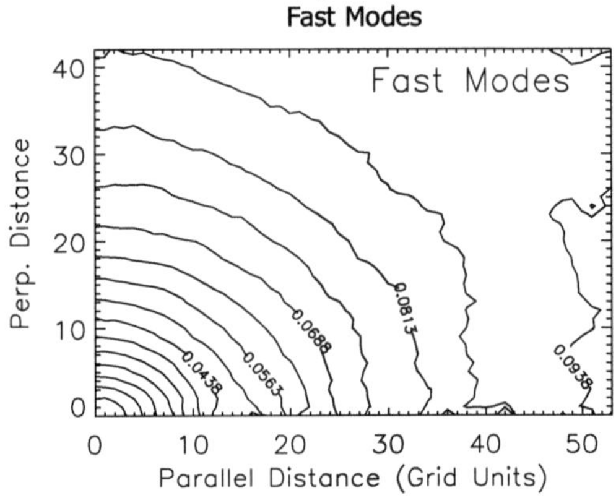

(f)

jection velocity is much less than the Alfven velocity. Although the weak turbulence picture corresponds to the early representation of MHD turbulence (see discussion in Sect. 2.1), one should keep in mind that the strength of Alfvenic interactions increases with the decrease of the scale along the cascade. Therefore the Alfvenic turbulence gets eventually strong, while both the inertial range and the astrophysical utility of the weak Alfvenic cascade are limited. The interaction of weak Alfvenic turbulence with fast modes has dependences on the angle between **B** and the wavevector, as was shown by Chandran (2005). In addition, as we discuss in Sect. 4, the absence of slab waves in MHD simulations does not mean that such modes do not appear in realistic astrophysical circumstances, e.g. when cosmic rays are present. Moreover, we show in Sect. 5 that in many circumstances, the way cascade is initiated and dissipates affect both the properties and implications of magnetic turbulence.

2.2 Imbalanced Turbulence

MHD turbulence in the presence of sources and sinks gets *imbalanced*, in the sense that the flow of energy in one direction is larger than the flow of energy in the opposite direction. Solar wind presents a vivid example of imbalanced turbulence with most waves near the Sun moving in the direction away from the Sun.

While theories of balanced MHD turbulence enjoyed much attention, the theory of imbalanced turbulence[7] attracted less work, unfortunately (see Biskamp 2003 and references therein). The analytical results were obtained for *weak imbalanced turbulence* (Galtier et al. 2002; Lithwick and Goldreich 2003) and they are applicable in a rather narrow range of imbalance ratios. Some earlier simulations of strong imbalanced turbulence were limited to rather idealized set ups (Maron and Goldreich 2001; Cho et al. 2002a), i.e. for the initial state the results of the simulations of strong balanced turbulence were used, but the amplitudes of waves moving in one direction were reduced. This did not allow making definitive conclusions about properties of imbalanced turbulence.

We think that the best experimental data on the imbalanced regime is currently available from observations of solar wind turbulence (e.g., Horbury 1999). This data, collected by spacecrafts, is consistent with Kolmogorov $-5/3$ spectrum, but does not provide sufficient insight into the anisotropy with respect to the local magnetic field. The imbalanced turbulence is not a rare exception, on the contrary, such processes as preferential decay of a weaker wave and the escape of turbulent energy from regions that generate perturbations make imbalanced turbulence widely spread in various astrophysical circumstances.

Attempts to construct the model of stationary *strong* imbalanced turbulence were done in Lithwick et al. (2007), Beresnyak and Lazarian (2008b), Chandran (2008). Below we discuss only the model in Beresnyak and Lazarian (2008b) as this is the only model that agrees with numerical simulations performed so far. In view of a big picture, this model can be viewed as an extension of GS95 model into the imbalanced regime.

While the classic formulation of the GS95 critical balance ($\delta v k_\perp \sim v_A k_\parallel$), based on causality, is unable to describe consistently the imbalanced case, Beresnyak and Lazarian (2008b) proposed a new way to introduce the balance between parallel and perpendicular modes. The new conditions for the critical balanced between the parallel and perpendicular modes of oppositely moving waves was obtained appealing to the process termed "propagation cascading". In the case of the balanced turbulence the "old" and "new" critical balance condition results in the same GS95 scaling. However, in the case of imbalanced turbulence the new formulation actually predicts *smaller* anisotropy for the stronger wave, which directly contradicts old causal critical balance, but is consistent with simulations.

We assume that waves have different anisotropies, i.e. the dependence of longitudinal scale Λ to transverse scale λ is different for each kind of wave. This situation is presented in Fig. 2, where some arbitrary longitudinal scale Λ^- corresponds to the *two different* transverse scales, λ_1 for weak wave w^- and λ_2 for strong wave w^+. Λ^+ is a longitudinal scale of w^+ wave having transverse scale λ_1. In case of strong turbulence, we expect that at least the w^- is being strongly cascaded by w^+. In this case the most effective mixing of w^- on scale λ_1 will be obtained through w^+ motions that are on the same scale. The longitudinal scale for w^- will be provided by causal critical balance, since its cascading is fast.

The cascading of w^+ is somewhat more complicated. Since the amplitude of w^- is not large enough to provide strong perturbations in w^+, the w^+ will be perturbed weakly, and the cascading timescale will be diminished according to the "strength" of the w^-, just like it does in weak turbulence. Moreover, now the w^- eddies will be cascading w^+ eddies with similar longitudinal scales, which is the generic feature of weak cascading.

The perturbations provided by w^- will have a transverse scale of λ_1. In other words, the energy of w^+ will be transferred between λ_2 and λ_1. The longitudinal scale for *cascaded* w^+ will be determined by the propagation critical balance in the following way. The wavepackets of w^- are strongly aligned to the mean field on scale λ_1 and therefore they are randomly

[7] Another name for imbalanced turbulence is a turbulence with non-zero cross-helicity.

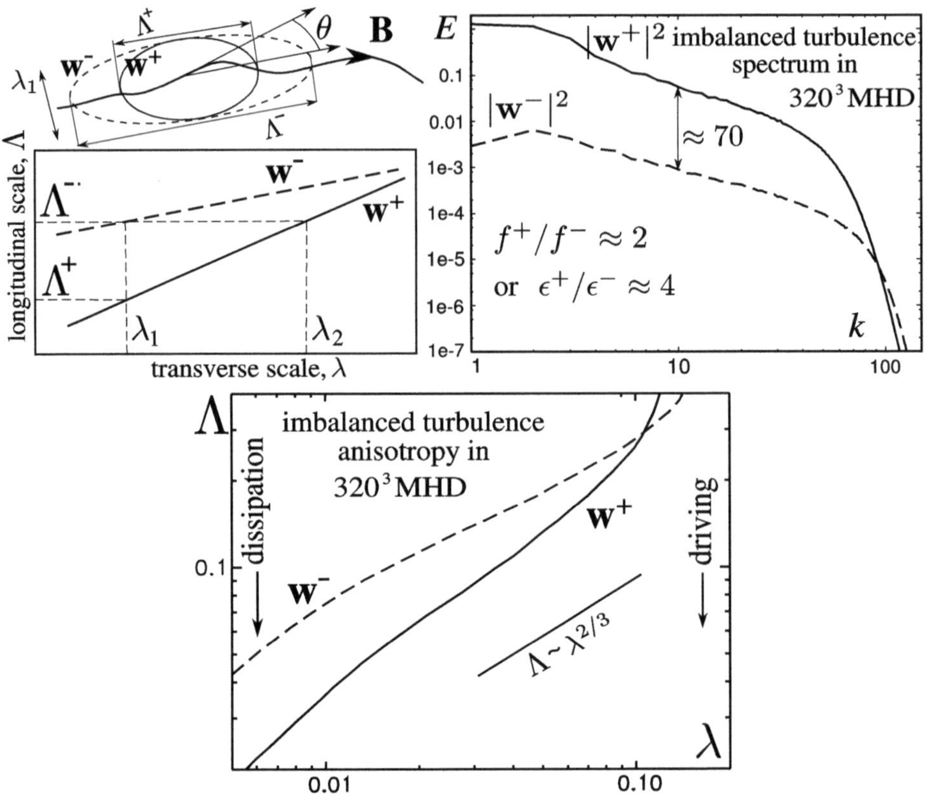

Fig. 2 *Left upper*: $\mathbf{w}^+$ wavepacket, produced by cascading by $\mathbf{w}^-$ wavepacket is aligned with respect to $\mathbf{w}^-$ wavepacket, but misaligned with respect to the local mean field on scale λ_1, by the angle θ. *Left lower*: the longitudinal scale Λ of the wavepackets, as a function of their transverse scale, λ; *Upper right*: the power spectrum of energies for both waves in an imbalanced forced incompressive 320^3 numerical simulation. *Bottom*: the $\Lambda(\lambda)$ dependence in the same simulation, the length scales are in the units of the cube size. From Beresnyak and Lazarian (2008b)

oriented with respect to the mean field at a larger scale λ_2. The RMS angle of wavevector of w^- eddies with respect to mean field on λ_2 will be around $\theta \approx \delta w^+(\lambda_2)/v_A$. This slant of w^- wavepackets will determine the increase of $k_\parallel$ for newly cascaded w^+ packets at λ_1 (see Fig. 2). The new interpretation of critical balance in the strongly imbalanced case is that the $k_\parallel$ of the weak wave increases due to the *finite lifetime* of the wave packet, while in the strong wave it increases due to the field wandering of the strong wave itself *on larger scales*. This effect does not contradict the exact MHD solution of the wave propagating in one direction, because it requires the oppositely propagating wave as an intermediary. The de-alignment of the cascaded strong wave is possible because the weak wave, acting as a cascading agent, is strongly aligned with the field lines on scale which is *different* (smaller) than the scale of the strong wave it is acting upon.

While the model in Beresnyak and Lazarian (2008b) is consistent with numerical simulations, higher resolution simulations are definitely required. Moreover, the three models of strong imbalanced turbulence, namely, Lithwick et al. (2007), Beresnyak and Lazarian 2008b and Chandran (2008), provide predictions to be compared with the solar wind observations. Naturally, additional processes should be accounted for in the stellar wind studies, e.g. parametric instabilities, reflections from preexisting density fluctuations (see Leroy 1980; Roberts et al. 1987; Horbury 1999; Del Zanna et al. 2001). This calls for further studies of imbalanced turbulence in compressible magnetized fluids.

2.3 Viscous Turbulence

When the magnetic turbulence takes place in viscous, but well conducting gas, its properties differ from those described above. The effect of the neutral gas can act in a number respects as the proxy of fluid viscosity[8]. Schekochihin et al. (2004) and Goldreich and Sridhar (2006) argued that the plasma viscosity parallel to magnetic field can act in the same way as the normal viscosity of unmagnetized fluids (cp Braginskii 1965). If this is true, the fluids with high magnetic Prandtl number $Pr_m = Re_m/Re$, i.e. with viscosity much larger than magnetic diffusivity, are widely spread. In particular, the properties of turbulence in high Pr_m fluid are relevant both to fully and partially ionized media of the Local Bubble.

For high Pr_m fluid Cho et al. (2002b) reported a new regime of MHD turbulence. Lazarian et al. (2004) showed that while the spectrum of volume-averaged magnetic fluctuations scales as $E_b(k) \sim k^{-1}$ (see Fig. 3), the pressure within intermittent magnetic structures increases with the decrease of the scale $\propto k$ and the filling factor decreases $\propto k^{-1}$. The magnetic pressure compresses the gas as demonstrated in Fig. 3. More importantly, extended current sheets that naturally emerge as magnetic field fluctuates in the plane perpendicular to the mean magnetic field (see Fig. 3). It was speculated in Lazarian (2007b) that these

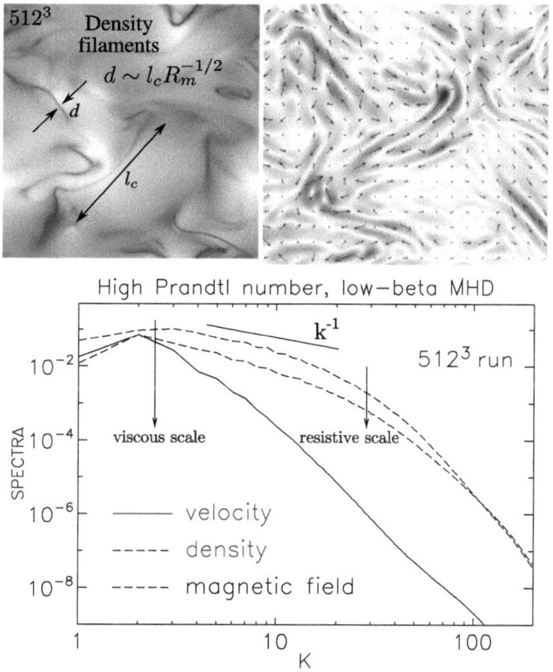

Fig. 3 *Top left*: Filaments of density created by magnetic compression of the gas in the viscosity-damped regime of MHD turbulence. Darker regions correspond to higher density. The viscous damping scale l_c is much larger than the current sheet thickness d. This creates large *observed* density contrasts. *Top right*: Magnetic reversals (in the plane $\perp$ to mean $\langle \mathbf{B} \rangle$) that create compressions of density. Darker regions correspond to higher magnetic field. *Arrows* show direction of the magnetic field. *Bottom*: Spectra of density and magnetic field are similar, while velocity is damped. The resistive scale in this regime is not L/Rm but $LRm^{-1/2}$. From Beresnyak & Lazarian, in prep

[8]We, however, warn our reader not to identify the viscosity and the effects of neutral gas on turbulence. The actual physics of neutral–ion interactions in turbulent gas goes beyond the effects of inducing the drag.

current sheets can account for the origin of the small ionized and neutral structures (SINS) on AU spatial scales (Dieter et al. 1976; Heiles 1997; Stanimirovic et al. 2004).

Goldreich and Sridhar (2006) appealed not to high Pr_m MHD turbulence per se, but to the generation of the magnetic field in the turbulent plasma (see Schekochihin et al. 2004) to account for the high amplitude, but small scale fluctuations of plasma density observed in the direction of the Galactic center. We believe that the regime of dynamo in Schekochihin et al. (2004) and the turbulence in Lazarian et al. (2004) have similarities in terms of the density enhancement that are created. Although in the case of magnetic turbulence with sufficiently strong mean magnetic field, global reversals, that Goldreich and Shridhar (2006) appeal to in compressing plasma, do not happen, the reversals of the magnetic field direction occur in the direction perpendicular to the mean magnetic field. As the mean magnetic field goes to zero, the two regimes become indistinguishable. A systematic study of the density enhancement within the Local Bubble, where more detailed knowledge of the plasma and magnetic field properties are potentially available, may test the theoretical constructions above.

3 Intermittency: Small Volumes with Extreme Conditions

An anisotropic spectrum alone, say $E(\mathbf{k})\,d\mathbf{k}$, cannot characterize MHD turbulence in all its complexity because it involves only the averaged energy in motions along a particular direction. To have a full statistical description, one needs to know not only the averaged spectrum of a physical variable but higher orders as well. The tendency of fluctuations to become relatively more violent but increasingly sparse in time and space as the scales get smaller, so that their influence remains appreciable, is called *intermittency*. The intermittency increases with the ratio of the size scales of injection and dissipation of energy, so the very limited range of scales within numerical simulations may fail to reflect the actual small scale processes.

Falgarone et al. (2005, 2006, 2007) and collaborators (Hily-Blant and Falgarone 2007; Hily-Blant et al. 2007 and refs. therein) attracted the attention of the interstellar community to the potential important implications of intermittency. A small and transient volume with high temperatures or violent turbulence can have significant effects on the net rates of processes within the ISM. For instance, many interstellar chemical reactions (e.g., the strongly endothermic formation of CH^+) might take place within very intensive intermittent vortices. The aforementioned authors claimed the existence of the observational evidence for such reactions and heating. If the effects of intermittency are as strong as Falgarone et al. believe, they should not be neglected when processes in the Local Bubble are considered.

To get insight into intermittency effects, it is necessary to study high moments of velocity fluctuations. Fortunately, both laboratory and numerical studies demonstrate that the higher moments of velocity fluctuations can be predicted remarkably well by the expression derived by She and Leveque (1994) (see discussion in Lazarian 2006a and ref. therein). The key parameter is the spatial dimension of the dissipation structures, D, $\sim$1 for filamentary vortices and ~ 2 for sheets in MHD (see Müller and Biskamp 2000; Cho et al. 2003a; Boldyrev et al. 2002). In some instances D can be a fractal dimension in between. The dimension can be obtained even with low resolution numerics. This opens an avenue of evaluating the effects of intermittencies, such as intense local heating or transient effects, for *Re* and *Rm* numbers that will probably never be achieved in numerical simulations.

The She and Leveque (1994) model describes the pth power of longitudinal velocity fluctuations at scale l, i.e. $(\delta V_l)^p \sim l^{\xi_p}$, where $\delta \mathbf{V_l} \equiv \delta V_l \mathbf{l}/l$. Here $\delta \mathbf{V_l}$ is the mean difference

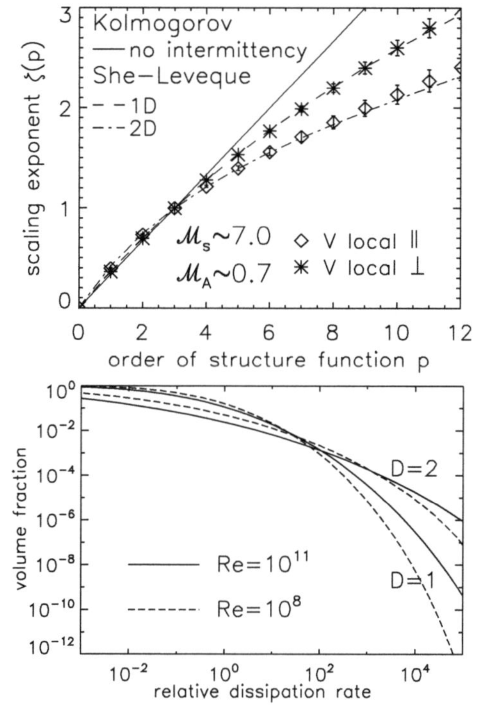

Fig. 4 *Top panel*: The intermittencies of velocities in our subAlfvenic, $M_A = 0.7$ supersonic $M_s = 7$ MHD simulations. From Kowal and Lazarian (2008, in preparation). *Bottom panel*: Volume fraction with the dissipation rate is higher than the mean rate for the She-Leveque model of intermittency with $D = 1$ and 2. From Beresnyak and Lazarian

of the local velocity from the average. For non-intermittent Kolmogorov turbulence, $\xi_p = p/3$ (recall the well-known result, $\delta V_l \sim l^{1/3}$). In the She and Leveque (1994) model of turbulence, ξ_p is a more complex function that depends on (a) the scaling of velocity $\delta V_l \sim l^\alpha$, (b) the energy cascading rate $t_l \sim l^{-\beta}$, and (c) the dimension of the dissipation structures, D. The She and Leveque (1994) expression is $\xi_p = \alpha\beta(1-\beta) + (3-D)(1-[1-\beta/(3-D)]^{\alpha\beta})$, with $\alpha \sim 1/3$ and $\beta \sim 2/3$ for MHD turbulence. The principal dependence of ξ_p is on D, between 1 and 2.

Dubrulle (1994) and She and Waymire (1995) have figured out that the scaling exponents ξ_p of the She-Leveque model correspond to so-called generalized log-Poisson distribution of the local dissipation rate ε_l. With this input, if we constrain the model parameters from low resolution numerical simulations we can calculate the probability of a fluctuation exceeding a given threshold of the local rate of deposition of energy, for arbitrary high values of Re and Rm. Very importantly, predictions can be made for the real ISM!

Figure 4 shows how to determine D. It involves a simulation (Mach numbers shown) for the component of **V** perpendicular and also parallel to the local **B**. The dot-dashed and dashed lines are the respective predictions of the She and Leveque (1994) expression, with the values of D as indicated. The simulation clearly provides appropriate values of D for the two components.

The bottom part of Fig. 4 shows calculations for the volume fractions of various dissipation rates (i.e., heating). While the temperatures achieved will depend upon the cooling functions, some important conclusions are available from the analysis of Fig. 4. Indeed, the model of chemical reactions by Falgarone et al. requires that a substantial part of the turbulent cascade energy dissipate in the very intermittent structures. Figure 4 shows that the bulk of the energy dissipates within structures where the dissipation rate is higher than the mean value less than the factor of 100, provided that the She-Leveque model is valid. This provides stringent constraints on what chemistry we could expect to be induced by intermittent turbulent heating.

Interestingly enough, the case of intermittency studies supports our point of the futility of the "brute force" numerical approach. For instance, for a typical ISM injection scale of 50 pc, the Reynolds number can be as high as $Re = 10^{11}$. In comparison, numerical simulations provide $\sim Re^3$ boxes for the present *record* resolution of a hydrodynamic simulation with 4096^3 boxes.

Even if the intermittency of turbulence is not as important as is in Falgarone et al. model, its implications for both ISM and Local Bubble plasma may be very important. Intermittency models, similar to ones presented in this section, show that for large Reynolds numbers, typical for ISM it is possible that up to 10% of the energy deposition can be localized in just 10^{-2}% of the volume! Such a concentration of energy dissipation in high Re can have many important consequences on inhomogeneous heating (see Sonnentrucker et al. 2006). It is clear that further studies in this direction are necessary.

4 Modification of Turbulence by Cosmic Rays

It is easy to argue that the interaction of turbulence and cosmic rays (CRs) is of the most important processes where properties of MHD turbulence are essential. Indeed, the interaction of turbulence with CRs is a cornerstone of CRs propagation and acceleration models (e.g., Ginzbirg 1966; Jokipii 1966; Wentzel 1969; Schlickeiser 2002 and ref. therein). To account for the interaction properly, one must know both the scaling of turbulence and the interactions of turbulence with various waves produced by CRs. As we mentioned above, slab Alfvenic modes, which are an essential part of many models of cosmic ray propagation (see Goldstein 1976; Bieber et al. 1994; Matthaeus et al. 2003; Shalchi et al. 2004; Shalchi 2005, 2006 and references therein), are not observed in direct MHD simulations. This does not mean that these modes are necessarily absent in realistic astrophysical environments. For instance, the influence of the CRs back onto the turbulence may be important.

The problem is that energetic particles or CRs being an important component in both the ISM and Local Bubble are not a part of the simulations of MHD turbulence. Nevertheless, CRs are dynamically important and well coupled to the rest of the ISM through magnetic fields and magnetic turbulence. The spectra of both CRs and interstellar turbulence show nice power laws, leading R. Jokipii (2001) to suggest a strong interrelation of the two. It has to be noted that the total pressure of CRs in the ISM is of the order of the kinetic pressure ρv^2 and, in many cases, exceed the thermal pressure by as much as a factor of 10. One cannot exclude that CRs may be as important for ISM simulations as dark matter is for the problems of galaxy formation and dynamics.

The direct application of the results of MHD turbulence theory to the problems of the CRs propagation and acceleration is well justified only when the modification of turbulence by CRs can be neglected. This has caused controversies since the classical work by Parker (1965). CRs can induce and are influenced by two instabilities: (a) streaming instability (Wentzel 1969, 1974), widely discussed in the CR literature and frequently referred to as the source of the postulated scattering "slab" , i.e. with $\mathbf{k}\|\mathbf{B}$ Alfvenic component, and (b) the anisotropic pressure kinetic gyroresonance instability[9] (see Mikhailovskii 1975; Gary 1993; Kulsrud 2004 and refs. therein), well known to the plasma community.

[9]These instabilities are due to the gyroresonance between a Cosmic Ray particle and one of the hydromagnetic oscillation (MHD mode).

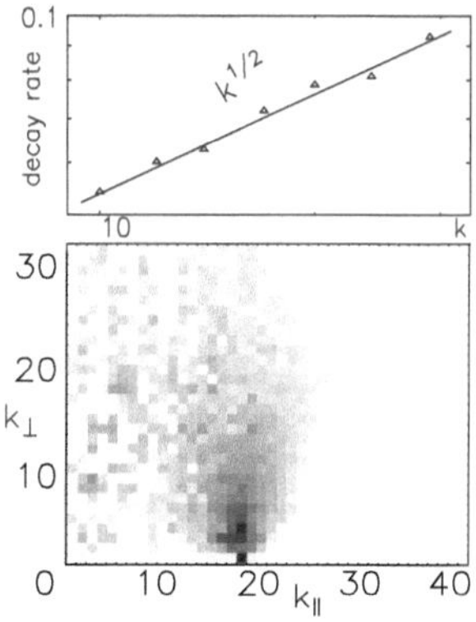

Fig. 5 Decorrelation of a plane, $k_\perp = 0$ Alfvén wave by turbulence. *Lower picture* shows the energy density of a wave in cylindrical k-space. Alfvén waves were injected at $k_\parallel = 17$. From Beresnyak and Lazarian (2008a)

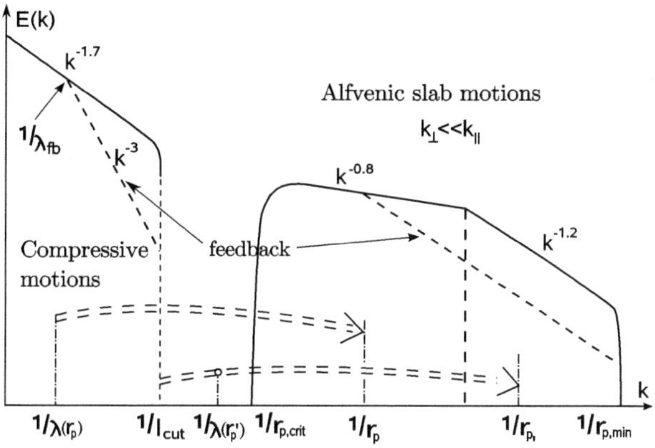

Fig. 6 Energy density of compressive modes and Alfvénic slab-type waves, induced by CRs. The energy is transferred from the mean free path scale to the CR Larmor radius scale. See more on feedback in Lazarian and Beresnyak (2006)

The streaming instability, as implied by its name, requires an anisotropic distribution of particles, while in the interstellar gas, if we judge from our local measured values, the distribution of CRs is close to isotropic. In addition, Yan and Lazarian (2002) noticed that the streaming instability can be non-linearly suppressed by ambient turbulence. Farmer and Goldreich (2004) estimated that galactic turbulence can efficiently suppress the formation of the "slab" Alfvenic waves for protons with relativistic factor $\gamma > 100$. Numerical calculations in Beresnyak and Lazarian (2008a) confirmed (see Fig. 5) those predictions.

Lazarian and Beresnyak (2006) showed that compressible MHD turbulence can generate kinetic gyroresonance instability that can couple efficiently CRs with turbulence and back-react on turbulence, modifying its spectrum (see Fig. 6). The instability arises from compressing magnetic fields with gyrating CRs. Indeed, the CR pressure becomes anisotropic as a consequence of adiabatic invariant conservation, thereby inducing the instability (Lazarian and Beresnyak 2006). The latter gives rise to "slab" Alfvenic perturbations that scatter and randomize the CR momenta via gyroresonance. As the "slab" mode grows, it increases CR scattering and coupling of CRs and magnetic fields.

Further research of the kinetic gyroresonance instabilities, as well as possibly other instabilities induced by CRs, should improve our understanding of the importance of the CRs-generated slab Alfven waves. Unlike earlier models which used somewhat arbitrary (see Sect. 2) prescriptions for the amplitude of the "slab" Alfvenic mode (see Bieber et al. 1994), future theories appealing to such modes should be able to evaluate their expected amplitudes (see first attempts to do this in Lazarian and Beresnyak 2006). We should also mention, that the slab mode will be subject to damping while interacting with the surrounding strong Alfvenic turbulence (Yan and Lazarian 2002; Farmer and Goldreich 2004; Lazarian and Beresnyak 2006; Beresnyak and Lazarian 2008b). Including this damping in Shalchi et al. (2008) improved the fit to the solar wind measurement by Ulysses (Gloeckler et al. 1995) and AMPTE spacecraft (Mobius et al. 1998).

5 Selected Implications of Magnetic Turbulence

The implications of magnetic turbulence for astrophysical fluids, e.g. interstellar medium, are numerous. For instance, turbulence can heat Diffuse Ionized Gas within the Milky Way (see Minter and Spangler 1997; Cho et al. 2002b), determine the evolution of molecular clouds (see McKee and Ostriker 2007), scatter and accelerate cosmic rays (see Schlickeiser 2002). Below we consider a couple of *selected* examples relevant to the research by the authors to illustrate the progress and problems on the way of evaluating the effects of turbulence.

5.1 Cosmic Ray Scattering by Fast Modes of MHD Turbulence

As we mentioned earlier, numerical simulations of MHD turbulence supported the GS95 model of turbulence, which does not have the "slab" Alfvenic modes that produced most of the scattering in the earlier models of CR propagation. Can the turbulence that does not appeal to CRs back-reaction (see Sect. 4) produce efficient scattering?

In most models of ISM turbulence, such as Armstrong et al. (1995), McKee and Ostriker (2007), the energy is injected at large scales. In this scenario, fast modes were identified as a scattering agent for cosmic rays (Yan and Lazarian 2002, 2004). These works made use of the quantitative description of turbulence obtained in CL02 to calculate the scattering rate of cosmic rays. The results are shown in Fig. 7. For instance, the scattering rate of relativistic protons by Alfvenic turbulence was shown to be nearly 10^{10} times lower than the generally accepted estimates obtained assuming the Kolmogorov scaling of turbulence. Although this estimate is 10^4 times larger than that obtained by Chandran (2000), who employed GS95 ideas of anisotropy, but lacked the quantitative description of the eddies, it is clear that for most interstellar circumstances the Alfvenic scattering is suppressed. The low efficiency of scattering by Alfvenic modes arises from the large anisotropy of the modes on scales of cosmic ray Larmor radius.

YL02 showed that the scattering by fast modes, which are isotropic (CL02), dominates (see Fig. 7). However, fast modes are subject to both collisional and collisionless damping[10], which was taken into account in Yan and Lazarian (2004). More recent studies of cosmic ray propagation and acceleration that explicitly appeal to the effect of the fast modes include Cassano and Brunetti (2005), Brunetti (2006), Brunetti and Lazarian (2007), Yan and Lazarian (2008) and Yan et al. (2008). Incidentally, fast modes have been also identified as primary agents for the acceleration of charged dust particles (Yan and Lazarian 2003; Yan et al. 2004).

[10] On the basis of weak turbulence theory, Chandran (2005) has argued that high-frequency fast waves, which move mostly parallel to magnetic field, generate Alfven waves also moving mostly parallel to magnetic field. We expect that the scattering by thus generated Alfven modes to be similar to the scattering by the fast modes created by them. Therefore we expect that the simplified approach adopted in Yan and Lazarian (2004) and the papers that followed to hold.

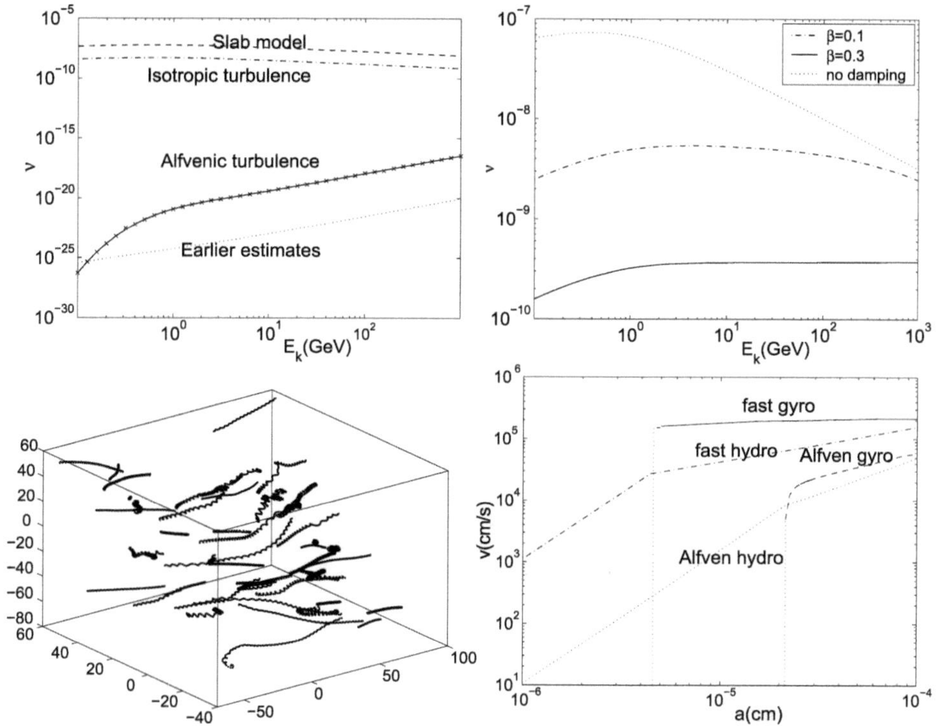

Fig. 7 Implications of interstellar turbulence for the cosmic rays and interstellar dust. *Upper left*: Rate of CR scattering by Alfven waves versus CR energy. The *lines at the top of the figure* are the accepted estimates obtained for Kolmogorov turbulence. The *dotted curve* is from Chandran (2000). The analytical calculations are given by the *solid line* with our numerical calculations given by crosses. *Upper right:* The rate of CR scattering (ν) by fast modes in magnetically dominated plasma. The rate of scattering depends on damping of the fast waves, which in turn depends on the ratio of gaseous to magnetic pressure ($\beta = P_{gas}/P_{mag}$). *Lower left*: Individual trajectories of CRs tracked by the Monte Carlo scattering code. **B** is obtained through 3-D simulations of MHD turbulence. These calculations provide estimates of CR diffusion. *Lower right*: Velocities of charged dust grains in cold neutral media (CNM). Gyroresonance with fast modes ("fast gyro") dominates for large grains, while hydrodynamic drag ("fast hydro") for small grains. The cutoffs for Alfven and fast gyro (*vertical lines*) are due to MHD turbulence damping caused by neutral–ion collisions. From Yan and Lazarian (2002, 2003, 2008)

5.2 Turbulent Reconnection and Cosmic Ray Acceleration

Magnetic reconnection can be associated with the ability of magnetic flux tubes to change their topology, while being submerged within conducting fluids (see Biskamp 1996). Due to high numerical diffusivity of present-day simulations, reconnection is fast there, which, for instance, means that magnetic fields in the ISM, Local Bubble and solar wind simulations change their topology fast. Is this true for the real astrophysical circumstances?

Recent years have been marked by a substantial progress in simulations of collisionless reconnection (see Shay and Drake 1998; Bhattacharjee et al. 2003; Drake et al. 2006a and references therein). However, while the researchers argue whether Hall MHD[11] or fully kinetic description (Daughton et al. 2006) is necessary, one statement is defi-

[11] A special form of MHD equations taking into account Hall effect in plasma.

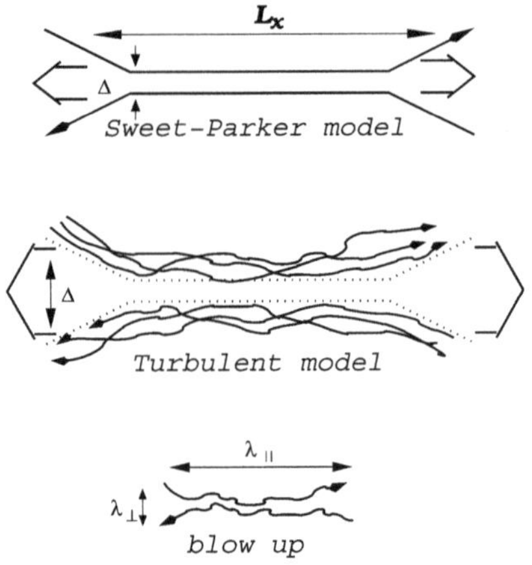

Fig. 8 *Upper plot*: Sweet-Parker model of reconnection. The outflow is limited by a thin slot Δ, which is determined by Ohmic diffusivity. The other scale is an astrophysical scale $L \gg \Delta$. *Middle plot*: Turbulent reconnection model that accounts for the stochasticity of magnetic field lines. The outflow is limited by the diffusion of magnetic field lines, which depends on field line stochasticity. *Low plot*: An individual small scale reconnection region. The reconnection over small patches of magnetic field determines the local reconnection rate. The global reconnection rate is substantially larger as many independent patches come together

nitely true. If magnetic reconnection is only fast in collisionless environments, most of the MHD simulations, e.g. of interstellar medium, accretion disks, stars, where the environment is collisional, are in error. We shall argue below that this radical conclusion may not be true and the reconnection is also fast in most astrophysical collisional environments.

Lazarian and Vishniac (1999, henceforth LV99) considered turbulence as the agent that makes magnetic reconnection fast. The scheme proposed by LV99 differs appreciably from the earlier attempts to enhance reconnection via turbulence (Speiser 1970; Jacobson and Moses 1984; Matthaeus and Lamkin 1985; Bhattacharjee and Hameiri 1986; Hameiri and Bhattacharjee 1987; Strauss 1988, see Lazarian et al. 2004 for a detailed comparison). The scheme proposed is a generalization of the Sweet-Parker scheme (see Fig. 8). The problem of the Sweet-Parker model is that the reconnection is negligibly slow for any realistic astrophysical conditions. However, astrophysical magnetic fields are generically turbulent.

LV99 consider the case in which there exists a large scale, well-ordered magnetic field, of the kind that is normally used as a starting point for discussions of reconnection. In addition, we expect that the field has some small scale 'wandering' of the field lines. On any given scale the typical angle by which field lines differ from their neighbors is $\phi \ll 1$, and this angle persists for a distance along the field lines $\lambda_{\parallel}$ with a correlation distance $\lambda_{\perp}$ across field lines (see Fig. 8).

The modification of the global constraint induced by mass conservation in the presence of a stochastic magnetic field component is self-evident. Instead of being squeezed from a layer whose width is determined by Ohmic diffusion, the plasma may diffuse through a much broader layer, $L_y \sim \langle y^2 \rangle^{1/2}$ (see Fig. 8), determined by the diffusion of magnetic field lines. This suggests an upper limit on the reconnection speed of $\sim V_A(\langle y^2 \rangle^{1/2}/L_x)$. This will be the actual speed of reconnection if the progress of reconnection in the current sheet does not impose a smaller limit. The value of $\langle y^2 \rangle^{1/2}$ can be determined once a particular model of turbulence is adopted, but it is obvious from the very beginning that this value is determined by field wandering rather than Ohmic diffusion as in the Sweet-Parker case.

What about limits on the speed of reconnection that arise from considering the structure of the current sheet? In the presence of a stochastic field component, magnetic reconnection dissipates field lines not over their entire length $\sim L_x$ but only over a scale $\lambda_{\|} \ll L_x$ (see Fig. 8), which is the scale over which magnetic field line deviates from its original direction by the thickness of the Ohmic diffusion layer $\lambda_{\perp}^{-1} \approx \eta / V_{\rm rec,local}$. If the angle ϕ of field deviation does not depend on the scale, the local reconnection velocity would be $\sim V_A \phi$ and would not depend on resistivity. In LV99 it is taken into account that ϕ does depend on scale (see Sect. 2). Therefore the *local* reconnection rate $V_{\rm rec,local}$ is given by the usual Sweet-Parker formula but with $\lambda_{\|}$ instead of L_x, i.e. $V_{\rm rec,local} \approx V_A (V_A \lambda_{\|}/\eta)^{-1/2}$. It is obvious from Fig. 8 that $\sim L_x/\lambda_{\|}$ magnetic field lines will undergo reconnection simultaneously (compared to a one by one line reconnection process for the Sweet-Parker scheme). Therefore the overall reconnection rate may be as large as $V_{\rm rec,global} \approx V_A (L_x/\lambda_{\|})(V_A \lambda_{\|}/\eta)^{-1/2}$. Whether or not this limit is important depends on the value of $\lambda_{\|}$.

The relevant values of $\lambda_{\|}$ and $\langle y^2 \rangle^{1/2}$ depend on the magnetic field statistics. This calculation was performed in LV99 using the GS95 model of MHD turbulence providing the upper limit on the reconnection speed:

$$V_{r,up} = V_A \min\left[\left(\frac{L_x}{l} \right)^{\frac{1}{2}}, \left(\frac{l}{L_x} \right)^{\frac{1}{2}} \right] \left(\frac{v_l}{V_A} \right)^2, \tag{1}$$

where l and v_l are the energy injection scale and turbulent velocity at this scale respectively. In LV99 other processes that can impede reconnection were found to be less restrictive. For instance, the tangle of reconnection field lines crossing the current sheet will need to reconnect repeatedly before individual flux elements can leave the current sheet behind. The rate at which this occurs can be estimated by assuming that it constitutes the real bottleneck in reconnection events, and then analyzing each flux element reconnection as part of a self-similar system of such events. This turns out to impede the reconnection. As the result, LV99 concludes that (1) is not only an upper limit, but is the best estimate of the speed of reconnection. The model has been recently tested numerically in Kowal et al. (2008) (see Fig. 9). The thick current sheets observed during the 2003 November 4 Coronal Mass Ejection reported in Ciaravella and Raymond (2008) are also consistent with the LV99 model.

The most interesting process is the first-order Fermi acceleration that is intrinsic to the turbulent reconnection. To understand it consider a particle entrained on a reconnected magnetic field line (see Fig. 8). This particle may bounce back and forth between magnetic mirrors formed by oppositely directed magnetic fluxes moving towards each other with the velocity V_R. Each of such bouncing will increase the energy of a particle in a way consistent with the requirements of the first-order Fermi process. The interesting property of this mechanism that potentially can be used to test the idea observationally is that the resulting spectrum is different from those arising from shocks. de Gouveia dal Pino and Lazarian (2001, 2003, 2005) used this mechanism of particle acceleration[12] to explain the synchrotron power-law spectrum arising from the flares of the microquasar GRS 1915+105. Note, that the mechanism acts in the Sweet-Parker scheme as well as in the scheme of turbulent reconnection. However, in the former the rates of reconnection and therefore the efficiency of acceleration are marginal in most cases.

[12] The mechanism has physical similarities to the acceleration mechanism that was proposed later for electrons by Drake et al. (2006b). In Drake's mechanism, similarly, to the Matthaeus et al. (1984) mechanism, however, the process of acceleration happens within 2D contracting loops. For LV99 model of reconnection the generic configuration of magnetic field are contracting spirals.

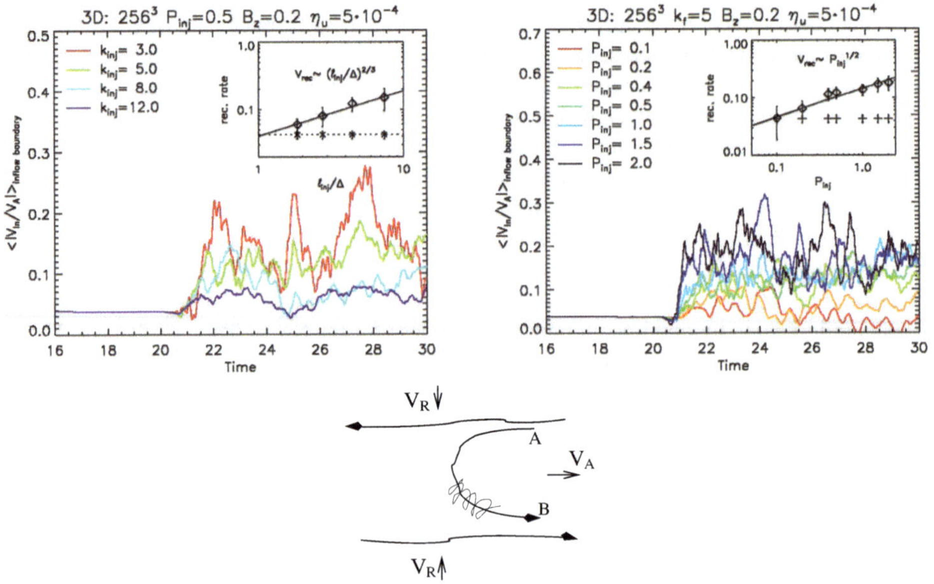

Fig. 9 *Top left panel*: Testing of the model of 3D turbulent reconnection with MHD simulations in Kowal et al. (2008). The results, consistent with the LV99 model, show an increase of the reconnection rate with the increase of the injection scale of the turbulence. The reconnection is bursty, as is expected for the turbulence model. The reconnection rate is normalized to the Alfvenic velocity, time is given in the units of Alfvenic crossing times. Initially the turbulence is slow, but its velocity increases as the turbulence is injected. *Top right*: The same as for the left panel, but the injection energy is being changed. The results are also consistent with the LV99 model. *Bottom panel*: Cosmic rays spiral about a reconnected magnetic field line and bounce back at points A and B. The reconnected regions move towards each other with the reconnection velocity V_R. The advection of cosmic rays entrained on magnetic field lines happens at the outflow velocity, which is in most cases of the order of V_A. Bouncing at points A and B happens because either of streaming instability or turbulence in the reconnection region

5.3 Perpendicular Diffusion and Subdiffusion

Modeling of clouds submerged in hot plasma present one of the challenges to the research of the Local Bubble physics. It is easy to see that the diffusion heat perpendicular to magnetic field is very slow if the field is laminar. Incidentally, the same is true in terms of CR diffusion, while the observations indicate that the diffusion of the CRs perpendicular to magnetic field in the Galaxy is reduced just by a factor ~ 3 compared to the parallel diffusion (Jokipii 1999). How could this be?

The reader may already guess that, similar to the reconnection problem discussed above, magnetic field wandering may allow the particles to diffuse (Jokipii and Parker 1969; Giacalone and Jokipii 1999). In analogy with the reconnection example, one may expect that the expression for the field wandering would depend on the model of turbulence accepted. Within the GS95 model of turbulence and assuming that the injection turbulent velocity is equal to the Alfven one, the calculations for heat diffusion were performed in Narayan and Medvedev (2001). A general case of arbitrary injection velocity was considered in Lazarian (2006). The heat diffusion in plasma was found to be a function of the Alfven Mach number M_A defined as the ratio of the turbulent velocity at the injection scale to the Alfven speed in the medium. For the case of strong mean magnetic field, i.e. for $M_A < 1$, and mean free path of a electron λ less than the injection scale of the turbulence L, the heat diffusion co-

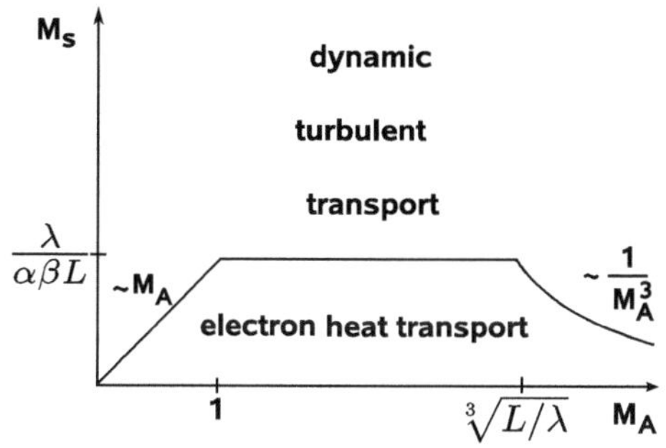

Fig. 10 The parameter spaces for the advection of heat by turbulent motions and heat conduction by plasma electrons. Sonic Mach number M_S is plotted against the Alfven Mach number M_A. The mean free path of electrons is λ, while α is the square root of the electron to proton mass ratio, i.e. $(m_e/m_p)^{1/2}$, β is a numerical factor ≈ 4. From Lazarian (2006, 2007a)

efficient was obtained to be $1/3M_A^4\kappa_{\mathrm{Spitzer}}$, where $\kappa_{\mathrm{Spitzer}} = \lambda v_{\mathrm{therm}}$ is the usual Spitzer heat diffusion coefficient of unmagnetized plasma. The factor 1/3 there reflects the 1D nature of diffusion along magnetic field lines, while the M_A^4 power[13] reflects the inefficiency of magnetic field line wandering in the presence of the strong mean magnetic field. Similarly, for superAlfvenic turbulence, i.e. for $M_A > 1$ one gets the heat diffusion coefficient $1/3\kappa_{\mathrm{Spitzer}}$ if the mean free path of an electron $\lambda < LM_A^{-3}$ and gets $1/3(LM_A^{-3}/\lambda)\kappa_{\mathrm{Spitzer}}$ in the opposite regime.

Cho et al. (2003b) noticed that in magnetized turbulence, the turbulence eddies should induce in plasma the heat advection with the effective diffusion coefficient $2/3LV_L$, where the factor of 1/3 is similar to the advective diffusion in hydrodynamics (Lesieur 1990) and the factor of 2 takes into account that both electrons and protons participate in the process. The relative importance of the two diffusion processes is exemplified in Fig. 10, where the parameter space for the dominance of the heat advection by turbulent motions and the dominance of the heat transfer by electrons are defined. These results for the fully ionized plasma can be used for parameterizing heat transport in the astrophysical codes. They also can easily be generalized for the partially ionized gas. Note, that following individual particles in simulations may be prohibitively expensive (see Tilley and Balsara 2006). Moreover, the actual structure of magnetic field is distorted by the limited numerical resolution.

The last point of magnetic field structure in numerical simulations is also important for the CRs diffusion perpendicular to magnetic field. Take, for instance, the issue of subdiffusion, which is the non-diffusive behavior (Δx slower than $\sqrt{t}$) arising from the CRs retracing their trajectories, as a result of the backward scattering (see Kóta and Jokipii 2000; Qin et al. 2002; Webb et al. 2006). In the case of the unrealistic turbulence with only one scale of turbulent motions l_{turb}, the retracing of particles stops as soon as the particles diffuse the so-called Rechester and Rosenbluth (1978) length, which is $l_{\mathrm{turb}} \ln(l_{\mathrm{turb}}/r_{\mathrm{Lar}})$, where r_{Lar} is the Larmor radius of a charged particle (see also Chandran and Cowley 1998). In the case of realistic turbulence with a range of scales one should use the dissipation scale instead of the t_{turb} (Narayan and Medvedev 2001; Lazarian 2006; Yan and Lazarian 2008). The corresponding scale for the Alfvenic turbulence in the ISM may be less than $l_{\mathrm{crit}} \approx 10^6$ km, thus one should expect subdiffusion only at scales less than this. For the solar wind turbulence l_{crit} is even smaller, of the order of 10^3 km. On the contrary, for the numerical simulations with a limited inertial range, the subdiffusion may be much more important (see Qin et al. 2002).

[13]There is a wrong power in the original Lazarian (2006) paper, which was corrected in Lazarian (2007a).

6 Studying Magnetic Turbulence in Interplanetary Medium Using Comets

As we illustrated with selected examples above, the modeling of turbulent astrophysical fluids is far from being simple. Therefore, observational testing is essential. We are fortunate to have in situ measurements of magnetic field in the interplanetary medium (see Opher et al. 2007). The advantage of direct studies of magnetic perturbations by spacecrafts has been explored through many important missions. Such studies, unlike numerical ones, may deliver information about the actual magnetic turbulence at high *Re* and *Rm* numbers. However, the spacecraft measurement are rather expensive. Are there any other cost-effective ways to study magnetic turbulence in interplanetary medium?

Below we present a new way to explore the turbulence in interplanetary medium by using the alignment of sodium atoms ejected from comets. Atomic alignment of atoms in their ground state was an effect studied in the middle of the previous century in laboratories, in relation to maser research (Hawkins 1955; Kastler 1957). The alignment of atoms is understood in terms of their non-equilibrium distribution of ground level substates. Therefore, to be aligned in the ground state the atom in question should have either fine or hyper-fine structure. The non-equilibrium distribution arises from radiative pumping, while the magnetic field realigns atoms inducing their Larmor precession. As the substates of the ground level are long-lived, even a small magnetic field modifies the distribution and therefore the polarization arising from the aligned atoms (see Yan and Lazarian 2007 for more details).

Comets are known to have Sodium tails and Sodium is an atom that can be aligned by radiation and realigned by solar wind magnetic fields. This opens an opportunity of studying magnetic fields in the solar wind from the ground, but tracing the polarization of the Sodium line. At the moment this is a suggestion supported by the synthetic ground based observations. For our synthetic observations we used a model of magnetic field in Liu et al. (2008). It employed the space weather model developed by University of Michigan, namely, The Space Weather Modeling Framework (SWMG) (Toth et al. 2005). More specifically, this is a solar corona and inner heliosphere model that extends the description of media from the solar surface to 1 AU.

The structure of the magnetic field in the heliosphere can be studied by the polarization of Sodium D2 emission in the comet's wake. Though the abundance of sodium in comets is very low, its high efficiency in scattering sunlight makes it a good tracer (Thomas 1992). As discussed in Yan and Lazarian (2007), the gaseous sodium atoms in the comet's tail acquire angular momentum from the solar radiation, i.e. they are aligned. Resonant scattering from these aligned atoms is polarized. Distant from comets, the Sun can be considered a point source. As shown in Fig. 11, the geometry of the scattering is well defined, i.e., the scattering angle θ_0 is known. The alignment is modulated by the local magnetic field. The polarization of the sodium emission thus provides exclusive information on the magnetic field in the interplanetary medium. We take the data cube from the spacecraft measurement as described above. Depending on its direction, the embedded magnetic field alters the degree of alignment and therefore the polarization of the light scattered by the aligned atoms. Figure 11 illustrates the trajectory of a comet along which the magnetic field varies and the polarization of Sodium D2 emission changes accordingly. By comparing observations with it, we can determine the structure of magnetic field in the heliosphere. For interplanetary studies, one can investigate not only spatial, but also temporal variations of magnetic fields. Since alignment happens at a time scale τ_R, magnetic field variations on this time scale will be reflected. This can allow for a cost-effective way of studying interplanetary magnetic turbulence at different scales. On the basis of the results above we expect that comets may become an important source of information about interplanetary magnetic fields and their variations.

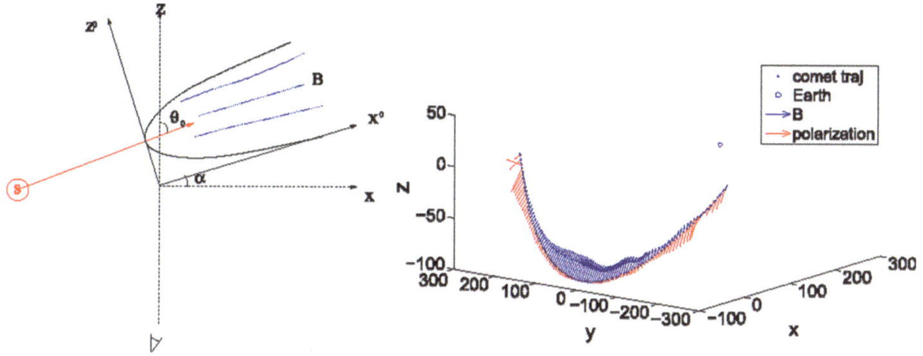

Fig. 11 *Left panel.* Schematic of the resonance scattering of sunlight by the sodium in a comet wake. The sodium tail points in the direction opposite to the Sun. The observer on the Earth sees the stream at the angle θ_0. The magnetic field realigns atoms via fast Larmor precession. Thus the polarization traces the interplanetary magnetic fields. *Right panel.* The magnetic field from the inner heliosphere model (*blue lines*) (Toth et al. 2005) and the predicted vectors of linear polarization of Sodium D2 emission (*red lines*) along the comet trajectory. From Yan et al. (2008)

7 Discussion and Summary

The goals of the above review were, first of all, to demonstrate the necessity of detailed studies of magnetic turbulence for modeling of various astrophysical environments, in particular, the environment of the Local Bubble and, secondly, to appeal for the necessity of new ways of studying magnetic fields in the solar wind. To address the former goal we discussed the properties of MHD turbulence in different environments (fully ionized and partially ionized gas) for different ways of excitation (balanced and imbalanced turbulence). The second goal was addressed by considering a new promising way of tracing magnetic field structure in the interplanetary medium.

The examples are representative of authors' interests, but they vividly show that the "brute force approach" to simulating Local Bubble and other astrophysical environments may easily fail. For instance, we showed that it would not be fruitful for calculating the intermittency of environmental heating and it is likely to fail in representing the effects of subdiffusion. It also may deliver unreliable results when simulating heat transfer in magnetized plasmas, etc. Therefore detailed quantitative modeling may require, first of all, the creation of the "tool box" of the particular recipes of how to parameterize particular properties of turbulent fluid. We claimed that studies of these properties were not only desirable, but were absolutely necessary. For instance, unless we reach a consensus on the rate of reconnection in collisional environments, the simulations of the interstellar gas will be highly suspicious (see the corresponding discussion in Sect. 5.2). All in all, we believe that the progress can be achieved via better understanding of fundamental properties of magnetized plasma and aggressive testing the results of modeling with observations. The fact that the properties of turbulent fluid are important for a wide range of astrophysical problems makes it easier to concentrate efforts and resources on such studies. This also calls for more vigorous scientific exchanges between the disciplines.

Our points of the review above can be briefly summarized in the following way:

1. Properties of astrophysical fluids, including those of solar wind, Local Bubble and interstellar turbulence are, to large extend, determined by magnetic turbulence. The "brute

force" approach aimed at detailed modeling of the aforementioned astrophysical environments keeping magnetic turbulence realistic is doomed. Instead, we feel that focused studies of particular physical phenomena in turbulent magnetized plasma can clarify when the simulations that take into account many processes at once provide a correct physical picture. These studies can help parameterize effects of turbulence in numerical codes.
2. By considering imbalanced turbulence, as this is the case of the Solar wind turbulence, and high-Pr number turbulence, that can approximate the turbulence in the partially ionized plasma, we demonstrated that magnetic turbulence can have many different properties. These properties affect the transport properties of the magnetized media, e.g. scattering of cosmic rays, the formation of density enhancements, including the formation of small ionized and neutral structures (SINS), the change of magnetic topology, i.e. magnetic reconnection, thermal conductivity.
3. We advocate the approach to simulating processes in the aforementioned environments that uses the synergy of focused studies of particular processes in turbulent magnetized plasmas, numerical simulations of astrophysical situations and observational studies. We also feel that collecting of new data on magnetic fields is essential. In the current situation, it is promising to explore new techniques of magnetic field studies, e.g. the ones that makes use of atoms aligned in their ground state by radiation and realigned by the external magnetic field. In particular, we discuss possible studies of interplanetary turbulence using Na aligned atoms in a comet wake.

Acknowledgements AL research is supported by the NSF Center for Magnetic Self Organization in Laboratory and Astrophysical Plasmas and NSF Grant AST-0808118. MO research is supported by NSF CAREER grant 201467. We thank our referee Jeff Linsky for a valuable input.

References

J.W. Armstrong, R. Woo, Astron. Astrophys. **103**, 415 (1981)
J.W. Armstrong, B.J. Rickett, S.R. Spangler, Astrophys. J. **443**, 209 (1995)
J. Ballesteros-Paredes, R. Klessen, M. Mac Low, E. Vasquez-Semadeni, in *Protostars and Planets*, ed. by V. Reipurth, D. Jewitt, K. Keil (University of Arizona Press, Tucson, 2006), pp. 63–80
A. Beresnyak, A. Lazarian, Astrophys. J. **640**, L175–L178 (2006)
A. Beresnyak, A. Lazarian, Astrophys. J. **678**, 961 (2008a)
A. Beresnyak, A. Lazarian, Astrophys. J. **682**, 1070 (2008b)
A. Beresnyak, A. Lazarian, J. Cho, Astrophys. J. **624**, L93–L96 (2006)
A. Bhattacharjee, E. Hameiri, Phys. Rev. Lett. **57**, 206 (1986)
A. Bhattacharjee, Z.W. Ma, X. Wang, Turbul. Magn. Fields Astrophys. **614**, 351 (2003)
J. Bieber, C. Smith, W. Matthaeus, Astrophys. J. **334**, 470 (1988)
J. Bieber, W. Matthaeus, C. Smith, Astrophys. J. **410**, 294 (1994)
D. Biskamp, Astrophys. Space Sci. **242**, 165 (1996)
D. Biskamp, Magnetohydrodynamic turbulence (CUP, 2003)
S. Boldyrev, Phys. Rev. Lett. **96**, 115002 (2006)
S. Boldyrev, Å. Nordlund, P., Padoan, Phys. Rev. Lett. **89**, 031102 (2002)
S.I. Braginskii, Rev. Plasma Phys. **1**, 205 (1965)
G. Brunetti, Astron. Nachr. **327**, 615 (2006)
G. Brunetti, A. Lazarian, Mon. Not. R. Astron. Soc. **378**, 245 (2007)
R. Cassano, G. Brunetti, Mon. Not. R. Astron. Soc. **357**, 1313 (2005)
B.D.G. Chandran, Phys. Rev. Lett. **85**(22), 4656–4659 (2000)
B.D.G. Chandran, Phys. Rev. Lett. **95**, 265004 (2005)
B.D.G. Chandran, Astrophys. J. **685**, 646 (2008)
B.D.G. Chandran, S.C. Cowley, Phys. Rev. Lett. **80**, 3077 (1998)
J. Cho, A. Lazarian, Phys. Rev. Lett. **88**(24), 5001-(1-4) (2002)
J. Cho, A. Lazarian, Mon. Not. R. Astron. Soc. **345**, 325–339 (2003)

J. Cho, A. Lazarian, Theor. Comput. Fluid Dyn. **19**, 127–157 (2005)
J. Cho, E. Vishniac, Astrophys. J. **539**, 273–282 (2000)
J. Cho, A. Lazarian, A. Honein, B. Knaepen, S. Kassinos, P. Moin, Astrophys. J. **589**, L77–L80 (2003)
J. Cho, A. Lazarian, E.T. Vishniac, Astrophys. J. **564**, 291 (2002a)
J. Cho, A. Lazarian, E. Vishniac, Astrophys. J. **566**, L49–L52 (2002b)
J. Cho, A. Lazarian, E. Vishniac, Astrophys. J. **595**, 812–823 (2003a)
J. Cho, A. Lazarian, E.T. Vishniac, *Turbulence and Magnetic Fields in Astrophysics*. Lecture Notes in Physics, vol. 614 (2003b), p. 56
A. Ciaravella, J. Raymond, Astrophys. J. **686**, 1372 (2008)
S.R. Cranmer, A.A. van Ballegooijen, Astrophys. J. Suppl. **156**, 265 (2005)
W. Daughton, J. Scudder, H. Karimabadi, Phys. Plasmas **13**, 2101 (2006)
E. de Gouveia dal Pino, A. Lazarian, Astrophys. J. **560**, 358–364 (2001)
E. de Gouveia dal Pino, A. Lazarian (2003). arXiv:astro-ph/0307054
E. de Gouveia dal Pino, A. Lazarian, Astron. Astrophys. **441**, 845–853 (2005)
L. Del Zanna, M. Velli, P. Londrillo, Astron. Astrophys. **367**, 705 (2001)
N.H. Dieter, W.J. Welch, J.D. Romney, Astrophys. J. **206**, L113–L115 (1976)
M. Dobrowolny, A. Mangeney, P. Veltri, Phys. Rev. Lett. **45**, 144 (1980)
J. Drake, M. Swisdak, K. Schoeffler, B. Rogers, S. Kobayashi, Geophys. Res. Lett. **33**, 13105 (2006a)
J. Drake, M. Swisdak, H. Che, M. Shay, Nature **443**, 05116 (2006b)
B. Dubrulle, Phys. Rev. Lett. **73**, 959–962 (1994)
B. Elmegreen, J. Scalo, Annu. Rev. Astron. Astrophys. **42**, 211 (2004)
E. Falgarone, P. Hily-Blant, J. Pety, G. Pineau Desforts, in *Magnetic Fields in the Universe*, ed. by E. de Gouveia Dal Pino, G. Lugones, A. Lazarian, AIP Conference Proceedings, vol. 784 (2005), p. 299
E. Falgarone, G. Pineau Des Forets, P. Hily-Blant, P. Schilke, Astron. Astrophys. **452**, 511 (2006)
E. Falgarone, P. Hily-Blant, J. Pety, G. Pineau Desforts, *ASP Conference Series*, vol. 365, ed. by M. Haverkorn, W.M. Goss (San Francisco, 2007), pp. 190–196
A. Farmer, P. Goldreich, Astrophys. J. **604**, 671 (2004)
S. Galtier, S.V. Nazarenko, A.C. Newel, A. Pouquet, J. Plasma Phys. **63**, 447 (2000)
S. Galtier, S. Nazarenko, A. Newell, A. Pouquet, Astrophys. J. **564**, L49 (2002)
S.P. Gary, *Theory of Space Plasma Microinstabilities* (CUP, Cambridge, 1993)
J. Giacalone, J.R. Jokipii, Astrophys. J. **520**, 204 (1999)
V. Ginzbirg, Sov. Astr. **9**, 877 (1966)
G. Gogoberidze, Phys. Plasma **14**, 022304 (2007)
P. Goldreich, S. Sridhar, Astrophys. J. **438**, 763–775 (1995)
P. Goldreich, S. Sridhar, Astrophys. J. **640**, L159 (2006)
M. Goldstein, Astrophys. J. **204**, 900 (1976)
G. Gloeckler, N. Schwadron, L. Fisk, J. Geiss, Geophys. Res. Lett. **22**, 2665 (1995)
E. Hameiri, A. Bhattacharjee, Phys. Fluids **30**, 1743 (1987)
W.B. Hawkins, Phys. Rev. **98**, 478 (1955)
C. Heiles, Astrophys. J. **481**, 193–204 (1997)
J. Higdon, Astrophys. J. **285**, 109 (1984)
P. Hily-Blant, E. Falgarone, Astron. Astrophys. **469**, 173–187 (2007)
P. Hily-Blant, J. Pety, E. Falgarone, ASP Conference Series, vol. 365, ed. by M. Haverkorn, W.M. Goss (San Francisco, 2007), pp. 184–190
T. Horbury, in *Plasma Turbulence and Energetic particles*, ed. by M. Ostrowski, R. Schlickeiser (Universytet Jagiellonski, Cracow, 1999), p. 28
A.R. Jacobson, R.W. Moses, Phys. Rev. A **29**, 3335 (1984)
R. Jokipii, Astrophys. J. **146**, 480 (1966)
R. Jokipii, in *Interstellar turbulence*, ed. by J. Franco, A. Carraminana, Proceedings of the 2nd Guillermo Haro Conference (Cambridge University Press, Cambridge, 1999), p. 70
R. Jokipii, in *The 7th Taipei Astrophysics Workshop on Cosmic Rays in the Universe*, ed. by C. Ko, ASP Conf. Proc., vol. 241 (Astronomical Society of the Pacific, San Francisco, 2001), p. 223
J.R. Jokipii, E.N. Parker, Astrophys. J. **155**, 777 (1969)
A. Kastler, J. Opt. Soc. Am. (1917–1983) **47**, 460 (1957)
A.P. Kazantsev, Sov. J. Exp. Theor. Phys. **26**, 1031 (1968)
J. Kóta, J.R. Jokipii, Astrophys. J. **531**, 1067 (2000)
G. Kowal, A. Lazarian (2006). astro-ph/0611396
G. Kowal, A. Lazarian, A. Beresnyak, Astrophys. J. **658**, 423–445 (2007)
G. Kowal, A. Lazarian, E. Vishniac, Astrophys. J. (2008, in preparation)
R. Kulsrud, *Plasma Physics for Astrophysics* (Princeton University Press, Princeton, 2004)
A. Lazarian, Astronomische Nachrichten **327**, 609 (2006a)

A. Lazarian, Astrophys. J. Lett. **645**, L25–L28 (2006)
A. Lazarian, Astrophys. J. Lett. **660**, L173 (2007a)
A. Lazarian, in *ASP Conference Series*, vol. 365, ed. by M. Haverkorn, W.M. Goss (San Francisco, 2007b), p. 324
A. Lazarian, A. Beresnyak, Mon. Not. R. Astron. Soc. **373**, 1195–1202 (2006)
A. Lazarian, E. Vishniac, Astrophys. J. **517**, 700–718 (1999)
A. Lazarian, E. Vishniac, J. Cho, Astrophys. J. **603**, 180–197 (2004)
B. Leroy, Astron. Astrophys. **91**, 136 (1980)
M. Lesieur, *Turbulence in Fluids* (Kluwer, Dordrecht, 1990)
Y. Lithwick, P. Goldreich, Astrophys. J. **562**, 279–296 (2001)
Y. Lithwick, P. Goldreich, Astrophys. J. **582**, 1220–1240 (2003)
Y. Lithwick, P. Goldreich, S. Sridhar, Astrophys. J. **655**, 269 (2007)
Y.C. Liu, M. Opher, O. Cohen, P. Liewer, T.I. Gombosi, Astrophys. J. **680**, 757 (2008)
M.M. Mac Low, Astrophys. J. **524**, 169–178 (1999)
J. Maron, P. Goldreich, Astrophys. J. **554**, 1175–1196 (2001)
W.H. Matthaeus, S.L. Lamkin, Phys. Fluids **28**, 303 (1985)
W. Matthaeus, M. Goldstein, D. Montgomery, Phys. Rev. Lett **51**, 1484 (1983)
W. Matthaeus, J. Ambrosiano, M. Goldstein, Phys. Rev. Lett. **53**, 1449 (1984)
W. Matthaeus, G. Qin, J. Bieber, G. Zank, Astrophys. J. **590**, L53 (2003)
C. McKee, E. Ostriker, Annu. Rev. Astron. Astrophys. **45**, 565–687 (2007)
A.B. Mikhailovskii, Rev. Plasma Phys. **2**, 201 (1975)
A.H. Minter, S.R. Spangler, Astrophys. J. **485**, 182 (1997)
E. Mobius, D. Rucinski, M. Lee, P. Isenberg, J. Geophys. Res. **10**, 257 (1998)
H.K. Moffatt, *Planetary and Space Science* (Cambridge, Cambridge University Press, 1978)
D. Montgomery, L. Turner, Phys. Fluids **24**, 815 (1981)
W.-C. Müller, D. Biskamp, Phys. Rev. Lett. **84**, 475 (2000)
R. Narayan, M.V. Medvedev, Astrophys. J. Lett. **562**, L129 (2001)
M. Opher, E.C. Stone, T.I. Gombosi, Science **316**, 875 (2007)
S. Ougthton, P. Dmitruk, W. Matthaeus, *Turbulence and Magnetic Fields in Astrophysics*, Lecture Notes in Physics, vol. 614 (2003), p. 56
E. Parker, Astrophys. J. **142**, 584–590 (1965)
G. Qin, W.H. Matthaeus, J.W. Bieber, Astrophys. J. Lett. **578**, L117 (2002)
A.B. Rechester, M.N. Rosenbluth, Phys. Rev. Lett. **40**, 38 (1978)
D. Roberts, M. Goldstein, L. Klein, W. Matthaeus, J. Geophys. Res. **92**, 12023 (1987)
A.A. Schekochihin, S.C. Cowley, S.F. Taylor, J.L. Maron, J.C. McWilliams, Astrophys. J. **612**, 276 (2004)
R. Schlickeiser, *Cosmic Ray Astrophysics* (Astronomy and Astrophysics Library, Springer, 2002)
A. Shalchi, Phys. Plasmas **12**, 052905 (2005)
A. Shalchi, Astron. Astrophys. **453**, L43 (2006)
A. Shalchi, J. Bieber, W. Matthaeus, G. Qin, Astrophys. J. **616**, 617 (2004)
A. Shalchi, A. Lazarian, R. Schlickeiser, Mon. Not. R. Astron. Soc. **383**, 803 (2008)
M.A. Shay, J.F. Drake, Geophys. Res. Lett. **25**, 3759 (1998)
Z.-S. She, E. Leveque, Phys. Rev. Lett. **72**, 336 (1994)
Z.-S. She, E.C. Waymire, Phys. Rev. Lett. **74**(2), 262–265 (1995)
J.V. Shebalin, W.H. Matthaeus, D. Montgomery, J. Plasma Phys. **29**, 525 (1983)
P. Sonnentrucker, E. Gonzlez-Alfonso, D.A. Neufeld, E.A. Bergin, G.J. Melnick, W.J. Forrest, J.L. Pipher, D.M. Watson, Astrophys. J. Lett. **650**, L71–L74 (2006)
T.W. Speiser, P&SS **18**, 613 (1970)
S. Stanimirovic, J.M. Weisberg, A. Hedden, K. Devine, T. Green, S.B. Anderson, Astrophys. Space Sci. **292**, 103 (2004)
H.R. Strauss, Astrophys. J. **326**, 412 (1988)
E.C. Stone, A.C. Cummings, F.B. McDonald, B.C. Heikkila, N. Lal, W.R. Webber, Nature **454**, 71 (2008)
N. Thomas, Survey Geophys. **13**, 91 (1992)
D.A. Tilley, D.S. Balsara, Astrophys. J. Lett. **645**, L49 (2006)
G. Toth et al., J. Geophys. Res. 110 (2005)
A. Verdini, M. Velli, Astrophys. J. **662**, 669 (2007)
J. Vestuto, E. Ostriker, J. Stone, Astrophys. J. **590**, 858 (2003)
E. Vishniac, A. Lazarian, J. Cho, *Turbulence and Magnetic Fields in Astrophysics*. Lecture Notes in Physics, vol. 614 (2003), p. 56
G.M. Webb, G.P. Zank, E.K. Kaghashvili, J.A. le Roux, Astrophys. J. **651**, 211 (2006)
D.G. Wentzel, Astrophys. J. **156**, 303 (1969)
D.G. Wentzel, in Annual Review of Astronomy and Astrophysics (Palo Alto, Calif., 1974), p. 12

H. Yan, A. Lazarian, Phys. Rev. Lett. **89**(28), 1102-(1-4) (2002)
H. Yan, A. Lazarian, Astrophys. J. **592**, L33 (2003)
H. Yan, A. Lazarian, Astrophys. J. **614**, 757–769 (2004)
H. Yan, A. Lazarian, Astrophys. J. **657**, 618 (2007)
H. Yan, A. Lazarian, Astrophys. J. **673**, 942 (2008)
H. Yan, A. Lazarian, B.T. Draine, Astrophys. J. **616**, 895 (2004)
H. Yan, A. Lazarian, V. Petrosian, Astrophys. J. **684**, 1461 (2008)
G. Zank, W.J. Matthaeus, Plasma Phys. **48**, 85 (1998)
E. Zweibel, F. Heitsch, Y. Fan, *Turbulence and Magnetic Fields in Astrophysics*. Lecture Notes in Physics, vol. 614 (2003), p. 101

Section VIII: The Emerging Broad Picture

The Heliosphere in Time

H.-R. Müller · P.C. Frisch · B.D. Fields · G.P. Zank

Originally published in the journal Space Science Reviews, Volume 143, Nos 1–4, 415–425.
DOI: 10.1007/s11214-008-9448-7

Abstract Because of the dynamic nature of the interstellar medium, the Sun should have encountered a variety of different interstellar environments in its lifetime. As the solar wind interacts with the surrounding interstellar medium to form a heliosphere, different heliosphere shapes, sizes, and particle contents result from the different environments. Some of the large possible interstellar parameter space (density, velocity, temperature) is explored here with the help of global heliosphere models, and the features in the resulting heliospheres are compared and discussed. The heliospheric size, expressed as distance of the nose of the heliopause to the Sun, is set by the solar wind-interstellar pressure balance, even for extreme cases. Other heliospheric boundary locations and neutral particle results correlate with the interstellar parameters as well. If the H^0 clouds identified in the Millennium Arecibo survey are typical of clouds encountered by the Sun, then the Sun spends $\sim$99.4% of the time in warm low density ISM, where the typical upwind heliosphere radii are up to two orders of magnitude larger than at present.

Keywords Global heliosphere modeling · Time-dependent interstellar conditions

H.-R. Müller (✉)
Department of Physics and Astronomy, Dartmouth College, Hanover, NH 03755, USA
e-mail: hans.mueller@dartmouth.edu

P.C. Frisch
Department of Astronomy and Astrophysics, University of Chicago, Chicago, IL 60637, USA
e-mail: frisch@oddjob.uchicago.edu

B.D. Fields
Center for Theoretical Astrophysics, Departments of Astronomy and Physics, University of Illinois, Urbana, IL 61801, USA
e-mail: bdfields@uiuc.edu

G.P. Zank
Center for Space Plasma and Aeronomic Research, University of Alabama in Huntsville, Huntsville, AL 35899, USA
e-mail: zank@cspar.uah.edu

1 Introduction

The expansion of the coronal solar wind is terminated by encountering the interstellar material surrounding the solar system, the circum-heliospheric interstellar medium (CHISM). The Sun is currently embedded in partially ionized material, with about three fourths of the material in neutral form. Absorption studies toward nearby stars reveal a very inhomogeneous local interstellar medium, with many distinct velocity components in most lines of sight. Also on bigger scales, the ISM is a dynamic medium, constantly replenished and mixed anew by the outflows of star forming regions and supernova explosions, for example. Shock passages create large regions of almost empty, high-temperature regions in their wake which may be unstable and hence short-lived.

On its journey around the Galaxy, the Sun is likely to encounter a variety of such interstellar environments, each of which is characterized by a set of physical quantities, including density, temperature, magnetic field, and velocity state. As the solar wind is often assumed to have been relatively stable over long periods of time, the question arises how the heliosphere has reacted to the differing interstellar environments encountered through time. We characterize in this paper the heliospheric response to some non-catastrophic changes in the interstellar environment ("galactic weather"). Studies like this, paired with cosmic ray calculations and investigations of the terrestrial consequences of changes in the interplanetary medium due to galactic weather events, will contribute to the understanding of long-ago events that left their mark in terrestrial records.

A concrete example is the evidence that the relatively warm and dense material of the contemporary local interstellar cloud (LIC) is thought to be embedded in a larger structure, the Local Bubble (LB). With some reasonable assumptions, it can be estimated that the Sun was exposed to the tenuous, very hot LB environment for a long time, has only entered the LIC some 40,000 years ago, and will exit it in 0–4000 years from now. In this sense, roughly steady interstellar conditions are short-lived, and changes will occur on different time scales, with cloud passages as short as 10^3 years.

The variety of possible heliosphere boundary conditions that follow from our understanding of interstellar clouds dictate that diverse models of the heliosphere are required. The models must accommodate the extremes that the Sun may have encountered over time. For most of the discussion below we use a heliospheric multifluid model that self-consistently incorporates the charge-exchange interactions between interstellar neutral atoms and the solar wind. The dynamics of the solar wind, and therefore the heliosphere characteristics, are altered substantially by mass-loading from interstellar H^0 atoms, so modeling both the pristine and charge-exchange H is a key aspect of understanding variations in the heliosphere as the Sun passes through different galactic environments. These models form the basis for the discussions in Sects. 2 and 4. For the extreme and short-lived interstellar environment experienced by the passage of an interstellar shock over the heliosphere, discussed in Sect. 3, we use a different model.

2 Basic Model Results

To investigate the influence of the interstellar conditions on the heliosphere, a detailed global heliospheric multifluid code is used, which self-consistently calculates plasma components and neutral hydrogen (Pauls et al. 1995; Zank et al. 1996; Müller et al. 2006a). Its strength lies in the detailed treatment of the neutral component, which originates in the ISM but is out of equilibrium when passing through the heliosphere. This is the consequence of charge exchange where neutral atoms are lost to the plasma, and new neutrals are inserted into the distribution with velocity characteristics that represent the underlying plasma protons. Because the mean free paths can be quite long, the multifluid approach treats the neutrals with three or four fluids, each representing the characteristic major different plasma regions where the respective secondary neutrals were born (e.g., Müller et al. 2008).

We first establish a detailed global heliospheric multifluid code that can accommodate the variable heliosphere boundary conditions that the Sun may have encountered over time. It assumes axisymmetry about the stagnation axis (the axis through the Sun that is parallel to the ISM flow) and an isotropic solar wind. The model grid is polar, with an inner inflow boundary at 1 AU (or even 0.7 AU for the smaller models of Sects. 2.3 and 2.4). There, the 1 AU solar wind parameters of 5.0 cm^{-3} for the plasma density, a temperature of 10^5 K, and a radial velocity of 400 km s^{-1} are used as standard values; they represent a typical in-ecliptic wind during solar minimum (slow solar wind). Each of the neutral fluids interacts with the plasma through resonant charge exchange, using the Fite et al. (1962) cross section, and all neutrals are subjected to photoionization which depends on the squared distance to the Sun. For simplicity, radiation pressure is assumed to balance gravity for neutral hydrogen, and heliospheric and interstellar magnetic fields are neglected. The interstellar medium is prescribed as inflow boundary condition at a suitably large distance from the interstellar bow shock (outer grid boundary at 500 AU for Sect. 2.3, 800 AU for Sect. 2.4, 1000 AU for Sect. 2.1, and 1500 AU for Sect. 2.2). The boundary parameters are the CHISM H^0 and H^+ number densities, and the (common) hydrogen velocity and temperature. Outflow boundary conditions are imposed on all fluids at the downwind outer boundary, and at the inner boundary for all neutral fluids. Photoionization already depletes the neutral density considerably there. Secondary neutrals are permitted to escape also through an outer outflow boundary condition in upwind directions.

2.1 Contemporary ISM (LIC)

As a proxy for the contemporary values of the CHISM boundary parameters we choose a model with $n(\text{H}^+) = 0.047\ \text{cm}^{-3}$, $n(\text{H}^0) = 0.216\ \text{cm}^{-3}$, $v = 26\ \text{km s}^{-1}$, and $T = 7000$ K (Slavin and Frisch 2008). Figure 1 shows the plasma temperature (top panel) and density (bottom panel) along the stagnation axis as solid lines, together with the neutral H density (dash-dotted line). The heliospheric boundaries appear as discontinuities. The interstellar plasma goes through a bow shock (BS); the accompanying decreased speed communicates via charge exchange to the neutrals and creates a hydrogen wall (hydrogen overdensity; Baranov and Malama 1993; Linsky and Wood 1996) in the post-bow shock region. On the sunward side of the heliopause, the supersonic solar wind undergoes a termination shock (TS) transition, and the hot heliosheath plasma is diverted tailwards. Solar and interstellar plasma are separated by a contact discontinuity, the heliopause (HP). The thermodynamically distinct plasma regions (supersonic solar wind; hot heliosheath; interstellar plasma) define the characteristic heliospheric regions that form the basis of the multifluid treatment of the neutrals in the model.

Figure 1 also displays the temperature and density profile of a plasma-only model, corresponding to the Local Bubble environment ("LB", dashed lines; presented in detail in Sect. 2.2 below). It demonstrates that charge exchange is the source of solar wind heating from about 10 AU all the way to the TS: The "LIC"-model plasma temperature is monotonically increasing in this range, while the plasma-only model follows a strict adiabatic cooling. At the HP, charge exchange organizes an anomalous heat transport from solar heliosheath plasma to interstellar plasma; such an energy transport is absent in the "LB" case without neutrals (HP at 300 AU in Fig. 1).

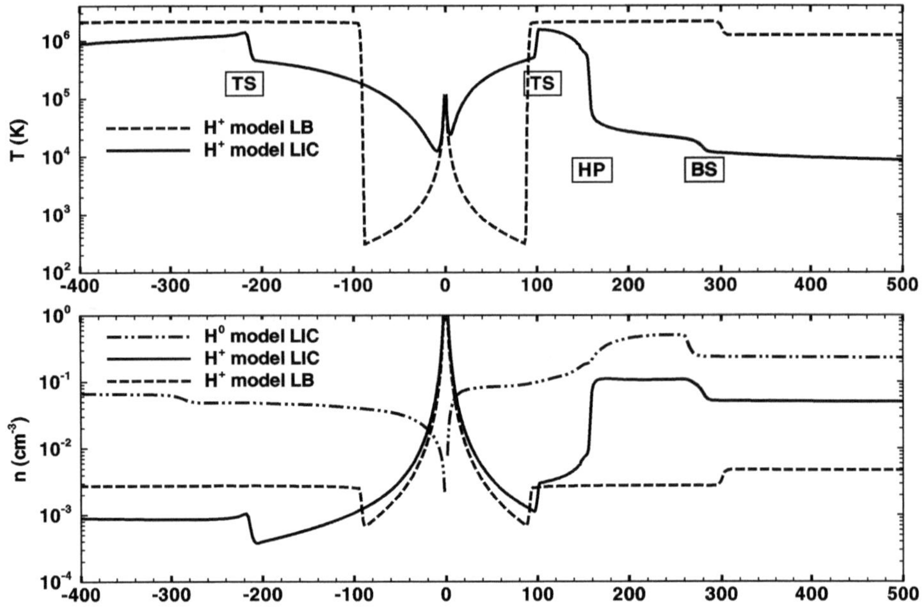

Fig. 1 One-dimensional profiles along the stagnation axis (distance in AU), with the Sun at center (0) and the CHISM coming from right. *Top*: plasma temperature of the Local Bubble case (*dashed*) and the contemporary conditions (*solid*). The heliospheric boundaries of the LIC model are marked in the plot. The *bottom panel* contains the corresponding densities (LB plasma model, *dashed*; LIC model: plasma, *solid*; neutral H, *dash-dot* pattern)

The neutral atom density at the upwind TS is 0.098 $\text{cm}^{-3} = 0.46\, n(\text{H}^0)$. The factor 0.46 is called filtration factor, based on the image that the pristine interstellar neutral flow gets processed ("filtered") while traversing the heliospheric interface between BS and TS (Wallis 1971). Close to the Sun, photoionization and increased charge exchange probability (due to increased solar wind density) deplete neutral H exponentially, creating a neutral H cavity. In the tail direction, the stagnation axis re-populates slowly with off-axis neutral H.

2.2 Hot Local Bubble

The relative motions of the Sun and surrounding interstellar gas indicate the Sun has emerged from the deepest void of the Local Bubble interior within the past $\sim$130,000 years. Such regions are common in the Milky Way Galaxy. Following this interpretation, the interior of this so-called Local Bubble is assumed hot and highly ionized but of low density. We adopt the ISM parameters of $n(\text{H}^+) = 0.005$ cm^{-3}, $n(\text{H}^0) = 0$, $v =$ 13.4 km s^{-1}, and $\log T(\text{K}) = 6.1$ as a proxy model "LB" for such an interstellar environment, with the velocity based on the Dehnen and Binney (1998) solar apex motion since the LB plasma is assumed to be at rest in the Local Standard of Rest (LSR). The realization that part of the soft X-ray emission could be due to heliospheric foreground composed of line emission of solar wind charge exchange products being left in excited states, has recently cast doubt whether the Local Bubble gas is in fact as hot as stated above, so that the LB model outlined here can be taken as an upper limit.

The speed of sound in such a LB plasma is high, so that the solar movement through it is decidedly subsonic. In this case, there is no bow shock, and interstellar plasma gets deceler-

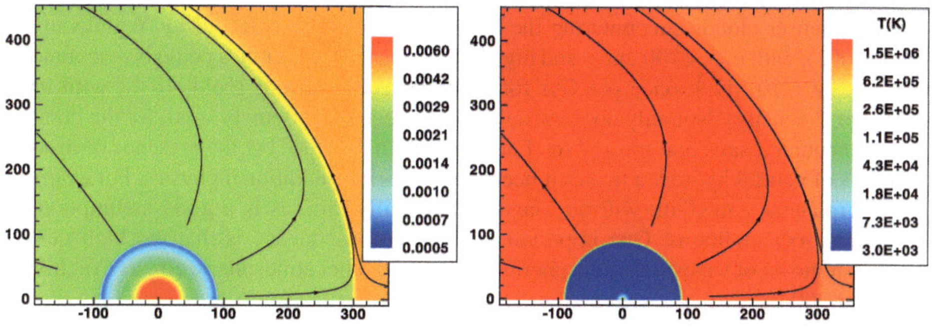

Fig. 2 Plasma density (*left*) and temperature (*right*) for the heliosphere surrounded by Local Bubble gas

ated adiabatically to match the zero velocity of the stagnation point. The isotropic thermal interstellar pressure dominates the ram pressure, and the termination shock is spherical at a distance of 90 AU from the Sun. This distance is comparable to that of the contemporary heliosphere. The distance to the nose of the heliopause is 300 AU, which makes the heliosheath very large in comparison to the one of the contemporary heliosphere. The temperature in the sheath reaches values as high as 2.2×10^6 K. Figure 2 shows 2D maps of density and temperature for the LB case, while Fig. 1 contains both variables along the stagnation axis, as dashed lines. In the termination shock transition, the density jumps by a factor of 3.8, and the wind speed decreases to 100 km s^{-1}.

In the LB model there are no neutral atoms in the entire combined system of the solar wind and ISM. The result is that a host of physical effects present if neutral gas surrounds the Sun were missing when the Sun was in the LB: There are no pickup ions (PUI) produced by charge exchange, there are no anomalous cosmic rays, and no slowdown or heating in the supersonic solar wind beyond the inner solar system takes place. As mentioned by Müller et al. (2006a), the plasma-only LB model exhibits a morphology and flow field that are similar to the gross properties of the hydrodynamic problem of a flow around a rigid sphere (excluding the flow detachment and vortices occurring in the latter case). They also point to the fact that absent the mitigating effects of neutrals, the magnetic field of both solar and interstellar origin takes on a more important role in the pressure balance, and corrections to the above results can be expected when magnetic fields are included realistically.

2.3 Dense Neutral ISM

Müller et al. (2006a) have presented results from models with an ISM density that is about two orders of magnitude higher than the contemporary value. The resultant heliospheres are small because of the increased interstellar ram pressure shifting the pressure balance. The presence of a higher neutral density makes charge exchange a more frequent occurrence, so that the plasma slowdown at the BS gets immediately communicated to the neutrals with a resulting hydrogen wall that is sharply defined post-BS, and has a high amplitude. Although filtration is very effective, enough neutral hydrogen enters the inner heliosphere inside the TS to lead to a pronounced solar wind slowdown (values down to 260 km s^{-1} upstream of the TS) and heating; the TS itself is therefore quite weak even in the nose direction. The BS compression ratio, on the other hand, is somewhat elevated. Atypically, the HP is not a sharp temperature transition as in most other models, but the temperature profile is more gradual, with the solar wind cooled close to the HP by frequent charge exchange, and the interstellar plasma heated close to the HP.

Even denser clouds encountering the heliosphere have been treated by Yeghikyan and Fahr (2003), with $n_{\text{tot}} = 100\ \text{cm}^{-3}$ and higher. Statistically such clouds could be encountered once every ~10 Myrs, once per Gyr for giant molecular clouds (cold, 10 K, with up to $1000\ \text{cm}^{-3}$ density). Not only does the heliosphere become extremely small for the duration of the encounter, such that most of the Earth orbit is in the ISM, but the presence of massive amounts of neutral hydrogen makes it necessary to include additional physics. For example, the interplanetary medium will no longer by optically thin, as is a good assumption for contemporary conditions. One important terrestrial consequence is that the high density leads to removal of terrestrial atmospheric oxygen, and other atmospheric effects (Yeghikyan and Fahr 2004).

2.4 High Velocity ISM

Similarly to density, it is also worthwhile to explore variations in the relative Sun-ISM speed. The interstellar velocity V is a key variable in understanding the ISM-heliosphere interactions because the ram pressure varies as V^2. As noted above, if the ISM is at rest with respect to the Local Standard of Rest, relative velocities of about 13 km s^{-1} result from the motion of the Sun in that frame of reference. In the contemporary case, the LIC moves with respect to the LSR, resulting in an overall 26 km s^{-1} relative motion. In addition, warm and cold H^0 clouds have radial velocities that vary from -80 to $+6$ km s^{-1}, with the possibility that 3D velocities are even larger. For the purpose of calculating astrospheres around stars in the solar neighborhood that are still embedded in a partially ionized ISM, Wood et al. (2003, 2005) determine relative Star-ISM velocities for selected objects. In their list there are entries with 68 and 86 km s^{-1}, and 40 Eri has a relative motion of 127 km s^{-1} (Wood et al. 2003). When applying such relative motions to the Sun, the resulting heliospheres can be described as "wind-swept," i.e., a narrow leading cavity and a long, drawn-out tail (Müller et al. 2006a). The high velocity generates a large ram pressure, making the resulting heliosphere smaller, similar to, but more elongated than, the high density cases. Such heliospheres also tend to have a triple point at the heliotail termination shock, necessitated by the heliosheath flow accelerating to supersonic velocities (e.g., Pauls and Zank 1996).

Not only are the high-speed heliospheres quite asymmetric, but the neutral mean free paths are now on the order of the heliosheath thickness or larger, meaning that the hydrogen wall is not very pronounced, and neutral filtration is very weak so that the interstellar density is not much higher than the neutral density entering the inner heliosphere through the termination shock. Such filtration ratios close to 1 seem only possible when a high interstellar velocity combines with a modest or low density so that the peak hydrogen wall occurs close to the HP without room for depletion of neutral H between peak and HP.

3 Supernova Remnant Encounters with the Heliosphere

The possibility that spikes in ^{10}Be isotopes in the Antarctic ice core samples were caused by a cosmic-ray enhancement due to a supernova shock passing over the heliosphere was evaluated by Sonett et al. (1987). Initial supernova remnant (SNR) expansion velocities are much larger than the high-velocity cases discussed in Sect. 2.4. Fields et al. (2008) have recently calculated the effect on the heliosphere of a supernova going off in the solar neighborhood. One of the primary motivations of such studies is the evidence of live radioisotopes in thin layers of ocean floors with lifetimes shorter than the age of the solar system (Knie et al. 1999, 2004). A plausible explanation of such data is that a SNR collision with Earth has deposited this material (^{60}Fe) in a short-duration event (2.8 ± 0.4) Myr ago.

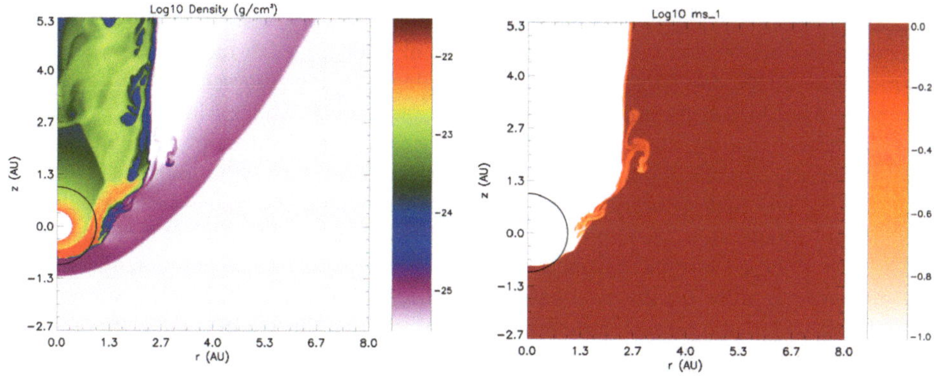

Fig. 3 (*Left*) Snapshot (logarithmic density map) of the heliosphere during arrival of a shock from a supernova explosion 8 pc away, for LIC-like ambient ISM conditions (the SNR velocity is 12,000 km s^{-1}). A *circle* is drawn at a radius of 1 AU indicating the Earth's orbit (modulo inclination effects). For a period of slow solar wind, a SNR under these conditions will engulf the Earth, and the Earth's orbit may carry it into the remnant where it will be directly exposed to supernova material. (*Right*) Corresponding contamination map ($C = 1$ supernova material, $C = 0$ solar material). Supernova material penetrates to 1 AU; solar wind and SNR plasmas mix also through instabilities at the heliopause

The effects of a SNR collision with the heliosphere depend on many factors, starting with the initial explosive energy, the density of the interstellar medium into which the supernova explodes, and the supernova distance. SNR shock front density, pressure, and velocity decrease with increasing distance from the supernova. Fields et al. (2008) have investigated a variety of ISM density parameters and distances, and modeled the SNR evolution from source to heliosphere with an AMR numerical code (FLASH; Fryxell et al. 2000); neutrals are neglected. Figure 3 displays an example of a heliosphere encountering a SNR from a 10^{51} erg supernova explosion at a distance of 8 pc, where the intervening ISM was assumed to have densities similar to the Local Bubble. This distance corresponds to the recent and conservative Gehrels et al. (2003) estimate for the limit at which the ionizing radiation from a supernova (prompt UV, X- and gamma-ray photons, as well as the later diffusive cosmic rays) inflicts damage on terrestrial stratospheric ozone at a level which can cause severe damage to the biosphere.

The passage of the supernova blast profile persists for thousands of years, with a slow secular decrease in pressure, velocity, and density. Consequently, high-speed, dense material comprises an interstellar wind on time scales of many solar cycles, such that on scales of the heliosphere the SNR blast is a basically steady, plane-parallel ISM wind. The snapshot in Fig. 3 is taken with solar wind conditions reflecting solar minimum. It can be seen that the heliosphere is severely compressed, and the Earth orbit (black circle) dips into the shocked interstellar medium for part of the year, providing an interplanetary medium rich in supernova ejecta that might eventually precipitate to Earth to form the mentioned ocean bottom sediment layer. We see that an event close enough to cause biological damage is also able to deliver supernova debris to the Earth, raising the possibility that isotopic signatures can be correlated with possible supernova-induced mass extinctions.

4 Sensitivity of the Heliosphere to ISM Conditions

Utilizing a broad set of plausible interstellar boundary conditions for the heliosphere, it is possible to establish some correlations between such parameters and the resultant heliosphere configuration. To establish these correlations, numerous multi-fluid models of the heliosphere and astrospheres have been analyzed together. The models span the low-to-high interstellar velocities and densities of Müller et al. (2006a), with most of them obeying $500 < P/k < 6000$ cm^{-3} K, and a study with values probing parameter space around the contemporary CHISM parameters with a similar range of P/k (see Müller et al. (2006b) for preliminary results). Regarding the latter study, the inferred values of the contemporary CHISM are $v = 26.3$ km s^{-1} and $T \sim 6300 \pm 340$ K (Witte et al. 1996; Möbius et al. 2004; Witte 2004). The contemporary interstellar proton and neutral H densities are not well constrained but should lie in the range from 0.04–0.14 cm^{-3} and 0.14–0.24 cm^{-3}, respectively (e.g. Zank 1999; Slavin and Frisch 2002). The setup for the systematic parameter study was therefore to probe the distinct densities of 0.05, 0.11, 0.17, and 0.23 cm^{-3} for both the interstellar neutral and plasma densities (16 combinations). The CHISM ionization fraction ranges from 18% to 82%. To explore the effects of temperature, it was decided to probe four temperatures 4000, 6000, 8000, and 10,000 K, for a total of 64 models. The velocity in this parameter study was set to 26.24 km s^{-1}.

One result from the parameter study as well as from the extension to a wider parameter space is the predictability of the distance of the upwind heliopause (stagnation point) as an expression of the pressure balance between interstellar and solar wind. The HP distance is calculated from the Rankine-Hugoniot termination shock transition conditions and treating the heliosheath and shocked interstellar flows as incompressible (Suess and Nerney 1990; Zank 1999; Müller et al. 2006a). A relation is obtained that links the solar wind ram pressure $P_1 = \rho_1 v_{SW}^2$ at 1 AU and its scaling with heliocentric distance, to the interstellar total pressure P_{ISM}. In a supersonic case, the latter is dominated by $\rho_{ISM} v_{ISM}^2$ with $\rho_{ISM} = m_p[n(\mathrm{HI}) + n(\mathrm{HII})]$. The resulting heliopause distance takes the form

$$r_{HP} = r_0 \sqrt{\frac{\rho_1 v_{SW}^2}{P_{ISM}}} \left(1 - \frac{v_{ISM}^2}{v_{SW}^2}\right)^{\frac{5}{4}} . \tag{1}$$

The constant r_0 is a product of factors from the theoretical calculation and from the empirical fact that neutral hydrogen does not fully participate in the pressure balance, but only weakly couples to the plasma through charge exchange. A fit between model results and r_{HP} values yields a value of $r_0 = 1.70$ AU, with about a 6% accuracy. A similar distance law was obeyed even by the high-velocity models of Fields et al. (2008) which represent SNR collisions with the heliosphere (Sect. 3). Baranov et al. (1979) have studied analytically the plasma-only case without neutrals and arrive at a distance relation similar to (1), without the last factor (typically close to unity) involving the velocity ratio. Because of the entire ram pressure $P_{ISM} = \rho_{p,ISM} v_{ISM}^2$ participating fully in the pressure balance in this case, they arrive at a lower $r_0 = 1.4$ AU, and the plasma-only heliospheric boundaries scale self-similarly with the factor $(\rho_1 v_{SW}^2 / \rho_{p,ISM} v_{ISM}^2)^{1/2}$.

The magnetic fields of solar and interstellar origin are neglected here; models that include interplanetary and interstellar magnetic fields are presented elsewhere (see for example the Opher contribution in this volume). It can be reasonably expected that for models with neutrals and magnetic fields, the inclusion of magnetic pressure in the pressure balance (1) will preserve its validity to some extent. However, most sensible interstellar magnetic field configurations render the heliosphere asymmetric, and the distances of the boundaries as a function of direction become more complicated than (1)–(3) suggest.

In all available (non-magnetic) models, the nose distance of the heliopause is accurately coupled to the termination shock distance,

$$r_{HP} = (1.40 \pm 0.03)\, r_{TS}. \tag{2}$$

Again, this relation holds for SNR-type velocities as well (Fields et al. 2008, Fig. 7). For the contemporary heliosphere, this result means that if the termination shock lies between 86 and 94 AU, the heliosheath in upwind direction is between 35 and 38 AU thick, for a heliopause distance between 121 and 132 AU.

The interstellar bow shock distance also correlates with the termination shock distance; however, there is an additional temperature dependence. For a higher interstellar temperature, the Mach number of the interstellar flow is lower, and hence the bow shock itself weaker. To still achieve a zero velocity at the stagnation point at the heliopause, the bow shock hence must be further out for higher temperatures, in spite of the higher interstellar pressure contributing to a slight shrinking of the overall heliospheric size. Multiplying a linear correlation with an ad-hoc term that grows for increasing temperatures gives a very good correlation between termination shock and bow shock distance,

$$r_{BS} = 1.5 r_{TS}\left(1 + \frac{1}{M-1}\right) + 23 \text{ AU} \tag{3}$$

(Müller et al. 2006b).

The shape of the termination shock is not spherical, but it is elongated in the tail directions. Typical values of the downwind/upwind ratio range between 2 and 3. The tail TS distance correlates with the upwind TS distance; however, there is again a temperature influence in this correlation. The higher the interstellar temperature, the lower the asymmetry of the TS between downwind and upwind direction. Empirically there is a correlation between $r_{TS}(180°)$ and the quantity $r_{TS}(0°)\sqrt{M}$, with a weak dependence of the proportionality factor on the interstellar neutral hydrogen density.

These results have been obtained from models where neutral hydrogen was present and influenced the plasma via charge exchange. The neutral hydrogen reaction to the different interstellar environments resists categorization more than is the case for the plasma results. For the 64-model systematic study, the height of the hydrogen wall (the peak neutral density between BS and HP) correlates with the interstellar Mach number. The higher the Mach number, the higher the resulting hydrogen wall. An expected anticorrelation between hydrogen wall height and amount of neutral hydrogen entering through the termination shock into the inner heliosphere could not be confirmed. There is, however, an anticorrelation of this filtration factor f with the interstellar plasma density, $f \propto 1/\sqrt[3]{n(\mathrm{HII})}$.

5 Paleo-Heliosphere

The Arecibo Millennium Survey of the radio H^0 sky (Heiles and Troland 2005) provides reliable statistics on the distribution of the column densities and velocities of warm and cold interstellar clouds seen from the tropical sky. About 60% of interstellar H^0 is warm neutral material (WNM). The median column density of the WNM, 1.3×10^{20} cm^{-2}, exceeds that of CNM, 0.5×10^{20} cm^{-2} (Frisch 2008). Although WNM is less dense than CNM, higher typical WNM velocities and column densities indicate that the Sun will spend more time in the WNM than the CNM. We calculate the length of time that the Sun would be in each H^0 component detected in the Arecibo survey, and compare that time with the heliopause radius calculated from (1). We use assumed densities of 0.27 cm^{-3} and 15 cm^{-3} for the WNM and CNM, respectively, and assume that the observed LSR radial velocity for each component is typical of the encounter velocity. These values then give us a good estimate of the total length of time the Sun is likely to spend in WNM versus CNM interstellar gas. From Fig. 4 we quickly see that the Sun spends significantly more time in WNM than CNM types of clouds. According to the tropical Arecibo sky, the Sun is in WNM 99.4% of the time, but

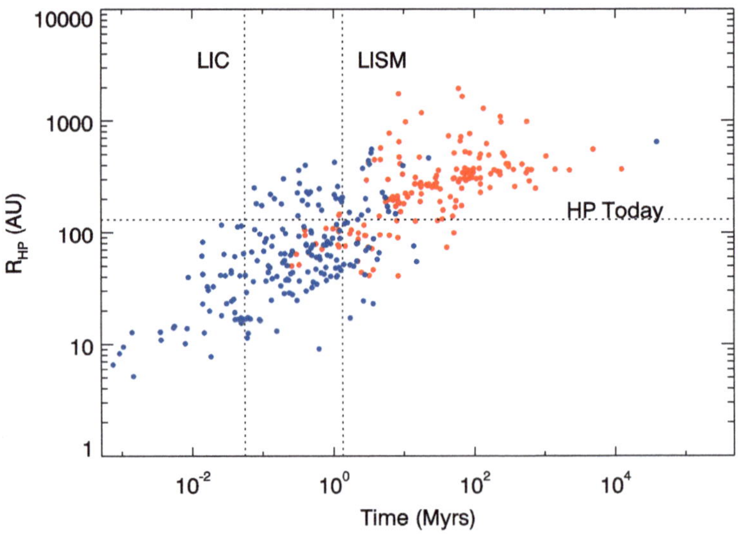

Fig. 4 The heliosphere in time, according to the tropical skies. The length of time, in Myrs, that the Sun would spend in the cold neutral (CNM) and warm neutral or partially ionized (WNM) interstellar clouds detected in the Arecibo Millennium Survey is plotted against the heliosphere radius predicted by (1) for those clouds. The CNM clouds (*blue dots*) assume densities of 15 cm^{-3}, and the WNM clouds (*red dots*) assume densities of 0.27 cm^{-3}. The length of times that the Sun spends in the LIC, and in the very local interstellar material (LISM) is indicated, as is the present heliopause distance in the upwind. Since the true three-dimensional velocity of these clouds is unknown, this comparison assumes that the relative Sun-cloud velocity is the LSR velocity of each component detected by Arecibo. This plot shows that if the Arecibo sky samples ISM that is typical of that encountered by the Sun, then the Earth and Sun spends 99.4% of the time in WNM clouds

this estimate does not include the times spent in fully ionized regions. The Sun has entered the LIC within the past ~ 0.056 Myrs, and the LISM within the past ~ 0.13 Myrs, according to UV absorption lines towards nearby stars (Frisch and Slavin 2006).

6 Conclusions

The ISM is a very dynamic environment. During its 5 Gyrs galactic trajectory, the solar system likely has been embedded in a wide variety of different interstellar environments. Some of these environments lead to a drastically reduced heliosphere size, allowing direct access of interstellar material to the solar system and Earth. Even without direct ISM access, the particle flux background like cosmic rays and dust is sensitive to the ISM environment. The passage of the solar system through an arm of the galaxy likely triggers a pronounced increase of both the rate of supernovae going off near the Sun, and the flux of galactic cosmic rays. Consequences for the terrestrial atmosphere/climate, and their geological records, are likely (e.g., Frisch 2006), but not part of this paper.

When calculating the heliosphere while including the interstellar neutrals self-consistently, and probing the vast interstellar parameter space, it can be confirmed that the overall heliospheric size is set by a pressure balance between solar wind and interstellar medium, with only weak modifications involving the interstellar Mach number and velocity (and most likely the magnetic pressure which was not included in this present analysis). The termination shock, heliopause, bow shock, and downwind termination shock locations exhibit simple correlations to the location of pressure balance. Neutral results like the filtration by the heliospheric interface, and the height of the hydrogen wall, also show a dependence on the interstellar parameters. For an encounter with dense ISM or very fast ISM, particle fluxes like neutrals, cosmic rays, and dust increase markedly at Earth orbit. A passage of a remnant of a nearby supernova blast has the ability to compress the heliosphere to a size that allows interstellar material including supernova ejecta direct access to Earth. In contrast, during the passage of the Sun through the Local Bubble, or through any almost completely ionized region, neutrals, pickup ions, and anomalous cosmic rays were entirely absent.

Acknowledgements HRM and PCF thank the International Space Science Institute (ISSI), Bern, Switzerland, for hosting the 2007 workshop "From the Heliosphere to the Local Bubble" and partially funding participation. HRM acknowledges partial funding through NSF grant AST-0607641 and NASA SHP grants NNG06GD48G and NNG06GD55G. PCF acknowledges funding through NASA grants NAG5-13107 and NNG05GD36G. The work of BDF was supported by NASA Exobiology grant EXB03–0000-0031.

References

V.B. Baranov, Y.G. Malama, J. Geophys. Res. **98**, 15157 (1993)
V.B. Baranov, M.G. Lebedev, M.S. Ruderman, Astrophys. Space Sci. **66**, 441–451 (1979)
W. Dehnen, J.J. Binney, Mon. Not. R. Astron. Soc. **298**, 387–394 (1998)
B.D. Fields, T. Athanassiadou, S.R. Johnson, Astrophys. J. **678**, 549–562 (2008)
W. Fite, A. Smith, R. Stebbings, Proc. R. Soc. Lond. Ser. A **268**, 527 (1962)
P.C. Frisch, *Solar Journey: The Significance of Our Galactic Environment for the Heliosphere and Earth* (Springer, Dordrecht, 2006)
P.C. Frisch, Space Sci. Rev. (2008, this issue). arXiv:0804.3798
P.C. Frisch, J.D. Slavin, Astrophys. Space Sci. Trans. **2**, 53–61 (2006)
B. Fryxell, K. Olson, P. Ricker, F.X. Timmes, M. Zingale, D.Q. Lamb, P. MacNeice, R. Rosner, J.W. Truran, H. Tufo, Astrophys. J. Suppl. **131**, 273 (2000)
N. Gehrels, C.M. Laird, C.H. Jackman, J.K. Cannizzo, B.J. Mattson, W. Chen, Astrophys. J. **585**, 1169 (2003)
C. Heiles, T.H. Troland, Astrophys. J. **624**, 773–793 (2005)
K. Knie, G. Korschinek, T. Faestermann, E.A. Dorfi, G. Rugel, C. Wallner, Phys. Rev. Lett. **93**, 171103 (2004)
K. Knie, G. Korschinek, T. Faestermann, C. Wallner, J. Scholten, W. Hillebrandt, Phys. Rev. Lett. **83**, 18 (1999)
J. Linsky, B. Wood, Astrophys. J. **463**, 254 (1996)
E. Möbius, M. Bzowski, S. Chalov, H.-J. Fahr, G. Gloeckler, V. Izmodenov, R. Kallenbach, R. Lallement, D. McMullin, H. Noda, M. Oka, A. Pauluhn, J. Raymond, D. Ruciński, R. Skoug, T. Terasawa, W. Thompson, J. Vallerga, R. von Steiger, M. Witte, Astron. Astrophys. **426**, 897–907 (2004)
H.-R. Müller, P.C. Frisch, V. Florinski, G.P. Zank, Astrophys. J. **647**, 1491 (2006a)
H.-R. Müller, G.P. Zank, P. Faurot-Pigeon, V. Jiwariyavej, R. Mutiso, F. Shafi, in *AIP Conf. Proc. 858: Physics of the Inner Heliosheath*, ed. by J. Heerikhuisen, V. Florinski, G.P. Zank, N.V. Pogorelov (2006b), pp. 33–38
H.-R. Müller, V. Florinski, J. Heerikhuisen, V.V. Izmodenov, K. Scherer, D.B. Alexashov, H.-J. Fahr, Astron. Astrophys. (2008, in press). doi:10.1051/0004-6361:20078708
H.L. Pauls, G.P. Zank, J. Geophys. Res. **101**, 17081–17092 (1996)
H.L. Pauls, G.P. Zank, L.L. Williams, J. Geophys. Res. **100**, 21595 (1995)
J.D. Slavin, P.C. Frisch, Astrophys. J. **565**, 364–379 (2002)
J.D. Slavin, P.C. Frisch, Astron. Astrophys. (2008, in press). doi:10.1051/0004-6361:20078101
C.P. Sonett, G.E. Morfill, J.R. Jokipii, Nature **330**, 458 (1987)
S.T. Suess, S. Nerney, J. Geophys. Res. **95**, 6403–6412 (1990)
M.K. Wallis, Nature **254**, 202–203 (1971)
M. Witte, Astron. Astrophys. **426**, 835–844 (2004)
M. Witte, M. Banaszkiewicz, H. Rosenbauer, Space Sci. Rev. **78**, 289 (1996)
B. Wood, J. Linsky, H.-R. Müller, G. Zank, Astrophys. J. **591**, 1210–1219 (2003)
B. Wood, H.-R. Müller, G. Zank, J. Linsky, S. Redfield, Astrophys. J. Lett. **628**, L143–L146 (2005)
A. Yeghikyan, H. Fahr, Ann. Geophys. **21**, 1263–1273 (2003)
A. Yeghikyan, H. Fahr, Astron. Astrophys. **425**, 1113–1118 (2004)
G.P. Zank, Space Sci. Rev. **89**, 413–688 (1999)
G.P. Zank, H.L. Pauls, L.L. Williams, D.T. Hall, J. Geophys. Res. **101**, 21639 (1996)

Some Observations Related to the Origin and Evolution of the Local Bubble/Local ISM

Rosine Lallement

Originally published in the journal Space Science Reviews, Volume 143, Nos 1–4, 427–436.
DOI: 10.1007/s11214-008-9428-y

Abstract I discuss some recent observations of the Local Interstellar medium (ISM) that are related to its history and temperature structure. I focus on three topics: (i) the abundance pattern of interstellar deuterium and metals, (ii) highly charged ion data, (iii) soft X-ray data.

Deuterium has been unambiguously shown to correlate almost linearly with refractory metals, confirming that interstellar grains are a "reservoir" of deuterium, and release it into the gaseous phase jointly with metals when the gas is shocked and heated. By interpreting the observed level of deuterium with respect to the abundance patterns of metals and oxygen, these data give some clues to the event, which gave rise to the expanding Gould belt. As a matter of fact abundance data seem to be linked to the belt, and the observed trends suggest an explosive origin, rather than a collision with an external cloud made of unastrated material.

X-rays and high ions trace hot gas and interfaces between hot and cool gas. However absorption lines of high ions show highly complex characteristics and no relationships have been established yet between their detected columns and the existence of hot-cool gas interfaces. Adding to the complexity, the X-ray emission through charge-exchange reactions between highly charged solar wind ions and neutrals plays a significant role, calling for modifications of the global picture of the LISM. In addition to the ubiquitous contamination of background data by locally emitted X-rays, there are also potential distant charge transfer (CX) X-ray emissions from clouds-hot gas interfaces.

There is a strong need for high quality, high spectral resolution X-ray data, because X-ray lines emitted after charge-transfer neutralization of helium-like ions bear a clear signature of the charge transfer process, if present, and allow to disentangle thermal and CX emission. More precise density and velocity distributions of the local ISM are also needed to take full advantage of the X-ray and high ion data and build a consistent picture of the Local Cavity (LC) and its surroundings. As an example of these requirements I discuss the case of the North Polar Spur for which there may be some evidence for CX emission.

Keywords ISM · X-ray background

R. Lallement (✉)
Service d'Aéronomie du CNRS, BP 3, 91371 Verrières-le-Buisson, France
e-mail: rosine.lallement@aerov.jussieu.fr

1 Introduction

Understanding the detailed structure of the Local Interstellar Medium (ISM) in our galactic neighborhood is becoming increasingly important. Firstly, it is a foreground that contaminates extragalactic diffuse background emission. A precise description of the gas distribution, resulting radiation field and dust emission is a mandatory tool for foreground corrections to the Cosmic Microwave Background (ESA Planck, Herschel forthcoming space missions). Second, the distribution of gas and dust influences the local gamma-ray radiation field, which will be the focus of the NASA GLAST mission. Third, it provides foregrounds and also helpful insights on the environment of a number of nearby galactic objects, and subsequently on potential interactions between these objects and the ambient medium. Examples are the astrospheres around cool stars detected in absorption at Lyman-alpha (Wood et al. 2008) or the cometary-shape envelopes recently discovered around evolved giant stars (e.g. Liebert et al. 2007). An extreme case is the spectacular trail around Mira which has been used to infer the date of entry of the star in the Local Bubble (Wareing et al. 2007). Last but not least, it provides a way to study and better understand the multi-phase structure of the local galactic ISM (LISM).

During this meeting, all of the local ISM aspects have been presented and debated by the best experts. A global picture emerges from these contributions that is at first sight consistent. The distribution of clouds and blown up, often reheated cavities as they come out from the simulations of stellar winds and supernovae (De Avillez and Breitschwerdt 2005; De Avillez 2008; Fuchs 2008) seems to represent very well the observations of dense and diffuse gas (Redfield 2008) and X-ray or radio data (Snowden et al. 2000). Intermediate ions (Shelton 2008) also roughly correspond to the theoretical interfaces between million degree gas and cool clouds and the estimated radiation field roughly explains the ionization states of the clouds (Slavin and Frisch 2002; Frisch 2008).

Still, a few contradictions remain, suggesting that the structure is still not entirely understood. Why are there very cold clouds and ultra-small scale structures within the hot gas of the Local Cavity (LC) (see Stanimirovic 2008), which should not happen in an old and repeatedly heated bubble? Why is the O VI distribution that traces interfaces between the hot gas and the clouds (inside and at the boundary) still defying a consistent picture, with many line-of-sight (LOS) devoid of O VI while other angularly close LOS have significant column densities? How to explain that warm (i.e. with thermally broadened lines) C IV and Si IV are absent along LOS going through the LC and nearby cavities (see Sect. 2)? Tangled magnetic fields around the clouds inhibiting the conduction can hardly be the unique solution. Finally, as stated by Jenkins (2008), the origin of the strong ionization of helium may be a problem if the properties of the local hot gas are going to be strongly revised due to the solar wind contamination of the X-ray background. New "ingredients" may be missing in the present picture. I discuss some observations potentially related to these "ingredients".

2 The Local ISM Deuterium Abundance Variability and its Link with the Local Bubble/Gould Belt Structure

2.1 The Deuterium Abundance within 500 pc

The analysis of the FUSE far UV spectra and the inferred D/H ratios have given rise to a strong debate. The situation is illustrated in Fig. 1, from Linsky et al. (2006). Within uncertainties the D/H ratio is about constant along lines-of-sight having a H column-density

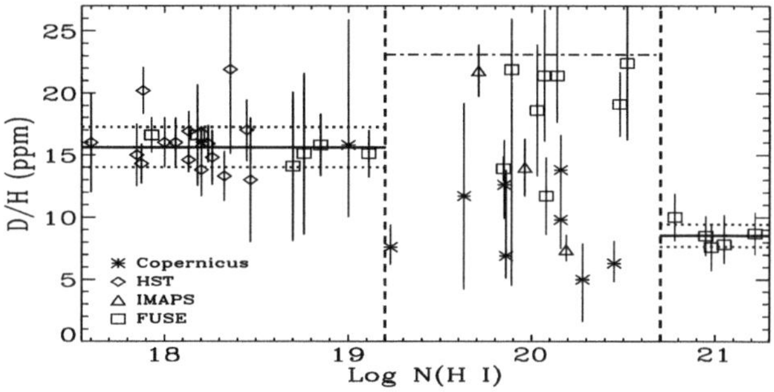

Fig. 1 Combined D/H measurements as a function of the total H I column-density to the target star. From Linsky et al. (2006)

less than 2×10^{19} cm^{-2}, which corresponds to distances of 50 to 150 pc according to the direction sampled. Then, beyond this value, the ratio is highly variable (by more than a factor of 4), while for columns above 10^{21} cm^{-2} it becomes significantly smaller and seems again about constant. Such a variability has been interpreted in two different ways. Linsky et al. (2006) attribute it to the effect of the preferential adsorption of deuterium at the surface of dust grains, and identify the low columns homogenous volume as the Local Cavity. Extreme deuteration of dust grains has been first proposed by Jura (1982), and the associated release of deuterium from grains into the gas following grain destruction has been suggested by Draine (2004) as the source of the strong variability of the D/H ratio in the local interstellar medium. Grain destruction and subsequent gas phase deuteration should in principle occur at shocks and in heated matter, i.e. especially within stellar wind cavities and supernovae remnants. Observational support to this scenario is brought by the observed correlations between deuterium and singly ionized titanium on one hand (Ellison et al. 2007), singly ionized iron and silicon on the other (Linsky et al. 2006). Grains are a "reservoir" of metals, and as a consequence such correlations naturally arise if they are a also "reservoir" of deuterium. Additional evidence comes from the apparent correlation between D/H and the excitation temperature of molecular hydrogen (Linsky et al. 2006).

At variance with this interpretation, Hébrard et al. (2005) invoked the very large uncertainties on the H I measurements as the source of the "apparent" abundance correlations. If one uses abundances relative to H for both D and the metal M, they are evidently both influenced by measurements errors on H. If the measured column NH_{obs} is in error by some factor X, both abundances are under- or over-estimated by the same factor X and the resulting set of data points for different lines-of-sight (and different X) is mimicking a linear correlation between the two abundances. Another argument used against the "depletion hypothesis" is the fact that the D/O ratio seemed to vary in a significantly less extent compared to D/H. Because the D and O absorption lines are less saturated than the H lines, O-related abundances are more reliable. Hébrard et al. (2005) suggest that D is locally overabundant due to some mixing processes, i.e. the gas within the first hundred parsecs is less processed than the gas at larger distance. This would be in agreement with Geiss et al. (2002) who suggest that the local gas is a mixture of "fresh" gas fallen from the halo down to the disk and processed galactic gas.

2.2 End of the Debate about Dust Release of Deuterium?

High resolution spectra of target stars whose D/H is known were recorded with the UVES spectrograph at the ESO Paranal Very Large Telescope and have been used to extend and reinforce the correlations with titanium. Figure 2 shows the data and the fitted linear relationship. For the first time titanium (and also iron) abundances relative to H have been

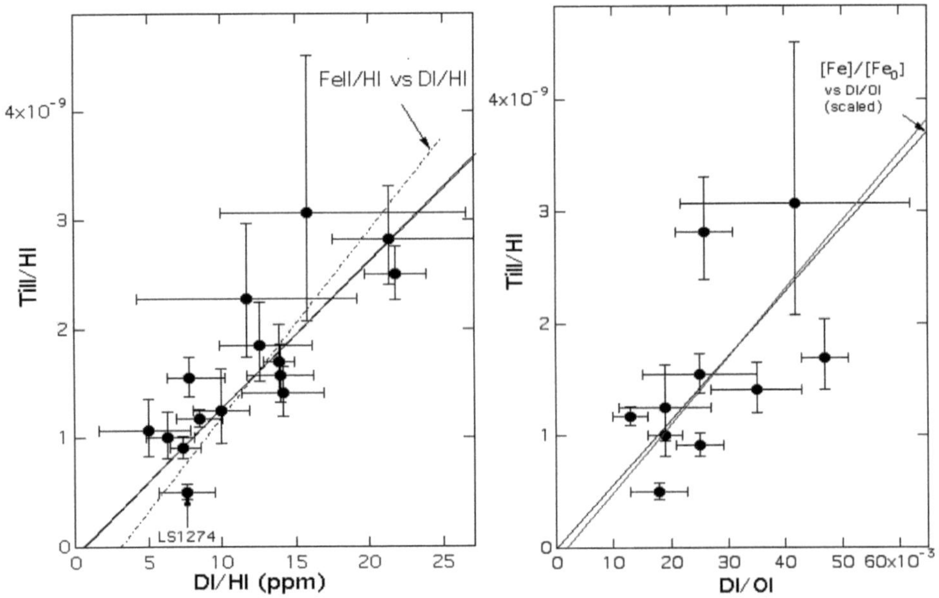

Fig. 2 Singly ionized titanium abundance and deuterium abundance. The data are consistent with linear relationships for Ti II/H vs both D/H and D/O (Lallement et al. 2008)

also correlated with D/O (in addition to D/H), convincingly revealing a new relationship and demonstrating that the uncertainty on H is definitely ruled out as the unique source of the deuterium-metals correlations (Lallement et al. 2008). Surprisingly, the deuterium abundance seems to vary linearly with the titanium abundance.

2.3 The Gould Belt

The deuterium depletion has some important consequences on its abundance and galactic evolutionary models, but it also opens a new perspective: deuterium can be used as a tracer of the ISM history. In particular it may bring some information on the local ISM, i.e. the local pattern shown in Fig. 1 must be related to past events having heated the dust.

Already noticed by Herschel in 1847, the peculiar distribution of the bright early-type stars in the solar neighborhood known as the Gould belt is still a subject of debate (for a review see Pöppel 1997). Hipparcos-based maps of early-type stars show the belt as a wavy band with a maximum at about −20 deg of galactic latitude in the direction close to the anticenter and another maximum at +20 deg in the opposite direction. The belt is close to an inclined plane made of stars, H I and H II regions as well as molecular clouds. The system is in expansion and rotation and is mainly composed of stars younger than 20–40 Myrs. The supernova rate is believed to be enhanced along the belt by comparison with the galactic average and has reached 20–27 SNe per million years within the belt (Grenier 2000). A representation of the associations along the belt is shown in Fig. 3 from Perrot and Grenier (2003). Potential sources of the belt include the impact of a giant cloud (Comerón and Torra 1994), which helps to explain the inclination, a strong explosion followed by a circular shock wave (Olano 1982) which helps to explain the expansion but not very well the inclination. Olano (2001) invokes the interaction between a super-cloud and the spiral arm. A gamma burst origin has also been proposed by Perrot and Grenier (2003), in parallel

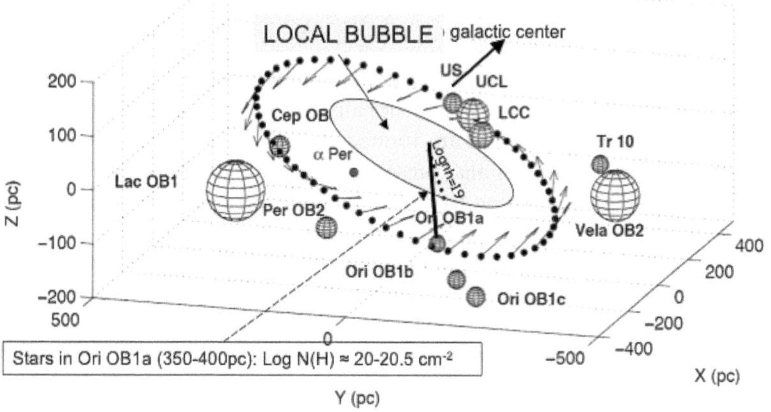

Fig. 3 A schematic view of the expanding belt, as modeled by Perrot and Grenier (2003). Despite the visual impression the longitude of the ascending order of the belt is here 296 deg (highest positive galactic latitudes at $l = 26$ deg). The LC (*solid oval*) corresponds to small H I columns, while the columns towards the OB associations defining the belt correspond to the high variability of D/H. The Sun is between the center of the belt and the Scorpius, Centaurus, Lupus associations. The "chimneys" of the LC opening to the southern and northern galactic halos are roughly perpendicular to the belt. The belt corresponds to the region of high variability of D/H

with the suggestion by Lallement et al. (2003) of a GRB origin for the Local Cavity and its inclined "chimneys" connecting the LC to the halo (Welsh et al. 1999). Both structures, the LC and the belt, may be related, and the inclination of the "chimney" axis at right angle to the may not be fortuitous belt (Sfeir et al. 1999).

2.4 A Link between the LC, the Gould Belt and the D/H Pattern?

As shown in Fig. 4, the LC, if defined by $N_H < 10^{19.3}$ cm^{-2}, is interior to the belt. This is also the region of almost constant D/H (see Fig. 1). On the other hand, belt associations like the Ori OBIa correspond to $N_H \approx 10^{20-20.5}$ cm^{-2}, a column that corresponds to the transition value between D/H constancy and high variability. In the light of the deuterium release by heated dust, the high variability transition region could be associated with the expanding front of the GB, where shocked and unshocked gas and dust do coexist (Lallement 2007).

Both a cloud impact and an explosive event can produce the Gould belt expansion and its associated stellar births and deaths, and the associated D/H variability. In the first case however two cumulative sources of deuterium enrichment are at work within the belt: dust evaporation on one hand, and mixing of galactic gas with D-rich extra-galactic gas on the second hand. Dust evaporation is accompanied by an equivalent increase of metals (Fig. 2) and a negligible or very small oxygen increase, while mixing with unastrated gas produces a strong decrease of metal and oxygen, which precludes simple proportionalities between metal abundances and both D/H and D/O. Thus, the observed linear relationships favor the first process more than the second. This is reinforced by the absence of significant O/H variations within the first 400 parsecs (Meyer et al. 1998; André et al. 2003; Oliveira et al. 2006)

3 Highly Charged Ions

The intermediate ions C IV and Si IV are preferentially found in photoionized gas within the ISM. N V and O VI are present in collisionally ionized gas found in interfaces between hot and cool gas (Slavin 2008), and are formed at temperatures of the order of 150,000–3,000,000 K. C IV and Si IV are also present in interfaces, but are formed at lower temperatures (around 80,000 K). These four ions will be found within blown up "bubbles" at locations where the gas has moderately cooled down.

The O VI ion has been detected with FUSE along a large number of target stars (Savage and Lehner 2006; Oegerle et al. 2005). The overall spatial distribution of local interstellar O VI absorption is found to be quite patchy, with both detections or non-detections for LOS of the same length or angularly close, which has been interpreted as being due to the presence of tangled or tangential magnetic fields which can quench the thermally conductive interfaces of cloud surfaces on which the O VI is thought to form (Cox and Helenius 2003). However, since there are numerous clouds around the Sun, one would expect the resulting columns to make a more homogeneous set.

C IV, Si IV, N V have been detected along numerous sight-lines >100 pc (e.g. Sembach et al. 1994; Savage et al. 2001; Gry and Jenkins 2001) with IUE and HST spectrographs, however often with too low resolution or with too many blends to allow accurate measurements of the line-widths. When there is enough resolution and the absorptions are well separated, some high ion lines are narrower than what one would expect for gas at 10^5 K or more (e.g. Knauth et al. 2003), and are likely to be due to photo-ionization.

Figure 4 shows part of HST/STIS measurements of C IV, Si IV and NV along two lines-of-sight that are known to cross both the near and far neutral interface boundaries to the Loop I cavity, in addition to intersecting the fragmented shell of neutral and partially ionized gas that defines the boundary to the Local Bubble. The LOS studied are superimposed on the low resolution dense gas distribution deduced from absorption studies (Lallement et al. 2003). While for one sampled direction there is no C IV detection in association with the distant dense gas (at heliocentric velocity around 0 km/s), for the second direction there is a clear detection, but the narrow line profile-widths, corresponding to temperatures of a few 10^4 K at maximum are found to be incompatible with theoretical models that predict high ion absorption due to the presence of evaporating cloud conduction interfaces (Welsh and Lallement 2005). Either the gas is very far from equilibrium, or these ions are produced by photo-ionization, as in the case of the Knauth et al. data. In any case, there is clearly a lack of "warm" ions from the expected interfaces dense gas and the Loop 1 cavity. On the other hand, the gas located between the LC and Loop 1 (at negative velocities) has some C IV counterpart for the 2 stars, but again when there is no blend (direction A) the line width is too narrow for a classical warm-hot interface. As a conclusion, the picture drawn from the high ions is still far from well understood and must be studied in details for each different context.

4 Charge eXchange (CX) X-Ray Emission

4.1 The Heliospheric CX Emission and its Impact on LC and Halo Hot Gas

Snowden et al. (1998, 2000) have used a large number of ROSAT $1/4$ keV shadows, to derive and map the unabsorbed fraction of the soft X-ray diffuse background, i.e. supposedly hot gas filling the LC. It is now clear that a non negligible fraction of this locally emitted soft

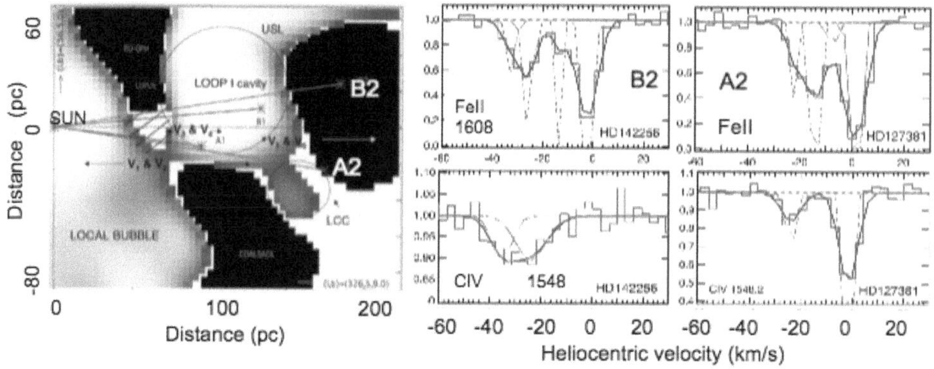

Fig. 4 High resolution HST/STIS spectra towards two target stars (*A2* (HD142256), *B2* (HD127381) on the *left graph*) within dense gas beyond Loop 1. This dense gas is at heliocentric velocity ≈0 km/s and is detected with Fe II (*top right*) for both *A2* and *B2*. Towards *B2* there is no corresponding C IV (nor Si IV) at the dense gas velocity. Towards *A2* C IV is detected, but the line-width is narrow and precludes collisional equilibrium at $T \approx 10^5$ K gas. Similarly, the negative velocity C IV detected towards target *A2*, which corresponds to the gas between the LC and Loop 1 is too narrow

X-ray emission is solar wind charge-exchange emission (or SWCX) (Cox 1998; Cravens et al. 2001; Lallement 2004; Snowden et al. 2004). From updated calculations using the best available atomic data, Koutroumpa et al. (2008) show that, while at high galactic latitude the major part of the measured brightness is definitely not from the heliosphere, at low galactic latitudes the emission generated throughout the heliosphere is potentially large enough to explain the observed signal. The whole picture of the hot gas distribution in the LC is thus to be revised, in the light of this strong CX contamination. In case the low galactic latitude emission is entirely solar, the existence of hot gas at small distances from the disk is no longer required, nor the pressure equilibrium between the local clouds and the pressure of the high latitude, because in this case the X-ray emitting hot gas and the disk gas would be physically disconnected. The actual location and characteristics of this high latitude hot gas would have to be revisited. Is this hot gas falling down towards the disk, as do the high velocity clouds (HVCs)? Is this infall a result of the large explosion having given rise to the Gould belt and created a low pressure region in the Sun vicinity? It is tempting to imagine such scenarios, but they are far too speculative at this stage and before new analyses of the hot gas and of the interfaces are performed.

4.2 Non-Solar Charge Transfer X-Ray Emission? The North Polar Spur?

The CX X-ray/EUV emission mechanism, which is extremely efficient, may be at work in other astrophysical situations, and in particular at interfaces between hot gas and cool clouds. In the same way neutral interstellar atoms enter the solar wind, while the ionized fraction is deviated and excluded from the heliosphere, neutral gas from partially ionized clouds may penetrate hot gas of stellar winds or SNR cavities. For an H atom entering a hot plasma, the mean free path against charge transfer is of the same order as the mean free path against collisional ionization for a large number of hot gas temperatures and neutral/hot gas relative velocities. The X-ray emission arising from charge transfer between neutrals and highly charged ions of the hot plasma occurs only within a very narrow layer of the order of this mean free path. However, despite this narrowness, the CX emission may be significant w.r.t. the thermal emission of hot gas, generated along large distances (Lallement 2004).

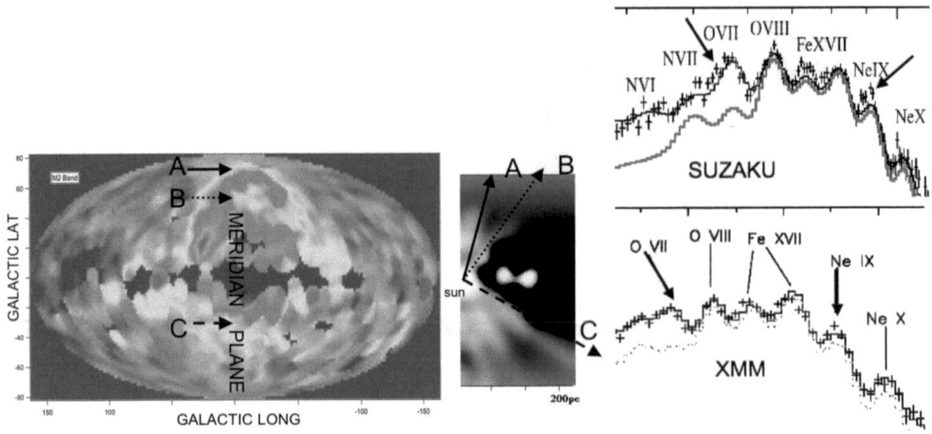

Fig. 5 (*Left*) The ROSAT 3/4 keV map reveals the conspicuous North Polar Spur (NPS) and Loop 1 emission (*grey* area between *A* and *C*). The emission discontinuities *A*, *B*, *C* correspond to lines-of-sight that are tangential to the dense gas, here shown in the meridian plane containing the galactic center (Lallement et al. 2003). (*Right*) XMM and Suzaku spectra of very bright NPS regions (Willingale et al. 2003; Miller et al. 2008). The O VIII and Ne IX emission lines seem in both cases to be shifted to lower energies, while O VIII or Ne X are not. A signature of charge-transfer emission?

Interestingly, crude estimates of the CX emission from high velocity clouds in a hot halo is of the same order as the emission detected with ROSAT (e.g. Kerp et al. 1998) towards some HVCs. In the case of stellar winds expanding within dense media, there may be CX emission lines from ion stages corresponding to the ionization degree within the shocked wind.

Is the diffuse X-ray background contaminated by the CX emission from such interfaces? We discuss here two (although weak) evidences for a positive answer, that call for further investigations.

Figure 5 shows the 3/4 keV ROSAT map, with the spectacular Loop 1/North Polar Spur (NPS) emission. While the NPS has been initially detected in radio, it corresponds to the edge of a very bright X-ray enhancement. We focus on three directions A, B, C defining three edges of the X-ray emission along a meridian line containing the galactic center. The same three directions are superimposed on a dense gas density cut in the corresponding meridian plane containing the Sun, the galactic poles and the galactic center. The density distribution comes from neutral sodium absorption data inversion (Lallement et al. 2003) and shows the shell of dense gas around Loop 1 (the Loop 1 cavity appears as a small white area within the dark region, its small size is due to the very coarse resolution of the distribution). Interestingly, the A, B and C bright edges correspond to directions that are tangential to the dense gas, as one would expect from emission arising at the periphery of the dense gas in potential agreement with a dense/hot gas interface CX emission. However, the limited precision of the density maps precludes a firm conclusion.

There is on the other hand a more direct diagnostic of the CX emission mechanism. After the electron transfer from a neutral to a high ion, and contrary to an electron impact recombination, the electron has a large probability of populating high energy shells, which strongly modifies the subsequent cascades. In the case of electron capture by helium-like ions, the de-excitation of the newly formed excited ions (e.g. N VI, O VIII, Ne IX) generates a resonance/inter-combination/forbidden line triplet considerably different in case of charge-transfer (i.e. if the electron is extracted from a neutral) compared to electron impact (i.e. for a capture of a free electron). E.g., Pepino et al. (2004) predict a 561 eV O VIII forbidden line about 4 times more intense than the 574 eV resonance line in the case of CX, while for electron impact it is twice fainter. Such line ratios, or the triplet wavelength shift in case of unresolved features, bear a clear signature of the charge transfer process, if present. E.g. the O VIII triplet is shifted by about -10 eV and the Ne IX triplet by about -17 eV for XMM and Suzaku spectral resolutions. Unfortunately, such diagnostics are difficult to obtain in case of extended sources and at low resolution.

In the case of the Loop 1, it is interesting to compare the highest quality NPS spectra recorded by XMM (Willingale et al. 2003) and recently Suzaku (Miller et al. 2008) with the model spectra. Figure 5 shows fractions of the recorded spectra at low energy. The O VIII and Ne IX triplets seem indeed to be shifted w.r.t. to the model, by about 10–15 eV, while the O VIII and Ne X lines are well centered. The two Fe XVII lines also seem to be shifted, as it should probably be in case of CX, but there are at present no calculations of the charge transfer shifts for these complex ions. These weak evidences for additional distant CX emission deserve more investigations. If present, interface CX emission should modify the overall picture of the hot gas distribution.

5 Data Need for a Better Global Picture of the LISM

Understanding the history of the local ISM, the origin of the expanding Gould belt, the link between the GB and the LC, the nature of the gas in the LC and the transition to the halo, and finally the ion and X-ray data is a difficult task, presently in progress as shown during this meeting. Is the LC a remnant of an explosive event that gave rise to the Gould belt? Is this region under-pressured, which induces halo gas down-flow, as suggested by high and intermediate velocity clouds negative velocities? Or is it on the contrary a typical, old, reheated SNR bubble? Answering these questions requires:

(i) to disentangle thermal vs. charge-exchange X-ray emission, which requires new generation high-resolution X-ray spectra (McCammon et al. 2002)
(ii) to better characterize the low and high ionization degree ions thanks to high resolution FUV/UV spectra
(iii) to build detailed 3D density and velocity distributions of dense and diffuse gas, thanks to combinations of absorption and emission data. This requires much larger data sets than available ones
(iv) to locate the high ions in the gas and velocity distribution and correlate them with shocks, interfaces, cavities
(iv) to compare abundance variations, such as deuterium, with the same features
(v) to compare the 3D picture with sophisticated models of ISM evolution, including magnetic fields

This is an ambitious program for a research field which is not at high priority, but its achievement would be useful in many ways as discussed in the introduction.

References

M.K. André et al., Astrophys. J. **591**, 1000 (2003)
F. Comerón, J. Torra, Astron. Astrophys. **281**, 35 (1994)
D.P. Cox, in *Lecture Notes in Physics*, vol. 506 (Springer, Berlin, 1998), p. 121
D.P. Cox, L. Helenius, Astrophys. J. **583**, 205 (2003)

T.E. Cravens, I.P. Robertson, S.L.J. Snowden, Geophys. Res. **106**(A11), 24883 (2001)
M. De Avillez, Space Sci. Rev. (2008), this issue
M. De Avillez, D. Breitschwerdt, Astron. Astrophys. **436**, 585 (2005)
B.T. Draine, in *Origin and Evolution of the Elements*, ed. by A. McWilliams, M. Rauch (Cambridge Univ. Press, Cambridge, 2004), p. 320
S.L. Ellison, J.X. Prochaska, S. Lopez, Mon. Not. R. Astron. Soc. **380**, 1245 (2007)
P.C. Frisch, Space Sci. Rev. (2008), this issue. doi: 10.1007/s11214-008-9394-4
B. Fuchs, Space Sci. Rev. (2008), this issue. doi:10.1007/s11214-008-9427-z
J. Geiss, G. Gloeckler, C. Charbonnel, Astrophys. J. **578**, 863 (2002)
I. Grenier, Astron. Astrophys. **364**, L93 (2000).
C. Gry, E.B. Jenkins, Astron. Astrophys. **367**, 617 (2001)
G. Hébrard, T.M. Tripp, P. Chayer et al., Astrophys. J. **635**, 1136 (2005)
E.B. Jenkins, Space Sci. Rev. (2008), this issue. doi:10.1007/s11214-008-9352-1
M. Jura, in *Advances in Ultraviolet Astronomy*, ed. by Y. Kondo. vol. CP-238 (NASA, 1982), p. 54
J. Kerp, J. Pietz, P.M.W. Kalberla et al., in *Lect. Notes in Phys.*, vol. 506 (1998), p. 457
D. Koutroumpa, R. Lallement, V. Kharchenko, A. Dalgarno, Space Sci. Rev. (2008), this issue
D. Knauth, C. Howk, K. Sembach, J. Lauroesch, D. Meyer, Astrophys. J. **592**, 964 (2003)
R. Lallement, Astron. Astrophys. **422**, 391 (2004)
R. Lallement, Space Sci. Rev. **130**, 341 (2007)
R. Lallement, B.Y. Welsh, J.L. Vergely, F. Crifo, D. Sfeir, Astron. Astrophys. **411**, 447 (2003)
R. Lallement, G. Hebrard, B.Y. Welsh, Astron. Astrophys. **481**, 381 (2008)
Y. Liebert, E. Gérard, T. Le Bertre, Mon. Not. R. Astron. Soc. **380**, 1161 (2007)
J. Linsky, B. Draine, W. Moos et al., Astrophys. J. **647**, 1106 (2006)
D. McCammon et al., Astrophys. J. **576**, 188 (2002)
D.M. Meyer, M. Jura, J.A. Cardelli, Astrophys. J. **493**, 222 (1998)
E.D. Miller, H. Tsunemi, M. Bautz et al., Publ. Astron. Soc. Jpn. **60**(SP1), S95 (2008)
W. Oegerle, E.B. Jenkins, R. Shelton, D. Bowen, P. Chayer, Astrophys. J. **622**, 377 (2005)
C.A. Olano, Astron. Astrophys. **112**, 195 (1982)
C.A. Olano, Astron. J. **121**, 295 (2001)
C.M. Oliveira, H.W. Moos, P. Chayer, J.W. Kruk, Astrophys. J. **642**, 283 (2006)
C. Perrot, I. Grenier, Astron. Astrophys. **404**, 519 (2003)
R. Pepino, V. Kharchenko, A. Dalgarno, R. Lallement, Astrophys. J. **617**, 1347 (2004)
W.G.L. Pöppel, Fundam. Cosm. Phys. **18**, 1 (1997)
S. Redfield, Space Sci. Rev. (2008), this issue
B. Savage, N. Lehner, Astrophys. J. Supp. Ser. **162**, 134 (2006)
B.D. Savage, M.R. Meade, K.R. Sembach, Astrophys. J. Supp. Ser. **136**, 631 (2001)
K.R. Sembach, B.D. Savage, E.B. Jenkins, Astrophys. J. **421**(2), 585 (1994)
D.M. Sfeir, R. Lallement, F. Crifo, B.Y. Welsh, Astron. Astrophys. **346**, 785 (1999)
R. Shelton, Space Sci. Rev. (2008), this issue
J.D. Slavin, Space Sci. Rev. (2008), this issue. doi:10.1007/s11214-008-9342-3
J.D. Slavin, P.C. Frisch, Astrophys. J. **565**, 364 (2002)
S.L. Snowden, R. Egger, D.P. Finkbeiner, M.J. Freyberg, P.P. Plucinsky, Astrophys. J. **493**, 715 (1998)
S.L. Snowden, M.J. Freyberg, K.D. Kuntz, W.T. Sanders, Astrophys. J. **128**, 171 (2000)
S.L. Snowden, M.R. Collier, K.D. Kuntz, Astrophys. J. **610**, 1182 (2004)
S. Stanimirovic, Space Sci. Rev. (2008), this issue. doi:10.1007/s11214-008-9363-y
J. Wareing, A.A. Zijlstra, T.J. O'Brien, M. Seibert, Astrophys. J. Lett. **670**, L125 (2007)
B.Y. Welsh, R. Lallement, Astron. Astrophys. **436**, 615 (2005)
B.Y. Welsh, D.M. Sfeir, M.M. Sirk, R. Lallement, Astron. Astrophys. **352**, 308 (1999)
R. Willingale, A.D.P. Hands, R.S. Warwick, S.L. Snowden, D.N. Burrows, Mon. Not. R. Astron. Soc. **343**(3), 995 (2003)
B.E. Wood, V.V. Izmodenov, Y.G. Malama, Space Sci. Rev. (2008), this issue. doi:10.1007/s11214-008-9369-5

Origin of the Local Bubble

**Burkhard Fuchs · Dieter Breitschwerdt ·
Miguel A. de Avillez · Christian Dettbarn**

Originally published in the journal Space Science Reviews, Volume 143, Nos 1–4, 437–448.
DOI: 10.1007/s11214-008-9427-z

Abstract We present a new unbiased search for OB associations in the Solar neighbourhood which have hosted the progenitor stars of the core collapse supernovae responsible for the Local Bubble in the interstellar gas. For this purpose we have analyzed a volume complete set (with a diameter of 400 pc) of B stars drawn from the Hipparcos catalogue and the Arivel data base, from which candidate members were selected by a kinematical criterion. After careful dereddening the star colours we have constructed a colour-magnitude diagram and confirmed that the Upper Scorpius, Upper Centaurus Lupus, and Lower Centaurus Crux subgroups of the Sco OB2 association are the youngest nearby OB associations. We dated their ages with theoretical isochrones in the range of 20–30 Myr, in agreement with previous work. We have traced backwards in time the paths of the stars and found that they entered the volume of the present bubble at 10 to 15 Myr ago. We argue that the Local Bubble began to form then and estimate that 14 to 20 supernovae have exploded since. The implied energy input into the ambient medium can be shown to be sufficient to excavate a bubble of the presently observed size.

Keywords Open clusters and associations: Sco OB2 · ISM: local bubble

1 Introduction

The Sun is located inside the Local Bubble (LB) in the interstellar medium (ISM), a low density X-ray emitting cavity which is deficient of HI. Sfeir et al. (1999) and Lallement et

B. Fuchs (✉) · C. Dettbarn
Astronomisches Rechen-Institut am Zentrum für Astronomie der Universität Heidelberg, Mönchhofstraße 12-14, 69120 Heidelberg, Germany
e-mail: fuchs@ari.uni-heidelberg.de

D. Breitschwerdt
Institut für Astronomie der Universität Wien, Türkenschanzstraße 17, 1180 Wien, Austria
e-mail: breitschwerdt@astro.univie.ac.at

M.A. de Avillez
Department of Mathematics, University of Evora, R. Romao Ramalho 59, 7000 Evora, Portugal

al. (2003) have delineated the outer boundaries of the LB by studying NaI absorption lines of the ISM in the spectra of stars for which trigonometric parallaxes obtained in the Hipparcos mission (ESA 1997) were available. The outer walls of the LB show up as a sudden increase of the equivalent widths of the NaI lines, when the lines of sight towards these stars are passing through the walls. According to these measurements the LB extends to about 200 pc in the Galactic plane, and 600 pc perpendicular to it, but is inclined about 20 degrees relative to the direction of the Galactic pole. Thus the LB has an orientation perpendicular to Gould's Belt.

The physical status of the ISM within the LB has been observed in X-rays and by EUV spectroscopy. However, the interpretation of the observations is controversial, all the more the X-ray observations are complicated by charge exchange reactions between solar wind ions (SWCE) and heliospheric gas (Lallement 2004). We do not go into any details here but refer to the accompanying paper by Breitschwerdt et al. (2008).

The LB is situated next to the Loop I superbubble, from which it is separated by a wall showing up in absorption of soft X-rays in ROSAT PSPC images (Egger and Aschenbach 1995).

It has been suggested early that the LB has been excavated by core collapse supernovae (Cox and Anderson 1982; Innes and Hartquist 1984; Smith and Cox 2001) even though the LB does not contain any OB associations which might have harboured the progenitors of the SNII.

The search for the "smoking gun" of the origin of the LB proved partially successful by discovering that moving groups of young stars in the solar neighbourhood could provide an adequate number of SN explosions while crossing the path of the LB. Berghöfer and Breitschwerdt (2002, henceforth BB02) calculated the trajectory of the Pleiades subgroup B1 backwards in time, and found that 19 SNe could have exploded between 10–20 Myr ago in the region that is occupied by the LB. The remaining stars of B1 are now part of the Scorpius Centaurus OB association. It could be shown that this is in good agreement with the size of the LB and the present soft X-ray emissivity. A similar analysis was carried out by Maíz-Apellániz (2001), who calculated backwards in time the trajectories of Sco Cen subgroups and claimed that about 6 SNe that went off in the Lower Centaurus Crux subgroup of the Sco OB2 association 7 to 9 Myr ago formed the LB.

However, the search for possible OB associations was not very systematical, mainly because their discovery history is unclear. Thus Fuchs et al. (2006) were prompted to carry out a fresh and unprejudiced search and scrutinized *all stars* that are within a volume of about 400 pc in diameter centered around the Sun, and performed a selection according to spectral and kinematical properties. The latter is based on three dimensional space velocities of the stars. Thus our approach is complementary to studies like by de Zeeuw et al. (1999) which are based on proper motions alone. Sartori et al. (2003) did include radial velocities when analyzing the subgroups of the Sco OB2 association, but work from a list of stars preselected by de Zeeuw et al. (1999). From the positions of B stars in the HR diagram and the main sequence's turnoff point we could reliably determine the age of the associations and the number of SNe within the volume occupied at present by the Local Bubble.

2 Search for Nearby OB Associations

Fuchs et al. (2006) tried to improve and harmonize previous studies. As a starting point we used the Hipparcos catalogue (ESA 1997) from which we selected all stars bluer than $(B - V) < -0.05$ with parallaxes larger than 5 milliarcseconds, because otherwise the distances would be to inaccurate. Drimmel et al. (2000) find by a comparison with the Tycho catalogue that the Hipparcos catalogue is about 97 percent complete down to $V = 7.5$ which corresponds at a distance of $1/(5\text{ mas}) = 200$ pc to an absolute magnitude of $M_V = 1$. Allowing for an extinction of $A_V \leq 0.2$ we have $M_V < 1.2$, and we have chosen the colour cut in $(B - V)$ accordingly (cf. Fig. 4). We have obtained in this way an unbiased complete sample of 762 B stars within a distance of 200 pc from the Sun. All stars in our list appear in the Arivel data base, which combines Hipparcos parallaxes and proper motions with radial velocity data collected from the literature (Wielen et al., in preparation). We find that for 610 stars radial velocities are available. These are accurate enough to be used for a kinematical analysis (cf. Fig. 2). We have tested if the availability of radial velocities is correlated with the colours of the stars and find that the 152 out of 762 stars for which no radial velocity is known are redder than $(B - V) > -0.1$. Fuchs et al. (2006) could show that this does not have any significant effect on the expected number of supernovae. For each star, spatial X, Y, and Z coordinates and the corresponding U, V, and W velocity components have been calculated. The coordinates are centered on the Sun with X pointing towards $l = 0°$, $b = 0°$, Y towards $l = 90°$, $b = 0°$, and Z towards $b = 90°$, respectively. Since OB associations disperse slowly on time scales of the order of several 10^7 years (Blaauw 1964), we have searched for kinematically coherent structures in our sample. For this purpose we have traced the paths of the stars backward in time over 3×10^7 years. Stellar orbits have been calculated using the epicyclic equations of motion for the stars in the sample (Lindblad 1959).

$$\begin{aligned}
X(t) &= X(0) - \frac{V(0)}{-2B}(1 - \cos(\kappa t)) + \frac{U(0)}{\kappa}\sin(\kappa t),\\
U(t) &= U(0)\cos(\kappa t) - \frac{\kappa}{-2B}V(0)\sin(\kappa t),\\
Y(t) &= Y(0) + 2A\left(X(0) - \frac{V(0)}{-2B}\right)t\\
&\quad + \frac{\Omega_0}{-B\kappa}V(0)\sin(\kappa t) + \frac{2\Omega_0}{\kappa^2}U(0)(1 - \cos(\kappa t)),\\
V(t) &= \frac{-2B}{\kappa}U(0)\sin(\kappa t) + V(0)\cos(\kappa t),\\
Z(t) &= \frac{W(0)}{\nu}\sin(\nu t) + Z(0)\cos(\nu t),\\
W(t) &= W(0)\cos(\nu t) - Z(0)\nu\sin(\nu t).
\end{aligned} \tag{1}$$

Here κ denotes the epicyclic frequency, $\kappa = \sqrt{-4\Omega_0 B}$. A and B are the Oort constants, and Ω_0 is the angular frequency of the rotation of the local standard of rest around the Galactic centre, $\Omega_0 = V_{LSR}/R_\odot$. ν denotes the vertical oscillation frequency which is related to the local density ρ_0 by the Poisson equation as $\nu = \sqrt{4\pi G\rho_0}$, where G is the constant of gravitation. For the angular velocity of the local standard of rest we have adopted a value of $\Omega_0 = 220$ km/s/8 kpc. The Oort constants describe in (1) the smooth Galactic gravitational potential, which is consistent with an essentially flat shape of the local Galactic rotation curve, $A = -B = \Omega_0/2$ (Feast and Whitelock 1997). For the local density we adopt a value of $\rho_0 = 0.1\ \mathcal{M}_\odot/\text{pc}^3$ (Holmberg and Flynn 2004). These parameter values imply $\kappa = 0.039$ km/s/pc $= 4 \times 10^{-8}\ \text{yr}^{-1}$ and $\nu = 0.074$ km/s/pc $= 7.5 \times 10^{-8}\ \text{yr}^{-1}$. In Fig. 1 we show the positions of the stars today and 3×10^7 years ago. Most of the 610 stars do not belong to the OB association, which hosted the SNe responsible for the origin of the

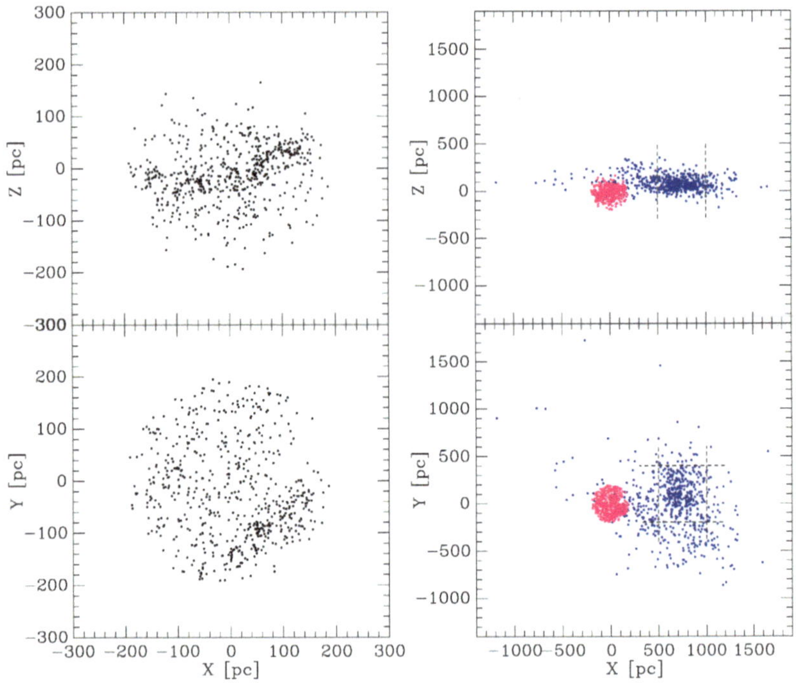

Fig. 1 *Left panel*: Positions of 610 stars drawn from the Hipparcos catalogue. The selected stars have colours $B - V < -0.05$ and for each star its radial velocity is known. The X-axis points towards the Galactic centre, Y into the direction of Galactic rotation, and Z towards the Galactic north-pole, respectively. *Right panel*: Positions of the originally selected stars today (*pink*) and 3×10^7 years ago (*blue*). The Sun is at rest in the diagrams. Stars lying in the windows indicated by *dashed lines* are identified as putative members of the searched for OB association

LB, and have space velocities different from the velocity of the association. Thus they are dispersed away into a wide cloud. However, the overdense regions in Fig. 1 indicate that there is a considerable number of stars which stayed together. The larger size of the overdense regions in the back projected sample compared to its size today is obviously due to the observational errors. The typical accuracy of Hipparcos proper motions is about 1 mas/yr which corresponds at a distance of 100 pc to a velocity of 0.5 km/s, whereas the accuracy of the radial velocities is several km/s, dominating the errors of the resulting space velocities. Assuming for the latter in each component an error of 5 km/s (cf. Fig. 2) we expect errors in the star positions 3×10^7 years ago of about 150 pc in each direction. Thus, if the overdensity of stars representing the kinematically homogeneous group is today occupying an area in X and Y of 200×200 pc (cf. Fig. 2), it will have been artificially spread out to a size of roughly 500×500 pc in X and Y. As can be seen from Fig. 1 there is an outer shroud of stars which lies at greater distances from the core of the overdensity. These must be stars with genuinely different space velocities from the kinematically homogenous group of stars. We identify this kinematically homogeneous group of stars as an OB association and select 302 stars lying in the windows indicated as dashed lines in Fig. 1. As expected these stars are more or less closely related to the Sco OB2 association.

In Fig. 2 we show the present day velocity distribution of the 302 selected stars. Since the velocity dispersion of an OB association is of the order of 10 km/s (Blaauw 1964) or even less (Kamaya 2004), we make a second selection indicated by windows drawn as dashed

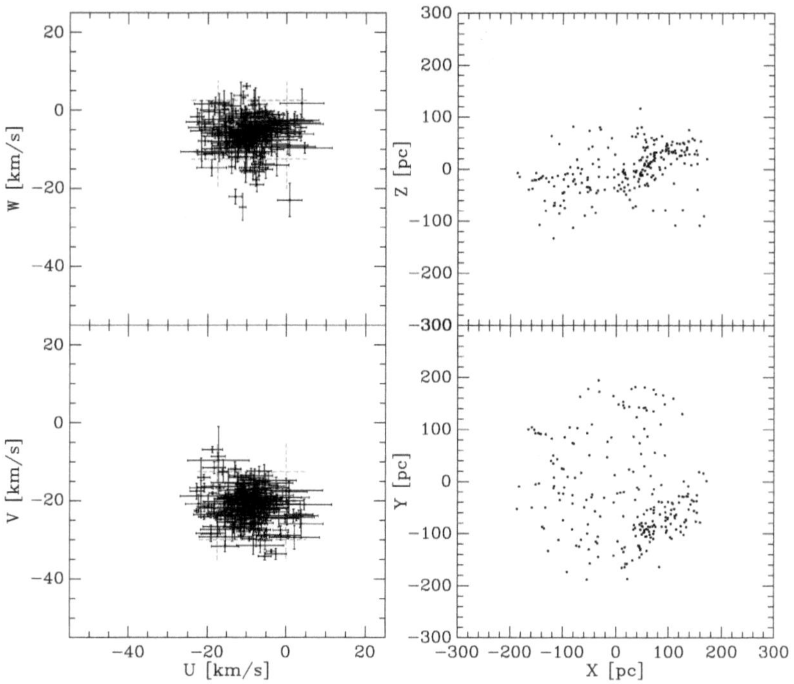

Fig. 2 *Left panel*: Present day velocity distribution of the 302 selected stars. A second selection is made of the stars lying in the windows indicated by *dashed lines*. *Right panel*: Spatial distribution of the finally 236 selected stars

lines in Fig. 2. This leaves a sample of 236 stars which we analyze in the following. Their spatial distribution is shown in the right panel of Fig. 2 where the Sco OB2 association is now clearly discernible. In Fig. 3 we reproduce a colour image of parts of the association.

The final sample is shown as a colour-magnitude diagram in Fig. 4. For this purpose we had to cross-identify the sample stars in the Geneva photometry data base (Mermilliod et al. 2000) and replaced the $(B-V)_T$ colours given in the Hipparcos catalogue by $(B-V)_J$ colours, because they can be then directly compared with theoretical isochrones available in the literature. In the colour range, which we consider here, $B-V$ given in the Tycho system can not be transformed directly to the Johnson system (ESA 1997). The absolute magnitudes have been determined from the visual magnitudes given in the Hipparcos catalogue in the Johnson system.

We have compared our sample with the extensive membership list of the Sco OB2 association compiled by de Zeeuw et al. (1999) who applied a combination of a modified convergent point method and the so called spaghetti method (Hoogerwerf and Aguilar 1999) to Hipparcos data. Of particular interest turned out the membership lists of the subgroups Upper Scorpius (US), Upper Centaurus Lupus (UCL), and Lower Centaurus Crux (LCC). With only very few exceptions all stars in the membership lists, which fulfill our colour selection criterion, appear also in our sample, which gives confidence in our selection procedure. A few stars from our final sample could be identified additionally in the membership list of de Geus et al. (1989) as members of the subgroups. The 79 stars common to both lists are colour coded in the right panel of Fig. 4. Of these stars 10 are belonging to the US, 42 to the UCL, and 27 stars belong to the LCC subgroup.

Fig. 3 Optical image of the Upper Scorpius (*left*) and Upper Centaurus Lupus (*right*) subgroup of the Sco OB2 association (copyright: T. Preibisch, Bonn)

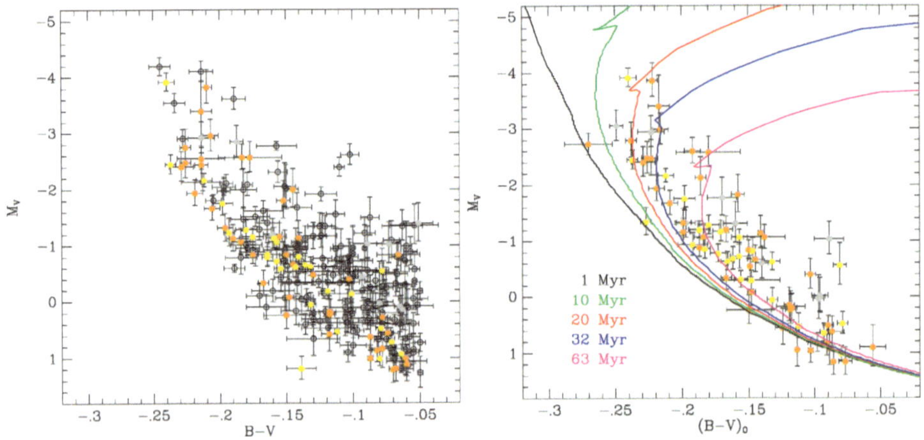

Fig. 4 *Left panel*: Colour-magnitude diagram of the final sample. Members of the Upper Centaurus Lupus subgroup of Sco OB2 are highlighted in *orange*, Lower Centaurus Crux in *yellow*, and Upper Scorpius in *grey*, respectively. *Right panel*: Dereddened colour-magnitude diagram of the members of the Upper Scorpius (*grey*), Upper Centaurus Lupus (*orange*) and Lower Centaurus Crux (*yellow*) subgroups. The *solid lines* are theoretical isochrones colour coded according to their ages

3 Results and Discussion

3.1 The Search for the "Smoking Gun"

The colour-magnitude diagram presented in the left panel of Fig. 4 shows a clearly discernible main sequence, which is particularly well delineated by the members of the UCL group. The turn-off point at the tip is defined by both the members of the UCL and the LCC subgroups. Apparently these together with the US subgroup are indeed the youngest OB associations in the solar neighbourhood (de Geus et al. 1989; Sartori et al. 2003). In order to determine the main sequence turnoff point we have compared the colour-magnitude diagram with theoretical isochrones calculated by Schaller et al. (1992) for solar metallicities. Fortunately de Bruijne (1999) and Sartori et al. (2003) have determined individually for most members of the US, UCL, and LCC subgroups, respectively, the extinction and colour excess by comparing the observed $(V - I)_C$ colours with the intrinsic colours of stars of the same spectral type and luminosity class. Dereddened data of the 79 stars are shown together with isochrones in the right panel of Fig. 4. We conclude from Fig. 4 that the ages of the UCL and LCC subgroups lie in the range of 20 to 30 Myr, whereas our data do not allow to date the age of the US subgroup. We note that this estimate of the ages of the subgroups is nearly twice of that of de Geus et al. (1989), who determined an age of 11–12 Myr of the LCC subgroup and 14–15 Myr of the UCL subgroup, respectively. These age estimates were revised by Sartori et al. (2003) to 16–20 Myr on the basis of the then newly available Padova isochrones (Bertelli et al. 1994). The Schaller et al. (1992) isochrones, which we used, are an upgrade of the Maeder (1981a, 1981b, 1981c) isochrones by the Geneva group. Moreover we note that Sartori et al. (2003) have adopted for the majority of their stars the spectral types given in the Hipparcos catalogue, which led to a very coarse binning in the $(B - V)_0$ colour. Given these uncertainties we conclude that our age datings of the LCC and UCL subgroups are consistent with Sartori et al.'s result. This agrees also well with the age of Pleiades subgroup B1, which was suggested to be responsible for the origin of the LB by BB02, but is significantly larger than assumed by Maíz-Apellániz (2001), especially for the LCC subgroup. Moreover, we have examined with the help of the Simbad data base each star of the subgroups lying not on the main sequence and found that practically all these stars are either binaries or peculiar in the sense that they are variable, emission line stars etc., so that their position off the main sequence in the colour-magnitude diagram shown in Fig. 4 can be explained in our interpretation by such effects.

In Fig. 5 we trace back the positions of the UCL and LCC subgroup members over the last 30 Myr using again the epicycle equations (1). However, we have not used the individual space velocities of the stars, but adopted for each star the mass-weighted mean velocity of the combined subgroups. This avoids any unphysical spread of the spatial distribution of the stars at earlier times due to the errors of the space velocity components of the stars. We estimate that the center of the combined UCL and LCC sample is defined with an accuracy of 17×17 pc in X, Y. To the mean velocity of the stars we have added the solar motion $(U, V, W)_\odot = (10, 5.3, 7.2)$ km/s (Dehnen and Binney 1998), so that the orbits are calculated in the reference frame of the local standard of rest. Since interstellar gas has usually only small peculiar motions, the local interstellar gas, and with it the LB which is indicated in Fig. 5 by the contour line taken from Lallement et al. (2003), will basically corotate with the local standard of rest around the Galactic centre. This means that the LB is at rest in the reference frame of Fig. 5. As can been seen from Fig. 5 the path of the association has aligned itself 15 Myr ago nearly parallel to the tangential Y-direction, the direction of Galactic rotation. Remnants of supernova explosions occurring during this period will have experienced very little shear due to the differential rotation of the Galaxy. The shear effect is described quantitatively by the term linearly proportional to time in the epicycle equation for $Y(t)$ (cf. (1)). The coefficient $X(0) - \frac{V(0)}{-2B}$ defines the mean guiding centre radius of an orbit. If the spread of these radii is small, as was the case in the last 15 Myr, the shear effect of the Galactic differential rotation is minimized. In our view this might well explain why all supernovae occurring during that time have combined together to form the LB, while supernova remnants formed at earlier times have drifted away (cf. Fig. 5).

Next we compare in the right panel of Fig. 5 the position of the Upper Centaurus Lupus and Lower Centaurus Crux associations relative to the LB today and at earlier times and also with the present day LB contours in meridional sections through the bubble. From the left panel of Fig. 5 we estimate the Galactic longitude in which direction we expect the associations to approach. Choosing then the appropriate meridional section through the bubble

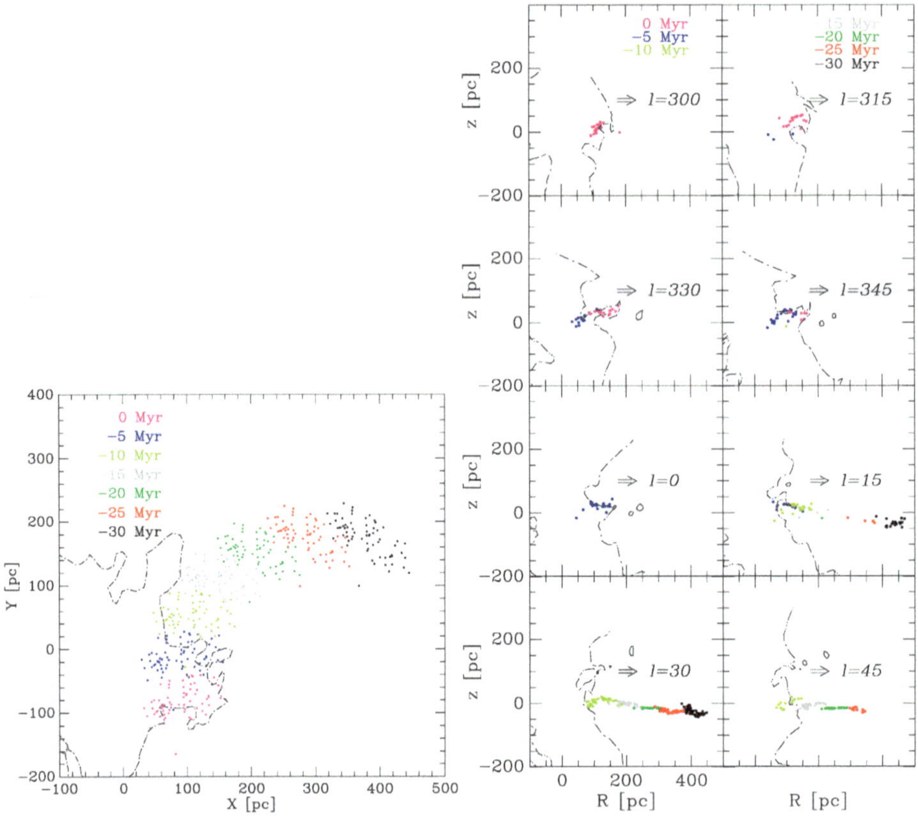

Fig. 5 *Left panel*: Path of the Upper Centaurus Lupus and Lower Centaurus Crux associations over the last 30 Myr projected onto the Galactic plane. The look-back time is colour coded. The orbits are calculated backwards in the reference frame of the local standard of rest assuming for each star the same mass-weighted mean velocity of the stars. The position of the Local Bubble is indicated by the *dash-dotted contour* line and is at rest in this reference frame. *Right panel*: Meridional sections of the contours delineating the outer boundary of the Local Bubble together with the positions of the stars in the Upper Centaurus Lupus and Lower Centaurus Crux associations. The *horizontal axis* in the *upper left panel* points into the direction $l = 300°$, in the *upper right* towards $l = 315°$ and so on. The vertical direction is always perpendicular to the Galactic midplane

from the paper by Lallement et al. (2003), we can determine immediately the positions of the stars in that longitude range relative to the LB. As can be seen from the upper right panels of Fig. 5 the associations are today just about to exit the bubble. Five and 10 Myr ago they were inside. The bottom right panel of Fig. 5 indicates that they entered 15 Myr ago the region occupied by the LB today. In this scenario the LB was starting to form about 15 Myr ago, which is consistent with the estimates of the age of the LB by Maíz-Apellániz (2001) and BB02. Lallement (2007, private comm.) has announced an upgrade of the contours of the LB's outer walls based on an extended set of target stars. It will be interesting to see how the new data will support our arguments.

One might wonder how realistic a scenario is, in which the SNe explode rather close to the edge of the present day bubble. As the high resolution simulation of Fuchs et al. (2006) shows, the location of the star cluster with respect to the centre of the bubble is not crucial. The bubble expands always fastest in the direction of the lowest ambient density and pressure. Since in the direction of the Galactic centre the Loop I superbubble was formed almost at the same time as the LB by SNe exploding within the Sco–Cen association, the pressure in this direction is very high. Hence the LB was forced to expand rather towards the anticentre direction and perpendicular to the plane, in agreement with the observations.

3.2 How Many Supernovae and Were They Enough to Create the LB?

Following BB02, the number of supernovae, which have exploded during the journey of the OB association through the LB, can be estimated with the aid of the initial mass function (IMF, cf. BB02). We have fitted an IMF of the form (Massey et al. 1995).

$$\frac{dN}{d\mathcal{M}} = \frac{dN}{d\mathcal{M}}\bigg|_0 \mathcal{M}^{\Gamma-1} \tag{2}$$

with an index $\Gamma = -1.1 \pm 0.1$ to the data. Masses are given in units of solar masses. The lower end of the main sequence at $M_V = 1$ mag corresponds to A0 stars with masses of $\mathcal{M}_l = 2.6\ \mathcal{M}_\odot$ and the upper tip at $M_V = -3.7$ mag to B0 stars with masses of $\mathcal{M}_u = 8.2\ \mathcal{M}_\odot$ (Schaller et al. 1992), respectively. The total number of stars in the UCL and LCC associations, respectively, allow the determination of the normalization constants,

$$N = \int_{2.6}^{8.2} \frac{dN}{d\mathcal{M}}\bigg|_0 \mathcal{M}^{-2.1}\, d\mathcal{M} = 0.228\, \frac{dN}{d\mathcal{M}}\bigg|_0, \tag{3}$$

implying $\frac{dN}{d\mathcal{M}}|_0 = 184$ for the 42 UCL and $\frac{dN}{d\mathcal{M}}|_0 = 118$ for the 27 LCC stars, respectively. As we have shown in the previous section, OB stars entered the LB region 10–15 Myr ago, setting the clock for its origin to $t = 0$. From a further fit to Schaller et al.'s isochrone data we estimate that the main sequence life time of such bright stars scales with mass as

$$\tau = \tau_0\, \mathcal{M}^{-\alpha} \quad (8\, \mathcal{M}_\odot \le \mathcal{M} \le 67\, \mathcal{M}_\odot) \tag{4}$$

with $\tau_0 = 1.6 \times 10^8$ yr and $\alpha = 0.932$. This means that the masses of the most massive stars $\mathcal{M}_{\Delta\tau}$ in the associations at a lookback time of $\Delta\tau$ years ago are given by

$$\mathcal{M}_{\Delta\tau} = \left(\mathcal{M}_u^{-\alpha} - \frac{\Delta\tau}{\tau_0}\right)^{-\frac{1}{\alpha}}, \tag{5}$$

implying $\mathcal{M}_{10} = 15.4\, \mathcal{M}_\odot$ if $\Delta\tau = 10$ Myr or $\mathcal{M}_{15} = 26.6\, \mathcal{M}_\odot$ if $\Delta\tau = 15$ Myr depending on the entry time of the associations into the volume occupied by the LB today. The expected number of supernovae, i.e. the number of 'missing' stars, is then calculated by

$$N_{SN} = \int_{8.2}^{\mathcal{M}_{\Delta\tau}} \frac{dN}{d\mathcal{M}}\bigg|_0 \mathcal{M}^{-2.1}\, d\mathcal{M}. \tag{6}$$

We thus obtain estimates of $N_{SN} = 8$–12 from the Upper Centaurus Lupus and $N_{SN} = 6$–8 from the Lower Centaurus Crux associations, respectively. The estimate of 14 to 20 supernovae, which created the LB, is in good agreement with the value determined by BB02. Extrapolating the IMF to masses beyond $\mathcal{M}_{\Delta\tau}$ we estimate that 12 to 5 SNe exploded before the associations entered the present LB volume.

In order to assess the question, whether the estimated number of supernovae would suffice to excavate the LB, we consider the energy input by the supernova explosions into the interstellar gas. According to the initial mass function (2) there are

$$dN = \frac{dN}{d\mathcal{M}}\bigg|_0 \mathcal{M}^{\Gamma-1}\, d\mathcal{M} \tag{7}$$

stars in the mass range $(\mathcal{M}, \mathcal{M} + d\mathcal{M})$ with main sequence life times $(\tau, \tau - d\tau)$. Thus

$$dN = \frac{dN}{d\mathcal{M}}\bigg|_0 \mathcal{M}^{\Gamma-1} \left(-\frac{d\mathcal{M}}{d\tau}\right) d\tau, \tag{8}$$

and the energy input rate is given by (cf. BB02)

$$\dot{\mathcal{E}}_{\rm SN} = \frac{d}{dt}\mathcal{E}_{\rm SN} N_{\rm SN} = \mathcal{E}_{\rm SN}\frac{dN_{\rm SN}}{dt} = \mathcal{E}_{\rm SN}\frac{dN}{d\mathcal{M}}\bigg|_0 \mathcal{M}^{\Gamma-1}(-1)\frac{d}{d\tau}\left(\frac{\tau}{\tau_0}\right)^{-\frac{1}{\alpha}}\frac{d\tau}{dt}, \tag{9}$$

where $\mathcal{E}_{\rm SN}$ denotes the energy released by a single supernova, $\mathcal{E}_{\rm SN} = 10^{51}$ ergs. According to the way we have set up (8) $\frac{d\tau}{dt}$ is equal to 1. Equation (9) describes the trade-off of the increasing number of supernova progenitors and their increasing main-sequence life times with decreasing mass. Inserting the age-to-mass relation (4) into (9) leads then to

$$\dot{\mathcal{E}}_{\rm SN} = \dot{\mathcal{E}}_{\rm SN0}\, t_7^{-\frac{\Gamma+\alpha}{\alpha}} \tag{10}$$

with t_7 defined as $t_7 = t/10^7$ yr. For the constant $\dot{\mathcal{E}}_{\rm SN0}$ we find

$$\dot{\mathcal{E}}_{\rm SN0} = \frac{\mathcal{E}_{\rm SN}}{\alpha\tau_0}\frac{dN}{d\mathcal{M}}\bigg|_0 \left(\frac{10^7 yr}{\tau_0}\right)^{-0.1803} = 3.5\times 10^{35}\frac{dN}{d\mathcal{M}}\bigg|_0 \text{ erg/s}. \tag{11}$$

Equation (10) shows a rather weak decline of the supernova energy input rate into the LB as a result of partial compensation between the increasing number of stars with decreasing mass and a corresponding increase in main sequence life time.

Winds from hot massive stars can potentially contribute significantly to the energy input into the interstellar gas (Lamers and Cassinelli 1999). However we do not expect this for the stars which we consider here. B stars with masses below 30 $\mathcal{M}_\odot$ have negligible main sequence winds. According to Maeder and Meynet (1987) a star with an initial mass of 20 $\mathcal{M}_\odot$ loses then during the supergiant phase with a duration of 1.3 Gyrs mass at a rate of about $10^{-5.5}$ $\mathcal{M}_\odot$ yr^{-1}, or in total 5.7 $\mathcal{M}_\odot$. Assuming as an upper limit a terminal speed of the stellar wind of 100 km/s this implies an energy input into the ISM of 5.7×10^{47} ergs which has to be compared with the energy release by a supernova of 10^{51} ergs.

BB02 have derived a bubble wind equation which describes the growth of the size of the bubble with time. The radius of the bubble is given by

$$R_{\rm b}(t) = R_{\rm b0}\, t^{\frac{2\alpha-\Gamma}{5\alpha}} \tag{12}$$

with the constant

$$R_{\rm b0} = \left[\frac{475\alpha}{(4\alpha-7\Gamma)(3\alpha-4\Gamma)}\right]^{\frac{1}{5}} \times \left[\frac{\alpha\dot{\mathcal{E}}_{\rm SN0}\,\tau_0^{1+\Gamma/\alpha}}{2\pi(2-\Gamma)\rho_0}\right]^{\frac{1}{5}}. \tag{13}$$

In (12) a constant density ρ_0 of the ambient interstellar gas is assumed for which we adopt a value of $\rho_0 = 2 \times 10^{-24}$ g/cm^3. The index in (12), $(2\alpha - \Gamma)/5\alpha = 0.564$, lies between the index of 0.4 of the Sedov equation, describing supernova remnants, and the index of 0.6 of the stellar wind/superbubble expansion law. For a LB age of 10 to 15 Myr (12) predicts a bubble radius of 78 pc to 100 pc, respectively. This in good agreement with the observed size of the LB in the Galactic disk, as determined by Lallement et al. (2003) (cf. also Fig. 5). For the determination of the expected LB size we have used the expected numbers of supernovae both from the LCC and UCL subgroups. Maíz-Apellániz (2001) has argued that the LB owes its existence only to the 6 SNe stemming from the LCC subgroup, because stars from this subgroup came closest to the Sun in the past. We find the same when tracing the orbits of the stars backwards in time. However, the members of the UCL subgroup did enter the region occupied by the LB today and SNe stemming from the UCL subgroup have to be taken into account, in our view, in the energy considerations as well. The energy input of 6 SNe would excavate a bubble with radius of only 65 pc, which is more difficult to reconcile with the fact that the walls of the LB have been blown out above and below the Galactic plane so that the LB has become effectively a chimney. In general, however, similarity solutions as applied here can only give a rough estimate of the LB age and size due to several severe restrictions. Firstly, the ambient medium has to be assumed to be either homogeneously distributed or to follow a power law distribution in density, and its pressure has to be small compared to the bubble pressure. Secondly, turbulent mixing and mass loading, which occur in real bubbles, are hard to incorporate without further assumptions (Dyson et al. 2002). Therefore the most realistic approach to model *existing bubbles* is to perform 3D high-resolution numerical simulations of their formation. A first simulation of this kind was carried out by Breitschwerdt and de Avillez (2006) which was based on the older and less detailed LB formation scenario of BB02, but which took into account the presence of the neighbouring Loop I. In Fuchs et al. (2006) we presented an upgrade of that simulation which is now based on the better understood supernova rate and the calculated paths of their progenitors through the LB. The simulations are described in detail in the contribution by Breitschwerdt et al. in this volume.

4 Conclusions

We have reported on a fresh, unbiased search for the OB associations in the solar vicinity which might have harboured the progenitors of the core collapse supernovae responsible for the creation of the LB. We analyzed a volume complete set of B stars in the Hipparcos catalogue and our Arivel data base. Tracing back the stars' paths in time we were able to isolate a kinematically homogenous group of stars which we identify with the OB association we were looking for. Following this strategy we confirm the rather robust result that besides the Upper Scorpius (US) subgroup the Upper Centaurus Lupus (UCL), and Lower Centaurus Crux (LCC) subgroups of the Sco OB2 association with ages of 20–30 Myr are the youngest stellar associations in the solar neighbourhood. However, we also found other B stars with similar kinematics. A tracing back of the stellar orbits showed that the association entered the region of the present LB 10 to 15 Myr ago. This estimate of the LB's age is in excellent agreement with an estimate obtained by numerical simulations of the ISM during the formation of LB and Loop I triggered by supernovae in a realistic background. Deriving O VI column densities from their simulations de Avillez and Breitschwerdt (2005), Breitschwerdt and de Avillez (2006), Fuchs et al. (2006) found that O VI absorption line data taken with FUSE (Oegerle et al. 2005;

Savage and Lehner 2006) could be well fitted adopting a LB age of about 14 Myr. We therefore conclude that the LB must have been excavated during this time. Modelling the IMFs of the LCC and UCL subgroups we find that about 14 to 20 SNe originated from the associations. The implied energy input into the ambient interstellar gas explains quantitatively the present size of the LB.

A future task will be to study the stars in Loop I in the same way as in the LB. We hope to develop then a self-consistent scenario of the simultaneous formation of the LB and Loop I.

Acknowledgements This research has made extensive use of the Simbad data base at CDS, Strasbourg, France. This work has been partially funded by the Portuguese Science Foundation under the project PESO/P/PRO/40149/2000 to MAdeA and DB.

References

T.W. Berghöfer, D. Breitschwerdt, Astron. Astrophys. **390**, 299 (2002) (BB02)
G. Bertelli, A. Bressan, C. Chiosi, F. Fagotto, E. Nasi, Astron. Astrophys. Suppl. Ser. **106**, 275 (1994)
A. Blaauw, Annu. Rev. Astron. Astrophys. **2**, 213 (1964)
D. Breitschwerdt et al., Space Sci. Rev. (2008, this issue)
D. Breitschwerdt, M.A. de Avillez, Astron. Astrophys. Lett. **452**, L1 (2006)
D.P. Cox, P.R. Anderson, Astrophys. J. **253**, 268 (1982)
M.A. de Avillez, D. Breitschwerdt, Astrophys. J. **634**, L65 (2005)
J.H.J. de Bruijne, Mon. Not. R. Astron. Soc. **310**, 585 (1999)
E.J. de Geus, P.T. de Zeeuw, J. Lub, Astron. Astrophys. **216**, 44 (1989)
W. Dehnen, J. Binney, Mon. Not. R. Astron. Soc. **298**, 387 (1998)
P.T. de Zeeuw, R. Hoogerwerf, J.H.J. de Bruijne, A.G.A. Brown, A. Blaauw, Astron. J. **117**, 354 (1999)
R. Drimmel, R.L. Smart, M.G. Lattanzi, Astron. Astrophys. **354**, 67 (2000)
J.E. Dyson, S.J. Arthur, T.W. Hartquist, Astron. Astrophys. **390**, 1063 (2002)
R. Egger, B. Aschenbach, Astron. Astrophys. **294**, L25 (1995)
M. Feast, P. Whitelock, Mon. Not. R. Astron. Soc. **291**, 683 (1997)
B. Fuchs, D. Breitschwerdt, M.A. de Avillez, C. Dettbarn, C. Flynn, Mon. Not. R. Astron. Soc. **373**, 993 (2006)
J. Holmberg, C. Flynn, Mon. Not. R. Astron. Soc. **352**, 440 (2004)
R. Hoogerwerf, L.A. Aguilar, Mon. Not. R. Astron. Soc. **306**, 394 (1999)
D.E. Innes, T.W. Hartquist, Mon. Not. R. Astron. Soc. **209**, 7 (1984)
H. Kamaya, Astron. J. **128**, 761 (2004)
R. Lallement, Astron. Astrophys. **422**, 391 (2004)
R. Lallement, B.Y. Welsh, J.L. Vergely, F. Crifo, D. Sfeir, Astron. Astrophys. **411**, 447 (2003)
H.J.G.L.M. Lamers, J.P. Cassinelli, *Introduction to Stellar Winds* (Cambridge Univ. Press, Cambridge, 1999)
B. Lindblad, *Handbuch der Physik*, vol. 53 (1959), p. 21
A. Maeder, Astron. Astrophys. **93**, 136 (1981a)
A. Maeder, Astron. Astrophys. **99**, 97 (1981b)
A. Maeder, Astron. Astrophys. **102**, 401 (1981c)
A. Maeder, G. Meynet, Astron. Astrophys. **182**, 243 (1987)
J. Maíz-Apellániz, Astrophys. J. **560**, L83 (2001)
P. Massey, K.E. Johnson, K. Degioia-Eastwood, Astrophys. J. **454**, 151 (1995)
J.C. Mermilliod, B. Hauck, M. Mermilliod, The general catalogue of photometric data (2000)
W.R. Oegerle, E.B. Jenkins, R.L. Shelton, D.V. Bowen, P. Chayer, Astrophys. J. **622**, 377 (2005)
M.J. Sartori, J.R.D. Lépine, W.S. Dias, Astron. Astrophys. **404**, 913 (2003)
B.D. Savage, N. Lehner, Astrophys. J. Suppl. Ser. **162**, 134 (2006)
G. Schaller, D. Schaerer, G. Meynet, A. Maeder, Astron. Astrophys. Suppl. Ser. **96**, 269 (1992)
D.M. Sfeir, R. Lallement, F. Crifo, B.Y. Welsh, Astron. Astrophys. **346**, 785 (1999)
R.K. Smith, D.P. Cox, Astrophys. J. Suppl. Ser. **134**, 283 (2001)

Kinetic-Fluid Perspective on Modeling the Heliospheric/Interstellar Medium Interface

V.B. Baranov

Originally published in the journal Space Science Reviews, Volume 143, Nos 1–4, 449–464.
DOI: 10.1007/s11214-008-9392-6

Abstract Content of the invited paper for Workshop "From the Outer Heliosphere to the Local Bubble: Comparison of New Observations with Theory" organized by International Space Science Institute (ISSI) is presented.

Main results of the kinetic-fluid model of the solar wind interaction with the partially ionized supersonic flow of the local interstellar medium (LISM) by Baranov and Malama (J. Geophys. Res. 98(A9):15,157–15,163, 1993) and its development by Moscow group to take into account effects of solar cycles, cosmic rays, non-equilibrium processes, magnetic field and so on are given. It is shown that multi-fluid models are not correct for problem considered due to Knudsen number $Kn = l_{ch}/L \sim 1$, where l_{ch} is the mean free path of H atoms for collisions connected with the resonance charge exchange, L is the characteristic length of the problem. Important problems connected with the physical properties of the heliospheric boundary and with magnetohydrodynamics (MHD) of the solar wind interaction with the LISM are discussed.

Keywords Heliosphere · Solar wind · Interstellar gas · Termination shock · Heliopause · Heliosheath · Bow shock · Charge exchange · Magnetic field

1 Introduction

Parker (1961) was the first who constructed an axis-symmetric quantitative model of the hypersonic stellar wind (including the solar wind) interaction with the subsonic interstellar gas flow. It was assumed by Parker (1961) that the Mach number of the interstellar gas flow is $M_\infty = V_\infty/a_\infty \ll 1$, where V_∞ and a_∞ are the constant bulk and thermal sound velocities, respectively. The stellar wind termination shock is approximately spherical in this case and the flow was considered as the interaction between incompressible potential fluids. An alternative model of the solar wind interaction with the local interstellar medium (LISM) flow

V.B. Baranov (✉)
Department of Mechanics and Mathematics, Institute for Problem in Mechanics, Moscow State University, Russian Academy of Sciences, Moscow, Russia
e-mail: baranov@ipmnet.ru

was constructed by Baranov et al. (1970, 1971). It was assumed in this model that LISM's flow is hypersonic ($M_\infty \gg 1$) as well as the solar wind at the Earth's orbit ($M_E \gg 1$). The interstellar gas was assumed to be a fully ionized plasma to use a hydrodynamic approach. In this case three strong discontinuities must be formed: a tangential discontinuity (heliopause) and two shocks. The bow shock (BS) in the interstellar plasma and the termination shock (TS) in the solar wind are formed due to decelerating the supersonic interstellar and solar wind plasma flows, respectively. The heliopause is a surface separating these flows. The thin layer approximation was used by Baranov et al. (1970, 1971) to construct the quantitative model. In this approximation the thickness of the region between the BS and TS (interface or heliosheath) is assumed to be small as compared with their heliocentric distance. It is interesting to note here that Baranov et al. (1970, 1971) proceeded from the concept of the Sun's motion relative to the nearest stars with a supersonic velocity of $V_\infty \sim 20$ km/s rather than from the real concept of the solar system motion relative to the local interstellar cloud (LIC) with a velocity of $V_\infty \sim 26$ km/s (Lallement and Bertin 1992).

First experiments on scattered solar radiation at wavelengths of 1216 Å and 584 Å and their interpretation proved (Bertaux and Blamont 1971; Thomas and Krassa 1971; Blum and Fahr 1970; Fahr 1974; Weller and Meier 1974) that the LISM's gas is moving with the supersonic velocity $V_\infty \sim 20$–25 km/s relative to the Sun. Therefore, the supersonic character of the LISM's gas flow relative to the Sun, suggested in the model by Baranov et al. (1970, 1971), was confirmed by these experiments. However, the assumption by authors, that the LISM is the fully ionized gas, is not real as it was shown on the basis of experiments on the scattered solar radiation mentioned above.

At present, there is no doubt that the LISM is a partially ionized gas where the main components are electrons and protons (plasma component below) and hydrogen atoms (neutral component below). Wallis (1975) was the first who qualitatively showed that the heliosheath introduced in the model by Baranov et al. (1970, 1971) can be a good "filter" for penetrating H atoms from the LISM to the solar system due to their charge exchange with interstellar protons decelerated in the BS. In doing so, plasma and neutral components can influence each other due to this process. In this case the equations of mass, momentum and energy conservation have the following form (one-fluid approximation for the plasma component)

$$\frac{\partial \rho}{\partial t} + \nabla \cdot \rho \mathbf{V} = 0, \qquad \frac{\partial \mathbf{V}}{\partial t} + (\mathbf{V}\cdot\nabla)\mathbf{V} + \frac{1}{\rho}\nabla p = \mathbf{F}_1(f_H, \rho, \mathbf{V}, p),$$
$$\frac{\partial}{\partial t}\left(\varepsilon + \frac{p}{\rho} + \frac{V^2}{2}\right) + \nabla\cdot\left[\rho\mathbf{V}\left(\varepsilon + \frac{p}{\rho} + \frac{V^2}{2}\right)\right] = F_2(f_H, \rho, \mathbf{V}, p), \quad p = (\gamma - 1)\rho\varepsilon, \tag{1}$$

where $p, \rho, \mathbf{V}$ and ε are pressure, mass density, bulk velocity and internal energy of the plasma component; γ is the ratio of specific heats; $\mathbf{F}_1$ and F_2 are the functionals ("source terms"), determining the change of momentum and energy of the plasma component due to collisions between H atoms and protons which are accompanied by the resonance charge exchange; and $f_H(\mathbf{r}, \mathbf{w}_H, t)$ is the H atom distribution function, depending on radius-vector $\mathbf{r}$, individual velocity vector $\mathbf{w}_H$ of H atoms and time t.

The self-consistent, axis-symmetric and stationary problem was solved by Baranov et al. (1981) for the following source terms (Holzer 1972)

$$\mathbf{F}_1 = \nu_c(\mathbf{V}_H - \mathbf{V}), \quad \nu_c = n_H\sigma U_*, \quad U_* = [(\mathbf{V}_H - \mathbf{V})^2 + 128k(T + T_H)/(9\pi m_H)]^{1/2},$$
$$F_1 = \rho\nu_c(V_H^2/2 - V^2/2 + 3kT_H/2m_H - 3kT/2m_H), \tag{2}$$

where $\mathbf{V}_H, n_H$ and T_H are the bulk velocity, number density and the temperature of H atoms, T is the temperature of the plasma component, σ and ν_c are the effective cross section and

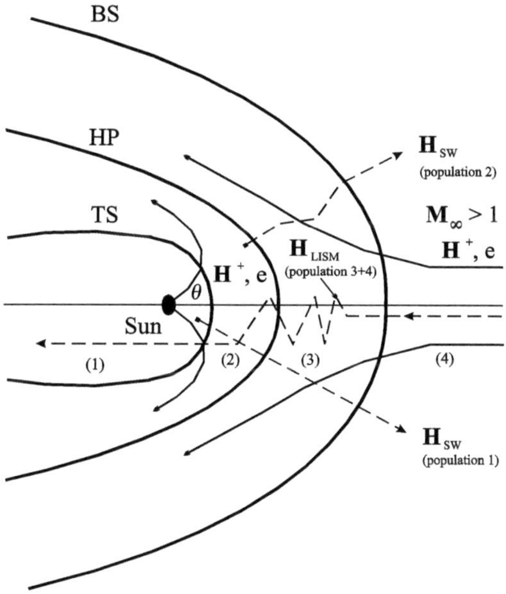

Fig. 1 General picture of the solar wind interaction with the supersonic flow of the partially ionized LISM gas

the frequency of charge exchange collisions, respectively, m_H is the proton (or hydrogen atom) mass, and k is the Boltzmann constant. Equations (2) were obtained at an assumption that the plasma and neutral components have the Maxwellian distribution function. Baranov et al. (1981) assumed also that parameters of H atoms are described by the following equations

$$\mathbf{V}_H = \mathbf{V}_\infty = \text{const}; \qquad T_H = T_\infty = \text{const}, \qquad \nabla \cdot n_H \mathbf{V}_H = -n_p \nu_c \quad (n_p = \rho/m_H). \quad (3)$$

Equations (3) were accepted because the hydrodynamic momentum and energy equations are not correct for H atoms due to their mean free path l_{ex} in collisions accompanying the resonance charge exchange is approximately on the same order as the characteristic length L of the problem considered (Knudsen number $Kn = l_{ex}/L \sim 1$). We also see from the last (3) that only the LISM's hydrogen atom disappearance is taken into account in the model by Baranov et al. (1981). Evidently, this model does not describe the distribution of the neutral component parameters since it takes into account only the primary H atoms which penetrate from the LISM to the solar system. However, the TS crossing by Voyager-1 in December 2004 and Voyager-2 in September 2007 at the heliocentric distances ~94 AU and ~86 AU, respectively, was predicted by Baranov et al. (1981) with 10% accuracy (see, in details, Baranov 2006).

In reality, a minimum four populations of H atoms are formed (see Fig. 1). Populations 1 and 2 are energetic H atoms formed due to charge exchanges of the LISM's hydrogen atoms with the supersonic solar wind protons (in the region 1) and with the protons in the inner heliosheath (region 2), respectively. Population 3 is H atoms born due to charge exchange of the LISM's hydrogen atoms with protons in the outer heliosheath (region 3) and population 4 is primary H atoms penetrating to the solar system from the undisturbed LISM (region 4 in Fig. 1) without charge exchanging.

Formation of these H atom populations is taken into account in the self-consistent, kinetic-fluid model by Baranov and Malama (1993). The numerical solution was obtained using the stationary version of (1) together with the Boltzmann equation for distribution function $f_H(\mathbf{r}, \mathbf{w}_H)$ of hydrogen atoms

$$\mathbf{w}_H \cdot \frac{\partial f_H(\mathbf{r}, \mathbf{w}_H)}{\partial \mathbf{r}} + \left[\frac{(\mathbf{F}_r + \mathbf{F}_g)}{m_H}\right] \cdot \frac{\partial f_H(\mathbf{r}, \mathbf{w}_H)}{\partial \mathbf{w}_H}$$
$$= f_p(\mathbf{r}, \mathbf{w}_H) \int |\mathbf{w}'_H - \mathbf{w}_H| \sigma f_H(\mathbf{r}, \mathbf{w}'_H) d\mathbf{w}'_H$$
$$- f_H(\mathbf{r}, \mathbf{w}_H) \int |\mathbf{w}_H - \mathbf{w}_p| \sigma f_p(\mathbf{r}, \mathbf{w}_p) d\mathbf{w}_p. \qquad (4)$$

Here $\mathbf{F}_r$ and $\mathbf{F}_g$ are the solar radiation pressure and the solar gravitational force, f_p is the locally-maxwellian distribution function of protons. Source terms of (1) were calculated on the basis of the following formula

$$\mathbf{F}_1 = \frac{1}{n_p} \int d\mathbf{w}_H \int d\mathbf{w}_p \sigma |\mathbf{w}_H - \mathbf{w}_p| (\mathbf{w}_H - \mathbf{w}_p) f_H(\mathbf{r}, \mathbf{w}_H) f_p(\mathbf{r}, \mathbf{w}_p),$$
$$F_2 = m_H \int d\mathbf{w}_H \int d\mathbf{w}_p \sigma |\mathbf{w}_H - \mathbf{w}_p| \left(\frac{w_H^2}{2} - \frac{w_p^2}{2}\right) f_H(\mathbf{r}, \mathbf{w}_H) f_p(r, w_p), \qquad (5)$$
$$n_H = \int d\mathbf{w}_H f_H(\mathbf{r}, \mathbf{w}_H), n_p = \int d\mathbf{w}_p f_p(\mathbf{r}, \mathbf{w}_p).$$

The numerical solution of the Boltzmann equation (4) was obtained by a Monte Carlo method with splitting trajectories (Malama 1991) which has a very high accuracy. Numerical results of the considered problem were obtained by the iteration method suggested by Baranov et al. (1991). The following boundary conditions were used

$$n_{pE} = 7\ \text{cm}^{-3}, \quad V_E = 450\ \text{km s}^{-1}, \quad M_E = 10,$$
$$n_{p\infty} = 0.07\ \text{cm}^{-3}, \quad n_{H\infty} = 0.14\ \text{cm}^{-3}, \quad V_\infty = 25\ \text{km s}^{-1}, \quad M_\infty = 2,$$

where indexes "E" and "∞" are at the Earth's orbit and at the undisturbed LISM, respectively. The ratio of the solar radiation pressure to the force of the solar gravitation was accepted to be equal of $\mu = F_r/F_g = 0.75$.

In Fig. 2 the total number density of populations 3 + 4 are given as a function of the heliocentric distance in the upwind, crosswind and downwind directions (curves 1, 2 and 3, respectively). We see a non-monotonic distribution with a maximum in the vicinity of the heliopause. This distribution was named by "hydrogen wall" and was experimentally discovered by Linsky and Wood (1996) who interpreted the Lyman-alpha absorption spectra obtained by the GHRS instrument onboard of Hubble Space Telescope.

The geometrical pattern of the interface obtained by Baranov and Malama (1993) are presented in Fig. 3. We see that heliocentric distances of the BS, HP and TS are much less at the real case of $n_{H\infty} \neq 0$ (solid lines) as compared with the case of $n_{H\infty} = 0$ (dashed lines), i.e. processes of the resonance charge exchange play a very important role. In doing so, as we see from Fig. 3, the TS location, detected by Voyager 1 and 2, was theoretically predicted more than 10 years ago with accuracy $\sim$5%. It is interesting to note here that a complicated structure of the tail region at $n_{H\infty} = 0$ with the triple point A, where Mach disc (MD), tangential discontinuity (TD) and reflected shock (RS) are crossing, disappears at $n_{H\infty} \neq 0$. In this case the sonic lines (dotted-dashed lines) are absent and the flow in the inner heliosheath is subsonic.

Fig. 2 H atom number density of populations 3 + 4 as a function of the heliocentric distance ("hydrogen wall") in upwind, crosswind and downwind directions (*curves 1*, *2*, and *3*, respectively)

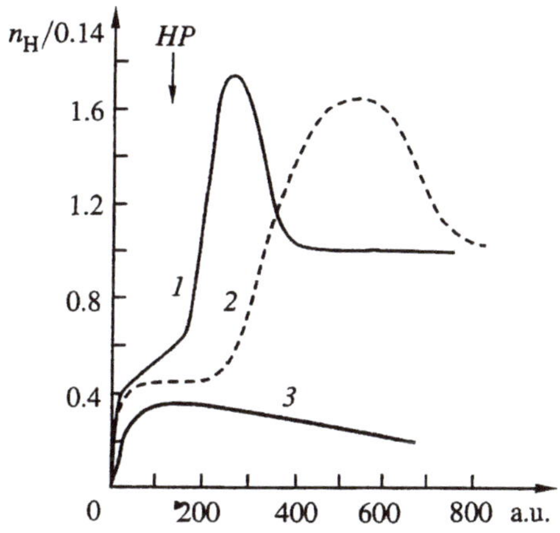

Fig. 3 Geometrical pattern of the flow calculated (*solid lines*). *Dashed lines* are the geometrical pattern without H atoms. *Dotted-dashed lines* are sonic lines

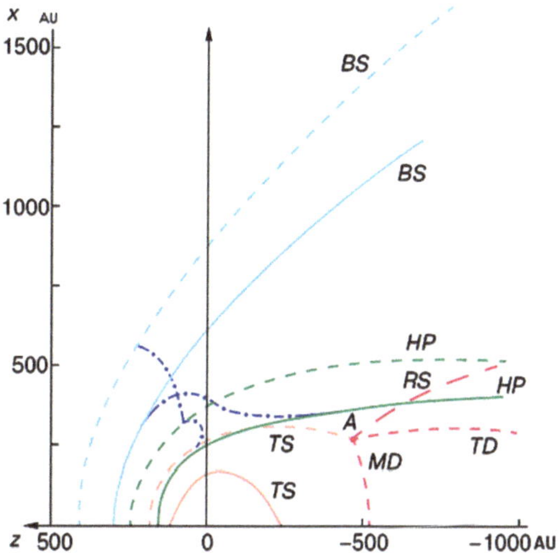

2 Development of the Kinetic-Fluid Model

Although some physical phenomena were predicted in the model by Baranov and Malama (1993), for example, the TS location, hydrogen wall, presence of the heliosheath region and so on it had several simplifying assumptions. Instantaneous relaxation of pick-up protons formed as a result of charge exchange and primary protons is assumed. One-fluid approach (1) for the plasma component is accepted in this model. It is axi-symmetric and stationary and does not take into account the effect of cosmic rays, magnetic fields and so on. A development of the kinetic-fluid model by Baranov and Malama (1993) is considered below (this development is described in details by Izmodenov and Baranov 2006 and Baranov and Izmodenov 2006).

2.1 Extent of the Heliosphere in the Tail Region

The main purpose for modelling the tail region of the heliospheric interface is to find the answers to two fundamental questions: (i) where is the heliospheric boundary in this region

Fig. 4 Lines of Mach number (M) are constant in the tail region of the heliosphere

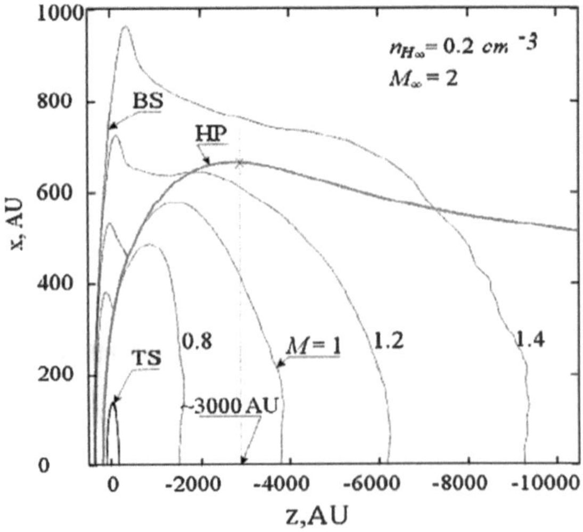

and (ii) what is the extent of the solar wind influence on the surrounding interstellar medium? As it is seen from Fig. 3 the tail region of the flow considered was calculated by Baranov and Malama (1993) only to the distances of about 1000 AU, where the solar wind parameters are strongly distinguished from the parameters of the undisturbed LISM flow. Therefore, this model can not answer the two questions given above.

A relaxation process of the solar wind and LISM parameters in the tail region of the heliosphere was calculated by Izmodenov and Alexashov (2003) on the basis of (1), (4) and (5). The results of their calculations in Fig. 4. are shown. We see that immediately after crossing the heliospheric termination shock (TS) the solar wind plasma acquires a subsonic velocity of about 100 km/s and a temperature of about 1.5×10^6 K ($M < 1$). Then the solar wind velocity continues to be decreasing due to "loading" by new protons born as a result of charge exchange of primary protons and gradually approaches a velocity which is typical for the undisturbed LISM flow ($V_\infty \sim 25$ km/s). Since the interstellar H atom temperature is much lower then the solar wind protons behind the TS, charge exchange leads to the effective cooling of the solar wind in the tail region. As a result of this cooling the solar wind Mach number increases, so that at a distance of about 4000 AU it again becomes supersonic ($M > 1$). With further increase of the heliocentric distance the plasma and H atom parameters approach their values in the undisturbed LISM ($M_\infty = 2$). Calculations of Izmodenov and Alexashov (2003) showed that at distances of about 40 to 50,000 AU the solar wind parameters are almost indistinguishable from those of the undisturbed LISM. These distances can be regarded as the heliospheric boundary in the tail region. It is also interesting to note that the jumps in density and tangential velocity at the heliopause HP almost vanish at considerably smaller heliocentric distances (of about 3000 AU).

Turbulence or heliopause instability could, of course, decrease the size of the heliospheric boundary. However, calculations by Izmodenov and Alexashov (2003), who taken into account processes of the resonance charge exchange only, give an upper limit of this boundary size.

2.2 Effect of the 11-Year Solar Activity Cycle

The solar wind parameters have already been measured onboard spacecraft over about four solar cycles (about 45 years). The measurements showed that the solar wind ram pressure

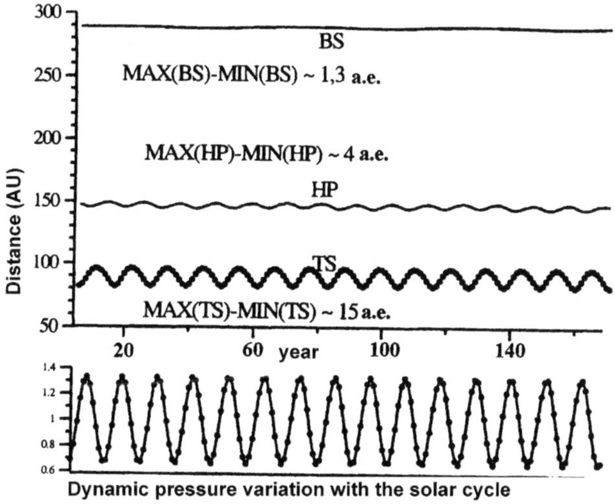

Fig. 5 Variations of the TS, HP and BS locations with the solar cycles

changes by a factor of about two on transition from maximum to minimum of the solar activity. Most global models of the solar cycle effects have ignored the interstellar H atom component or took this component into account by using simplified fluid approximation. This simplifications were made because it is difficult to solve a 6D (time, two dimensions in space and three dimensions in velocity space) kinetic equation for the H atom distribution function.

Development of the kinetic-fluid model by Baranov and Malama (1993) to take into account these non-stationary effects was made by Izmodenov and Malama (2004) and Izmodenov et al. (2005b). The 11-year periodic variations of the termination (TS) and bow (BS) shocks and the heliopause (HP) locations in upwind direction are shown in Fig. 5 obtained by Izmodenov et al. (2005b). We see from Fig. 5 that the oscillation amplitudes of the TS, HP and BS are equal to 7.5 AU, 2 AU and less than 0.7 AU, respectively. It is interesting to note that the values of the solar-cycle-averaged heliocentric stand-off distances of the shocks and the heliopause, as well as the plasma parameter distributions, are similar to those obtained within the framework of the stationary model by Baranov and Malama (1993).

2.3 Effect of Cosmic Rays

On the hydrodynamic level the cosmic ray effect on the flow is described by the cosmic ray pressure gradient ∇p_{cr} and by the energy transport $\mathbf{V} \cdot \nabla p_{cr}$ in the momentum and energy equations, respectively (second and third equations of (1)). In so doing (1) must be solved together with (4), (5) and an equation for the cosmic ray pressure. It was shown by Myasnikov et al. (2000) that the effect of galactic cosmic rays (GCR) is negligible as compared with the effect of the resonance charge exchange. However, the effect of the anomalous cosmic rays (ACR) is not negligible in the vicinity of the termination shock as it was shown by Alexashov et al. (2004) on the basis of numerical calculations.

The distribution of the solar wind bulk velocity in the vicinity of the termination shock in Fig. 6 is presented at diffusion coefficients $k = 3.75 \times 10^{20}$; 3.75×10^{21} and 3.75×10^{22} cm s^{-1} (curves 1, 2 and 3, respectively). Curve 4 was calculated for the case when the ACR is absent. We see that the ACR effect gives rise to a smooth deceleration of the solar wind in the so-called "precursor". In so doing, the intensity of the termination shock

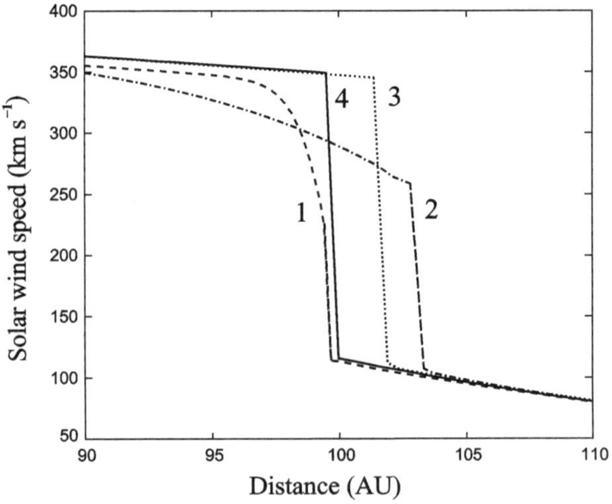

Fig. 6 Effect of ACR on the solar wind velocity in the vicinity of the TS at different values of the diffusion coefficients. Precursor is pronounced at small and medium values of the diffusion coefficients (*curve 1* and *2*, respectively). *Curve 4* shows the TS position without ACR

decreases and, as a result, the temperature in the inner heliosheath also decreases. This effect is important for interpreting the measurements of H atom fluxes of population 2 at 1 AU which the USA plans to make onboard IBEX spacecraft in 2008.

2.4 Effect of Nonequilibrium of the Pickup and Solar Wind Protons

As we noted above the one-fluid approximation is used in the model by Baranov and Malama (1993), where the instantaneous relaxation of the pickup and solar wind protons is assumed. However, measurements of the pickup proton distribution function onboard Ulysses and ACE spacecraft showed that this distribution is non-maxwellian, though isotropic. These data pointed to the absence of thermodynamic equilibrium between the pickup and solar origin protons, although their mean velocities are equal. This effect was taken into account by Malama et al. (2006) in the framework of the stationary and axi-symmetric model by Baranov and Malama (1993), but (1), (4) and (5) were numerically solved together with an equation for pickup proton distribution function. The fluxes of H atoms of population 2 at 1 AU in Fig. 7 are shown. We see from this figure that the model by Malama et al. (2006) predicts smaller fluxes for particles with energies less than 1 keV and larger fluxes for particles with energies greater than 1 keV as compared with the model by Baranov and Malama (1993). This result can be also interesting for the IBEX project (see Sect. 2.3).

2.5 Interstellar Magnetic Field Effect

The interstellar magnetic field effect on the heliosheath structure was studied by Izmodenov et al. (2005a) in the general three-dimensional case. The electromagnetic force $1/4\pi(\nabla \times \mathbf{B}) \times \mathbf{B}$ was included in the momentum equation (second of equations (1)). Equations (1), (4) and (5) were numerically solved together with the equation for the magnetic field induction $\mathbf{B}$ (Reynolds number $R_m \gg 1$)

$$\nabla \times (\mathbf{V} \times \mathbf{B}) = 0.$$

The following magnitudes of parameters in the undisturbed LISM and in the solar wind at the Earth's orbit were used

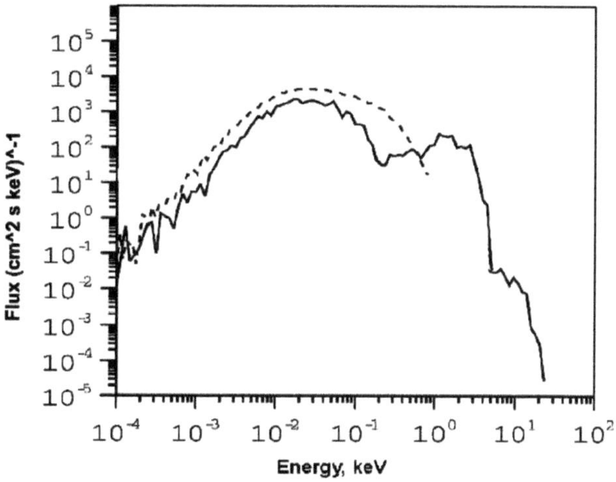

Fig. 7 Fluxes of population 2 H atoms at 1 AU (*solid line*). *Dashed line* is the fluxes in the model by Baranov and Malama (1993)

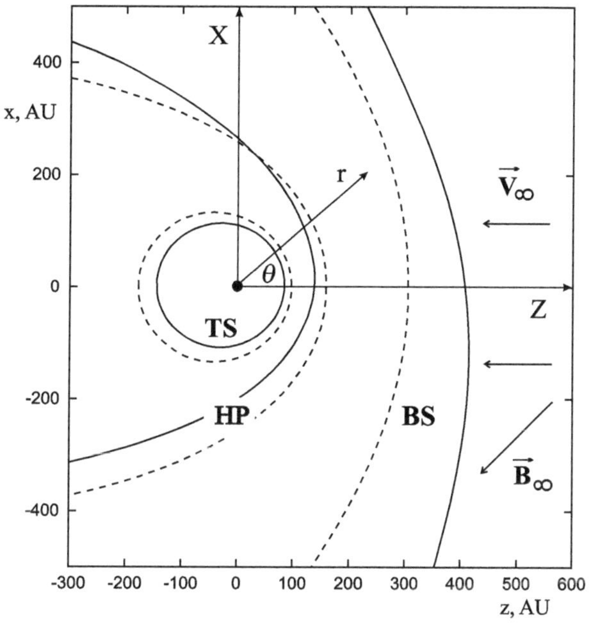

Fig. 8 Effect of the interstellar magnetic field on the geometrical pattern. *Dashed lines* are the BS, HP and TS positions without magnetic field

$$V_\infty = 26.4 \text{ km/s}, \quad M_\infty = 2, \quad n_{H\infty} = 0.18 \text{ cm}^{-3}, \quad n_{p\infty} = 0.06 \text{ cm}^{-3}, \quad B_\infty = 2.5 \text{ µGs},$$

$$V_E = 432 \text{ km/s}, \quad M_E = 10, \quad n_E = 7.39 \text{ cm}^{-3}.$$

Figure 8 demonstrates the geometrical pattern of the flow considered in the xOz plane determined by the vectors $\mathbf{V}_\infty$ and $\mathbf{B}_\infty$ (the angle between these vectors is assumed to be equal of 45°). Figure 8 clearly indicates the flow asymmetry about the Oz axis directed opposite to the vector $\mathbf{V}_\infty$.

Taking the interstellar magnetic field into account causes the TS to approach and the BS to recede from the Sun. The heliocentric stand-off distance of the HP depends on the magnetic pressure to magnetic tension ratio. In the regions where the magnetic field tension is greater than magnetic pressure the heliopause recedes from the Sun. The motion of H atoms, of course, depends on the motion of the plasma component. The deviation of the LISM hydrogen atom velocity direction inside of the heliopause from the undisturbed LISM velocity direction was detected from measurements of the backscattered solar Lyman- α radiation by Lallement et al. (2005). onboard SOHO spacecraft. This deviation is explained by Izmodenov et al. (2005a) as the effect of the interstellar magnetic field.

3 Some Critical Remarks

The kinetic-fluid model by Baranov and Malama (1993) and its development were considered in previous sections. However, multi-fluid models were developed in the last time due to their simplicity as compared with kinetic-fluid models although, as it was noted above, they are not correct due to the Knudsen number $Kn \sim 1$ for hydrogen atoms of all populations. Locally-maxwellian distribution function is required to use hydrodynamic equations. Figure 9 taken from a paper by Baranov et al. (1998) demonstrates a strong anisotropy of the population 2 temperature (moment of H atom distribution function) which must be isotropic in hydrodynamic approach. A direct calculation of H atom distribution function by Monte Carlo method also shows (Izmodenov et al. 2001) that all populations of H atoms are not locally-maxwellian, i.e. the fluid approach is not correct in this case.

> "In the "West", fluid codes for both neutrals and plasma remained the new theoretical works, which in post-talk discussions during meetings had the effect to stimulate quick tempered reactions from Vladimir Baranov, for the pleasure of those who knew him (and the surprise of the others)". (Rosine Lallement 2001, Century for Space Science, vol. II, p. 1200)

A tempered reaction, presented by R. Lallement in this sentence, could be connected with an unpleasant tendency in the heliospheric theoretical community to ignore results of other authors instead of comparisons with the results of previously published papers. Below we present several examples of this tendency.

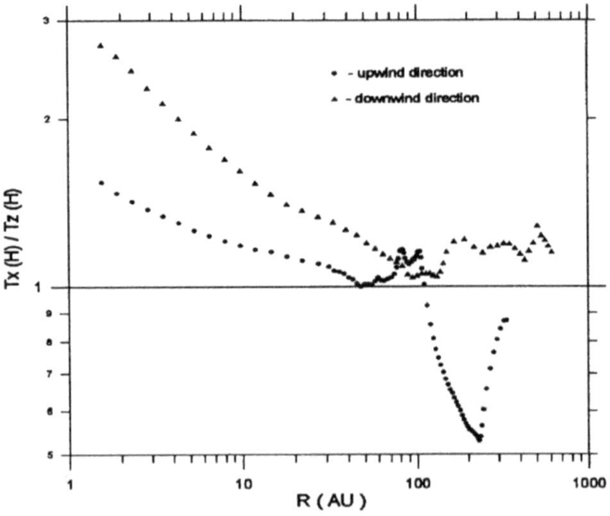

Fig. 9 Ratio of tangential and radial "temperatures" of population 2 H atoms as a function of the heliocentric distance

1. Zank et al. (1996) used a simplified multi-fluid and axi-symmetric model of the solar wind interaction with the partially ionized LISM gas without comparisons with the kinetic-fluid model by Baranov and Malama (1993) considered in Sect. 1. These comparisons were made later by Baranov et al. (1998) who showed that the results of the first model are erroneous (see also Fig. 9). There is no a reaction on this criticism up to now although authors of the multi-fluid model by Zank et al. (1996) continue to use results of their paper.
2. Williams et al. (1997) claimed that H–H and H–proton elastic collisions are important in the model by Baranov and Malama (1993). However, Izmodenov et al. (2000) have shown that estimations of Williams et al. (1997) are not correct. There is no a reaction in literature up to now.
3. Zank and Pauls (1996), Florinskii et al. (2004) calculated three-fluid model, where H atom of populations 1 and 2 were considered as different fluids and populations 3 and 4 as one fluid. Authors of these papers did not compare their results with the results of the kinetic-fluid model by Baranov and Malama (1993). It is interesting to note here that Alexashov and Izmodenov (2005) have compared the results of kinetic-fluid and multi-fluid models. They have demonstrated large differences between results of these models. Especially large differences were presented by Alexashov and Izmodenov (2005) for H atom parameters.
4. Magnetohydrodynamic (MHD) models taking into account an effect of the interstellar magnetic field are important for studying the heliosheath structure. Most of MHD models do not take into account neutral H atoms. Such models are not real for problem of the solar wind interaction with the partially ionized LISM flow, although have a theoretical interest. However, authors of published papers do not compare their results. For example, different geometrical forms of the bow shock splitting (appearance of different shock branches) were obtained by Myasnikov (1997), Ratkiewicz et al. (1998) and Pogorelov and Matsuda (2000) in the simplest axi-symmetric MHD problem ($\mathbf{B}_\infty || \mathbf{V}_\infty$) and at the same values of boundary conditions. However, numerical results of these authors were not compared neither each other nor with the results by Baranov and Zaitsev (1995) who did not obtain the bow shock splitting.

In conclusion of this section we would like to present a brilliant aphorism by Blaise Pascal:

> "Truth, we know, is so delicate that, if we make the slightest deviation from it, we fall into error; but this alleged error is so extremely fine-spun that, if we diverge from it in the slightest degree, we fall back upon the truth". (Blaise Pascal, "The Provincial Letters", Letter III)

4 To the Problem of the Heliopause Properties

The heliopause is a boundary separating the solar wind and the plasma component of the LISM. In the framework of the ideal fluid approach the heliopause is the tangential discontinuity which in reality can have a structure or can be unstable. At present Voyager 1 and Voyager 2 crossed the termination shock, entered to the inner heliosheath and are moving to this boundary. That is why the problem of the real physical properties of the heliopause is of great importance for observers.

4.1 To the Problem of the Heliopause Instability

The problem of the heliopause instability relative to the small perturbations without taking into account the H atom component was considered by Baranov et al. (1992). The heliopause is treated as a classical tangential planar discontinuity in compressible gas. This approximation is locally correct for the flanks of the heliopause and for perturbations with characteristic length scale which is much less than the curvature radius of the heliopause. Figure 10 demonstrates the qualitative picture of the disturbed surface of the heliopause. The shaded regions correspond to directions of exponentially growing perturbations.

We see that in the subsonic region out of the nose one the wave vectors of the all directions correspond to unstable disturbances (shaded circle), whereas in the supersonic region the heliopause is unstable only with respect to the meridional surface perturbation modes forming a typical "Dorian column" surface structure. However, the interstellar magnetic field as well as the magnetic field of the solar origin could give rise to the stabilization of this structure (Ruderman and Fahr 1993, 1995).

The results by Baranov et al. (1992) are not correct in the vicinity of the stagnation point (in the nose region) because the basic flow in this region is, as minimum, two-dimensional. A study of the stability problem in the nose region was first made by Belov (1997a) analytically on the basis of the following equations for the plane background flow of the incompressible fluid

$$V_{0x1} = \beta x, \quad V_{0z1} = -\beta z, \quad V_{0x2} = (\rho_1/\rho_2)^{1/2}\beta x, \quad V_{0z2} = -(\rho_1/\rho_2)^{1/2}\beta z \quad (\beta = \text{const} > 0), \tag{6}$$

formulated by Baranov et al. (1992). Here V_{0ij} $(i = x, z;\ j = 1, 2)$ are the components of the bulk velocity along (Ox) and normal (Oz) to the heliopause plane in the vicinity of the stagnation point, ρ is the mass density, indexes 1 and 2 are for the shocked interstellar

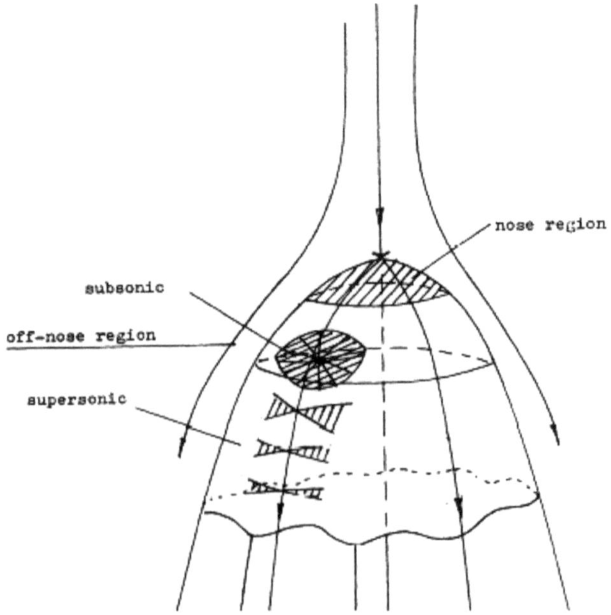

Fig. 10 Qualitative picture of disturbed flow in vicinity of the heliopause. *Shaded* angular sectors along one of streamlines indicate wave vector directions of growing disturbances. In supersonic region far from nose one a typical "Dorian column" surface is presented

plasma and for the shocked solar wind. Axi-symmetric analogy of (6) was considered by Belov (1997b). These studies together with a numerical solution of the flow considered by Belov and Myasnikov (1999) showed that amplitudes of small perturbations, formed in the nose region of the heliosphere, increase with their transfer to the wings, i.e. the plasma instability in the nose region have a convective character.

These investigations did not take into account the H atom component and, therefore, important processes of the resonance charge exchange. Numerical investigations of the heliopause instability with taking into account charge exchange processes were first made by Liewer et al. (1996) using simplified hydrodynamic approach for H atoms determined by (2) and (3). However, following comments by Pauls and Zank (1997) and a reply by Liewer and Brackbill (1997) show that the problem of the heliopause instability is not numerically solved.

An attempt to study the linear heliopause instability in the vicinity of the stagnation point (in the nose region) taking into account H atoms was analytically made by Florinskii et al. (2005) in the framework of (2) and (3) for H atoms. However, they used incorrect the energy equation

$$2\gamma \nabla \cdot \mathbf{V} = -\nu$$

instead of the correct equation

$$\frac{\partial p}{\partial t} + \mathbf{V} \cdot \nabla p + \gamma p \nabla \cdot \mathbf{V} = \nu\rho \left[\frac{\gamma - 1}{2} (V - V_H)^2 + \frac{kT}{m_p} - \frac{p}{2\rho} \right].$$

Besides, the one-dimensional background flow is considered by Florinskii et al. (2005) in the vicinity of the stagnation point, although the number density of the basic flow is unbounded on the heliopause in this case. The linearized equations, which were used by Florinskii et al. (2005), for the unbounded solution of the background flow become mathematically incorrect.

It is possible to conclude this subsection by statement that the problem of the real heliopause instability is not yet solved.

4.2 What Is the Role of the Magnetic Field?

The heliopause properties can be determined by the magnetic fields of the solar and interstellar origin. In particular, they can stabilize the heliopause (Ruderman and Fahr 1993, 1995) if it is a tangential discontinuity (the normal component of the magnetic field $B_n = 0$), as it was noted in Sect. 4.1. In this case magnetic field lines of the solar and interstellar origin are not related to each other. However, these lines can be connected either due to processes of reconnection or in the case when the magnetic field is not freezed in plasma. The first 3D MHD numerical model taking into account the H atom component in the framework of the MHD version of (1)–(3) was constructed by Linde et al. (1998). In so doing the solar as well as interstellar magnetic fields were taken into account. Accurate results were obtained due to excellent numerical scheme. However, their numerical results connected with a transition from the magnetic field of the solar origin to the interstellar magnetic field in the vicinity of the heliopause could not be interpreted in this paper as it was noted by authors. Evidently, this negative result was connected with a problem of the numerical resolution. Therefore, the problem concerning the MHD structure of the heliopause was not solved in this paper.

5 Whether the MHD Outer Heliosheath Is Bounded

Numerical difficulties of the kinetic description of the H atom motion together with MHD description of the plasma component leaded to many publications where the motion of all H atom populations is described by the multi-fluid equations (Pogorelov and Zank 2006; Pogorelov et al. 2006). The Ohm's law is not used in this approach because it is a result of mathematical manipulations with the multi-fluid equations. Evidently, the results of the interstellar magnetic field effect obtained in the framework of the multi-fluid models, and the results which could be obtained in the framework of the classical one-fluid MHD equations (see, for example, Cowling 1976), must be almost coinciding. An analysis of the generalized Ohm's law for partially ionized gas made by Baranov and Fahr (2003a) showed that the magnetic field distribution in the outer heliosheath depends significantly on the proton currents determined by the formula

$$\mathbf{j}_p = en_p(\mathbf{V}_p - \mathbf{V}) = \frac{2e\tau_{pH}(1-\alpha)^2}{m_p}\left[-\frac{\alpha}{1+\alpha}\nabla p + \frac{1}{4\pi}(\nabla \times \mathbf{B}) \times \mathbf{B}\right], \tag{7}$$

where $\mathbf{V}_p$ and $\mathbf{V}$ are the proton and mean bulk velocities, respectively, $\alpha = n_e/(n_e + n_H)$ is the ionization degree, τ_{pH} is the mean free time of protons in process of the resonance charge exchange. In this case the equation of the magnetic field induction will have the form (Baranov and Fahr 2003a, 2006)

$$\frac{\partial \mathbf{B}}{\partial t} = \nabla \times (\mathbf{V} \times \mathbf{B}) + \left\{\nabla \times \frac{(1-\alpha)^2\tau_{pH}}{n_p m_p}\left[\frac{\alpha}{1+\alpha}\nabla p \times \mathbf{B} - \frac{1}{4\pi}\mathbf{B} \times (\nabla \times \mathbf{B}) \times \mathbf{B}\right]\right\}.$$

The second term of the RHS of this equation is much larger than the first one for the problem of the outer interface structure. This term contains the Laplasian $\nabla^2\mathbf{B}$ due to

$$\left[(\nabla \times \mathbf{B}) \times \mathbf{B}\right] \times \mathbf{B} = -\mathbf{B} \times \left[(\nabla \times \mathbf{B}) \times \mathbf{B}\right] = \left[-B^2\nabla \times \mathbf{B} + \mathbf{B}(\mathbf{B} \cdot \nabla \times \mathbf{B})\right]$$

and

$$\nabla \times (B^2\nabla \times \mathbf{B}) = -B^2\nabla^2\mathbf{B} + \nabla(B^2) \times \mathbf{B},$$

i.e. there is a magnetic field diffusion rather than the magnetic field freezing in the outer heliosheath for the partially ionized gas of the LISM. MHD equations with second derivatives have no solutions with strong discontinuities. Thus, the bow shock cannot be formed in the models based on the classical one-fluid MHD equations. Evidently, the formation of the bow shock in the models based on the multi-fluid MHD equations is questionable. This effect could be interpreted on the basis of results obtained by Izmodenov et al. (2005a) in the framework of the kinetic-MHD model. Really, Fig. 11 (see Baranov and Fahr 2006) demonstrates the plasma component number density as a function of the heliocentric distance in the upwind direction. We see that the bow shock disappears at $B_\infty \neq 0$. This result is in contradiction with the results by Pogorelov and Zank (2006) and Pogorelov et al. (2006) who obtained the formation of the bow shock.

We would like to note in conclusion of this section that the problem of the magnetic field freezing in the outer heliosheath was discussed by Baranov and Fahr (2003b) and Florinskii and Zank (2003). It is evidently from (7) that the proton current increases with increasing the magnetic field as B^2 (Baranov and Fahr 2003b) rather than decreases as B^{-2} (Florinskii and Zank 2003). The increase of the proton current is physically connected with the decrease of the charged particle velocity relative to the H atom velocity due to the magnetic field effect.

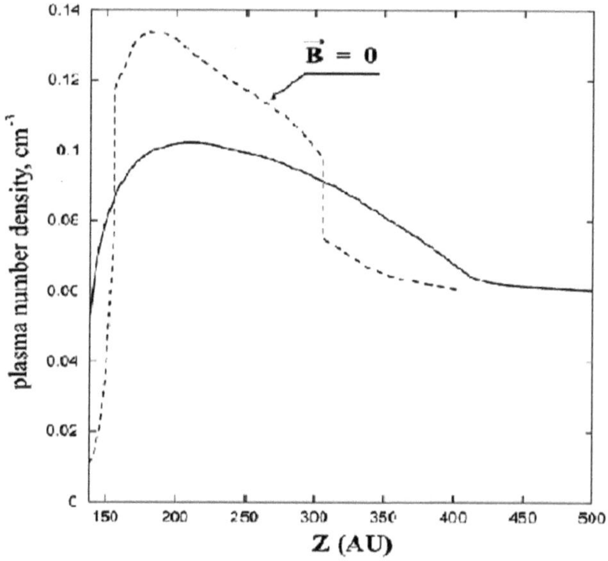

Fig. 11 Effect of the interstellar magnetic field on the distribution of the plasma number density in the outer heliosheath in the upwind direction (*solid line*). *Dashed line* is for vanishing interstellar magnetic field

Acknowledgements I thank V.V. Izmodenov, D.B. Alexashov and S.V. Chalov for their significant contribution in a preparation of this review paper. This work was supported by RFBR grants 07-01-00291; 07-02-01101; and 06-02-72557 and Programm of Fundamental Investigations of RAS.

References

D.B. Alexashov, V.V. Izmodenov, Astron. Astrophys. **439**, 1171–1181 (2005). doi:10.1051/0004-6361:20052821

D.B. Alexashov, S.V. Chalov, A.V. Myasnikov, V.V. Izmodenov, R. Kallenbach, J. Astron. Astrophys. **420**, 729–736 (2004). doi:10.1051/0004-6361:20041056

V.B. Baranov, in *The Physics of the Heliospheric Boundaries*, ed. by V. Izmodenov and R. Kallenbach. ISSI Scientific Report, No 5, pp. 27–44, ESA-ESTEC, Paris 2006

V.B. Baranov, H.-J. Fahr, J. Geophys. Res. **108**(A3), 1110 (2003a). doi:10.1029/2001JA009221

V.B. Baranov, H.-J. Fahr, J. Geophys. Res. **108**(A12), 1439 (2003b). doi:10.1029/2003JA010118

V.B. Baranov, H.-J. Fahr, Astrophys. Space Sci. Trans. **2**, 45–51 (2006)

V.B. Baranov, V.V. Izmodenov, Fluid Dyn. **41**(5), 689–707 (2006). doi:10.1007/s10697-006-0089-9

V.B. Baranov, Yu.G. Malama, J. Geophys. Res. **98**(A9), 15,157–15,163 (1993)

V.B. Baranov, N.A. Zaitsev, Astron. Astrophys. **304**, 631–637 (1995)

V.B. Baranov, K.V. Krasnobaev, A.G. Kulikovskii, Dokl. Akad. Nauk SSSR **194**, 41–43 (1970)

V.B. Baranov, K.V. Krasnobaev, A.G. Kulikovskii, Sov. Phys. Dokl. **15**(9), 791–793 (1971). (Engl. Translation)

V.B. Baranov, M.K. Ermakov, M.G. Lebedev, Sov. Astron. Lett. **7**, 206–210 (1981)

V.B. Baranov, M.G. Lebedev, Y.G. Malama, Astrophys. J. **375**, 347–351 (1991). doi:10.1086/170194

V.B. Baranov, H.-J. Fahr, M.S. Ruderman, Astron. Astrophys. **261**, 341–351 (1992)

V.B. Baranov, V.V. Izmodenov, Y.G. Malama, J. Geophys. Res. **103**(A5), 9575–9586 (1998). doi:10.1029/97JA03662

N.A. Belov, Fluid Dyn. **32**(2), 78–82 (1997a)

N.A. Belov, Fluid Dyn. **32**(6), 25–29 (1997b)

N.A. Belov, A.V. Myasnikov, Fluid Dyn. **34**(3), 379–387 (1999)

J.-L. Bertaux, J. Blamont, Astron. Astrophys. **11**, 200–217 (1971)

P. Blum, H.-J. Fahr, Astron. Astrophys. **4**, 280–290 (1970)

T.G. Cowling, *Magnetohydrodynamics* (Adam Hilger, London, 1976)

H.-J. Fahr, Space Sci. Rev. **15**, 483–540 (1974). doi:10.1007/BF00178217

V. Florinskii, G.P. Zank, J. Geophys. Res. **108**(A12), 1438 (2003). doi:10.1029/2003JA009950

V. Florinskii, N. Pogorelov, G. Zank, B. Wood, D.P. Cox, Astrophys. J. **604**, 700–715 (2004). doi:10.1086/382017
V. Florinskii, G.P. Zank, N.V. Pogorelov, J. Geophys. Res. **110**(A7) (2005). doi:10.1029/2004JA010879
T.E. Holzer, J. Geophys. Res. **77**, 5407–5431 (1972). doi:10.1029/JA077i028p05407
V.V. Izmodenov, D.B. Alexashov, Astron. Lett. **29**(1), 58–63 (2003). doi:10.1134/1.1537379
V.V. Izmodenov, Y.G. Malama, Adv. Space Res. **34**, 74–78 (2004). doi:10.1016/j.asr.2003.05.047
V.V. Izmodenov, V.B. Baranov, in *The Physics of the Heliospheric Boundaries*, ed. by V. Izmodenov, R. Kallenbach. ISSI Scientific Report, No 5, pp. 67–136, ESA-ESTEC, Paris, 2006
V.V. Izmodenov, Yu.G. Malama, A.P. Kalinin, M.A. Gruntman, R. Lallement, I.P. Rodionova, Astrophys. Space Sci. **274**(1–2), 71–76 (2000)
V.V. Izmodenov, M.A. Gruntman, Y.G. Malama, J. Geophys. Res. **106**, 10,681–10,690 (2001). doi:10.1029/2000JA000273
V.V. Izmodenov, D.B. Alexashov, A.V. Myasnikov, Astron. Astrophys. **437**, L35–L38 (2005a). doi:10.1051/0004-6361:200500132
V.V. Izmodenov, Y.G. Malama, M.S. Ruderman, Astron. Astrophys. **429**, 1069–1080 (2005b). doi:10.1051/0004-6361:20041348
R. Lallement, P. Bertin, Astron. Astrophys. **266**, 479–485 (1992)
R. Lallement, E. Quemerais, J.-L. Bertaux, S. Ferron, D. Kotroumpa, R. Pellinen, Science **307**(5714), 1447–1449 (2005). doi:10.1126/science.1107953 Medline
P.C. Liewer, J.U. Brackbill, J. Geophys. Res. **102**(A5), 9839–9842 (1997). doi:10.1029/97JA00153
P.C. Liewer, S.R. Karmesin, J.U. Brackbill, J. Geophys. Res. **101**(A8), 17,119–17,127 (1996). doi:10.1029/96JA00606
T.J. Linde, T.I. Gombosi, P.L. Roe, K.G. Powell, D.L. DeZeëuw, J. Geophys. Res. **103**, 1889–1904 (1998). doi:10.1029/97JA02144
J. Linsky, B. Wood, Astrophys. J. **463**, 254–270 (1996). doi:10.1086/177238
Y.G. Malama, Astrophys. Space Sci. **176**, 21–46 (1991). doi:10.1007/BF00643074
Y.G. Malama, V.V. Izmodenov, S.V. Chalov, Astron. Astrophys. **445**, 693–701 (2006). doi:10.1051/0004-6361:20053646
A.V. Myasnikov, Preprint of Institute for Problems in Mechanics RAS, No 585, 1997
A.V. Myasnikov, D.B. Alexashov, V.V. Izmodenov, S.V. Chalov, J. Geophys. Res. **105**(A3), 5167–5178 (2000)
E. Parker, Astrophys. J. **134**, 20–46 (1961). doi:10.1086/147124
H.L. Pauls, G.P. Zank, J. Geophys. Res. **102**(A5), 9835–9837 (1997). doi:10.1029/97JA00152
N.V. Pogorelov, T. Matsuda, Astron. Astrophys. **313**, 697–705 (2000)
N.V. Pogorelov, G.P. Zank, Astrophys. J. **636**, L161–L164 (2006). doi:10.1086/500087
N.V. Pogorelov, G.P. Zank, T. Ogino, Astrophys. J. **644**, 1299–1316 (2006). doi:10.1086/503703
R. Ratkiewicz, A. Barnes, G.A. Molvik et al., Astron. Astrophys. **335**, 363–375 (1998)
M.S. Ruderman, H.-J. Fahr, Astron. Astrophys. **275**, 635–644 (1993)
M.S. Ruderman, H.-J. Fahr, Astron. Astrophys. **299**, 258–266 (1995)
G. Thomas, R. Krassa, Astron. Astrophys. **11**, 218–233 (1971)
M. Wallis, Nature **254**, 207–208 (1975). doi:10.1038/254202a0
C. Weller, R. Meier, Astrophys. J. **193**, 471–476 (1974). doi:10.1086/153182
L.L. Williams, D.T. Hall, L. Pauls, G.P. Zank, Astrophys. J. **476**, 366 (1997)
G.P. Zank, H.P. Pauls, Space Sci. Rev. **78**, 95–106 (1996). doi:10.1007/BF00170796
G.P. Zank, H.L. Pauls, L.L. Williams, D.T. Hall, J. Geophys. Res. **101**, 21,639 (1996). doi:10.1029/96JA02127

Section IX: Summary

From the Heliosphere to the Local Bubble—What Have We Learned?

Eberhard Möbius

Originally published in the journal Space Science Reviews, Volume 143, Nos 1–4, 465–473.
DOI: 10.1007/s11214-008-9477-2 © Springer Science+Business Media B.V. 2008

Abstract The ISSI Workshop "From the Heliosphere to the Local Bubble" has brought together space physicists who work on the *in-situ* study of the interaction of the heliosphere with the surrounding circum-heliospheric interstellar medium (CHISM) and astrophysicists interested in the interstellar medium itself and in the wider neighborhood known as the Local Bubble. This paper contains a summary of the workshop except for the big picture presentations given at the last day. It contains highlights as viewed by the author and is organized to lead the reader from the heliosphere, via the immediate solar neighborhood into the Local Bubble, with the attempt of a brief outlook into the future.

Keywords Heliosphere · Local interstellar medium · Local bubble

1 Purpose and Philosophy of Summary

The ISSI Workshop "From the Heliosphere to the Local Bubble" was held at a very interesting juncture in our exploration of space, in October 2007, 50 years after the launch of Sputnik 1 as the first Earth orbiting satellite. Back then during the International Geophysical Year (IGY) in 1957/8, humans just started to stick their head out of the Earth's atmosphere. Now, celebrating the International Heliophysical Year (IHY) in 2007/8, we have watched Voyager 1 and 2, the most distant man-made space probes, crossing the termination shock (TS) of the solar wind in 2004 and 2007, respectively. This foray into the heliospheric boundary regions is complemented by an impressive heliospheric network of spacecraft, distributed over distance from the Sun, longitude and latitude, which has just been joined by the Interstellar Boundary Explorer (IBEX) on October 19, 2008, providing the missing global imaging. In a nutshell, 50 years after the dawn of the space age, humans have begun to stick their head outside the heliosphere.

This paper loosely follows the concluding summary presentation at the ISSI Workshop "From the Heliosphere to the Local Bubble", which was organized around the key objectives set for the workshop. At the same time it will probably be a bit more selective because

E. Möbius (✉)
Space Science Center & Department of Physics, University of New Hampshire, Morse Hall, 8 College Road, Durham, NH 03824, USA
e-mail: Eberhard.Moebius@unh.edu

of space constraints and the need to place key issues into context in the written form, which could be assumed as still present with the audience at the workshop. Starting with recent results and puzzles in the outer heliosphere and at its boundary, we will discuss multi-pronged approaches to infer the state of the CHISM from within the heliosphere. We will then discuss the overall structure and pressure balance in the Local Bubble (LB), followed by small-scale structure and turbulence in the LB. We will let the big picture talks of the last day speak for themselves and conclude with some remarks on future perspectives for the next 50 years of the space age.

2 Outer Heliosphere and Its Boundary Regions

Clearly, the recent TS crossings of the Voyager spacecraft and their results provided the majority of the highlights concerning our emerging knowledge of the outer heliosphere. After years of anticipation and ever more sophisticated predictions, Voyager 1 finally crossed the TS at 94 AU in December 2004 and V2 at 83 AU in August 2007. However, their findings were quite different from what had been expected.

Highly collimated and beaming energetic particle distributions in the foreshock region provided clear evidence that the overall shape of the TS must be blunt. It thus cuts into the regular spiral pattern of the interplanetary magnetic field (IMF) where it is closest to the Sun in the apex direction of the Sun's motion relative to the CHISM. As a result, the beaming points anti-sunward along the spiral on V1, while sunward on V2. After first causing some confusion, this is the natural consequence of V1 and V2 exiting the heliosphere on either side of the apex in longitude and a blunt shock, as is illustrated in Fig. 5 by Richardson and Stone (2008). Together with MHD modeling of the solar wind, as it changes with solar activity, the distances and timing of the TS crossings of the two probes are a clear indication of a distinct asymmetry in the heliospheric shape relative to the CHISM-heliosphere flow axis in latitude and longitude (Richardson and Stone 2008).

A complementary remote sensing method to infer the heliospheric shape makes use of high-resolution absorption spectroscopy of the Ly α line from nearby stars, which senses the slowdown of the interstellar material in the outer heliosheath. Wood et al. (2008) presented new results with a significant blueshift in the line for stars within $\approx 20°$ of the heliotail direction, which provide constraints on the width of the heliotail. The observed strength of the Ly α absorption feature due to H atoms in the heliotail suggests that the energetic neutral atom (ENA) signal reported from Mars Express observations by Galli et al. (2006) may be inconsistent with these results, i.e. too high. The future resolution of this apparent disagreement will be a task for IBEX (McComas 2008).

Another major puzzle provided by data from both Voyager probes is that the almost unanimously predicted anomalous cosmic ray (ACR) source was not found at the TS in either crossing. The turnover of the ACR spectra at low energies, a sign for substantial modulation of these particles in the injection range of the spectra and thus a remote source location, did not go away when the Voyagers crossed the TS (Richardson and Stone 2008). Several models have been suggested to explain this observation. McComas and Schwadron (2006) suggested an increasing effectiveness of the acceleration towards the flanks of the heliosphere where the IMF encounters the TS at angles significantly less than 90° relative to the shock normal and subsequent transport of the ACRs along the magnetic field in the heliosheath. Fisk and Gloeckler (2006) have proposed turbulent acceleration in the heliosheath as a solution, and Florinski (2008) has presented arguments that the formation of anisotropic pitch angle distributions in the acceleration process at the TS could explain the observed dips in the spectra at low energies. At this point the jury is out concerning these competing explanation attempts, with the expectation that continuous Voyager observations and IBEX will provide clues.

With both Voyagers roaming the heliosheath, their next highly anticipated objective will be the heliopause. The questions arising from this "horse race" are which of the two probes may make it first and when this is likely to happen. Given the speed and the locations of the probes, the answers to these questions are tied to the magnitude of the heliospheric asymmetry and the thickness of the heliosheath. The local plasma distribution $f(v)$ in the heliosheath can now be deduced from Voyager observations, and the interstellar neutral H density n_H is known from pickup ion and solar wind slowdown measurements. The flux of ENAs $j_{ENA}(v)$ from the heliosheath is the line of sight integral of the ENA production from the TS to the heliopause (HP) according to

$$j_{\mathrm{ENA}}(v) = n_{\mathrm{H}} \cdot \sigma_{\mathrm{chex}} \int_{\mathrm{TS}}^{\mathrm{HP}} f(v) v dx \approx n_{\mathrm{H}} \cdot \sigma_{\mathrm{chex}} \cdot \bar{f}(v) v \cdot L, \tag{2.1}$$

where σ_{chex} is the cross section for charge exchange (taken as the average for simplicity in the argument). Assuming an approximately homogenous distribution of the heliosheath plasma, the thickness L can be deduced from the ENA fluxes that will be obtained with IBEX (McComas 2008).

With its global ENA maps of the heliosheath region, IBEX will provide the currently missing overview, thus ideally complementing the detailed *in situ* observations, at two locations at the TS and in the heliosheath. It will allow us to assess the strength of the TS across the entire sky and provide insight into the heated pickup ion distributions in the heliosheath. IBEX will also be the ideal tool to draw a detailed picture of the heliospheric asymmetry. Current global heliospheric models already show hints of a potentially rich structure in the anticipated ENA images (Heerikhuisen et al. 2008). It should be noted that the abovementioned puzzles and the complexity of the questions emphasize the need for detailed realistic global simulations and basic theory along with the full complement of observations.

3 Inferring the CHISM from within

The Sun is moving relative to the interstellar medium, which is only partially ionized. As a consequence an interstellar wind of neutral interstellar gas is blowing through the heliosphere. Ionization rates of 5×10^{-8}–5×10^{-7} s^{-1} for neutral species with high ionization potential (H, He, N, O, and Ne) at 1 AU and the inflow speed of 26.3 km/s (Witte 2004) conspire to penetration distances to the Sun in the upwind direction of 0.3–3 AU, i.e. where the original interstellar density is reduced to $1/e$. This situation enables *in situ* diagnostics of the neutral interstellar component in the inner heliosphere through backscattering of solar resonance lines, pickup ion and direct neutral gas observations. In the framework of a scientific Team at ISSI the physical parameters of He and its flow vector, as derived from these three methods, have recently been consolidated into a consensus set (Möbius et al. 2004). Because the heliospheric boundary region is transparent for He the derived values represent the pristine CHISM values for He, and thus by inference the dynamic parameters of the medium.

Because other neutrals, notably H and O, are greatly affected by charge exchange, they are substantially depleted during their passage through the outer heliosheath, partially replacing the removed fraction of atoms with a hotter and slower secondary component that stems from the plasma in the outer heliosheath. Therefore, the interstellar densities of these species must be reconstructed by detailed modeling of the boundary interaction from the observed densities in the heliosphere. Another ISSI Team has attempted to consolidate observations of the interstellar neutral H density at the TS, which is the basis for the derivation of the majority component in the CHISM. The density has been derived from pickup ion observations by Ulysses as 0.087 ± 0.022 cm^{-3} (Bzowski et al. 2008a, 2008b), from the slowdown of the solar wind due to the implantation of H pickup ions in the outer heliosphere as 0.09 ± 0.02 cm^{-3} (Richardson et al. 2008), and from the damping of activity region related modulation of the backscattered Ly α signal with distance from the Sun as 0.09 ± 0.025 cm^{-3} (Pryor et al. 2008). The combination of these results leads to a neutral H density $n_{\mathrm{H}} = 0.089 \pm 0.022$ cm^{-3}, compiled as the weighted mean analogous to the consensus values of the CHISM He parameters (Möbius et al. 2004). At this workshop Gloeckler et al. (2008) have presented another complementary method to determine the densities of neutral interstellar species at the TS. They use the spectra of the suprathermal tails of pickup ions and ACRs as observed with the Low Energy Charged Particle (LECP) instrument on Voyager in the heliosheath to derive abundance ratios. They are normalized to the interstellar He density as derived with the Solar Wind Ion Composition Spectrometer (SWICS) on Ulysses from He^{2+} pickup ions (Gloeckler and Geiss 2001), which rests on the relative accuracy measuring the same species in the solar wind and as pickup ions with the same sensor and thus, in principle, can be more precise than observations that require absolute calibration of sensors. However, its derivation involves a number of modeling steps and parameters that are obtained by empirical modeling. This is reflected in the conservative uncertainty of heir value $n_{\mathrm{H}} = 0.055 \pm 0.021$ cm^{-3}, which is lower than the other results. Including this new result by Gloeckler et al. (2008) and a reduced value from their Ulysses pickup ion evaluations ($n_{\mathrm{H}} = 0.067 \pm 0.023$ cm^{-3}, including estimated systematic geometric factor uncertainties) in the weighted average, a combined value for the H density at the termination shock $n_{\mathrm{H,Mean}} = 0.083 \pm 0.023$ cm^{-3} is found, which is now somewhat lower than previously favored values, but within the range of several published results with different observation methods. However, it has become more and more evident that direct observation of the interstellar neutral gas flow for species affected by the heliospheric interface, such as H and O, will be very important to reduce these mostly systematic and modeling uncertainties as has been done successfully for interstellar He (Witte 2004). The IBEX mission will provide this opportunity at least for O (McComas 2008; Möbius et al. 2008).

A potential question mark was thrown into the modeling of the H densities from Ly α backscattering by the recent observation of an unexplained increase of the Ly α intensities, as seen with Voyager 1 outside 85 AU (Quemerais et al. 2008). Attempts to attribute this increase to an asymmetry in the heliosphere as inferred from other observations have not led to conclusive results.

On the other hand the distances of the two Voyager TS crossings from the Sun can be directly tied to the total outside pressure on the heliosphere and thus constrain the interstellar plasma density and magnetic field strength. In a parameter study, Izmodenov et al. (2003) showed that these observations leave a rather small two-dimensional box for the neutral and plasma density in the CHISM proper. Furthermore, Slavin and Frisch (2008) and Slavin (2008) show that the combination of the He parameters and the composition in the CHISM as derived from ACR and pickup ion observations severely constrain the ionization fraction and thus the UV and X-ray radiation environment.

These results leave the interstellar magnetic field (IsMF) the least known quantity both in strength and direction. Observations with the Solar Wind ANisotropy (SWAN) instrument on the SOlar and Heliospheric Observer (SOHO) show that the direction of the

H flow vector in the heliosphere deviates significantly from that of the He flow (Lallement et al. 2005). This deviation is evidence for a deflection of the plasma flow around the heliosphere, which is imprinted onto the secondary component of neutral H. Shortly thereafter, these observations were explained in terms of a magnetic-field-induced asymmetry of the heliosphere (Izmodenov et al. 2005). These results clearly complement the asymmetry deduced from the Voyager TS crossings. Kurth and Gurnett (2003) had previously interpreted pointing observations of the heliospheric radio bursts at 2–3 kHz that followed the strongest solar ejections of the past few solar cycles as an indication for the IsMF direction. It was interesting to see at this workshop that several modelers presented results of their global heliospheric MHD models that included magnetic fields. With moderate differences they came to similar conclusions, i.e. their models could simultaneously reproduce the Voyager and H flow observations (Izmodenov 2008; Opher 2008; Pogorelov 2008). In addition, Opher (2008) could show that the resulting most likely IsMF direction is also consistent with the observations by Kurth and Gurnett (2003).

Frisch (2008) linked the CHISM parameters as inferred from inner heliosphere observations through detailed modeling to the wider solar neighborhood and concluded that the physical parameters are those of a warm cloud where the Sun is located close to an interface between clouds. In addition, she made the point that the currently inferred interstellar magnetic field direction is consistent with average fields obtained for the adjacent few hundred pc from Faraday rotation observations. It will be important though to narrow the uncertainties for the magnetic field. Directly probing the decelerated and deflected flow of interstellar O with the IBEX-Lo sensor for its interaction with the heliospheric interface will provide an opportunity to achieve this goal (McComas 2008; Möbius et al. 2008).

Another important, yet often neglected, component of the interstellar medium is dust. It typically carries about 1% of the interstellar material and the lion share of the refractory elements. *In-situ* observations currently present two puzzles: Firstly, when matching the directly observable portion of the dust mass spectrum, as obtained with the Ulysses and Galileo probes, the inferred integral mass exceeds the local budget for dust by a wide margin (Draine 2008). Secondly, Krüger and Grün (2008) showed that the most recent interstellar dust observations in the inner heliosphere between 2005 and 2008 appear to indicate a substantial shift in arrival direction by 30° compared to previous observations. An explanation for such a dramatic shift is not in sight.

Could it be that the multitude of puzzles presented at this workshop point to a very special location for the Sun, i.e. at the edge of a cloud? One of the conclusions drawn by Redfield (2008) from his absorption spectroscopy survey of neighboring interstellar clouds appears to support this view; the relative velocity of the Sun and the temperature of the CHISM as derived from He neutral gas observations in the heliosphere (Witte 2004) are remarkably close to the mean of the values attributed to the Local Interstellar Cloud (LIC) and the G-Cloud, implying that the heliosphere may be located in a transition region between the two clouds. The G-cloud is visible in the ram direction, while the LIC appears around the periphery when looking in that direction.

4 Structure and Pressure Balance in the Local Bubble

As a corollary to the more local conclusion mentioned in the previous section, Redfield (2008) also presented convincing evidence that in its wider neighborhood, the Sun is located in a flow of interstellar clouds, apparently approaching from the direction of the Scorpius-Centaurus Association in the sky. The theater in which this flow plays out is the Local Bubble (LB), our county in the universe so to speak, an irregular sphere of $\approx$200 pc in diameter filled with hot dilute gas with $\approx 10^6$ K in temperature and many denser warm clouds (Jenkins 2008). A rather detailed picture of the LB has been obtained with NaI absorption line spectroscopy (e.g. Lallement et al. 2003). The kinematics of the flow of clouds is attributed to the neighboring Loop I superbubble pushing into the LB. The interaction of the two clouds is evident in soft X-ray images (e.g. Egger 1998). Both the LB and the Loop I superbubble have been formed by a succession of supernova explosions. While the Loop I formation still enshrouds the responsible Sco-Cen Association, the Pleiades cluster has been identified as the past producer of the LB, which are now separated from Loop I by their relative motion (Breitschwerdt et al. 2008). Modeling of the Pleiades cluster evolution using a typical mass distribution of its stars and assuming supernova deaths of stars larger than 10 solar masses leads to results consistent with the current size and location of the LB.

While a pressure $P/k \approx 15{,}000\ \mathrm{cm}^{-3}$ K is inferred inside the LB to maintain pressure balance at the outer edge of the LB, the LIC only provides $P/k \approx 2{,}300\ \mathrm{cm}^{-3}$ K as deduced from its measured physical parameters (Jenkins 2008). This conspicuous mismatch makes it very interesting to actually measure the internal pressure of the LB, but at a temperature of 10^6 K diagnostics are rather challenging. The diffuse X-ray background was thought to be the best indicator for this hot plasma. Attributing the observed background completely to hot bubble plasma leads to pressures consistent with the LB boundary condition or to even higher values.

However, the observation of strong solar wind charge exchange generated X-ray emission from comet Hale-Bopp and the realization that the solar wind—interstellar gas interaction could produce similar X-ray emission threw a monkey wrench into the soft X-ray diagnostics of the hot LB gas. Koutroumpa et al. (2008) presented a careful evaluation of potential foreground contributions to the soft X-ray background originating from solar wind charge exchange, using the comparison of observations with very local interstellar clouds and with no clouds in the field-of-view. The conclusion from these observations is that likely a large fraction, if not all, of the soft X-ray background intensity could be explained by heliospheric contributions to the diffuse X-ray background. If this were the final answer, the internal LB pressure would fall way short of the required pressure balance. A complementary view was presented by Shelton (2008), who pointed out that there are still substantial 1/4 keV X-ray intensities after solar wind charge exchange subtraction and that the observed column densities in the OVI band cannot be explained by local emission. Perhaps there is room for both, a local foreground and a LB component. In addition, Welsh (2008) showed a wide range of ultraviolet line observations, most of which reemphasized the challenge of such observations to identify hot LB gas.

The discussion of this technique gave the impression of an attempt to use the infamous "elephant method": Determine the weight of a sheet of paper by subtracting the weight of an elephant from the combination of both. This analogy may sound harsh, but it illuminates the challenge. Clearly, new tools are needed to reduce current uncertainties enough to allow decisive observations. Snowden (2008) showed that high spectral resolution in combination with good angular resolution should separate foreground from background components by selection of specific lines and object identification. Of course, this requires investment in new instrumentation, which already appears to be on the horizon.

In addition to these new X-ray techniques, the importance of high spectral resolution UV spectroscopy with the Space Telescope imaging Spectrograph (STIS) camera of Hubble Space Telescope (HST) was emphasized. This instrument has provided amazing details of the heliosphere-CHISM interaction and has enabled the study of the interaction with nearby astrospheres (Wood et al. 2005). This requires a repair of the Hubble Wide Field camera, which is currently part of the Hubble repair mission with the Shuttle launch planned for the near future. Together with interstellar clouds embedded in the LB, astrospheres provide excellent probes of the surrounding medium. Coupling the detailed study of the heliosphere-CHISM interaction with surveys of astrospheres and ISM clouds in the LB will lead to a strong synergism. This situation, with the emergence of new more detailed observation techniques, appears to be a *déjà vu* of the situation in the heliosphere-CHISM studies in the 1980s when new *in-situ* techniques, such as pickup ion and neutral gas observations, led to a dramatic reduction of the observational uncertainties.

5 Small Scale Structure and Turbulence

Another method to observe interstellar clouds in the LB and the motion of the interstellar gas is the use of radio scintillations in pulsar signals. Stanimirowic (2008) presented this method for the diagnostics of the interstellar cloud population, and Spangler (2008) used it to identify turbulent motion in the ISM. Stanimirowic (2008) showed a variety of cloud examples from the typical cold and warm neutral clouds that appear to be in pressure equilibrium with each other at $P/k \approx 3000\ \mathrm{cm}^{-3}$ K, although there are also more exotic examples with much higher pressures, yet much smaller size. Such clouds may survive several million years, but then they are destined to evaporate. Such cloud diagnostics point to a much more dynamic picture than simple static pressure equilibrium.

Breitschwerdt et al. (2008) expanded upon this topic and presented MHD simulations of the LB structure. These simulations indicate that magnetic fields play an important role in the formation and maintenance of cloud structures and that the situation is highly dynamic. Thus gas pressure alone cannot be taken as a safe indicator for pressure balance and stability.

This twist focused the attention towards two universal themes touched upon in this workshop, which closed the loop from the large-scale structures to the heliospheric boundary structures, i.e. magnetic fields and turbulence. Magnetic fields play an enormous role in the formation and stabilization of the individual interstellar cloud structures, and the strength and orientation of the interstellar magnetic field determines the shape and asymmetry of the heliosphere. Likewise, we cannot expect a smooth and orderly boundary of the heliosphere, as Jokipii and Balogh (2008) pointed out. Both magnetic field and plasma density show substantial fluctuations, which can be described more naturally in terms of turbulent interactions. On the large scale, substantial motions within the LB must include turbulence as indicated by radio scintillation observations (Spangler 2008). The observed spectrum of density fluctuations showed the clear presence of a Kolmogorov spectrum (Lazarian 2008).

6 Future Perspectives

Given recent developments in the fields covered by this workshop, what will likely be the expected progress in the coming years and where might the field be in 50 years? The two Voyagers have just made it across the termination shock in an amazing journey over 30 years. They are turning up more puzzles as they go and are very likely to cross the heliopause in tact, if funding is maintained. IBEX, which has been launched in October 2008, will open a completely new window for local astrophysics with its ENA observations. Complemented by the fleet of the other heliospheric spacecraft, the Voyagers and IBEX lead us into a golden

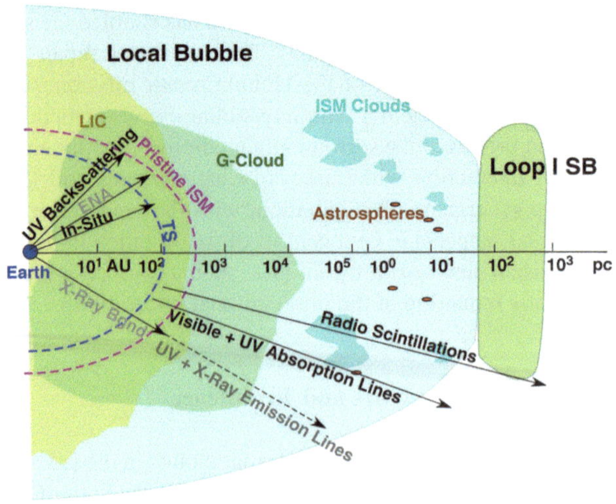

Fig. 1 Schematic view of the local environment of the solar system up to the Loop I structure in logarithmic scale, including sketches of typical structural elements. The reach of diagnostic methods is indicated by arrows pointing away from the Earth (local ones in the *upper half*; existing techniques, named in *black*, emerging ones in *grey*)

age for studies of the heliospheric boundary and the CHISM. The availability of these local tools calls for restoring an important survey tool for our interstellar neighborhood: The HST STIS instrument, which had been collecting high-resolution absorption spectra and was our most important tool for gleaning the cloud structure of the surrounding medium. Appropriate priority on the HST service mission will ensure STIS availability for the coming 5–10 years that are critical to these synergistic studies. To unravel the larger structure and pressure balance in the LB, high-resolution soft X-ray spectra appear to hold the most promise. Technical capabilities have been developed; implementing them appears to be a high priority for the near future. Figure 1 provides an overview from the Earth via the heliosphere, the CHISM and the LB to the neighboring Loop I structure on a logarithmic scale along with the typical reach of the diagnostic techniques presented in this workshop. Emerging and new techniques are named in grey.

Right on time, 50 years after the beginning of the space age we are now sticking our head out of the heliosphere. Perhaps we may be about to stick our head out of the Local Bubble 50 years from now? Not so fast! This would require us to develop Warp-Drive capability now, which is clearly not in sight. As a more modest foray, we may be sticking our head into the warm clouds of the local interstellar medium. There is some important new evidence in the observations of the local cloud structure that the solar system may be in a transition region between the LIC and the G cloud (Redfield 2008). We look forward to new observations that will test this and other recent results concerning the heliosphere and CHISM discussed at this notable ISSI Workshop.

Acknowledgements This work was partly supported by NASA under Explorer Program Grant NNG05EC85C and SR&T Grant NNG06GD55G. The author thanks all ISSI Workshop participants for making their presentation materials readily available.

References

D. Breitschwerdt, M.A. deAvillez, B. Fuchs, C. Dettbarn, Space Sci. Rev. (2008, this issue)
M. Bzowski, E. Möbius, S. Tarnopolski, V. Izmodenov, G. Gloeckler, Astron. Astrophys. **491**, 7–19 (2008a)
M. Bzowski et al., Space Sci. Rev. (2008b, this issue)
B.T. Draine, Space Sci. Rev. (2008, this issue)

R. Egger, in *The Local Bubble and Beyond*, ed. by D. Breitschwerdt, M.J. Freyberg, J. Truemper. Proc. IAU Coll., vol. 166 (1998), pp. 287–296
L.A. Fisk, G. Gloeckler, Astrophys. J. **640**, L79–L82 (2006)
V. Florinski, Space Sci. Rev. (2008, this issue)
P.C. Frisch, Space Sci. Rev. (2008, this issue)
A. Galli, P. Wurz, S. Barabash et al., Astrophys. J. **644**, 1317–1325 (2006)
G. Gloeckler, L.A. Fisk, J. Geiss, M.E. Hill, D.C. Hamilton, R.B. Decker, S.A. Krimigis, Space Sci. Rev. (2008, this issue)
G. Gloeckler, J. Geiss, Space Sci. Rev. **97**, 169 (2001)
J. Heerikhuisen, N.V. Pogorelov, V. Florinski, G.P. Zank, in Numerical Modeling of Space Plasma Flows: Astronum 2007, ed. by N.V. Pogorelov, E. AUdit, G.P. Zank, ASP Conf. Ser., vol. 385, p. 204 (2008)
V.V. Izmodenov, Yu.G. Malama, G. Gloeckler, J. Geiss, Astrophys. J. **594**, L59–L62 (2003)
V.V. Izmodenov, D.B. Alexashov, A.V. Myasnikov, Astron. Astrophys. **437**, L35–L38 (2005)
V.V. Izmodenov, Space Sci. Rev. (2008, this issue)
E.B. Jenkins, Space Sci. Rev. (2008, this issue)
R. Jokipii, A. Balogh, Space Sci. Rev. (2008, this issue)
D. Koutroumpa, R. Lallement, V. Kharchenko, A. Dalgarno, Space Sci. Rev. (2008, this issue)
H. Krüger, E. Grün, Space Sci. Rev. (2008, this issue)
W.S. Kurth, D.A. Gurnett, J. Geophys. Res. **108**, 8027 (2003)
R. Lallement et al., Astron. Astrophys. **411**, 447 (2003)
R. Lallement, E. Quemerais, J.L. Bertaux et al., Science **307**, 1447–1449 (2005)
A. Lazarian, Space Sci. Rev. (2008, this issue)
D.J. McComas, Space Sci. Rev. (2008, this issue)
D.J. McComas, N.A. Schwadron, Geophys. Res. Lett. **33**, L04102 (2006)
E. Möbius, M. Bzowski, S. Chalov et al., Astron. Astrophys. **426**, 897 (2004)
E. Möbius, H. Kucharek, L. Petersen et al., Space Sci. Rev. (2008, submitted)
M. Opher, Space Sci. Rev. (2008, this issue)
N. Pogorelov, Space Sci. Rev. (2008, this issue)
W. Pryor et al., Astron. Astrophys. **491**, 21–18 (2008)
E. Quemerais, R. Lallement, B.R. Sandel, J.T. Clarke, Interplanetary Lyman α Observations: Intensities from Voyagers and line profiles from HST/STIS (2008, this issue)
S. Redfield, Space Sci. Rev. (2008, this issue)
J.D. Richardson, Y. Liu, C. Wang, D.J. McComas, Astron. Astrophys. **491**, 1–5 (2008)
J.D. Richardson, E.C. Stone, Space Sci. Rev. (2008, this issue)
R.L. Shelton, Space Sci. Rev. (2008, this issue)
J. Slavin, Space Sci. Rev. (2008, this issue)
J. Slavin, P.C. Frisch, Astron. Astrophys. **491**, 53–68 (2008)
S. Snowden, Space Sci. Rev. (2008, this issue)
S.R. Spangler, Space Sci. Rev. (2008, this issue)
S. Stanimirowic, Space Sci. Rev. (2008, this issue)
B.Y. Welsh, Space Sci. Rev. (2008, this issue)
M. Witte, Astron. Astrophys. **426**, 835 (2004)
B. Wood et al., Astrophys. J. **628**, L143 (2005)
B.E. Wood, V.V. Izmodenov, Yu.G. Malama, Space Sci. Rev. (2008, this issue)

Space Science Series of ISSI

1. R. von Steiger, R. Lallement and M.A. Lee (eds.): *The Heliosphere in the Local Interstellar Medium*. 1996 ISBN 0-7923-4320-4
2. B. Hultqvist and M. Øieroset (eds.): *Transport Across the Boundaries of the Magnetosphere*. 1997 ISBN 0-7923-4788-9
3. L.A. Fisk, J.R. Jokipii, G.M. Simnett, R. von Steiger and K.-P. Wenzel (eds.): *Cosmic Rays in the Heliosphere*. 1998 ISBN 0-7923-5069-3
4. N. Prantzos, M. Tosi and R. von Steiger (eds.): *Primordial Nuclei and Their Galactic Evolution*. 1998 ISBN 0-7923-5114-2
5. C. Fröhlich, M.C.E. Huber, S.K. Solanki and R. von Steiger (eds.): *Solar Composition and its Evolution – From Core to Corona*. 1998 ISBN 0-7923-5496-6
6. B. Hultqvist, M. Øieroset, Goetz Paschmann and R. Treumann (eds.): *Magnetospheric Plasma Sources and Losses*. 1999 ISBN 0-7923-5846-5
7. A. Balogh, J.T. Gosling, J.R. Jokipii, R. Kallenbach and H. Kunow (eds.): *Co-rotating Interaction Regions*. 1999 ISBN 0-7923-6080-X
8. K. Altwegg, P. Ehrenfreund, J. Geiss and W. Huebner (eds.): *Composition and Origin of Cometary Materials*. 1999 ISBN 0-7923-6154-7
9. W. Benz, R. Kallenbach and G.W. Lugmair (eds.): *From Dust to Terrestrial Planets*. 2000 ISBN 0-7923-6467-8
10. J.W. Bieber, E. Eroshenko, P. Evenson, E.O. Flückiger and R. Kallenbach (eds.): *Cosmic Rays and Earth*. 2000 ISBN 0-7923-6712-X
11. E. Friis-Christensen, C. Fröhlich, J.D. Haigh, M. Schüssler and R. von Steiger (eds.): *Solar Variability and Climate*. 2000 ISBN 0-7923-6741-3
12. R. Kallenbach, J. Geiss and W.K. Hartmann (eds.): *Chronology and Evolution of Mars*. 2001 ISBN 0-7923-7051-1
13. R. Diehl, E. Parizot, R. Kallenbach and R. von Steiger (eds.): *The Astrophysics of Galactic Cosmic Rays*. 2001 ISBN 0-7923-7051-1
14. Ph. Jetzer, K. Pretzl and R. von Steiger (eds.): *Matter in the Universe*. 2001 ISBN 1-4020-0666-7
15. G. Paschmann, S. Haaland and R. Treumann (eds.): *Auroral Plasma Physics*. 2002 ISBN 1-4020-0963-1
16. R. Kallenbach, T. Encrenaz, J. Geiss, K. Mauersberger, T.C. Owen and F. Robert (eds.): *Solar System History from Isotopic Signatures of Volatile Elements*. 2003 ISBN 1-4020-1177-6
17. G. Beutler, M.R. Drinkwater, R. Rummel and R. von Steiger (eds.): *Earth Gravity Field from Space – from Sensors to Earth Sciences*. 2003 ISBN 1-4020-1408-2
18. D. Winterhalter, M. Acuña and A. Zakharov (eds.): *"Mars" Magnetism and its Interaction with the Solar Wind*. 2004 ISBN 1-4020-2048-1
19. T. Encrenaz, R. Kallenbach, T.C. Owen and C. Sotin: *The Outer Planets and their Moons* ISBN 1-4020-3362-1
20. G. Paschmann, S.J. Schwartz, C.P. Escoubet and S. Haaland (eds.): *Outer Magnetospheric Boundaries: Cluster Results* ISBN 1-4020-3488-1
21. H. Kunow, N.U. Crooker, J.A. Linker, R. Schwenn and R. von Steiger (eds.): *Coronal Mass Ejections* ISBN 978-0-387-45086-5

22. D.N. Baker, B. Klecker, S.J. Schwartz, R. Schwenn and R. von Steiger (eds.): *Solar Dynamics and its Effects on the Heliosphere and Earth* ISBN 978-0-387-69531-0
23. Y. Calisesi, R.-M. Bonnet, L. Gray, J. Langen and M. Lockwood (eds.): *Solar Variability and Planetary Climates* ISBN 978-0-387-48339-9
24. K.E. Fishbaugh, P. Lognonné, F. Raulin, D.J. Des Marais, O. Korablev (eds.): *Geology and Habitability of Terrestrial Planets* ISBN 978-0-387-74287-8
25. O. Botta, J.L. Bada, J. Gomez-Elvira, E. Javaux, F. Selsis, R. Summons (eds.): *Strategies of Life Detection* ISBN 978-0-387-77515-9
26. A. Balogh, L. Ksanfomality, R. von Steiger (eds.): *Mercury* ISBN 978-0-387-77538-8
27. R. von Steiger, G. Gloeckler, G.M. Mason (eds.): *The Composition of Matter* ISBN 978-0-387-74183-3
28. H. Balsiger, K. Altwegg, W. Huebner, T.C. Owen, R. Schulz (eds.): *Origin and Early Evolution of Comet Nuclei, Workshop honouring Johannes Geiss on the occasion of his 80th birthday* ISBN 978-0-387-85454-0
29. A.F. Nagy, A. Balogh, T.E. Cravens, M. Mendillo, I. Mueller-Wodarg (eds.): *Comparative Aeronomy* ISBN 978-0-387-87824-9
30. F. Leblanc, K.L. Aplin, Y. Yair, R.G. Harrison, J.P. Lebreton and M. Blanc (eds.): *Planetary Atmospheric Electricity* ISBN 987-0-387-87663-4
31. J.L. Linsky, V. Izmodenov, E. Möbius, R. von Steiger (eds.): *From the Outer Heliosphere to the Local Bubble: Comparison of New Observations with Theory* ISBN 978-1-4419-0246-7

Springer – Dordrecht / Boston / London

MIX
Papier aus verantwortungsvollen Quellen
Paper from responsible sources
FSC® C105338

If you have any concerns about our products,
you can contact us on
ProductSafety@springernature.com

In case Publisher is established outside the EU,
the EU authorized representative is:
Springer Nature Customer Service Center GmbH
Europaplatz 3, 69115 Heidelberg, Germany

Printed by Libri Plureos GmbH
in Hamburg, Germany